Catalytic Antibodies
Edited by Ehud Keinan

Further Titles of Interest

C. Niemeyer, C. Mirkin (eds.)

Nanobiotechnology – Concepts, Applications and Perspectives

2004
ISBN 3-527-30658-7

A. S. Bommarius, B. Riebel

Biocatalysis – Fundamentals and Applications

2004
ISBN 3-527-30344-8

K. Drauz, H. Waldmann (eds.)

Enzyme Catalysis in Organic Synthesis (2nd Edition)

2002
ISBN 3-527-29949-1

S. Brakmann, K. Johnsson (eds.)

Directed Molecular Evolution of Proteins

2002
ISBN 3-527-30423-1

Catalytic Antibodies

Edited by Ehud Keinan

WILEY-VCH Verlag GmbH & Co. KGaA

Prof. Dr. Ehud Keinan
Department of Chemistry
Technion-Israel Institute of Technology
Technion City
Haifa 32000
Israel
keinan@techunix.technion.ac.il

and
The Scripps Research Institute
Department of Molecular Biology
and
The Skaggs Institute for Chemical Biology
10550 North Torrey Pines Road
La Jolla, CA 92037
USA
keinan@scripps.edu

■ All books published by Wiley-VCH are carefully produced. Nevertheless, authors, editors and publisher do not warrant the information contained in these books, including this book, to be free of errors. Readers are advised to keep in mind that statements, data, illustrations, procedural details or other items may inadvertently be inaccurate.

Library of Congress Card No.:
applied for

British Library Cataloguing-in-Publication Data:
A catalogue record for this book is available from the British Library.

Bibliographic information published by Die Deutsche Bibliothek
Die Deutsche Bibliothek lists this publication in the Deutsche Nationalbibliografie; detailed bibliographic data is available in the Internet at http://dnb.ddb.de

© 2005 WILEY-VCH Verlag GmbH & Co KGaA, Weinheim

Printed on acid-free paper

All rights reserved (including those of translation into other languages). No part of this book may be reproduced in any form – by photoprinting, microfilm, or any other means – nor transmitted or translated into a machine language without written permission from the publishers. Registered names, trademarks, etc. used in this book, even when not specifically marked as such, are not to be considered unprotected by law.

Composition Steingraeber Satztechnik GmbH, Dossenheim
Printing Strauss GmbH, Mörlenbach
Bookbinding Litges & Dopf Buchbinderei GmbH, Heppenheim

Printed in the Federal Republic of Germany
ISBN 3-527-30688-9

"וְכִתְּתוּ חַרְבוֹתָם לְאִתִּים,
וַחֲנִיתוֹתֵיהֶם לְמַזְמֵרוֹת ..."
(ישעיהו ב,ד)

"And they shall beat their swords
into plowshares, and their spears
into pruning-hooks ..."

Isaiah (2:4)

Antibodies represent the major weapon our body utilizes to attack foreign invaders. The idea of chemists using antibodies to catalyze their non-biological reactions is reminiscent of the biblical metaphor of a military weapon being converted to a working tool. This phrase from Isaiah inspired Mordecai Ardon (1896–1992), whose monumental stained glass windows adorn the Jewish National and University Library on the campus of the Hebrew University of Jerusalem. Photo courtesy of the Hebrew University.

Foreword

It is not our intent here to review the many accomplishments of the field of antibody catalysis. The chapters contained in this volume amply attest to the vigor and accomplishments of the field. Rather we will attempt to extract some of the general principles that have and continue to guide the field.

The first issues concern why one would attempt to make enzymes since there are already some 4000 enzymes known that cover many chemical transformations that one might desire to accomplish. The simple answer is that there are many transformations that existing enzymes do not cover and one might wish to have enzymes for these transformations. However, there is a much deeper answer to this question that goes to the heart of the field and, indeed, is the centerpiece of chemistry. Chemists make things in order to understand the rules of chemistry. For instance, the object of natural product chemistry is not to obtain carloads of material, but rather to understand the general principles that govern how atoms and molecules interact. One learns from both the successes and failures of attempts to accomplish a given transformation. Likewise, if we are to understand the complex chemical engines that are protein enzymes we must ultimately make them. As with natural products, we can learn from both successes and failures. In every case when the binding energy of antigen-antibody union is converted to catalysis, we can state that we know a route by which the transformation under consideration can occur. The initial problem – that one didn't know in detail exactly how the transition state is bound – has largely been obviated by the many crystal structures of catalytic antibodies bound to their transition state analogues. Of course, a catalytic antibody tells one *a* way that a given transformation can proceed, but not necessarily *the* way that a natural enzyme might accomplish the same transformation. The fact that proteins can accomplish the same transformation by a variety of routes is in of itself very interesting and one can only wonder why certain roads were not taken by evolution especially if they are isoenergetic.

So what have we learned? The first very large lesson is that it is not very difficult to generate protein catalysts so long as one correctly understands some of the general chemical principles of the reaction. In stating that it is easy to obtain catalysis we are not making a statement about overall efficiency, but for now are only concerned about starting with binding energy and winding up with catalysis. The lessons that

Catalytic Antibodies. Edited by Ehud Keinan
Copyright © 2005 WILEY-VCH Verlag GmbH & Co. KGaA, Weinheim
ISBN: 3-527-30688-9

one derives from the experimental conversion of binding energy to catalysis in real time speaks most generally to the large subject of enzyme evolution. This is because evolution can improve things so long as the function is useful and, thus, the fact that it is rather easy to get catalysis started in the confines of a generic protein-binding pocket gives a selectable function. One of the main insights upon which the field of antibody catalysis was founded is that even a combinatorial library as large as the antibody repertoire was, when used randomly, not large enough to achieve catalysis in real time. Thus, relative to the evolutionary time scale, the process was shortened by millions to billions of years by programming the binding with detailed chemical instructions. These instructions were given, of course, by use of transition state analogues as antigens. The process by which the early antibody catalysts were generated is not precisely identical to evolution for two reasons. First, there is presumably no process in evolution in which a concert of chemical instructions is given at once. Second, screening rather than selection was used to find the early catalysts. Thus, unlike the real time induction of antibody catalysts, evolution does not need instruction because it has time and selection on its side. However, one can assume that immunization with a transition state analogue is simply a device that shortens the time to reach a point that evolution would come to anyway. From this vantage point, antibody catalysis can teach us much about the evolution of enzymes. In the simplest of terms, immunization with a transition state analogue does select for a concert of binding parameters such as the size of the binding pocket, the geometrical arrangement, of protein functionalities around the transition state, and the distribution of charges. Presumably evolution would have to do the same and, in this sense, these parameters are like state functions in that their overall energetics are independent of the pathway by which they were formed. Because we understand that proteins are complex and dynamic entities, we would not expect one-step perfection even though a proper concert of favorable binding interactions has been achieved. In this sense it is paradoxical that we can learn the most about enzyme evolution from the imperfection of antibody catalysts. This is because we can isolate binding parameters and learn what they worth in terms of rate acceleration. Further, because of the fact that antibodies are easy to engineer we can make stepwise improvements and study how they affect catalysis. Finally, the focus here on imperfect enzymes is not to say that, as we will see later, enzymes that rival natural catalysts have not been achieved.

The extraction of information that relates binding to catalysis depends on how many different binding motifs can be generated because this becomes the database from which the general principles are extracted. Fortunately, the antibody repertoire is sufficiently large that a sizeable database of binding interactions is achieved. Another feature that needs to be recognized derives from the fact that the induction of an immune is, itself, an evolutionary process that selects for binding energy. Thus, in making a database on the best way to bind to organic functionalities, we are also learning about the permitted and most successful ways that substrates and transition states might be bound. One of the early difficulties of the field was that the fine details of how binding occurred were not totally controlled by the experimenter. While this is still somewhat the case, this has largely been obviated by the bait and switch strategy and covalent immunization.

One of the most startling lessons from the field concerns how precisely a complex enzyme mechanism can be copied into another protein with a completely different fold and presumably dynamic repertoire. This was seen in the ease by which aldolase antibody catalysts that proceed by an enamine mechanism were generated by immunization with a simple 1,3-diketone antigen to select for a reactive lysine in the antibody active site. The mechanism and rate acceleration of these antibody aldolases are nearly identical to nature's own enzymes. One interesting side point that came out of these studies concerns the vigor with which immunological evolution drives toward binding energy even to the point of taking covalent options when they are available. Thus, even though a diketone antigen "offered" the opportunity for covalent binding if a lysine with a perturbed pK_a appeared by mutation, there were many other ways that binding energy could be achieved without resorting to the covalent option (for instance by simple binding to the benzene ring that was also present in the hapten). The fact that most antibodies mutated to express a rare lysine that is not present in the germ line speaks to the fact that antibody evolution is, as expected, a chemistry driven process but also one whose chemistry is only limited by opportunity so long as the reaction in question increases binding energy beyond that which can be achieved by a concert of non-covalent interactions. In other words, any chemical reaction that can happen will.

One of the largest lessons about proteins and catalysis came from an antibody that is not a catalyst and, thus, is not covered in this volume. Several of us reasoned that the time had come to use the binding energy of antibodies to perturb electronically excited states in the same way that thermal transition states have been studied in the field so far. To attempt this, antibodies were made to stilbene. Remarkably, even though these antibodies were not "taught" about the excited state, when irradiated with U.V light they adapted to and perturbed the excited state of stilbene and emitted intense blue light. These antibodies were called "blue fluorescent antibodies". This effect means that proteins that bind to ground states can adapt to high-energy states of the ligand with sufficient energy to perturb the excited state surface. This is probably a general property of proteins and may explain how enzymes bind to ground state substrates while also maintaining the property of adapting to transition states. Again we see a lesson from antibodies that goes to the centerpiece of protein catalysis.

In terms of breaking news, recent events have demonstrated that all antibodies have the catalytic potential to generate highly reactive oxygen species including ozone and a masked form of the hydroxyl radical. This may be the most potent effector function of antibodies ever discovered. Thus, this preface ends with a remarkable irony. The able workers whose beautiful experiments are detailed in this volume turned antibodies into enzymes to learn more about natural enzymes only to ultimately find out that antibodies were natural enzymes all along!

Richard A. Lerner and Peter G. Schultz

The Scripps Research Institute and
The Skaggs Institute for Chemical Biology,
Department of Chemistry,
10550 N. Torrey Pines Road, La Jolla, CA 92037

Table of Contents

1		**Immunological Evolution of Catalysis** *1*
		Jun Yin, Peter G. Schultz
1.1		Introduction *1*
1.2		Parallels between Antibody and Enzyme Evolution *1*
1.3		Evolution of Catalytic Antibodies *3*
1.4		Ferrochelatase Antibody 7G12 – Evolution of the Strain Mechanism *3*
1.5		Esterase Antibody 48G7 – Effect of Distant Mutations on Catalysis *15*
1.6		Sulfur Oxidase Antibody 28B4 – Incremental Changes in Evolution *18*
1.7		Oxy-Cope Antibody AZ28 – Evolution of Conformational Diversity in Catalysis *22*
1.8		Diels-Alderase Antibody 39A11 – Evolution of a Polyspecific Antibody combining Site *25*
1.9		Conclusions *26*
		References *28*
2		**Critical Analysis of Antibody Catalysis** *30*
		Donald Hilvert
2.1		Introduction *30*
2.2		Exploiting Antibodies as Catalysts *30*
2.3		Catalytic Efficiency *31*
2.4		Hapten Design *33*
2.5		Representative Catalytic Antibodies *34*
2.5.1		Proximity Effects *34*
2.5.1.1		Sigmatropic Rearrangements *34*
2.5.1.2		Cycloadditions *38*
2.5.2		Strain *43*
2.5.2.1		Ferrochelatase Mimics *43*
2.5.2.2		Other Systems *46*
2.5.3		Electrostatic Catalysis *46*
2.5.3.1		Acyl Transfer Reactions *46*
2.5.4		Functional Groups *53*

Catalytic Antibodies. Edited by Ehud Keinan
Copyright © 2005 WILEY-VCH Verlag GmbH & Co. KGaA, Weinheim
ISBN: 3-527-30688-9

2.5.4.1	Aldolases *54*
2.6	Perspectives *57*
2.6.1	General Lessons from Comparisons of Enzymes and Antibodies *57*
2.6.2	How efficient does catalysis need to be? *58*
2.6.3	Strategies for Optimizing Efficiency *59*
2.6.3.1	Better Haptens *60*
2.6.3.2	Screening *61*
2.6.3.3	Engineering *61*
2.6.3.4	Selection *62*
2.6.3.5	Other Scaffolds *62*
2.7	Conclusions *63*
	References *65*

3 Theoretical Studies of Antibody Catalysis *72*

Dean J. Tantillo, Andrew G. Leach, Xiyun Zhang, K. N. Houk

3.1	Introduction *72*
3.2	Questions Subject to Theoretical Elucidation *73*
3.2.1	Predicting Antibody Structure from Sequence *73*
3.2.2	Predicting Binding Modes and Binding Energies *74*
3.2.3	Understanding Antibody Catalysis [14] *75*
3.2.4	General Considerations *76*
3.3	Hydrolytic Antibodies *76*
3.3.1	Gas and Solution Phase Hydrolysis of Aryl Esters *76*
3.3.2	Hapten Fidelity *78*
3.3.3	Theoretical Exploration of Antibody Catalysis *80*
3.3.3.1	16G3 *81*
3.3.3.2	6D9 *82*
3.3.3.3	43C9 *83*
3.3.3.4	CNJ206 *84*
3.3.3.5	48G7 *85*
3.3.3.6	17E8 and 29G11 *86*
3.4	Cationic Cyclizations *88*
3.4.1	Antibody Catalysis of Solvolysis *88*
3.4.2	Antibody-Catalyzed Hydroxyepoxide Cyclization *89*
3.5	Antibody-Catalyzed Diels-Alder and *retro*-Diels-Alder Reactions *90*
3.5.1	The Most Efficient *endo*-Diels-Alderase 1E9 *91*
3.5.2	*endo*-Diels-Alderase 39A11 and its Germline Precursor *93*
3.5.3	*exo*-Diels-Alderase 13G5 *96*
3.5.4	*retro*-Diels-Alderase 10F11 *100*
3.6	Other Antibody-Catalyzed Pericyclic Reactions *101*
3.6.1	Oxy-Cope Rearrangement Catalyzed by Antibody AZ-28 *101*
3.6.2	1,3-Dipolar Cycloaddition Catalyzed by Antibody 29G12 *105*
3.6.3	Chorismate-Prephenate Claisen Rearrangement Catalyzed by Antibody 1F7 *106*

3.7	Antibody-Catalyzed Carboxybenzisoxazole Decarboxylation	107
3.8	Summary	110
	References	111

4 The Enterprise of Catalytic Antibodies: A Historical Perspective *118*
Michael Ben-Chaim

4.1	Introduction	118
4.2	Methods	120
4.3	Results	122
4.3.1	The Conceptual Origins of Catalytic Antibodies	122
4.3.2	Tapping the Immune System for Catalysts	125
4.4	Conclusions	128
	References	130

5 Catalytic Antibodies in Natural Products Synthesis *132*
Ashraf Brik, Ehud Keinan

5.1	Introduction	132
5.2	Total Synthesis of α-Multistriatin via Antibody-Catalyzed Asymmetric Protonolysis of an Enol Ether	134
5.3	Total Synthesis of Epothilones Using Aldolase Antibodies	136
5.4	Total Synthesis of Brevicomins Using Aldolase Antibody 38C2	142
5.5	Synthesis of 1-Deoxy-L-Xylose Using 38C2 Antibody	145
5.6	Synthesis of (+)-Frontalin and Mevalonolactone via Resolution of Tertiary Aldols with 38C2	145
5.7	Wieland-Miescher Ketone via 38C2-Catalyzed Robinson Annulation	147
5.8	Formation of Steroid A and B Rings via Cationic Cyclization	147
5.9	Synthesis of Naproxen via Antibody-Catalyzed Ester Hydrolysis	148
5.10	Conclusions	148
	References	150

6 Structure and Function of Catalytic Antibodies *153*
Nicholas A. Larsen, Ian A. Wilson

6.1	Introduction	153
6.2	Electrostatic Complementarity	155
6.2.1	Anionic Binding Motifs and Ester Hydrolysis	155
6.2.2	Structural Motifs	156
6.2.3	Mechanistic Considerations	157
6.2.4	Mechanistic Pitfalls	158
6.2.5	Structural and Functional Considerations for Hapten Design	158
6.2.6	Anionic Binding Motifs and Control of the Microenvironment in Decarboxylation	159

6.2.7	Anionic Binding Motifs and the Periodate Cofactor in Sulfide Oxygenation *160*
6.2.8	Cation Stabilization and Allylic Isomerization *161*
6.3	Shape Complementarity and Approximation *162*
6.3.1	Unimolecular Rearrangements *163*
6.3.1.1	Antibody 1F7 *163*
6.3.1.2	Antibody AZ-28 *164*
6.3.2	Bimolecular Rearrangements and the Diels-Alder Reaction *165*
6.3.2.1	Retro-Diels-Alder Reaction *165*
6.3.2.2	Disfavored *exo* Diels-Alder Reaction *166*
6.3.2.3	*Endo* Diels-Alder Reactions *167*
6.4	Shape Complementarity and Control of the Reaction Coordinate *169*
6.4.1	*Syn*-Elimination *169*
6.4.2	Disfavored Ring Closure *171*
6.4.3	Selective Control of a Reactive Carbocation: Antibodies 4C6 and 19A4 *172*
6.5	Shape Complementarity and Substrate Strain *174*
6.6	Reactive Amino Acids and the Possibility of Covalent Catalysis *175*
6.7	New Challenges *178*
	References *179*

7	**Antibody Catalysis of Disfavored Chemical Reactions** *184*
	Jonathan E. McDunn, Tobin J. Dickerson, Kim D. Janda
7.1	Introduction *184*
7.2	Formal Violation of Baldwin's Rules for Ring Closure: the 6-*endo*-tet Ring Closure *185*
7.3	Cationic Cyclization *191*
7.3.1	Initial Studies *191*
7.3.2	Tandem Cationic Cyclization Mediated by Antibody *197*
7.4	*exo*-Diels-Alder Reactions *203*
7.5	Miscellaneous Disfavored Processes *206*
7.5.1	*Syn*-Elimination to a *cis* Olefin *206*
7.5.2	Aryl Carbamate Hydrolysis via $B_{Ac}2$ Mechanism *207*
7.5.3	Transesterification *208*
7.5.4	Ketal Formation *210*
7.5.5	Controlling Photoprocesses *211*
7.6	Summary *212*
	References *214*

8	**Screening Methods for Catalytic Antibodies** *217*
	Jean-Louis Reymond
8.1	Introduction *217*
8.2	Theoretical Consideration: What to look for during screening *218*
8.2.1	Converting Antibody Binding to Catalysis (Quantitatively) *218*

8.2.2	Screening and the Theoretical Detection Limits	219
8.2.3	The Proof of Selective Antibody Catalysis	220
8.3	A Practical Perspective: Screening in the Published Literature	221
8.3.1	Statistical Overview	221
8.3.2	HPLC Analysis	223
8.3.3	Chromogenic and Fluorogenic Reactions	225
8.4	High-Throughput Screening Methods	226
8.4.1	Enzyme-Coupled Assays	227
8.4.2	cat-ELISA	228
8.4.3	Thin-Layer Chromatography	230
8.4.4	Chromogenic and Fluorogenic Substrates	230
8.4.5	Isotopic Labeling	235
8.5	Examples with Fluorogenic Substrates and Antibodies from Hybridoma	235
8.5.1	Retro-Diels-Alderase Antibodies	236
8.5.2	Pivalase Catalytic Antibodies	237
8.6	Conclusion	239
	References	240

9	***In vitro* Evolution of Catalytic Antibodies and Other Proteins via Combinatorial Libraries**	**243**
	Ron Piran, Ehud Keinan	
9.1	Introduction	243
9.2	Technologies and Constructs – Basic Principles	244
9.2.1	Phage Display	244
9.2.2	The Fab Construct	246
9.2.3	The Fv Construct	248
9.2.4	*In vitro* Immunization	250
9.3	Screening of Libraries	251
9.4	Directed Evolution	262
9.4.1	Mutagenesis and Selection	263
9.4.2	DNA Shuffling	267
9.5	*In vitro* Libraries	270
9.5.1	Ribosomal Display	270
9.5.2	*In vitro* Compartmentalization	271
9.5.3	Plasmid Display	272
9.6	*In vivo* Libraries	273
9.6.1	Protein Libraries Presented on an *E. coli* Surface	273
9.6.2	Protein Libraries in Yeast	274
9.6.3	Protein Libraries in Other Organisms	276
9.7	Conclusions and Outlook	279
	References	280

10	**Medicinal Potential of Catalytic Antibodies** *284*
	Roey Amir, Doron Shabat
10.1	Introduction *284*
10.2	Prodrug Activation *284*
10.2.1	Catalytic Antibodies Designed for Carbamate/Ester Hydrolysis *284*
10.2.2	Catalytic Antibody Designed for *retro*-Diels-Alder Reaction *287*
10.2.3	Catalytic Antibody 38C2 (*retro*-Aldol-*retro*-Michael Cleavage Reaction) *288*
10.3	Cocaine Inactivation *300*
10.4	Conclusions *301*
	References *302*

11	**Reactive Immunization: A Unique Approach to Aldolase Antibodies** *304*
	Fujie Tanaka, Carlos F. Barbas, III
11.1	Introduction *304*
11.2	Generation of Aldolase Antibodies by Reactive Immunization *305*
11.2.1	Development of the Concept of Reactive Immunization *305*
11.2.2	Combining Reactive Immunization with Transition State Analog for Aldolase Antibodies *309*
11.3	Structural Insight into Aldolase Antibodies *310*
11.4	Aldolase Antibody-catalyzed Reactions *313*
11.4.1	Broad Scope and High Enantioselectivity *313*
11.4.2	Aldol Reactions Catalyzed by Aldolase Antibodies 38C2 and 33F12 *314*
11.4.3	Antibody 38C2 and 33F12-catalyzed *retro*-Aldol Reactions and Their Application to Kinetic Resolutions *315*
11.4.4	Aldol and *retro*-Aldol Reactions Catalyzed by Aldolase Antibodies 93F3 and 84G3 *316*
11.4.5	Preparative scale kinetic resolutions using aldolase antibodies in a biphasic aqueous/organic solvent system *318*
11.4.6	Aldolase Antibody-catalyzed Reactions in Natural Product Syntheses *318*
11.4.7	Aldolase Antibodies in Reactions Involving Imine and Enamine Mechanisms and Exploitation of the Nucleophilic Lysine ε-Amino Group *319*
11.4.8	Concise Catalytic Assays for Aldolase Antibody-catalyzed Reactions *321*
11.4.9	Prodrug Activation by an Aldolase Antibody *322*
11.5	Evolution of Aldolase Antibodies *in vitro* *325*
11.5.1	Selection of Aldolase Antibodies with Diketones using Phage Display *325*
11.5.2	Selection Systems for Aldolase Antibodies *328*
11.6	Other Catalytic Antibodies Selected with Reactive Compounds *in vitro* *328*
11.7	Summary *330*
11.8	Acknowledgments *331*
	References *332*

12	**The Antibody-Catalyzed Water Oxidation Pathway** *336*
	Cindy Takeuchi, Paul Wentworth Jr.
	References *348*

13	**Photoenzymes and Photoabzymes** *350*
	Sigal Saphier, Ron Piran, Ehud Keinan
13.1	Introduction *350*
13.2	Photoenzymes *351*
13.2.1	Protochlorophyllide Reductase *351*
13.2.2	DNA Photolyase *352*
13.2.3	[6-4] Photoproduct Lyase *355*
13.2.4	General Considerations *356*
13.3	Photocatalytic Antibodies *357*
13.4	Conclusions *366*
13.5	Acknowledgement *366*
	References *367*

14	**Selectivity with Catalytic Antibodies – What Can Be Achieved?** *370*
	Veronique Gouverneur
14.1	Introduction *370*
14.2	Acyl Transfer Reactions: Ester Hydrolysis, Transacylations, and Amide Hydrolysis *371*
14.3	Glycosyl and Phosphoryl Group Transfer *383*
14.4	Pericyclic Reactions *385*
14.4.1	Sigmatropic Rearrangement *386*
14.4.2	Pericyclic Eliminations *387*
14.4.3	Cycloaddition *388*
14.5	Aldol Reactions *392*
14.6	Cyclization *398*
14.6.1	Reduction and Oxidation *402*
14.7	Additions and Eliminations *405*
14.8	Conclusion *408*
	References *410*

15	**Catalytic Antibodies as Mechanistic and Structural Models of Hydrolytic Enzymes** *418*
	Ariel B. Lindner, Zelig Eshhar, Dan S. Tawfik
15.1	Introduction *418*
15.2	Chapter Overview *421*
15.2.1	Catalysis by Oxyanion Stabilization (OAS) *421*
15.2.1.1	Fidelity of TSA design *422*
15.2.1.2	Esterolytic Antibodies Based Solely on Oxyanion Stabilization *422*

15.2.1.3 Antibody Oxanion Holes *425*
15.2.1.4 Oxyanion holes – Antibodies vs. Enzymes *426*
15.2.1.5 Antibody Affinity and Rate Acceleration Limitation *429*
15.2.1.6 Nucleophilic Catalysis *430*
15.2.2 Endogenous Nucleophiles in Hydrolytic Antibodies *431*
15.2.2.1 Reactive immunization (RI) *431*
15.2.2.2 Serendipity and 43C9 Antibody *432*
15.2.3 Exogenous Nucleophiles and Chemical Rescue in Hydrolytic Antibodies *435*
15.2.4 General Acid/Base Mechanisms in Hydrolytic Antibodies *439*
15.2.5 Metal-Activated Catalytic Antibodies *441*
15.2.6 Substrate Destabilization *442*
15.3 The Role of Conformational Changes in Catalytic Antibodies *443*
15.3.1 Conformational changes in catalytic antibodies *444*
15.3.2 The contribution of structural isomerization to catalysis *445*
15.4 Conclusions *447*
References *449*

16 Transition State Analogs – Archetype Antigens for Catalytic Antibody Generation *454*
Anita D. Wentworth, Paul Wentworth, Jr., G. Michael Blackburn
References *467*

17 Polyclonal Catalytic Antibodies *470*
Marina Resmini, Elizabeth L. Ostler, Keith Brocklehurst, Gerard Gallacher
17.1 Introduction *470*
17.2 The Importance of Polyclonal Catalytic Antibodies *471*
17.3 Demonstration of Polyclonal Antibody Catalysis *472*
17.4 Hapten Design and Catalytic Antibody Activity *475*
17.5 Kinetic Activity, Homogeneity, and Variability *477*
17.6 Assessment of Contamination *478*
17.7 Mechanistic Studies *480*
17.8 Investigations of Active-Site Availability *480*
17.9 Therapeutic Applications *481*
17.10 Antiidiotypic Antibodies *484*
17.11 Naturally-Occurring Catalytic Antibodies *484*
17.12 Conclusions *485*
References *487*

18 Production of Monoclonal Catalytic Antibodies: Principles and Practice *491*
Diane Kubitz, Ehud Keinan
18.1 Introduction *491*

18.2	Immunization *492*
18.3	Hybridoma Production and Screening *494*
18.4	Large-Scale Antibody Production *497*
18.5	Antibody Purification *499*
18.6	Testing for Catalytic Activity *500*
18.7	Preparation of Fab, F(ab')$_2$, and Fab' Fragments *500*
18.8	Conclusion *503*
	References *504*

19	**Natural Catalytic Antibodies – Abzymes** *505*
	Georgy A. Nevinsky, Valentina N. Buneva
19.1	Abbreviations *505*
19.2	Introduction *505*
19.3	Natural Catalytic Antibodies *506*
19.4	Peculiarities of the Immune Status of Patients with Various Autoimmune Diseases *508*
19.5	The Origin of Natural Autoimmune Abzymes *510*
19.6	Peculiarities of the Immune Status of Pregnant and Lactating Women and the Origin of Natural Abzymes from Human Milk *513*
19.7	Purification of Natural Abzymes *519*
19.8	Criteria to Establish that Catalytic Activity is Intrinsic to Antibodies *525*
19.9	Catalytic Antibodies Catalyzing Transformations of Water and Oxygen Radicals *530*
19.10	Antibodies with Proteolytic Activities *531*
19.10.1	Antibodies Hydrolyzing Vasoactive Intestinal Peptide *532*
19.10.2	Abzymes Hydrolyzing Thyroglobulin, Prothrombin, and Factor VIII *533*
19.11	Human and Animal RNase and DNase Abzymes *534*
19.12	Human Milk Abzymes with Various Activities *542*
19.12.1	Abzymes with Protein Kinase Activity *542*
19.12.2	Abzymes with Lipid Kinase Activity *545*
19.12.3	Abzymes with Oligosaccharide Kinase Activity *549*
19.12.4	Nucleotide-Hydrolyzing Abzymes of Human Milk *551*
19.12.5	Human Blood Sera and Milk Abzymes Hydrolyzing Polysaccharides *552*
19.12.6	Peculiarities of Milk Abzymes *554*
19.13	Biological Roles of Abzymes *555*
19.14	Natural Abzymes as Tools for Investigating RNA Structure *557*
19.15	Abzymes as Diagnostic Tools and Tools for Biological Manipulations *557*
19.16	Conclusion *561*
	References *562*

Index *571*

"What I cannot create I do not understand."
Richard P. Feynman

Preface

It is now 18 years since the first antibody catalysts elicited against transition-state analogs were induced. At this age of maturation we found it appropriate to comprehensively cover the field of antibody catalysis in one volume. The catalytic antibody technology merges the combinatorial diversity of the immune system with a programmable design by the experimenter. In fact, no other area of bioorganic chemistry has taught us so much about the use of large libraries of molecules in the service of chemistry. We have learned from the immune system that natural evolution does not necessarily require billions of years; it may be completed on a laboratory timescale. This lesson of applying the three major components of evolution – diversity, selection and amplification – has inspired biologists and chemists alike on their quest to discover new functional biopolymers, new catalysts, new drugs, and even new solid-state materials with desired physical properties.

There are many conceptual steps on the way towards the realization of a new antibody catalyst, including mechanistic understanding of the specific reaction to be catalyzed, scholarly prediction of the transition state of highest energy, creative design of a chemically stable transition state analog (TSA), and the planning of synthetic schemes for haptens and substrates. Yet, as is appropriately expressed by the above-cited dictum of Feynman, antibody catalysis is primarily an experimental science where the keys to success reside in the details of the experimental procedures, including organic synthesis, immunization protocols, screening procedures, production of monoclonal antibodies, kinetic experiments, and crystallographic studies.

Many facets of biocatalysis are not yet fully understood, particularly those related to the dynamics of the protein catalyst along the reaction coordinate. Even when some of that can be envisaged, we do not yet know how to design the antibody active site in order to achieve the desired dynamic properties. Therefore, the use of a TSA, even an optimal one, which is often impossible to make, represents only a "snapshot" of a continuous, dynamic process and only a general guideline for the immune system to produce the appropriate catalyst. The resultant, broadly diverse population of relevant antibodies reflects the variety of ways by which the immune system can respond to the given TSA. The beauty of this approach is the element of serendipity, allowing us to find valuable items that were not looked for. Consequently, the importance of efficient screening strategies can never be overestimated.

Catalytic Antibodies. Edited by Ehud Keinan
Copyright © 2005 WILEY-VCH Verlag GmbH & Co. KGaA, Weinheim
ISBN: 3-527-30688-9

One of the most significant trends in modern science is the collapse of traditional barriers between well-defined scientific disciplines. Antibody catalysis has illustrated, probably more than any other field, the rewards of working at the interface of biology and chemistry. The diverse chapters of this book reflect the broad spectrum of activities in this highly interdisciplinary field.

Chapter 1 describes structural and functional studies of the immunological evolution of catalytic antibodies, which have provided important insights into the evolution of binding energy and catalysis in biological systems, including antibodies and enzymes. These studies suggest that there is an intrinsic conformational flexibility within the germline antibody combining site. The binding of ligands to the germline antibody induces conformational rearrangements that result in enhanced complementarity between the antibody and the ligand. Consequently, structural plasticity in the combining site allows a limited number of protein scaffolds to bind a broad array of substrates with moderate affinity. Then, somatic mutations and selection processes optimize the complementarity of the combining site with its hapten by fixing the conformation of the CDR loops. Interestingly, these mutations are introduced not only at the hapten combining site but also at the peripheries of that site.

Chapter 2 analyzes the main achievements in the field and defines the key challenges ahead of us. The chapter examines the structure and mechanism of representative catalytic antibodies that were generated by various strategies for different reaction types. It concludes that the field has progressed rapidly from simple model reactions to complex multistep processes, thus defining the scope and limitations of this technology. Now that the approach is well established, attention must be paid to strategies for optimizing catalytic efficiency and for promoting more demanding transformations. Development of improved transition-state analogues, refinements in immunization and screening protocols, and elaboration of general strategies for enhancing the efficiency of first-generation catalytic antibodies are identified as evident, but difficult, challenges for this field. In addition, learning how to create, manipulate and evolve large combinatorial libraries of proteins outside the immune system should help to automate the processes of catalyst discovery and optimization.

Chapter 3 reviews the theoretical studies that have been performed on antibody catalysis to date. Various tools have been applied to the computational investigations of different types of reactions that have been catalyzed by antibodies. Many insights about both quantitative and qualitative aspects of antibody catalysis have been obtained. It may be concluded that quantum mechanical computations on model systems can now be carried out with high accuracy, affording good qualitative determinations of the mode of binding of haptens, substrates, and transition states into antibody binding sites. Yet, the major challenges to be met on the way to a full understanding of antibody catalysis in particular, and biocatalysis, in general, include the development of practical quantitative computation methods for solvation energies and protein-ligand interaction energies.

Chapter 4, which is written by a science historian, brings in to this book a unique historical perspective of the field, drawing special attention to the initial endeavors to establish antibody catalysis as a scientific discipline at the crossroads of chemistry and immunology. The chapter is neither a comprehensive coverage of the history of the field and the chronological development nor a presentation of the key individuals who contributed to the field. It treats the emergence of this field as a case study and a model for what does it take for a new scientific discipline to be born. Dr. Ben Chaim's thesis is based on many hours of recorded interviews with almost all the active scientists in the field. I believe that this rather unusual, thought-provoking contribution blends well with the scientific chapters of this book.

Chapter 5 covers the use of catalytic antibodies in the synthesis of natural products. One of the main goals of the field of antibody catalysis has been to learn how to design catalysts to improve the overall yield of existing synthetic routs, thus allowing practical construction of more totally synthetic drugs and other important natural products. The examples covered by this chapter, including (–)-α-multistriatin, epothilones, brevicomis, 1-deoxy-l-xylose, (+)-frontalin, the formal synthesis of (–)- and (+)-mevalonolactone, partial synthesis of the steroid skeleton, and naproxen, testify for the important role catalytic antibodies may play in asymmetric synthesis. The remarkable ability of these biocatalysts to control the rate, stereo-, regio-, chemo-, and enantioselectivity of many reactions, including highly disfavored chemical processes, and sometimes even with high substrates promiscuity, render these agents valuable tools for the total synthesis of pharmaceuticals, fine chemicals, and complex natural products.

Chapter 6 presents an impressive number of X-ray crystallographic studies of catalytic antibodies, which have provided valuable descriptions of the specific interactions with substrates and TSAs within the antibody combining sites. The antibodies studied catalyze a variety of chemical transformations, including ester hydrolysis, Diels-Alder reaction, cationic cyclization, elimination, decarboxylation, aldol reactions, sulfide oxidation, rearrangements, and metal chelation. This chapter catalogues and critically assesses the contributions that structural studies have made to the understanding of the catalytic mechanism. Comparisons are made to natural enzymes that catalyze similar reactions.

Chapter 7 reviews several remarkable cases of antibody-catalyzed disfavored chemical transformations. Catalysis of such reactions is a most compelling aspect in both enzymes and catalytic antibodies not only because the reactions are synthetically useful but also because these biocatalysts reveal how nature solves particularly difficult problems. Biochemical and structural studies of the enzymes and antibodies that catalyze disfavored reactions provide chemists with insights into important mechanistic considerations as well as a starting point in the development of *de novo* catalysts for these and other disfavored processes. The examples covered in this chapter include formal violation of the Baldwin's rules for ring closure, exo-Diels-Alder reactions, formation of a ketal in water, *syn*-elimination to produce a cis olefin, and selective formation of C-C bonds via cationic cyclization reactions.

Chapter 8 underscores the importance of screening methods for catalytic antibodies. Selective transition state binding is just one of many elements of catalysis, one that can be well designed by immunization with an appropriate TSA. Therefore, a good TSA-binder may not necessarily be a good catalyst. Although screening for binding is easily performed by the well-established immunological methods, one should keep in mind that this approach is reminiscent of searching for a lost coin under the street lamp. Certainly, the optimal approach would be the screening for actual catalysis rather than for binding.

Chapter 9 highlights the strengths of directed evolution for the design of novel enzymes with emphasis on catalytic antibodies. The representative techniques of displayed combinatorial protein libraries that are described in this chapter do not cover the entire field of directed evolution. They demonstrate, however, the great opportunities in the employment of such techniques for the development of functional proteins, and of catalytic antibodies in particular. The various strategies harness a number of important technologies, including the methods of cloning and expression of Fab and scFv in different organisms, the ways of physically linking the phenotype to its genotype, the means of creating large libraries in very small volumes, methods of clone selection, and techniques of amplification of clones having the desired properties.

Chapter 10 reviews the medicinal applications of catalytic antibodies, particularly for prodrug activation. The addition of a targeting device to the relevant catalytic antibody renders this treatment highly selective for tumor cells, thus allowing the conversion of a nontoxic prodrug into a toxic drug in high local doses at the tumor site. The concept of using catalytic antibodies as therapeutic agents has become even more appealing when it was shown that most of the amino acids in a mouse antibody molecule could be replaced with human sequences, thereby making it compatible for *in vivo* treatment in humans. Furthermore, since antibodies can catalyze specific reactions that are not catalyzed by natural enzymes they can be used *in vivo*, avoiding specific catalytic competition with natural proteins.

Chapter 11 discusses the development of the concept of reactive immunization and its application to the creation of aldolase antibodies *in vivo*. The *in vitro* application of this strategy can be achieved using phage display selections with reactive compounds. Reactive immunization *in vivo* and reactive selection *in vitro* utilize designed reactive compounds that covalently react with antibodies during their induction and selection in such a way as to effectively program a reaction mechanism into the selected protein. The direct consequence of this strategy is the development of efficient catalysts that operate via experimenter-defined mechanisms. This approach has provided highly proficient aldolase antibodies that had not been accessible by traditional immunization with transition state analogs. Broad scope, enhanced catalytic activity, and defined chemical mechanism are three features that distinguish antibodies derived from reactive immunization from those obtained by immunization with transition state analogs.

Chapter 12 is devoted to the newly discovered antibody-catalyzed water oxidation pathway. A central concept within immunology is that antibodies are the key molecular link between recognition and destruction of antigens/pathogens. The antibody catalysis field has demonstrated that the antibody molecule is capable of performing sophisticated chemistry, but there was no compelling evidence that antibodies use this catalytic potential in their normal immune function. It was found recently that all antibodies, regardless of source or antigenic specificity, can catalyze the oxidation of water by singlet oxygen via a pathway, which is postulated to include trioxygen species such as dihydrogen trioxide and ozone, to the ultimate product, hydrogen peroxide. It has been shown that oxidants generated by antibodies and by activated human neutrophils, which are present in inflammatory tissues, can kill bacteria.

Chapter 13 describes the known enzymes and antibodies that catalyze photochemical reactions. Biocatalysis is an attractive and useful strategy by which mechanistic manifolds can be restrained and a reactive intermediate such as excited species in photochemical reactions can be channeled into a single product. The lesson we learn from the very few known natural photoenzymes shows that catalysis originates mainly from entropic stabilization of a productive conformer and from the involvement of a photoactive group. Such groups may be available in the form of either a cofactor or an amino acid residue. The few reported examples of photocatalytic antibodies have demonstrated that this approach can be utilized successfully for the design of novel photocatalysts and this opportunity should be considered seriously, particularly because natural evolution has voted against the development of photocatalytic enzymes.

Chapter 14 highlights the high selectivity of catalytic antibodies, including chemo-, regio-, stereoselectivity, and, in particular, enantioselectivity. The ability of antibodies to discriminate between closely related isomeric transition states have been utilized in asymmetric synthesis and other selective transformations, some of which have no enzymatic equivalent.

Chapter 15 analyzes various examples of catalytic antibodies as mechanistic and structural models of hydrolytic enzymes, emphasizing the ability of antibodies to mimic multiple aspects of enzyme catalysis. Undoubtedly, these research efforts have created unique opportunities to learn about enzyme structure and function, about antibody diversity, and about evolution. Although hydrolytic catalytic antibodies have shown high selectivity, they are still inferior to natural enzymes in terms of catalytic efficiency. It is relatively easy to install in antibodies individual parameters of a catalytic machinery but it is still very difficult to design an optimal catalyst having all catalytic elements operating in concert, including the relative positioning of catalytic residues and the appropriate protein dynamics. The obvious consequence is that the keys for obtaining better catalysts are screening methods for catalysis and techniques of directed evolution.

Chapter 16 focuses on the use of transition state analogs as archetype antigens. Many catalytic antibodies elicited against TSAs have rarely approached the catalytic efficiency of enzymes. Yet, from a pragmatic perspective, the achievement of enzyme-like efficiency is not the ultimate condition for a satisfactory catalytic activity either *in vitro* or *in vivo*. From a chemist's point of view, the ability to program an antibody to catalyze reactions for which there is no enzymatic counterpart, such as the Diels-Alder cycloaddition, the 1,3-dipolar cycloaddition, or the Bergman cycloaromatization reaction, with reasonable rate enhancement and with high regio-, chemo-, stereo-, and enantioselectivity, is a good reason to continue exploring the scope of these remarkable bioctalysts.

Chapter 17 highlights the advantages of using polyclonal catalytic antibodies, which, in contrast with antibodies produced by the hybridoma technology, represent the entirety of the immune response. The study of polyclonal catalytic antibodies has already contributed significantly to the understanding of how the immune system works and of the requirements for successful hapten design. Recently this has led to interesting developments in therapeutic applications, both by active or passive immunization.

Chapter 18 is devoted to the experimental work that is needed for the general production and purification of catalytic monoclonal antibodies via the hybridoma technology. Since antibody catalysis is primarily an experimental science, the keys to success reside in the details of the experimental procedures. Although hybridoma technology is a mature and well-established method, it involves a variety of independent variables and steps. Each step can be carried out in many different ways, and the diversity of the published approaches reflects specific biological problems and experimental tradition that characterize any given laboratory. The literature offers a broad variety of methods, which differ from one another in speed, convenience, reproducibility, and cost. The immunization protocols and the methods of producing monoclonal antibodies described in this chapter have been used continuously over the past two decades in our Scripps laboratories, without claiming that these methods are superior to others.

Finally, **Chapter 19** focuses on naturally occurring catalytic antibodies, which have been surprisingly discovered in the immune system. Such catalysts could function as important mediators of the immunological defense, regulation, and autoimmune dysfunction.

Due to the page limit of this volume, these nineteen chapters describe only some of the most important aspects but certainly do not cover fully the entire science of catalytic antibodies. Likewise, not all the active researchers in the field could be represented in this book. I thank the authors who agreed to participate in this endeavor for their contribution and insight and apologize to the many other scientists who could not be included in the project. It is my hope that this volume will stimulate further research and attract young scientists to join the intriguing field of catalytic antibodies.

Haifa, August 2004   *Ehud Keinan*

List of Contributors

Roey Amir
Department of Organic Chemistry
Faculty of Exact Sciences
Tel-Aviv University
Tel-Aviv, 69978
Israel

Carlos F. Barbas, III
Department of Molecular Biology
The Scripps Research Institute
10550 North Torrey Pines Road
La Jolla, CA 92037
USA
carlos@scripps.edu

Michael Ben-Chaim
48 McKeel Avenue
Tarrytown, NY 10591
USA

G. Michael Blackburn
Department of Chemistry
University of Sheffield
Sheffield S3 7HF
United Kingdom

Ashraf Brik
Department of Chemistry
Technion-Israel Institute of Technology
Technion City
Haifa 32000
Israel

Keith Brocklehurst
School of Biological Sciences
Queen Mary, University of London
Mile End Road
London E1 4NS
United Kingdom

Valentina N. Buneva
Institute of Chemical Biology and Fundamental Medicine
Siberian Division of Russian Academy of Sciences
8, Lavrentiev Ave.
Novosibirsk 630090
Russia

Tobin J. Dickerson
Department of Chemistry
The Scripps Research Institute
10550 North Torrey Pines Road
La Jolla, CA 92037
USA

Zelig Eshhar
Department of Immunology
Weizmann Institute of Science
POB 26

Catalytic Antibodies. Edited by Ehud Keinan
Copyright © 2005 WILEY-VCH Verlag GmbH & Co. KGaA, Weinheim
ISBN: 3-527-30688-9

Rehovot 76100
Israel

Gerard Gallacher
School of Pharmacy and Biomolecular Sciences
University of Brighton
Cockcroft Building, Moulsecoomb
Brighton BN2 4GJ
United Kingdom

Veronique Gouverneur
University of Oxford
The Central Chemistry Laboratory
South Parks Road
OX1 3QH Oxford
United Kingdom
veronique.gouverneur@chem.ox.ac.uk

Donald Hilvert
Laboratory of Organic Chemistry
Swiss Federal Institute of Technology
Universitätsstrasse 16
CH-8092 Zürich
Switzerland
hilvert@org.chem.ethz.ch

Kendall N. Houk
Department of Chemistry and Biochemistry
University of California, Los Angeles
405 Hilgard Ave.
Los Angeles, CA 90095-1569
USA
houk@chem.ucla.edu

Kim D. Janda
Department of Chemistry
The Scripps Research Institute
10550 North Torrey Pines Road
La Jolla, CA 92037
USA
kdjanda@scripps.edu

Ehud Keinan
Department of Chemistry
Technion-Israel Institute of Technology
Technion City
Haifa 32000
Israel
keinan@techunix.technion.ac.il

Diane Kubitz
Department of Chemistry
The Scripps Research Institute
10550 North Torrey Pines Road
La Jolla, CA 92037
USA
dkubitz@scripps.edu

Nicholas A. Larsen
Harvard Medical School
250 Longwood Ave. SGM123
Boston, MA 02115
USA

Andrew G. Leach
Astrazeneca Pharmaceuticals
Mereside Alderley Park
Macclesfield SK10 4TG
United Kingdom

Richard A. Lerner
Department of Chemistry
The Scripps Research Institute
10550 North Torrey Pines Road
La Jolla, CA 92037
USA
rlerner@scripps.edu

Ariel B. Lindner
Department of Immunology
Weizmann Institute of Science
POB 26
Rehovot 76100
Israel

Jonathan E. McDunn
Department of Chemistry

The Scripps Research Institute
10550 North Torrey Pines Road
La Jolla, CA 92037
USA

Georgy A. Nevinsky
Institute of Chemical Biology and Fundamental Medicine
Siberian Division of Russian Academy of Sciences
8, Lavrentiev Ave.
Novosibirsk 630090
Russia
nevinsky@niboch.nsc.ru

Elizabeth L. Ostler
School of Pharmacy and Biomolecular Sciences
University of Brighton
Cockcroft Building, Moulsecoomb
Brighton BN2 4GJ
United Kingdom

Ron Piran
Department of Chemistry
Technion-Israel Institute of Technology
Technion City
Haifa 32000
Israel

Marina Resmini
Department of Chemistry
University of London
Mile End Road
London E1 4NS
United Kingdom
m.resmini@qmul.ac.uk

Jean-Louis Reymond
Department of Chemistry & Biochemistry
University of Berne
Freiestrasse 3
CH-3012 Bern
Switzerland
jean-louis.reymond@ioc.unibe.ch

Sigal Saphier
Department of Chemistry
Technion-Israel Institute of Technology
Technion City
Haifa 32000
Israel

Peter G. Schultz
Department of Chemistry
The Scripps Research Institute
10550 North Torrey Pines Road
La Jolla, CA 92037
USA
schultz@scripps.edu

Doron Shabat
School of Chemistry
Tel-Aviv University
Tel-Aviv, 69978
Israel
chdoron@post.tau.ac.il

Cindy Takeuchi
Department of Immunology
The Scripps Research Institute
10550 North Torrey Pines Road
La Jolla, CA 92037
USA

Fujie Tanaka
Department of Molecular Biology
The Scripps Research Institute
10550 North Torrey Pines Road
La Jolla, CA 92037
USA

Dean J. Tantillo
Department of Chemistry
University of California, Davis
One Shields Ave.
Davis, CA 95616
USA

Dan Tawfik
Department of Biological Chemistry
Weizmann Institute of Science
POB 26
Rehovot 76100
Israel
dan.tawfik@weizmann.ac.il

Anita D. Wentworth
Department of Chemistry
The Scripps Research Institute
10550 North Torrey Pines Road
La Jolla, CA 92037
USA
anitad@scripps.edu

Paul Wentworth, Jr.
Department of Chemistry
The Scripps Research Institute
10550 North Torrey Pines Road
La Jolla, CA 92037
USA
paulw@scripps.edu

Ian A. Wilson
Department of Molecular Biology
The Scripps Research Institute
10550 North Torrey Pines Road
La Jolla, CA 92037
wilson@scripps.edu

Jun Yin
Department of Chemistry
University of California, Berkeley
Berkeley, CA 94720
USA

Xiyun Zhang
Department of Chemistry and Biochemistry
University of California, Los Angeles
405 Hilgard Ave.
Los Angeles, CA 90095-1569
USA

1
Immunological Evolution of Catalysis
Jun Yin, Peter G. Schultz

1.1
Introduction

Both antibodies and enzymes are able to bind a large number of ligands, ranging from small molecules to macromolecules, with high affinity and specificity. Pauling was the first to note that the fundamental difference between the two is that enzymes evolve to selectively bind high-energy transition states and are selected based on catalytic efficiency, whereas antibodies evolve to maximize the affinity for molecules in their ground state [1, 2]. A logical consequence of this comparison is that an antibody generated to a stable analog of a rate-limiting transition state for a particular reaction should act as a selective enzyme-like catalyst. Indeed, there are now many examples in which, with the appropriate chemical instruction, antibodies have been generated that catalyze a large number of reactions, ranging from pericyclic reactions to carbonium ion rearrangements [3]. Importantly, detailed structural and functional studies of the immunological evolution of catalytic antibodies have provided important insights into the evolution of binding energy and catalysis in biological systems. Here we focus on a number of such studies that have been carried out in this laboratory.

1.2
Parallels between Antibody and Enzyme Evolution

There are many parallels between the immunological evolution of antibodies and the natural evolution of enzymes (Table 1.1). Both processes involve genetic recombination and point mutation coupled with a selection process during which molecules with desired function are identified from a large diverse library of proteins. In the case of enzymes, gene duplication and exon shuffling provide proteins with new structures and functions [4, 5]. Random mutations further increase molecular diversity and refine biological function. Similar mechanisms are responsible for the diversity of the antibody molecule (Fig. 1.1) [6–8]. Antibody genes are segmented and exist as groups with multiple members. For an example, in the murine genome there are four functional joining (J) segments for the light chain variable region and four for

Catalytic Antibodies. Edited by Ehud Keinan
Copyright © 2005 WILEY-VCH Verlag GmbH & Co. KGaA, Weinheim
ISBN: 3-527-30688-9

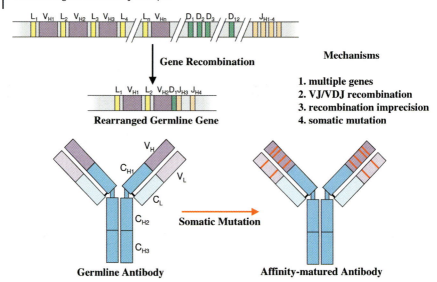

Fig. 1.1 The generation of immunological diversity by genetic recombination and somatic mutation.

the heavy chain, at least 12 diversity (D) segments, hundreds of light chain variable (V_L) segments and heavy chain variable (V_H) segments. Genes for the antibody light and heavy chain variable regions are assembled by the random recombination of V, (D) and J segments; random association of light and heavy chain genes give rise to the germline antibody repertoire. Moreover, the exact site of the joining between segments during gene recombination is imprecise and leads to junctional diversity at the V-(D)-J junctions. Nucleotides can be deleted from or added randomly to both ends of the D gene segments, leading to CDR H3 length variations and even greater variability at the joining region. After the combinatorial assembly of the antibody genes, the germline antibody undergoes affinity maturation during which somatic mutations are introduced throughout the antibody variable domain, further expanding the sequence diversity of the germline antibody repertoire. As a result, the immune system has the potential to generate more than 10^{11} unique antibodies [9] that possess high affinity and specificity toward virtually any ligand.

Despite the similarities in the mechanism used to generate molecular diversity, the two evolutionary processes operate on very different time scales: enzymes typically evolve over millions of years; in contrast, antibody binding energy evolves over a period of weeks. The immunological evolution of a catalytic antibody in which immunological evolution is programmed with chemical information about the rate-limiting transition state of a reaction bridges the gap between these two processes. Indeed, antibodies have been generated that recapitulate the catalytic efficiencies, selectivities, and even mechanisms of enzymes in a number of instances [10, 11]. In other cases, catalytic antibodies have been generated for reactions that are not known to exist in Nature [12, 13].

Because the immune response occurs over a time scale of weeks, one has the opportunity to characterize the entire evolutionary process – from the germline precursor to the affinity-matured catalytic antibody. Such an analysis would allow us to address some fundamental questions regarding the evolution of binding energy and catalysis both in immune response and in the evolution of enzymes. For example, (1) How does the immunoglobulin fold bind so many different chemical structures with high affinity and specificity? Is the binding potential of the antibody molecule a result of sequence diversity alone or are some other mechanisms also operating? Similarly, since more than a hundred reactions have been shown to be catalyzed by antibodies, how does the antibody combining site manage to accommodate such a wide array of catalytic mechanisms? (2) Are there any differences between the binding mode of the germline and that of the affinity-matured antibodies to antigens and what are the functional roles of the somatic mutations, both in the antibody combining site and distal to it, that lead to high affinity and high specificity binding in the affinity-matured antibodies? (3) How do binding energy and catalysis coevolve during the process of immunological evolution of catalytic antibodies, and does this process provide insights into the mechanisms of enzyme evolution?

1.3
Evolution of Catalytic Antibodies

In order to characterize the immunological evolution of catalytic antibodies, both the germline and the affinity-matured antibodies are cloned, sequenced and expressed as the Fab (Fragment, antigen binding) [14]. The sites and identities of the somatic mutations are determined by sequence alignments, and their functional roles can be analyzed by site-directed mutagenesis of the germline and affinity-matured Fab. High resolution X-ray crystal structures of the germline Fab and the affinity-matured Fab with and without hapten bound provide an opportunity to analyze the underlying structural basis for the evolution of binding energy and catalysis. To date, such detailed studies of the immunological evolution of catalytic antibodies have been carried out with five catalytic antibody systems: esterase antibody 48G7 [15–18], Diels-Alderase antibody 39A11 [19–21]oxy-Cope antibody AZ28 [12, 22, 23], sulfur oxidase antibody 28B4 [24–26], and ferrochelatase antibody 7G12 [27–30]. These antibodies were raised against haptens of distinct structures and catalyze a variety of reactions.

1.4
Ferrochelatase Antibody 7G12 – Evolution of the Strain Mechanism

The enzyme ferrochelatase catalyzes the insertion of Fe^{2+} into protoporphyrin IX as the last step in heme biosynthetic pathway [31]. N-alkylporphyrins, in which the porphyrin macrocycle is distorted because of alkylation at one of the pyrrole nitrogen atoms, are strong inhibitors of this enzyme [32]. Based on this observation, it was thus proposed that the enzyme catalyzes the porphyrin metalation reaction by distorting

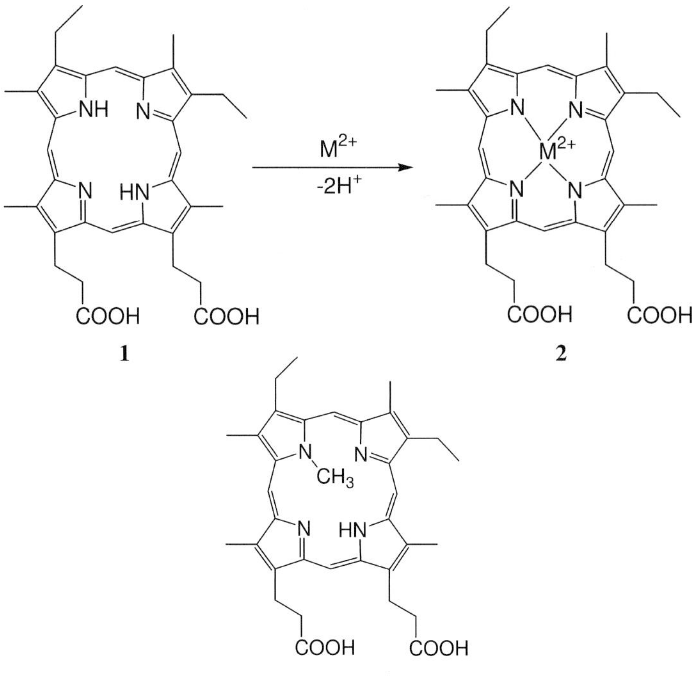

Scheme 1.1

the porphyrin substrate out of planarity so that the pyrrole nitrogen lone pairs are more accessible for metal chelation [33]. To test this notion, antibody 7G12 was generated against an analog of the strained substrate, N-methylmesoporphyrin (**3**) (NMP), and was found to catalyze the metalation of mesoporphyrin (**1**) (MP) by Zn^{2+} with rates comparable to that of the natural enzyme (Scheme 1.1) [27]. The same antibody also catalyzes the insertion of Cu^{2+} into mesoporphyrin at a lower rate. Resonance Raman spectroscopy shows that antibody 7G12 induces distortion in the bound mesoporphyrin substrate corresponding to an alternative up-and-down tilting of the two opposite pyrrole rings [34]. In contrast, the enzyme ferrochelatase induces tilting of all four pyrrole rings in the same direction (doming) [34].

In order to determine the precise conformation of the porphyrin substrate in the antibody combining site, we solved the X-ray crystal structure of 7G12 Fab-MP Michaelis complex to 2.6 Å resolution [29]. Crystals of 7G12 Fab-MP complex give intense red fluorescence upon irradiation with green excitation light around 546 nm (Fig. 1.2). The addition of copper acetate to the crystallization drop leads to a rapid loss of fluorescence due to the formation of the non-fluorescent Cu(II)MP complex, demonstrating that the antibody-MP cocrystal is catalytically active. The F_o–F_c electron density map clearly shows that the substrate MP molecule adopts a non-planar conformation (Fig. 1.3); the out-of-plane displacements for the substrate MP and the hapten NMP are shown in Fig. 1.4. All pyrroles in MP show significant displacements away from the porphyrin least-squares (PLS) plane. Pyrrole A in MP adopts a similar an-

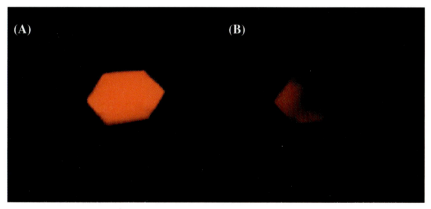

Fig. 1.2 Fluorescent micrographs of a single crystal of 7G12 Fab-MP complex (**A**) before and (**B**) after the addition of copper acetate.

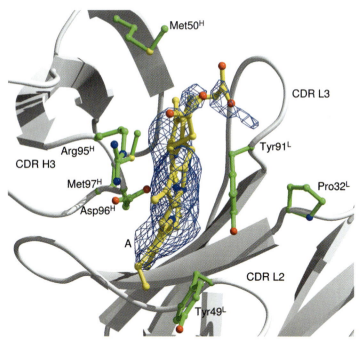

Fig. 1.3 F_o–F_c electron density map contoured at 2.0σ for the substrate MP molecule (colored in yellow) bound with 7G12 Fab. Residues making important packing interactions with MP in the antibody combining site are also shown.

gle (26°) relative to the PLS plane to that observed in NMP [28], however, the other pyrrole rings form larger angles to the PLS plane than do their NMP counterparts (16°, 17°, and 25° for rings B, C, and D, respectively). The porphyrin conformation observed in the crystal structure of Fab-MP complex agrees with prior resonance Ra-

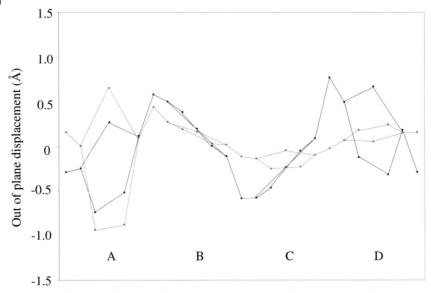

Fig. 1.4 Out-of-plane displacement of porphyrin ring atoms from the porphyrin least-squares (PLS) plane for MP (blue) and NMP (pink) bound to antibody 7G12. The porphyrin atoms that are involved in the same pyrrole ring are connected to give a pentagon shape. **A, B, C,** and **D** denote the porphyrin pyrroles.

man spectroscopy data, which indicates an up-and-down tilting of the pyrrole rings, based on the observation of specific out-of-plane vibration mode [34]. A normal mode decomposition (NCD) analysis, which deconstructs into low-frequency normal coordinate displacements [35], shows a moderate doming (A2u) deformation, as well as strong saddling (B2u) and ruffling (B1u) deformations for the antibody-bound MP.

An analysis of the crystal structure of the 7G12 Fab-substrate MP complex reveals those interactions between the residues in the antibody combining site and MP that lead to substrate distortion (Fig. 1.3). The porphyrin molecule is bound in a cleft, with CDR L2 and CDR L3 forming one side of the cleft and CDR H3 forming the other side. Part of the CDR H3 loop, composed of Arg95^H, Asp96^H and Met97^H, packs on the macrocyclic ring of the porphyrin. The carboxylic oxygen of Asp96^H points toward the center of the porphyrin ring and is within hydrogen-bonding distance of all pyrrole nitrogen atoms. The guanidino group of Arg95^H forms salt bridges with both carboxylates of the propionic acid side chains of the porphyrin ring. The aromatic side chains of Tyr49^L and Tyr91^L π stack on pyrrole rings A and B of the substrate. While the π stacking interaction on one face of pyrrole ring B is balanced by the packing of Asp96^H on the other face, the π stacking interaction with Tyr49^L pushes pyrrole ring A out of plane because of the absence of heavy chain residue contacts on the opposite face. In the crystal structure of Fab-hapten NMP complex, Tyr49^L also packs against N-methyl pyrrole, which is distorted out of plane of the other pyrrole rings because of the N-methyl substitution [28]. Thus, the distortion

of the MP substrate induced by the antibody combining site is a direct result of the
distortion in NMP induced by N-methyl substitution on the free MP. The active site
of antibody 7G12 reflects the "instruction" from the hapten NMP during the process
of immune response *in vivo* and acts to strain the MP substrates *in vitro* and catalyze
porphyrin metalation.

Modern theories of biological catalysis date from Haldane's 1930 treatise "Enzymes", in which he first proposed the strain theory: "using Fischer's lock-and-key simile, the key does not fit the lock perfectly but exercises a certain strain on it" [36]. In other words, a degree of misfit between the enzyme and its bound substrate is needed to distort the substrate toward the transition state conformation. Although the notion that enzymes use binding energy to strain or distort substrates is a fundamental theory of enzyme catalysis, it has proven difficult to validate experimentally [37]. The crystal structure of the 7G12 Fab-MP Michaelis complex provides unequivocal structural evidence for the strain theory proposed by Haldane more than seventy years ago. Thus, the study on the evolution of catalytic antibody not only yields new biological catalysts, but also tests and validates fundamental principles of enzyme catalysis. In addition, the detailed structural and biophysical characterization of the germline and affinity-matured antibodies should provide a detailed mechanistic picture of the evolution of this strain mechanism.

The germline precursor of antibody 7G12 accumulates five somatic mutations during affinity maturation: two in the V_L chain: Ser14^{L}Thr and Ala32^{L}Pro; three in V_H: Arg50^{H}Met, Ser76^{H}Asn and Ser97^{H}Met [28]. The X-ray crystal structures of the Fab fragment of the germline and affinity-matured antibody both unbound and hapten NMP-bound were solved to high resolution [29]. Superposition of the crystal structures of the unliganded and NMP-bound germline Fab shows that there are significant conformational changes in the loop of CDR H3 upon hapten binding (Fig. 1.5A). In the unliganded germline Fab, CDR H3 adopts a relaxed and extended conformation. The side chains of Arg95^H and Asp96^H, which are at the tip of

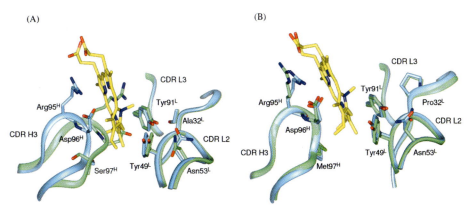

Fig. 1.5 (A) Overlay of the unliganded germline Fab (green) of antibody 7G12 and the germline Fab-hapten NMP 3 complex (blue). (B) Overlay of the unliganded affinity-matured Fab (green) of antibody 7G12 and the affinity-matured Fab-hapten NMP 3 complex (blue). Hapten 3 is in yellow.

(A) 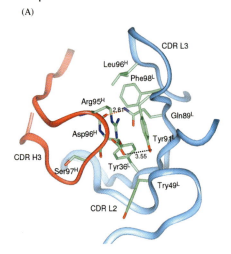 (B)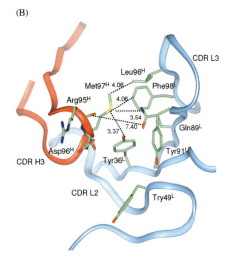

Fig. 1.6 (**A**) In the 7G12 germline Fab without NMP bound, CDR H3 extends into the antibody combining site, and a hydrogen bond (red dotted lines) is formed between the backbone CO of Arg95^H and carboxamide NH of Gln89^L. (**B**) In the 7G12 affinity-matured Fab without NMP bound, CDR H3 is kinked because of Ser97^{H}Met somatic mutation, and the antibody combining site is preorganized for NMP binding. Dotted lines with numbers show the distance (Å) between the two atoms. NMP is in yellow.

CDR H3 loop, move into the hapten-binding site, occupying the space taken by the hapten in the germline Fab-NMP complex. There is a hydrogen bond between the backbone carbonyl oxygen of Asp96^H and carboxamide NH of Gln89^L (Fig. 1.6A). Upon NMP binding, the backbone of CDR H3 loop is pushed backward toward the side of the heavy chain by roughly 4 Å (at Cα of Arg95^H) because of the insertion of NMP molecule into the hapten-binding pocket. There is even larger movement for the side chains of Arg95^H (7.0 Å at the guanidino C) and Asp96^H (5.5 Å at the carboxamide C) upon NMP binding. In the germline Fab-NMP complex, Arg95^H packs onto the macrocyclic ring of NMP and the carboxylate group of Asp96^H is positioned equidistant from the four pyrrole nitrogen atoms of NMP (2.6–3.0 Å) and may form hydrogen bonds with one or more of the pyrrole NH groups. The structural flexibility of germline antibody is also manifested by comparing the electrostatic surface of the antibody combining site before and after the binding of NMP (Fig. 1.7). In the hapten-free form, the germline antibody combining site is flat with minor electrostatic charges distributed on the surface. However, upon hapten binding, the movement of the CDR H3 loop creates a cavity on the surface of the germline antibody that complements the shape of the distorted porphyrin molecule. Exposure of residues Arg95^H and Asp96^H also introduces negative charges inside the cavity and positive charges at the rim, which complement the charge distribution of NMP. These results suggest the germline antibody of 7G12 adopts an "induced fit" binding mode and the conformational changes in the germline antibody upon hapten binding lead to enhanced complementarity between the antibody combining site and the hapten.

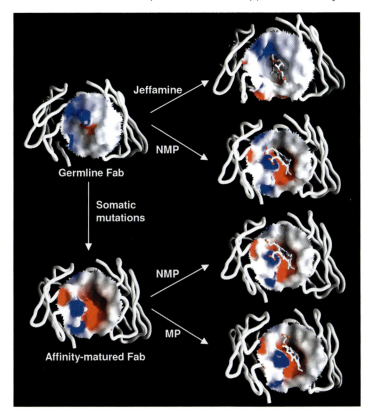

Fig. 1.7 Electrostatic surface potential of the antibody combining site in the germline and affinity-matured Fab either unliganded or bound with NMP, MP or jeffamine. The red and blue correspond to negative and positive surface potential, respectively.

The conformational flexibility of the germline antibody is further confirmed by a crystal structure of the germline Fab-jeffamine complex [30].

Jeffamine $(CH_3OCH_2CH_2O(CH(CH_3)CH_2O)_nCH_2CH(NH_2)CH_3, n \approx 8)$, which is structurally very distinct from the hapten NMP, was found to bind the germline antibody of 7G12 but not the affinity-matured antibody. The crystal structure of the germline Fab-jeffamine complex shows that five isopropoxy units of the jeffamine molecule fold into a U-shaped conformation and are deeply embedded in the antibody combining site consisting of mainly hydrophobic residues including $Tyr36^L$, $Leu46^L$, $Tyr49^L$, $Tyr91^L$, $Leu96^L$ and $Phe98^L$ from the light chain and $Thr93^H$ and $Trp100^H$ from the heavy chain (Fig. 1.8). A hydrogen bond is formed between O6 of jeffamine and the indole $N\epsilon1$ of $Trp100^H$ (2.8 Å). $His35^H$ and $Arg95^H$ also pack on the poly(isopropoxy) backbone of the jeffamine molecule. Superposition of the unliganded germline Fab and the germline Fab with bound jeffamine or bound hapten NMP reveals that the heavy chain CDR loops undergo an even larger rearrangement upon binding jeffamine than the binding of NMP (Fig. 1.9A). The Cα of $Asp96^H$ of CDRH3 moves 10.9 Å away from the light chain in order to accommodate jeffamine.

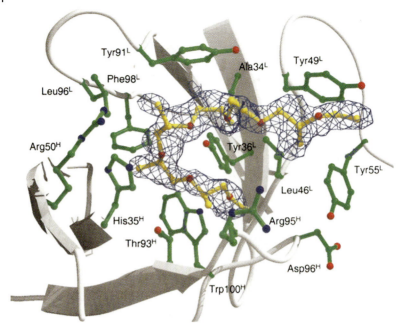

Fig. 1.8 $F_o–F_c$ electron density map contoured at 2.0σ for the jeffamine molecule (colored in yellow) bound to the germline Fab of antibody 7G12. Residues making important packing interactions with jeffamine in the antibody combining site are also shown.

This forces a conformational change of CDRH1 to avoid steric clashes between the phenol residue of Tyr27^H and the carboxyl group of Asp96^H. As a result of antibody combining site reorganization, two different sets of residues are used by the germline antibody to interact with either NMP or jeffamine (Fig. 1.10). The charge distribution and the shape of the antibody combining site in the germline Fab-jeffamine complex is very different from the that of the same antibody when there is no ligand bound or with its hapten NMP bound (Fig. 1.7). Thus the germline antibody of 7G12 has an intrinsic conformational flexibility in the heavy chain CDR loops; binding of either hapten or structurally distinct ligands induces CDR loop rearrangements that increase complementarity between the antibody and the ligands. Importantly, the conformational diversity of the germline antibody gives rise to binding polyspecificity and plays a significant role in expanding the germline-binding repertoire.

In contrast to the germline antibody, a comparison of the crystal structures of the unliganded and NMP-bound 7G12 Fab indicates that minimal changes occur upon hapten binding in the affinity-matured antibody (Fig. 1.5B). Neither the shape nor the electrostatic characteristics of the antibody combining site of the affinity-matured antibody change significantly upon the binding of the hapten or the substrate molecule (Fig. 1.7), suggesting that the antibody combining site in the affinity-matured Fab is rigid and preorganized for the binding of distorted porphyrins. This "lock-and-key" binding mode [38] of the affinity-matured antibody versus "induced fit" binding [39]

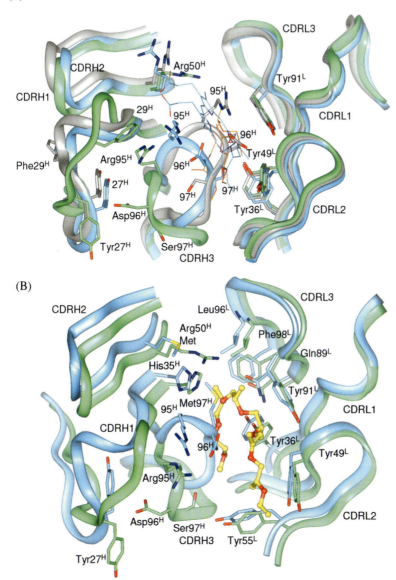

Fig. 1.9 (A) Overlay of the unbound germline Fab (gray) of antibody 7G12, the germline Fab-NMP complex (blue), and the germline Fab-jeffamine complex (green). The heavy chain CDR loops of the germline antibody undergoes significant conformational changes upon the binding of NMP (blue lines) or jeffamine (orange). (B) Overlay of the germline Fab-jeffamine complex (green) and the unliganded affinity-matured Fab (blue). Jeffamine is drawn in yellow and in ball and stick.

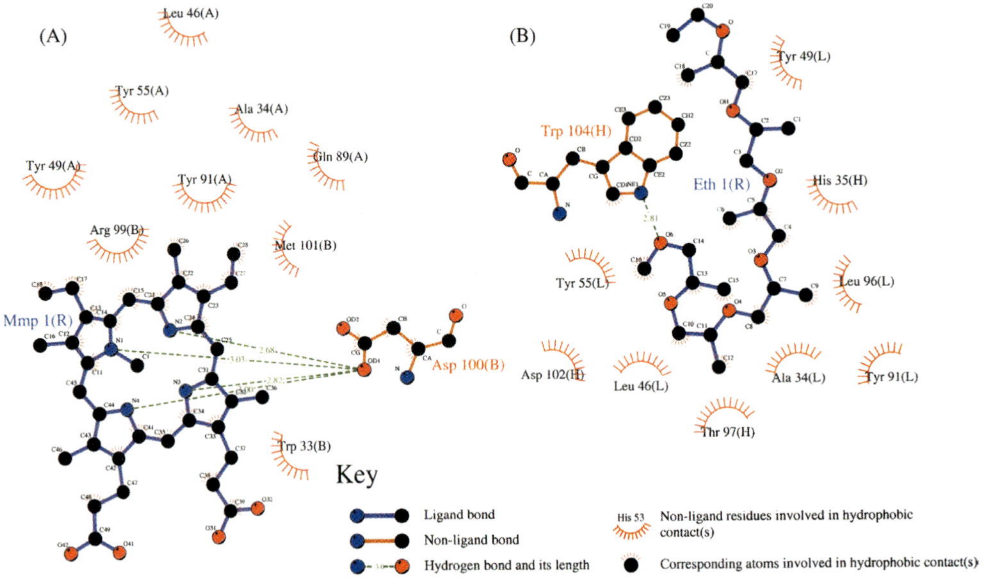

Fig. 1.10 Active-site residues responsible for the binding with **(A)** NMP and **(B)** jeffamine in the antibody combining site of the 7G12 germline Fab.

in the germline antibody was previously observed in the antibodies 48G7 [17], AZ28 [23] and 28B4 [26]. In all cases, the affinity-matured antibodies show between 40 (AZ28) and more than 30 000 (48G7) times higher hapten-binding affinity relative to the germline antibody. The decrease in K_d and the correspondingly more negative free energy of binding (ΔG) partially originate from a more favorable entropy term ($-T\Delta S$) due to a preorganized combining site. The conformational flexibility of the germline antibody leads to a higher entropic penalty for binding of ligand, since side-chain motion must be restricted upon complex formation.

The structural rigidity in the affinity-matured antibody of 7G12 is also manifested in catalysis [29]. When the substrate MP binds to the affinity-matured antibody 7G12, it is forced to adopt a strained conformation in order to fit the antibody combining site (Fig. 1.7), rather than the antibody combining site reorganizing to accommodate the substrate in its planar conformation. Analysis of the crystal structures of the germline and affinity-matured Fab also provides insights into the mechanism by which this strain mechanism evolved. The somatic mutation Ser97^{H}Met is located in CDR H3 and leads to a sharp turn in the backbone of CDR H3 in the affinity-matured Fab resulting from the packing interactions of Met97^H with the side chains of Tyr36^L, Gln89^L, Leu96^L and Phe98^L (Fig. 1.6B). This maintains the position of CDR H3 loop out of the active site and places residues Arg95^H and Asp96^H in the proper positions to interact with the NMP molecule upon its binding to the affinity-matured antibody. Thus, somatic mutation Ser97^{H}Met fixes the conformation of CDR H3 for NMP binding; in its absence in the germline antibody the CDR H3 loop is flexible.

The same somatic mutation also renders the affinity-matured antibody with higher binding specificity: the side chain of Met97^H occupies the position that C10 to C14 of jeffamine binds in the germline antibody and at the same time fixes the conformation of CDR H3 for specific NMP recognition (Fig. 1.9B). As a result, the affinity-matured antibody does not bind jeffamine to any measurable degree, but has almost 100-fold higher affinity for NMP than the germline antibody. The methionine side chain introduced by the Arg50^{H}Met somatic mutation packs against the imidazole ring of His35^H and orients its side chain to pack against Met97^H, which plays an important role in organizing the conformation of CDR H3 for hapten binding. Thus, somatic mutation Arg50^{H}Met in combination with the Ser97^{H}Met somatic mutation helps to fix the conformation of CDR H3. Finally, the Ala32^{L}Pro somatic mutation increases the steric bulk of the residue at this site in order to reinforce the packing interactions of the antibody with the hapten (Fig. 1.5). In the affinity-matured Fab, Pro32^L packs on Tyr49^L, which packs directly on pyrrole ring A of NMP bound to the antibody and plays a crucial role in forcing pyrrole A out of the plane of the porphyrin ring in the Fab-MP complex (Fig. 1.3).

The hapten binding and catalytic properties of the germline Fab, affinity-matured Fab and a number of somatic mutants were measured to determine how the structural changes associated with affinity maturation affect binding and catalysis [29]. The 7G12 Fab binds NMP with a dissociation constant (K_d) of 20.7 nM, 95-fold lower than the K_d of the germline Fab (1.96 μM). The value of k_{cat}/K_m of the antibody-catalyzed porphyrin metalation reaction also increases 92-fold from the germline Fab to the affinity-matured Fab. Somatic mutation Ser97^{H}Met fixes the conformation of CDR H3 and preorganizes the antibody combining site for the binding of a strained porphyrin molecule. The Met97^{H}Ser mutant of the affinity-matured Fab reverses this somatic mutation and results in a more flexible antibody combining site that should exert less strain on a porphyrin substrate. Consistent with the structural analysis, the Met97^{H}Ser mutant has a K_d of 373 nM for the binding of NMP, an 18-fold decrease in binding compared to the affinity-matured Fab. Also, this mutant catalyzes the insertion of Cu^{2+} into MP with a K_m of 191 μM and a k_{cat} of 2.2 h^{-1}, corresponding to a virtually unchanged K_m and a tenfold reduced k_{cat} relative to the affinity-matured Fab (K_m = 150 μM and k_{cat} = 24.2 h^{-1}). This suggests that the structural flexibility introduced into the antibody combining site by Met97^{H}Ser mutation does not affect the binding of the Fab to the substrate in the planar ground state (K_m unchanged), but decreases the ability of the antibody combining site to distort the substrate toward the non-planar transition state. This is manifested by an additional 1.4 kcal/mol free energy of activation for the Met97^{H}Ser-catalyzed reaction compared to the affinity-matured antibody. Thus, the catalytic activity for porphyrin metalation appears to have evolved as a consequence of binding affinity for a distorted porphyrin (NMP). Somatic mutations lead to a rigid antibody combining site preorganized to bind the substrate MP in a strained, nonplanar conformation. However, the ability to bind porphyrin in a distorted conformation is compromised in the germline Fab because of the structural flexibility in the antibody combining site, resulting in a correspondingly lower catalytic activity. In summary, the affinity maturation of antibody 7G12 in response to the strained substrate mimetic NMP resembles the evolution of enzymatic function, in which binding energy is evolved to lower the activation energy of a reaction, in this case by straining the substrate.

Similar examples of conformational plasticity can be found in other proteins, allowing them to bind a large number of different protein and small molecule ligands using the same molecular surface. For an example, human growth hormone (hGH) not only binds and activates human growth hormone receptor (hGHR) but also binds and activates prolactin receptor despite the low sequence identity between hGH and prolactin (23%) as well as between their receptors (28%) [40]. Structural analysis of hGH-hGHR complex and hGH-prolactin receptor complex shows that hGH uses virtually the same set of contact residues to bind both partners. Recently it was shown that the hinge region on the Fc fragment of human immunoglobulin G interacts with four different protein scaffolds that bind at a common site between the C_{H2} and C_{H3} domains [41]. Moreover, some enzymes such as hexokinase [42, 43] and triosephosphate isomerase [44] function by an induced fit mechanism in which binding of substrate induces a conformational change in the active site that leads to enhanced catalytic rate. Perhaps this structural plasticity is a remnant of early proteins. The ability of a receptor to alter conformation to bind multiple ligands would have allowed a limited number of proteins to bind a large number of ligands or substrates. Point mutations, like somatic mutations during the affinity-maturation of antibodies, coupled with the proper selection pressure, would then fix a particular active site conformation to bind a specific ligand or catalyze a particular reaction. Indeed the ability to evolve so many different binding and catalytic activities from the antibody framework reinforces this hypothesis for the evolution of protein function.

Not only do studies of catalytic antibodies provide insights into the evolution of the binding and catalytic function of enzymes, they also provide fundamental insights into the molecular basis of the immune response itself. Over half a century ago there was considerable debate over the mechanisms by which the immune system is able to evolve selective, high affinity receptors for an almost infinite number of ligands. Once it was established that the immune system produces a large number of antibodies with different sequences through recombination and somatic mutation, sequence diversity was widely accepted as the basis for the tremendous binding potential of antibody repertoire. However, Haurowitz, Pauling, indexPauling, L.and others argued that conformational diversity could also account for the virtually infinite binding potential of the antibody molecule [45, 46]. Just as a human hand can bind and adapt its shape to a large number of structures, so could an antibody active site change its shape to complement a virtually infinite number of ligands. This theory was termed the chemical instruction theory. Our studies on the process of antibody affinity maturation reveals that conformational diversity does indeed play a key role in expanding the structural diversity of the germline antibody repertoire, allowing the germline antibody to adopt many different structures (and ligand binding modes). The somatic mutations acquired by the affinity-matured antibody act to both fix the conformation of the CDR loops for specific hapten binding and to introduce side chains that interfere with the binding of non-hapten ligands so as to render the affinity-matured antibody more specific. This may be a general strategy used by the immune system to achieve both highly diverse germline antibody binding repertoire while at the same time allowing rapid selection for antibodies with high affinity and specificity.

1.5
Esterase Antibody 48G7 – Effect of Distant Mutations on Catalysis

Antibody 48G7 was raised against *p*-nitrophenyl phosphonate transition state analog (**8**) and catalyzes the hydrolysis of the corresponding *p*-nitrophenyl ester and carbonate (**4**) with rate accelerations exceeding 10^4-fold (Scheme 1.2) [15]. Nine somatic mutations were accumulated during the affinity maturation process of antibody 48G7, six in the heavy chain (Glu42^{H}Lys, Gly55^{H}Val, Asn56^{H}Asp, Gly65^{H}Asp, Asn76^{H}Lys, and Ala78^{H}Thr) and three in the light chain (Ser30^{L}Asn, Ser34^{L}Gly, and Asp55^{L}His) [15]. The affinity-matured Fab of 48G7 binds hapten with a K_d of 4.5 nM, whereas its germline precursor Fab has a K_d of 135 μM, 30 000 times higher than the affinity-matured Fab. For the antibody-catalyzed ester hydrolysis reaction, the affinity-matured Fab has a k_{cat}/K_m value 100-fold higher than the germline Fab [15].

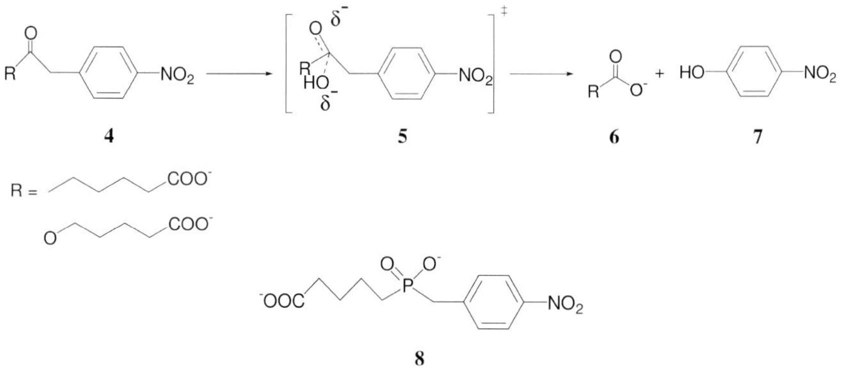

Scheme 1.2

Comparison of the X-ray crystal structures of the unliganded and hapten **8** complexed 48G7 Fab reveals that very few structural changes occur in the affinity-matured antibody upon hapten binding (Fig. 1.11B). In contrast, binding of hapten **8** to the germline Fab again leads to significant conformational changes in the antibody combining site, especially in CDR H3 [17] (Fig. 1.11A): the side chain of Tyr99^H on CDR H3 moves 6 Å away from the hapten binding site at phenyl OH to make room for the incoming hapten, and the side chain of Tyr98^H moves 8.3 Å at phenyl OH and inserts between Tyr99^H and Tyr33^H, which also moves toward the phosphonate group of the hapten. These movements establish a π-cation interaction between the side chains of Arg46^L and Tyr99^H, a π–π interaction between the aryl groups of Tyr99^H and Tyr98^H, and a T-stack interaction between the aryl rings of Tyr98^H and Tyr33^H. All these interactions help to stabilize the conformation of the CDR loops in the germline antibody for hapten binding.

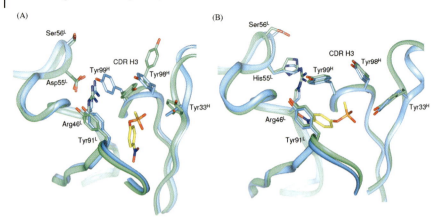

Fig. 1.11 (A) Overlay of the unliganded germline Fab (green) of antibody 48G7 and the germline Fab-hapten **8** complex (blue). (B) Overlay of the unliganded affinity-matured Fab (green) of antibody 48G7 and the affinity-matured Fab-hapten **8** complex (blue). Hapten **8** is in yellow.

However, in the affinity-matured antibody, the same CDR loop conformations are fixed by somatic mutations, and, as a result, the antibody combining site is preorganized for hapten binding. Two somatic mutations, Gly55^{H}Val and Asn56^{H}Asp, are mainly responsible for fixing the conformation of CDR H1, CDR H2 and CDR H3 in the affinity-matured antibody, despite the fact that both of them are some 15 to 20 Å from the bound hapten (Fig. 1.12). Somatic mutation Gly55^H Val changes the back-

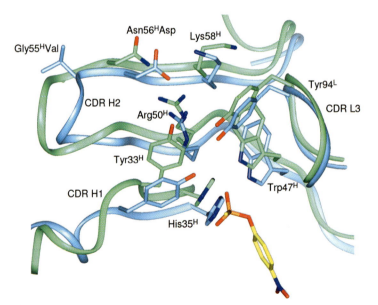

Fig. 1.12 Superposition of the structures of the 48G7 germline Fab-hapten **8** complex (green) and the affinity-matured Fab-hapten **8** complex (blue), illustrating the changes that occur as results of somatic mutations Gly55^{H}Val and Asn56^{H}Asp. Hapten **8** is in yellow.

bone conformation of CDR H2 loop and leads to two salt bridges formed between the ε-amino group of Lys58^H and the carboxyl group of Asp56^H which is introduced by Asn56^{H}Asp somatic mutation. These interactions stabilize the conformation of Asp56^H which is part of the hydrogen bond network involving Asp56^H, Arg50^H, Tyr33^H, His35^H, Trp47^H, and Tyr94^L that fix the conformation of heavy chain CDR loops for optimized hapten binding. Other somatic mutations including Ser30^{L}Asn, Glu42^{H}Lys, Gly65^{H}Asp, Asn76^{H}Lys, and Ala78^{H}Thr are also distant from the bound hapten, yet they reconfigure active site residues involved in binding interactions with hapten by reorganizing networks of hydrogen bonding, electrostatic and Van der Waals interactions between variable region residues over distances as long as 20 Å.

Thus, the structural studies on antibody 48G7 and its germline precursor again suggest that the germline antibody has an intrinsic structural flexibility and may undergo significant conformational changes to achieve better complementarity with the hapten. Somatic mutations serve to remove the CDR loop flexibility and preorganize the antibody combining site for hapten binding. These studies have also shown that somatic mutation can be either in the active site or significantly removed in distance, affecting ligand binding through coupled secondary sphere interactions such as sophisticated hydrogen bond networks. Similarly, mutational studies of enzymes have shown that one can significantly affect the binding and catalytic properties of enzymes through mutations outside the active site [47, 48], in much the same way as somatic mutations throughout the antibody variable region affect the antibody binding affinity. For an example, it was found in dihydrofolate reductase that two mutations, Arg44Leu and His45Gln, perturb the structure of the protein and elevate the pK_a of a folate binding site residue Asp27 (25 Å away from the site of the mutations) by one and two pH units, respectively [49]. In the case of β-lactamase, mutation Met182Thr, which is located 17 Å from the enzyme active site, is found to be responsible for the 500-fold increase in antibiotic resistance in combination with the two other mutations [50]. Structural studies suggest that the threonine hydroxy group introduced by this mutation forms two new hydrogen bonds with the backbone carbonyl groups of Glu63 and Glu64 and helps to fix the position of the catalytic residues [51]. The long-range effects of mutations in antibody and enzyme active sites suggest that distal regions of protein structures can be highly interconnected. This may be the structural basis for the evolution of allosteric binding sites in enzymes, in which changes at one binding site affect the binding of ligands at other sites that are spatially removed from each other.

The somatic mutations in antibody 48G7 have also been found to effect hapten binding in a context-dependent and cooperative manner. For example, analysis of the 48G7 crystal structure suggested that the somatic mutation Gly55^{H}Val may interact cooperatively with the somatic mutation Asn56^{H}Asp, since changes in the conformation of the turn in which residue 56^H resides appear to be induced by mutations at position 55^H (Fig. 1.12). The single reversion mutations Val55^{H}Gly and Asp56^{H}Asn were made in the affinity-matured antibody, as well as the double mutation in which both somatically introduced residues at 55^H and 56^H were switched back to their germline identities. Simple additive effects of the two single reversion mutations predicted a 0.4 kcal/mol loss in free energy of binding. However, the actual loss of binding energy observed for 48G7 Val55^{H}Gly/Asp56^{H}Asn double mutant is 1.0 kcal/mol, with an additional 0.6 kcal/mol in binding free energy due to cooperativity between the sites of 55^H and 56^H. Cooperativity was also found between other proximal pairs of somatic mutations (30^L/34^L and 76^H/78^H), as well as between a non-proximal pair (55^L/76^H). Thus, the sources of a global gain in function may not always be delineated through changes in structure produced by individual mutations; cooperativity amongst somatic mutations may be another mechanism by which the immune system produces large increases in binding affinity for the affinity-matured antibody. This is well illustrated by the affinity 28B4.

1.6
Sulfur Oxidase Antibody 28B4 – Incremental Changes in Evolution

Antibody 28B4 catalyzes the periodate-dependent oxidation of sulfide **9** to sulfoxide **11** (Scheme 1.3) [25]. Hapten **12** was designed to mimic the transition state of the oxidation reaction using periodate as the cofactor. A total of nine replacement mutations, two in the light chain and seven in the heavy chain, occurred during the evolution of this antibody [25]. The X-ray crystal structures of the unliganded and hapten-bound germline Fab of antibody 28B4 were determined [26]. Comparison with the corresponding structures of the affinity-matured Fab of 28B4 [25] reveals the site of somatic mutations relative to the antibody combining site. Three of the mutations, Ser35^{H}Asn, Asn53^{H}Lys, and Asp95^{H}Trp, occur at the hapten-binding site; four mutations, Ser25^{L}Phe, Met34^{H}Phe, Val37^{H}Ala, and Ser76^{H}Gly, are one shell removed from the residues that form the hapten-binding pocket. The other two, Pro40^{L}Ser and Val12^{H}Gly, are close to the constant region of the antibody and far away from the bound hapten.

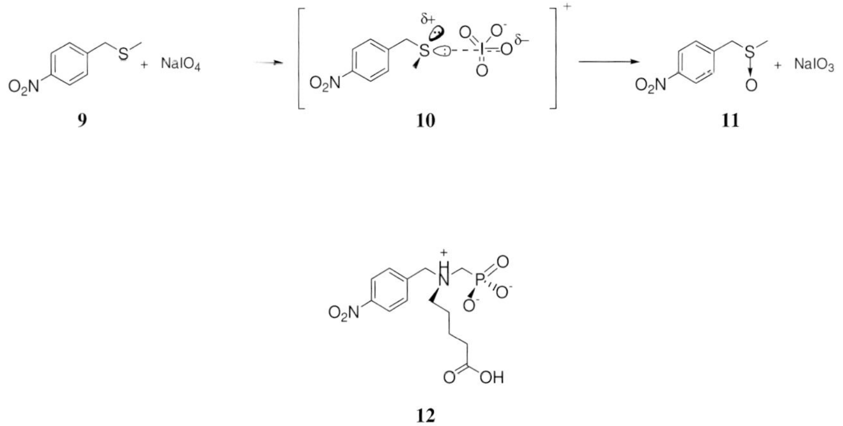

Scheme 1.3

Comparison of unbound and hapten 12-bound germline Fab structures again shows that there are significant changes in the loops of CDR H3 and CDR L1 upon hapten binding. In the hapten-bound structure, residues H95 to H99 of CDR H3 are shifted away from hapten 12. In the absence of such a CDR H3 conformational change, the backbone of Tyr98^H would sterically clash with the *p*-nitro group of the bound hapten. The backward movement of the backbone of CDR H3 (a maximum of 3.7 Å between the Cα positions of Tyr98^H in the two structures) removes this unfavorable steric interaction and at the same time introduces a new hydrogen bond between the backbone amide of Ala99^H and the *p*-nitro oxygen atom of hapten 12. Asp95^H in this loop also makes hydrophobic contacts to the *p*-nitrophenyl ring of hapten 12. There are also significant movements of residues L27c to L32 in the CDR L1 loop. Thus, a flexible, induced-fit type of binding mode is adopted by the germline antibody of 28B4 as with the other germline antibodies discussed so far. Comparison of the unliganded and hapten 12-bound affinity-matured 28B4 structures shows that minimal changes occur upon hapten binding. This is again consistent with a lock-and-key mechanism of binding.

A comparison of the structures of the hapten-bound germline and affinity-matured Fab of antibody 28B4 reveals that hapten 12 is bound in different orientations in the two antibodies (Fig. 1.13). The somatic mutations at the antigen combining site of the germline antibody, Ser35^{H}Asn, Asn53^{H}Lys, and Asp95^{H}Trp, are mainly responsible for the new orientation adopted by the bound hapten in the affinity-matured antibody. While the phosphonate group of hapten 12 is bound in the same orientation in the two Fabs, the *p*-nitrophenyl ring of hapten 12 in 28B4 Fab is rotated relative to its orientation in the germline Fab and forms a parallel π-stacking interaction with the indole ring of Trp95^H which was introduced by somatic mutation. The *p*-nitro group of the hapten is rotated away from Glu34^L and forms a new hydrogen bond with the carboxamide group of somatically incorporated Asn35^H. The Asn53^{H}Lys somatic mutation introduces two new hydrogen bonds between the ε-amino group of Lys53^H and the phosphonate moiety of hapten 12. Thus, as a result of the active site mutations, hapten 12 has gained two additional hydrogen bonds to the phosphonate group and a higher degree of packing complementarity with residues Trp95^H, Phe50^H, Tyr32^L, and Tyr36^L. This is consistent with the 700-fold increase in hapten binding affinity of the affinity-matured antibody relative to the germline antibody.

An analysis of the binding affinities of pair-wise germline mutations reveals that their effects on hapten binding are coupled (Fig. 1.14). The single Asp95^{H}Trp mutation on the germline Fab results in a K_d of 450 nM, a 55-fold net gain in binding affinity relative to the germline Fab, whereas the K_d values of the Ser35^{H}Asn and Asn53^{H}Lys mutants are 31 µM and 27 µM, respectively, virtually unchanged from the germline Fab. This result suggests that the Asp95^{H}Trp mutation is largely responsible for switching the hapten from the germline to the affinity-matured binding orientation (Fig. 1.13). The π-stacking interactions involving Trp95^H and Tyr50^H, which result from the Asp95^{H}Trp mutation, are clearly responsible for the altered binding geometry of the hapten in the affinity-matured antibody. In the germline Fab in which residue 95^H is Asp, the hapten is bound in a geometry far away from residues 35^H and 53^H, such that the somatically mutated residues Asn35^H and Lys53^H are too dis-

(A)

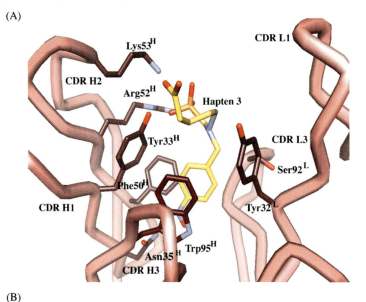

(B)

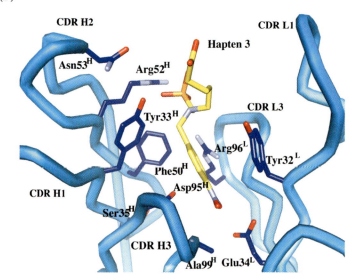

Fig. 1.13 Active sites of (**A**) the affinity-matured 28B4 Fab-hapten **12** complex (red) and (**B**) the germline Fab-hapten **12** complex (blue). Hapten **12** is in yellow.

tant to have any positive effect on hapten binding (Fig. 1.13). Thus, single somatic mutations at these sites alone do not increase the binding affinity of the germline antibody for the hapten. Indeed, the crystal structure of the germline antibody-hapten **12** complex shows that the closest distances between hapten **12** and Ser35^H and Asn53^H are 9.2 Å and 7.2 Å, respectively. However, the Asp95^{H}Trp mutation leads

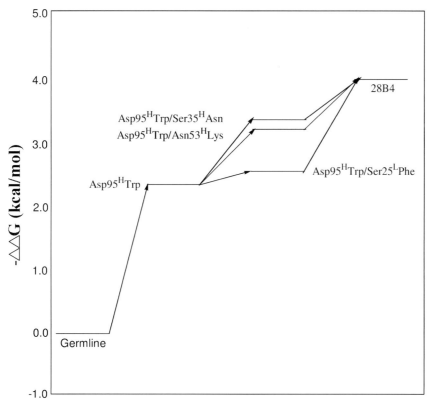

Fig. 1.14 Potential energy diagram showing that stepwise acquisition of somatic mutations by 28B4 germline antibody is accompanied by stepwise increase in binding affinity between the antibody and hapten **12**.

to a different orientation of the bound hapten, which allows Asn35^H and Lys53^H to make direct interactions with hapten **12**. The double mutants Ser35^{H}Asn/Asp95^{H}Trp and Asn53^{H}Lys/Asp95^{H}Trp have K_d values of 94 nM and 110 nM, respectively, corresponding to a roughly five-fold gain in binding affinity over the single somatic mutant Asp95^{H}Trp (K_d = 450 nM).

These observations suggest that the somatic mutations Ser35^{H}Asn and Asn53^{H}Lys must have been introduced into an intermediate which had already acquired the Asp95^{H}Trp mutation during previous rounds of mutation and selection. Thus, there is a stepwise acquisition of functional mutations by the germline antibody that concomitantly results in a stepwise increase in hapten-binding affinity (Fig. 1.14). This is most likely accomplished through an iterative cycle of mutation, affinity selection, and clonal expansion. A similar context-dependent effect has also been seen with the 48G7 antibody, [18] in which significant was found between pairs of somatically mutated residues: together, the mutations have a much stronger favorable effect on hapten binding than a simple sum of the two individual effects.

One of Darwin's fundamental conclusions with regard to evolution is gradualism [52]. This theory appears to be applicable to the evolution of binding and catalysis in proteins as suggested by the evolution of catalytic antibody 28B4. Somatic mutations improve binding and/or catalytic function *incrementally* rather then in discontinuous jumps. In this way beneficial mutations from the previous round of selection are retained and incorporated in the next round of evolution. This realization is significantly changing the current mutagenesis strategies for the *in vitro* evolution of enzymes to better reflect those used by natural evolution [50].

1.7
Oxy-Cope Antibody AZ28 – Evolution of Conformational Diversity in Catalysis

The oxy-Cope reaction is a unimolecular [3,3] sigmatropic rearrangement that is widely used in organic synthesis but is not catalyzed by any known enzyme. Immunization with hapten **16**, a chair-like analog of the putative pericyclic transition state, led to the generation of catalytic antibody AZ28 (Scheme 1.4) [12]. This antibody catalyzes the unimolecular oxy-Cope rearrangement of substrate **13** to product **15** with a rate acceleration (k_{cat}/k_{uncat}) of 5300, a K_m of 74 µM, and a K_d of 17 nM for hapten **16** [22]. During affinity maturation, the germline antibody acquired two somatic mutations in the light chain (Ser34^{L}Asn and Ala51^{L}Thr) and four in the heavy chain (Tyr32^{H}Phe, Ser56^{H}Gly, Asn58^{H}His, and Thr73^{H}Lys). As expected, the germline antibody has a lower binding affinity for hapten (K_d = 670 nM) than AZ-28, but surprisingly a higher catalytic rate: k_{cat}/k_{uncat} = 163 000 and K_m = 73 µM [22]. These values are close to that of chorismate mutase, which catalyzes the related [3, 3]-sigmatropic rearrangement of chorismate to prephenate [53].

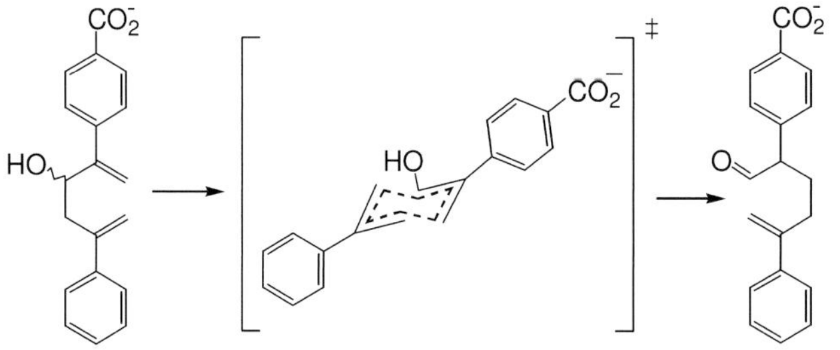

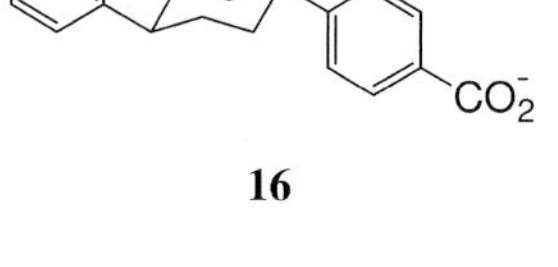

Scheme 1.4

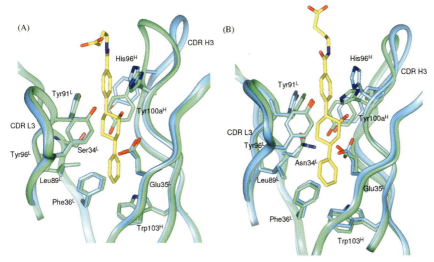

Fig. 1.15 (A) Overlay of the unliganded germline Fab (green) of antibody AZ28 and the germline Fab-hapten **16** complex (blue). (B) Overlay of the unliganded affinity-matured Fab (green) of antibody AZ28 and the affinity-matured Fab-hapten **16** complex (blue). Hapten **16** is in yellow.

To assess the structural basis for the decreased affinity but increased catalytic efficiency of the germline antibody, the X-ray crystal structures of the germline Fab and affinity-matured AZ28 Fab with and without hapten **16** bound were determined (Fig. 1.15) [23]. In the affinity-matured antibody, hapten **16** again binds in a lock-and-key mode with packing interactions between the antibody and 2,5-aryl substituents locking the substrate in the desired cyclic conformation. This is confirmed by NMR studies with TRNOE (transferred nuclear Overhauser effects) measurements showing that AZ28 preorganizes the normally extended substrate into a cyclic conformation so that its termini are in close proximity [54]. This leads to a cyclic alignment of the 4π + 2σ orbital system of the hexadiene core upon its binding to the antibody, lowering the overall entropy of activation ($\Delta S^{\neq}$) of the oxy-Cope reaction.

In addition, electronic effects arising from the 3-hydroxyl and 2,5-diphenyl substituents also affect the energetics of 3,3-sigmatropic rearrangement reactions. It is well known that anionic substituent effects accelerate the oxy-Cope rearrangement through hyperconjugation of electron density on oxygen [55]. Thus, His96^H which hydrogen bonds to the 3-hydroxyl group of the substrate might act to enhance the rate of the rearrangement by increasing the electron density on the oxygen substituent of substrate **13**. Upon binding to the affinity-matured antibody, the cyclohexyl ring of hapten **16**, which mimics the cyclic 4π + 2σ transition state, is rotated out of the planes of the 5- and 2-phenyl rings by 81° and 83°, respectively. The two phenyl substituents are rotated with respect to each other by a dihedral angle of 19° and their conformations are fixed by the contacts with active-site residues, especially the π-stacking interactions between the 2-aryl substituent and His96^H, Tyr91^L, and Tyr100a^H. The aryl substituents at the 2 and 5 positions of the 1,5-hexadiene have been shown to lower the activation energy by 5–10 kcal/mol by stabilization of a biradicaloid-like transition state [56]. However, the crystal structure of AZ28 Fab-hapten **16** complex shows that the aryl substituents of hapten **16** are rotated out of planarity with the cyclohexyl ring, leading to decreased-orbital overlap. Thus it appears that the substrate is fixed in a catalytically unfavorable conformation upon its binding to the affinity-matured Fab.

The structures of the hapten-bound and free forms of the germline antibody provide an explanation for the increased rate of this antibody despite its lower affinity for hapten **16** (Fig. 1.15A). In the unliganded germline antibody structure, CDR H3 (residues $His96^H$ to $Asp101^H$) has a different conformation than that observed in the germline Fab-hapten **16** complex. Residue $Phe99^H$ (C_α) at the top of the loop is shifted 4.9 Å away from the hapten in the germline Fab-hapten complex relative to the unliganded structure. Such conformational flexibility in the germline antibody combining site would allow the 2-phenyl ring to rotate into planarity, increasing π overlap, and, as a result, lowering the activation energy.

The static snapshots of the oxy-Cope catalytic antibodies described here strongly suggest that the conformational diversity of the germline repertoire can also play a dynamic role in catalysis, much as (as is now being realized) side-chain dynamics play a key role in enzymatic catalysis. Indeed, a number of enzymes undergo conformational change upon binding of substrate and lead to enhanced catalytic rate, including hexokinase [42, 43] and triosephosphate isomerase [44]. Recent single-molecule kinetic studies of enzyme-catalyzed reactions also suggest that different conformational states of proteins are characterized by different catalytic rates [57].

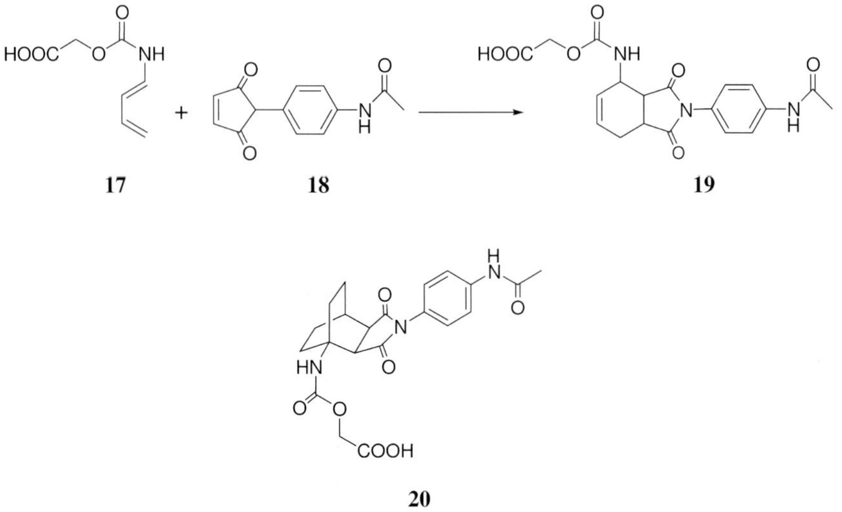

Scheme 1.5

1.8
Diels-Alderase Antibody 39A11 – Evolution of a Polyspecific Antibody combining Site

Antibody 39A11 was generated to the bicyclo[2.2.2]octene hapten **20**, a mimic of the boat-like transition state of the Diels-Alder reaction (Scheme 1.5) [19]. This antibody catalyzes the cycloaddition reaction of diene **17** and dienophile **18** to give the Diels-Alder adduct **19**. The affinity maturation of antibody 39A11 results in only two somatic mutations: Val27c^{L}Leu in CDR L1 and Ser91^{L}Val in CDR L3 [20]. Mutagenesis studies suggest that somatic mutation Ser91^{L}Val is largely responsible for the 40-fold increase in binding affinity and 4-fold increase in k_{cat} for the affinity-matured Fab over the germline Fab. The X-ray crystal structures of the affinity-matured Fab with hapten **20** bound and the germline Fab unliganded were solved [20]. A comparison of the two structures indicates that neither somatic mutation nor ligand binding results in substantial structural or conformational changes in the active site. Thus the germline precursor to antibody 39A11 appears to be a good start point for the evolution of high-affinity combining site for hapten **20** – only one somatic mutation in the combining site is required to bind hapten with nano-molar affinity. It was further found that the germline antibody of 39A11 is polyspecific and shows binding for a panel of structurally very different ligands with affinity within an order of magnitude of that for its own hapten [20]. This polyspecificity may be general to several germline-encoded antibodies and may have been selected for by the immune system to provide a mechanism for rapid generation of antibodies of moderate to high affinity for a broad range of antigens.

Three other antibodies, DB3, TE33, and IE9 were raised against progesterone, a 16-amino acid peptide, and a hexachloronorbornene derivative [58–60], respectively, and are found to use V_H and V_L chains highly homologous to those of 39A11. Both DB3 and IE9 show some cross reactivity, and all four antibodies including 39A11 use a light chain variable region encoded by the Vκ1 gene, which is common to a relatively large population of antibodies that bind a large number of antigens including proteins, DNA, steroids, peptides, and small haptens [61]. However, all four antibodies have quite different CDR H3 loops. This suggests that certain combinations of Vκ–Jκ and V_H give rise to CDRs L1-3 and CDRs H1-2 that are responsible for the assembly of a partial antigen-combining site, which is polyspecific in nature. CDR H3, which is encoded by highly diverse D–J$_H$ joining genes in each antibody, is responsible for the ultimate specificity of the fully assembled antibody. This notion is supported by a comparison of the X-ray crystal structures of DB3, TE33, and 39A11 (Fig. 1.16) [20, 58, 59]. Superposition of those structures reveals that the CDR H3 and CDR L3 loops, together with Trp50^H, form a deep hydrophobic binding pocket as the antibody combining site. In antibodies DB3 and 39A11, Trp50^H and residue 100^H in the CDR H3 loop sandwich the hapten, providing critical hydrogen-bonding or hydrophobic contacts that define opposite walls of the deep binding pocket – Trp100^H in DB3 packs with the central nonpolar region of the steroid, and Arg100^H and Trp50^H in 39A11 provide key hydrophobic and hydrogen-bonding interactions with hapten **20**. CDR H3 of antibody TE33 also packs on the C terminus of the peptide antigen. Thus, in these antibodies, CDR H3 is mainly responsible for the introduction of specific interactions into the antibody combining site that render the fully affinity-matured antibodies specific for their hapten. Such a strategy for variable region gene assembly would benefit the evolution of highly specific antibodies in two ways: (i) by increasing the diversity of the germline antibody binding repertoire, since a single combination of Vκ–Jκ and V_H can be used to construct combining sites for structurally very different antigens, and (ii) by accelerating the immunological evolution process, in that a single combination of Vκ–Jκ and V_H is polyspecific and can be tested in conjunction with all possible D–J_H joining genes for optimal antigen binding specificity and affinity. A similar strategy has also been used for the gene assembly of the germline antibody of 28B4 [26].

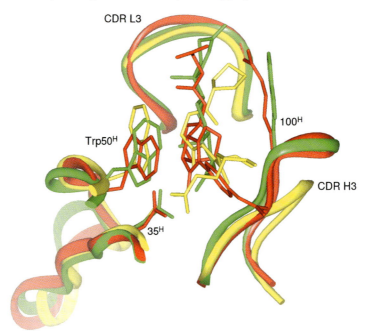

Fig. 1.16 Superposition of the CDR L3 and CDR H3 loops of antibodies DB3, TE33 and 39A11 with bound steroid (green), peptide (yellow) and hapten 20 (red), respectively.

1.9
Conclusions

The immune system is able to generate high-affinity receptors for virtually any chemical structure through its ability to generate a large library of antibodies and to select members of the library based on the affinity of antibody with antigen. There are many parallels between this process and the natural evolution of enzymes. Consequently, studies of the immunological evolution of catalytic antibodies, as a prototypical example of enzyme evolution, have provided a number of insights into the mechanisms of both antibody and enzyme evolution:

1. There is an intrinsic conformational flexibility within the germline antibody combining site. The binding of ligands to the germline antibody induces CDR loop conformational rearrangements that result in enhanced complementarity between the antibody and the ligand. The conformational diversity in the germline antibodies allows the immune system to evolve different antigen specificity using the same germline antibody scaffold, thus greatly expanding the germline binding repertoire. Such an "induced fit" binding mode in the germline antibody may have also existed in primitive proteins. Structural plasticity in the combining sites would have allowed a limited number of protein scaffolds to bind a broad array of substrates with moderate affinity. This plasticity might have allowed these proteins to catalyze a number of distinct reactions, albeit with low rate accelerations. Mutation and selection processes would optimize the complementarity of the protein active site with either a ligand (in the case of a simple receptor) or with a transition state (in the case of an enzyme), leading to enhanced catalytic efficiency and specificity. This process would provide an efficient way to allow a limited number of protein frameworks to evolve many distinct binding and catalytic functions.

2. Certain combinations of germline V genes (Vκ, Jκ and V_H) are polyspecific in nature and can be used to construct antibody combining sites for structurally very distinct ligands. CDR H3 (encoded by D_H–J_H joining) and later somatic mutations play a key role in defining the ultimate specificity of the antibody by introducing specific interactions that strengthen the binding of hapten or interfere with the binding of non-hapten ligands. Germline antibody polyspecificity further expands the binding potential of the germline repertoire.

3. Somatic mutations acquired by the affinity-matured antibody act to fix the conformation of the CDR loops and preorganize the antibody combining site to increase the complementarity with the hapten and/or introduce specific interactions that diminish the binding affinity for non-hapten ligands. This results in an antibody combining site with high specificity and affinity for the hapten in the affinity-matured antibody.

4. Somatic mutations can be introduced into the germline antibody at the peripheries of the hapten combining site that affect hapten binding through secondary sphere interactions, or they can occur at the hapten combining site and affect hapten binding by direct contact. Similarly, residues distal from the enzyme active site have been shown to have a long-range effect on enzyme catalysis.

5. The stepwise acquisition of functional mutations by the germline antibody genes is coupled with stepwise increases in binding affinity between antibody and antigen, often involving cooperative interactions.

This is manuscript 15490-CH of The Scripps Research Institute.

References

1 PAULING, L., *Chem. Eng. News* 24 (**1946**), p. 1375
2 PAULING, L., *Am. Scient.* 36 (**1948**), p. 51
3 SCHULTZ, P. G., YIN, J., LERNER, R. A., *Angew. Chem. Int. Ed.* 41 (**2002**), p. 4427–4437
4 GILBERT, W., *Nature* 271 (**1978**), p. 501
5 EIGEN, M., *Steps Towards Life, a Perspective on Evolution*, Oxford University Press, Oxford **1992**
6 BURNET, F. M., *The Clonal Selection Theory of Aquired Immunity*, Cambridge University Press, Cambridge **1959**
7 TALMAGE, D. W., *Science* 129 (**1959**), p. 1643
8 TONEGAWA, S., *Nature* 302 (**1983**), p. 575–81
9 JANEWAY JR, C. A., TRAVERS, P., *Immunobiology, the Immunosystem in Health and Disease*, Current Biology Ltd., London **1997**
10 JACOBSEN, J. R., PRUDENT, J. R., KOCHERSPERGER, L., YONKOVICH, S., SCHULTZ, P. G., *Science* 256 (**1992**), p. 365–7
11 WIRSCHING, P., ASHLEY, J. A., BENKOVIC, S. J., JANDA, K. D., LERNER, R. A., *Science* 252 (**1991**), p. 680
12 BRAISTED, A. C., SCHULTZ, P. G., *J. Am. Chem. Soc.* 116 (**1994**), p. 2211
13 MA, L. F., SWEET, E. H., SCHULTZ, P. G., *J. Am. Chem. Soc.* 121 (**1999**), p. 10227–10228
14 ULRICH, H. D., PATTEN, P. A., YANG, P. L., ROMESBERG, F. E., SCHULTZ, P. G., *Proc. Natl. Acad. Sci. USA* 92 (**1995**), p. 11907–11
15 PATTEN, P. A., GRAY, N. S., YANG, P. L., MARKS, C. B., WEDEMAYER, G. J., BONIFACE, J. J., STEVENS, R. C., SCHULTZ, P. G., *Science* 271 (**1996**), p. 1086–91
16 WEDEMAYER, G. J., WANG, L. H., PATTEN, P. A., SCHULTZ, P. G., STEVENS, R. C., *J. Mol. Biol.* 268 (**1997**), p. 390–400
17 WEDEMAYER, G. J., PATTEN, P. A., WANG, L. H., SCHULTZ, P. G., STEVENS, R. C., *Science* 276 (**1997**), p. 1665–9
18 YANG, P. L., SCHULTZ, P. G., *J. Mol. Biol.* 294 (**1999**), p. 1191–201
19 BRAISTED, A. C., SCHULTZ, P. G., *J. Am. Chem. Soc.* 112 (**1990**), p. 7430
20 ROMESBERG, F. E., SPILLER, B., SCHULTZ, P. G., STEVENS, R. C. *Science* 279 (**1998**), p. 1929–33
21 ROMESBERG, F. E., SCHULTZ, P. G., *Bioorg. Med. Chem. Lett.* 9 (**1999**), p. 1741–4
22 ULRICH, H. D., MUNDORFF, E., SANTARSIERO, B. D., DRIGGERS, E. M., STEVENS, R. C., SCHULTZ, P. G., *Nature* 389 (**1997**), p. 271–5
23 MUNDORFF, E. C., HANSON, M. A., VARVAK, A., ULRICH, H., SCHULTZ, P. G., AND STEVENS, R. C., *Biochemistry* 39 (**2000**), p. 627–32
24 HSIEH, L. C., STEPHANS, J. C., SCHULTZ, P. G., *J. Am. Chem. Soc.* 116 (**1994**), p. 2167–2168

25 Hsieh-Wilson, L. C., Schultz, P. G., Stevens, R. C., *Proc. Natl. Acad. Sci. USA* 93 (**1996**), p. 5363–7
26 Yin, J., Mundorff, E. C., Yang, P. L., Wendt, K. U., Hanway, D., Stevens, R. C., Schultz, P. G., *Biochemistry* 40 (**2001**), p. 10764–73
27 Cochran, A. G., Schultz, P. G., *Science* 249 (**1990**), p. 781–3
28 Romesberg, F. E., Santarsiero, B. D., Spiller, B., Yin, J., Barnes, D., Schultz, P. G., Stevens, R. C., *Biochemistry* 37 (**1998**), p. 14404–9
29 Yin, J., Andryski, S., Beuscher, A. E., IV, Stevens, R. C., Schultz, P. G., (2003) *Proc. Natl. Aca. Sci.* (in press)
30 Yin, J., Beuscher, A. E., IV, Andryski, S., Stevens, R. C., Schultz, P. G., *J. Mol. Biol.* 330 (**2003**), p. 51–56
31 Dailey, H. A., Dailey, T. A., Wu, C. K., Medlock, A. E., Wang, K. F., Rose, J. P., Wang, B. C., *Cell Mol. Life Sci.* 57 (**2000**), p. 1909–26
32 Dailey, H. A., Fleming, J. E., *J. Biol. Chem.* 258 (**1983**), p. 11453–9
33 McLaughlin, G. M., *J. Chem. Soc., Perkin Trans.* 2 (**1974**), p. 136
34 Blackwood, M. E., Jr, Rush, T. S., 3rd, Romesberg, F., Schultz, P. G., Spiro, T. G., *Biochemistry* 37 (**1998**), p. 779–82
35 Jentzen, W., Song, X. Z., Shelnutt, J. A., *J. Phys. Chem. B* 101 (**1997**), p. 1684–1699.
36 Haldane, J. B. S., *Enzymes*, Longmans Green, London **1930**, p. 182
37 Jencks, W. P., *Catalysis in Chemistry, and Enzymology*, McGraw-Hill, New York **1969**, p. 263
38 Fischer, E., *Ber.* 27 (**1894**), p. 3189
39 Koshland, D. E., Jr, *Proc. Natl. Acad. Sci. U. S. A.* 44 (**1958**), p. 98–104
40 Wells, J. A., de Vos, A. M., *Annu. Rev. Biochem.* 65 (**1996**), p. 609–34
41 DeLano, W. L., Ultsch, M. H., de Vos, A. M., Wells, J. A., *Science* 287 (**2000**), p. 1279–83.
42 Anderson, W. F., Steitz, T. A., *J Mol Biol* 92 (**1975**), p. 279–87
43 DelaFuente, G., Lagunas, R., Sols, A., *Eur. J. Biochem.* 16 (**1970**), p. 226–33
44 Knowles, J. R., *Nature* 350 (**1991**), p. 121–4
45 Breinl, F., Haurowitz, F., *Z. Physiol. Chem.* 192 (**1930**), p. 45
46 Pauling, L., *J. Am. Chem. Soc.* 62 (**1940**), p. 2643
47 Baker, P. J., Waugh, M. L., Wang, X. G., Stillman, T. J., Turnbull, A. P., Engel, P. C., Rice, D. W., *Biochemistry* 36 (**1997**), p. 16109–15
48 Oue, S., Okamoto, A., Yano, T., Kagamiyama, H., *J. Biol. Chem.* 274 (**1999**), p. 2344–9
49 Adams, J., Johnson, K., Matthews, R., Benkovic, S. J. *Biochemistry* 28 (**1989**), p. 6611–6618
50 Stemmer, W. P., *Nature* 370 (**1994**), p. 389–91
51 Orencia, M. C., Yoon, J. S., Ness, J. E., Stemmer, W. P., Stevens, R. C., *Nat Struct Biol* 8 (**2001**), p. 238–42
52 Koshland, D. E., Jr. *Cold Spring Harb. Symp. Quant. Biol.* 52 (**1987**), p. 1–7
53 Andrews, P. R., Smith, G. D., Young, I. G., *Biochemistry* 12 (**1973**), p. 3492–8
54 Driggers, E. M., Cho, H. S., Liu, C. W., Katzka, C. P., Braisted, A. C., Ulrich, H. D., Wemmer, D. E., Schultz, P. G., *J. Am. Chem. Soc.* 120 (**1998**), p. 1945
55 Steigerwald, M. J., Goddard, W. A., Evans, D. A., *J. Am. Chem. Soc.* 101 (**1979**), p. 1994–1997
56 Dewar, M. J. S., Wade, L. E., *J. Am. Chem. Soc.* 99 (**1977**), p. 4417
57 Xie, X. S., Lu, H. P., *J. Biol. Chem.* 274 (**1999**), p. 15967–70
58 Arevalo, J. H., Taussig, M. J., Wilson, I. A., *Nature* 365 (**1993**), p. 859–63
59 Scherf, T., Hiller, R., Naider, F., Levitt, M., Anglister, J., *Biochemistry* 31 (**1992**), p. 6884–97
60 Haynes, M. et al., *Israel J. Chem.* 136 (**1996**), p. 151
61 Kim, H. et al., *J. Immunol.* 143 (**1989**), p. 638

2
Critical Analysis of Antibody Catalysis*
Donald Hilvert

2.1
Introduction

Antibody molecules elicited with rationally designed transition-state analogs catalyze numerous reactions, including many that cannot be achieved by standard chemical methods. Although relatively primitive when compared with natural enzymes, these catalysts are valuable tools for probing the origins and evolution of biological catalysis. Mechanistic and structural analyses of representative antibody catalysts, generated with a variety of strategies for several different reaction types, suggest that their modest efficiency is a consequence of imperfect hapten design and indirect selection. Development of improved transition-state analogs, refinements in immunization and screening protocols, and elaboration of general strategies for augmenting the efficiency of first-generation catalytic antibodies are identified as evident, but difficult, challenges for this field. Rising to these challenges and more successfully integrating programmable design with the selective forces of biology will enhance our understanding of enzymatic catalysis. Further, it should yield useful protein catalysts for an enhanced range of practical applications in chemistry and biology.

2.2
Exploiting Antibodies as Catalysts

Some three decades ago, Jencks suggested that stable molecules resembling the transition state of a reaction might be used as haptens to elicit antibodies with tailored catalytic activities and selectivities [1]. Implementation of this clever idea was made possible by the development of monoclonal antibodies, viable transition-state analogs, and versatile screening assays, and more than 100 reactions have now been successfully accelerated by antibodies [2, 3]. These include pericyclic processes, group transfer reactions, additions and eliminations, oxidations and reductions, aldol condensations, and miscellaneous cofactor-dependent transformations. Because selectivities

---

* This article originally appeared in *Annu. Rev. Biochem.* 2000. 69:751–93 and is reprinted here with permission

Catalytic Antibodies. Edited by Ehud Keinan
Copyright © 2005 WILEY-VCH Verlag GmbH & Co. KGaA, Weinheim
ISBN: 3-527-30688-9

in these systems generally reflect the structure of the hapten and can rival those of natural enzymes, transformations that cannot be achieved efficiently or selectively via more traditional chemical methods are the subject of much current research [4].

Total synthesis of the natural product epothilone from a chiral intermediate prepared by antibody catalysis [5] and activation of a prodrug *in vivo* [6] are two recent accomplishments that illustrate the potential of this technology. Nevertheless, practical applications are still the exception rather than the rule. Low catalytic efficiency, in particular, appears to be a significant limitation. While modest rate accelerations are easily achieved, enzyme-like activity remains elusive. The goal of this review is to address this problem in the light of recent structural and mechanistic work. After first considering the intertwined issues of catalytic efficiency and hapten design, a few well-characterized examples of antibodies promoting a diverse set of reactions are used to illustrate how antibody binding energy is exploited for catalysis and to identify factors that limit overall efficiency. Finally, possible strategies for improving these systems are discussed.

2.3
Catalytic Efficiency

By almost any criterion, natural enzymes are incredibly efficient catalysts. The fastest enzymes are limited by the rate at which they encounter substrate, and even those that have not achieved this level of evolutionary perfection typically have apparent bimolecular rate constants (k_{cat}/K_m) between 10^6 and 10^8 $M^{-1}s^{-1}$, irrespective of the rate of the corresponding uncatalyzed reaction [7]. Rate accelerations over background (k_{cat}/k_{uncat}) are also very high, in the range 10^6 to 10^{12}, and can even reach 10^{17} in some special cases [7, 8].

These extraordinary effects have been explained by an enzyme's ability to bind transition states more tightly than ground states [9]. When enzymatic and non-enzymatic reactions occur by the same mechanism and chemistry is rate determining, a simple thermodynamic cycle based on transition-state theory can be used to show that $k_{cat}/k_{uncat} = K_m/K_{TS}$, where K_m and K_{TS} are the dissociation constants for substrate and transition state from the enzyme [10, 11]. The chemical proficiency of an enzyme, as defined by Wolfenden [7], is then given as the ratio $(k_{cat}/K_m)/k_{uncat}$ or $1/K_{TS}$. This term represents the lower limit of a protein's affinity for the transition state and varies from 10^8 to 10^{23} M^{-1} for a series of natural enzymes [7, 8]. Although true transition-state affinity may be underestimated if chemistry is not clearly rate limiting or if the enzyme uses a different mechanism than the uncatalyzed reaction, this term provides a useful measure of catalytic power for the purpose of discussion.

One practical consequence of the application of transition-state theory to enzymes has been the design of potent inhibitor molecules through mimicry of structural and electronic features of otherwise ephemeral transition states (for recent reviews, see [12–14]). Another has been the use of such compounds to generate catalytic antibodies [2, 3]. The latter strategy has been found to be broadly useful, as any chemical transformation compatible with a biological milieu is potentially amenable to antibody catalysis so long as an appropriate transition-state analog can be devised.

Broad scope and high, programmable catalyst selectivity certainly make this route to tailored enzymes one of the most promising to emerge in the last two decades. In terms of efficiency, however, catalytic antibodies do not yet match their natural counterparts. Published k_{cat}/K_m values for the best catalysts are only in the range 10^2 to 10^4 M^{-1} s^{-1}, well below the limit for diffusion-controlled processes, and rate accelerations are usually between 10^3- to 10^5-fold over background [15, 16]. Smaller effects are often found, but higher efficiencies are extremely rare.

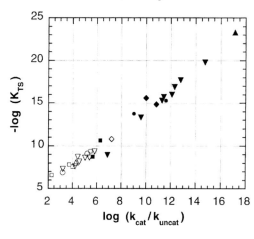

Fig. 2.1 Comparison of chemical proficiency, $-\log (K_{TS})$, and rate acceleration, k_{cat}/k_{uncat}, for a series of reactions catalyzed by enzymes (filled symbols) and antibodies (open symbols). Rearrangements (■,□), hydrolyses (▼, ▽), decarboxylations (▲, △), deprotonations (●, ○) and retroaldol reactions (◆, ◇) are included. Enzyme data are from [7, 112, 131, 132] and antibody data were taken from [25, 26, 38, 75, 91, 110, 113, 115, 121, 127, 200–202].

Differences between antibodies and enzymes are readily apparent in Fig. 2.1, where rate acceleration is plotted against chemical proficiency for a representative set of reactions. Antibody-catalyzed reactions cluster in the lower left quadrant of the plot, corresponding to the lowest activities, whereas enzymatic reactions lie above and to the right, spanning a much greater range of efficiency. Because K_m values are roughly comparable in all of these systems, an excellent correlation between rate acceleration and proficiency is observed. In essence, better transition-state binding translates directly into higher activity. Although the best antibodies approach the efficiency of the least efficient enzymes, it should be noted that the corresponding reactions often involve conversions of relatively activated substrates (e.g., the hydrolysis of aryl esters). In contrast, enzymes specialize in accelerating extremely slow reactions like the cleavage of amides ($t_{1/2}$ = 7.3 years) and phosphate diesters ($t_{1/2}$ = 130 000 years), few of which have yielded to significant antibody catalysis.

In a complementary analysis, Stewart and Benkovic [15] found a weak correlation between rate acceleration and the ratio of the equilibrium binding constants of the reaction substrate and the hapten: $k_{cat}/k_{uncat} = K_m/K_{TS} \approx K_m/K_i$. In other words, affinity for the transition-state analog roughly approximates affinity for the true transition state, in accord with the basic premise underlying the production of catalytic antibodies. Practically speaking, this means that the search for high-performance catalysts can often be reduced to a search for the best hapten binders – provided the hapten is a reasonably good transition-state mimic. However, association constants for low molecular weight haptens ($1/K_i = 10^6$–10^{10} M^{-1}) are generally much smaller than chemical proficiencies of the most effective enzymes. Given K_m values of ca. 10^{-4} M, these affinities suggest that k_{cat}/k_{uncat} will seldom exceed ca. 10^6 for first generation catalytic antibodies, as found experimentally. For reactions with modest activation barriers, effects of this magnitude may be useful. If product inhibition and protein production pose no further technical hurdles, kinetic resolutions can be achieved, the fate of reactive intermediates controlled and so on. For truly difficult reactions, however, such effects are almost certainly inadequate.

2.4
Hapten Design

Before turning to a description of specific antibodies, a few words about hapten design are in order. Transition states themselves have fleeting lifetimes and cannot be isolated. Synthesis of effective analogs must therefore draw on our chemical intuition about the conformational, stereochemical, and electronic properties of the reaction under study. Because no stable molecule can reproduce all the characteristics of an actual transition state, design efforts have tended to focus on salient features distinguishing transition state from ground state [12–14]. For instance, changes in hybridization or charge occurring as a reaction proceeds can be mimicked by incorporating different elements or charged groups at appropriate sites within the analog. Conformational constraint can help approximate the reactive geometry of a flexible substrate, while multisubstrate analogs are used to imitate the relative disposition of reactants in a bimolecular process.

For reactions that require catalytic functionality within the antibody pocket, more sophisticated strategies appear to be needed. In the "bait-and-switch" approach, charge complementarity between hapten and antibody is exploited to induce appropriately positioned acids, bases and nucleophiles. Alternatively, catalytic residues can be selected directly by irreversible chemical modification when mechanism-based inhibitors [17, 18] are employed as haptens. The latter strategy, dubbed "reactive immunization" [19], has the virtue of allowing rational engineering of covalent catalysis.

For each new reaction, hapten design must be optimized to maximize the probability of finding an antibody catalyst. Because subtle differences between even the best transition-state analogs and actual transition states almost certainly contribute to lower efficiencies of antibodies compared with enzymes, it is important to understand how instructions implicit in any given hapten design are realized in the complementary immunoglobulin binding pocket. Characterization of successful antibody catalysts at the atomic level currently provides the most useful insights into how binding energy is exploited for catalysis.

2.5
Representative Catalytic Antibodies

2.5.1
Proximity Effects

Utilization of binding energy to constrain flexible molecules into reactive conformations or to preorganize reactants for bimolecular reaction is a potentially powerful strategy for accelerating reactions with unfavorable entropies of activation [20, 21]. To test whether antibodies might serve as entropy traps [21], concerted pericyclic reactions requiring neither nucleophilic nor acid-base catalysis have been investigated.

2.5.1.1
Sigmatropic Rearrangements

The biologically important Claisen rearrangement of chorismate to prephenate (Fig. 2.2) and the abiological oxy-Cope rearrangement (Fig. 2.3) are typical [3,3]-sigmatropic processes. They proceed via highly ordered, entropically unfavorable, cyclic transition states involving the simultaneous formation of a carbon-carbon bond and cleavage of either a carbon-oxygen or another carbon-carbon bond.

For the chorismate rearrangement, the conformationally locked oxabicyclic dicarboxylic acid **1** [22] proved to be a successful hapten (Fig. 2.2). It has a chair-like

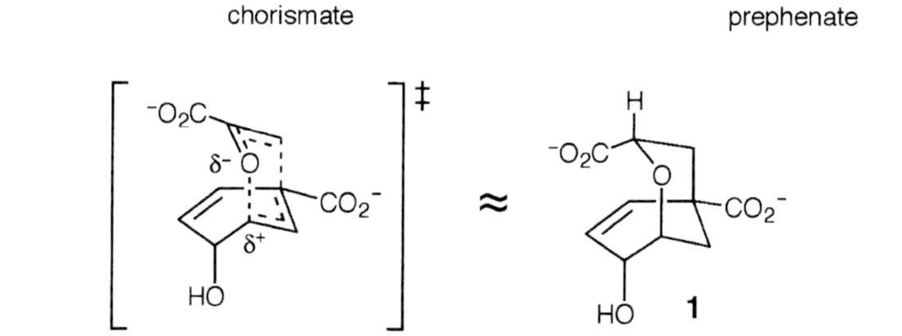

Fig. 2.2 (*Top*) The Claisen rearrangement of chorismate to prephenate catalyzed by antibody 1F7; the flexible chorismate adopts an extended pseudo-diequatorial conformation in solution and must undergo a conformational change to populate the less stable pseudo-diaxial conformer in order for reaction to proceed. (*Bottom*) The conformationally restricted dicarboxylic acid **1** [22] mimics the transition state of the Claisen rearrangement and was used to elicit catalytic antibody 1F7 [25].

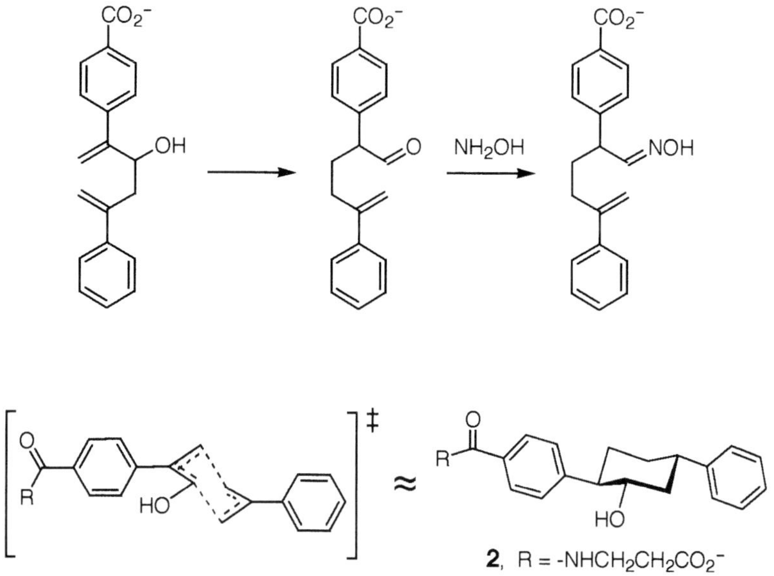

Fig. 2.3 Antibody-catalyzed oxy-Cope rearrangement [36]. The aldehyde product is trapped with hydroxylamine to give an oxime to prevent time-dependent inactivation of the catalyst. The 2,5-diaryl cyclohexanol derivative **2** was used to imitate the structure of the pericyclic transition state.

geometry very similar to that of the presumed transition state and is an effective inhibitor of natural chorismate mutases [22, 23]. Antibodies that bind **1** catalyze the chorismate rearrangement enantioselectively with rate accelerations (k_{cat}/k_{uncat}) of 10^2 to 10^4 over background [24–27]. For comparison, enzymes accelerate this reaction by a factor of ca. 10^6 [28]. Spectroscopic and X-ray studies of 1F7 (which achieves a 200-fold rate enhancement) have provided insights into the origins of catalysis in these systems.

Transferred nuclear Overhauser effects show that 1F7 binds the flexible chorismate molecule in the diaxial conformation specified by the transition-state analog [29]. Crystallographic data confirm that the induced binding pocket faithfully reflects hapten design [30]. Compound **1** is deeply buried in the complex, and the active site's overall shape and charge are complementary to a single hapten enantiomer (Fig. 2.4a). Consequently, only the corresponding (–)-isomer of chorismate binds in a conformation appropriate for reaction. The subsequent rearrangement of bound substrate then occurs by the same concerted mechanism as that deduced for the uncatalyzed reaction and for natural chorismate mutases.

Nevertheless, 1F7 is likely to be a much poorer entropy trap than mutase enzymes. It exploits many fewer hydrogen bonds and electrostatic interactions for ligand recognition (Fig. 2.4a,b). It also appears to accommodate charge separation in the transition state less effectively. Isotope effects show that the transition state for the rearrangement is highly polarized, with C–O bond cleavage preceding C–C bond formation [31, 32]. The enzymes stabilize this species electrostatically by placing a

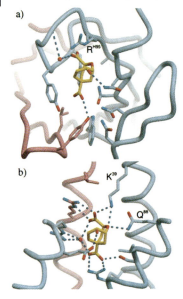

Fig. 2.4 (a) Active site of antibody 1F7 [30], showing interactions with hapten 1. The light and heavy chains are pink and blue, respectively; the hapten is yellow. Note that Arg^{H95} forms a salt bridge with the ligand's secondary carboxylate but is too far away to form a hydrogen bond with the ether oxygen. **(b)** Active site of *E. coli* chorismate mutase [203]. Bound **1** is completely inaccessible to solvent and makes numerous contacts with protein residues; hydrogen bonds from Lys^{39} and Gln^{88} to the ligand's ether oxygen are essential for high activity [28]. Graphics were prepared with the programs BobScript [204] and Raster3D [205].

cationic residue (either Arg or Lys) near the partially negatively-charged ether oxygen of the breaking C–O bond [28, 33, 34], but 1F7 lacks an analogous feature (Fig. 2.4a,b). These differences presumably explain the antibody's 10^4-fold lower efficiency. They can be attributed, in large part, to shortcomings in hapten design. While **1** reproduces the geometry of the actual transition state reasonably well, it mimics the polarized character of this high energy species poorly [35].

Shortcomings in hapten design are also likely to account for the modest activity of antibody AZ-28. This antibody was raised against cyclohexanol derivative **2** (K_d = 17 nM) and catalyzes the oxy-Cope rearrangement of the corresponding 2,5-diaryl-3-hydroxy-1,5-hexadiene with a k_{cat}/k_{uncat} of 5300 (Fig. 2.3) [36]. In this case, disposition of the aryl substituents in the transition state is imitated imperfectly in the stable hapten.

Like the chorismate mutase antibody, AZ-28 has been shown by TRNOE measurements to preorganize the normally extended hexadiene substrate into a cyclic conformation so that its termini are in close proximity [37]. In this case, ligand recognition is mediated by extensive van der Waals contacts, π-stacking interactions with the aromatic rings, and hydrogen bonding interactions with the alcohol, all evident in the X-ray structure of the antibody-hapten complex (Fig. 2.5) [38]. However, the conformation adopted by the substrate at the AZ-28 active site is unlikely to be optimal for reaction. The two aryl substituents at C2 and C5 are key recognition elements (Fig. 2.5) but are oriented very differently in the hapten, where they are sp^3 hybridized and equatorial to the plane of the cyclohexane ring, and the transition state, where they are sp^2 hybridized and conjugated with the reacting olefins (Fig. 2.3). Thus, even though the hapten-induced pocket brings together the ends of the hexadiene substrate, binding energy directed to the peripheral aryl groups almost certainly imposes physical constraints that preclude effective alignment of the reacting [4π+ 2σ]

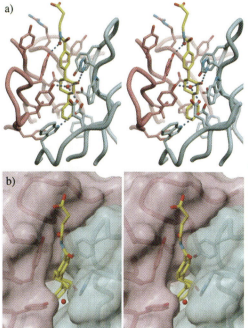

Fig. 2.5 (a) Stereoview of the AZ-28 active site showing hapten-contacting residues [38]. The 5-phenyl ring of **2** sits deep in the hydrophobic pocket, while the 2-phenyl substituent and linker are near the surface. The hapten's alcohol forms a hydrogen bond with His^{H96}. Two water molecules in the pocket are depicted as red spheres. **(b)** GRASP [206] surface representation of AZ-28 showing the high degree of shape complementarity between ligand and protein.

orbitals. Alterations in antibody structure leading to improved orbital overlap should therefore result in significant increases in catalytic efficiency.

This inference is supported by studies of AZ-28's germline precursor. This antibody binds hapten **2** with 40-fold lower affinity than the mature AZ-28 but is a substantially better catalyst, achieving a 163 000-fold rate acceleration over background [38]. Mutagenesis experiments showed that replacement of Ser^{L34} in the germline sequence with Asn is responsible for both effects [38]. In the crystal structure of AZ-28, Asn^{L34} interacts directly with the cyclohexyl ring of the hapten and is therefore in a position to influence the conformation of the substrate at the active site.

Although designed as entropy traps, and despite evident restriction in the conformational freedom of their substrates, neither 1F7 nor AZ-28 lowers the entropy of activation for its reaction. Both have $\Delta S^{\ddagger}$ values that are 10–20 cal K^{-1} mol^{-1} *less* favorable than the corresponding uncatalyzed reactions [25, 37]. This contrasts with some natural chorismate mutases which do reduce the entropy barrier to reaction significantly [39]. Mechanistic interpretations of activation parameters are necessarily uncertain [40], but the unfavorable $\Delta S^{\ddagger}$ values are consistent with the need for substantial conformational change in the bound substrate as the reaction proceeds. However, other factors, including changes in solvation or conformation associated with the antibody, cannot be excluded.

More generally, these two examples show how the chemical instructions implicit in hapten structure, including deficiencies with respect to transition-state mimicry, are accurately imprinted on an antibody binding site. Improved transition-state analogs should therefore yield much better catalysts. To obtain more efficient chorismate mutase antibodies, for example, haptens containing additional negative charges might be used to elicit the catalytically essential cation in the vicinity of the substrate's ether oxygen. Similarly, haptens in which the aryl substituents are coplanar with the cyclohexyl ring should increase the probability of identifying faster catalysts for the oxy-Cope rearrangement of Fig. 2.3. The sensitivity of the latter reaction to anionic substituent effects [41] could also be drawn on. Haptens containing an appropriately positioned ammonium group might induce an antibody residue capable of deprotonating the substrate alcohol.

Fine-tuning of the first-generation antibodies is also likely to yield substantially better catalysts. Identification of a second chorismate mutase antibody possessing significantly higher activity than 1F7 [26] supports the feasibility of such an undertaking. Plausible strategies for optimizing activity include site-directed mutagenesis or random mutagenesis coupled with *in vivo* selection. The 1F7 Fab fragment's ability to replace the missing enzyme in a chorismate mutase-deficient yeast cell line [42] is a promising indication that it will be amenable to directed evolution in the laboratory [43].

Overall, there are many ways to bind any given transition-state analog, only some of which will be effective for catalysis. Indeed, a significant fraction of hapten binders in any given experiment is usually found to be inactive. Hapten affinity rather than catalytic activity drives maturation of the immune response, so mutations can arise that favor tighter hapten binding but are deleterious for catalysis, as seen for AZ-28. Broad screening of antibodies raised to each hapten is therefore necessary to guarantee a representative sampling of the immune response. In the present instance, antibodies other than 1F7 and AZ-28 might be obtained that ultimately prove to be better starting points for optimization.

2.5.1.2
Cycloadditions

Loss of both translational and rotational degrees of freedom should make bimolecular reactions particularly sensitive to proximity effects [20]. Diels-Alder reactions between dienes and dienophiles have been used to test this notion. They typically have high activation entropies in the range –30 to –40 cal K^{-1} mol^{-1} [44], reflecting the low probability of bringing together two substrates in an orientation optimal for reaction. The transition state for these concerted cycloadditions is highly ordered and resembles the boat form of the cyclohexene product more closely than is does the starting materials. Antibodies raised against bicyclic compounds that mimic the transition-state geometry have displayed a range of useful catalytic effects, including control over reaction pathway and absolute stereochemistry [45–49].

The general approach to catalysis is exemplified by antibody 1E9, which promotes the [4+2] cycloaddition cycloadditionand N-ethylmaleimide (Fig. 2.6) with multiple turnovers [45]. The initially formed adduct **3** spontaneously eliminates sulfur dioxide to give *N*-ethyl tetrachlorophthalimide as the final product after oxidation *in situ*. The *endo* hexachloronorbornene derivative **4**, an excellent mimic of the intermediate and

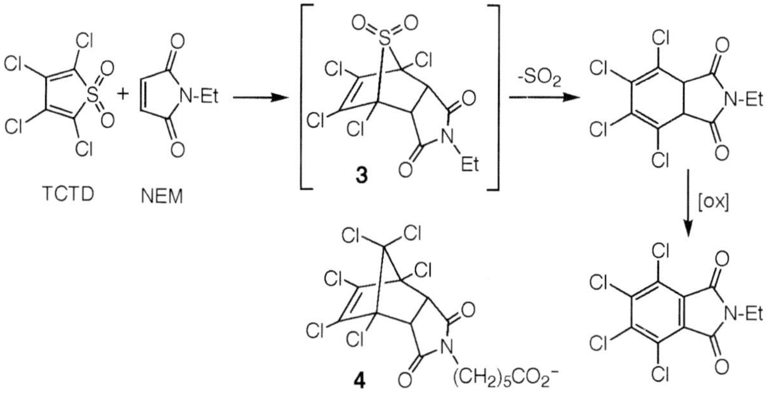

Fig. 2.6 Diels-Alder condensation of tetrachlorothiophene dioxide (TCTD) and N-ethyl maleimide (NEM) yields a high-energy intermediate (**3**) which eliminates sulfur dioxide. The initially formed product is subsequently oxidized *in situ*. The transition state for cycloaddition and chelotropic elimination of SO_2 closely resemble the hexachloronorbornene derivative **4** used as a hapten to elicit antibody 1E9 [45].

its flanking transition states, served as the hapten (K_d = 2 nM). Because the planar product is structurally so different from **4**, it binds 10^5-fold less tightly to the induced antibody, effectively minimizing product inhibition.

In this case, catalytic efficiency can be estimated as an effective molarity (EM) [50]. EM is the ratio of the pseudo-first-order rate constant for the antibody reaction (k_{cat}) to the second-order rate constant for the uncatalyzed process (k_{uncat}). This ratio gives the nominal concentration of one reactant needed to convert the spontaneous bimolecular reaction into a pseudo-first-order process with a rate equivalent to that achieved in the antibody ternary complex. It is usually interpreted as the entropic advantage of a unimolecular over a bimolecular process, with an upper limit of about 10^8 M for 1 M standard states [51]. For 1E9, the EM is ca. 10^3 M [52]. Although much lower than the theoretical limit, this value is significantly higher than EMs reported for other antibody Diels-Alderases, which rarely exceed 10 M [46–49], making 1E9 the most efficient such catalyst described to date.

Recent structural work has shown that the 1E9 active site is exactly what one might expect of an antibody that functions as an entropy trap [52]. Extensive van der Waals contacts, π-stacking interactions, and a strategically-placed hydrogen bond to one of the succinimide carbonyl groups create a pocket that is highly complementary to the hapten (Fig. 2.7a). When complexed, the ligand (excluding the hexanoate linker) is 86% buried. Its fit to the protein is so snug that no interfacial cavities are detectable, even when a probe of 1.2 Å radius is used. Thus, the 1E9 binding pocket appears ideally suited to the task of preorganizing its diene and dienophile substrates in a reactive complex that closely approximates the transition-state geometry. Nevertheless, a simple entropy trap mechanism does not appear to be operative. The temperature dependence of k_{cat} and k_{uncat} shows that catalysis by 1E9 is achieved entirely by reducing the enthalpy of activation; the solution and the antibody processes are equally

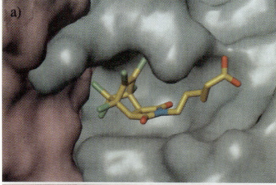

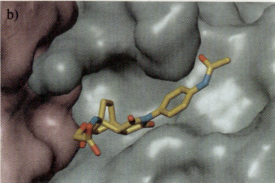

Fig. 2.7 (a) The active site of Diels-Alderase 1E9 provides a snug, complementary binding surface for its hexachloronorbornene hapten [52]. With the exception of a hydrogen bond between the deeply buried carbonyl group of the ligand and the side chain of Asn^{H35} (not shown), most of the contacts are hydrophobic in nature. (b) Antibody 39-A11 (see Fig. 2.8) binds its hapten much more loosely. Note that the "dienophile-like" N-aryl succinimide is significantly better packed than the bicyclo[2.2.2]octene moiety, which serves as a diene surrogate.

unfavorable entropically, with $\Delta S^{\ddagger}$ values of -22 cal K^{-1} mol^{-1} [52]. Nor is the rate acceleration due to a simple medium effect associated with the apolar binding cavity, since the uncatalyzed reaction is 10 times slower in acetonitrile than in water.

Catalysis by 1E9 can be explained by enthalpic stabilization of the transition state through an unusually close fit to the apolar binding surface of the antibody active site and a strong hydrogen bond between the side chain of Asn^{H35} and the maleimide carbonyl. Quantum mechanical calculations indicate that the numerous, energetically favorable van der Waals interactions provide the driving force for binding [52]. Although these interactions distinguish poorly between ground and transition state, they hold the substrates against a relatively unfavorable electrostatic field that becomes substantially more favorable as the transition state is approached because of the increased strength of the hydrogen bond to the dienophile carbonyl.

Comparison of 1E9 with another Diels-Alderase, 39-A11 [47], dramatically illustrates the importance of close packing for high efficiency. Antibody 39-A11 was generated with the substituted bicyclo[2.2.2]octene derivative 5, which contains an ethano bridge locking the cyclohexene ring into the requisite boat conformation (Fig. 2.8). It catalyzes the Diels-Alder reaction between an electron-rich acyclic diene and an N-aryl maleimide to give a cyclohexene derivative, albeit with a relatively low effective molarity of 0.35 M.

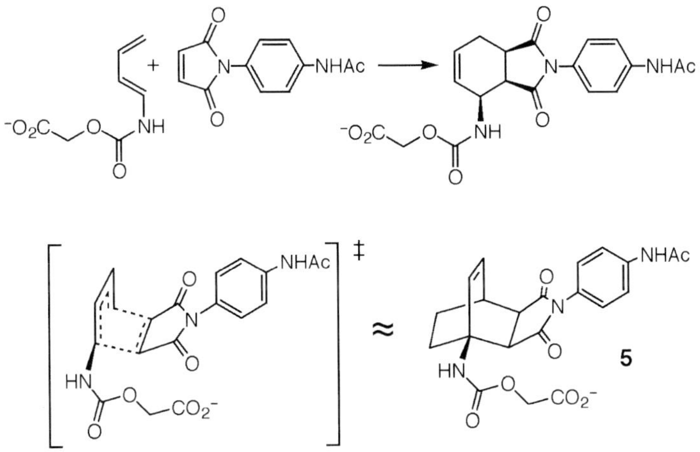

Fig. 2.8 Antibody 39-A11 catalyzes a Diels-Alder reaction between an electron-rich acylic diene and an N-aryl maleimide. It was elicited with the bicyclo[2.2.2]octene hapten **5**. The ethano bridge in **5** locks the cyclohexane ring into a boat conformation but has no counterpart in the substrate itself.

Considering the low dissociation constant reported for the antibody-hapten complex ($K_d = 10$ nM) [47], the fit of the bicyclo[2.2.2]octene to 39-A11 is surprisingly loose (Fig. 2.7b) [53]. Only 66% of the hapten surface area is buried in the complex. Poor complementarity is indicated by the large cavity volume of 117 Å^3 detected between ligand and antibody. The portion of the hapten corresponding to the reacting [4+2] system is particularly poorly packed, whereas peripheral substituents of **5**, especially the aryl side chain, appear to be important recognition elements. Moreover, the hapten's ethano bridge, which has no counterpart in the substrates or the transition state, carves out additional unwanted space within the pocket. Consequently, the bound substrates – particularly the diene, which must bind in the least complementary region of the pocket – are likely to retain considerable degrees of freedom. Low catalytic efficiency is therefore unsurprising. Consistent with this idea, introducing large aromatic groups at positions L91 and L96 to improve packing interactions with the kinetically favored *endo* transition state results in 5 to 10-fold higher k_{cat} values [54].

Relatively non-polar as it is, the environment of the 39-A11 active site may further erode catalytic efficiency. Like 1E9, 39-A11 provides a hydrogen bond (also from Asn^{H35}) to the dienophile carbonyl, but its reaction involves a strong donor diene and acceptor dienophile rather than two electron-deficient addends. The corresponding transition state should therefore be more polar and hence more sensitive to transfer from water than that of the 1E9-catalyzed reaction. Unfortunately, mutagenesis experiments to augment activity by providing additional hydrogen bonds to the dienophile have not been successful [54].

Structurally distinct haptens notwithstanding, 1E9 and 39-A11 are unexpectedly closely related in primary sequence and tertiary structure [52, 53, 55]. Both belong to a family of polyspecific antibodies that exhibit extensive cross-reactivity for hydrophobic ligands containing one or two polar groups. For example, 39-A11 and its germline precursor accommodate a range of structurally diverse compounds [53], while 1E9 and the related progesterone-binding antibody DB3 [56, 57] bind each other's ligands with affinities only 25-fold to 50-fold lower than their own [55]. The side chain of Asn^{H35} and conserved hydrophobic interactions seem to be particularly important for achieving recognition. However, docking experiments [52] with 1E9 and DB3 suggest that non-cognate molecules bind randomly in the apolar cavity, whereas specific ligands adopt a single, well-defined binding mode similar to that seen in crystal structures of the corresponding antibody-hapten complexes. Specificity in these systems appears to be conferred by a small number of mutations to the shared scaffold. In the case of 39A11, for instance, substituting Val for Ser at position L91 in the germline sequence accounts for almost all the 40-fold increase in hapten affinity achieved during affinity maturation [53]. Similarly, a somatic mutation in the L3 CDR loop (Ser^{L89} → Phe) and a rare mutation in the antibody framework region (Trp^{H47} → Leu) substantially alter the shape of the 1E9 combining pocket compared to that of 39-A11 or DB3 [52]. These changes are primarily responsible for its virtually perfect shape complementarity to the transition-state analog. This complementarity appears crucial for reactivity, since DB3 does not detectably accelerate the 1E9 reaction despite its affinity for hapten 4.

Given the immune system's enormous combinatorial diversity (more than 10^8 antibodies are available in the primary response [58]), it is surprising that similar antibodies were generated in separate immunization experiments with 4, 5 and progesterone. Nor, as will be shown below, is this finding unique. To what extent does utilization of a few restricted sets of antibody germline genes limit the catalytic potential of the immune system? Do these frequently selected scaffolds represent local minima from which it will be difficult to evolve more active catalysts? Poor shape complementarity between 39-A11 and the bicyclo[2.2.2]octene is a case in point. However, site-directed mutagenesis experiments demonstrate that the mature antibody is not an evolutionary dead end. The same Val^{L91}Tyr change that improves catalysis 10-fold also increases hapten affinity 2.5 times [54], and it should be possible to find additional mutations that further tighten the structure. Because only ten antibodies were screened for activity, we cannot know whether analogous mutations were present in the antibody population induced in response to hapten 5. It is conceivable that they never emerged: a 10 nM dissociation constant for the 39-A11-hapten complex is more than adequate for the immune system's purposes, so there may be little or no selection pressure to increase affinity beyond that point.

As shown by 1E9, the combining pocket can be molded remarkably well under favorable circumstances to achieve nearly perfect shape complementarity with a ligand. Superior fit is not necessarily manifest in tighter binding, since 1E9 and 39-A11 have similar hapten affinities (K_d = 2 nM versus 10 nM). Compound 4 is more highly optimized than 5 with respect to transition-state mimicry, however, and important binding interactions in 1E9 are concentrated where they are needed for catalysis, rather than loosely dispersed as in 39-A11. A high degree of complementarity at the site of reaction thus appears to pay off in this case in terms of a more efficient catalytic outcome.

Even in the case of 1E9, though, much higher efficiency should be attainable. An analogous enzyme is unavailable for direct comparison (see [59] for evidence regarding possible Diels-Alderases in Nature), but 1E9's chemical proficiency, defined as $(k_{cat}/K_{diene}K_{dienophile})/k_{uncat} = 1.4 \times 10^7$, is far from what is expected of a fully evolved catalyst. Given an already excellent fit between protein and ligand, it is unlikely that mutation of residues lining the binding cavity will substantially improve complementarity. To optimize electrostatic interactions with the transition state and reduce any remaining degrees of freedom available to the bound substrates, residues distant from the active site will have to be modified. Because such mutations will be difficult to identify by inspection of the protein structure, combinatorial mutagenesis and an efficient screening assay [60] or selection protocol will be needed if 1E9 variants with enhanced properties are to be developed.

2.5.2
Strain

Substrate destabilization through strain has been proposed as another mechanism for achieving rate accelerations with enzymes [20, 61]. Binding energy can be used to strain molecules in various ways. Destabilization can involve geometric distortion of the substrate, electrostatic repulsion between groups of like charge, or desolvation effects. If substrate destabilization is relieved at the transition state, significant reductions can result in the free energy of activation for reaction. Rate accelerations obtained in this way are potentially quite large, limited only by the amount of binding energy available to force the substrate into the destabilizing environment. As with entropic effects, it was predicted that strain mechanisms might be readily exploited for antibody catalysis.

2.5.2.1
Ferrochelatase Mimics

Ferrochelatase is an example of an enzyme that is believed to exploit geometric distortion for catalysis. As the terminal enzyme in heme biosynthesis, it promotes complexation of Fe^{2+} by protoporphyrin IX [62, 63]. Early work suggested that ferrochelatase functions by distorting the substrate porphyrin from its preferred planar conformation into a bent structure [62]. This distortion exposes the nitrogen of one of the pyrroles to solvent, thereby facilitating metal ion complexation. In fact, non-planar N-methylated porphyrins are known to chelate metal ions 3 to 5 orders of magnitude faster than their non-alkylated counterparts [62]. They are also potent inhibitors of ferrochelatase [64]. When used as haptens, they have yielded antibodies [65] that catalyze insertion of divalent metal ions into mesoporphyrins with k_{cat} values approaching those of the enzyme (Fig. 2.9).

Ferrochelatase antibody 7G12 has been characterized in some detail. It promotes incorporation of Zn^{2+} and Cu^{2+} into mesoporphyrin IX with k_{cat} values of 0.022 s^{-1} and 0.0069 s^{-1}, respectively [65]. By way of comparison, the k_{cat} value for recombinant *Bacillus subtilis* ferrochelatase is ca. 0.4 s^{-1} for metalation with Fe^{2+}, Zn^{2+} and

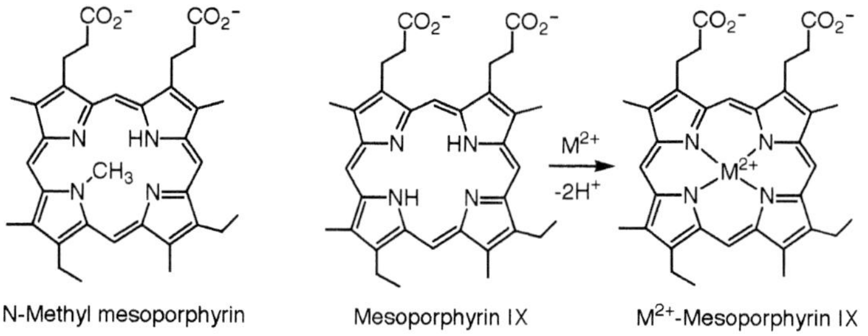

Fig. 2.9 N-methyl mesoporphyrin and the mesoporphyrin metalation reaction catalyzed by antibody 7G12 [65].

Cu^{2+} [66]. K_m values for the porphyrin are also comparable (50 µM for the antibody and 8 µM for the enzyme). However, whereas substrate metal ions bind tightly to the enzyme (20–170 µM), no evidence for saturation of the antibody has been observed up to concentrations up to 2.5 mM, suggesting that binding of metal ions to the antibody does not contribute to catalysis. Severe inhibition of the antibody-catalyzed reaction by the metalloporphyrin product is a further point of contrast.

Mechanistic and structural studies have clarified how the antibody chelatase functions. The crystal structure of 7G12 complexed with its hapten has been solved [67]. The N-methylated porphyrin binds at the junction of the heavy and light chains (Fig. 2.10). One face of the ligand makes extensive contacts with V_H, while the other is relatively exposed to solvent. Packing interactions from light chain Tyr residues may reinforce the distortion from planarity of pyrrole ring A, which bears the N-methyl group. Replacement of these residues with Ala caused large increases in K_m and 10-fold to 40-fold decreases in k_{cat}/K_m [67], suggesting that they may play a similar role with the non-methylated substrate. In analogy with the enhanced reaction rates achieved by porphyrin alkylation, a catalytic mechanism involving binding and distortion of the porphyrin by the protein, followed by direct chelation of metal ions from solution, seems plausible. Although its precise role is still unclear, amino acid Asp^{H96} is evidently required for this process [67]. Its carboxylate side chain is directed from the V_H domain toward the center of the porphyrin ring with one oxygen roughly equidistant from the porphyrin's four pyrrole nitrogens (Fig. 2.10). The other oxygen is fixed in place through a hydrogen bond to Arg^{H95}. It is conceivable that the carboxylate acts as a base which shuttles protons from the porphyrin during metal ion exchange. This residue is also probably responsible, at least in part, for product inhibition, since axial coordination to the metal ion will anchor the metaloporphyrin to the antibody.

Direct experimental evidence for porphyrin distortion by 7G12 has been obtained with resonance Raman spectroscopy [68]. Spectral data show that the antibody induces an alternating up-and-down tilting of the pyrrole rings very similar to the distortion produced by porphyrin alkylation. In contrast, yeast ferrochelatase apparently causes all four pyrrole rings to tilt in the same direction in a domed fashion. The enzymatic reaction is regulated allosterically by a metal-dependent protein con-

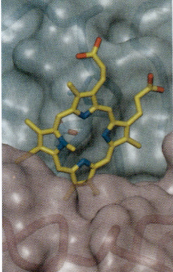 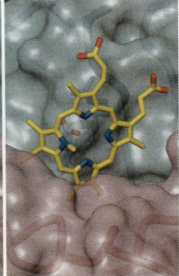

Fig. 2.10 Stereoview of the 7G12 binding site [67], showing how bound porphyrin packs against the heavy chain (blue) and exposes one face to solvent. The side chains of heavy chain residues Asp^{H96} and Arg^{H96} are visible under the surface behind the porphyrin. Interactions with the light chain (pink) are largely restricted to contacts with the methyl and ethyl substituents of the A and B pyrrole rings of the porphyrin. Although a mixture of N-alkylated porphyrins was used, only the derivative with the A ring methylated appears to bind to the antibody.

formational change. Since the antibody has no metal binding site, the distortion it induces must be brought about entirely by binding interactions between porphyrin and protein. An atomic-level explanation of this effect will require elucidation of the antibody-substrate complex structure.

The broad lesson to be derived from these experiments is that substrate destabilization can be a very successful approach to antibody catalysis. Although 7G12 and ferrochelatase perturb the porphyrin structure in different ways, both kinds of distortion appear to be effective for metal ion chelation, yielding k_{cat} values within a factor of 10 of each other. That said, destabilization mechanisms are expected to have little to no effect on k_{cat}/K_m [69], the steady-state parameter generally optimized through evolution. By this criterion, 7G12 is substantially more primitive than its natural counterpart. Because chelatases catalyze a bimolecular reaction, flux through the catalyst is limited by the least favorably processed substrate. For both enzyme and antibody, this is the metal ion. Hence, rate enhancements are given by $[(k_{cat}/K_{M^{2+}})/k_{uncat}]$. Since $K_{M^{2+}}$ values for 7G12 are at least 10^3 times larger than those for natural ferrochelatases (assuming the metal ion binds at all), the antibody suffers an equivalent rate disadvantage under practical operating conditions, despite its favorable k_{cat}. Similar considerations apply to the chemical proficiencies of the two catalysts [($k_{cat}/K_{M^{2+}} K_{porphyrin})/k_{uncat}$]. The antibody's transition-state affinity is at least four orders of magnitude lower than that of the enzyme.

To improve 7G12, a suitable binding site for metal ions should be constructed, perhaps by extending light chain CDR loops that are near the exposed face of the bound porphyrin. The challenge will be to bring the metal ion into close proximity with the porphyrin without further increasing the active site's affinity for product. As for the chorismate mutase antibody discussed above, an *in vivo* selection strategy is feasible. Ferrochelatase-deficient yeast auxotrophs have been reported [70]. Complementing this metabolic deficiency with the antibody catalyst would provide a means of identifying more efficient variants.

2.5.2.2
Other Systems

Large rate accelerations are also expected when charged reactants are enthalpically destabilized relative to a charge delocalized transition state by desolvation [20]. Such a mechanism may contribute to the efficacy of enzymes that promote biologically relevant decarboxylations [71–74]. Antibody catalysis of model reactions, such as the solvent-sensitive decarboxylation of 3-carboxy-benzisoxazoles to salicylonitriles [75, 76] and the difficult decarboxylation of orotate to uracil [77], has been used to probe the role of medium effects in a tailored binding pocket. Significant rate enhancements have been achieved, and structural work characterizing the properties of successful active sites will be instructive. Given the well-established and often dramatic sensitivity to solvent change of many reaction types, medium effects are likely to be pervasive in antibody and enzymatic catalysis. Reactions that display large changes in charge localization, including decarboxylations, nucleophilic substitutions, and aldol condensations, should be especially amenable to such effects.

2.5.3
Electrostatic Catalysis

2.5.3.1
Acyl Transfer Reactions

Enzymes frequently utilize hydrogen bonding and charged groups to stabilize polar transition states electrostatically [69]. When haptens containing positive and negative charges are used, electrostatic interactions can also be exploited for antibody catalysis. For example, anionic phosphonates and phosphonamidates, originally designed as potent inhibitors of hydrolytic enzymes [12, 78], have been useful in the production of antibodies that hydrolyze esters, carbonates and (more rarely) amides [79]. Such analogs resemble the transition state for hydrolysis in a number of ways, including tetrahedral geometry, negative charge, and increased bond lengths. How these features are reflected in the induced binding pockets has been deduced through structural studies of six different esterase families [80–90].

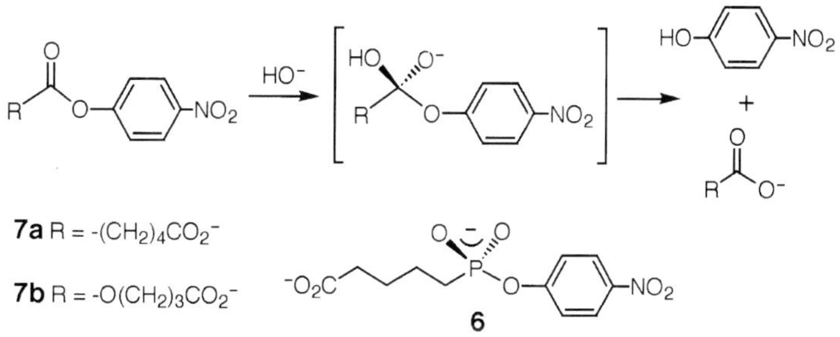

Fig. 2.11 Hydrolysis of activated aryl esters **7a** and carbonates **7b** proceeds via an anionic and tetrahedral intermediate (in brackets). Hydrolytic antibody 48G7 was elicited with an aryl phosphonate derivative (**6**) that mimics this high-energy species and its flanking transition states [91].

Antibody 48G7 is typical of this class of catalyst [91]. It was generated against p-nitrophenyl phosphonate (**6**) and accelerates the hydrolysis of the corresponding activated ester **7a** and carbonate **7b** by factors of 1.6×10^4 and 4×10^4, respectively (Fig. 2.11). Both the antibody and its germline precursor, with and without bound hapten, have been characterized to provide insight into the origins and evolution of its catalytic effects [85–87].

The mature antibody has a deep, well-defined combining site rich in aromatic residues (Fig. 2.12a). It binds the hapten in an extended conformation with the aryl group at the bottom of a hydrophobic cleft formed between CDRs L1 and H3. The negatively charged phosphonate moiety lies near the pocket entrance, where it forms multiple interactions with charged and neutral antibody residues. The pro-R phosphonyl oxygen hydrogen bonds with the Tyr^{H33} phenolic hydroxyl group and forms a salt bridge with the Arg^{L96} guanidinium group, while the pro-S oxygen hydrogen bonds to the His^{H35} ε-imino group and the Tyr^{H96} backbone NH. Comparison of the mature antibody with and without the transition-state analog shows no major conformational changes [86], suggesting a simple lock-and-key mechanism for hapten binding (see, however, [92]).

Antibody 48G7 has a 30 000-fold higher affinity for the phosphonate hapten and 20-fold greater catalytic efficiency than its germline precursor. Nine somatic mutations, all lying outside the combining site, are responsible for these effects [85, 87, 93]. Although not in direct contact with bound ligand, the mutated residues appear to preorganize the pocket for binding and catalysis. Improved packing and secondary hydrogen-bonding interactions help limit side-chain and backbone flexibility inherent in the germline protein. In fact, germline 48G7 (Fig. 2.12b) is conformationally much more flexible than the mature antibody and undergoes significant reorganization upon hapten binding [87]. The hapten itself binds quite differently in the two complexes (Fig. 2.12a,b). The phosphonate moiety occupies essentially the same location in both, but it cannot form a hydrogen bond with Tyr^{H33} in the germline complex because of an altered conformation of the CDR H1 loop. Further, the p-nitrophenyl group is rotated away from the position it adopts in the mature antibody to occupy a hydrophobic cleft constructed from framework residues. This second apolar pocket is

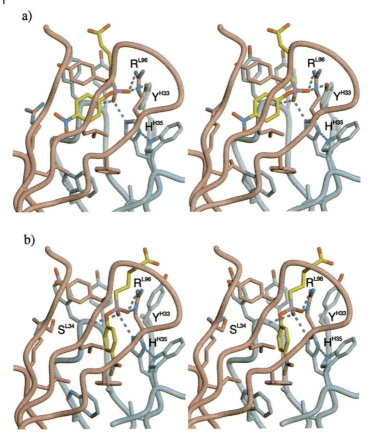

Fig. 2.12 Stereoview of the active site of mature 48G7 (**a**) and its germline precursor (**b**), showing the different orientation of the hapten in the two complexes [85–87]. The phosphonate moiety of **6** binds in roughly the same location in both, although the network of hydrogen-bonding interactions that constitute the oxyanion hole is slightly different. The deeply bound aryl group binds in one of two available hydrophobic clefts at the bottom of the pocket depending on whether the residue at position L34 is Ser (germline) or Gly (mature).

present, but empty, in mature 48G7. The alternative binding mode is made possible by removal of an otherwise repulsive interaction with the nitro group of the hapten in the mature antibody by the somatic mutation Ser^{L34}Gly.

Concordant with hapten design, the crystallographic data suggest that 48G7 is a relatively simple catalyst which promotes ester hydrolysis by direct attack of hydroxide on the scissile carbonyl. Although the two hapten binding modes seen in the germline and mature 48G7 complexes create some ambiguity about the substrate's preferred orientation, the side chains of Tyr^{H33}, His^{H35} and Arg^{L96} and the backbone amide of Tyr^{H96} apparently stabilize the tetrahedral and anionic transition states electrostatically through hydrogen-bonding and ionic interactions (Fig. 2.12). An analogy between this anion binding site and the well-characterized "oxyanion hole" of serine proteases is apparent. Individually, however, the antibody residues are not very effective oxyanion stabilizers. Mutations at positions H33, H35 and L96 cause only 3 to 30-fold reductions in k_{cat} [85]. Loss of the hydrogen bond between Tyr^{H33} and the transition state probably contributes to the germline's 20-fold lower efficiency, as well. In contrast, replacement of a single asparagine in the oxyanion hole of the protease subtilisin results in 10^2 to 10^3-fold losses in specific activity [94, 95]. High solvent accessibility and/or conformational mobility of the antibody residues may account for their limited efficacy.

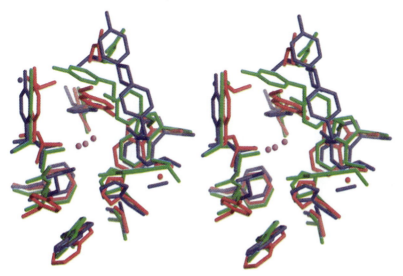

Fig. 2.13 Overlay of three hydrolytic antibodies, 48G7 (green), 17E8 (blue), and CNJ206 (red), shows remarkable structural convergence in these independently-generated active sites [96]. Only hapten-contacting side chains are illustrated; the purple balls indicate the position of the phosphorus of the respective bound hapten.

The 48G7 active-site structure and hapten-recognition properties are strikingly similar to those of other, independently-derived *anti*-phosphonate antibodies [96]. These include three catalysts (CNJ206 [80, 81], 17E8 [89] and 29G11 [90]) for the cleavage of *p*-nitrophenyl esters and three (D2.3, D2.4 and D2.5 [83, 84]) for the energetically more demanding hydrolysis of *p*-nitrobenzyl esters. Although many of these esterolytic antibodies derive from different germline sequences, they appear to have in common a deep hydrophobic pocket in the framework region into which the leaving group's aryl/benzyl ring binds (Fig. 2.13). All exploit multiple hydrogen bonds and salt bridges near the mouth of the cavity for recognition of the phosphonate moiety. Consequently, it is likely that all employ the same basic hydrolytic mechanism as 43G7 [96]. Given this, their comparative efficiency, which ranges over two orders of magnitude, must reflect differing abilities to stabilize the hydrolytic transition state relative to the bound ground state. Examination of the hapten complexes suggests that the rate enhancement ($k_{cat}/k_{uncat} = 10^3 - 10^5$) roughly parallels the number of hydrogen bonds to the phosphonate. Flexibility in the active site (as seen in the 48G7 germline antibody) also correlates with low efficiency [80, 81, 87]. In some cases, the antibodies may exploit binding energy to distort the substrate ester from its thermodynamically favorable Z-conformation, making it easier to reach the tetrahedral transition state during catalysis [82, 85]. For example, this factor may come into play in 48G7, depending on whether the substrate binds like the hapten in the mature or the germline complex (Fig. 2.12).

The remarkable degree of structural convergence observed in antibodies selected for tight binding to aryl phosphonate transition-state analog finds parallels in a number of other systems. For example, the immune responses to phosphorylcholine [97], *p*-azophenylarsonates [98], and 2-phenyloxazolones [99] are dominated by specific combinations of heavy and light chain variable regions. Interestingly, when a phenyloxazolone-binding antibody was found with a unique V_H domain, its structure showed conservation of important antigen binding residues [100]. These results point to the immune system's utilizing a relatively limited number of mechanisms to recognize any given type of antigen. Apparently, strong selective pressure reduces the broad diversity initially present in the primary repertoire to a small set of "best" solutions that can be further optimized by somatic mutation. As discussed above for the Diels-Alderases, structural convergence is even evident in antibodies raised against unrelated haptens. Rather than possessing an infinite variety of differently configured active sites, the immune system appears to play with a limited deck.

Variations within the general theme do arise. Another esterase that has been structurally characterized, antibody 6D6, catalyzes the hydrolysis of a chloramphenical monoester with relatively low efficiency (k_{cat}/k_{uncat} = 900) [101]. It shows specific differences from the other antibodies [82]. In particular, the hapten is bound more shallowly, although the stacked aromatic rings of the leaving group and the acyl side chain are the most deeply buried portions of the molecule. The tetrahedral phosphonate is highly solvent exposed, forming only one hydrogen bond to the antibody, presumably explaining 6D6's relatively low efficiency.

At this point, it is not clear how much activity can be obtained from *anti*-phosphonate antibodies. Over 50 esterolytic antibodies have been shown to have properties roughly comparable to those discussed here: simple hydrolytic mechanisms, rate accelerations up to 10^5 over background, and chemical proficiencies of $10^7 - 10^8 M^{-1}$ [15, 16]. While notable, particularly considering concomitant high and predictable selectivity [3], such effects are still orders of magnitude lower than those achieved by analogous enzymes. The immunological approach appears to have hit a ceiling in catalytic efficiency. It seems likely that these activities reflect what can be achieved with an "oxyanion hole" mechanism alone. Recent attempts to augment activity by rational mutagenesis or affinity selection with phage-displayed antibody fragments have met with only limited success [102–105]. Negative results should not be overinterpreted, but they do raise concerns that the aryl phosphonate binding pocket common to these catalysts is not an intermediate that can be further refined but an evolutionary dead end. Alternative antigen presentation strategies and *in vitro* selection methods may provide access to different subsets of the immune repertoire more amenable to optimization. For example, antibodies that bind tetrahedral anions not at the entrance to the combining site but deep within their active site, as seen for enzymes that promote hydrolyses, would be of interest.

Of course, highly evolved hydrolytic enzymes are more than simple oxyanion holes. They exploit arrays of catalytic groups (and, often, metal cofactors) to catalyze energetically demanding reactions such as amide hydrolysis. Induction of several precisely aligned functional groups in a single immunization step is extremely improbable, however. Such arrays are unlikely to be present in the primary repertoire of the immune system, and, depending on the basic immunoglobulin scaffold, may not be accessible through somatic hypermutation. Occasionally, though, serendipitous mutations that open up new opportunities for catalysis can and do occur.

Although phosphonate transition-state analogs specify a simple hydrolytic mechanism, some *anti*-phosphonate antibodies have been identified that exploit more complex mechanisms. For example, the lipase-like antibody 21H3, generated against a typical benzyl phosphonate ester [106], unexpectedly accelerates ester hydrolysis by a two-step mechanism involving transient acylation of an amino acid within the binding pocket [107]. Its efficiency at hydrolyzing esters is no greater than that of the esterolytic antibodies discussed above, but use of covalent catalysis makes possible stereoselective transesterification reactions that cannot be carried out in water without the antibody [107, 108]. Similarly, considerable biochemical evidence has been adduced for a two-step sequence involving an acyl-antibody intermediate in hydrolytic reactions catalyzed by antibody 43C9 [109]. The latter was generated against phosphonamidate **8** [110], rather than a phosphonate, and it is unique in its ability to promote the hydrolysis of activated amides as well as esters (Fig. 2.14). A 2.5×10^5-fold rate enhancement (chemical proficiency = 5×10^8 M^{-1}) achieved in the cleavage of *p*-nitroanilide **9** makes 43C9 one of the most efficient hydrolytic antibodies known.

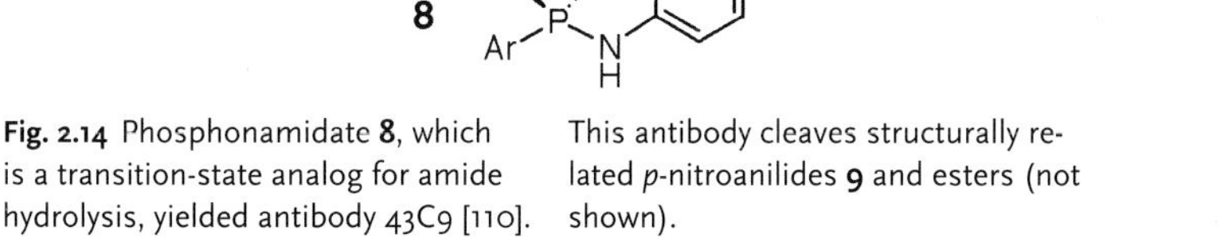

Fig. 2.14 Phosphonamidate **8**, which is a transition-state analog for amide hydrolysis, yielded antibody 43C9 [110]. This antibody cleaves structurally related *p*-nitroanilides **9** and esters (not shown).

The crystal structures of free 43C9 and its complex with *p*-nitrophenol were recently solved [88]. Although detailed understanding of interactions specific to ligand recognition and catalysis await characterization of the hapten complex, likely participants in the reaction are identifiable upon inspection of the binding pocket (Fig. 2.15). As with other hydrolytic antibodies, several residues are available for stabilizing the negative charge that develops in the transition state, many of which are shared with

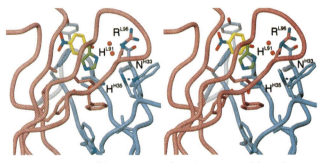

Fig. 2.15 Stereoview of the active site of amidase 43C9 with bound p-nitrophenol [88]. The imidazole side chain of His^{L91} (green) points into the pocket toward the region normally occupied by the phosphonate in other hydrolytic antibodies. Asn^{H33}, His^{H35} and Arg^{L96} are potential oxyanion stabilizers. Two active-site water molecules are shown as red spheres.

48G7. These include His^{H35} and Arg^{L96}, whose importance for catalysis is supported by the results of site-directed mutagenesis experiments [103]. Residue Asn^{H33}, like 48G7's Tyr^{H33}, may also play a role in transition-state stabilization. What distinguishes 43C9 from other esterolytic antibodies, though, is the presence of a second histidine at position L91. His^{L91} is directed into the pocket toward the region where substrate must bind (Fig. 2.15). Docking experiments have suggested a plausible orientation of the substrate, placing its scissile carbonyl in an excellent position for nucleophilic attack by this residue's imidazole side chain [88]. Detailed mechanistic inferences should be considered tentative in the light of likely perturbations to the binding pocket caused by packing interactions between the active sites of adjacent molecules in the crystal [88]. Nevertheless, mutagenesis of His^{L91} to Gln decreases catalytic efficiency >50-fold with little effect on ligand binding, supporting its role as the nucleophile that is transiently acylated during catalysis [103]. Consistent with this possibility, the acyl intermediate detected by electrospray mass spectrometry at pH 5.9 is not observed with the His^{L91}Gln variant [111].

Covalent catalysis by imidazole is well-established in non-enzymatic reactions of carboxylic acid derivatives [1]. Its efficiency is a consequence of imidazole's high nucleophilicity and the relative instability of the acyl-imidazole intermediate. The presence of His^{L91} can thus explain why 43C9, but not 48G7 or other esterolytic antibodies, cleaves an amide. Context is clearly relevant, since lipase 21H3 lacks amidase activity though it also exploits nucleophilic catalysis [106, 107]. The fact that 43C9's mechanism is unspecified by its hapten's design also underscores the importance of serendipity in these experiments. The rarity of such occurrences is reflected in many failed experiments to generate amidases with phosphonamidate haptens.

Despite its relative mechanistic sophistication, 43C9 is still a primitive amidase when compared with a typical protease. Subtilisin, which also exploits nucleophilic catalysis, catalyzes the cleavage of succinyl-Ala-Ala-Pro-Phe-p-nitroanilide 10^4 times more efficiently than 43C9. Its rate acceleration is 3.9×10^9 over background and its chemical proficiency is 2.2×10^{13} M^{-1} [112]. Moreover, subtilisin can cleave unactivated amides, whereas 43C9 is restricted to substrates with leaving groups of $pK_a <$ ca. 12 [113].

The high efficiency of covalent catalysis in serine proteases like subtilisin derives from cooperative action of several functional groups in addition to the active site nucleophile. Acid-base chemistry, in particular, is used to activate the serine nucleophile, facilitate proton transfers, and stabilize the amide leaving group. Removal of any single component of the protease's catalytic triad results in 10^4 to 10^6-fold decreases in activity [112]. Unsurprisingly, 43C9 lacks an analogously complex catalytic machinery. Although it has proven difficult to install additional functional groups by mutagenesis [103], more productive changes may become obvious when the structure of the hapten complex is known. Lessons from the immunological evolution of 48G7 [53, 87] suggest that mutagenesis of residues outside the binding pocket will be required to fine-tune critical synergies between participants in catalysis.

2.5.4
Functional Groups

Functional groups – acids, bases, and even exogenous cofactors – can clearly extend the capabilities of catalytic antibodies. They will certainly be needed if reactions with large activation barriers are to be significantly accelerated. As we have seen in the case of amidase 43C9, unplanned but useful residues can appear by chance in the immunoglobulin pocket during affinity maturation. To increase the probability of eliciting such functional groups where they are needed, several strategies have been devised.

Charge complementarity between an antibody and its ligand is the easiest principle to exploit in generating functionalized binding pockets. Cationic haptens have been used, for instance, to elicit negatively charged carboxylates that can serve as bases or nucleophiles or as general acids when protonated. Elimination reactions, epoxide ring openings, cationic cyclizations, and hydrolyses of esters, ketals, and enol ethers have been successfully catalyzed by this approach [114]. In favorable cases, very large catalytic effects have been achieved. For example, antibodies raised against a protonated benzimidazolium derivative use an active-site Asp or Glu to deprotonate substituted benzisoxazoles with effective molarities of 40 000 M and rate accelerations in excess of 10^8 over the acetate-promoted background reaction [115].

Nevertheless, it is unlikely that a single hapten will ever elicit arrays of residues as sophisticated as those present in highly evolved enzymes. Heterologous immunization has been explored as a method of circumventing this limitation [116]. In this procedure, two molecules, each containing different functional groups, are serially used as haptens to elicit the immune response. Ideally, a subset of the resulting antibodies will possess multiple catalytic groups induced in response to both templating molecules and have, as a result, enhanced activities. An important advantage of this strategy is that simplified haptens can be used, reducing the need for laborious synthesis. Although generality must still be established, results from initial studies indicate that antibody esterases generated by heterologous immunization are more efficient than those generated in response to the individual haptens [116, 117].

Mechanism-based enzyme inhibitors [17, 18] are potentially of even greater utility than haptens [19, 118–122]. Such molecules exploit a protein's ability to initiate a cascade of events, ultimately leading to its own covalent modification. This irreversible chemical reaction thus provides a means of selecting immunoglobulins *in vivo* and/or *in vitro* on the basis of their activity. When the selected antibody is challenged with substrate rather than hapten, the same group(s) responsible for protein modification can be used to promote the desired chemical transformation. Covalent catalysis can thus be specified through hapten design. The potential of reactive immunization is perhaps best illustrated by the production of aldolase antibodies [120]. These catalysts not only utilize a complex chemical mechanism but are among the most efficient catalytic antibodies described to date.

2.5.4.1
Aldolases

Aldol condensations are broadly useful carbon-carbon bond-forming reactions in organic synthesis. They pose special difficulties for biocatalysts, however, because they proceed via a series of consecutive transition states, each requiring acid-base catalysis. Class I aldolases solve these problems by using a reactive lysine to activate the ketone donor through Schiff base formation at the active site [123–125]. Deprotonation of the Schiff base yields an enamine that then adds stereoselectively to the acceptor aldehyde to form a new carbon-carbon bond. Subsequent hydrolysis of the Schiff base releases product and regenerates the active catalyst.

β-Diketones inhibit class I adolases by forming a Schiff base with the active-site lysine and then rearranging to a more stable vinylogous amide. When β-diketone **10** was used as a hapten instead of a more conventional transition-state analog [120], two analogously modifiable antibodies (33F12 and 38C2) were obtained (Fig. 2.16). The reactive group on the antibodies is a lysine with an anomalously low pK_a (5.5 for 33F12 and 6.0 for 38C2 [123]). The vinylogous amide it forms with the hapten (λ_{max} 316 nm, ε 15 000 M^{-1} cm^{-1}) can be irreversibly trapped by reduction with sodium cyanoborohydride. These same antibodies also mimic the activity of class I aldolases [120]. Their reactive lysine reacts with a wide range of ketones to form enamine adducts. These condense with diverse aldehydes to form aldol products.

The structure of antibody aldolase 33F12 [123] shows the catalytic lysine (Lys^{H93}) buried at the bottom of a hydrophobic pocket (Fig. 2.17). It is not hydrogen bonded with other amino acids and there are no charged residues within 7Å. The absence of such interactions must be responsible for this group's unusual reactivity: the hydrophobic microenvironment lowers the amine's pK_a by disfavoring the protonated state. In contrast, class I aldolases are believed to increase the acidity of their catalytically essential lysine through proximity to several positively charged residues [126].

Unfortunately, the unliganded antibody provides few clues about the interactions that stabilize the transition states for formation and breakdown of the carbinol amine, deprotonation to afford the enamine, and creation of the new carbon-carbon bond. Sequestered water molecules or the hydroxyl groups of Tyr or Ser residues that con-

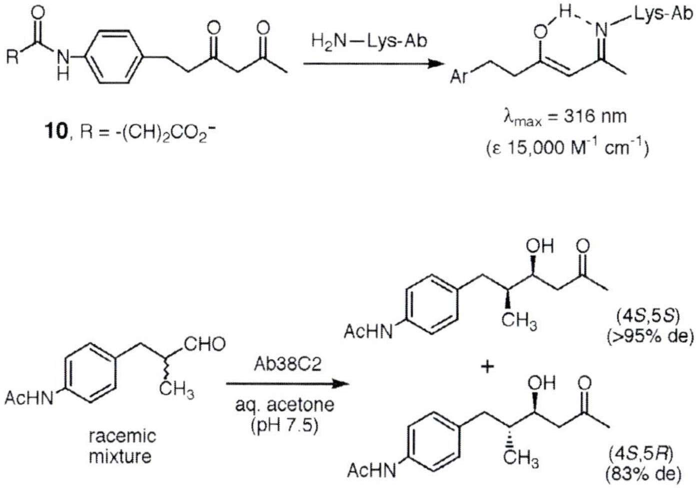

2.5 Representative Catalytic Antibodies

[chemical scheme showing compound 10, R = -(CH)$_2$CO$_2^-$, reacting with H$_2$N–Lys-Ab to form vinylogous amide product with Lys-Ab; λ_{max} = 316 nm (ϵ 15,000 M^{-1} cm^{-1})]

[scheme showing racemic mixture of AcHN-aryl-CH(CH$_3$)-CHO reacting via Ab38C2 in aq. acetone (pH 7.5) to give two aldol products: (4S,5S) >95% de and (4S,5R) 83% de]

Fig. 2.16 β-Diketone **10**, used as a hapten to raise antibodies 38C2 and 33F12, traps a reactive, active-site amine to form a stable, chromophoric vinylogous amide [120]. These antibodies promote diverse aldol condensations. In the example shown (bottom), the two products are formed in a ca. 1:1 ratio with the indicated diastereoselectivities.

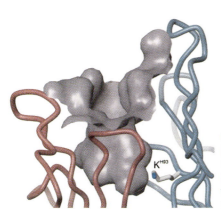

Fig. 2.17 The binding pocket of antibody 33F12 [123] is seen through a slice in the molecular surface calculated with a sphere of 1.4 Å radius [206]. Only the ε-amino group of Lys^{H93} contacts the molecular surface at the bottom of the antigen binding site.

stitute the pocket walls may be involved. For instance, a simple rotation of the Lys side chain from its position in the unliganded antibody would bring its amino group near Ser^{H100}, which lies on the opposite side of the pocket. Structures of enamine adducts will be important for clarifying these points.

An unanticipated feature of the aldolase antibodies is their promiscuity. Over 100 different reactions, including aldehyde-aldehyde, ketone-aldehyde and ketone-ketone condensations, are subject to catalysis [123]. Because the 33F12 pocket is relatively hydrophobic, polyhydroxylated aldehydes, such as glyceraldehyde, glucose, and ribose, which are good acceptors for natural aldolases, are not substrates for the antibodies. Aside from this restriction, a wide range of donors and acceptors is tolerated. Non-specific van der Waals interactions likely provide the driving force for sequestering the first substrate. Once bound, it encounters the reactive lysine and forms the nucleophilic enamine. Provided there are no steric clashes, similar interactions should allow the aldehyde acceptor to bind and undergo aldol addition. The binding site is 11 Å deep and quite capacious, much larger in fact than the β-diketone that induced it, accounting for this broad specificity. These properties have been rationalized as a consequence of the reactive immunization process itself [123]: capture of the antibody by a covalent chemical event early in the process of affinity maturation may obviate the need for further refinement of the binding pocket by somatic mutation.

Although induced with an achiral hapten and despite their broad substrate specificity, these aldolase antibodies are surprisingly stereoselective (Fig. 2.16). When acetone is the donor substrate, addition preferentially occurs on the si-face of the aldehyde acceptor; with hydroxy acetone, attack occurs on the re-face. In most cases, enantiomeric excesses >95% are found [127]. Enantioselective Robinson annulations [128], resolution of tertiary aldols [129], and preparation of chiral intermediates for the total synthesis of brevicomins [130] and epothilones [5] illustrate the synthetic utility of these catalysts. Kinetic resolution of the epothilone precursor was carried out on a gram scale using 0.06 mol% of antibody 38C2 (corresponding to 0.5 g of IgG). The reaction proceeded with good yield (37%) and high enantiomeric excess (90%). More recently, antibody 38C2 has been used to activate prodrugs [6]. It catalyzes the selective removal of generic drug-masking groups via sequential retro-aldol and retro-Michael reactions. This advance could prove useful in the development of selective chemotherapeutic strategies.

How do these remarkable antibody aldolases compare to natural aldolases? To address this question, it is easiest to consider representative retroaldol reactions. For such transformations, the antibodies achieve turnover numbers ranging from 0.0003 to 0.08 s^{-1} [127]. One of the best antibody substrates is 4-(4'-isobutyramidophenyl)-4-hydroxy-2-butanone. In the presence of 38C2, it undergoes retroaldolization with k_{cat} and k_{cat}/K_m values of 0.083 min^{-1} and 3.3×10^3 M^{-1} s^{-1}, respectively. The rate constant for the background reaction is 1.4×10^{-9} s^{-1} (aqueous buffer, pH 7 and 25 °C). The antibody thus accelerates this reaction by a factor of 2×10^7-fold; its chemical proficiency is 6×10^{10} M^{-1}. These are impressive effects for a catalytic antibody, and may be contrasted with activity accruing to FDP aldolase and KDPG aldolase, typical class I enzymes involved in sugar metabolism.

FDP aldolase catalyzes the interconversion of fructose-1,6-diphosphate (FDP) to give dihydroxyacetone phosphate and glyceraldehyde-3-phosphate. Its steady state kinetic parameters in the cleavage direction are k_{cat} = 48 s^{-1} and k_{cat}/K_m = 1.6×10^7 M^{-1} s^{-1} [131]. KDPG aldolase, which cleaves 2-keto-3-deoxy-6-phosphogluconate (KDPG) into pyruvate and glyceraldehyde-3-phosphate, is even more active, having k_{cat} and k_{cat}/K_m values of 290 s^{-1} and 4.0×10^6 M^{-1} s^{-1}, respectively [132]. Although rate constants for the corresponding uncatalyzed reactions are unavailable, taking 1.4×10^{-9} s^{-1} as a conservative estimate [127] yields rate enhancements of ca. 10^{10}–10^{11} and chemical proficiencies of ca. 10^{15}–10^{16} M^{-1}. By either criterion, the natural enzymes are several orders of magnitude more efficient than the antibody aldolases.

Natural aldolases are more restrictive in their substrate requirements than the antibodies, though they could lose considerable activity and still be competitive. For example, removal of a single phosphate in FDP causes a 50-fold loss in rate with FDP aldolase [131]. The extensive interactions that confer such specificity are almost certainly coupled to the enzyme's high efficiency. By further refining the antibody pocket to the precise steric and electronic demands of a particular aldol reaction, while maintaining the active-site lysine's high reactivity, enzyme levels of activity may be attainable. Narrowed scope may be the unavoidable cost of truly high efficiency, however.

2.6
Perspectives

2.6.1
General Lessons from Comparisons of Enzymes and Antibodies

Structural and mechanistic work reviewed here reveals many notable parallels between antibodies and their more highly evolved counterparts. Not only are the sizes and shapes of their active sites comparable, but antibodies and enzymes utilize the same set of molecular interactions to bind their respective ligands and stabilize transition states. Although antibodies tend to be less extensively functionalized than enzymes, the basic mechanistic strategies they employ to lower kinetic barriers are strikingly similar. As more primitive catalysts, however, they provide an alternative vantage point for examining the relationship between binding energy and catalysis. In this regard, simplicity is a virtue. Rather than working backwards from a fully evolved enzyme, uncomplicated, tailored model systems can be constructed to illuminate specific mechanistic questions. As multiple mechanistic strategies are combined to augment efficiency, valuable insight into the evolution of catalytic function can be gained. Functional analysis of the antibody intermediates that arise during affinity maturation [38, 53, 85, 87] also sheds light on these issues.

Currently, antibodies appear less successful than enzymes in their ability to achieve the fine level of recognition required for optimal discrimination between transition states and ground states. Their modest efficiencies appear to be a direct consequence of the simple strategy used to generate them. While the process of natural selection optimizes enzymes on the basis of their catalytic activity, the immune system's microevolutionary mechanisms select antibodies for increased affinity to an imperfect transition-state analog. It is unrealistic to expect that proteins engineered to recognize such haptens will provide an ideal steric and electrostatic environment for chemical transformation. Even with a perfect transition-state analog, the chances of obtaining a fully evolved catalyst through immunization would be low. As noted above, there is generally insufficient selection pressure to attain the high binding energies that characterize complexes between true enzymes and their transition states. On both micro and macro levels, mechanistic improvements arise as a function of time, so differences in time scales for the evolution of enzymes and antibodies – millions of years versus weeks or months – also come into play.

Although Nature uses a wide variety of different protein scaffolds to build enzyme active sites [133], she does not seem to have adopted the immunoglobulin fold. It is therefore conceivable that antibody structure itself places intrinsic limitations on the kind of reactions amenable to catalysis and on attainable efficiencies. In general, though, structural studies show excellent shape and chemical complementarity between antibodies and their ligands. Depending on the hapten, deep pockets, clefts, grooves, and flatter, more undulating surfaces can be created [134, 135]. Because certain classes of haptens tend to be recognized in the same way [52, 53, 83, 96], structural diversity must be considerably more restricted than might have been expected given 10^8 variants available in the primary immune repertoire [136, 137], but whether these consensus sites significantly restrict the catalytic capabilities of antibodies is still unclear.

Conformational flexibility is another potential concern. Protein conformational changes in enzymes provide a means of excluding water from the active site and enable the catalyst to adjust to changes in substrate as the reaction coordinate is traversed [138, 139]. Antibodies are known to undergo a comparable range of ligand-induced conformational changes, including alterations in side chain rotamers, segmental movements of hypervariable loops and changes in the relative disposition of the V_H and V_L domains [134]. Without direct selection for activity, however, these dynamic effects will be difficult to exploit deliberately for catalysis. In fact, conformational flexibility in catalytic antibodies, when observed [80, 81, 87], usually results in lower rather than higher efficiency.

2.6.2
How efficient does catalysis need to be?

Enzymes represent an extraordinarily high standard against which to judge new catalysts that are rationally designed from simple principles. From the perspective of the chemist, one exciting aspect of catalytic antibody technology is its ability to deliver tailored catalysts for reactions which are difficult to carry out selectively using existing methods or for which natural enzymes do not exist. Must such systems attain enzyme-like efficiency to be useful?

Because antibodies are biocompatible and have long serum half-lives, many *in vivo* applications would be conceivable if sufficient activity were available. In fact, existing catalytic antibodies have already achieved significant effects in biological systems. When expressed at high levels, they have been shown to be competent (if inefficient) catalysts in metabolism, replacing essential enzymes in amino acid [42] and pyrimidine biosynthesis [77] pathways. Therapeutically relevant concentrations of the aldolase antibodies discussed earlier have been used to activate prodrugs and kill colon and prostate cell lines [6]. Similarly, an esterolytic antibody has been employed as a cocaine antagonist, protecting rats from cocaine-induced seizures and sudden death [140, 141].

Many chemical reactions cannot proceed in the absence of catalysts because competing pathways have lower energies. In several instances, antibody binding energy has been successfully utilized to alter the course of such reactions by selectively stabilizing the less favorable transition state. For example, antibody catalysts have been developed for normally disfavored *syn*-eliminations [142], *exo* rather than *endo* Diels-Alder cycloadditions [46, 48], and 6-*endo*-tet ring closures of epoxy alcohols [143, 144]. Antibodies have also been used to control the fate of high-energy intermediates, allowing them to partition along only one of several possible pathways, as in the case of conversion of an enol ether to a cyclic ketal in water [145]. Formation of a strained cyclopropane derivative in a cationic olefin cyclization is another such example [146]. Binding energies up to 5.5 kcal/mol are typically available for achieving such discrimination, and even more energy may be available in favorable cases [15]. Such selectivities could be of great utility in organic synthesis.

Assuming an antibody is available for any given transformation, its turnover and cost will ultimately determine whether it is used in practice. Presently, the steady state parameters of typical catalysts necessitate high antibody concentrations (≥ 10 µM = 1.6 mg/ml) and long reaction times to achieve useful conversions [12]. Preparative applications of several antibodies show that gram-scale reactions are feasible, particularly if antibody selectivity is high and competing reactions are substantially slower than the desired transformation [5, 147, 148]. Costs associated with high-volume antibody production are certainly an issue, but some antibodies are now produced on a large scale for diagnostic and medical applications. They are readily obtained in good yield through ascites production [149] or by fermentation in hollow fiber reactors [150, 151]. Their Fab and Fv fragments can often be produced efficiently in plants [152, 153] or microorganisms [154]. Technical advances in microbial fermentation can be expected to make antibody production even more economically favorable in the future.

In short, current levels of activity may be adequate for some laboratory applications, but higher efficiencies would certainly be beneficial. A 10^3-fold increase in turnover would mean that 10^3-fold less catalyst is needed to achieve useful levels of performance. Enhanced proficiency will certainly be necessary if energetically more demanding reactions are to be tackled. The creation of antibody equivalents of site-specific proteases, glycosidases, and nucleases, for example, remains a significant yet unrealized goal. The use of antibodies to synthesize or modify structurally complex and biologically important macromolecules will depend on solving this basic problem.

2.6.3
Strategies for Optimizing Efficiency

If imperfect design and indirect selection for binding rather than function are the primary reasons for low catalytic efficiency, creation of substantially better antibody catalysts will be feasible. Conceptually, two approaches can be envisaged: (1) refining methods for producing first generation catalysts, and (2) developing new strategies to optimize existing active sites. Improved transition-state analogs and more effective screening of the immune response address the first point. Rational reengineering and directed evolution methods are relevant to the second. These strategies have already been discussed in the context of specific antibody catalysts but are summarized in more general terms below.

2.6.3.1
Better Haptens

The rewards of good – and the penalties of deficient – design are evident in the properties of catalytic antibodies characterized to date. In general, however, it is not clear *which* structural features of a transition state are most important to mimic to elicit maximally effective catalysts. Statistically meaningful correlation of different hapten types and the properties of their complementary active sites are needed to optimize analogs for each type of reaction.

Incorporating design features that maximize transition-state affinity while minimizing ground-state stabilization remains a major challenge. For example, aryl groups are constituents of many haptens. Binding energy directed toward them, while increasing hapten affinity, may be useless or even harmful for catalysis, since these elements are common to both ground and transition state. Inefficient utilization of intrinsic binding energy in this way may help to explain the modest activities seen in the oxy-Cope [38] and Diels-Alder reactions [53] discussed above. The finding that catalytic activity for a series of polyclonal esterases correlates inversely with the size and hydrophobicity of the haptenic aryl phosphates [155] also illustrates this problem. When binding energy is used to recognize parts of the substrate distant from the site of reaction, product inhibition becomes another concern. Strategies to facilitate product release must therefore be considered integral to hapten design. Antigen presentation is a further issue. Small molecules are not immunogenic and must be coupled to carrier proteins. Variation of the tether site may be a useful means of focusing immunorecognition to a hapten's catalytically relevant epitopes [141].

Linkage of chemical reactivity with the immune system's selection and amplification processes may mitigate some limitations in design. For this reason, the reactive immunization strategy [19, 118–122] merits increased attention. Many additional examples will be needed to establish its true scope and limitations. Given their importance in natural enzymes [156], metal ions and exogenous organic cofactors should considerably extend the properties of antibody catalysts, as well. Although versatile hybrid catalysts that combine the intrinsic reactivity of the metal ion or coenzyme with the tailored binding specificity of the antibody can be readily envisaged, this strategy has been surprisingly underutilized. Metal ion binding sites have been engineered into antibody binding sites [157], creating a sensitive $Zn(II)$ sensor in one case [158], but catalysis has not been realized. An alternative, seemingly promising strategy using metal chelates for peptide cleavage has received little attention since first reported [159]. Modest activities have been described for other miscellaneous cofactor-dependent reactions [158, 160–163], but more work is obviously needed. Such strategies will be very important for promoting reactions with high kinetic barriers and reactions that cannot be carried out with protein residues alone.

2.6.3.2
Screening

Because unusual germline sequences or fortuitous mutations may be necessary for high activity, the best antibody catalysts are also potentially the rarest. For this reason, more extensive screening of the immune response may dramatically increase the probability of finding highly active clones. Usually, small panels of antibodies chosen for their ability to bind the transition-state analog are purified and tested individually for catalytic activity. This procedure is necessarily indirect and slow. Sensitive chemical [164], biological [42, 77, 91, 165] and immunological assays [60, 166, 167] can facilitate the screening of thousands of candidates directly for catalysis and thereby accelerate the preliminary evaluation process.

One practical problem associated with broad screening is that some antigens yield many hapten binders, others relatively few. To increase the size and diversity of the antibody population available for testing, multiple fusions can be performed and several mouse strains utilized for immunization. Mice prone to autoimmunity have been shown to yield unusually large numbers of esterolytic antibodies and may prove more generally useful for expanding the repertoire of catalytic clones elicited by a single transition-state analog [168].

Significant progress has also been made in copying the combinatorial processes of the immune system *in vitro*. Libraries of antibody fragments containing more than 10^6 members can be constructed and produced in microorganisms or displayed on phage particles [169–172]. These systems are attractive vehicles for exploring the catalytic potential of different subsets of the primary immunological repertoire. Binders can be selected from these libraries on the basis of hapten affinity and subsequently screened for catalytic activity [169]. Clever strategies for capturing active clones based on their activity should be even more effective [173–177]. Alternatively, catalysts can be obtained directly by selection *in vivo* using yeast or bacterial auxotrophs [42, 77, 91]. In these approaches, iterative rounds of mutagenesis and reselection replace somatic mutation as a means of refining initial hits [170]. Domain swapping [178] and powerful DNA shuffling methods [179, 180] have been developed to speed up this process.

2.6.3.3
Engineering

The upper limit on activity that can be achieved with antibodies is unknown and may be reaction dependent. It is therefore important to push several test cases as far as possible. Site-directed mutagenesis is an attractive tool for improving catalytic power, particularly given the availability of increasing numbers of high-resolution structures. In general, it will probably be easiest to reengineer the poorest catalysts, since changes that improve packing or provide missing but critical interactions may be relatively obvious. The mutational study leading to an order of magnitude increase in activity of the Diels-Alderase antibody 39-A11 is a case in point [54]. The fact that relatively few changes are needed to tailor the properties of germline structure during affinity maturation [53, 85, 93] is also encouraging. Pinpointing subtle changes needed to optimize more active clones is likely to be more difficult, however. The obstacles encountered in augmenting the activity of hydrolytic antibodies sound a cautionary note [103–105]. Ultimately, our understanding of will determine what can be achieved in this way.

2.6.3.4
Selection

Enzymes have been brought to peak efficiency over millions of years by the process of natural selection. An analogous process in the laboratory, involving recursive cycles of mutagenesis and genetic selection for function, may provide the ultimate test of the capabilities of antibody catalysts. Evolution of antibodies can be accomplished perhaps most directly by complementation of auxotrophic yeast or bacterial strains. The chorismate mutase antibody 1F7 [42] and an orotate decarboxylase antibody [77] have been shown to confer a significant growth advantage under selective conditions to host cells lacking the corresponding enzymes. Though the experiments are technically difficult because of poor antibody expression in microorganisms, preliminary results with the chorismate mutase antibody have demonstrated the feasibility of selecting antibodies with novel properties [43]. Recent work showing that cytoplasmic production of antibody fragments is optimizable by selection [181] augers well for these efforts. Selection systems are available for many transformations, including the ferrochelatase and metabolic reactions already mentioned [42, 70, 77]. A generalized selection scheme for hydrolytic reactions has also been reported [91]; analogous assays could be developed to exploit the ability of a catalyst to synthesize or destroy nutrients, drugs, hormones, or toxins. In addition to providing clues about the perfectibility of catalytic antibodies, such experiments may yield fundamental insights into structure-activity relationships and the evolution of molecular function.

2.6.3.5
Other Scaffolds

The immune system was originally tapped as a source of catalysts as a matter of convenience: it is unrivaled in its ability to fashion high affinity protein receptors – the antibodies – to virtually any natural or synthetic molecule, essentially on demand [58, 136, 137]. However, now that the combinatorial processes of the immune system can be mimicked *in vitro* and libraries of macromolecules can be generated relatively easily using the tools of molecular biology, there is no compelling need to restrict Jencks's original strategy to a single protein fold or even to a single class of macromolecule.

Indeed, in analogy to catalytic antibody experiments, catalytic RNAs and DNAs have been obtained from large libraries of nucleic acids by selection for binding to transition-state analogs. Catalysts for porphyrin metalation obtained in this way [182–184] have activity comparable to that of the ferrochelatase antibody 7G12 discussed above, but an RNA rotamase is 30 times less effective than its antibody counterpart [185]. The lower activity of the latter largely reflects the RNA's lower affinity for the hapten used (K_d = 7 µM, compared with 0.21 µM for the antibody). A similar explanation has been invoked for RNAs that bind the 1E9 hapten (compound 4, Fig. 2.6) but fail to catalyze the corresponding Diels-Alder reaction [186]. Nucleic acids are likely to be intrinsically more limited than proteins in their capacity for high affinity molecular recognition of structurally diverse ligands as well as for catalysis [187].

Direct selection for function rather than transition-state analog binding has proven to be a much more powerful approach for obtaining nucleic acids with novel catalytic properties (for a recent comprehensive review, see [188]). RNA and have been prepared in this way for a variety of reactions, including a Diels-Alder cycloaddition [189]. Although the resulting catalysts often have relatively modest efficiency, phosphoryl transfers, which are difficult to achieve with antibodies because of the dearth of stable analogs of the pentacoordinate transition state, have been particularly amenable to catalysis. One of the most impressive accomplishments in this regard is the selection of highly efficient ribozymes capable of a self-ligation reaction from large pools of random sequence [190]. The number of starting molecules in these experiments was huge (ca. 10^{15}), dwarfing the diversity of the primary immune repertoire (ca. 10^8 molecules), allowing even extremely rare catalysts to be found. One of the ribozymes obtained by selection was subsequently reengineered to function as a true catalyst; it promotes an intermolecular ligation with multiple turnovers and a rate acceleration approaching 10^9 [191]. This impressive activity is higher than any seen for most antibody catalysts, and provides an impressive demonstration of the power of direct selection.

In principle, *in vitro* selection of peptides and proteins from vast combinatorial libraries is now possible as well using ribosome display [192–194], mRNA-protein fusion methods [195, 196] and more established phage display formats [172]. Coupled with efficient ways of linking genotype with phenotype [174–176], these methods can be expected to facilitate the production of proteins with novel properties and functions. As such, they powerfully complement other efforts to harness the power of evolution to redesign the structures and activities of existing enzymes [197–199, 207, 208].

2.7
Conclusions

Catalytic antibody technology combines programmable design with the combinatorial diversity of the immune system. This fusion has allowed the field to progress in relatively short order from simple model reactions to complex multistep processes, but much remains to be learned. Early efforts focused largely on defining the scope and limitations of this technology. Now that the approach is well established, attention must be paid to strategies for optimizing catalytic efficiency and for promoting more demanding transformations. In many ways, these are far greater challenges than identifying first-generation catalysts with modest activity. Continued mechanistic and structural analysis of these systems will inform such endeavors. In addition, learning how to create, manipulate and evolve large combinatorial libraries of proteins outside the immune system should help to automate the processes of catalyst discovery and optimization.

Acknowledgements

The author is indebted to Kinya Hotta for creating the graphics for this article and to the National Institutes of Health, the ETH Zurich, the Swiss National Science Foundation, and Novartis Pharma for generous support.

References

1 JENCKS, W. P., *Catalysis in Chemistry and Enzymology*, McGraw Hill, New York **1969**
2 SCHULTZ, P. G., LERNER, R. A., *Science* 269 (**1995**), p. 1835–1842
3 HILVERT, D., In *Topics in Stereochemistry*, ed. S. E. Denmark 22, John Wiley & Sons, Inc., New York **1999**, p. 83–135
4 SCHULTZ, P. G., LERNER, R. A., *Acc. Chem. Res.* 26 (**1993**), p. 391–395
5 SINHA, S. C., BARBAS III, C. F., LERNER, R. A., *Proc. Natl. Acad. Sci. USA* 95 (**1998**), p. 14603–14608
6 SHABAT, D., RADER, C., LIST, B., LERNER, R. A., BARBAS III, C. F., *Proc. Natl. Acad. Sci. USA* 96 (**1999**), p. 6925–6930
7 RADZICKA, A., WOLFENDEN, R., *Science* 267 (**1995**), p. 90–93
8 WOLFENDEN, R., LU, X. D., YOUNG, G., *J. Am. Chem. Soc.* 120 (**1998**), p. 6814–6815
9 PAULING, L., *Am. Sci.* 36 (**1948**), p. 51–58
10 KURZ, J. L., *J. Am. Chem. Soc.* 85 (**1963**), p. 987–991
11 WOLFENDEN, R., *Nature* 223 (**1969**), p. 704–705
12 MADER, M. M., BARTLETT, P. A., *Chem. Rev.* 97 (**1997**), p. 1281–1301
13 RADZICKA, A., WOLFENDEN, R., *Methods Enzymol.* 249 (**1995**), p. 284–312
14 LOLIS, E., PETSKO, G. A., *Annu. Rev. Biochem.* 59 (**1990**), p. 597–630
15 STEWART, J. D., BENKOVIC, S. J., *Nature* 375 (**1995**), p. 388–391
16 THOMAS, N. R., *Appl. Biochem. Biotechnol.* 47: (**1994**), p. 345–372
17 ABELES, R. H., MAYCOCK, A. L., *Acc. Chem. Res.* 9 (**1976**), p. 313–319
18 WALSH, C., *Tetrahedron* 38 (**1982**), p. 871–909
19 WIRSCHING, P., ASHLEY, J. A., LO, C.-H. L., JANDA, K. D., LERNER, R. A., *Science* 270 (**1995**), p. 1775–1782
20 JENCKS, W. P., *Adv. Enzymol.* 43 (**1975**), p. 219–410
21 WESTHEIMER, F. H., *Adv. Enzymol.* 24 (**1962**), p. 441–482
22 BARTLETT, P. A., NAKAGAWA, Y., JOHNSON, C. R., REICH, S. H., LUIS, A. *J. Org. Chem.* 53 (**1988**), p. 3195–3210
23 GRAY, J. V., EREN, D., KNOWLES, J. R., *Biochemistry* 29 (**1990**), p. 8872–8878
24 HILVERT, D., NARED, K. D., *J. Am. Chem. Soc.* 110 (**1988**), p. 5593–5594
25 HILVERT, D., CARPENTER, S. H., NARED, K. D., AUDITOR, M.-T. M., *Proc. Natl. Acad. Sci. USA* 85 (**1988**), p. 4953–4955
26 JACKSON, D. Y., JACOBS, J. W., SUGASAWARA, R., REICH, S. H., BARTLETT, P. A., SCHULTZ, P. G., *J. Am. Chem. Soc.* 110 (**1988**), p. 4841–4842
27 JACKSON, D. Y., LIANG, M. N., BARTLETT, P. A., SCHULTZ, P. G., *Angew. Chem. Int. Ed. Engl.* 31 (**1992**), p. 182–183
28 LEE, A. Y., STEWART, J. D., CLARDY, J., GANEM, B., *Chem. Biol.* 2 (**1995**), p. 195–203
29 CAMPBELL, P. A., TARASOW, T. M., MASSEFSKI, W., WRIGHT, P. E., HIL-

vert, D., *Proc. Natl. Acad. Sci. USA* 90 (**1993**), p. 8663–8667
30 Haynes, M. R., Stura, E. A., Hilvert, D., Wilson, I. A., *Science* 263 (**1994**), p. 646–652
31 Addadi, L., Jaffe, E. K., Knowles, J. R., *Biochemistry* 22 (**1983**), p. 4494–4501
32 Gustin, D. J., Mattei, P., Kast, P., Wiest, O., Lee, L., Cleland, W. W., Hilvert, D., *J. Am. Chem. Soc.* 121 (**1999**), p. 1756–1757
33 Kast, P., Asif-Ullah, M., Jiang, N., Hilvert, D., *Proc. Natl. Acad. Sci. USA* 93 (**1996**), p. 5043–5048
34 Liu, D. R., Cload, S. T., Pastor, R. M., Schultz, P. G., *J. Am. Chem. Soc.* 118 (**1996**), p. 1789–1790
35 Wiest, O., Houk, K. N., *J. Org. Chem.* 59 (**1994**), p. 7582–7584
36 Braisted, A. C., Schultz, P. G., *J. Am. Chem. Soc.* 116 (**1994**), p. 2211–2212
37 Driggers, E. M., Cho, H. S., Liu, C. W., Katzka, C. W., Braisted, A. C., Ulrich, H. D., Wemmer, D. E., Schultz, P. G., *J. Am. Chem. Soc.* 120 (**1998**), p. 1945–1958
38 Ulrich, H. D., Mundorff, E., Santarsiero, B. D., Driggers, E. M., Stevens, R. C., Schultz, P. G., *Nature* 389 (**1997**), p. 271–275
39 Görisch, J., *Biochemistry* 17 (**1978**), p. 3700–3705
40 Kast, P., Asif-Ullah, M., Hilvert, D., *Tetrahedron Lett.* 37 (**1996**), p. 2691–2694
41 Steigerwald, M. J., Goddard, W. A., Evans, D. A., *J. Am. Chem. Soc.* 101 (**1979**), p. 1994–1997
42 Tang, Y., Hicks, J. B., Hilvert, D., *Proc. Natl. Acad. Sci. USA* 88 (**1991**), p. 8784–8786
43 Tang, Y., *Evolutionary studies with a catalytic antibody*, Ph.D., The Scripps Research Institute **1996**
44 Sauer, J., Sustman, R., *Angew. Chem. Int. Ed. Engl.* 19 (**1980**), p. 779–807
45 Hilvert, D., Hill, K. W., Nared, K. D., Auditor, M.-T. M., *J. Am. Chem. Soc.* 111 (**1989**), p. 9261–9262
46 Gouverneur, V. E., Houk, K. N., Pascual-Teresa, B., Beno, B., Janda, K. D., Lerner, R. A., *Science* 262 (**1993**), p. 204–208
47 Braisted, A. C., Schultz, P. G., *J. Am. Chem. Soc.* 112 (**1990**), p. 7430–7431
48 Yli-Kauhaluoma, J. T., Ashley, J. A., Lo, C.-H., Tucker, L., Wolfe, M. M., Janda, K. D., *J. Am. Chem. Soc.* 117 (**1995**), p. 7041–7047
49 Resmini, M., Meekel, A. A. P., Pandit, U. K., *Pure Appl. Chem.* 68 (**1996**), p. 2025–2028
50 Kirby, A. J., *Adv. Phys. Org. Chem.* 17 (**1980**), p. 183–278
51 Page, M. I., Jencks, W. P., *Proc. Natl. Acad. Sci. USA* 68 (**1971**), p. 1678–1683
52 Xu, J., Deng, Q., Chen, J., Houk, K. N., Bartek, J. D. H., Wilson, I. A., *Science* (**1999**), in press
53 Romesberg, F. E., Spiller, B., Schultz, P. G., Stevens, R. C., *Science* 279 (**1998**), p. 1929–1933
54 Romesberg, F. E., Schultz, P. G., *Bioorg. Med. Chem. Lett.* 9 (**1999**), p. 1741–1744
55 Haynes, M. R., Lenz, M., Taussig, M. J., Wilson, I. A., Hilvert, D., *Israel J. Chem.* 36 (**1996**), p. 151–159
56 Arevalo, J. H., Taussig, M. J., Wilson, I. A., *Nature* 365 (**1993**), p. 859–863
57 Arevalo, J. H., Hassig, C. A., Stura, E. A., Sims, M. J., Taussig, M. J., Wilson, I. A., *J. Mol. Biol.* 241 (**1994**), p. 663–690
58 Kabat, E. A., *Structural Concepts in Immunology and Immunochemistry*. Holt, Rinehart, and Winston, Inc., New York **1976**
59 Laschat, S., *Angew. Chem. Int. Ed. Engl.* 35 (**1996**), p. 289–291
60 MacBeath, G., Hilvert, D., *J. Am. Chem. Soc.* 116 (**1994**), p. 6101–6106
61 Haldane, J. B. S., *Enzymes*. Longmans, Green and Co., London **1930**
62 Lavalee, D. K., *Mol. Struct. Energ.* 9 (**1988**), p. 279–314
63 Al-Karadaghi, S., Hansson, M., Stanislav, N., Johnson, B., Hederstedt, L., *Structure* 5 (**1997**), p. 1501–1510
64 Dailey, H. A., Fleming, J. E., *J. Biol. Chem.* 258 (**1983**), p. 11453–11459

65 COCHRAN, A. G., SCHULTZ, P. G., *Science* 249 (**1990**), p. 781–783
66 HANSSON, M., HEDERSTEDT, L. *Eur. J. Biochem.* 220 (**1994**), p. 201–208
67 ROMESBERG, F. E., SANTARSIERO, B. D., SPILLER, B., YIN, J., BARNES, D., SCHULTZ, P. G., STEVENS, R. C., *Biochemistry* 37 (**1998**), p. 14404–14409
68 BLACKWOOD JR., M. E., RUSH III, T. S., ROMESBERG, F., SCHULTZ, P. G., SPIRO, T. G., *Biochemistry* 37 (**1998**), p. 779–782
69 WARSHEL, A., *J. Biol. Chem.* 273 (**1998**), p. 27035–27038
70 GORA, M., GRZYBOWSKA, E., RYTKA, J., LABBE-BOIS, R., *J. Biol. Chem.* 271 (**1996**), p. 11810–11816
71 CROSBY, J., STONE, R., LIENHARD, G. E., *J. Am. Chem. Soc.* 92 (**1970**), p. 2891–2900
72 O'LEARY, M. H., PIAZZA, G. J., *Biochemistry* 20 (**1981**), p. 2743–2748
73 ALSTON, T. A., ABELES, R. H., *Biochemistry* 26 (**1987**), p. 4082–4085
74 SUN, S. X., DUGGLEBY, R. G., SCHOWEN, R. L., *J. Am. Chem. Soc.* 117 (**1995**), p. 7317–7322
75 LEWIS, C., KRÄMER, T., ROBINSON, S., HILVERT, D., *Science* 253 (**1991**), p. 1019–1022
76 TARASOW, T. M., LEWIS, C., HILVERT, D., *J. Am. Chem. Soc.* 116 (**1994**), p. 7959–7963
77 SMILEY, J. A., BENKOVIC, S. J., *Proc. Natl. Acad. Sci. USA* 91 (**1994**), p. 8319–8323
78 BARTLETT, P. A., MARLOW, C. K., *Biochemistry* 22 (**1983**), p. 4618–4624
79 STEWART, J. D., LIOTTA, L. J., BENKOVIC, S. J., *Acc. Chem. Res.* 26 (**1993**), p. 396–404
80 CHARBONNIER, J.-B., CARPENTER, E., GIGANT, B., GOLINELLI-PIMPANEAU, B., ESHHAR, Z., GREEN, B. S., KNOSSOW, M., *Proc. Natl. Acad. Sci. USA* 92 (**1995**), p. 11721–11725
81 GOLINELLI-PIMPANEAU, B., GIGANT, B., BIZEBARD, T., NAVAZA, J., SALUDJIAN, P., ZEMEL, R., TAWFIK, D. S., ESHHAR, Z., GREEN, B. S., KNOSSOW, M., *Structure* 2 (**1994**), p. 175–183
82 KRISTENSEN, O., VASSYLYEV, D. G., TANAKA, F., MORIKAWA, K., FUJII, I., *J. Mol. Biol.* 281 (**1998**), p. 501–511
83 CHARBONNIER, J.-B., GOLINELLI-PIMPANEU, B., GIGANT, B., TAWFIK, D. S., CHAP, R., SCHINDLER, D. G., KIM, S.-H., GREEN, B. S., ESHHAR, Z., KNOSSOW, M., *Science* 275 (**1997**), p. 1140–1142
84 GIGANT, B., CHARBONNIER, J.-B., ESHHAR, Z., GREEN, B. S., KNOSSOW, M., *Proc. Natl. Acad. Sci. USA* 94 (**1997**), p. 7857–7861
85 PATTEN, P. A., GRAY, N. S., YANG, P. L., MARKS, C. B., WEDEMAYER, G. J., BONIFACE, J. J., STEVENS, R. C., SCHULTZ, P. G., *Science* 271 (**1996**), p. 1086–1091
86 WEDEMAYER, G. J., WANG, L. H., PATTEN, P. A., SCHULTZ, P. G., STEVENS, R. C., *J. Mol. Biol.* 268 (**1997**), p. 390–400
87 WEDEMAYER, G. J., PATTEN, P. A., WANG, L. H., SCHULTZ, P. G., STEVENS, R. C., *Science* 276 (**1997**), p. 1665–1669
88 THAYER, M. M., OLENDER, E. H., ARVAI, A. S., KOIKE, C. K., CANESTRELLI, I. L., STEWART, J. D., BENKOVIC, S. J., GETZOFF, E. D., ROBERTS, V. A., *J. Mol. Biol.* 291 (**1999**), p. 329–345
89 ZHOU, G. W., GUO, J., HUANG, W., FLETTERICK, R. J., SCANLAN, T. S., *Science* 265 (**1994**), p. 1059–1064
90 GUO, J., HUANG, W., ZHOU, G. W., FLETTERICK, R. J., SCANLAN, T. S., *Proc. Natl. Acad. Sci. USA* 92 (**1995**), p. 1694–1698
91 LESLEY, S. A., PATTEN, P. A., SCHULTZ, P. G., *Proc. Natl. Acad. Sci. USA* 90 (**1993**), p. 1160–1165
92 LINDNER, A. B., ESHHAR, Z., TAWFIK, D. S., *J. Mol. Biol.* 285 (**1999**), p. 421–430
93 TOMLINSON, I. M., WALTER, G., JONES, P. T., DEAR, P. H., SONNHAMMER, E. L., WINTER, G., *J. Mol. Biol.* 256 (**1996**), p. 813–817
94 WELLS, J. A., CUNNINGHAM, B. C., GRAYCAR, T. P., ESTELL, D. A., *Philos. Trans. R. Soc. Lond. (A)* 317 (**1986**), p. 415–423
95 BRYAN, P., PANTOLIANO, M. W., QUILL, S. G., HSIAO, H.-Y., POULOS, T., *Proc. Natl. Acad. Sci. USA* 83 (**1986**), p. 3743–3745

96 MacBeath, G., Hilvert, D., *Chem. Biol.* 3 (**1996**), p. 433–445

97 Padlan, E. A., Cohen, G. H., Davies, D. R., *Ann. Inst. Pasteur Immunol.* C 136 (**1985**), p. 271–276

98 Strong, R. K., Petsko, G. A., Sharon, J., Margolies, M. N., *Biochemistry* 30 (**1991**), p. 3749–3757

99 Griffiths, G. M., Berek, C., Kaartinen, M., Milstein, C., *Nature* 312 (**1984**), p. 271–275

100 Alzari, P. M., Spinelli, S., Mariuzza, R. A., Boulot, G., Poljak, R. J., Jarvis, J. M., Milstein, C., *EMBO J.* 9 (**1990**), p. 3807–3814

101 Miyashita, H., Karaki, Y., Kikuchi, M., Fujii, I., *Proc. Natl. Acad. Sci. USA* 90 (**1993**), p. 5337–5340

102 Roberts, V. A., Stewart, J., Benkovic, S. J., Getzoff, E. D., *J. Mol. Biol.* 235 (**1994**), p. 1098–1116

103 Stewart, J. D., Roberts, V. A., Thomas, N. R., Getzoff, E. D., Benkovic, S. J., *Biochemistry* 33 (**1994**), p. 1994–2003

104 Baca, M., Scanlan, T. S., Stephenson, R. C., Wells, J. A., *Proc. Natl. Acad. Sci. USA* 94 (**1997**), p. 10063–10068

105 Arkin, M. R., Wells, J. A., *J. Mol. Biol.* 284 (**1998**), p. 1083–1094

106 Janda, K. D., Benkovic, S. J., Lerner, R. A., *Science* 244 (**1989**), p. 437–440

107 Benkovic, S. J., Adams, J. A., Borders, C. L., Janda, K. D., Lerner, R. A., *Science* 250 (**1990**), p. 1135–1139

108 Fernholz, E., Schloeder, D., Liu, K. K.-C., Bradshaw, C. W., Huang, H., Janda, K. D., Lerner, R. A., Wong, C.-H., *J. Org. Chem.* 57 (**1991**), p. 4756–4761

109 Stewart, J. D., Krebs, J. F., Siuzdak, G., Berdis, A. J., Smithrud, D. B., Benkovic, S. J., *Proc. Natl. Acad. Sci. USA* 91 (**1994**), p. 7404–7409

110 Janda, K. D., Schloeder, D., Benkovic, S. J., Lerner, R. A., *Science* 241 (**1988**), p. 1188–1191

111 Krebs, J. F., Siuzdak, G., Dyson, H. J., Stewart, J. D., Benkovic, S. J., *Biochemistry* 34 (**1995**), p. 720–723

112 Carter, P., Wells, J. A., *Nature* 332 (**1988**), p. 564–568

113 Gibbs, R. A., Benkovic, P. A., Janda, K. D., Lerner, R. A., Benkovic, S. J., *J. Am. Chem. Soc.* 114 (**1992**), p. 3528–3534

114 Lerner, R. A., Benkovic, S. J., Schultz, P. G., *Science* 252 (**1991**), p. 659–667

115 Thorn, S. N., Daniels, R. G., Auditor, M.-T. M., Hilvert, D., *Nature* 373 (**1995**), p. 228–230

116 Suga, H., Ersoy, O., Williams, S. F., Tsumuraya, T., Margolies, M. N., Sinskey, A. J., Masamune, S., *J. Am. Chem. Soc.* 116 (**1994**), p. 6025–6026

117 Tsumuraya, T., Suga, H., Meguro, S., Tsunakawa, A., Masamune, S., *J. Am. Chem. Soc.* 117 (**1995**), p. 11390–11396

118 Janda, K. D., Lo, C.-H. L., Li, T., Barbas, C. F., Wirsching, P., Lerner, R. A. *Proc. Natl. Acad. Sci. USA* 91 (**1994**), p. 2532–2536

119 Janda, K. D., Lo, L.-C., Lo, C.-H. L., Sim, M.-M., Wang, R., Wong, C.-H., Lerner, R. A., *Science* 275 (**1997**), p. 945–948

120 Wagner, J., Lerner, R. A., Barbas, C. F., *Science* 270 (**1995**), p. 1797–1800

121 Lo, C.-H. L., Wentworth Jr., P., Jung, K. W., Yoon, J., Ashley, J. A., Janda, K. D., *J. Am. Chem. Soc.* 119 (**1997**), p. 10251–10252

122 Gao, Cs., Lavey, B. J., Lo, C.-H. L., Datta, A., Wentworth Jr., P., Janda, K. D., *J. Am. Chem. Soc.* 120 (**1998**), p. 2211–2217

123 Barbas III, C. F., Heine, A., Zhong, G., Hoffmann, T., Gramatikova, S., Björnestedt, R., List, B., Anderson, J., Stura, E. A., Wilson, I. A., Lerner, R. A., *Science* 278 (**1997**), p. 2085–2092

124 Horecker, B. L., Tsolas, O., Lai, C.-Y., In *The Enzymes*, ed. PD Boyer, 7, Academic Press, New York **1975**, p. 213–258

125 Marsh, J. J., Lebherz, H. G., *Trends Biochem. Sci.* 17 (**1992**), p. 110–113

126 Highbarger, L. A., Gerlt, J. A., Kenyon, G. L., *Biochemistry* 35 (**1996**), p. 41–46

127 Hoffmann, T., Zhong, G., List, B., Shabat, D., Anderson, J., Gramatikova, S., Lerner, R. A., Bar-

bas III, C. F., *J. Am. Chem. Soc.* 120 (**1998**), p. 2768–2779
128 Zhong, G., Hoffmann, T., Lerner, R. A., Danishefsky, S., Barbas III, C. F., *J. Am. Chem. Soc.* 119 (**1997**), p. 8131–8132
129 List, B., Shabat, D., Zhong, G., Turner, J. M., Li, A., Bui, T., Anderson, J., Lerner, R. A., Barbas III, C. F., *J. Am. Chem. Soc.* 121 (**1999**), p. 7283–7291
130 List, B., Shabat, D., Barbas III, C. F., Lerner, R. A., *Chem. Eur. J.* 4 (**1998**), p. 881–885
131 Pentoet E. E., Kochman, M., Rutter, W. J., *Biochemistry* 8 (**1969**), p. 4396–4402
132 Hammerstedt, R. H., Möhler, H., Decker, K. A., Ersfeld, D. E., Wood, W. A., *Methods Enzymol.* 42C (**1975**), p. 258–264
133 Branden, C., Tooze, J., *Introduction to Protein Structure*, Garland Publishing, Inc., New York **1991**
134 Wilson, I. A., Stanfield, R. L., *Curr. Opin. Struct. Biol.* 3 (**1993**), p. 113–318
135 Davies, D. R., Padlan, E. A., Sheriff, S., *Annu. Rev. Biochem.* 59 (**1990**), p. 439–473
136 Alt, F. W., Blackwell, T. K., Yancopoulos, G. D., *Science* 238 (**1987**), p. 1079–1087
137 Rajewsky, K., Förester, I., Cumang, A., *Science* 238 (**1987**), p. 1088–1094
138 Tsou, C. L., *Ann. N. Y. Acad. Sci.* 864 (**1998**), p. 1–8
139 Post, C. B., Ray, W. J., Jr, *Biochemistry* 34 (**1995**), p. 15881–15885
140 Mets, B., Winger, G., Cabrera, C., Seo, S., Jamdar, S., Yang, G., Zhao, K., Briscoe, R. J., Almonte, R., Woods, J. H., Landry, D. W., *Proc. Natl. Acad. Sci. USA* 95 (**1998**), p. 10176–10181
141 Yang, G., Chun, J., Arakawa-Uramoto, H., Wang, X., Gawinowicz, M. A., Zhao, K., Landry, D. W., *J. Am. Chem. Soc.* 118 (**1996**), p. 5881–5890
142 Cravatt, B. F., Ashley, J. A., Janda, K. D., Boger, D. L., Lerner, R. A., *J. Am. Chem. Soc.* 116 (**1994**), p. 6013–6014
143 Janda, K. D., Shevlin, C. G., Lerner, R. A., *Science* 259 (**1993**), p. 490–493
144 Janda, K. D., Shevlin, C. G., Lerner, R. A., *J. Am. Chem. Soc.* 117 (**1995**), p. 2659–2660
145 Shabat, D., Itzhaky, H., Reymond, J.-L., Keinan, E., *Nature* 374 (**1995**), p. 143–146
146 Li, T., Janda, K. D., Lerner, R. A., *Nature* 379 (**1996**), p. 326–327
147 Reymond, J.-L., Reber, J.-L., Lerner, R. A., *Angew. Chem. Int. Ed. Engl.* 33 (**1994**), p. 475–477
148 Sinha, S. C., Keinan, E., *J. Am. Chem. Soc.* 117 (**1995**), p. 3653–3654
149 Harlow, E., Lane, D., *Antibodies: A Laboratory Manual*, Cold Spring Harbor Laboratory **1988**
150 Lowry, D., Murphy, S., Goffe, R. A., *J. Biotechnol.* 36 (**1994**), p. 35–38
151 Jackson, L. R., Trudel, L. J., Fox, J. G., Lipman, N. S., *J. Immunol. Methods* 189 (**1996**), p. 217–231
152 Ma, J. K. C., Hiatt, A., Hein, M., Vine, N. D., Wang, F., Stabila, P., Van Dolleweerd, C., Mostov, K., Lehner, T., *Science* 268 (**1995**), p. 716–719
153 Hiatt, A., Ma, J. K. C., *FEBS Lett.* 307 (**1992**), p. 71–75
154 Carter, P., Kelley, R. F., Rodrigues, M. L., Snedecor, B., Covarrubias, M., Velligan, M. D., Wong, W. L. T., Rowland, A. M., Kotts, C. E., Carver, M. E., Yang, M., Bourell, J. H., Shephard, H. M., Henner, D., *BioTechnology* 10 (**1992**), p. 163–167
155 Wallace, M. B., Iverson, B. L., *J. Am. Chem. Soc.* 118 (**1996**), p. 251–252
156 Walsh, C., *Enzymatic Reaction Mechanisms*, W.H. Freeman and Company, New York **1979**
157 Iverson, B. L., Iverson, S. A., Roberts, V. A., Getzhoff, E. D., Tainer, J. A., Benkovic, S. J., Lerner, R. A., *Science* 249 (**1990**), p. 659–662
158 Stewart, J. D., Roberts, V. A., Crowder, M. W., Getzoff, E. D., Benkovic, S. J., *J. Am. Chem. Soc.* 116 (**1994**), p. 415–416
159 Iverson, B. L., Lerner, R. A., *Science* 243 (**1989**), p. 1184–1188
160 Shokat, K. M., Leumann, C. J., Sugasawara, R., Schultz, P. G., *Angew.*

Chem. Int. Ed. Engl. 27 (**1988**), p. 1172–1174

161 NIMRI, S., KEINAN, E., *J. Am. Chem. Soc.* 121 (**1999**), in press

162 COCHRAN, A. G., SCHULTZ, P. G., *J. Am. Chem. Soc.* 112 (**1990**), p. 9414–9415

163 COCHRAN, A. G., PHAM, T., SUGASAWARA, R., SCHULTZ, P. G., *J. Am. Chem. Soc.* 113 (**1991**), p. 6670–6672

164 EISENTHAL, R., DANSON, M. J., *Enzyme Assays – A Practical Approach*. IRL Press, Oxford **1993**

165 FENNIRI, H., JANDA, K. D, LERNER, R. A., *Proc. Natl. Acad. Sci. USA* 92 (**1995**), p. 2278–2282

166 LEE, I., STEWART, J. D., ZHONG, W., BENKOVIC, S. J., *Anal. Biochem.* 230 (**1995**), p. 62–67

167 TAWFIK, D. S., GREEN, B. S., CHAP, R., SELA, M., ESHHAR, Z., *Proc. Natl. Acad. Sci. USA* 90 (**1993**), p. 373–377

168 TAWFIK, D. S., CHAP, R., GREEN, B. S., SELA, M., ESHHAR, Z., *Proc. Natl. Acad. Sci. USA* 92 (**1995**), p. 2145–2149

169 CHISWELL, D. J., MCCAFFERTY, J., *Trends Biotechnol.* 10 (**1992**), p. 80–84

170 MARKS, J. D., HOOGENBOOM, H. R., GRIFFITHS, A. D., WINTER, G., *J. Biol. Chem.* 267 (**1992**), p. 16007–16010

171 HUSE, W. D., SASTRY, L., IVERSON, S. A., KANG, A. S., ALTING-MEES, M., BURTON, D. R., BENKOVIC, S. J., LERNER, R. A., *Science* 246 (**1989**), p. 1275–1281

172 SMITH, G. P., PETRENKO, V. A., *Chem. Rev.* 97 (**1997**), p. 391–410

173 JESTIN, J. L., KRISTENSEN, P., WINTER, G., *Angew. Chem. Int. Ed. Engl.* 38 (**1999**), p. 1124–1127

174 GAO, C., LIN, C. H., LO, C.-H. L., MAO, S., WIRSCHING, P., LERNER, R. A., JANDA, K. D., *Proc. Natl. Acad. Sci. USA* 94 (**1997**), p. 11777–11782

175 PEDERSEN, H., HOLDER, S., SUTHERLIN, D. P., SCHWITTER, U., KING, D. S., SCHULTZ, P. G., *Proc. Natl. Acad. Sci. USA* 95 (**1998**), p. 10523–10528

176 TAWFIK, D. S., GRIFFITHS, A. D., *Nat. Biotechnol.* 16 (**1998**), p. 652–656

177 SOUMILLION, P., JESPERS, L., BOUCHET, M., MARCHAND-BRYNAERT, J., WINTER, G., FASTREZ, J., *J. Mol. Biol.* 237 (**1994**), p. 415–422

178 MILLER, G. P., POSNER, B. A., BENKOVIC, S. J., *Bioorg. Med. Chem.* 5 (**1997**), p. 581–590

179 STEMMER, W. P. C., *Nature* 370 (**1994**), p. 389–391

180 CRAMERI, A., CWIRLA, S., STEMMER, W. P. C., *Nat. Med.* 2 (**1996**), p. 100–102

181 MARTINEAU, P., JONES, P., WINTER, G., *J. Mol. Biol.* 280 (**1998**), p. 117–127

182 CONN, M. M., PRUDENT, J. R., SCHULTZ, P. G., *J. Am. Chem. Soc.* 118 (**1996**), p. 7012–7013

183 LI, Y. F., SEN, D., *Nat. Struct. Biol.* 3 (**1996**), p. 743–747

184 LI, Y. F., SEN, D., *Biochemistry* 36 (**1997**), p. 5589–5599

185 PRUDENT, J. R., UNO, T., SCHULTZ, P. G., *Science* 264 (**1994**), p. 1924–1927.

186 MORRIS, K. N., TARASOW, T. M., JULIN, C. M., SIMONS, S., HILVERT, D., GOLD, L., *Proc. Natl. Acad. Sci. USA* 91 (**1994**), p. 13028–13032

187 NARLIKAR, G. J., HERSCHLAG, D., *Annu. Rev. Biochem.* 66 (**1997**), p. 19–59

188 WILSON, D. S., SZOSTAK, J. W., *Annu. Rev. Biochem.* 68 (**1999**), p. 611–647

189 TARASOW, T. M., TARASOW, S. L., EATON, B. E., *Nature* 389 (**1997**), p. 54–57

190 BARTEL, D. P., SZOSTAK, J. W., *Science* 261 (**1993**), p. 1411–1418

191 EKLAND, E. H., SZOSTAK, J. W., BARTEL, D. P., *Science* 269 (**1995**), p. 364–370

192 HANES, J., PLÜCKTHUN, A., *Proc. Natl. Acad. Sci. USA* 94 (**1997**), p. 4937–4942

193 HANES, J., JERMUTUS, L., WEBER-BORNHAUSER, S., BOSSHARD, H. R., PLÜCKTHUN, A., *Proc. Natl. Acad. Sci. USA* 95 (**1998**), p. 14130–14135

194 HE, M. Y., TAUSSIG, M. J., *Nucleic Acids Res.* 25 (**1997**), p. 5132–5134

195 ROBERTS, R. W., SZOSTAK, J. W., *Proc. Natl. Acad. Sci. USA* 94 (**1997**), p. 12297–12302

196 NEMOTO, N., MIYAMOTO-SATO, E., HUSIMI, Y., YANAGAWA, H., *FEBS Lett.* 414 (**1997**), p. 405–408

197 MACBEATH, G., KAST, P., HILVERT, D., *Science* 279 (**1998**), p. 1958–1961

198 Minshull, J., Stemmer, W. P. C., *Curr. Opin. Chem. Biol.* 3 **(1999)**, p. 284–290

199 Arnold, F. H., *Acc. Chem. Res.* 31 **(1998)**, p. 125–131

200 Shokat, K. M., Leumann, C. J., Sugasawara, R., Schultz, P. G., *Nature* 338 **(1989)**, p. 269–271

201 Tawfik, D. S., Lindner, A. B., Chap, R., Eshhar, Z., Green, B. S., *Eur. J. Biochem.* 244 **(1997)**, p. 619–626

202 Zemel, R., Schindler, D. G., Tawfik, D. S., Eshhar, Z., Green, B. S., *Mol. Immunol.* 31 **(1994)**, p. 127–137

203 Lee, A. Y., Karplus, A. P., Ganem, B., Clardy, J., *J. Am. Chem. Soc.* 117 **(1995)**, p. 3627–3628

204 Esnouf, R. M., *J. Mol. Graph. Model.* 15 **(1997)**, p. 132–134, 112-113

205 Merritt, E. A., Murphy, M. E. P., *Acta Crystallogr.* D 50 **(1994)**, p. 869–873

206 Nicholls, A., Sharp, K. A., Honig, B., *Proteins Struct. Funct. Genet.* 11 **(1991)**, p. 281–296

207 Joo, H., Lin, Z. L., Arnold, F. H., *Nature* 399 **(1999)**, p. 670–673

208 Taylor, S. V., Kast, P., Hilvert, D., *Angew. Chem. Int. Ed.* 40 **(2001)**, 3310–3335

3
Theoretical Studies of Antibody Catalysis
Dean J. Tantillo, Andrew G. Leach, Xiyun Zhang, K. N. Houk

3.1
Introduction

Catalytic antibodies have the potential to supply an abundant source of catalysts for a variety of reactions, limited only by our creativity in mimicking transition states and producing effective antigenic derivatives. Nevertheless, a comparison of the effectiveness of various catalytic antibodies to typical enzymes shows a stark contrast in catalytic efficiency.

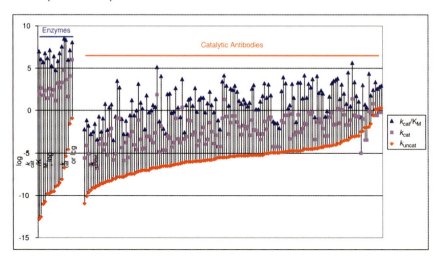

Fig. 3.1 Catalytic efficiencies of various enzymes and catalytic antibodies [1, 2].

Fig. 3.1 shows the catalytic efficiencies of a number of enzymes (compiled by Wolfenden) [1], along with data compiled for a range of catalytic antibodies [2]. The common measures of catalytic efficiency, k_{cat}/k_{uncat}, are often in the range of 10^8–10^{12} for enzymes, but 10^3–10^6 is more common for catalytic antibodies. Since K_M for both enzymes and antibodies tend to be similar, 10^3–10^6 M, the "proficiency", $1/k_{TS}$ = $k_{cat}/K_M/k_{uncat}$, for enzymes exceeds those for catalytic antibodies by 10^5–10^9. Over-

Catalytic Antibodies. Edited by Ehud Keinan
Copyright © 2005 WILEY-VCH Verlag GmbH & Co. KGaA, Weinheim
ISBN: 3-527-30688-9

arching goals of theoretical research in this field are to understand the details of antibody catalysis and why it often falls short of enzyme catalysis, and to propose approaches to create more proficient catalytic antibodies.

From the beginning, our UCLA group has eagerly explored these new protein catalysts. Quantum mechanics has been used to explore the mechanisms of the uncatalyzed reactions, while docking and molecular dynamics have been employed to determine how the binding of substrates and transition states contributes to acceleration. Modern tools of theory and computation are used to understand antibody catalysis and to predict means by which the scope and effectiveness of catalytic antibody technology may be extended.

This chapter reviews the theoretical studies that have been performed on antibody catalysis to date. After an introduction to the methods used and questions explored, the chapter is organized around different types of reactions that have been catalyzed by antibodies.

3.2
Questions Subject to Theoretical Elucidation

There are many aspects of antibody structure, binding, and reactivity [3, 4] that can be modeled using computational methods. For various reasons – their highly conserved structural motifs, the short time scale for their evolution, their ability to function as catalysts for non-metabolic processes, the fact that their combining sites tend to contain fewer key catalytic residues than the active sites of enzymes – antibodies have served as testing grounds for the application of various computational methods to biological problems. Nearly all antibody modeling studies have focused on one or more of three major problems of chemical and biological significance: (1) predicting structure from sequence, (2) modeling antigen/substrate binding (predicting binding modes and binding energies), (3) understanding catalysis (correlating relative rates and selectivities with noncovalent or covalent interactions).

3.2.1
Predicting Antibody Structure from Sequence

There have been various approaches to modeling the structure of antibody combining sites based on amino acid sequences [4]. The problem is simpler than the general protein-folding problem due to the fact that the large framework portion of immunoglobulin Gs is conserved. Modeling of the mobile regions relies heavily on sequence homology between the target antibody and antibodies whose three-dimensional structures have been solved by X-ray crystallography. The modeling process therefore usually begins with fragments of known antibody structures that are used as templates. These fragments can include the peptide backbones of the complementarity-determining regions (CDRs), which comprise antibody combining sites as well as the framework regions on which the CDRs are displayed, and may or may not include the exact positions of atoms in amino acid side chains.

A representative suite of homology-driven structure prediction programs – which was designed specifically to model antibody structures – is AbM [5]. The AbM modeling protocol has been tested, for example, on several hydrolytic antibodies whose structures were known but not contained in the AbM structural database [6]. The AbM procedure, a fairly typical knowledge-based procedure, produced reasonable orientations for canonical loops [7] and some non-canonical loops, but unreasonable structures for others – particularly CDR-H3, a loop whose length and corresponding variability makes it especially difficult to model. Root mean square deviations between modeled and experimental side chain positions were typically 2 to 5 times higher than those for backbone atoms alone, emphasizing that improvement is necessary in procedures for modeling side chain packing. While binding residues were generally modeled in reasonable orientations, binding cavities with appropriate dimensions were generally not produced. Nonetheless, programs like AbM can provide reasonable starting points for more detailed theoretical and experimental studies. The AbM protocol has been updated and made available as a web-based source [5b]. An improved antibody modeling algorithm, WAM (for Web Antibody Modeling), has also been developed for antibody structure prediction [5b].

3.2.2
Predicting Binding Modes and Binding Energies

There are two major goals in modeling antibody–antigen binding (which are representative of host–guest chemistry in general): predicting reasonable binding modes and predicting reasonable binding energies.

The problem of predicting binding modes has usually been approached through computational docking procedures. Such procedures generally involve sampling of potential binding modes coupled to rapid evaluation of interaction energies between the antibody and antigen. There are many docking protocols, which differ in their sampling algorithms, the nature and accuracy of the functions used for evaluating interaction energies, and the degree of flexibility allowed for the antibody and antigen during the docking process [8]. Many of the available docking programs perform well for antibody-hapten binding problems [4, 8b]. Recently, this approach has also been used to predict binding modes for transition structures, as will be described later in this chapter.

The problem of predicting binding energies quantitatively is a major focus of current computational research [9, 10]. This is a difficult problem in that experimental binding energies reflect differences between interactions of host and guest with each other and with solvent in the bound and unbound states. Measured binding constants are also directly related to free energies of binding and therefore include entropic as well as enthalpic effects. Consensus has not yet been reached on the best methodologies for modeling solvation or entropic effects.

State-of-the-art methods for computing binding free energies typically involve molecular dynamics (MD) simulations [4, 10, 11] coupled with free-energy perturbation (FEP) [12] or linear interaction energy (LIE) [13] schemes. MD simulations are used to produce ensembles of low-energy binding modes and conformational states and evaluate their average energies by allowing the relative positions of the atoms in the antibody and its antigen/substrate to evolve under the influence of empirical force fields that are based on classical mechanics and electrostatics. While these methods cannot yet consistently and quantitatively predict binding constants with high accuracy, they can often reproduce trends in binding free energy for related substrates and hosts [10].

3.2.3
Understanding Antibody Catalysis [14]

Understanding the effects of biocatalysts on rates and selectivities requires, first, an understanding of the uncatalyzed reaction in solution and/or in the gas phase, and second, an understanding of the effects of the environment surrounding the reactants, intermediates (if any) and transition structure(s) in the catalyst – i.e., the effects of specific interactions (steric, electrostatic, hydrogen bonding) with active-site residues and/or the non-specific effects of a particular (homogenous or heterogeneous) dielectric environment of an antibody interior or surrounding solvent.

Modeling the geometries and energetics of interactions between transition state structures and catalytic antibody combining sites is in many ways similar to modeling antibody-antigen interactions. However, transition states involve bond-making and bond-breaking processes, and most empirical force fields are not parameterized to treat these processes [15]. In addition, some antibody-catalyzed reactions involve major changes to transition state geometries and mechanistic pathways relative to those in the gas phase or solution. Since the accurate description of transition states requires quantum rather than classical mechanics, modeling of biological reaction coordinates has often involved the integration of quantum mechanical and force field methods. Combined quantum mechanics/molecular mechanics (QM/MM) techniques [16], in which bond-making and bond-breaking are treated with quantum mechanics and the remainder of the system – protein and solvent – is treated with molecular mechanics, have been applied to problems of biological catalysis with varied degrees of success [16]. Such treatments are computationally demanding and often employ relatively low levels of theory for their quantum mechanical parts.

Several alternative methods have been proposed, including the theozyme [17] and transition state docking [18] approaches. The theozyme approach involves the construction of antibody-reactant and antibody–transition structure complexes by computing optimal geometries for transition state stabilization by small models of particular combining site functional groups. In this strategy, as much of the combining site as is computationally feasible is treated with high-level quantum mechanical methods. Transition state docking is, in a sense, a variation on the typical QM/MM strategy in which the transition structure is first determined using quantum mechanical methods and then computationally docked into an antibody combining site using the methods described above. Another alternative involves MD studies (without quantum mechanics) using relatively stable molecules as models of transition structures (e.g., tetrahedral intermediates as models of transition structures for the hydrolysis of carboxylic acid derivatives; see Section 3.3), although this methodology is only appropriate in the very few cases where a stable molecule or a reactive intermediate can be found that reasonably mimics the steric and electrostatic properties of a transition structure.

3.2.4
General Considerations

Given the size and complexity of most biological systems, calibration of computational methods is often difficult. In ideal cases, experimental information such as structures, binding constants, and the effects of mutations on reaction rates and selectivities will be available to compare with predictions arrived at using theory. In some cases, several different computational methods will be applied to the same problem, allowing for comparisons of the effectiveness of these different methods.

The remainder of this chapter will focus on studies that have employed the methods described above in efforts to determine the origins of rate accelerations and selectivities in antibody-catalyzed reactions.

3.3
Hydrolytic Antibodies

Ester and amide hydrolysis reactions and closely related reactions such as transesterification and aminolysis have most often been the target of antibody catalysis [14]. Studies of such reactions have been prevalent for several reasons: (1) efficient and selective catalysis of peptide bond hydrolysis has long been a goal of organic chemists, bio-organic chemists, and biochemists, because of the myriad applications such a process would have in molecular biology, chemical biology, biochemistry, and organic synthesis [19]; (2) the use of aryl (particularly *para*-nitrophenyl, PNP) esters and amides facilitates the study of mechanisms for these antibody-catalyzed reactions, since well-precedented analytical methods are available for monitoring reactions that involve aryloxy leaving groups [20, 21]. Because the barriers for hydrolysis of aryl esters are generally low in the absence of any catalyst, the widespread use of such substrates has occasionally encouraged criticism of the merits of catalytic antibody research [21a]. Nonetheless, the fundamental insights garnered from studies of aryl ester and amide hydrolysis have implications throughout the catalytic antibody field and beyond into other areas of catalysis and immunology.

3.3.1
Gas and Solution Phase Hydrolysis of Aryl Esters

The generalized mechanism for the alkaline hydrolysis of ester derivatives involves a tetrahedral intermediate flanked by two negatively charged *pseudo*-tetrahedral transition structures (Scheme 3.1). The details of a given hydrolytic mechanism depend, of course, on the nature of the substituents R, R', and X, and the environment in which the reaction is carried out. Since phenyl (R = alkyl, R'= phenyl, X = O) and

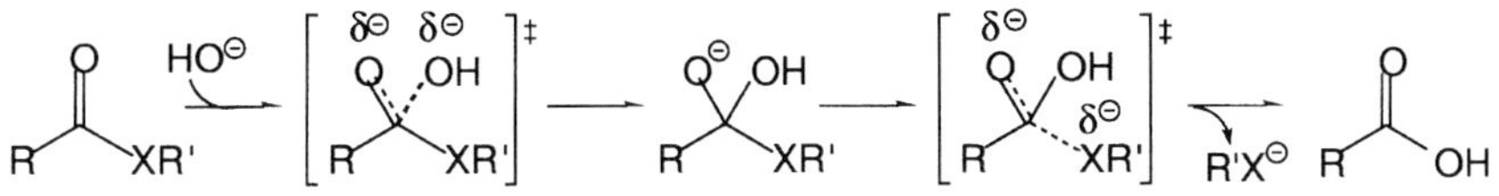

Scheme 3.1 A general mechanism for the alkaline hydrolysis of ester derivatives.

para-nitrophenyl (R = alkyl, R′= PNP, X = O) esters are, by far, the most common substrates for antibody-catalyzed hydrolyses, the following mechanistic discussion will be confined to the reactions of these species.

Hydrolytic mechanisms for phenyl acetate (PhOAc) and para-nitrophenyl acetate (PNPOAc) have been explored at several different levels of theory, in both the gas phase and aqueous solution [20, 22, 23]. In 1999, Tantillo and Houk reported a study in which gas phase geometries of stationary points were optimized at the RHF/6–31+G(d) level, and single point energy calculations were carried out on these geometries at both the RHF/6–31+G(d) and MP2/6–31+G(d) levels [20]. The effects of solvation on the relative energies of these stationary points were estimated using two related but different methods, and SCI-PCM (Self-Consistent Isodensity PCM), which compute stabilization energies associated with moving a structure from the gas phase to a homogeneous continuum dielectric environment (in this case with a dielectric constant of 78.5 for water) [24]. Although such calculations do not account for explicit solvent-solute interactions such as hydrogen bonding, they nonetheless seem to perform well in this and related reactions, especially in terms of assessing the effects of aqueous solvation on activation barriers [20, 23, 24].

The computed energetics of PhOAc hydrolysis in the gas phase are shown in Fig. 3.2 [20]. The overall energetics of this process agree well with the experimentally determined exothermicity [25]. Although a tetrahedral intermediate and two pseudo-tetrahedral transition structures were located in the gas phase, the energy of the transition structure for the expulsion of PhO⁻ is predicted to be comparable to that of the tetrahedral intermediate at the RHF/6–31+G(d) level and considerably lower than that of the tetrahedral intermediate at the MP2/6–31+G(d) level (due to the inclusion of zero point energy and electron correlation effects). This suggests that the gas phase attack of hydroxide on PhOAc leads directly to PhO⁻ and HOAc products, without formation of a tetrahedral intermediate; analogous calculations on PNPOAc [20] also failed to find a tetrahedral intermediate. Similar results were also obtained using B3LYP/6–31G(d) [23]. This situation – concerted hydrolysis – differs from that for alkyl esters, in which tetrahedral intermediates are predicted to exist with barriers of ca. 5 kcal/mol or larger for alkoxide expulsion [26], consistent with the fact that PhO⁻ and PNPO⁻ are much better leaving groups than simple alkoxides.

The results of solvation on the energetics of PhOAc hydrolysis [20] are also shown in Fig. 3.2. The presence of a polar environment is predicted to have two main effects. First, as expected, it selectively stabilizes the isolated reactants to a large extent, especially hydroxide because of its concentrated charge, making the ion molecule complex irrelevant to the energetics of the solution reaction. Second, the polar environment interacts slightly more favorably with the tetrahedral intermediate than with the more delocalized transition structure for phenoxide expulsion, thereby leading to

3 Theoretical Studies of Antibody Catalysis

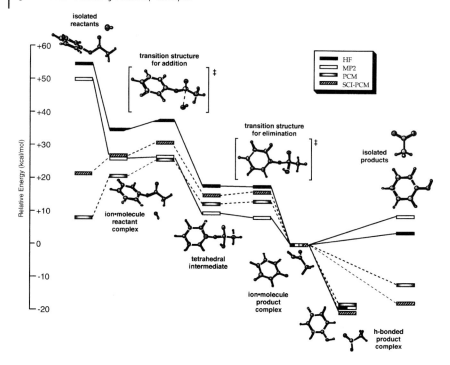

Fig. 3.2 Computed energetics of PhOAc hydrolysis.

a small barrier for PhO⁻ loss. DFT calculations, which incorporated solvation via the COSMO (Conductor-like Screening Model) method, gave qualitatively similar results [23].

Whether the alkaline hydrolyses of PhOAc and PNPOAc are stepwise with very small barriers to decomposition of the tetrahedral intermediate or truly concerted does not affect the strategy for catalyzing this reaction – this being to provide an environment which interacts most favorably with the rate-determining transition structures for the addition of hydroxide.

3.3.2
Hapten Fidelity

Transition state analog (TSA) haptens [27] for hydrolysis tend to contain structures with tetrahedrally disposed oxygen atoms that mimic the charge delocalization and geometry of putative hydrolytic transition states (Chart 3.1). The most common haptens used to elicit hydrolytic antibodies have been the phosphorus-containing haptens shown in Chart 3.1, although all of the functional groups shown there have been used (occasionally in combination) to successfully elicit antibodies which catalyze the hydrolysis of certain esters, amides, carbonates, and carbamates, as well as some which catalyze transesterification and ester aminolysis reactions.

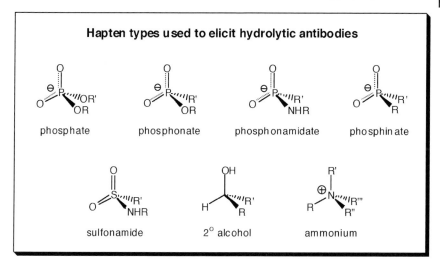

Chart 3.1 Transition state analog (TSA) haptens.

Several computational studies have compared the structures of TSAs with those of various tetrahedral intermediates and transition states for hydrolysis reactions. Teraishi et al. and Ohkubo et al. compared Hartree-Fock and MP2 structures of tetrahedral intermediates and phosphonate TSAs for methyl acetate hydrolysis [28]. Shields and coworkers used *ab initio* calculations to compare the geometries of transition states and tetrahedral intermediates involved in the hydrolysis of methyl acetate and methyl benzoate with those of the corresponding phosphonate haptens [29], and PM3 and SM3 calculated geometries and electrostatic potential surfaces of transition states for hydroxide addition to cocaine with those of phosphonate and thiophosphonate TSAs [30]. In related studies, Curley and Pratt examined various TSA inhibitors of a β-lactamase using force field computations [31], and Houk and coworkers examined TSA inhibitors of various proteases using *ab initio* calculations and analyses of electrostatic potential surfaces [32].

Tantillo and Houk compared the computed geometries and charge distributions of transition states for aryl ester hydrolysis (see above) with those of phosphonate and phosphinate TSA haptens (Fig. 3.3) [20]. This study showed that, in terms of molecular volumes, bond lengths, and bond angles, these haptens are more similar to elimination transition structures and tetrahedral intermediates (when located) than to the rate-determining transition structures for hydroxide addition. The P–O bond distances in the haptens do not mimic the C–O⁻ and C–OH distances in the addition transition states very closely either (Fig. 3.3). In terms of charge distributions, although the oxygen atoms attached to both phosphorus and carbon are quite negative, the fact that the oxygen of the attacking hydroxide bears a proton induces an asymmetry in the transition structures that is not well represented in the haptens.

These studies indicate that phosphonates and their relatives are reasonable isosteric mimics of transition states involved in ester hydrolysis, although some signif-

3 Theoretical Studies of Antibody Catalysis

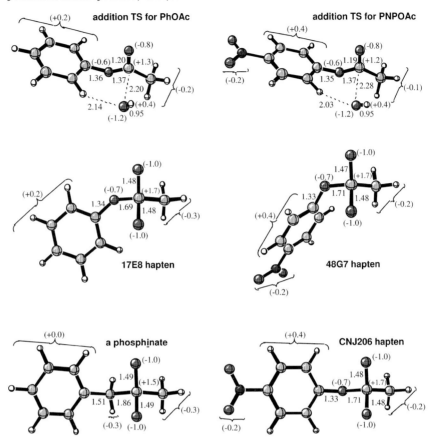

Fig. 3.3 Computed geometries and charge distributions of transition states with TSA haptens. Distances are shown in Ångstroms; partial atomic charges are shown in parantheses.

icant features of the charge distributions and hydrogen bonding patterns of these transition states are not well mimicked.

3.3.3
Theoretical Exploration of Antibody Catalysis

Various modeling strategies – from homology modeling to molecular dynamics and to detailed quantum mechanical studies – have been utilized to uncover the origins of rate acceleration and selectivity for a number of antibodies that promote hydrolysis or closely related reactions of aryl and benzyl esters. These studies are described in brief below.

3.3.3.1
16G3

Antibody 16G3 raised against a highly substituted *para*-nitrobenzyl phosphonate hapten, catalyzes peptide bond formation via the attack of an amino group in one amino acid onto a PNP-ester derivative of the acid group on another (Scheme 3.2) [33]. This catalyst has been shown to promote the formation of di-, tri-, and tetra-peptides with rate accelerations of up to 10^4 over the corresponding background reactions, without competing epimerization or hydrolysis. Recently, this catalyst has also been used to promote the cyclization of PNP-esters to form cyclic hexapeptides (Scheme 3.2). In the absence of a crystal structure for 16G3, a model of the antibody combining site was constructed based on homology to related structures. The hapten was then manually docked into this binding site, and the resulting structure was relaxed using a molecular mechanics-based energy minimization procedure. Models of complexes between 16G3 and several substrates were similarly constructed, using the hapten complex as a guide. Inspection of the resulting models led to explanations for several experimentally observed behaviors of 16G3 that involved specific structural features

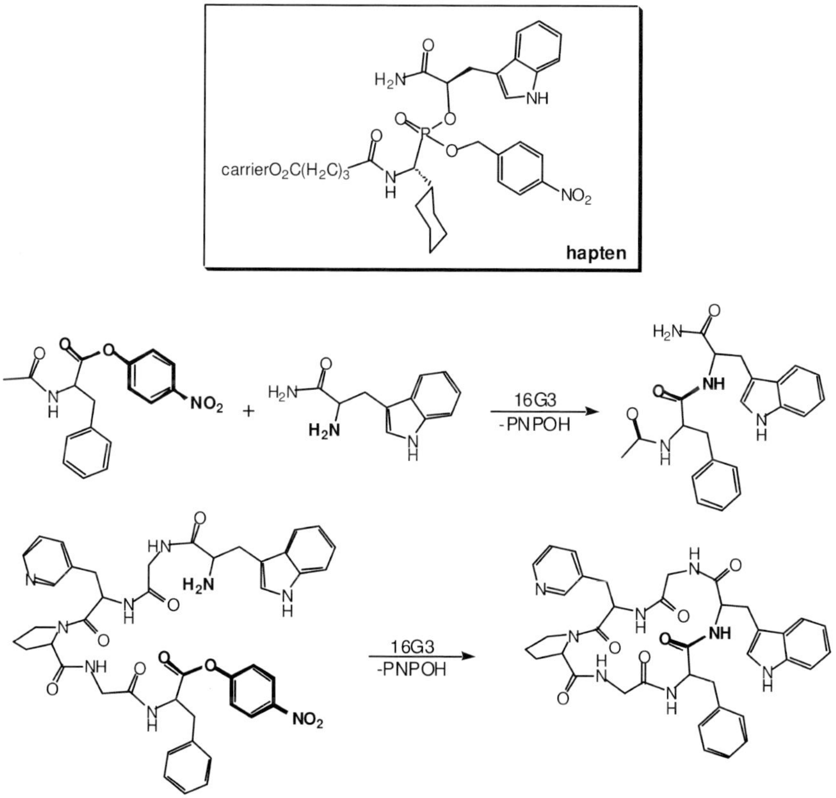

Scheme 3.2 Peptide bond formation catalyzed by antibody 16G3.

of the combining site. For example, the selectivity of 16G3 for substrates with certain amino acid side chains and certain stereochemistries, as well as low levels of product inhibition, were explained using arguments based on substrate pre-organization and shape/electrostatic complementarity. While the accuracy of these models awaits validation by more complex theoretical and/or experimental structural studies, their usefulness in guiding experiments has already been demonstrated [33c].

3.3.3.2
6D9

Antibody 6D9 catalyzes the hydrolysis of a chloramphenicol prodrug, releasing active chloramphenicol (Scheme 3.3) [34]. This antibody is the best of six catalytic antibodies obtained from a single immunization against a *para*-nitrobenzyl phosphonate TSA hapten (Scheme 3.3), displaying a k_{cat}/k_{uncat} of ~ 10^3. Based on a variety of biochemical experiments, a direct attack by hydroxide was proposed as the mechanism employed by 6D9 for prodrug hydrolysis. Before the crystal structure of this antibody became available, a model of its combining site was constructed by the Fujii group based on homology [34b]. Another antibody (an *anti*-peptide antibody), whose structure was available and whose amino acid sequence was quite similar to that of 6D9, despite the absence of any functional relationship between the two, was chosen as a template

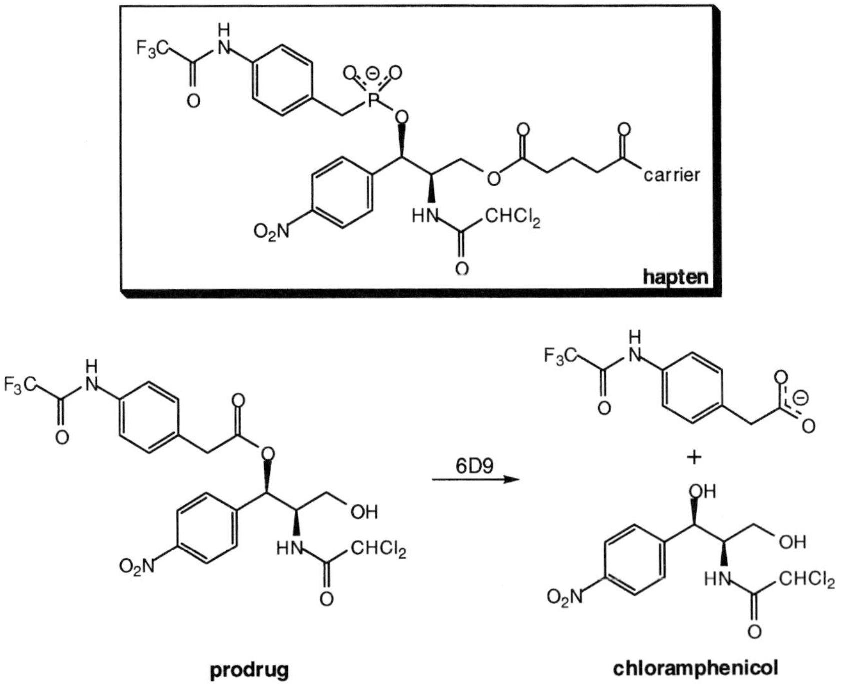

Scheme 3.3 Hydrolysis of a chloramphenicol prodrug catalyzed by antibody 6D9.

onto which the CDR loops of 6D9 were grafted. The structure was then allowed to relax through a series of constrained energy minimizations. The hapten was then manually docked into the modified binding site in an orientation consistent with the results from the battery of biochemical experiments that had been performed on 6D9. Based on the resulting model, a combining site histidine residue that was close to the phosphonate group was assigned a role of transition state stabilization. While the orientation of the hapten and the exact positions of the amino acids in this model differ somewhat from the crystal structure obtained later [34d], the histidine residue implicated in the modeling was indeed found to be within hydrogen-bonding distance of a phosphonate oxygen of the bound hapten, and so presumably also interacts with the corresponding oxygen atom of the transition structure for hydroxide attack.

3.3.3.3
43C9

Antibody 43C9 was raised against a *para*-nitrophenyl phosphonamidate, and promotes the hydrolysis of both *para*-nitrophenyl amides ($k_{cat}/k_{uncat} \approx 10^5$) and esters ($k_{cat}/k_{uncat} \approx 10^4$; Scheme 3.4) [35]. Before the crystal structure of this antibody became available, Benkovic et al. constructed a model of its variable fragment based on homology [35d]. Various portions of the antibody were constructed by cutting analogous portions from antibodies with high homology to each structure and manually grafting them together. The hapten was then manually docked into the resulting combining site in an orientation that allowed for formation of a salt bridge between the phosphonamidate group and the side chain of an arginine residue (Arg L96) that was exposed to the binding site, and a series of energy minimizations and molecular dynamics runs on various portions of the model 43C9–hapten complex were then used to allow the structure to relax. Based on the 43C9–hapten contacts observed in this model, Arg L96 was assigned key roles in both hapten binding and transition state stabilization, and these were then probed experimentally by site-directed

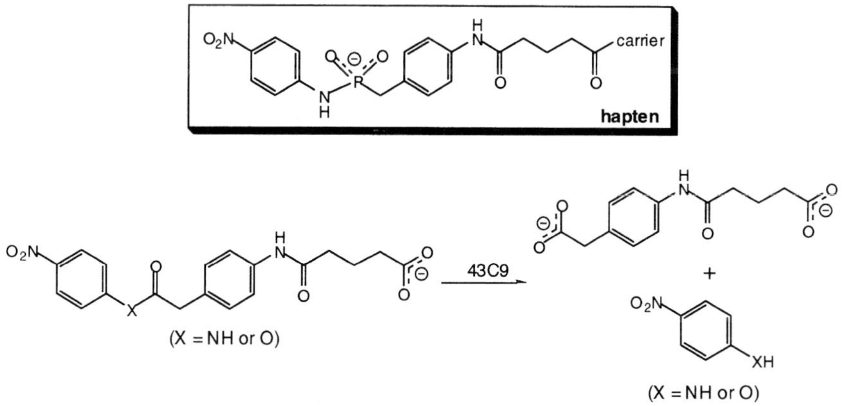

Scheme 3.4 Hydrolysis of *para*-nitrophenyl amides and esters catalyzed by antibody 43C9.

mutagenesis of this residue to glutamine – a transformation that diminished the hapten-binding ability of 43C9 by a factor of 20 and obliterated its ability to catalyze hydrolysis. In addition, previous biochemical studies had suggested that 43C9 accomplishes hydrolysis by initial acylation of a combining site histidine, but which histidine residue plays this role was not known. Based on the 43C9–hapten model, this role was assigned to His L91. Although subsequent solution of a crystal structure of 43C9 [35j] revealed some deficiencies in the homology model resulting from incorrect positioning of CDR H3, the proximity of both Arg L96 and His L91 to the phosphonamidate binding region was confirmed. In addition, the model was successfully used to design several variants of 43C9 containing metal ion binding sites [35e–g].

The crystal structure of 43C9 was solved with *para*-nitrophenol bound rather than with the phosphonamidate hapten [35j], so models of 43C9–hapten and 43C9–substrate complexes were constructed based on the 43C9–PNPOH structure via manual docking and energy minimizations. This simple modeling showed that His L91 is oriented in a reasonable geometry for attack on the substrate carbonyl, at a distance of approximately 4 Å. Recently, based on density functional theory calculations of activation barriers for hydroxide and deprotonated imidazole attack on substrate models, as well as molecular dynamics computations on substrate and transition state binding, Chong et al. suggested that direct hydroxide attack is also a plausible mechanism for hydrolysis catalyzed by 43C9, and may at least compete with a mechanism involving antibody acylation [23].

3.3.3.4
CNJ206

Antibody CNJ206 was raised against a *para*-nitrophenyl phosphonate hapten and catalyzes the hydrolysis of *para*-nitrophenyl acetate with a k_{cat}/k_{uncat} of ca. 10^3 (Scheme 3.5) [36]. A crystal structure of CNJ206 has been reported [36b]. Based on biochemical experiments and the structure of the combining site, it seems clear that this antibody promotes direct attack of hydroxide onto its aryl ester substrate. The

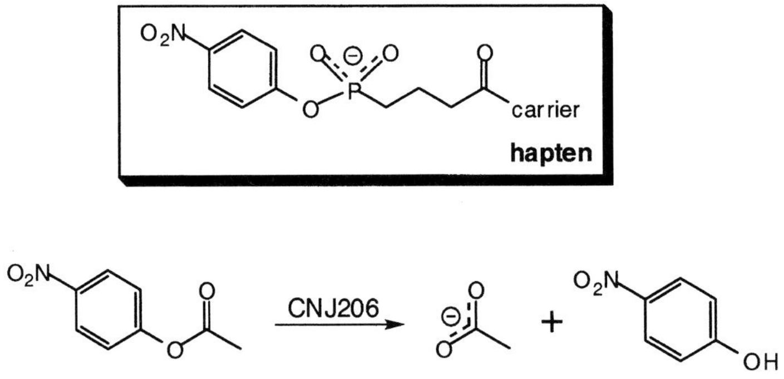

Scheme 3.5 Hydrolysis of *para*-nitrophenyl acetate catalyzed by antibody CNJ206.

specific antibody–transition structure interactions that lead to transition state stabilization were analyzed by Tantillo and Houk using the transition state docking method [18]. This procedure involves docking transition structures into an active site – in this case, conformationally unconstrained docking of structures optimized at an *ab initio* level (phenyl acetate was used as a model substrate) into the static, crystallographically determined CNJ206 combining site using AutoDock [37] with the assumption that the transition state structure is not significantly altered upon complexation (an assumption that appears to be valid for most antibody-catalyzed hydrolyses). In principle, either face of the *para*-nitrophenyl acetate carbonyl can be attacked by hydroxide, so enantiomeric transition structures for both *si* and *re* attack were docked into the binding site. The AutoDock calculations predict that the *si* transition structure binds more tightly to CNJ206 by several kcal/mol, demonstrating how stereoselective catalysis can arise from immunization with an achiral hapten just because the natural binding sites displayed to an antigen are constructed from inherently asymmetric amino acids. Based on the computational models, this selectivity was ascribed to stronger hydrogen bonding to the *si* transition structure than to the *re* transition structure [18]. A mutation was also proposed that might improve catalysis: Tyr H102 → Gln. A modified binding site was modeled and the *si* and *re* transition structures were re-docked. While the complex of the *re* transition structure does not benefit from this mutation, the complex of the *si* transition structure boasts an additional hydrogen bond, as hoped, and an improved computed binding energy. This computational prediction awaits experimental verification.

3.3.3.5
48G7

Antibody 48G7 catalyzes the hydrolysis of simple *para*-nitrophenyl esters and was raised against a *para*-nitrophenyl phosphonate TSA hapten (Scheme 3.6) [38]. The crystal structures of both 48G7 and its germline precursor, with and without hapten bound, have been reported by Schultz and Stevens, allowing the structural changes that occurred during the micro-evolutionary maturation process to be revealed [38c,d]. Based on these crystal structures, it was proposed that the 48G7 combining site

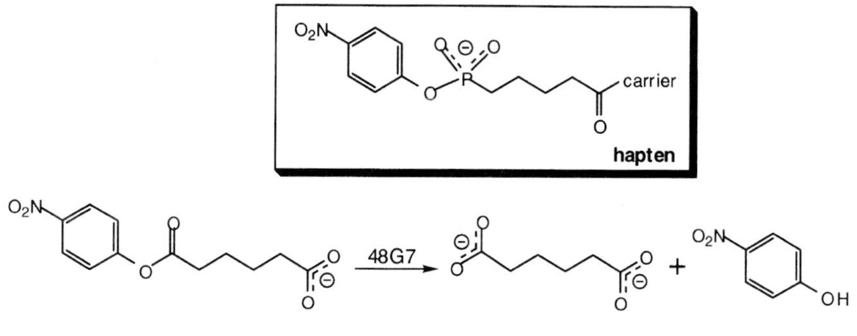

Scheme 3.6 Hydrolysis of *para*-nitrophenyl esters catalyzed by antibody 48G7.

loses conformational flexibility upon maturation, becoming rigidly pre-organized for hapten binding. This concept was explored by Kollman and coworkers using molecular dynamics simulations with explicit solvent combined with calculations using continuum solvent models [38e], which showed that the germline progenitor of 48G7 was indeed more flexible than its micro-evolved progeny. Hapten binding energies were also predicted in this study, and good qualitative agreement between the predicted and experimentally determined binding energies was achieved. Analysis of binding energies indicated that electrostatic interactions are unfavorable for all cases relative to water, but these interactions become more favorable upon maturation, leading to a net increase in hapten affinity. These electrostatic variations were shown to be intimately connected to conformational differences in the 48G7 combining site upon maturation, beyond just changes in the identities of specific residues.

The increase in hydrolysis rate constant (as well as hapten and product affinities) upon maturation ($k_{cat}/k_{uncat} \approx 10^4$ for 48G7 and $k_{cat}/k_{uncat} \approx 10^2$ for its germline precursor) was also correlated with specific structural changes using the transition state docking approach [18]. Like CNJ206, mature 48G7 has a preference for one of the two possible enantiomeric transition structures for hydroxide attack, but in this case the *re* transition structure is predicted to bind more tightly. Interestingly, 48G7's germline precursor displays no preference for either of the modes of attack; stereoselectivity arose during the affinity maturation process. In the germline case, the partially negatively charged hydroxyl and carbonyl oxygens are both stabilized in the *si* and *re* transition structure complexes, but in the case of mature 48G7, only the *re* transition structure is predicted to enjoy direct stabilizing interactions with both its hydroxyl and carbonyl oxygens. The structural changes that accompany maturation increase both the absolute stabilization of the *re* transition structure and its relative stabilization compared to the *si* structure. Additionally, the availability of kinetic data on various combining site mutants of mature 48G7 provided a true test of the limits of the transition state docking approach for modeling antibody-catalyzed hydrolyses. While quantitative predictions of the effects of these mutations on hydrolysis rates were not possible with this method, qualitative correlations were observed between computed transition state stabilization energies and k_{cat} and k_{cat}/K_m values, and these rate variations could be connected to changes in structure observed for modeled mutant combining sites.

3.3.3.6
17E8 and 29G11

Antibody 17E8 promotes the hydrolysis of several amino acid phenyl esters with k_{cat}/k_{uncat} up to 10^4 and was elicited in response to a phenyl phosphonate TSA hapten (Scheme 3.7) [39]. Antibody 29G11, obtained from the same immunization as 17E8, also catalyzes the hydrolysis of norleucine phenyl esters (X = $CH_2CH_2CH_2CH_3$) with $k_{cat}/k_{uncat} \approx 10^3$ [39c]. While it seemed clear from the outset that the hydrolytic mechanism employed by 29G11 involves direct attack of hydroxide on the substrate, whether 17E8 actually utilized a stepwise antibody acylation mechanism was the

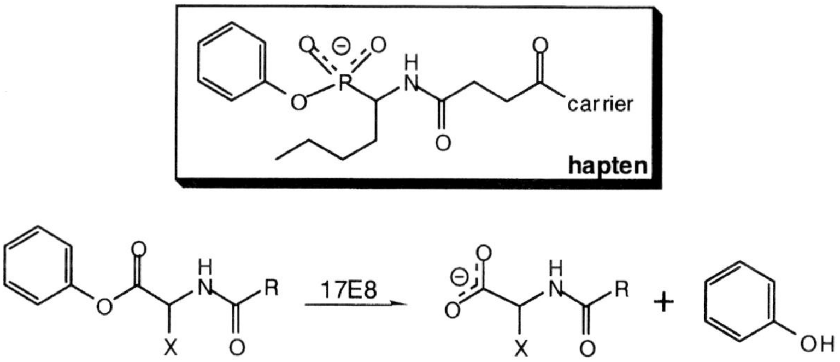

Scheme 3.7 Hydrolysis of several amino acid phenyl esters catalyzed by antibodies 17E8.

subject of some debate. The details of these catalytic mechanisms were explored using various computational techniques.

Binding of various substrates to 17E8 was examined using free-energy perturbation methods based on molecular dynamics simulations [39e]. The preference for norleucine over methionine substrates was accurately reproduced by these calculations and shown to be the result of better solvation of the methionine side chain in water.

Antibody–transition state interactions were also examined computationally. While X-ray structures of both 17E8 [39b] and 29G11 [39f] have now been reported, the gap of several years between the two prompted the construction of a computational model of 29G11 using the X-ray structure of 17E8, followed by manual replacements of key active site residues and energy minimization [39c]. Given the high sequence similarity of the 17E8 and 29G11 combining sites, it is not surprising that the active site structure in this model was extremely similar to that subsequently determined by X-ray. It had been proposed from the outset that an active site serine residue in 17E8 served as the nucleophile involved in antibody acylation, and the absence of this residue in 29G11 led to the suggestion that 29G11 promoted hydrolysis by a different mechanism – direct hydroxide attack.

The transition state docking method was also applied to both 29G11 and 17E8 [18]. Like CNJ206 and 48G7, antibody 29G11 is predicted to function through direct stabilization of the transition structure for hydroxide attack (in this case the *re* transition structure is preferred). In the case of 17E8, several variants of the combining site were modeled to explore the different roles for the active site serine and histidine; in all cases, complexes with the *re* transition structure were preferred. In one example, His H35, Arg L96, Lys H97, and a backbone NH group stabilize the partially negative hydroxyl and carbonyl oxygens of the transition structure for hydroxide attack. This is a model that is consistent with available experimental results but does not require antibody acylation. Subtle conformational differences in 29G11 prevent His H35 from hydrogen bonding to a transition state oxygen, thereby explaining 17E8's slightly better k_{cat}, and mutation of this histidine in 17E8 to alanine has been shown to greatly diminish catalysis [39f]. It has also been shown experimentally that mutation of the active site serine in 17E8 (Ser H99) to alanine does not diminish catalysis

[39f]. This observation is certainly the biggest strike against the antibody acylation mechanism. Consistent with this observation, replacement of the serine hydroxyl by a hydrogen atom and re-docking of the transition structure led to an extremely similar antibody–transition structure complex and predicted binding energy. Subsequent quantum mechanical and molecular dynamics computations also support a direct hydroxide attack mechanism [23].

3.4
Cationic Cyclizations

3.4.1
Antibody Catalysis of Solvolysis

Carbocation cyclizations are of central importance in biosynthesis, producing the cyclic terpenes, steroids, and related hormones. In analogy to biosynthetic polycyclizations, several different antibodies that catalyze the cyclization of one or two rings upon solvolysis have been reported by Lerner, Janda, and coworkers [40–42]. Interestingly, these antibody-catalyzed reactions often proceed with unusual regioselectivity or stereoselectivity, whereas the uncatalyzed reactions are very inefficient and have not been observed.

The solvolysis/cyclization reactions promoted by antibody TM1–87D7, which was raised against an ammonium hapten (Scheme 3.8), were studied computationally by Lee and Houk [43]. The goals of this work were to predict the product distributions that should be expected from the uncatalyzed background reactions (since the experimental product distribution in the absence of antibody could not be measured directly) and to develop a model for how these reactions might be accelerated upon binding to an antibody catalyst. Quantum mechanical calculations (MP2/6–311G(d)) indicated

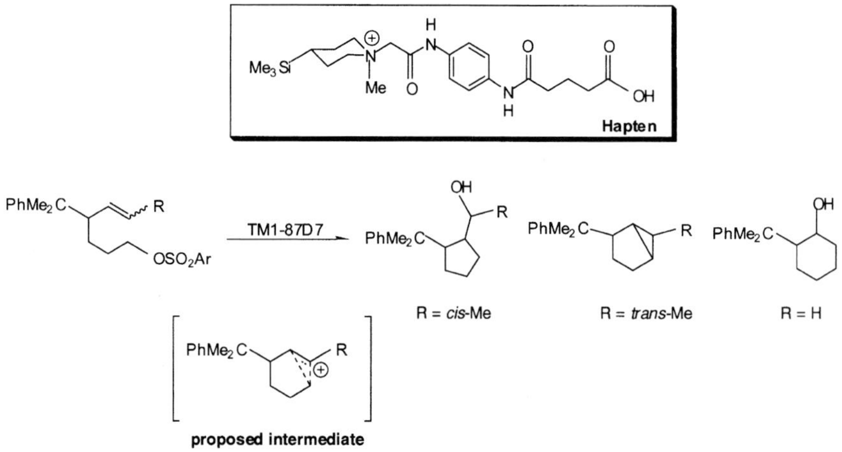

Scheme 3.8 Solvolysis/cyclization reactions catalyzed by antibody TM1–87D7.

that non-classical protonated cyclopropane intermediates (Scheme 3.8) were likely involved in the cyclization reactions, and these intermediates were docked into the binding site of McPC603, a well-characterized antibody binder of phosphorylcholine; phosphorylcholine has tetramethylammonium and phosphate groups that resemble the carbocation/sulfonate ion pair that is presumably formed in the solvation reactions catalyzed by antibody TM1–87D7. This modeling showed that the carbocation intermediate could fit into the McPC603 binding site in an orientation that allows for favorable electrostatic interactions between the carbocation and a combining site carboxylate. This carboxylate may also position a water molecule for trapping of the carbocation. The quantum mechanical calculations of Lee and Houk also showed that the position of the methyl group in the substrate alters the geometry of the non-classical cation intermediates toward those of the product structures found for each of the antibody-catalyzed reactions shown in Scheme 3.8; thus it seems that the substituent, rather than the antibody, directs the selectivity of product formation [43].

3.4.2
Antibody-Catalyzed Hydroxyepoxide Cyclization

The hydroxyepoxide cyclization shown in Scheme 3.9 could, in principle, form either a tetrahydropyran or tetrahydrofuran product. Janda et al. found that antibody 26D9 catalyzes the preferential formation of the tetrahydropyran, whereas normal acid catalysis in the absence of the antibody leads to the tetrahydrofuran product, as expected based on Baldwin's rules for cyclization reactions [44]. This hydroxyepoxide cyclization was one of the earliest antibody-catalyzed reactions studied using *ab initio* computations [45–46]. In fact, these pioneering studies by Na and Houk marked the origin of the "theozyme" concept [17].

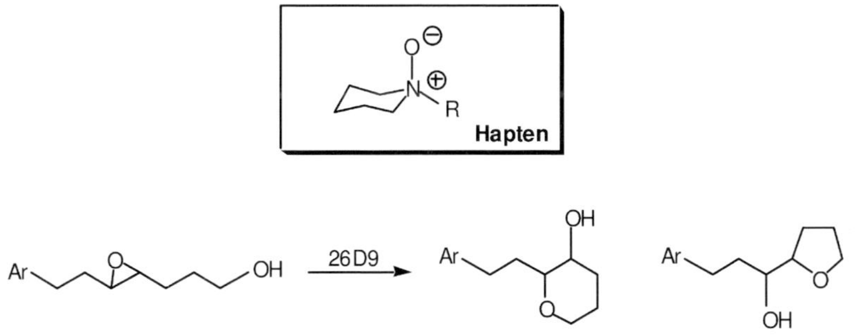

Scheme 3.9 Hydroxyepoxide cyclization catalyzed by antibody 26D9.

Na and Houk's calculations showed that the transition state for proton-catalyzed formation of the tetrahydrofuran product was favored over that for formation of the tetrahydropyran product by 1.8 kcal/mol, consistent with the experimental results for the solution reaction [45]. Given that no tetrahydrofuran product was observed in the 26D9-catalyzed reaction, this established that in the antibody-catalyzed reaction, the six-membered transition state must be favored by at least 3.6 kcal/mol over the five-membered transition state. Differences in the geometries and electrostatic profiles of these transition states were noted, and Na and Houk then used computational means to build – in the absence of any experimental information on what the binding site of 26D9 looked like – a theoretical model for a catalyst that could exploit these subtle differences to favor formation of the six-membered transition state. First, the positions of a formic acid and a formate molecule (reasonable models for typical active-site functional groups) were optimized around a model of the piperidine N-oxide hapten shown in Scheme 3.9. This gave an acid/base pair that was ideally situated to bind to the hapten; these two groups, fixed in space, constituted the "theozyme" or theoretical enzyme. Next, this theozyme was allowed to interact with the transition states for tetrahydrofuran and tetrahydropyran formation, and it was found that the complex with the six-membered transition state was in fact 4 kcal/mol more stable than that with the five-membered transition state.

Further computational studies on the cyclization of hydroxyepoxide reactants with an additional CH_2 group led to the prediction that this same theozyme would also favor the formation of ethers containing seven-membered rings rather than six-membered rings [46]. Janda et al. then tested this prediction and verified the formation of the seven-membered ring product in the antibody-catalyzed reaction [47–48].

Subsequently, Gruber et al. obtained the crystal structure of an antibody, 5C8, which has a very similar sequence to that of 26D9 and also catalyzes the selective formation of the tetrahydropyran product [49]. Remarkably, the binding site of this antibody contains an aspartate/aspartic acid pair oriented similarly to the formate/formic acid pair in the theozyme!

Thus, the concept has been validated as a reasonable approach to elucidating the origins of catalysis, predicting the products of reactions with altered substrates, and even predicting functional group arrays that are well suited for promoting particular reaction pathways [17].

3.5
Antibody-Catalyzed Diels-Alder and *retro*-Diels-Alder Reactions

The Diels-Alder reaction plays an important role in organic synthesis by providing a general and stereoselective route to cyclohexenes [50]. This reaction is facilitated by the presence of electron-donating groups on the diene and electron-withdrawing groups on the dienophile [51].

In spite of the significance of the Diels-Alder reaction in synthesis, the evidence for natural Diels-Alderases is only circumstantial [52]. However, much success has been achieved in developing catalysts for such reactions, and there are many artificial "Diels-Alderases", e.g., Lewis acids [53], antibodies [54], micelles [55], albumin [56], Baker's yeast [57], cyclodextrins [58], Rebek's tennis ball [59] and synthetic RNA fragments [60]. Antibody Diels-Alderases are the best characterized and most potent of these catalysts [2b]. Antibodies catalyzing four representative Diels-Alder reactions have been explored theoretically: (1) antibody 1E9, the most efficient antibody Diels-

Alderase discovered so far [61]; (2) antibody 39A11 and its germline precursor [62]; (3) an *exo*-Diels-Alderase, antibody 13G5 and related antibodies, which demonstrate stereocontrol [63]; (4) a *retro*-Diels-Alderase, antibody 10F11 [64]. Theoretical studies of these antibodies have led to proposals about how binding, catalysis, and selectivity by antibody Diels-Alderases are achieved.

3.5.1
The Most Efficient *endo*-Diels-Alderase 1E9

The first Diels-Alderase, antibody 1E9, was discovered by Hilvert and coworkers [61]. It catalyzes the cycloaddition between tetrachlorothiophene dioxide (**1**) and N-ethylmaleimide (**2**) (Scheme 3.10). The intermediate, **3**, eliminates sulfur dioxide spontaneously to give product **4**, which is subsequently oxidized to give N-ethyl tetrachlorophthalimide (**5**). Antibody 1E9 was raised against the hexachloronorbornene derivative **6**, a structural analog of **3** and the transition state leading to **3**. As judged by the k_{cat}/k_{uncat} value of 1000 M [61a], 1E9 is significantly more efficient than other antibodies in catalyzing Diels-Alder reactions [2b]. The crystal structure of the 1E9-6 complex was reported by Xu et al. [61b]. Structural analysis indicates that the key catalytic residues in the binding pocket were Asn H35, which forms a hydrogen bond to the carbonyl group on the succinimido moiety of **6**, and Trp H50, which provides π-stacking interactions with the succinimide ring.

The reaction pathway was explored quantum mechanically [65]. The cycloaddition between **1** and **2** was found to be the rate-determining step. Chen et al. located the *endo* and *exo* transition states for this cycloaddition step in the absence of catalyst and the *endo* transition state is 6.9 kcal/mol lower in energy than the *exo*.

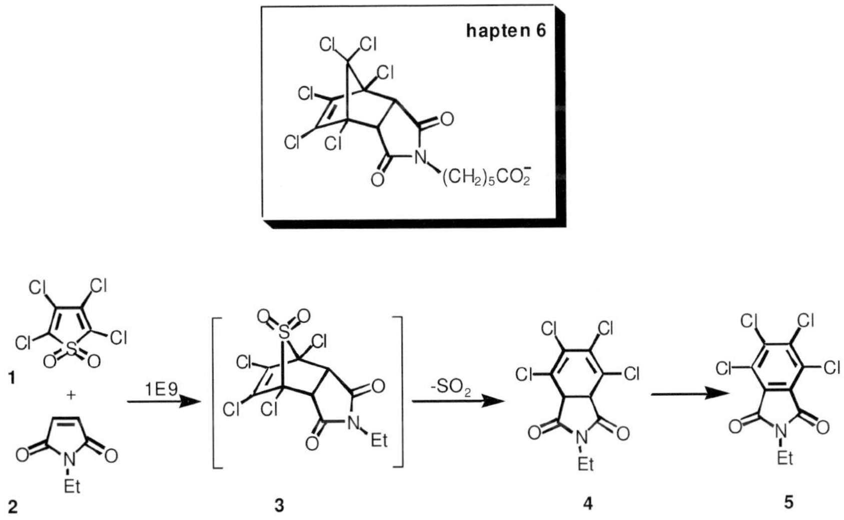

Scheme 3.10 *Endo*-cycloaddition catalyzed by antibody 1E9.

While the stereoselectivity of 1E9 could not be determined experimentally because of the spontaneous elimination of SO_2, docking experiments with the *endo* and *exo* hapten stereoisomers indicated that the binding pocket accommodates the *endo* conformation better than the *exo*. For the *endo* hapten, AutoDock [37] reproduces the binding mode revealed in the crystal structure.

The *endo* and *exo* transition states were docked in the binding pocket. Both transition states bind to antibody 1E9 in a similar mode, where a hydrogen bond forms between the carbonyl group of the dienophile and the amide group of Asn H35 (Fig. 3.4). Therefore, Chen et al. constructed a quantum mechanical theozyme [17] using a molecule of water hydrogen bonding to the carbonyl group of the *endo* transition state. Calculations showed that the theozyme could stabilize the *endo* transition state by 0.6 kcal/mol. Based on the inherent 6.5 kcal/mol energy difference between the *endo* and *exo* transition states, they predicted that the natural preference for the *endo* pathway will be maintained by 1E9 catalysis.

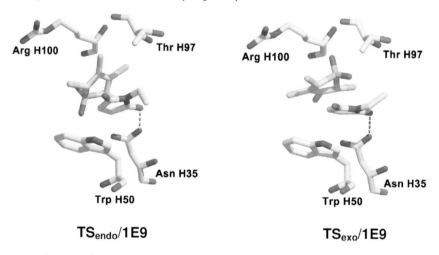

Fig. 3.4 *Exo* and *endo* transition states docked in the binding site of antibody 1E9.

Chen et al. also carried out MD [10] calculations and used the linear interaction method [13] to predict that the barrier for *endo* addition is lowered in energy by 5.7 kcal/mol in 1E9 compared to that of the uncatalyzed reaction [65]. The experimental lowering is 4.0 kcal/mol [61a]. These calculations show that favorable van der Waals interactions dominate the binding along the reaction coordinate, but provide little discrimination between the reactant complex and transition state. In contrast, electrostatic interactions are less favorable for reactant binding than for the transition state. This was attributed to the increased strength of the hydrogen bond from Asn H35 to the carbonyl oxygen of the dienophile in the transition state.

These calculations showed that the high efficiency of 1E9 catalysis arises from enthalpic stabilization of the transition state by a close fit with the hydrophobic binding pocket and a strategically placed hydrogen bond [65]. Recently, Zhang et al. reexamined antibody 1E9 by computing the electrostatic potential surface for its binding

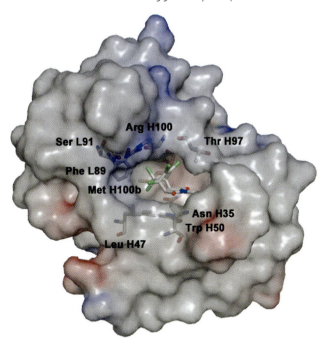

Fig. 3.5 Electrostatic potential surface of 1E9 binding site with the bound hapten (blue represents positive electrostatic potential and red represents negative electrostatic potential).

pocket resolved from the X-ray experiment (Fig. 3.5) [66]. Remarkable shape complementarity could be observed between the hapten and the antibody binding site. Gap indices – defined as the gap volume between antibody and hapten (in Å^3) divided by the buried surface of the hapten (in Å^2) – were computed to evaluate shape complementarity between catalytic antibodies and their corresponding haptens semi-quantitatively. Not surprisingly, antibody 1E9 with the hapten gave the smallest value (0.44) compared to other highly homologous antibodies (e.g., DB3: 0.90; 39A11: 1.69).

3.5.2
endo-Diels-Alderase 39A11 and its Germline Precursor

Antibody 39A11 catalyzes the cycloaddition of diene **7** and dienophile **8** to give Diels-Alder adduct **9** (Scheme 3.11) [62a]. It was generated against racemic hapten **10**, a mimic of the boat-like transition state of the cycloaddition. The crystal structures of the 39A11–**10** complex and its unliganded germline antibody have been reported [62b]. Structure analysis indicates that the important residues for binding are Asn H35, which hydrogen bonds to the carbonyl oxygen of **10**, and Trp H50, which π-stacks with the succinimide ring. However, the absolute stereoselectivity of 39A11 and the origins of 39A11-catalysis were still unknown when these structures were determined.

For the Diels-Alder reaction between *N*-formylaminobutadiene and *N*-methyl maleimide, DFT calculations predict that the *endo* pathway is favored by 5.2 kcal/mol

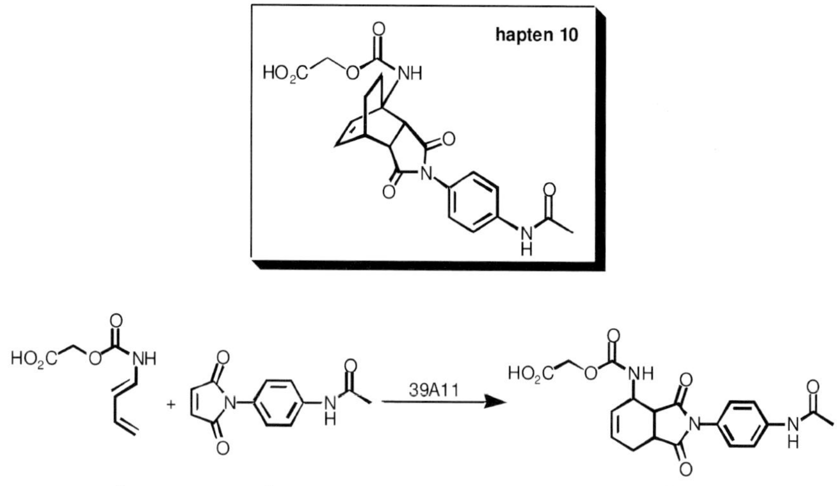

Scheme 3.11 *Endo*-cycloaddition catalyzed by antibody 39A11.

over the *exo* pathway in the gas phase. This preference decreases to 3.4 kcal/mol in water [66].

To understand the binding and stereoselectivity of 39A11, docking simulations were performed on hapten stereoisomers, hap1 (*exo*hap1, *endo*hap1 and sathap1) and hap2 (*exo*hap2, *endo*hap2 and sathap2) inside the binding pockets of 39A11 and the germline (Scheme 3.12) [66]. These calculations showed that: (1) the antibodies can accommodate all the haptens; (2) hapten stereoisomers belonging to the same series (1 or 2) have essentially identical binding modes in 39A11 or the germline; (3) the antibodies recognize the hap1 series better than the hap2 series. The binding mode of the hap1 series in antibody 39A11 actually reproduces the binding mode revealed in the crystal structure of 39A11-10 complex. The important residues for binding of the hap1 series are Asn H35 and Trp H50. Based on these observations, Zhang et al. concluded that the helical arrangement of the polar –NHR and succinimide substituents decides the binding mode of a given hapten stereoisomer [66].

They also docked the transition state stereoisomers into the binding pockets of 39A11 and the germline (Scheme 3.12 and Fig. 3.6). Not surprisingly, they found that the ts1 series (*exo*ts1 and *endo*ts1) fits better than the ts2 series (*exo*ts2 and *endo*ts2). The ts1 series adopts similar binding modes to the hap1 series with noncovalent interactions with Asn H35 and Trp H50. While the *endo*ts1 and *exo*ts1 bind equally well in the germline, the mature antibody 39A11 shows an overall preference for *endo*ts1 because of an additional hydrogen bond with Arg H97. Based on the docking results, quantum mechanical theozymes [17] for both *endo* and *exo* transition states were constructed, where two molecules of formamide were used to imitate the functionalities of Asn H35 and Arg H97. While the calculated preference for the *endo* pathway is 3.4 kcal/mol in water, the results of theozyme calculations show that the preference increases to 5.0 kcal/mol and 6.0 kcal/mol upon catalysis by the germline and 39A11,

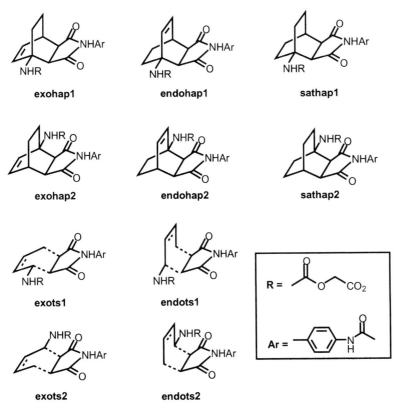

Scheme 3.12 Hapten stereoisomers, their saturated analogs, and transition state stereoisomers.

respectively. The results indicate that the stereoselectivity pre-exists in the germline and gets increased through the maturation process. Zhang et al. predicted that the Diels-Alder reaction would give a (3S,4R,5R) major product when catalyzed by either the germline or mature 39A11. This prediction awaits experimental tests [66].

Compared to 1E9, antibody 39A11 is much less efficient, with a k_{cat}/k_{uncat} value of 0.35 M [62a]. While the high efficiency of antibody 1E9 is largely attributed to the ideal shape complementarity between the hapten and the binding pocket [65], this does not apply to antibody 39A11. Electrostatic potential surfaces were built to visualize the binding pockets of the germline and antibody 39A11 (Fig. 3.7). A few key residue mutations in the maturation process result in a comparatively more polar and loosely fitting binding pocket of antibody 39A11, which consequently leads to its poor catalytic ability. The results agree with mutation studies by Romesberg and Schultz, which show that hapten packing is correlated with catalytic ability [67].

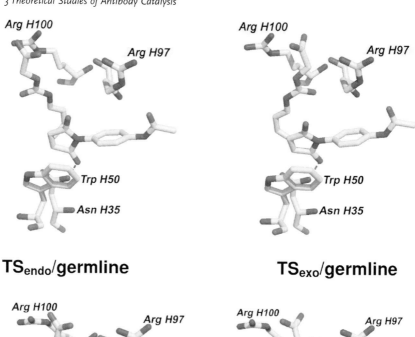

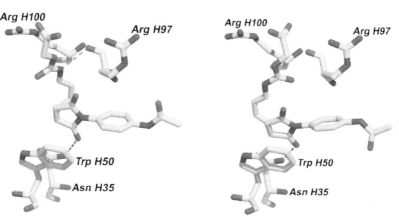

Fig. 3.6 *Exo* and *endo* transition states docked in the binding sites of the germline and antibody 39A11.

3.5.3
exo-Diels-Alderase 13G5

An excellent demonstration of the potential of antibody Diels-Alderases is the catalysis of the formation of a product stereoisomer that is disfavored in solution. In 1993, the first *exo* Diels-Alderase, antibody 22C8, was reported by Gouverneur et al. [63a]. The target reaction was the cycloaddition between *trans*-1-*N*-acylamino-1,3-butadiene

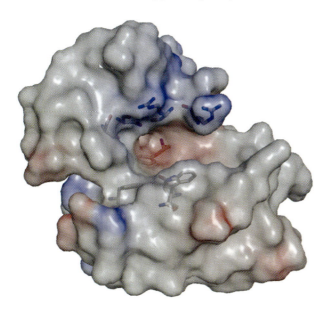

germline

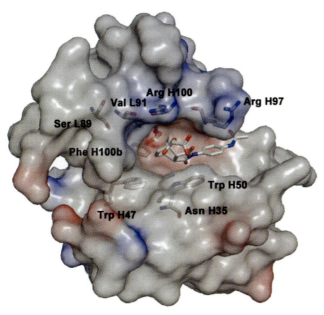

39A11

Fig. 3.7 Electrostatic potential surfaces of the germline binding site, and antibody 39A11 binding site with the bound hapten.

11 and N,N-dimethylacrylamide **12** (Scheme 3.13). The uncatalyzed reaction is highly regioselective but gives an 85:15 mixture of the *endo* and *exo* adducts. Model calculations with RHF theory on a system composed of acrylamide and N-(1-butadienyl)-carbamic acid showed that the two *ortho* transition states for cycloaddition are more favored than the two *meta* transition states by 2 to 4 kcal/mol, and the energy barrier for the *ortho endo* is lower than that of the *ortho exo* by 1.9 kcal/mol. Haptens **13** and **14** were selected to mimic the *ortho endo* and *ortho exo* transition states, respectively. Antibodies 7D4 and 22C8 were generated, which catalyze almost exclusive *endo* and *exo* addition, respectively.

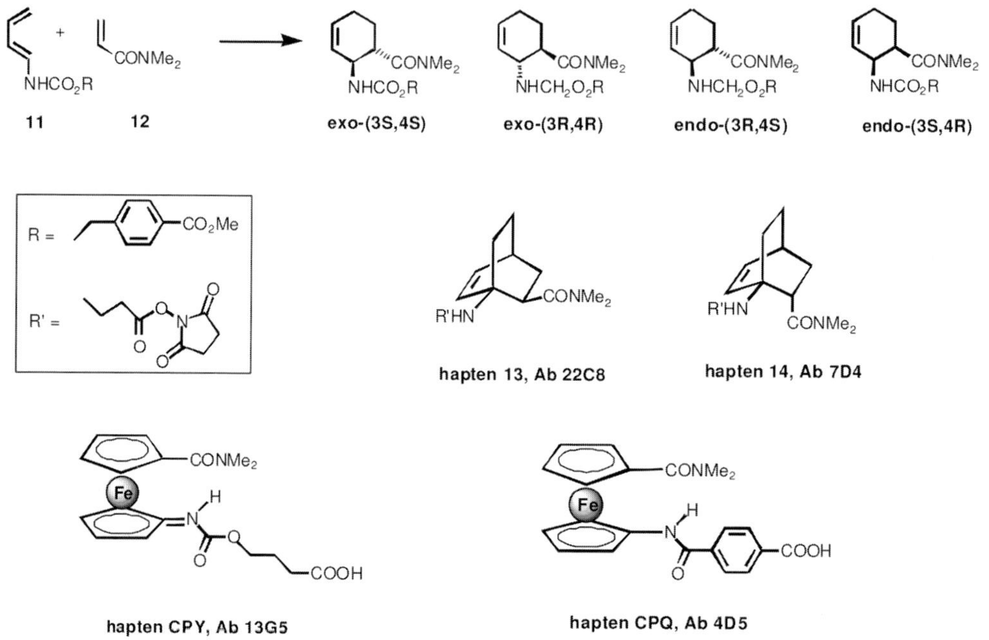

Scheme 3.13 Exo-cycloaddition catalyzed by antibodies elicited from various haptens.

Later, flexible and unconventional ferrocene haptens, **CPY** and **CPQ** (Scheme 3.13), were designed by Heine et al. for the same reaction [63b]. *Endo* Diels-Alderase 4D5 and *exo* Diels-Alderase 13G5 were generated in response to these haptens. Heine et al. reported the crystal structure of the 13G5–**CPY** complex and constructed the electrostatic surface for the binding pocket [63b]. They found that **CPY** is bound in a deep pocket with a narrow entrance (Fig. 3.8). However, the absolute stereochemistry of the *exo*-Diels-Alderase was not clear at that time. Docking experiments showed that the *exo* transition states dock better than the *endo* with several strong hydrogen bonds, and the (R,R) adduct appeared to be favored somewhat compared to the (S,S) [63b].

Recently, however, Cannizzaro et al. investigated the stereoselectivity of these *exo* and *endo* antibody Diels-Alderases experimentally [68]. They found that all the *exo* Diels-Alderases (13G5 as well as all other related antibodies from the haptens shown

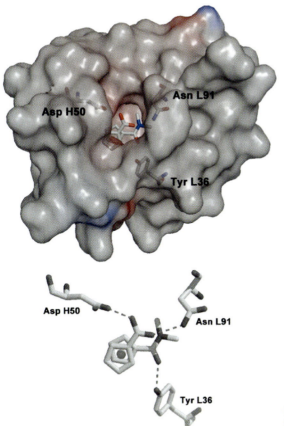

Fig. 3.8 13G5 binding site with the bound hapten.

in Scheme 3.13) catalyze the formation of the *exo*-(*S,S*) adduct. They further carried out detailed theoretical studies on the most efficient *exo* Diels-Alderase, antibody 13G5 (% *ee* = 98 ± 2 %), to elucidate the origins of its significant selectivity.

Cannizzaro et al. first carried out theozyme calculations [17] for the *exo* transition state at the B3LYP/6–311++G(d,p)//B3LYP/6–31+G(d,p) level [68]. Structural analysis of the crystal structure has indicated that the important residues that hydrogen bond with **CPY** are Tyr L36, Asn L91 and Asp H50 (Fig. 3.8) [65b]. Consequently, the theozymes consisted of a water molecule representing Tyr L36, a formamide molecule for Asn L91 and a formate molecule for Asp H50 (Scheme 3.14). The calculations show that while a water molecule (Tyr L36) and a formate molecule (Asp H50) lower the reaction barrier, a formamide molecule (Asn L91) does not. They further explored the synergistic stabilizing effects of both catalytic residues, Tyr L36, and Asp H50, on both *endo* and *exo* transition states, and in this case the *exo* transition state was found to be 3.0 kcal/mol lower in energy than the *endo*. Cannizzaro et al. explained the calculated preference as a consequence of electrostatic repulsion between the formate oxygen and the carbonyl oxygen of the dienophile, which is minimized in the *exo* transition state.

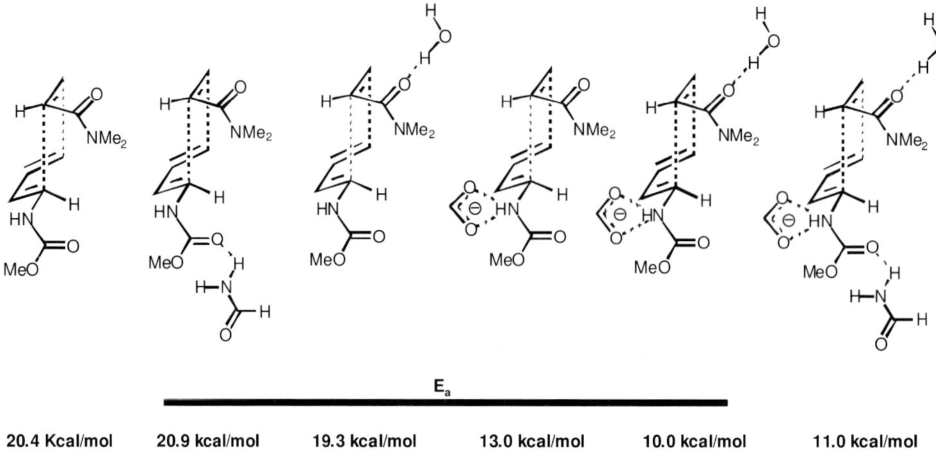

| 20.4 Kcal/mol | 20.9 kcal/mol | 19.3 kcal/mol | 13.0 kcal/mol | 10.0 kcal/mol | 11.0 kcal/mol |

Scheme 3.14 Theozyme calculations for the *exo* transition state.

The question remaining is why the *exo*-(S,S) adduct is the observed major product rather than the *exo*-(R,R) adduct. Cannizzaro et al. solved this question by docking and molecular dynamics simulations of the *exo*-(S,S) and *exo*-(R,R) transition states. The *exo*-(S,S) transition state fits well in the binding pocket of antibody 13G5 with a productive orientation. In contrast, in order for the *exo*-(R,R) transition state to fit in a productive orientation, a conformational change of the diene chain is necessary, which costs an additional 1.6 kcal/mol according to B3LYP calculations. Moreover, the results from molecular dynamics simulations suggest that the *exo*-(S,S) product is obtained when the diene is recognized by Asp H50 and Trp H33, while the *exo*-(R,R) product is obtained when the diene is recognized simultaneously by Asp H50, Trp H33 and Asn L91, which is energetically disfavored compared to the former because the hydrogen bond with Asp H50 is *anti*-catalytic [68].

3.5.4
retro-Diels-Alderase 10F11

The *retro*-Diels-Alder antibody 9D9 was found by Reymond and coworkers to catalyze the *retro*-Diels-Alder reaction of **15** (Scheme 3.15), a reaction attracting much attention because of its liberation of the biologically relevant HNO [64a]. Through a screen for catalysis rather than hapten binding, Reymond et al. obtained an improved antibody, 10F11 [64b], which was found to selectively catalyze the *retro*-Diels-Alder reaction of **15** compared to its regioisomer **16** (the k_{cat}/k_{uncat} is 13 times larger for **15** than for **16**). They also determined the crystal structure of 10F11 [64b].

Leach and Houk performed theoretical studies on the origins of catalysis and regioselectivity of antibody 10F11 [69]. Their docking experiments showed that while **15** binds in a specific way to the binding pocket, the regioisomeric substrate **16** does not, and there is less shape complementarity between the binding pocket and **16**. A

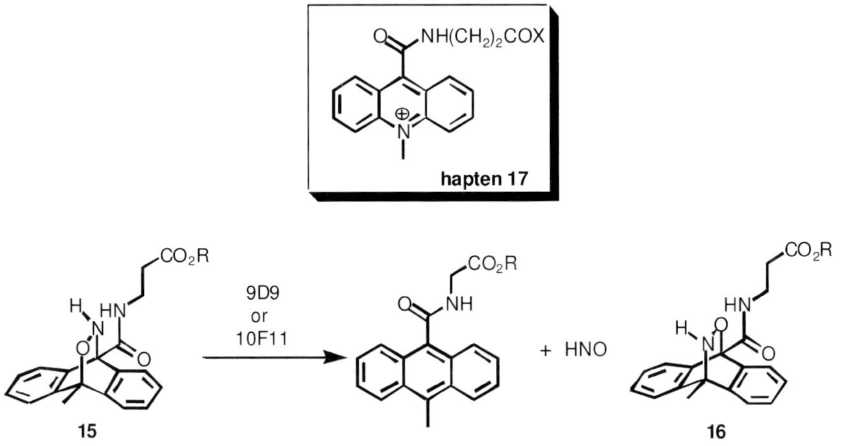

Scheme 3.15 *Retro*-Diels-Alder reaction catalyzed by antibodies 9D9 and 10F11.

tryptophan residue in the binding pocket contributes to catalysis by π-stacking more effectively with the flatter TS than with the substrate. These preliminary results have led to a comprehensive view of how the antibody *retro*-Diels-Alderases function, and further studies are under way [69].

3.6
Other Antibody-Catalyzed Pericyclic Reactions

3.6.1
Oxy-Cope Rearrangement Catalyzed by Antibody AZ-28

AZ-28 is found to catalyze the oxy-Cope rearrangement of **17** (Scheme 3.16) [70]. The S enantiomer of the substrate is found to react faster than the R by a factor of 15 [71], although this stereochemistry is not reflected in the achiral product. NMR studies indicate that the Michaelis complex has the two olefin termini in close contact. The activation enthalpies of the catalyzed and uncatalyzed reactions were measured and found to be 15.4 and 27.4 kcal/mol respectively. The entropies of activation were −23 and −3 e.u., so the $\Delta G^{\ddagger}$ is reduced from 28.3 to 22.0 kcal/mol by the catalyst at room temperature. The germline precursor to the antibody was also studied and found to be a better catalyst for the reaction, even though it binds the hapten less tightly [72].

An approach to computing the means by which AZ28 effects catalysis was adopted by Asada et al. [73]. They first used B3LYP to calculate the geometries and energies of the reactant, transition state and product of the uncatalyzed reaction. The stereoselectivity of the reaction was not considered – the transition state with the same stereochemical arrangement as the hapten in the crystal structure was used. Their B3LYP/6–311++G(d,p)//B3LYP/6–31G(d) calculations predicted an activation energy of 25.1 kcal/mol, in reasonable agreement with the barrier measured for the uncat-

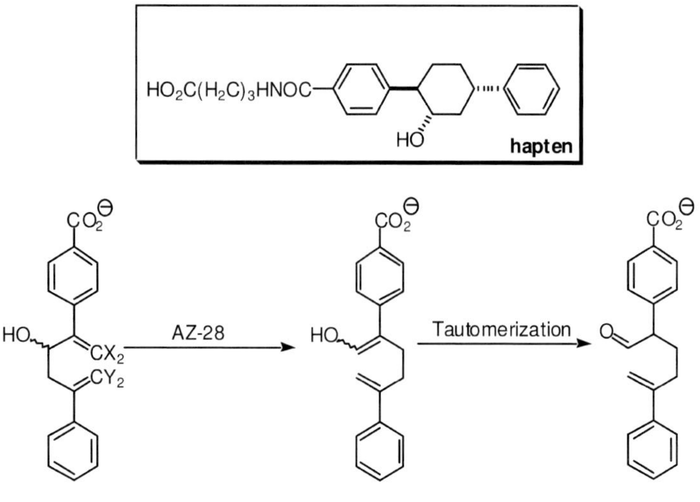

Scheme 3.16 Oxy-Cope rearrangement catalyzed by antibody AZ-28.

alyzed reaction. The reaction is 6.8 kcal/mol exothermic with the product being in its enolic form.

The B3LYP structures were then utilized to prepare parameters for the AMBER force field for use in subsequent molecular dynamics simulations [10]. These involved docking the three structures (substrate **17**, transition state, and product) into both the germline and mature antibody AZ28. Initial investigations revealed that His H96 makes intimate contact with the hapten; thus, its protonation state is of critical importance (Fig. 3.9). Interaction energy analysis revealed that the His is most likely protonated at the δ position. Using this protonation state, MD calculations revealed that when bound in the binding site of the germline, all three of the species showed greater motion during the course of the 1 ns trajectories than when the same species were bound in the mature antibody. The simulation also revealed that the side chain of the hapten forms two hydrogen bonds with Arg L50 in the mature antibody but only one in the germline structure. In line with experimental observations, they calculated that the hapten is bound more tightly by the mature antibody than the germline (ΔG_b = −15.4 kcal/mol for the germline and −17.4 kcal/mol for the mature antibody). Their calculations also show that the germline binds the transition state of the reaction more tightly than does the mature antibody (ΔG_b = −16.2 kcal/mol compared to − 12.5 kcal/mol by the mature antibody), and they ascribe this principally to the greater flexibility possible in the germline than in the mature antibody. This reflects the structural divergence between the actual transition state and the antigenic transition state analog. While the transition state is bound more tightly by the germline, the substrate is also bound more tightly (ΔG_b = −8.8 kcal/mol compared to −7.3 kcal/mol),

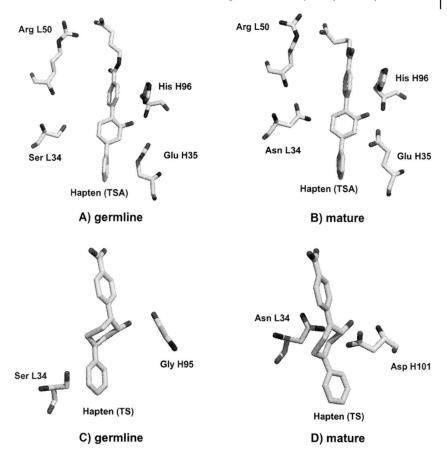

Fig. 3.9 Hapten(TSA) and hapten(TS) in the binding sites of the germline and antibody AZ-28.

but this difference is smaller, and hence the germline is a better catalyst than the mature antibody.

The final aspect of this study involved the calculation of the effect of mutation of positions L34, L51, H32, H58 and H73 to alanine and mutation of position H56 to glycine – these are the six positions affected by the somatic mutations that occur during maturation. These calculations revealed that L34 is the position that causes the most significant changes to the binding affinity. Mutation of Asn L34 to Ala (in the mature antibody) is calculated to diminish transition state binding by 7.6 kcal/mol and hapten binding by 6.8 kcal/mol. The effect is less pronounced in the germline, and mutating Ser L34 to Ala increases transition state binding by 0.3 kcal/mol and increases hapten binding by 1.8 kcal/mol.

A more detailed study by Black et al. is still in progress [74]. This study involved, as the first step, DFT calculations on the concerted reaction and stepwise reaction through radicals [74]. It was found that the reaction is preferentially concerted, but surprisingly, the favored transition state places the hydroxyl in an axial position. Kinetic isotope effects had been measured for the *di-* and *tetra*-deuterated substrates **18** and **19** in the catalyzed reaction (0.72 and 0.61 respectively). These were compared to those for the uncatalyzed Cope reaction of tetradeuterated 2,5-diphenylhexa-1,5-diene, measured experimentally (0.57) [76] and computed using B3LYP/6–31G(d) (0.60) [75]. Calculated results for the full system, including the hydroxyl and carboxylate groups were also in accord with these values (0.69 was calculated, also with B3LYP/6–31G(d)).

Docking of the hapten showed that the binding geometry observed in the crystal structure could be reproduced for both the protonated and unprotonated forms of the hapten. Docking of the carboxylate-substituted transition structure and substrate into AZ28 and its germline precursor, each in two protonation states, was also performed. All four transition states (*R*-axial, *R*-equatorial, *S*-axial and *S*-equatorial) were found to dock with similar energies and in one or the other of two modes – making an H-bond to either Glu H35 or to Asp H101 (Fig. 3.10). In the crystal structure with the hapten bound, Tyr H100a (adjacent to Asp H101) was involved in H-bonding but is buried slightly deeper, such that when the shorter transition state binds, the hydrogen bond shifts to Asp H101.

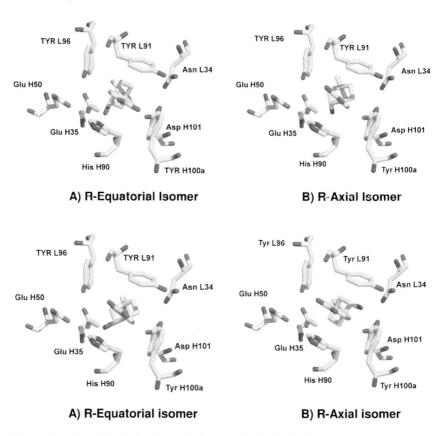

Fig. 3.10 Transition states docked in the binding site of antibody AZ-28.

More insights into the possible source of selectivity came with the docking of the substrates. The S substrate readily adopted conformations in the binding site that placed the two reactive termini in close proximity, as found in the NMR studies, whereas fewer binding modes of this type were found for the R substrate. It was hypothesized that the antibody may steer selectivity toward the S substrate by more readily permitting it to adopt a reactive conformation.

A theozyme model [17] was developed to test the hypothesis by Schultz that hydrogen bonding to the OH group is important in catalysis. An NH_3 was used to mimic a presumed H-bond acceptor in the active site, and as much as 2.1 kcal/mol lowering of the barrier was calculated. Aqueous solvation calculations performed using the PCM model [24, 77] revealed that water lowers the barrier compared to the reaction in the gas phase. The antibody is likely to provide a less polar environment than water, and the general medium effect of the antibody should thus diminish the rate of the reaction.

Schultz also proposed that the phenyls might be oriented in non-optimal arrangements in the binding site. Model calculations showed that altering the conformational positioning of the phenyl rings had significant effects on the energetics of the reaction. For example, rotating the two phenyl rings to positions orthogonal to the six-membered transition state ring cost 28.3 kcal/mol for the carboxylic acid form and 19.9 kcal/mol for the carboxylate form. This corresponds to 10–14 kcal/mol for each phenyl ring.

3.6.2
1,3-Dipolar Cycloaddition Catalyzed by Antibody 29G12

Catalytic antibody 29G12, elicited against hapten **20**, was found to bind its two substrates **21** and **22** with K_M values of 3.4 mM and 5.8 mM, respectively (Scheme 3.17)

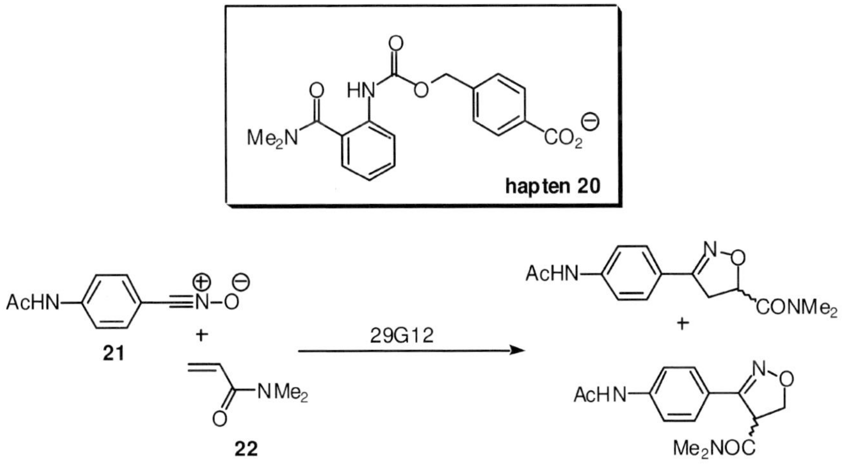

Scheme 3.17 1,3-Dipolar cycloaddition catalyzed by antibody 29G12.

[78]. $\Delta H^{\ddagger}$ and $\Delta S^{\ddagger}$ measured for the catalyzed and uncatalyzed reactions revealed that $\Delta H^{\ddagger}$ diminished by 6.8 kcal/mol in the presence of the antibody while $\Delta S^{\ddagger}$ goes from −28.1 to −38.1 e.u. The transition state of the reaction was found to be bound, with a dissociation constant of 0.7 µM.

Theoretical calculations were undertaken to quantitate the regioselectivity and to compare the transition state with the hapten. B3LYP/6−31G(d) modeling revealed significant changes in the atomic charges on going from reactants to the transition state. The computed dipole moment of the transition state (4.5 D) is less than the sum of the dipole moments of the reactants (11.2 D). This suggests that the antibody could stabilize the low-polarity transition state more than the two reactants with their dipoles aligned as must occur to form the observed product. This could provide a significant enthalpic advantage while adding to the entropic cost by requiring a very specific orientation to maximize this electrostatic stabilization. Further studies are under way as the crystal structure of 29G12 has now been solved [79].

3.6.3
Chorismate-Prephenate Claisen Rearrangement Catalyzed by Antibody 1F7

Wiest and Houk [80] performed a thorough study of model systems related to the chorismate-prephenate rearrangement catalyzed by antibody 1F7 [81] and also the enzyme, chorismate mutase (Scheme 3.18) [82]. Experimental studies of the uncatalyzed reaction reveal that bond breaking is much more advanced than bond making in the transition state (based on kinetic isotope effect evidence) [83] and that the transition state is more polar than the reactant [84]. Experimental studies of the enzyme catalyzed reaction show that the same pericyclic mechanism observed for the uncatalyzed reaction holds and that the $\Delta H^{\ddagger}$ is lowered by 4.6 kcal/mol while $\Delta S^{\ddagger}$ goes from −12.9 to 0.0 e.u., leading to an overall reduction in $\Delta G^{\ddagger}$ of 8.3 kcal/mol at room temperature [85]. Antibody 1F7 catalyzes the same reaction as chorismate mutase and lowers $\Delta H^{\ddagger}$ by approximately the same amount as the enzyme (5.5 kcal/mol) but

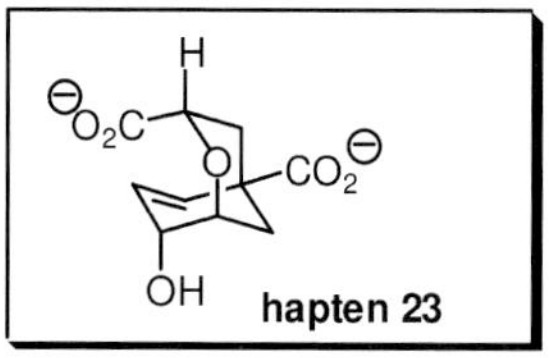

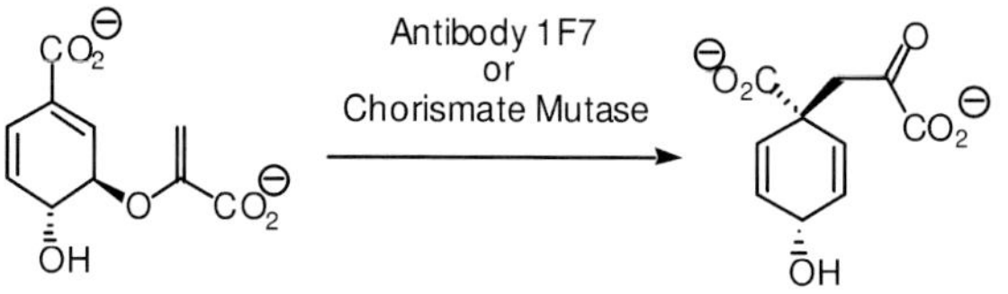

Scheme 3.18 Chorismate-prephenate Claisen rearrangement catalyzed by antibody 1F7.

incurs a larger entropic barrier (−22 e.u.), leading to an overall reduction in $\Delta G^\ddagger$ of 2.9 kcal/mol compared to the uncatalyzed reaction [81].

The crystal structures of the antibody-23 and enzyme-23 complexes revealed that there are less charged and polar residues in the antibody binding cavity than in the enzyme active site. Consequently, much fewer interactions between the antibody and bound 23 were observed than those between the enzyme and bound 23.

Computational studies of model systems revealed the effects of H-bonding to various fragments of the transition state, which facilitated the understanding of both the enzyme and antibody catalysis [80]. Deprotonation of either of the substrate carboxylic acids was found to lower the activation barrier substantially. This was attributed to a favorable electrostatic interaction between the carboxylate and the pericyclic transition state. Deprotonation of both acids, however, was expected to raise the barrier because of electrostatic repulsion in the transition state, although this could be offset by appropriately placed positively charged residues.

By comparing the theozyme to the crystal structure of the enzyme, Wiest and Houk concluded that the enzyme active site provides charge complementarity via Arg 7 interacting with the transition state carboxylate and Arg 90 interacting with the ether linkage. However, stabilization of the partial positive charge of the six-membered transition state ring was found to make a negligible contribution to the enzyme catalysis.

In the antibody, fewer electrostatic interactions are present. The carboxylate and the ether interact only with backbone amide NHs, and the potentially useful Asn H33, Asp H99 and Arg H95 side chains are all folded away from the binding site such that there are no suitable side chain functional groups present in the binding site to assist in catalysis in the manner suggested by the theozyme model.

3.7
Antibody-Catalyzed Carboxybenzisoxazole Decarboxylation

Antibody 21D8, raised against naphthalenedisulfonate hapten **25**, catalyzes the decarboxylation of 5-nitro-3-carboxybenzisoxazole **26** (Scheme 3.19) [86]. The rate of this reaction, in the absence of antibody, has been shown to vary by up to 8 orders of magnitude simply because of changes in solvent, with the fastest rates observed for nonpolar and aprotic solvents [87]. Consequently, the question arose as to whether the catalysis by 21D8 was merely the result of nonspecific medium effects. Houk and coworkers performed a detailed study of this decarboxylation reaction in 21D8 in an effort to resolve this mechanistic question [88].

A variety of computational methods were brought to bear on this problem. First, the gas phase reaction coordinate was explored using density functional theory (B3LYP/6-31+G(d)). Theozymes [17] were then constructed to explore the effects of hydrogen bond donors on the barrier of decarboxylation. The results of these computations are summarized in Scheme 3.20. The barrier in the absence of any theozyme was predicted to be 9.3 kcal/mol. When a single ammonium cation was placed near the substrate carboxylate group (**T1**), the barrier increased dramatically, as expected, given

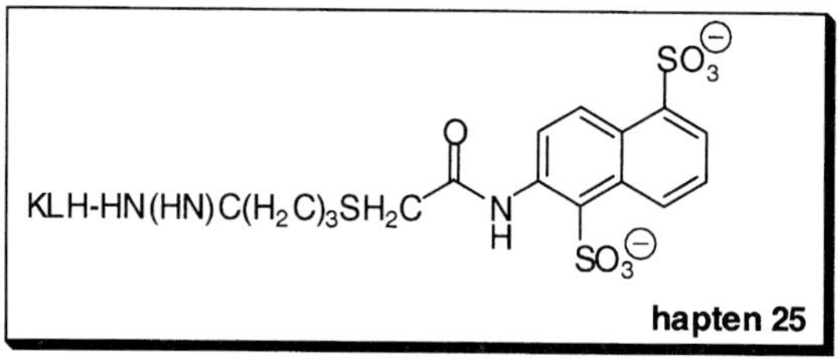
KLH-HN(HN)C(H2C)3SH2C
hapten 25

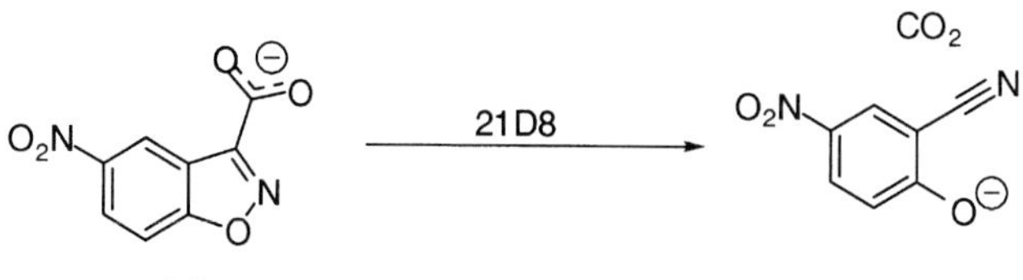
O2N
21D8
CO2

26

Scheme 3.19 Carboxybenzisoxazole decarboxylation catalyzed by antibody 21D8.

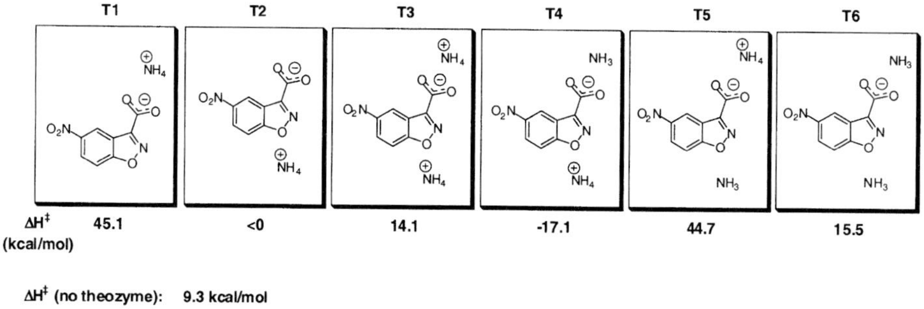
T1
T2
T3
T4
T5
T6
ΔH‡ (kcal/mol)
45.1
<0
14.1
-17.1
44.7
15.5
ΔH‡ (no theozyme): 9.3 kcal/mol

Scheme 3.20 Theozyme calculations to quantitate hydrogen bond stabilization.

that charge must flow off of this group during the decarboxylation. When a single ammonium cation was placed near the oxygen of the isoxazole ring (**T2**), however, no transition state could be found; instead, optimization led directly to products. This result indicates that strong hydrogen bond donors positioned near this oxygen atom can abolish the decarboxylation barrier, presumably by stabilizing the negative charge that accumulates at this site during the reaction. When ammonium cations were placed in the vicinity of both the carboxylate group and isoxazole oxygen (**T3**), the barrier was increased slightly over that without theozyme. Alternative protonation states of **T3** were also examined (**T4**–**T6**) (unlike **T1**–**T3**, the geometries of these complexes were not optimized; instead, protons were simply removed from **T3** and the relative energies of reactant and transition structure were calculated). These computations showed that barrier lowering requires a stronger hydrogen bond donor near the isoxazole oxygen than near the carboxylate. Thus, if the antibody binding site presented this sort of functional group array to the substrate, rate acceleration could arise.

Several approaches were used to model the interactions between the 21D8 binding site as determined by X-ray crystallography and the reactant and transition structure for decarboxylation. First, the gas phase structures were docked into 21D8 using several charge models and versions of AutoDock [37]. These calculations allowed four low-energy binding modes to be determined (modes **A–D**, Fig. 3.11). In all of these binding modes, the carboxylate and nitro groups interact with either Arg L96 or the polar pocket comprising Arg L46, Ala H101 and Trp H103. These four binding modes were predicted to be extremely close in energy, therefore more quantitative calculations (using free energy perturbation [12] and linear interaction energy [13] methods) were pursued.

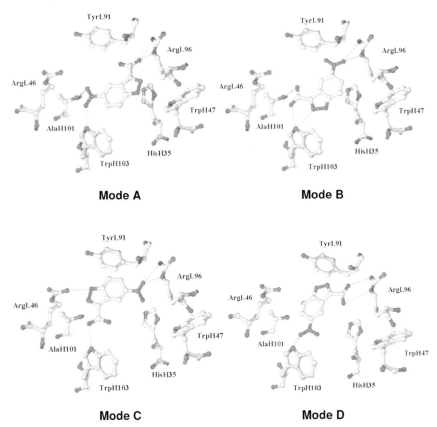

Fig. 3.11 Four favorable binding modes determined by AutoDock.

The free-energy perturbation studies indicated that all four binding modes should provide acceleration compared to water. The linear interaction energy calculations indicated that the substrate binds most tightly in modes **A** and **B**, resulting principally from better electrostatic complementarity between the transition state and binding site in these two modes. After these results were analyzed in great detail, the following model of catalysis by 21D8 was proposed. First, the substrate is bound out of solution into mode **A**, in which its carboxylate group remains partially solvated by water and its less polar portion is buried in the binding site. As the transition state is reached, the carboxylate interacts less favorably with solvent, but a stabilizing interaction between the isoxazole oxygen and Arg L96 develops (an interaction that is particularly strong in the relatively hydrophobic binding site). This model was termed "catalysis on the coastline" since the substrate seemed to have one (nonpolar) foot on land (the antibody) and the other (polar) foot in the sea (the solvent water).

This study is a *tour de force*, employing a battery of cutting edge techniques such as docking, free-energy perturbation, linear interaction energy, and quantum mechanical theozymes. The result is a very comprehensive description of how antibody 21D8 functions and a demonstration of how a delicate balance of maintaining solvation in some regions but excluding water in others can be critical for catalysis. The study paves the way for further studies of antibodies such as 33F12 and 38C2, which also catalyze decarboxylation reactions [89], but which provide a greater challenge for theory since they involve covalent linkages between the substrate and the catalyst – a challenge that also must be met in the study of many enzyme-catalyzed reactions.

3.8
Summary

A wide array of tools have been applied to the computational investigations of a range of catalytic antibodies. Many insights about both quantitative and qualitative aspects of antibody catalysis have been obtained. Quantum mechanics on model systems can now be carried out with high accuracy. Good qualitative determinations of the mode of binding of haptens, substrates, and transition states into antibody binding sites are also possible. The quantitative computation of solvation energies and protein-ligand interaction energies is still an area of active investigation; practical quantitative methods for such calculations are necessary before a full understanding of antibody catalysis – indeed, biological catalysis in general – is possible.

Acknowledgement

The UCLA research effort in catalytic antibodies has been generously supported by the National Science Foundation, the National Institutes of General Medical Sciences, National Institutes of Health, and the National Center for Supercomputing Applications/National Computational Science Alliance. We were attracted to this field by the encouragement, collaboration and stimulating discussions with the pioneers of catalytic antibodies at the Scripps Research Institute, some of whom are now spread around the world: Richard Lerner, Donald Hilvert, Ian Wilson, Kim Janda, Udi Keinan, and Jean-Louis Reymond.

References

1 (a) RADZICKA, A., WOLFENDEN, R., *J. Am. Chem. Soc.* 118 (**1996**), p. 6105–6109. (b) WOLFENDEN, R., LU, X., YOUNG, G., *J. Am. Chem. Soc.* 120 (**1998**), p. 6814–6815. (c) RADZICKA, A., WOLFENDEN, R., *Science* 267 (**1995**), p. 90–93. (d) TAYLOR, E. A., PALMER, D. R. J., GERLT, J. A., *J. Am. Chem. Soc.* 123 (**2001**), p. 5824–5825. (e) BRYANT, R. A. R., HANSEN, D. E., *J. Am. Chem. Soc.* 118 (**1996**), p. 5498–5499

2 (a) HOUK, K. N., LEACH, A. G., KIM, S. P., J., ZHANG, X., *Angew. Chem. Int. Ed.* 42 (**2003**), p. 4872–4897. (b) KIM, S. P., LEACH, A. G., HOUK, K. N., *J. Org. Chem.* 67 (**2002**), p. 4250–4260. (c) BLACKBURN, G. M., DATTA, A., DENHAM, H., WENTWORTH, P., JR., *Advances in Physical Organic Chemistry* 31 (**1998**), p. 249–392

3 (a) BURTON, D. R., *Acc. Chem. Res.* 26 (**1993**), p. 405–411 and references therein. (b) For a comprehensive introduction to antibody biology and other aspects of the immune system see: KUBY, J., *Immunology*, 3rd Ed., W. H. Freeman, New York **1997**

4 For reviews see: (a) NOVOTNY, J., BAJORATH, J., *Adv. Protein Chem.* 49 (**1996**), p. 149–260. (b) BOLGER, M. B., SHERMAN, M. A., *Methods Enzymol.* 203 (**1991**), p. 21–45. (c) PADLAN, E. A., KABAT, E. A., *Methods Enzymol.* 203 (**1991**), p. 3–21. (d) MARTIN, A. C. R., CHEETHAM, J. C., REES, A. R., *Methods Enzymol.* 203 (**1991**), p. 121–153

5 (a) *AbM V2.03 and V2.03a*, Oxford Molecular Ltd., Magdalen Centre, Oxford Science Park, Sandford-on-Thames, Oxford, OX4 4GA, England. (b) WHITELEGG, N. R. J., REES, A. R., *Protein Eng.* 13 (**2000**), p. 819–824

6 TANTILLO, D. J., HOUK, K. N., unpublished results.

7 CHOTHIA, C., LESK, A. M., *J. Mol. Biol.* 196 (**1987**), p. 901–917

8 (a) For a general review of docking algorithms see: HALPERIN, I., MA, B., WOLFSON, H., NUSSINOV, R., *Proteins: Struct. Funct. Genet.* 47 (**2002**), p. 409–443. (b) For a review on docking methods applied to antibody-antigen complexes see: SOTRIFFER, C. A., FLADER, W., WINGER, R. H., RODE, B. M., LIEDL, K. R., VARGA, J. M., *Methods — A Companion to Methods Enzymol.* 20 (**2000**), p. 280–291

9 For leading references and several examples involving antibodies, see: (a) COZZINI, P., FORNABAIO, M., MARABOTTI, A., ABRAHAM, D. J., KELLOGG, G. E., MOZZARELLI, A. *J. Med. Chem.* 45 (**2002**), p. 2469–2483. (b) HANDSCHUH, S., GOLDFUSS, B., CHEN, J., GASTEIGER, J., HOUK, K. N. *J. Comput. Aided Mol. Design* 14 (**2000**), p. 611–629. (c) LEE, F. S., CHU, Z. T., BOLGER, M. B., WARSHEL, A., *Protein Eng.* 5 (**1992**), p. 215–228

10 WANG, W., DONINI, O., REYES, C. M., KOLLMAN, P. A., *Annu. Rev. Biophys. Biomol. Struct.* 30 (**2001**), p. 211–243

11 For leading references and recent examples, see: (a) CHONG, L. T.,

Duan, Y., Wang, L., Massova, I., Kollman, P. A., *Proc. Natl. Acad. Sci. USA* 96 (**1999**), p. 14330–14335. (b) Viswanathan, M., Linthicum, D. S., Subramaniam, S., *Methods* 20 (**2000**), p. 362–371

12 For leading references and an example of free-energy perturbation techniques applied to antibody-antigen binding, see: (a) Kollman, P.A, Massova, I., Reyes, C., Kuhn, B., Huo, S., Chong, L., Lee, M., Lee, T., Duan, Y., Wang, W., Donini, O., Cieplak, P., Srinivasan, J., Case, D. A., Cheatham, T. E., III, *Acc. Chem. Res.* 33 (**2000**), p. 889–897. (b) Fox, T., Scanlan, T. S., Kollman, P. A., *J. Am. Chem. Soc.* 119 (**1997**), p. 11571–11577

13 (a) Aqvist, J., Medine, C., Samuelson, J. E., *Protein Eng.* 7 (**1994**), p. 385–391. (b) Aqvist, J., Luzhkov, V. B., Brandsdal, B. O., *Acc. Chem. Res.* 35 (**2002**), p. 358–365. (c) For a recent application to steroid binding antibodies, see: Chen, J., Wang, R., Taussig, M., Houk, K. N., *J. Org. Chem.* 66 (**2001**), p. 3021–3026

14 Reviews on antibody catalysis include, aside from other contributions to this volume: (a) Stevenson, J. D., Thomas, N. R., *Nat. Prod. Rep.* 17 (**2000**), p. 535–577. (b) Hilvert, D., In *Topics in Stereochemistry*, Denmark, S. E., Ed., John Wiley & Sons, New York **1999**, p. 83–135. (c) Reymond, J.-L., In *Topics in Current Chemistry*, Fessner, W.-D., Ed., Springer-Verlag, Heidelberg **1999**, p. 59–93. (d) Liu, D. R., Schultz, P. G., *Angew. Chem. Int. Ed. Engl.* 38 **1999**, p. 36–54. (e) Hilvert, D., MacBeath, G., Shin, J. A., In *Bioorganic Chemistry: Peptides and Proteins*, Hecht, S. M., Ed., Oxford University Press, New York **1998**, p. 335–366. (f) Blackburn, G. M., Datta, A., Denham, H., Wentworth, P., Jr., *Adv. Phys. Org. Chem.* 31 (**1998**), p. 249–392. (g) Smithrud, D. B., Benkovic, S. J., *Curr. Opin. Chem. Biol.* 8 (**1997**), p. 459–466. (h) Thomas, N. R., *Nat. Prod. Rep.* (**1996**), p. 479–511. (i) Kirby, A. J., *Acta Chem. Scand.* 50 (**1996**), p. 203–210. (j) Schultz, P. G., Lerner, R. A., *Science* 269, (**1995**), p. 1835–1842. (k) Lerner, R. A., Benkovic, S. J., Schultz, P. G., *Science* 252 (**1991**), p. 659–667

15 Some force fields have been parameterized for predictions of transition state structures and energies: Eksterowicz, J. E., Houk. K. N., *Chem. Rev.* 93 (**1993**), p. 2439–2461

16 Recent reviews of QM/MM methods applied to biological systems include: (a) Lee, T.-S., Massova, I., Kuhn, B., Kollman, P. A., *J. Chem. Soc., Perkin Trans. 2* (**2000**), p. 409–415. (b) Monard, G., Merz, K. M., Jr., *Acc. Chem. Res.* 32 (**1999**), p. 904–911. (c) Åqvist, J., Warshel, A., *Chem. Rev.* 93 (**1993**), p. 2523–2544. (d) Warshel, A., *Computer Modeling of Chemical Reactions in Enzymes and Solutions*, John Wiley & Sons, New York **1991**

17 (a) Tantillo, D. J., Chen, J., Houk, K. N., *Curr. Opin. Chem. Biol.* 2 (**1998**), p. 743–750. (b) Tantillo, D. J., Houk, K. N.: "Theozymes and Catalyst Design" In *Stimulating Concepts in Chemistry*, Wiley-VCH, Weinheim, Germany **2000**, p. 79–88. (c) Na, J., Houk, K. N., Shevlin, C. G., Janda, K. D., Lerner, R. A., *J. Am. Chem. Soc.* 115 (**1993**), p. 8453–8454

18 Tantillo, D. J., Houk, K. N., *J. Comp. Chem.* 23 (**2002**), p. 84–95

19 (a) Liotta, L. J., Gibbs, R. A., Taylor, S. D., Benkovic, P. A., Benkovic, S. J., *J. Am. Chem. Soc.* 117 (**1995**), p. 4729–4741. (b) Reymond, J.-L., *Topics Curr. Chem.* 200 (**1999**), p. 59–93. (c) Hasserodt, J. *Synlett* (**1999**), p. 2007–2022. (d) Hiratake, J., Oda, J., *J. Synth. Org. Chem. Jpn.* 55 (**1997**), p. 452–459. (e) Hirschmann, R., Smith III, A. B., Taylor, C. M., Benkovic, P. A., Taylor, S. D., Yager, K. M., Sprengler, P. A., Benkovic, S. J., *Science* 265 (**1994**), p. 234–237. (f) Smithrud, D. B., Benkovic, P. A., Benkovic, S. J., Taylor, C. M., Yager, K. M., Witherington, J., Philips, B. W., Sprengler, P. A., Smith III, A. B., Hirschmann, R., *J. Am. Chem. Soc.* 119 (**1997**), p. 278–282

20 Tantillo, D. J., Houk, K. N., *J. Org. Chem.* 64 (1999), p. 3066–3076

21 (a) Menger, F. M., Ladika, M., *J. Am. Chem. Soc.* 109 (1987), p. 3145–3146. (b) Many crystal structures of antibodies that hydrolyze aryl and benzyl esters have been reported, and the similarities and differences between the combining sites of these catalysts have been the subject of several reviews, which have led to the "transition state epitope" and "canonical binding array" concepts. (c) MacBeath, G., Hilvert, D., *Chem. Biol.* 3 (1996), p. 433–445. (d) Charbonni J. B., Gigant, B., Golinelli-Pimpaneau, B., Knossow, M., *Biochimie* 79 (1997), p. 653–660. (e) Tantillo, D. J., Houk, K. N., *Chem. Biol.* 8 (2001), p. 535–545

22 Luzhkov, V. B., Venanzi, C. A., *J. Phys. Chem.* 99 (1995), p. 2312–2323

23 Chong, L. T., Bandyopadhyay, P., Scanlan, T. S., Kuntz, I. D., Kollman, P. A., *J. Comp. Chem.* 24 (2003), p. 1371–1377

24 (a) Miertus, S., Tomasi, J., *Chem. Phys.* 65 (1982), p. 239–245. (b) Miertus, S., Scrocco, E., Tomasi, J. *Chem. Phys.* 55 (1981), p. 117–129. (b) Wiberg, K. B., Keith, T. A., Frisch, M. J., Murcko, M. *J. Phys. Chem.* 99 (1995), p. 9072–9079. (c) Wiberg, K. B., Castejon, H., *J. Comp. Chem.* 17 (1996), p. 185–190. (d) Zheng, Y.-J., Ornstein, R. L., *J. Mol. Struc. (Theochem)* 429 (1998), p. 41–48

25 (a) Pedley, J. B., Naylor, R. D., Kirby, S. P., *Thermochemical Data of Organic Compounds*, Chapman and Hall, New York 1986. (b) Chase, M. W., Jr., Davies, C. A., Downey, J. R., Jr., Frurip, D. J., McDonald, R. A., Syverud, A. N., *J. Phys. Chem. Ref. Data, Suppl.* 1 14 (1985), p. 1–1856. (c) Cox, J. D., *Pure Appl. Chem.* 2 (1961), p. 125–128. (d) Wenthold, P. G., Squires, R. R., *J. Am. Chem. Soc.* 116 (1994), p. 11890–11897.

26 (a) Zhan, C.-G., Landry, D. W., Ornstein, R. L., *J. Am. Chem. Soc.* 122 (2000), p. 1522–2530. (b) Pranata, J., *J. Phys. Chem.* 98 (1994), 1180–1184. (c) Hori, K., *J. Chem. Soc., Perkin Trans.* 1 (1992), p. 1629–1633. (d) Teraishi, K., Saito, M., Fujii, I., Nakamura, H., *Tetrahedron Lett.* 33 (1992), p. 7153–7156. (e) Zhan, C.-G., Landry, D. W., *J. Phys. Chem. A* 105 (2001), p. 1296–1301. (f) Pliego, J. R., Jr., Riveros, J. M., *Chem. Eur. J.* 7 (2001), p. 169–175 and references therein

27 (a) Jencks, W. P., *Catalysis in Chemistry and Enzymology*, McGraw Hill, New York 1969, p. 288. (b) Mader, M. M., Bartlett, P. A., *Chem. Rev.* 97 (1997), p. 1281–1301

28 (a) Teraishi, K., Saito, M., Fujii, I., Nakamura, H., *Tetrahedron Lett.* 33 (1992), p. 7153–7156. (b) Ohkubo, K., Urata, Y., Seri, K., Ishida, H., Sagawa, T., Nakashima, T., Imagawa, Y., *J. Mol. Catal.* 90 (1994), p. 355–365

29 (a) Sherer, E. C., Turner, G. M., Shields, G. C., *Int. J. Quantum Chem., Quantum Biol. Symp.* 22 (1995), p. 83–93. (b) Turner, G. M., Sherer, E. C., Shields, G. C., *Int. J. Quantum Chem., Quantum Biol. Symp.* 22 (1995), p. 103–112

30 (a) Sherer, E. C., Turner, G. M., Lively, T. N., Landry, D. W., Shields, G. C., *J. Mol. Modeling* 2 (1996), p. 62–69. (b) Sherer, E. C., Yang, G., Turner, G. M., Shields, G. C., Landry, D. W., *J. Phys. Chem. A* 101 (1997), p. 8526–8529

31 Curley, K., Pratt, R. F., *J. Am. Chem. Soc.* 119 (1997), p. 1529–1538

32 Radkiewicz, J. L., McAllister, M., Goldstein, E., Houk, K. N., *J. Org. Chem.* 63 (1998), p. 1419–1428

33 (a) Smithrud, D. B., Benkovic, P. A., Benkovic, S. J., Roberts, V., Liu, J., Neagu, I., Iwama, S., Phillips, B. W., Smith, A. B., III, Hirschmann, R., *Proc. Natl. Acad. Sci. USA* 97 (2000), p. 1953–1958. (b) Hirschmann, R., Smith, A. B., III, Taylor, C. M., Benkovic, P. A., Taylor, S. D., Yager, K. M., Sprengeler, P. A., Benkovic, S. J. *Science* 265 (1994), p. 234–237. (c) Smithrud, D. B., Benkovic, P. A., Benkovic, S. J., Taylor, C. M., Yager, K. M., Witherington, J. Philips, B. W., Sprengeler, P. A., Smith, A. B.,

III, Hirschmann, R., *J. Am. Chem. Soc.* 119 (**1997**), p. 278–282

34 (a) Miyashita, H., Hara, T., Tanimura, R., Tanaka, F., Kikuchi, M., Fujii, I., *Proc. Natl. Acad. Sci. USA* 91 (**1994**), p. 6045–6049. (b) Fujii, I., Tanaka, F., Miyashita, H., Tanimura, R., Kinoshita, K., *J. Am. Chem. Soc.* 117 (**1995**), p. 6199–6209. (c) Miyashita, H., Hara, T., Tanimura, R., Fukuyama, S., Cagnon, C., Kohara, A., Fujii, I., *J. Mol. Biol.* 267 (**1997**), p. 1247–1257. (d) Kristensen, O., Vassylyev, D. G., Tanaka, F., Morikawa, K., Fujii, I. *J. Mol. Biol.* 281 (**1998**), p. 501–511. (e) Gigant, B., Tsumuraya, T., Fujii, I., Knossow, M. *Structure* 7 (**1999**), p. 1385–1393

35 (a) Benkovic, S. J., Adams, J. A., Borders, C. L., Jr., Janda, K. D., Lerner, R. A., *Science* 250 (**1990**), p. 1135–1139. (b) Gibbs, R. A., Benkovic, P. A., Janda, K. D., Lerner, R. A., Benkovic, S. J., *J. Am. Chem. Soc.* 114 (**1992**), p. 3528–3534. (c) Stewart, J. D., Liotta, L. J., Benkovic, S. J., *Acc. Chem. Res.* 26 (**1993**), p. 396–404. (d) Roberts, V. A., Stewart, J., Benkovic, S. J., Getzoff, E. D., *J. Mol. Biol.* (**1994**), 235, p. 1098–1116. (e) Stewart, J. D., Roberts, V. A., Crowder, M. W., Getzoff, E. D., Benkovic, S. J., *J. Am. Chem. Soc.* 116 (**1994**), p. 415–416. (f) Crowder, M. W., Stewart, J. D., Roberts, V. A., Bender, C. J., Tevelrakh, E., Peisach, J., Getzoff, E. D., Gaffney, B. J., Benkovic, S. J., *J. Am. Chem. Soc.* 117 (**1995**), p. 5627–5634. (g) Roberts, E. A., Getzoff, E. D. *FASEB J.* 9 (**1995**), p. 94–100. (h) Miller, G. P., Posner, B. A., Benkovic, S. J. *Bioorg. Med. Chem. Lett.* 5 (1997), p. 581–590. (i) Janda, K. D., Schloeder, D., Benkovic, S. J., Lerner, R. A., *Science* 241 (**1988**), p. 1188–1191. (j) Thayer, M. M., Olender, E. H., Arvai, A. S., Koike, C. K., Canestrelli, I. L., Stewart, J. D., Benkovic, S. J., Getzoff, E. D., Roberts, V. A. *J. Mol. Biol.* 291 (**1999**), p. 329–345

36 (a) Zemel, R., Schindler, D. G., Tawfik, D. S., Eshhar, Z., Green, B. S., *Mol. Immunol.* 31 (**1994**), p. 127–137. (b) Golinelli-Pimpaneau, B., Gigant, B., Bizebard, T., Navaza, J., Saludjian, P., Zemel, R., Tawfik, D. S., Eshhar, Z., Green, B. S., Knossow, M. *Structure* 2 (**1994**), p. 175–183. (c) Charbonnier, J.-B., Golinelli-Pimpaneau, B., Gigant, B., Green, B. S., Knossow, M. *Isr. J. Chem.* 36 (**1996**), p. 143–149. (d) Gigant, B., Charbonnier, J.-B., Golinelli-Pimpaneau, B., Zemel, R. R., Eshhar, Z., Green, B. S., Knossow, M., *Eur. J. Biochem.* 246 (**1997**), p. 471–476

37 (a) Morris, G. M., Goodsell, D. S., Huey, R., Olson, A. J., *AutoDock V. 2.4*, The Scripps Research Institute, 10666 North Torrey Pines Road, La Jolla, CA 92037–5025, 1995. (b) Morris, G. M., Goodsell, D. S., Huey, R., Hart, W. E., Halliday, S., Belew, R., Olson, A. J. *AutoDock V. 3.0*, The Scripps Research Institute, 10666 North Torrey Pines Road, La Jolla, CA 92037–5025, 1999. (c) Morris, G. M., Goodsell, D. S., Huey, R., Olson, A. J., *J. Comput.-Aided Mol. Design*, 10 (**1996**) p. 293–304. (d) Goodsell, D. S., Morris, G. M., Olson, A. J., *J. Mol. Recognition* 9 (**1996**), p. 1–5. (e) Goodsell, D. S., Olson, A. J., *Proteins: Struct. Funct. Genet.* 8 (**1990**), p. 195–202. (f) For an application to antibody-peptide complexes, see: Friedman, A. R., Roberts, V. A., Tainer, J. A., *Proteins: Struct. Funct. Genet.* 20 (**1994**), p. 15–24. (g) For an application to antibody-small molecule complexes, see: Sotriffer, C. A., Liedl, K. R., Winger, R. H. Gamper, A. M., Kroemer, R. T., Linthicum, D. S., Rode, B.-M., Varga, J. M., *Mol. Immunol.* 33 (**1996**), p. 129–144, and Sotriffer, C. A., Winger, R. H., Liedl, K. R., Rode, B. M., Varga, J. M., *J. Comput.-Aided Mol. Design* 10 (**1996**), p. 305–320

38 (a) Lesley, S. A., Patten, P. A., Schultz, P. G., *Proc. Natl. Acad. Sci. USA* 90 (**1993**), p. 1160–1165. (b) Patten, P. A., Gray, N. S., Yang,

P. L., Marks, C. B., Wedemayer, G. J., Boniface, J. J., Stevens, R. C., Schultz, P. G., *Science* 271 (**1996**), p. 1086–1091. (c) Wedemayer, G. J., Patten, P. A., Wang, L. H., Schultz, P. G., Stevens, R. C., *Science* 276 (**1997**), p. 1665–1669. (d) Wedemayer, G. J., Wang, L. H., Patten, P. A., Schultz, P. G., Stevens, R. C., *J. Mol. Biol.* 268 (**1997**), p. 390–400. (e) Chong, L. T., Duan, Y., Wang, L., Massova, I., Kollman, P. A. *Proc. Natl. Acad. Sci. USA* 96 (**1999**), p. 14330–14335

39 (a) Guo, J., Huang, W., Scanlan, T. S., *J. Am. Chem. Soc.* 116 (**1994**), p. 6062–6069. (b) Zhou, G. W., Guo, J., Huang, W., Fletterick, R. J., Scanlan, T. S., *Science* 265 (**1994**), p. 1059–1064. (c) Guo, J., Huang, W., Zhou, G. W., Fletterick, R. J., Scanlan, T. S., *Proc. Natl. Acad. Sci. USA* 92 (**1995**), p. 1694–1698. (d) Wade, H., Scanlan, T. S., *J. Am. Chem. Soc.* 118 (**1996**), p. 6510–6511. (e) Fox, T., Scanlan, T. S., Kollman, P. A., *J. Am. Chem. Soc.* 119 (**1997**), p. 11571–11577. (f) Baca, M., Scanlan, T. S., Stephensen, R. C., Wells, J. A., *Proc. Natl. Acad. Sci. USA* 94 (**1997**), p. 10063–10068. (g) Buchbinder, J. L., Stephenson, R. C., Scanlan, T. S., Fletterick, R. J., *J. Mol. Biol.* 282 (**1998**), p. 1033–1041. (h) Wade, H., Scanlan, T. S., *J. Am. Chem. Soc.* 121 (**1999**), p. 1434–1443

40 Li, T., Janda, K. D., Lerner, R. A., *Nature* 379 (**1996**), p. 326–327

41 Li, T., Janda, K. D., Ashley, J. A., Lerner, R. A., *Science* 264 (**1994**), p. 1289–1293

42 Li, T., Janda, K. D., Hilton, S., Lerner, R. A., *J. Am. Chem. Soc.* 117 (**1995**), p. 2367–2368

43 Lee, J. K., Houk, K. N., *Angew. Chem. Int. Ed. Engl.* 36 (**1997**), p. 1003–1005

44 (a) Janda, K. D., Shevlin, C. G., Lerner, R. A. *Science* 259 (**1993**), p. 490–493. (b) Danishefsky, S., *Science* 259 (**1993**), p. 469–470

45 Na, J., Houk, K. N., Shevlin, C. G., Janda, K. D., Lerner, R. A., *J. Am. Chem. Soc.* 115 (**1993**), p. 8453–8454

46 Na, J., Houk, K. N., *J. Am. Chem. Soc.* 118 (**1996**), p. 9204–9205

47 Janda, K. D., Shevlin, C. G., Lerner, R. A., *J. Am. Chem. Soc.* 117 (**1995**), p. 2659–2660

48 This system has recently been reinvestigated computationally: Coxon, J. M., Thorpe, A. J., *J. Am. Chem. Soc.* 121 (**1999**), p. 10955–10957

49 Gruber, K., Zhou, B., Houk, K. N., Lerner, R. A., Shevlin, C. G., Wilson, I. A. *Biochemistry* (**1999**), 30, p. 7002–7014.

50 Pindur, U., Lutz, G., Otto, C., *Chem. Rev.* 93 (**1993**), p. 741–761

51 (a) Sauer, J., Sustmann, R., *Angew. Chem. Int. Ed. Engl.* 19 (**1980**), p. 779–807. (b) Li, Y., Houk, K. N., *J. Am. Chem. Soc.* 115 (**1993**), p. 7478–7485

52 (a) Laschat, S., *Angew. Chem. Int. Ed. Engl.* 35 (**1996**), p. 289–291. (b) Auclair, K., Sutherland, A., Kennedy, J., Witter, D. J., Van den Heever, J. P., Hutchinson, C. R., Vederas, J. C., *J. Am. Chem. Soc.* 122 (**2000**), p. 11519–11520

53 Recent reviews of enantioselective Lewis acid-catalyzed Diels-Alder reactions include: (a) Togni, A., Venanzi, L. M., *Angew. Chem. Int. Ed. Engl.* 33 (**1994**), p. 497–526. (b) Maruoka, K., Yamamoto, H., *Catalytic Asymmetric Synthesis* Ed: Ojima, I., VCH, New York **1993**. (c) Pindur, U., Lutz, G., Otto, C., *Chem. Rev.* 93 (**1993**), p. 741–761. (d) Delox, L., Srebnik, M. *Chem. Rev.* 93 (**1993**), p. 763. (e) Kagan, H. B., Riant, O. *Chem. Rev,* 92 **1992**, p. 1007–1019

54 Lerner, R. A., Benkovic, S. J., Schultz, P. G., *Science* 252 (**1991**), p. 659–667. (b) Schultz, P. G., Lerner, R. A., *Science* 269 (**1995**), p. 1835–1842. (c) Hilvert, D., *Annu. Rev. Biochem.* 69 (**2000**), p. 751–793. (d) Hilvert, D., *Top. Stereochem.* 22 (**1999**), p. 83–135. (e) Oikawa, H., Kobayashi, T., Katayama, K., Suzuki, Y., Ichihara, A., *J. Org. Chem.* 63 (**1998**), p. 8748–8756

55 Braun, R., Schuster, F., Sauer, J., *Tetrahedron Lett.* 27 (**1986**), p. 1285–1288

56 Colonna, S., Manfredi, A., Annunziata, R., *Tetrahedron Lett.* 29 (**1988**), p. 3347–3350

57 Rao, K. R., Srinivasan, T. N., Bhanumathi, N., *Tetrahedron Lett.* 31 (**1990**), p. 5959–5960

58 (a) Hudlicky, T., Butora, G., Fearnley, S. P., Gum, A. G., Persichini, P. J., III, Stabile, M. R., Merola, J. S., *J. Chem. Soc., Perkin Trans. I* 19 (**1995**), p. 2393–2398. (b) Alvira, E., Cativiela, C., Garcia, J. I., Mayoral, J. A., *Tetrahedron Lett.* 36 (**1995**), p. 2129–2132. (d) Sternbach, D. D., Possana, D. M., *J. Am. Chem. Soc.* 104 (**1982**), p. 5853–5854

59 Kang, J., Santamaria, J., Hilmersson, G., Rebek, J., *J. Am. Chem. Soc.* 120 (**1998**), p. 7389–7390

60 (a) Morris, K. N., Tarasow, T. M., Julin, C. M., Simons, S. L., Hilvert, D., Gold, L., *Proc. Natl. Acad. Sci. U.S.A.* 91 (**1994**), p. 13028–13032. (b) Seelig, B., Jaschke, A., *Chem. Biol.* 6 (**1999**), p. 167–176. (c) Tarasow, T. M., Eaton, B. E., *J. Am. Chem. Soc.* 122 (**2000**), p. 1015–1021. (d) Seelig, B., Keiper, S., Stuhlmann, F., Jaschke, A., *Angew. Chem. Int. Ed.* 39 (**2000**), p. 4576–4579

61 (a) Hilvert, D., Hill, K. W., Nared, K. D., Auditor, M.-T., M., *J. Am. Chem. Soc.* 111 (**1989**), p. 9261–9262. (b) Xu, J., Deng, Q., Chen, J., Houk, K. N., Bartek, J., Hilvert, D., Wilson, I. A., *Science* 286 (**1999**), p. 2345–2348

62 (a) Braisted, A. C., Schultz, P. G., *J. Am. Chem. Soc.* 112 (**1990**), p. 7430–7431. (b) Romesberg, F. E., Spiller, B., Schultz, P. G., Stevens, R. C., *Science* 279 (**1998**), p. 1929–1933

63 (a) Gouverneur, V. E., Houk, K. N., Pascual-Teresa, B., Beno, B., Janda, K. D., Lerner, R. A., *Science* 262 (**1993**), p. 204–208. (b) Heine, A., Stura, E. A., Yli-Kauhaluoma, J. T., Gao, C., Deng, Q., Beno, B., Houk, K. N., Janda, K. D., Wilson, I. A., *Science* 279 (**1998**), p. 1934–1940

64 (a) Bahr, N., Güller, R., Reymond, J.-L., Lerner, R. A., *J. Am. Chem. Soc.* 118 (**1996**), p. 3550–3555. (b) Bensel, N., Bahr, N., Reymond, M. T., Schenkels, C., Reymond, J.-L., *Helv. Chim. Acta* 82 (**1999**), p. 44–52

65 Chen, J., Deng, Q., Wang, R., Houk, K. N., Hilvert, D., *ChemBioChem* 1 (**2000**), p. 255–261 and references therein

66 Zhang, X., Deng, Q., Yoo, S., Houk, K. N., *J. Org. Chem.* 67 (**2002**), p. 9043–9053, and references therein

67 Romesberg, F. E., Schultz, P. G., *Bioorg. Med. Chem. Lett.* 9 (**1999**), p. 1741–1744

68 Cannizzaro, C. E., Ashley, J. A., Janda, K. D., Houk, K. N., *J. Am. Chem. Soc.* 125 (**2003**), p. 2489–2506 and references therein

69 Leach, A. G., Houk, K. N., manuscripts in preparation

70 Braisted, A. C., Schultz, P. G., *J. Am. Chem. Soc.* 116 (**1994**), p. 2211–2212

71 Driggers, E. M., Cho, H. S., Liu, C. W., Katzka, C. P., Braisted, A. C., Ulrich, H. D., Wemmer, D. E., Schultz, P. G., *J. Am. Chem. Soc.* 120 (**1998**), p. 1945–1958

72 Mundorff, E. C., Hanson, M. A., Varvak, A., Ulrich, H., Schultz, P. G., Stevens, R. C., *Biochemistry* 39 (**2000**), p. 627–632

73 Asada, T., Gouda, H., Kollman, P. A., *J. Am. Chem. Soc.* 124 (**2002**), p. 12535–12542

74 Black, K. A., Kalani, M. Y. S., Leach, A. G., Houk, K. N., *J. Am. Chem. Soc.*, in press

75 Hrovat, D. A., Chen, J., Houk, K. N., Borden, W. T., *J. Am. Chem. Soc.* 122 (**2000**), p. 7456–7460

76 Gajewski, J. J., Conrad, N. D., *J. Am. Chem. Soc.* 101 (**1979**), p. 6693–6704

77 Barone, V., Cossi, M., *J. Phys. Chem. A.* 102 (**1998**), p. 1995–1999

78 Toker, J. D., Wentworth, Jr., P., Hu, Y., Houk, K. N., Janda, K. D., *J. Am. Chem. Soc.* 122 (**2000**), p. 3244–3245

79 Heine, A., Hu, Y., unpublished results

80 Wiest, O., Houk, K. N., *J. Am. Chem. Soc.* 117 (**1995**), p. 11628–11639

81 Haynes, M. R., Stura, E. A., Hilvert, D., Wilson, I. A., *Science* 263 (**1994**), p. 646–652

82 ANDREWS, P. R., SMITH, G. D., YOUNG, I. G., *Biochemistry* 12 (**1973**), p. 3492–3498

83 ADDADI, L., JAFFE, E. K., KNOWLES, J. R., *Biochemistry* 22 (**1983**), p. 4494–4501

84 (a) COPLEY, S. D., KNOWLES, J. R., *J. Am. Chem. Soc.* 109 (**1987**), p. 5308–5313. (b) GAJEWSKI, J. J., JURAYJ, J., KIMBROUGH, D. R., GANDE, M. E., GANEM, B., CARPENTER, B. E., *J. Am. Chem. Soc.* 109 (**1987**), p. 1170–1186. (c) COATES, R. M., ROGERS, B. D., HOBBS, S. J., PECK, D. R., CURRAN, D. P., *J. Am. Chem. Soc.* 109 (**1987**), p. 1160–1170

85 GÖRISCH, H., *Biochemistry* 17 (**1978**), p. 3700–3705

86 (a) HOTTA, K. WILSON, I. A., HILVERT, D., *Biochemistry* 41 (**2002**), p. 772–779. (b) HOTTA, K., LANGE, H., TANTILLO, D. J., HOUK, K. N., HILVERT, D., WILSON, I. A., *J. Mol. Biol.* 302 (**2000**), p. 1213–1225. (c) TARASOW, T. M., LEWIS, C., HILVERT, D., *J. Am. Chem. Soc.* 116 (**1994**), p. 7959–7963. (d) LEWIS, C., PANETH, P., O'LEARY, M. H., HILVERT, D., *J. Am. Chem., Soc.* 115 (**1993**), p. 1410–1413 and references therein. (e) LEWIS, C., KRAMER, T., ROBINSON, S., HILVERT, D., *Science*, 253 (**1991**), p. 1019–1022

87 KEMP, D. S., PAUL, K. G., *J. Am. Chem. Soc.* 97 (**1975**), p. 7305–7312

88 UJAQUE, G., TANTILLO, D. J., HU, Y., HOUK, K. N., HOTTA, K., HILVERT D., *J. Comp. Chem.* 24 (**2003**), p. 98–110

89 BJORNESTEDT, R., ZHONG, G., LERNER, R. A., BARBAS, C. F., III., *J. Am. Chem. Soc.* 118 (**1996**), p. 11720–11724

4
The Enterprise of Catalytic Antibodies: A Historical Perspective
Michael Ben-Chaim

4.1
Introduction

Scientific work makes history. Traditionally, the history of science has been portrayed as a chronology of discrete discoveries. Each discovery comprised the proposal of novel theoretical ideas and their unambiguous corroboration by empirical evidence. Scientific growth was accordingly depicted as an emblem of the triumph of human reason, set apart from the non-linear, multi-faceted, and controversial changes that so often characterize human history as a whole. More recently, however, historians have convincingly demonstrated that traditional accounts misleadingly simplify and idealize the nuts and bolts of scientific change. Important discoveries, it now appears, involve a considerable reorganization of scientific practice. Far from being confined to novel theoretical ideas, innovations in science often include changes in methods, research instruments, social division of scientific work, disciplinary boundaries, and relations between science and its markets [1]. As a result of the changes in the historical perspective, interest in the history of science and interest in the evolution of human societies in general have become more closely related. Anthropologists have been primarily concerned with "archaic", or "pre-scientific" societies, whose culture is permeated with anthropomorphic beliefs about the physical universe. These beliefs form an important aspect of local religious communities and cults of magic. Scientific research, by contrast, is associated with objective knowledge and its rational acquisition. Given this apparent dichotomy, the natural sciences have long been considered to stretch beyond the scope of anthropological research. But do archaic cultures and modern science have *nothing* in common which is valuable to our understanding of the historical evolution of human behavior?

Unlike the authors of the rest of the chapters in this volume, I was initially attracted to catalytic antibodies for anthropological reasons, as I listened to a recording of a radio broadcast on the subject in 1996. It was a popular lecture, delivered by Ehud Keinan, a chemistry professor at the Technion – Israel Institute of Technology and adjunct professor at The Scripps Research Institute in La Jolla. The immune system, I learned, was considered by chemists as an enormous treasure, a natural repository of an almost infinite number of potential catalytic agents for desirable chemical

Catalytic Antibodies. Edited by Ehud Keinan
Copyright © 2005 WILEY-VCH Verlag GmbH & Co. KGaA, Weinheim
ISBN: 3-527-30688-9

reactions. Chemists have always been interested in making valuable compounds, and they now found in the immune system a new partner with which they might achieve this goal. The naturally available repertoire of antibody molecules presented an invaluable opportunity, and in 1997, the first catalytic antibody, labeled 38C2, made its way into the commercial market. This antibody was produced by a research team led by Richard A. Lerner, the president of The Scripps Research Institute. The commercialization of catalytic antibodies, according to Lerner, reaffirmed the age-old image of nature as a horn of plenty [2]. The goods it offered, as evinced by catalytic antibodies, were at once intellectual and practical.

This notion of specific natural properties affording gifts for mankind pervade the ethnographic literature of anthropologists. It appears that such gifts are perceived by potential beneficiaries as comprising specific constraints and opportunities that govern human action in specific domains of everyday life. For example, the Cree Indians of northeastern Canada believe that the caribou, a food staple in their diet, offers itself up to the hunter in a spirit of goodwill and compassion. When pursuing caribou, the Cree claim that there often comes a critical moment when a particular animal becomes aware of the hunter. The caribou stands stock still, turns its head, and stares you squarely in the face. The art of hunting involves learning to anticipate this moment and to be ready to take advantage of the opportunity it offers. As with many other hunting people around the world, the Cree accordingly believe that the nourishing substance of the animal is gratefully accepted by the skilled hunter. Gifted hunters thus demonstrate their virtuous role in the greater "circle of life" [3]. These hunters may seem to have little in common with the culture of modern scientific research, yet there is a common trait: the entrepreneurial activity which links learning with the cultivation of practical skills necessary for the transformation of resources into goods. The enterprise, whether in science or in hunting, integrates the quest for truth with the search for goods, and consists at its core in matching specific interests with the opportunities afforded by the environment. The following is a report on the origins of the enterprise of catalytic antibodies.

By virtue of its historical character, my report is inevitably and indelibly selective with respect to the evidence upon which it is based. It draws special attention to initial endeavors to establish antibody catalysis as a scientific discipline at the crossroads of chemistry and immunology. To put these endeavors in the appropriate historical perspective, the report examines antecedent scientific contributions and trends of research in the light of which the rise of the discipline of antibody catalysis can be understood within a framework of historical continuity and change. The report, on the other hand, does not focus on the diverse research projects that comprise the current state of the discipline. As far as the historian is concerned, the present can only serve as a point of reference, rather than the subject matter, of inquiry. I have therefore deliberately refrained from offering in what follows a comprehensive view of the research on catalytic antibodies. Such a view is more appropriately obtained by professional reviews, such as the collection of articles that make up this volume as a whole [4]. It is only in the light of future innovations that more advanced historical accounts of catalytic antibodies might be elaborated.

4.2 Methods

The scientific enterprise has become an important engine of cultural, political, and economic changes in the modern era. Notions of the scientific method, however, originated in the writings of ancient Greek philosophers who sought to differentiate their intellectual pursuits from the practical disciplines that structured human conduct in everyday life. Philosophy as a distinctive vocation was predicated on a particular notion of human agents as rational beings. Aristotle thus explained that the human capacity to reason could be cultivated in relation to two mutually exclusive goals, namely, the practical and the theoretical. Humans ordinarily exercise their intellect in an attempt to solve the practical problems of everyday life. According to Aristotle, this was the distinctively human expression of the desire to survive. Human reason, however, was not only crucial to the sheer survival of individual agents. He thus taught that it ought to be cultivated for the purpose of realizing as fully as possible the distinctively human capacity for abstract thought and desire to understand [5]. The rationality of theoretical thought was accordingly distinguished from practical reasoning. While the former characterized the quest for genuine knowledge, the latter characterized the search for more tangible goods.

The separation of philosophy from the realm of labor and industry in western civilization has gradually been undermined, however, since the late Middle Ages. The changing role of the natural sciences in society formed part of the entire transformation of western civilization during the Renaissance, the swell of humanism, the Christian Reformation and Counter-reformation, and the aggrandizement of the middle classes in the early modern period. A new perspective on natural philosophy, its goals, and methods was proposed during the second half of the 18th century by Adam Smith, the Enlightenment Scottish scholar acclaimed for his contributions to modern theories of the market economy. Smith endeavored to show that theoretical thinking partook in the production of wealth. He thus suggested, for example, that all forms of human learning were governed by the general principles of social division of labor and the growth of specialization. Specialized knowledge, "as in every other business, improves dexterity and saves time", he explained [6]. Moreover, although Smith accepted the traditional account of philosophy as a vocation that differed from practical endeavors that were directly involved in the production of goods, he sought to explain how learning on the one hand and industry on the other were nevertheless interdependent. Theoretical studies, in his view, did not imply the renunciation of practical aspirations. Quite to the contrary, the former comprised a highly efficient method to satisfy the latter. Precisely because theoretical thinkers were not entangled in the routine life of ordinary laborers, they could, from their privileged position in the matrix of social life, identify new opportunities for the deployment of already approved skills and methods. The natural philosopher, Smith claimed, was "one of those men who, though they work at nothing themselves, yet by observing all are enabled by this extended way of thinking to apply things together to produce effects to which they seem no way adapted". As an example, Smith suggested that "the man who first thought of applying a stream of water and still more the blast of the wind to turn, by an outer wheel in place of a crank, was neither a millar nor a mill-wright but a philosopher" [7]. Thus, the natural philosopher was an entrepreneur who specialized in the discovery of novel applications of existing methods and techniques. By augmenting the repertoire of methodical work, human learning contributed to the growth of public wealth.

Smith's views on the social role of natural philosophers suggest a model of the entrepreneurial "logic" of scientific discoveries. This model assumes that the variety of observable properties of a given environment affords a practically infinite repertoire of potential opportunities for the novel application of available skills and methods. Learning consists in the systematic endeavor to develop the appropriate methodology with which the value of a specific opportunity can be demonstrated. The enterprise of research is initiated when a particular opportunity is identified and selected for this purpose. Selection is then followed by searching for techniques and methods that can be adapted to realize the opportunity and amplify its value. Successful demonstrations thereby offer users of the novel methodology the opportunity to reproduce valuable natural properties. Variation, selection, and reproduction thus comprise, in a nutshell, the three components of the enterprise in which learning is unified with the transformation of resources into goods. This simple model, I suggest, is applicable to the Cree hunters who developed the art of caribou hunting as well as to the researchers who developed the innovative means to tap the immune system for the purpose of the catalysis of desirable chemical reactions.

In the light of this model, I set out to investigate the contexts and circumstances in which scientists originally selected antibodies as potentially valuable biocatalysts and invested a "critical mass" of resources to explore and develop methodologies that could demonstrate this value. The published scientific literature provided one important source of information on the origins of antibody catalysis, and I consulted it whenever possible. Another major source of information was provided through interviews with scientists who participated in research projects on catalytic antibodies or were familiar with initiatives to develop such projects. During the late 1990s, I conducted over two dozen interviews with individual senior researchers, research students, and laboratory workers affiliated to about a dozen different research institutes in several countries, including the USA, Great Britain, Israel, and Switzerland. Most of the interviews were tape-recorded with the permission of the interviewee. Interviews averaged 45 minutes, and formed open conversations in which interviewees related in chronological order their main research experiences and contacts in connection with antibody catalysis [8].

4.3
Results

4.3.1
The Conceptual Origins of Catalytic Antibodies

Since scientific work makes history, scientists are concerned with reconstructing the historical background of their own work, and this is especially manifested in the writing of professional reviews. A certain chronology has been reiterated in reviews on antibody catalysis [9]. Current work in this field has been traced to the seminal ideas of Linus Pauling on the chemical binding of enzymes and antibodies to respective substances, which he communicated on several occasions during the 1940s [10]. The second landmark in review narratives is William Jencks's *Catalysis in Chemistry and Enzymology*, which was originally published in 1969. In this book, Jencks, a leading physical organic chemist of enzyme catalysis, noted the possibility of synthesizing an enzyme by way of preparing an antibody to a hapten that resembles the transition state of a chosen reaction. The binding sites of such antibodies would be complementary to the transition state, Jencks pointed out, "and should cause an acceleration [in the rate of reaction] by forcing bound substrates to resemble the transition state" [11]. It is commonly acknowledged, however, that antibody catalysis made its way as a scientific field of international acclaim only in the late 1980s, following the publication of experimental results by two research groups from California, led by Richard Lerner from the Scripps Clinic and Research Foundation, and Peter Schultz from the University of California at Berkeley, respectively [12].

This chronology reflects a common image of scientific progress, according to which novel ideas and their empirical demonstration comprise the core of the history of science. But it does not take into consideration the historical context in which the testing of ideas may or may not be considered as sufficiently valuable to the scientific community. Moreover, it overlooks the development of the entrepreneurial effort to establish the intellectual and institutional framework within which the empirical examination of certain ideas becomes a collectively thriving, continuous project. Ironically, the existence of such a project justifies the writing of professional reviews; yet the reviews often ignore the conditions that made the project happen in the first place. These historiographic considerations, as will be shown in what follows, are particularly relevant to explaining the time-lag between the communication of Pauling's ideas in the 1940s on the one hand and the emergence of antibody catalysis as a promising research field in the late 1980s on the other.

During the late 19th century and the early part of the 20th century, research in the life sciences was considerably independent of physics and chemistry. While physicists were probing into the atomic structure of matter, advances in biochemistry barely addressed the molecular structure of the physiological activity of specific substances. In 1930, for example, John Haldane professed that scientists would eventually be able to synthesize enzymes, though at the time they were not yet able to decisively identify the molecular composition of these catalysts [13]. A year later, Pauling published his new theory of the nature of chemical bond [14]. Further contributions followed soon after, the result of Pauling's systematic endeavor to apply the knowledge of quantum physics to the various disciplines of classical chemistry. One of his fundamental achievements consisted in the elucidation of the secondary structure of proteins, which led to a series of studies of the physical chemistry of various physiological processes [15]. These studies were guided, to an important extent, by using the knowledge of the molecular structure of proteins to elucidate the specificity of their interactions with other substances. He thus extended the work of Karl Landsteiner on the specificity of binding between the antibody molecule and the molecules of antigens and haptens. With respect to enzyme catalysis, Pauling's work was based on formulations of a general treatment of reaction rates that were proposed in the early 1930s, notably by Henry Eyring and Michael Polanyi. An important aspect of these formulations was the notion that a reaction rate is proportional to the concentration of the activated complexes of reactants. Later known as "transition-state" theory, it provided a convenient means of calculating chemical rates while ignoring the details of what happens before the activated complex – denoting the no-return point at which reactants are transformed into products – is reached [16]. In 1948, Pauling presented an overview of the contribution of the chemical analysis of the conformation of proteins to the study of their physiological functions and their value to medical research. He proposed, more particularly, that chemotherapeutic agents could be synthesized in accordance with the specificity of protein binding to compete against harmful substances for binding with proteins such as enzymes and antibodies. In this context, he compared the complementary configuration of the enzyme to the transient state of the substrate on the one hand, with that of the antibody to the relatively stable antigen molecules on the other [17]. Pauling, however, divided his experimental work between enzymatic and immunochemical mechanisms, and did not report an attempt to combine them into a unified research project. His paper from 1948 did not offer special reasons that could lead other scientists to seriously consider such an attempt.

The post World War II era witnessed the rapid expansion of the scientific establishment as a whole. Enzymology and immunology grew into two large areas of specialized research. Knowledge of the reactivity and kinetics of many enzymes was greatly enhanced by advances in methods for the determination of the amino acid sequence of peptide chains and the study of the three-dimensional structure of globular proteins. Similar methods were applied to the study of the antibody molecule, yet immunological research predominantly focused on the various aspects of the regulatory mechanisms underlying the immune system as a whole. It is important to take this context into account in assessing Jencks's concept of the catalytic antibody from 1969. Jencks did not intend to suggest a novel research project at the crossroads of chemistry and immunology. As he pointed out later, he had not been aware, at the time, of Pauling's aforementioned comments on the analogy between antibodies and enzymes [18]. Jencks conceived the idea of a catalytic antibody to elucidate the chemical considerations that could account for the specific rate enhancements that are brought about by enzymes. His concept originated from the received view that the catalysis of reactions by the protein enzyme involves the stabilization of the transition-state of the reaction. Since the transition state is the most unstable configuration in the reaction path, no products can be formed from reactants unless the transition state is formed. As Jencks recently noted, his concept of antibody catalysis was originally meant to state a tautology: given the basic thermodynamic and kinetic principles of catalysis, enzymes were proteins whose affinity for the transition state must be greater than their affinity for either substrate or product. Hence, any protein that binds to the transition state reduces its free energy and thereby increases the rate of the reaction [19]. Moreover, as Jencks suggested in his aforementioned book, complementarity between the active site and the transition state did not provide a complete account of the mechanism of enzyme catalysis. An important challenge, in this respect, was to understand the chemical factors by virtue of which the conformation of the intact enzyme is modified upon binding to the substrate to induce the formation of the transition state complex [20].

Jencks did not pursue his idea of antibody catalysis. During the 1970s, however, several empirical demonstrations of his concept were published. One research project was carried out by Victor Raso, a graduate student at Tufts University under the supervision of David Stollar from the Department of Biochemistry and Pharmacology [21]. Raso's studies partook in an innovative program at the Tufts Medical School, where graduate students were encouraged to acquire a variety of research skills by taking part in the research activity of different laboratories. Their training thus crossed disciplinary boundaries and enabled them, in addition, to become highly independent in defining their research problems [22]. This setting explains, in part, how Raso could set up an original research project that was predicated on an intriguing idea that had not earlier been put to scrutiny by more established researchers. Experimental results of antibody catalysis were also reported, in the late 1970s, by a larger team of researchers from the Weizmann Institute of Science, Israel, who collaborated in part with a research team from the Ames Research Laboratory in Iowa [23]. The work of the Weizmann team was initiated by Bernard Green, a bioorganic chemist who had already submitted, by the mid 1970s, several detailed research proposals that were based on considerations of the enzyme-antibody analogy. These proposals failed, however, to yield sufficient financial resources for conducting long-term research, arguably because there was not on the first place sufficient evidence that could justify their plausibility and worth. It was the collaboration with other researchers, including several immunologists from the Weizmann Institute, whose assistance included the provision of monoclonal antibodies, that eventually enabled Green to realize some of his earlier proposals [24].

The reports of antibody catalysis from the 1970s did not spark off further explorations of the enzyme-antibody analogy. More particularly, they offered neither concrete results that indicated catalysis in an unambiguous manner nor a clear program for enhancing the potential of antibodies as biocatalysts. As a result, they did not persuasively show that the scientific interest in catalytic antibodies could successfully compete with the classical interest in enzymes. The catalytic power of enzymes is awesome. Many known enzymes catalyze complex reactions by enhancing their rate 10^{12}–10^{17} fold [25]. In the aforementioned reports, by contrast, the relatively few antibodies that were tested for catalysis performed somewhat poorly, enhancing the rate of reaction only 10^2–10^4 fold. Understanding the mechanisms of enzyme catalysis has always been one of the most prominent challenges facing scientists, and so there seemed to be no special reason for chemists to devote greater attention to the relatively poor performance of a few catalyst-like antibodies. The early demonstrations of antibody catalysis were therefore somewhat discouraging, and did not offer an attractive alternative to the study of the archetypes of biocatalysis. Apparently, the ideas of Pauling and Jencks did not suffice to stimulate the curiosity of scientists and encourage long-term investments in the study of the enzyme-antibody analogy. As will be shown in what follows, the emergence of antibody catalysis as a research field in the late 1980s was predicated upon new and broader interests, in the context of which Jencks's original concept was only the bait, so to speak.

4.3.2
Tapping the Immune System for Catalysts

Clearly, there is a difference between hunting a caribou and developing the art of caribou hunting. The Cree suggest that the caribou is given to them as a gift of nature, but they do not claim that for this reason all their hunting expeditions must always be successful. Rather, by disclosing useful information about the behavior of the caribou, the "theory" of the gift forms an integral part of the training of Cree hunters, as noted earlier. There is arguably a comparable difference between "hunting" for an antibody that catalyzes a reaction on the one hand, and, on the other, establishing the procedures by which the immune system as a whole can be gradually but systematically harnessed to yield antibodies that catalyze selected reactions. As shown above, the latter approach was not seriously considered by either Pauling or Jencks, since they both construed the enzyme-antibody analogy primarily in the context of kinetic and structural aspects of protein binding. As will be shown in what follows, the proliferation of studies of catalytic antibodies was predicated on a different approach, the articulation of which was stimulated by new opportunities that accompanied advancements in cellular and molecular studies of immunoregulation and their implications for medicine and pharmacology.

During the first half of the 20th century, serological and structural studies revealed that the immune system of vertebrates was biologically highly unique in its capacity to generate, on demand, numerous large molecules with diverse binding properties. This diversity was manifested in different variable (V) region structures of antibodies, and was shown to form an integral aspect of their physiological function. The aforementioned antigen-enzyme analogy notwithstanding, from a biological standpoint there was clearly an important difference between the two protein molecules. The diversity and specificity of antibodies is critically dependent on the immune response which the organism develops against a "foreign" agent. The diversity and specificity of enzymes, by contrast, remains relatively stable during the life-cycle of the organism. The production of enzymes in the organism is relatively "immune" to specific external stimuli. The immune system by contrast, is a mechanism that partakes in the organism's adaptation to environmental change by producing large molecules of diverse binding specifications.

Since the 1950s, rapid advancements in the study of the cellular mechanisms of the immune response, in conjunction with developments in molecular biology, enabled scientists to elaborate sophisticated strategies to understand and combat disease based on the analysis of pathogenic agents and their interaction with antibodies, as well as the study of the biological regulation of the immune response as a whole [26]. By learning how the information contained in the large protein structure of antibodies is encoded, and how molecular recognition induces the immune response, scientists could develop more rigorous methodologies for intervening in the generation of antibodies of specific chemical properties. In this context, the immune system could be treated as a robust manufacturer of proteins, endowed with a feedback mechanism that operated at the relatively short time scale of the immune response. The immune system embodied a mechanism for the *in vitro* evolution of proteins of specific chemical functions. By instructing the immune system to respond to more rigorously devised stimuli, the chemical structure of antibodies could be gradually modified to perfect desirable chemical functions. These considerations gave rise to a new concept of antibody catalysis. The old concept suggested that antibodies could mimic enzymes. The new concept suggested that the immune system could be manipulated to mimic the evolution of enzymes, or, more precisely, that it could be utilized to transform, in a gradual and systematic manner, relatively poor biocatalysts into more effective biocatalysts. The old concept was not, however, rejected. It was rather embraced as a key element in a new enterprise that was now more firmly established at the crossroads of chemistry and immunology.

A tentative outline of the new concept of catalytic antibodies was originally proposed by Richard Lerner, in 1982–3, at the Annual Conference of the American Chemical Society and the 18th Solvay Conference on Chemistry, which was devoted to discussions on molecular recognition in biological processes [27]. Lerner's proposal followed the presentation of experimental findings of recent work in immunochemistry that was carried out by his research team at Scripps. These findings demonstrated that short chains of amino acids, taken from different regions of the antigen protein, induced a humoral immune response as long as they were situated in a region which the antibody could easily access. It appeared, as Lerner pointed out, that "the immunogenecity of an intact protein is less than the sum of the immunogenecity of its pieces" [28]. Lerner then related these findings to the nascent hybridoma technology for the generation of monoclonal antibodies, which was developed in the mid 1970s by Georges Köhler and César Milstein. Coupled with hybridoma technology, the immunogenecity of small peptides afforded the opportunity to secrete tailor-made proteins that could then be utilized to perform tasks that were not necessarily related to immunization. Pointing out the enzyme-antibody analogy, Lerner suggested that "since the effectiveness of enzymes depends upon the stabilization of minor equilibrium states we might expect antibodies recognizing these same states to carry out catalytic functions". At this stage, however, he could only consider the likelihood of the project and note its potential value. Unfamiliar with the moderate results mentioned earlier, Lerner thus proclaimed that it "was a wonderful possibility since one could fish in the immunological repertoire for any kind of enzyme so long as the substrate was sufficiently large to be immunogenic" [29].

Back at Scripps, the pursuit of the new objective, alongside ongoing work in experimental pathology, was accompanied by major institutional changes which contributed decisively to its progress. In 1983, the Department of Molecular Biology was founded at Scripps, and Lerner was appointed to chair it. During the late 1970s, the relatively small research department of the Scripps Clinic, as it was then named, headed by the medical scientist Frank Dixon, was adjusting its infrastructure to incorporate new advances in molecular biology. Chairing the new Department, Lerner began to consolidate its physics and chemistry, recruiting X-ray crystallographers, experts in nuclear magnetic resonance (NMR), protein chemists, and synthetic organic chemists. Recruitment was a necessary, yet insufficient condition for growth. As he noted in his Annual Report a year later, "a new sociology has come into effect in the Molecular Biology Department" to overcome "the difficult problem...that different disciplines are like different cultures with different time frames". Accordingly, the new organization stressed the importance of establishing a working environment in which the various laboratories could effectively collaborate. By the mid 1980s, Lerner's laboratory was strategically set for long-term studies of antibody catalysis, and by the early 1990s Scripps became a worldwide center in this field. In the early 1980s, a unit specializing in hybridoma technology was established at Scripps, and the newly recruited organic and synthetic chemists who specialized in the preparation of transition-state analogs coordinated their work with the process of immunization stage by stage. This division and coordination of labor enabled several research teams at Scripps to simultaneously develop different strategies to refine the screening of antibodies for catalysis, thereby simulating *in vitro* evolutionary modifications in their catalytic properties.

Lerner's initiative coincided with the research interests of Peter Schultz, a newly appointed Chemistry Professor at the University of California in Berkeley. Studying at Caltech, Schultz specialized in bioorganic chemistry, and his early research pertained to the recognition of nucleic acids by proteins and restriction enzymes, as well as the application of genetic engineering methods to the study of the active site of enzymes [30]. His research on catalytic antibodies began at Berkeley, and soon after he pioneered research efforts to develop evolutionary models which combine the methods of chemistry with those of molecular biology and genetic engineering to identify and reproduce organic molecules with interesting biological functions.

Unlike Lerner, Schultz had to accommodate his initial trials to the relatively modest conditions offered by the academic structure of laboratory work. His research team, composed of graduate students and postdoctoral fellows, began to acquire the specialized experience in the highly interdisciplinary work that encompassed the chemistry of protein catalysis as well as the preparation of monoclonal antibodies. The two laboratories from Scripps and Berkeley have worked in parallel since the mid 1980s, and in 1986 it was decided that their reports on catalytic antibodies would be submitted to *Science* for publication in tandem [31]. Reinforcing one another, the two publications clearly aimed to establish the bridgehead for more advanced and diversified studies on catalytic antibodies. Both teams prepared reports on relatively simple reactions of the hydrolysis of esters, for which detailed mechanistic analysis had already appeared in the literature. Centered upon the approach of transition-state stabilization proposed by Jencks, their reports were designed to illuminate the potential of catalytic antibodies. As Schultz's team noted, their work "is a first step toward defining the rules and strategies whereby catalytic activity can be introduced into antibodies... for use as tools in biology, chemical synthesis, and medicine" [32].

As the rest of the chapters in this volume notably attest, transition-state stabilization has become one among a variety of strategies that were developed since the mid 1980s to augment the catalytic performance of antibodies. These developments were not only aimed at testing Jencks's original concept. Rather, they were commonly set to turn the regulative mechanisms of the immune response into a laboratory system for the directed reproduction of a diversity of proteins that can be screened for specific chemical and biological functions. These developments thus may seem to have eclipsed the inaugural research projects mentioned earlier. However, as this volume as whole amply suggests, the latter is currently still widely considered as one of the principal frames of reference for the former. As I noted in the introduction to this report, viewing recent contributions from the distance of historical time may provide, in the future, the adequate historical perspective on the development of antibody catalysis beyond the emergence of this discipline in the mid 1980s.

4.4
Conclusions

The emergence of antibody catalysis as a new research discipline at the crossroads of chemistry and immunology exemplifies the importance of the entrepreneurial moment in scientific innovations, which interlocks the social and the intellectual components in the scientists' role. Common images of scientific discoveries highlight the curious mind, but fail to elucidate the entrepreneurial effort that is involved in identifying new opportunities to expand upon prevailing skills and methods. From the standpoint of the enterprising agent, the natural world is not a mere object of human curiosity, but also inspires the belief in finding in it novel means to augment the human lot. As noted earlier in this chapter, this belief – dressed in numerous different cultural idioms – underlies systems of gift-exchange in many human societies. The methodical pursuit of this belief explains, as Adam Smith suggested, how scientific learning becomes a crucial aspect of the human endeavor to transform promising resources into valuable goods. The study of the entrepreneurial moment in scientific innovations contributes to our understanding of science as an engine of economic growth, and, more broadly, suggests an additional frame of reference for interfacing the natural sciences with the humanities.

Acknowledgements

The research for this chapter was supported by the Technion V.P.R. Fund for the Promotion of Research. I am grateful to the following scientists and research workers for their cooperation in granting interviews for my research (indicating institutional affiliations at the time of the interview): Carlos Barbas III, Scripps; Eleanor Butz, *Science* Editorial Board; Fortuna Cohen, Weizmann; Albert Eschenmoser, ETH; David Givol, Weizmann; Bernard Green, Hebrew University; Flavio Grynzpan, Scripps; Donald Hilvert, Scripps; Kim Janda, Scripps; William Jencks, Brandeis University; Ehud Keinan, Technion and Scripps; Anthony Kirby, Cambridge University; Richard Lerner, Scripps; Erling Norrby, the Royal Swedish Academy of Sciences; Victor Raso, Boston Biomedical Research Institute; Marina Resmini, Queen Mary College, University of London; Jean-Louis Reymond, University of Berne; Peter Schultz, University of California, Berkeley; Michael Sela, Weizmann; Doron Shabat, Scripps; Avidor Shulman, Technion; David Stollar, Tufts University; Dan Tawfik, Cambridge University; Paul Wentworth, Scripps; Ian Wilson, Scripps; Chi-Huey Wong, Scripps. I am especially thankful to Ehud Keinan, Tom Yule, and Karen-Beth Scholthof for helpful discussions and comments on earlier drafts.

References

1 GOLINSKI, J. V., *Making Natural Knowledge. Constructivism and the History of Science*, Cambridge University Press, **1998**
2 Personal communication with R. Lerner, September **1997**
3 For detailed ethnographic accounts of Cree attitudes to animals, see FEIT, H., "The ethnoecology of the Waswanipi Cree: or how hunters can manage their resources", in COX, B. (ed.), *Cultural Ecology: Readings on the Canadian Indians and Eskimos*, McClelland and Stewart, Toronto **1973**, p. 115–125; SCOTT, C., "Knowledge construction among Cree hunters: metaphors and literal understanding", *Journal de la Société des Américanistes* 75 (**1989**), p. 193–208. For general anthropological accounts of the cultural utilization of natural resources, INGOLD, T., *The Perception of the Environment. Essays on livelihood, dwelling and skill*, Routledge, London **2000**
4 For some earlier reviews, see [9].
5 For recent studies of Aristotle's philosophy of human cognition and learning, see LEAR, J., *Aristotle: the desire to understand*, Cambridge University Press, Cambridge **1988**; NUSSBAUM, M. C., OKSENBERG RORTY, A. (eds.), *Essays on Aristotle's De anima*. Clarendon Press, Oxford **1995**. Oksenberg Rorty **1992**
6 SMITH, A., *Inquiry into the Nature and Causes of the Wealth of Nations* Penguin Books, Harmondsworth **1986**, p. 115

7 SMITH, A., *Lectures on Jurisprudence*, ed. by MEEK, R. L., RAPHAEL, D. D., STEIN, P. G., Clarendon Press, Oxford **1978**, p. 347–348
8 The list of interviewees and their institutional affiliation is included in the Acknowledgements at the end of this chapter
9 Keinan, E., Lerner, R. A., "The first Decade of Antibody Catalysis: Perspective and Prospects", *Israel Journal of Chemistry* 36 (**1996**), p. 113–119; GREEN, B. S., "Monoclonal Antibodies as Catalysts and Templates for organic Chemical Reactions", in MIZRAHI A. (ed.), *Monoclonal Antibodies: Production and Application* Alan R. Liss, New York **1989**, p. 359–393; MAYFORTH, R. D., "Catalytic Antibodies", in *Designing Antibodies* Academic Press, San Diego **1993**, p. 167–19
10 PAULING, L., "A Theory of the structure and process of formation of antibodies", *J.A.C.S.*, (**1940**), p. 2643–2657; "Molecular architecture and biological reactions", *Chem. Eng. News* 24 (**1946**), p. 1375–1377; "Chemical achievement and hopes for the future", *Am. Sci.* 36 (**1948**), p. 51–58
11 JENCKS, W. P., *Catalysis in Chemistry and Enzymology*, McGraw-Hill, New York **1969**, p. 288
12 TRAMONTANO, A., JANDA, K. D., LERNER, R. A., "Catalytic Antibodies", *Science* 234 (**1986**), p. 1566–1570; POLLACK, S. J., JACOBS, J. W., SCHULTZ, P. G., "Selective Chemical Catalysis

by an Antibody", *Science* 234 (1986), p. 1570–1573
13 HALDANE, J. B. S., *Enzymes*, Longmans, Green and Company, London 1930
14 PAULING, L., "The nature of the chemical bond", *Journal of the American Chemical Society*, 53 (1931), p. 1367–1400
15 For a biographical sketch of these achievements, see BERNAL, J. D., "The pattern of Linus Pauling's work in relation to molecular biology", in RICH, A., DAVIDSON, N., *Structural Chemistry and Molecular Biology*, W.H. Freeman, San Francisco 1968, p. 370–379
16 LAIDLER K. J., KING, M. C., "The development of transition-state theory", *Journal of Physical Chemistry* 87 (1983), p. 2657–2664
17 Pauling (1948), p. 57–8
18 CHADWICK D. J., MARSH, J. (eds.), *Catalytic Antibodies*, John Wiley & Sons, Chichester 1991, p. 2
19 Personal communication, February 11, 1997
20 JENCKS (1969), p. 282–7
21 RASO, V., *The Antibody-Enzyme Analogy. Antibodies Designed to Simulate Pyridoxal Phosphate Dependent Tyrosine Utilizing Enzymes* (unpublished Ph.D. Thesis, 1973); RASO, V., STOLLAR B. D., "Antibodies specific for Conformationally Distinct Coenzyme-substrate Transition State Analogs", *J.A.C.S.* 95 (1973), p. 1621–1628; RASO, V., STOLLAR, B. D., "The Antibody-Enzyme Analogy", *Biochemistry* 14 (1975), p. 584–599
22 Personal communication, February 11, 1997
23 KOHEN, F., HOLLANDER, Z., BURD, J. F., BOGUSLASKI, R. C., "A Steroid Immunoassay Based on Antibody-Enhanced Hydrolysis", *FEBS Letters* 100 (1979), p. 137–140; KOHEN, F., KIM, J. B., LINDNER, H. R., ESHHAR, Z., GREEN, B., "Monoclonal Immunoglobulin Augments Hydrolysis", *FEBS Lett.* 111, p. 427–431
24 Personal communication, May 1998
25 FERSHT, A., *Structure and Mechanism in Protein Science. A Guide to Enzyme Catalysis and Protein Folding*. W.H. Freeman and Company, New York 1999, p. 60–1
26 SILVERSTEIN, A. M., *A History of Immunology*, Academic Press, 1989; TAUBER, A. I., PODOLSKY, S. H., *Clonal Selection Theory and the Rise of Molecular Immunology*, Harvard University Press, Cambridge 1997
27 Of Lerner's two conference presentations, one was reprinted in LERNER, R. A., "Antibodies of Predetermined Specificity in Biology and Medicine", in VAN BINST, G. (ed.), *Design and Synthesis of Organic Molecules Based on Molecular Recognition. Proceedings of the XVIIIth Solvay Conference on Chemistry* Springer-Verlag, Berlin 1983, p. 43–49
28 Lerner (1983), p. 45
29 Lerner (1983), p. 46
30 See, for example, TAYLOR, J. S., SCHULTZ, P. G., DERVAN, P. B., *Tetrahedron Symposium*, 40 (1984), p. 457–465; SCHULTZ, P. G., AU, K. G., WALSH, C. T., *Biochemistry*, 24 (1985), p. 6837–6840
31 TRAMONTANO ET AL. (1986), POLLACK ET AL. (1986); personal communications with Lerner and Schultz, February 21 and 26, 1997, respectively
32 POLLACK ET AL. (1986), p. 1572

5
Catalytic Antibodies in Natural Products Synthesis
Ashraf Brik, Ehud Keinan

5.1
Introduction

Natural product synthesis is the centerpiece of organic chemistry because it has always been the ultimate testing ground for new concepts and new synthetic methods. In fact, much of what we know about organic reaction mechanisms has come from attempts to carry out selective chemical transformations on the journey to the construction of these often very complex naturally occurring molecules. In spite of the tremendous successes in the fields of organic synthesis and synthetic methodology [1], there is a continual search for methods to improve organic synthesis and provide practical routes to drugs, natural products, and other important chemicals.

Enzymes and other biocatalysts have been attracting ever-increasing attention as useful tools in organic synthesis, mainly because of their high selectivity, specificity, and phenomenal rate acceleration. Enzymes may offer solutions to many synthetic problems that are difficult or even impossible to solve efficiently by other chemical means [2]. Biocatalysis has been successfully used not only in small-scale synthesis but also in the pharmaceutical industry for the preparation of enantiomerically pure intermediates and products [3]. Furthermore, recently developed biological tools, such as directed evolution of enzymes [4, 5, 6], phage display selection methods [7, 8], techniques of molecular genetics [9], as well as traditional screening methods [10], have increased the scope of opportunities of using enzymes in organic synthesis. Yet, for many chemical transformations there is no known natural enzyme, and in many cases the relevant enzyme is either difficult to isolate from its natural source or is too unstable for synthetic applications. Moreover, the high specificity and narrow substrate range of most enzymes are major drawbacks for general application in organic synthesis, where catalysts that are more promiscuous with respect to their range of substrates are desirable. These limitations have led to increased search, particularly over the last two decades, for new biocatalysts. Undoubtedly, of the various strategies to generate new biocatalysts, antibody catalysis has been the most successful and practical approach.

Challenging the immune system with stable analogs of the transition state of a given reaction has proven to be a useful approach to achieve monoclonal antibodies

Catalytic Antibodies. Edited by Ehud Keinan
Copyright © 2005 WILEY-VCH Verlag GmbH & Co. KGaA, Weinheim
ISBN: 3-527-30688-9

that catalyze the reaction. Since the demonstration of this idea by Lerner [11] and Schultz [12], its implementation coupled with versatile screening assays has yielded an abundance of catalytic antibodies for more than 100 reactions. Most of these reactions represent useful transformations for organic synthesis [13], including pericyclic processes, group transfer reactions, additions and eliminations, oxidations and reductions, aldol condensations, and miscellaneous cofactor-dependent transformations. In many cases, the selectivities of these catalysts rival those of natural enzymes, and transformations that could not be achieved efficiently or selectively via more traditional chemical methods were shown to be possible via antibody catalysis [14, 15].

One of the main goals of the field of antibody catalysis [11, 12, 16] has been the achievement of custom catalysts for synthetic schemes in order to open routes that are otherwise inaccessible or cumbersome. In this way, one hopes to improve the overall yield and allow practical construction of synthetic drugs and important natural products. In general, in addition to being highly chemoselective, most antibody-catalyzed reactions are diastereoselective, enantioselective [17], and regioselective [18], even reversal of chemoselectivity having been observed [19]. The relevance of these catalysts to synthetic organic chemistry has been demonstrated by the possibility of running reactions with gram scale quantities [20].

Since many organic synthetic methods require the manipulation of highly reactive, water-sensitive intermediates, the handling of such intermediates along with catalytic antibodies is an important issue for synthetic applications. It has been shown that such chemistry is accessible using catalytic antibodies in aqueous media [21]. Antibody 14D9 catalyzed the protonation of a prochiral enol ether with complete enantioselectivity to form a highly water-sensitive intermediate. In the hydrophobic environment of the antibody active site, this short-lived intermediate reacts intramolecularly with a neighboring hydroxyl group to form an enantiomerically pure ketal. This reaction is normally not possible in aqueous medium because the oxocarbonium ion is trapped by a water molecule to give the corresponding ketone. The ability to promote reaction pathways that are normally disfavored or have low probability represents one of the greatest advantages of antibody catalysis [14].

This review will not survey all the useful transformations catalyzed so far by antibodies, as other chapters in this book adequately cover these reactions. This chapter covers mainly work that resulted in either total synthesis or formal total synthesis of a natural product. It is important to emphasize that the key point in all of these examples is not simply that one can make specific target molecules, such as multistriatin, epothilones, brevicomins, etc., or even that the methods used are now the best available to synthesize the compounds, but rather that catalytic antibodies perform competitively in the important testing ground of natural product synthesis.

5.2
Total Synthesis of α-Multistriatin
via Antibody-Catalyzed Asymmetric Protonolysis of an Enol Ether

The compound (–)-α-multistriatin, **1**, is one of the three essential components of the aggregation pheromone of the European elm bark beetle, *Scolytus multistriatus* (Marsham), the principal vector of Dutch elm disease in Europe and North America [22]. The severe devastation of the elm population in Northeastern USA has resulted in extensive studies of the synthesis and field utilization of **1** [23]. Since the discovery of this pheromone by Silverstein in 1975 [24], it has been the subject of numerous synthetic efforts [25]. Field experiments showed that the inactive, (+)-enantiomer of **1** inhibits the biological activity of the naturally occurring, (–)-enantiomer [26]. Therefore, the high enantiomeric purity is a crucial issue in the synthesis of this pheromone. The enantioselective synthesis of **1** by Sinha and Keinan represents the first example of a natural product total synthesis that involves antibody catalysis [27].

The retrosynthetic analysis of **1** took advantage of the opportunity to obtain α-branched ketones with high enantiomeric purity via antibody 14D9-catalyzed enantioselective protonolysis of the appropriate enol ether [28]. The ability to carry out this reaction under mild conditions was of particular importance because high enantiomeric purity of a tertiary carbon center adjacent to a carbonyl function is difficult to generate and retain. Antibody 14D9 has already been proven to be an effective catalyst for the hydrolysis of a variety of oxygen functions, including acetals [29], ketals [30], epoxides [17], and enol ethers [28, 31]. Since all these reactions are characterized by a positively charged transition state, it is conceivable that antibody 14D9, which was elicited against the positively charged piperidinium cation **2**, catalyzes the reactions by stabilizing the positive charge that is created along the reaction coordinate [29]. Indeed, mechanistic studies with this antibody had suggested that an ionizable side chain in the active site combined with hydrophobic interactions directly participate in transition-state stabilization to achieve its catalysis [28].

Thus, substrate **3** (Scheme 5.1) was chosen for the antibody-catalyzed preparation of the desired α-branched ketone, **4**, the key asymmetric building block of the entire synthesis [27]. The Z enol ether **3** together with its E stereoisomer, along with two other regioisomers, were prepared from 3-pentanone and methyl 4-bromomethylbenzoate in a four-step sequence of chemical transformations followed by chromatographic separation. In acidic, aqueous media, all isomers are hydrolyzed to the racemic ketone **4**. Antibody 14D9 catalyzed this reaction under mildly acidic conditions, with hydrolysis of the Z enol ether, **3**, being much more effective (K_m = 230 µM, k_{cat} = 0.36 min^{-1} at pH 6.5, k_{cat}/k_{uncat} = 65 000) than that of the E isomer (K_m = 310 µM, k_{cat} = 0.044 min^{-1} at pH 6.0 (k_{cat}/k_{uncat} = 5000). These observations enabled a mixture of all four isomeric enol ethers to be used in the antibody-catalyzed hydrolysis, resulting in complete, selective consumption of **3**, and leaving the other three isomers essentially intact. Ketone **4** was thus obtained with configuration (S) in greater than 99% ee.

Another practical advantage of this process originated from the fact that there is no known enzyme or other biological component which can catalyze the enol

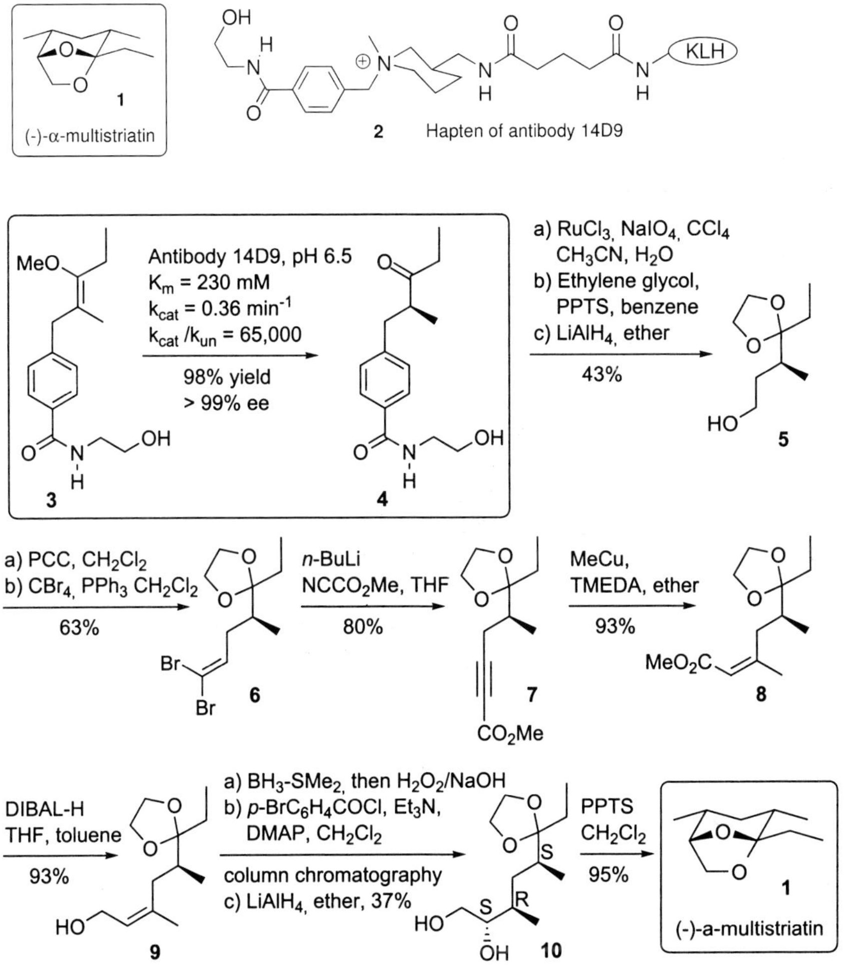

5.2 Total Synthesis of α-Multistriatin via Antibody-Catalyzed Asymmetric Protonolysis of an Enol Ether

Scheme 5.1

ether hydrolysis. Consequently, there was no need to use a purified antibody to catalyze this reaction. Usually, catalytic antibodies that are produced from murine hybridoma cell lines are obtained from ascites fluid followed by ammonium sulfate (SAS) precipitation, anion exchange, and affinity chromatography with immobilized protein G [32]. Efficient catalysis was thus achieved with a partially purified antibody 14D9, which was precipitated from the ascites fluid by SAS. Catalytic activity of this crude antibody was completely inhibited by the hapten, 2, thereby ruling out any non-specific catalysis by other components in the crude SAS-fraction.

The antibody-catalyzed reaction was carried out on a preparative scale using very simple organic-laboratory equipment. In each catalytic cycle, a solution of the enol ether 3 (180 mg, 0.65 mmol) in DMF (1 mL) was added to a solution of a crude SAS fraction of antibody 14D9 (22.5 mL containing 225 mg protein, 0.0015 mmol) in bis-tris buffer (50 mM, pH 6.5), and the mixture was stirred at 24 °C. Progress of the reaction could be observed visually, as the starting mixture was a turbid white (because of lower solubility of the starting material relative to that of the product) and became clear as the reaction reached completion. The reaction was transferred into a cellulose dialysis bag and dialyzed into the same buffer. The recovered antibody solution was taken to the next catalytic cycle with a fresh solution of **3**. Only minor deterioration of catalytic activity could be observed over five cycles of the reaction.

The crucial antibody-catalyzed step was followed by several chemical steps to complete the synthesis of **1** (Scheme 5.1). Ketone **4** was treated with $RuCl_3$ and sodium periodate. The resultant keto-acid was then converted to a ketal-ester by reaction with ethylene glycol and catalytic amounts of pyridinium p-toluenesulfonate in benzene. Reduction of the resultant carboxylic ester with $LiAlH_4$ in ether afforded alcohol **5**, which was oxidized with pyridinium chlorochromate in dichloromethane to the corresponding aldehyde. The latter was treated with a solution of triphenylphosphine and carbon tetrabromide in dichloromethane to give the dibromoalkene **6**. Treatment of **6** with n-butyllithium and methyl cyanoformate in hexane-THF produced the substituted methyl propargylate **7**. Reaction of the latter with methylcopper afforded geometrically pure (Z) α,β-unsaturated ester **8**. Reduction of this ester with diisobutylaluminum hydride in toluene-THF afforded the corresponding allylic alcohol, **9**, with retention of the (Z) geometry. Treatment of **9** with a solution of borane-dimethyl sulfide complex in THF followed by oxidation with basic hydrogen peroxide produced a 70:30 mixture of two diastereomeric products, **10**, which had the desired 2S,3R,5S configuration, and its 2R,3S,5S-diastereomer, respectively. The chromatographically purified diol **10** was treated with catalytic amounts of pyridinium p-toluenesulfonate in dichloromethane followed by kugelrohr distillation at 110 °C to produce (−)-α-multistriatin, **1**, in the form of a colorless oil. The synthetic pheromone **1** has been checked in field experiments and found to be as active as the naturally occurring compound in attracting the European elm bark beetles into traps loaded with a mixture of **1** with (−)-α-cubebene (−)-α-cubebene (a host-produced component) and (−)-4-methylheptan-3-ol [27].

To summarize this achievement, all four asymmetric centers of the target molecule had originated from the chirality that was accomplished in the antibody catalyzed step, i.e., enantioselective protonolysis of an enol ether. That specific step is a unique example of a chemical transformation that is difficult to achieve either by an available synthetic methodology or via catalysis with a known enzyme.

5.3
Total Synthesis of Epothilones Using Aldolase Antibodies

The total synthesis of many epothilone derivatives by Sinha and coworker represents the most advanced example of natural product synthesis assisted by antibody catalysis [33]. Several aldolase antibodies were employed to catalyze both aldol and retroaldol reactions in order to achieve the required enantioselectivity in the synthesis of various epothilone derivatives.

5.3 Total Synthesis of Epothilones Using Aldolase Antibodies

An active site nucleophilic lysine residue with a highly perturbed pK_a is an essential element of the catalytic machinery that is available to both the natural type I aldolase enzymes and the aldolase antibodies elicited by reactive immunization [34, 35]. Unlike normal immunization, in reactive immunization [36], chemically reactive haptens are used as immunogens, so that a chemical reaction occurs in the binding site of an antibody during its induction. This strategy creates antibodies with a broad substrate scope, which is the result of the fact that the chemical event is covalent and occurs early in the process of the antibody binding site refinement. After formation of a covalent bond between the hapten and antibody, further improvement of the binding site via non-covalent bonds cannot meaningfully increase the binding energy, and thus mutations are no longer selectable [37]. Thus, reactive immunizations often yield antibody binding sites into which an efficient chemical mechanism has been installed but are not otherwise highly refined. Such catalysts can be expected to be both efficient and promiscuous in both the aldol and retroaldol reactions [38]. Chapter 11 of this book comprehensively covers these issues [39].

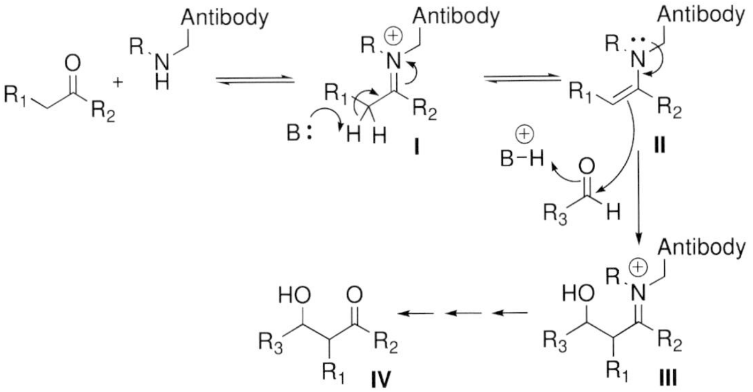

Scheme 5.2

In the aldolase antibodies, such as 38C2 and 33F12 [34], 24H6 [35] and 84G3, 85H6 and 93F3 [40], which were generated against β-diketone haptens, a hydrophobic microenvironment accounts for tuning the pK_a of the ε-amino group of this lysine residue [41]. This nucleophilic lysine reacts with ketones and aldehydes to form an imine (or iminium ion, **I**) and enamine (**II**) intermediates (Scheme 5.2). These intermediates may lead to many carbonyl transformations, such as carbonyl condensation reactions, alkylation, decarboxylation, etc. The aldol reaction [42], for example, involves a nucleophilic addition of the enamine intermediate, **II** (Scheme 5.2), to a carbonyl acceptor, e.g., an aldehyde, to form the iminium intermediate **III** and eventually produce the aldol product **IV**. The synthetic advantages of these aldolase antibodies, which share both the characteristics of the natural class I aldolases and broad substrate scope, have been demonstrated by various examples [43, 44].

Epothilones are molecules of current interest because of their medical promise and synthetic challenges [45, 33]. Epothilone A (**11**) and epothilone B (**12**) are powerful cytotoxic agents isolated from myxobacteria (*Sorangium cellulosum* strain 90). They possess a taxol-like mode of action, functioning through stabilization of cellular microtubules, and exhibit cytotoxicity even in taxol-resistant cell lines. Epothilone B has been reported to be about 3400 times more active than taxol against the resistant human leukemic cell line CCRF-CEM/VBL in cell-culture cytotoxicity studies. Desoxy precursors of **11** and **12**, epothilones C (**13**) and D (**14**) also possess comparable biological properties, particularly the tubulin polymerization activity. Besides epothilones A–E, many analogs of these compounds have been synthesized and studied for their effects on tubulin polymerization *in vitro* and *in vivo*.

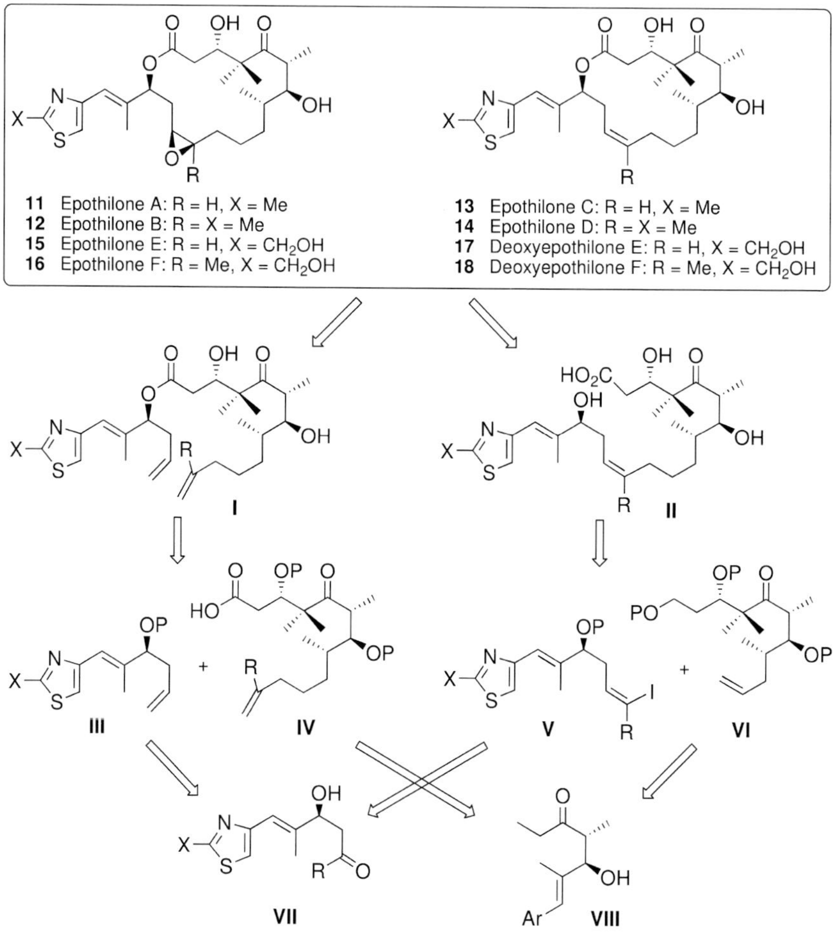

11 Epothilone A: R = H, X = Me
12 Epothilone B: R = X = Me
15 Epothilone E: R = H, X = CH$_2$OH
16 Epothilone F: R = Me, X = CH$_2$OH

13 Epothilone C: R = H, X = Me
14 Epothilone D: R = X = Me
17 Deoxyepothilone E: R = H, X = CH$_2$OH
18 Deoxyepothilone F: R = Me, X = CH$_2$OH

Scheme 5.3

Retrosynthetic analysis of the epothilone skeleton suggested that compounds **11–18** could be obtained either from **I** via metathesis or from **II** via macrolactonization (Scheme 5.3). Each of the intermediates, **I** and **II**, could be constructed from two major building blocks, **I** from **III** and **IV**, and **II** from **V** and **VI**. Both intermediates **III** and **V** could be obtained from a common aldol **VII**. Similarly, both intermediates **IV** and **VI** could be obtained from the common aldol **VIII**.

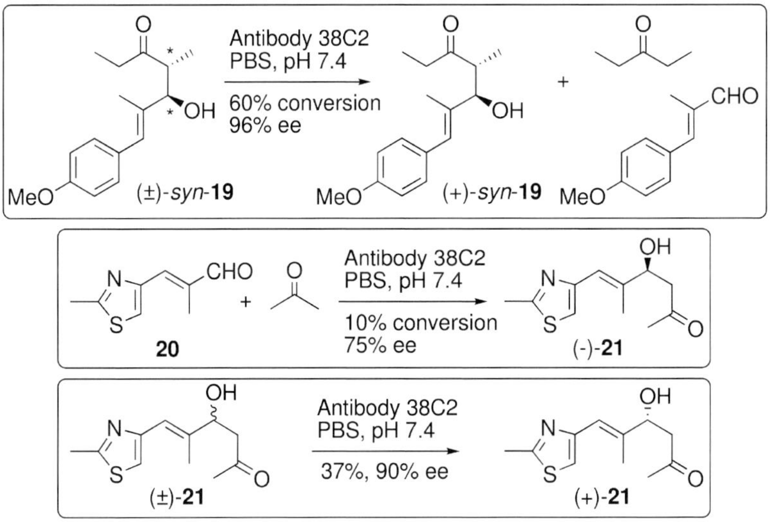

5.3 Total Synthesis of Epothilones Using Aldolase Antibodies

Scheme 5.4

Preliminary investigations have indicated that both enantiomers of each of the required aldol precursors, **VII** and **VIII**, could be accessible by antibody 38C2-catalyzed aldol and retroaldol reactions [46]. Indeed, aldol (±)-syn-**19** underwent a retro-aldol reaction to produce 4-methoxy-α-methylcinnamaldehyde and pentan-3-one (Scheme 5.4). At pH 5.4 (PBS) with 0.125 mol% of 38C2 at 60% conversion, the recovered unreacted aldol, (+)-syn-**19**, was obtained with 96% ee. An important point regarding this step is that the reaction could be carried out on gram scale. Typically, 0.75 g of compound (±)-syn-**19** was resolved to afford enantiomerically pure (+)-syn-**19** (0.3 g). Similarly, as expected on the basis of previous results [34, 38], antibody 38C2 also catalyzed the aldol condensation between aldehyde **20** and acetone to produce aldol (−)-**21** with 75% ee (Scheme 5.4). The reaction was carried out with 0.06 mol% catalyst and was interrupted at 10% conversion and 51% yield. At higher conversions the retro-aldol reaction became more significant, leading to diminished enantiomeric purity of **21**. An alternative approach to (−)-**21** used enantioselective 38C2-catalyzed retro-aldol reaction with (±)-**21** to produce (+)-**21** followed by inversion of the hydroxyl configuration by the Mitsunobu reaction to give (−)-**21** (Scheme 5.4).

With the enantiomerically enriched building blocks (+)-syn-**19** and (−)-**21** in hand, synthesis of the naturally occurring epothilone A (**11**) was achived via its desoxy precursor epothilone C (**13**) using either the metathesis (Schemes 5 and 6) or the macrolactonization approach. The use of building block (+)-syn-**19** secured the two stereogenic centers at C-6 and C-7 that allowed for the transfer of chirality to the remaining centers. Hydrogenation of (+)-syn-**19** afforded an easily separable 1:1 mixture of compounds **22** and epi-**22**. The free alcohol in **22** was protected in the form of a TBS ether, **22a**, before undergoing monomethylation to give **23**. Aldol reaction with 3-tert-butyldimethylsilyloxy propanal afforded a mixture of diastereomeric aldol products, with the desired stereoisomer, **24a**, being the major product. The latter was

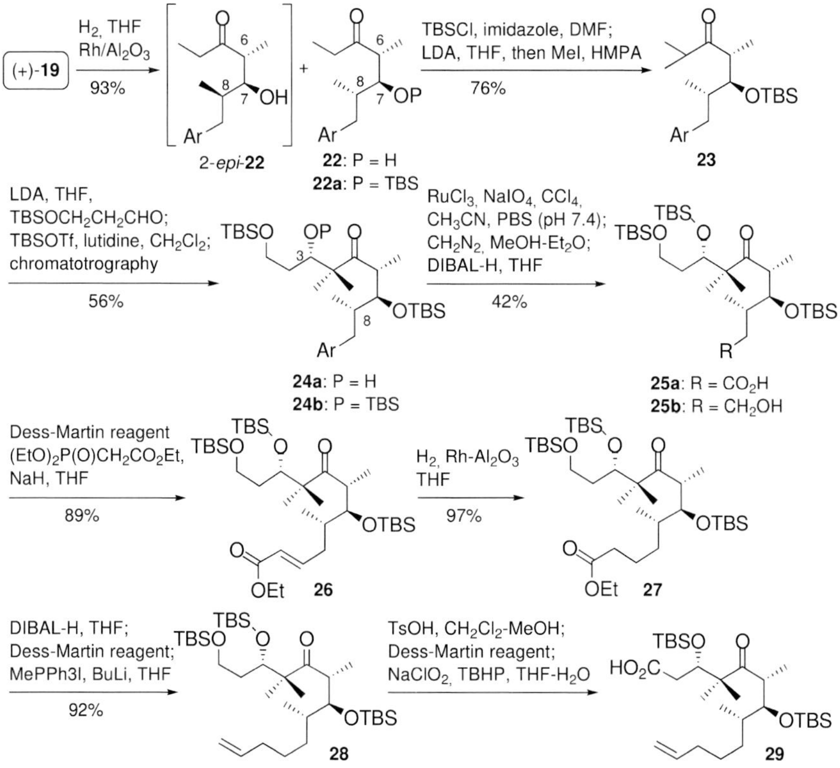

Scheme 5.5

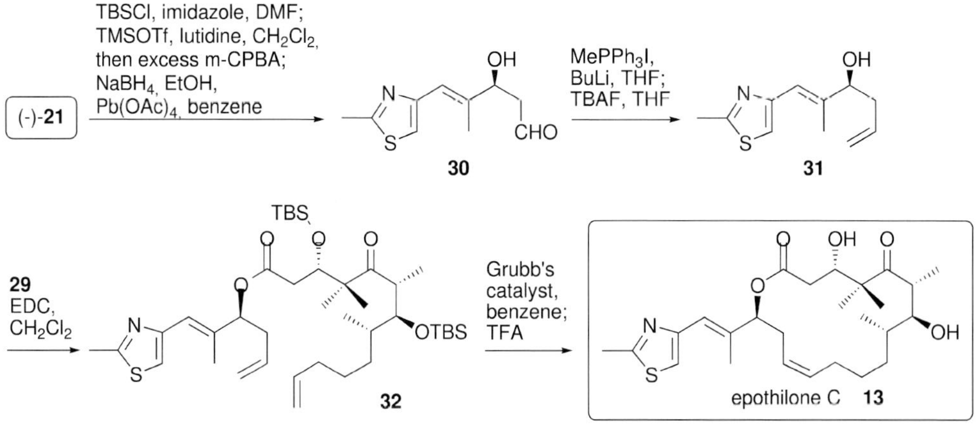

Scheme 5.6

silylated to give **24b**. Exhaustive degradation of the aromatic ring with RuCl$_3$-NaIO$_4$ afforded a carboxylic acid, **25a**, which was transformed to the corresponding alcohol **25b**. The latter was oxidized to the corresponding aldehyde, which was subjected to a Wittig olefination reaction to produce an α,β-unsaturated ester, **26**. Hydrogenation of the alkene gave the saturated ester **27**. Reduction of the ester to a primary alcohol and oxidation with afforded the corresponding aldehyde, which was reacted with methyltriphenylphosphorane to produce a terminal alkene, **28**. Deprotection of the primary alcohol followed by oxidation afforded the carboxylic acid **29**.

The other aldol building block, (−)-**21**, was first protected in the form of a TBS ether (Scheme 5.6). The latter was converted to the trimethylsilyl enol ether, which was then reacted with *m*-CPBA to produce the primary hydroxyketone. Reduction of the ketone to the vicinal diol followed by cleavage with Pb(OAc)$_4$ gave aldehyde **30**. The latter was subjected to a Wittig reaction with methylphosphorane followed by hydrolysis of the TBS ether to yield alcohol **31**. Esterification of the carboxylic acid **29** with alcohol **31** afforded ester **32**, which was converted to a macrocyclic lactone (a 3:2 mixture of the *cis*:*trans* isomers) via metathesis reaction with Grubb's catalyst Deprotection of the alcohols using TFA, followed by chromatographic separation, afforded epothilone C (**13**). Epoxidation of **13** with methyl(trifluoromethyl)dioxirane afforded epothilone A (**11**). An alternative approach to compound **13** employed the macrolactonization strategy, using precursor (−)-**21** [33].

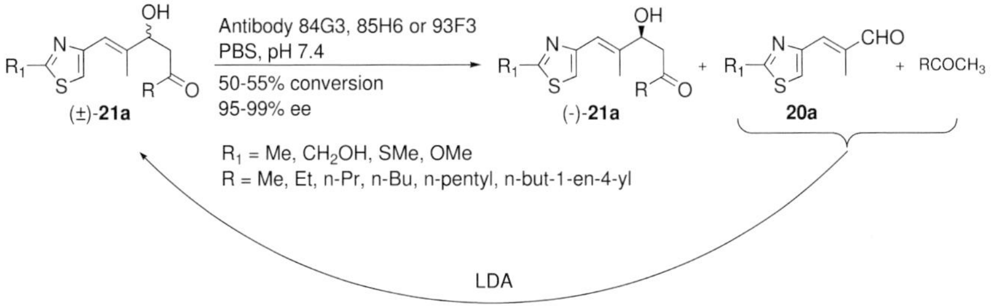

Scheme 5.7

The above-described syntheses were based on the aldol products (+)*syn*-**19** and (−)-**21**. However, the antibody 38C2-catalyzed aldol reaction afforded compound (−)-**21** in modest enantiomeric purity (75% ee) at 10% conversion, and these yields decreased even further as the reaction progressed as a result of retroaldol reaction. This problem was satisfactorily solved by using the more recently discovered antibodies, 84G3, 85H6, and 93F3 [40]. These aldolase antibodies exhibited antipodal reactivities in comparison with 38C2, allowing for the resolution of compound (±)-**21** and its derivatives, (±)-**21a**, in very high enantiomeric purity (Scheme 5.7). These aldol derivatives were practically resolved in multigram (> 16 g) quantities and essentially enantiomerically pure form using 0.003, 0.005, and 0.0004 mol % of antibodies 84G3, 85H6 and 93F3, respectively [47, 48]. Since these reactions and their work-up did not require harsh conditions, it was possible to recycle the thiazol aldehydes (**20**) for the synthesis of the racemic aldol substrate, (±)-**21a**, using LDA (Scheme 5.7).

With these chiral building blocks in hand, a sequence of chemical steps previously published by Schinzer et al. allowed the completion of the total synthesis of the above-mentioned epothilones, **11**, **12**, **15**, and **16**, via their desoxyepothilones, **13**, **14**, **17**, and **18** (Scheme 5.3). Recently, preliminary results were published in the synthesis of 13-alkyl analogs of epothilones A– E and fluoroepothilones using the same strategy [47, 49]. Other naturally occurring epothilones as well as many non-natural analogs, such as **33–38** (Scheme 5.8), were synthesized from the key building blocks, including (+)-**19** and (–)-**21** and similar precursors, all generated via antibody-catalyzed aldol and retroaldol reactions.

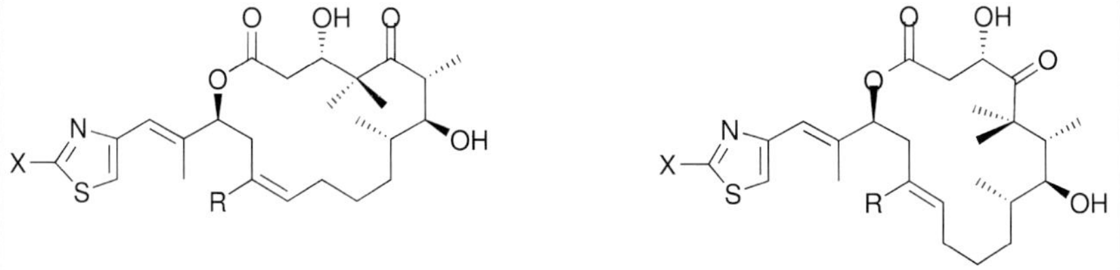

33: R = X = Me
34: R = Et, X = Me
35: R = Me, X = SMe
36: R = Et, X = SMe
37: R = Me, X = CH$_2$OH
38: R = Et, X = CH$_2$OH

(*E*)-**33**: R = X = Me
(*E*)-**34**: R = Et, X = Me
(*E*)-**35**: R = Me, X = SMe
(*E*)-**36**: R = Et, X = SMe
(*E*)-**37**: R = Me, X = CH$_2$OH
(*E*)-**38**: R = Et, X = CH$_2$OH

Scheme 5.8

In summary, the two antibody-catalyzed steps have created the required chirality in the total syntheses of epothilones A–F and their analogs. Three stereogenic centers were thus incorporated into the epothilone skeleton.

5.4
Total Synthesis of Brevicomins Using Aldolase Antibody 38C2

The brevicomins are derivatives of 6,8,-dioxabicyclo[3.2.1]octanes, which are known pheromones of a variety of bark beetle species [50]. Extensive outbreaks of bark beetles may result in the destruction of millions of trees per year, causing great ecological and economic damage [51]. Aldolase antibody 38C2 was successfully used by Barbas and coworkers for the enantioselective total syntheses of several brevicomins [52]. The key steps in these syntheses were achieved via either an antibody-catalyzed aldol addition or retroaldol reaction. For example, 38C2-catalyzes aldol reactions between hydroxyacetone as donor, and various aldol aldehyde and ketone acceptors afforded α,β-dihydroxyketones with an α(2*R*,3*S*) configuration in high regio- and stereoselectivity [37]. The usefulness of this reaction was demonstrated by the short syntheses of hydroxybrevicomins *ent*-**43** and *ent*-**44** in very high enantiomeric purity (Scheme 5.9). The synthesis was carried out by reacting aldehyde **39** and hydroxyacetone on a prepar-

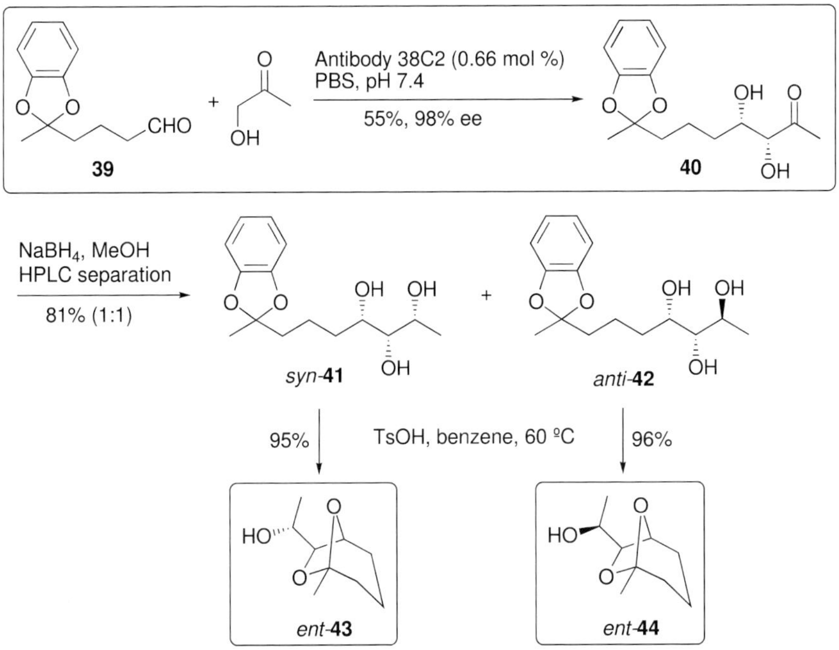

5.4 Total Synthesis of Brevicomins Using Aldolase Antibody 38C2

Scheme 5.9

ative scale in the presence of antibody 38C2 (0.66 mol %) to give diol **40** in 55% yield and 98% ee, along with the *anti*-diastereomer (4:1). Subsequent non-selective reduction with NaBH$_4$ followed by separation of diastereomers afforded triols *syn*-**41** and *syn*-**42**. Hydrolysis of the ketals and *in situ* ketalization produced the desired brevicomins *ent*-**43** and *ent*-**44**, respectively.

Antibody 38C2-catalyzed aldol reaction between aldehyde **39** and 1-hydroxybutan-2-one yielded the aldol product **45** in 99% ee (Scheme 5.10). As described above, reduction and separation of diastereomers produced triols *syn*-**46** and *syn*-**47**. Deprotection and intramolecular ketalization afforded brevicomins **48** and **49**, respectively. Brevicomin **48** can be easily converted to brevicomin **50** by reductive dehydroxylation, as described previously [53].

Since the aldolase antibodies are able to catalyze both the aldol addition and the retroaldol reaction, these catalysts are useful in the kinetic resolution of aldol products [46]. A single antibody catalyst can therefore be used for the preparation of both aldol enantiomers. This concept was applied in the synthesis of hydroxybrevicomins **43** and **44**. Antibody 38C2 catalyzed the retro-aldol reaction of the racemic mixture of **51** to produce (3S,4R)-**51** in 99% ee after 55% conversion of racemate (Scheme 5.11). Using the same procedures described in Scheme 5.9, diol (3S,4R)-**51** can be converted to the natural products **43** and **44**. These examples demonstrated that catalytic antibodies could be used to decrease the total number of synthetic steps and to increase enantioselectivity of natural product synthesis.

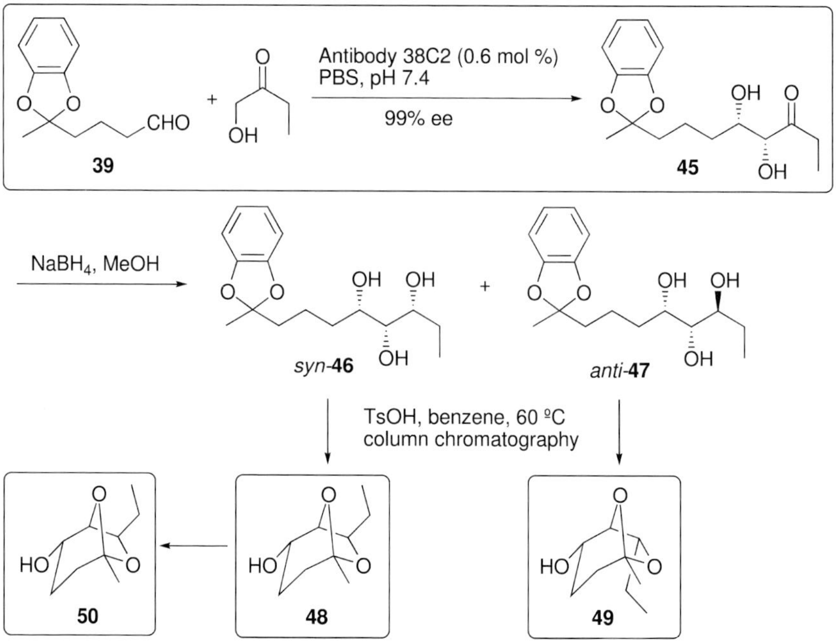

144 | 5 Catalytic Antibodies in Natural Products Synthesis

Scheme 5.10

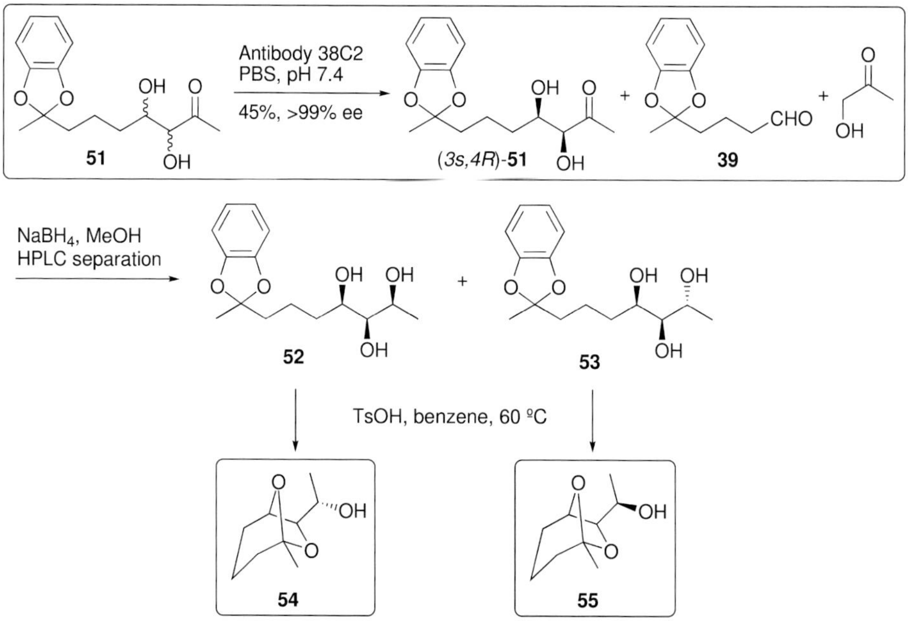

Scheme 5.11

5.5
Synthesis of 1-Deoxy-L-Xylose Using 38C2 Antibody

1-Deoxy-L-xylose (56), which was isolated from *Streptomyces hygroscopicuus*, is a key intermediate in the biosynthesis of thiamin (vitamin B_1) [54] and of pyridoxal (vitamin B_6) [55], as well as an alternative non-mevalonate biosynthetic precursor of terpenoid building blocks. Using 38C2, the total synthesis of this natural product by Shabat et al. was achieved via a two-step synthesis, which is considered to be the shortest synthesis known to date for this molecule [56].

While all the natural aldolases use hydroxyacetone in its phosphate-protected form, aldolase 38C2 antibody is known to use unprotected hydroxyacetone in the aldol addition reaction. As shown in Scheme 5.12, the key step of the antibody 38C2-catalyzed synthesis of 56 is an aldol reaction between unprotected hydroxyacetone and aldehyde 54. Using only 0.04 mol% of 38C2 afforded dihydroxyketone 55 in 32% yield and 97% ee after 56% conversion of aldehyde 54. Hydrogenolysis of the benzyl protecting group produced the target product, 56 in 81% yield.

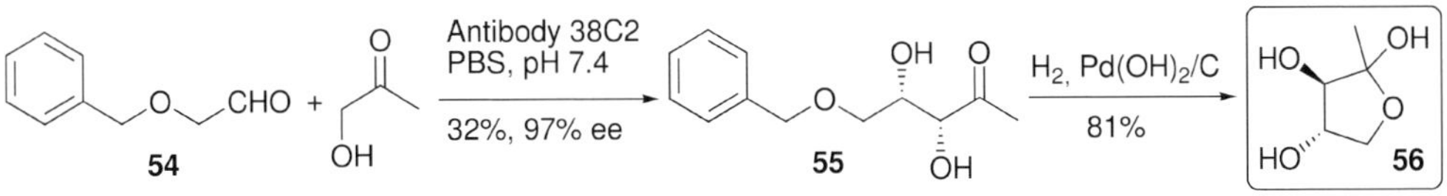

Scheme 5.12

5.6
Synthesis of (+)-Frontalin and Mevalonolactone via Resolution of Tertiary Aldols with 38C2

Both chemical and enzymatic approaches have been applied almost exclusively to the synthesis of secondary β-hydroxy carbonyl compounds (secondary aldols). General methods, either chemical or enzymatic, for the preparation of enantiomerically enriched tertiary aldols have not been developed. Tertiary aldols, which contain a heteroatom-substituted quaternary carbon stereocenter, constitute a challenge in synthetic chemistry. This is particularly true when this problem is approached through aldol chemistry. Interestingly, although tertiary aldols represent a common structural motif in many bioactive natural products, natural enzymes cannot be applied to this problem because no known natural aldolase catalyzes the synthesis of tertiary aldols.

Antibody 38C2 was found by List et al. to be a practical and highly enantioselective catalyst for the formation of tertiary aldols, accepting a broad variety of substrates [57]. This antibody catalyzes the retroaldol reaction of various tertiary aldols, thus providing a rapid entry to highly enantiomerically enriched tertiary aldols (typically >95%) via kinetic resolutions. The utility of this approach has been demonstrated in the synthesis of (+)-frontalin (59), a sex pheromone, which was found in several beetle species [58] as well as in the temporal gland secretion of the male Asian elephant [59]. The enantioselective synthesis of (+)-frontalin via kinetic resolution with 38C2 is shown in Scheme 5.13. Racemic aldol **57** (50 mg) was resolved with antibody 38C2 to give aldol product (R)-**57** (22 mg, 44%) in 95 % ee. LiOH-mediated Horner-Wadsworth-Emmons reaction of aldehyde (R)-**57** with diethyl (2-oxopropylphosphonate) afforded enone **58**, which was subjected to Pd-catalyzed hydrogenolysis and spontaneous cyclization to produce (+)-frontalin (**59**).

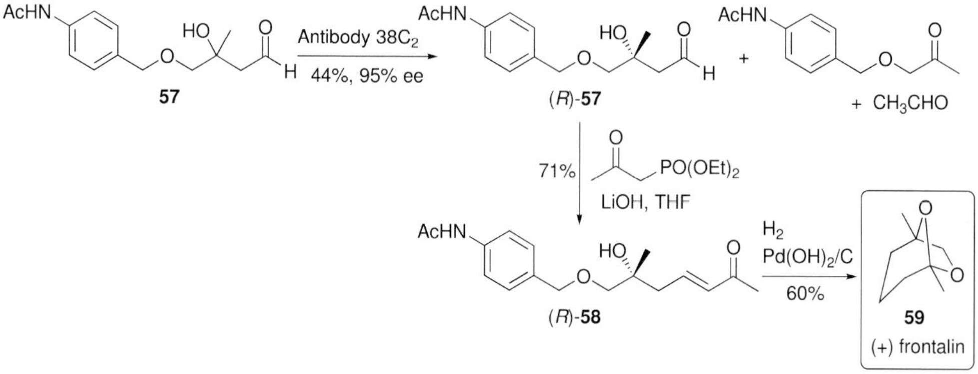

Scheme 5.13

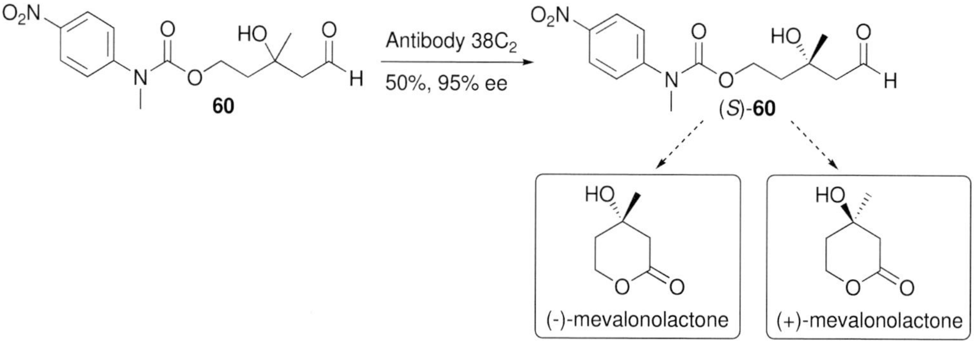

Scheme 5.14

Similarly, the formal synthesis of (–)- and (+)-mevalonolactone (Scheme 5.14) was achieved by the kinetic resolution of the racemic tertiary aldol **60**. The enantiomerically pure (S)-**60** could be converted to (–)-mevalonolactone by oxidation, basic hydrolysis, and lactonization. It has already been shown that (S)-**60** can also be used for the synthesis of (+)-mevalonolactone [60]. Since there are many other natural products containing the tertiary aldol moiety, such as saframycin H [61], the above-described examples represent a general approach to the synthesis of such compounds [57].

5.7
Wieland-Miescher Ketone via 38C2-Catalyzed Robinson Annulation

The remarkable substrate scope of antibody 38C2 allows it to catalyze a broad variety of enantioselective aldol-type reactions, including the Robinson annulation (Scheme 5.15) [62]. Zhong et al. reported that antibody 38C2 catalyzes not only the intramolecular cyclization of triketone **61**, but also the following dehydration of the resultant aldol product (**62**) with >95% ee. Aldol **62**, known as Wieland-Miescher ketone, represents a useful precursor in the total synthesis of various steroid derivatives.

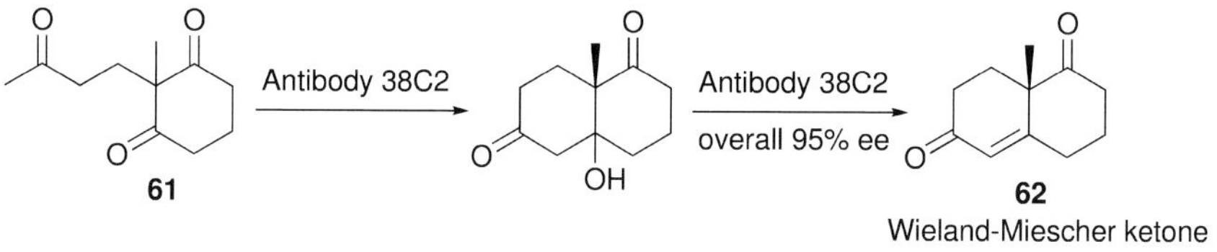

Scheme 5.15

5.8
Formation of Steroid A and B Rings via Cationic Cyclization

In efforts to mimic the natural enzyme oxidosqualene cyclase, several catalytic antibodies were elicited against charged transition state analogs. For example, Hasserodt et al. employed the bicyclic bridge-methylated decahydroquinoline N-oxide **63** as a hapten to elicit antibody HA5-19A4, which catalyzed the cationic cyclization of the dienol sulfonate **64** to produce the closely related decalin systems, **65–67**, with an average enantiomeric excess of 53 % (Scheme 5.16), which represent rings A and B of the steroid nucleus [63]. An extension of this work from the same group presented

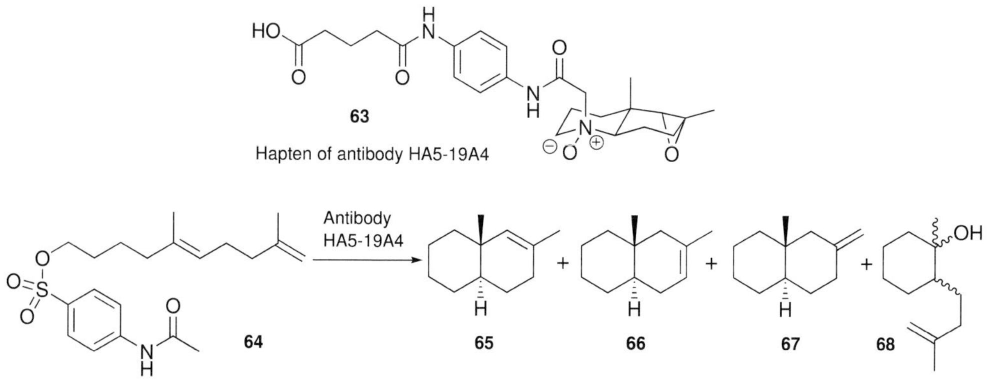

Scheme 5.16

cationic cyclization that focused on substrates analogous to those seen in triterpene biosynthesis. Three antibodies, 15D6, 20C7, and 25A10, which have been elicited against a 4-aza-steroid aminoxide hapten, **69**, initiated the cationic cyclization of an oxidosqualene derivative and catalyzed the formation of compound **71** from polyene **70** at neutral pH (Scheme 5.17). The latter represents ring A of the lanosterol nucleus [64].

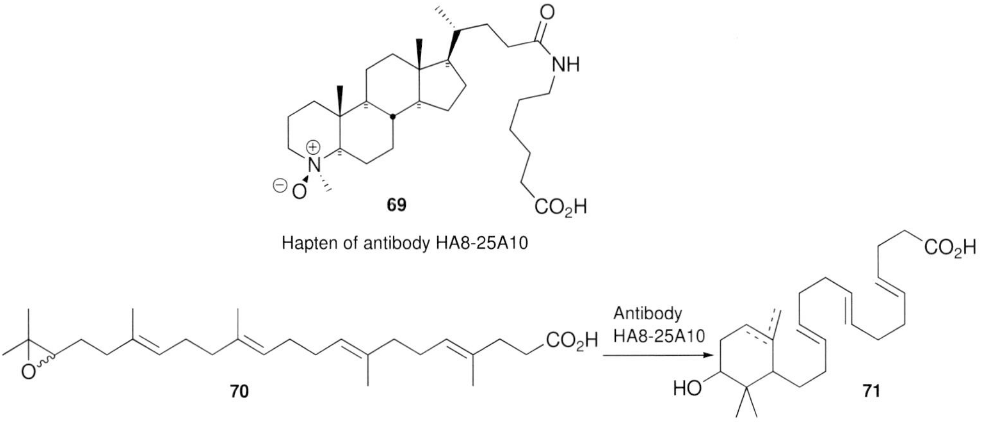

Scheme 5.17

5.9
Synthesis of Naproxen via Antibody-Catalyzed Ester Hydrolysis

Considerable efforts have been devoted to improving the preparation of the enantiomerically pure S-(+)-naproxen, **72**, a widely prescribed non-steroidal *anti*-inflammatory drug (NSAID), which has been made industrially through diastereomeric crystallization. Antibodies 5A9, 6C7, and 15G12, which were elicited against the phosphonate diester **73** in a reactive immunization approach, hydrolyze S-(+)-naproxen *p*-methylsulfonylphenyl ester **75** with rate enhancement as high as 6.6×10^5 and a kinetic resolution of *rac*–**75**, leading to S-(+)-**72** in 90% *ee* for 35% conversion (Scheme 5.18) [65]. Improved catalysts were obtained by employment of phosphonate monoester **74**, leading to a library of catalysts, 6G6, 12C8, and 12D9, with excellent turnover numbers, k_{cat}/k_{uncat} as high as 1.9×10^6 and a useful kinetic resolution of *rac*-**75**, generating S-(+)-**72** in >98% *ee* with up to 50% conversion [66].

5.10
Conclusions

The synthesis of natural products remains the ultimate testing ground for new concepts in organic chemistry. One of the main goals of the field of antibody catalysis

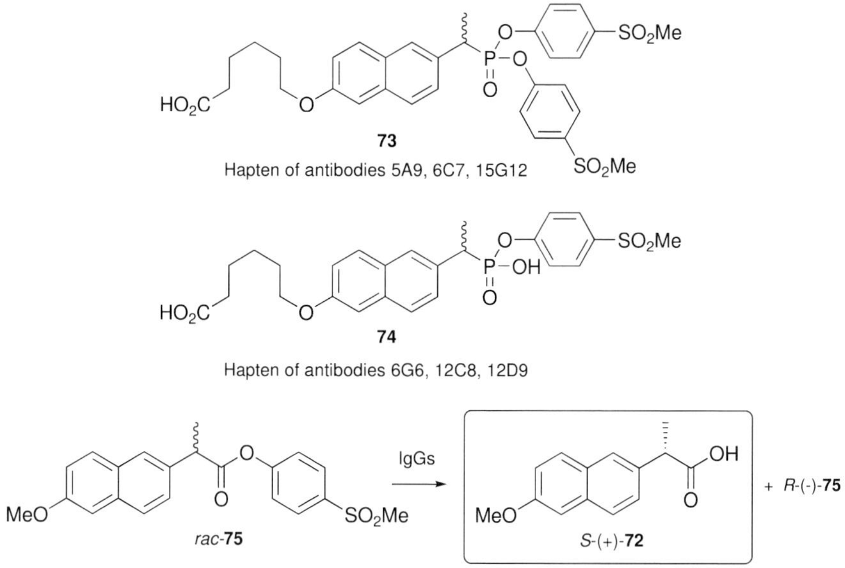

Scheme 5.18

has been to learn how to design catalysts to improve the overall yield of existing synthetic routes, thus allowing practical construction of more totally synthetic drugs and other important natural products. The examples of natural products synthesis covered by this chapter, including (−)-α-multistriatin, epothilones, brevicomins, 1-deoxy-L-xylose, (+)-frontalin, the formal synthesis of (−)- and (+)-mevanolactone, partial synthesis of the steroid skeleton, and naproxen, testify to the important role catalytic antibodies may play in asymmetric synthesis. The remarkable ability of these biocatalysts to control the rate, stereo-, regio-, chemo-, and enantioselectivity of many reactions, including highly disfavored chemical processes and sometimes even those with high substrates promiscuity, will undoubtedly place these agents in a central position on the map of total synthesis of pharmaceuticals, fine chemicals, and complex natural products.

To reach a wider use of catalytic antibodies in synthesis, an obvious need is the reduction of costs. This may be achieved through less expensive production and/or facile means of catalyst recovery. Much work has been done to reduce the cost of antibody production, including over-expression in bacteria, plants, seeds, and algae [67]. Production in seeds is particularly attractive because it may afford a highly concentrated source of stable protein that is easily stored in large quantities. Thus, the seed becomes a dry, well-protected storage device for the catalyst. A number of methods have been reported for immobilization and recovery of antibodies to allow operation in continuous-flow reactors. These methods include attachment to insoluble solid supports or soluble polymer, or entrapment in sol-gel matrices.

References

1 Nicolaou, K. C., Sorensen, E. J., *Classics in Total Synthesis*, Willey-VCH, Weinheim **1996**
2 (a) Wong, C.-H., Koeller, K. M., *Nature*, 409 (**2001**), p. 232. (b) Wong, C.-H., Whitesides, G. M., *Enzymes in synthetic organic chemistry*, Pergamon, Oxford **1994**. (c) Drauz, K., Waldmann, H., *Enzyme catalysis in organic synthesis* VCH, Weinheim **1995**
3 Schmid, A., Dordick, J. S., Hauer, B., Kiener, A., Wubbolts, M., Witholt B., *Nature* 409 (**2001**), p. 258
4 Arnold, F. H., *Nature* 409 (**2001**), p. 253
5 Arnold, F, H., Volkov, A. A., *Curr. Opin. Chem. Biol.* 3 (**1999**), p. 54
6 Reetz, M.T., Jaeger, K. E., *Chem. Eur. J.* 6 (**2000**), p. 407
7 . (a) Jestin, J.-L., Kristensen, P., Winter, G. A., *Angew. Chem. Int. Edn Engl.* 38 (**1999**) p. 1124
8 Janda, K. D, Lo, L. C., Lo, C. H., Sim, M. M., Wang, R., Wong, C. H., Lerner, R. A., *Science* 275 (**1997**), p. 945–948
9 Cane, D. E., Walsh, C. H., Khosla, C., *Science* 282 (**1998**), p. 63
10 Shimizu, S., Ogawa, J., Kataoka, M., Kobayashi, M., *Adv. Biochem. Eng. Biotechnnol.* 58 (**1997**), p. 45
11 Tramontano, A., Janda, K. D., Lerner, R.A., *Science* 234 (**1986**), p. 1566
12 Pollack, S. J., Jacobs, J. W., Schultz, P. G., *Science* 234 (**1986**), p. 1570
13 Schultz, P. G., Lerner, R. A., *Science* 269 (**1995**), p. 1835–42. (b) Hilvert, D. In *Topics in Stereochemistry*, Denmark, S. E. ed. 22 (**1995**), p. 83–135. Wiley, New York. (c) Hilvert, D., *Annu. Rev. Biochem.* 69 (**2000**), p. 751–793. (d) Reymond, J.-L., *Top. Curr. chem.* 200 (**1999**), p. 59. (e) Thomas, N. R., *Nat. Prod. Rep.* 13 (**1996**), p. 479
14 Schultz, P. G., Lerner, R. A., *Acc. Chem. Res.* 26 (**1993**), 391–95
15 McDunn, J. E., Dickerson, T. J., Janda, K. D., "Antibody catalysis of disfavored chemical reactions", Chapter xx in this book
16 Keinan, E., Lerner, R. A., *Israel J. Chem.* 36 (**1996**), p. 113–119
17 Sinha, S. C., Keinan, E., Reymond, J.-L., *J. Am. Chem. Soc.* 115 (**1993**), p. 4893–4894
18 Hsieh, L. C., Yonkovich, S., Kochersperger, L., Schultz, P. G., *Science* 260 (**1993**), p. 337–339
19 Sinha, S. C., Keinan, E., Reymond, J.-L., *Proc. Natl. Acad. Sci. USA* 90 (**1993**), p. 11910–11913
20 a) Reymond, J.-L., Reber, J.-L., Lerner, R. A., *Angew. Chem. Int. Ed. Engl.* 33 (**1994**), p. 475. b) Shevlin, C. G., Hilton, S., Janda, K. D., *Bioorg. Med. Chem. Lett.* 4 (**1994**), p. 297
21 Shabat, D., Itzhaky, H., Reymond, J.-L., Keinan, E., *Nature*, 374 (**1995**), p. 143
22 Lanier, G. N., Silverstein, R. M., Peacock, J. W., in Anderson, J. F., Kaya, H. K. Eds., *Perspectives of Forest Entomology*, Academic Press, New York **1976**, chapter 12

23 LANIER, G. N., GORE, W. E., PEARCE, G. T., PEACOCK, J. W., SILVERSTEIN, R. M., *J. Chem. Ecol.* 3 (**1977**), p. 1

24 PEARCE, G. T., GORE, W. E., SILVERSTEIN, R. M., PEACOCK, J. W., CUTHBERT, P. A., LANIER, G. N., SIMEONE, J. B., *J. Chem. Ecol.* 1 (**1975**), p. 115

25 a) GORE, W. E., PEARCE, G. T., SILVERSTEIN, R. M., *J. Org. Chem.* 40 (**1975**), p. 1705. b) PEARCE, G. T., GORE, W. E., SILVERSTEIN, R. M., *J. Org. Chem.* 41 (**1976**), p. 2797. c) ELLIOT, W. J., FRIED, J., *J. Org. Chem.* 41 (**1976**), p. 2475. d) MORI, K., *Tetrahedron* 32 (**1976**), p. 1979. e) CERNIGLIARO, G. J., KOCIENSKI, P. J., *J. Org. Chem.* 42 (**1977**), p. 3622. f) SUM. P.-E., WEILER, L., *Can. J. Chem.* 56 (**1978**), p. 2700. g) BARTLETT, P. A., MYERSON, J., *J. Org. Chem.* 44 (**1979**), p. 1625. h) FITZSIMMONS, B. J., PLAUMANN, D. E., FRASER-REID, B., *Tetrahedron Lett.* (**1979**), p. 3925. i) MARINO, J. P., ABE, H., *J. Org. Chem.* 46 (**1981**), p. 5379. j) SUM. P.-E., WEILER, L., *Can. J. Chem.* 60 (**1982**), p. 327. k) PLAUMANN, D. E., FITZSIMMONS, B. J., RITCHIE, B. M., FRASER-REID, B., *J. Org. Chem.* 47 (**1982**), p. 941. l) WALBA, D. M., WAND, M. D., *Tetrahedron Lett.* 23 (**1982**), p. 4995. m) LAGRANGE, A., OLESKER, A., SOARES COSTA, S., LUKACS, G., THANG, T. T., *Carbohydrate, Res.* 110 (**1982**), p. 159. n) HELBIG, W., *Liebigs Ann. Chem.* (**1984**), p. 1165. o) LARCHEVEQUE, M., HENROT, S., *Tetrahedron* 43 (**1987**), p. 2303. p) MORI, K., SEU, Y.-B., *Tetrahedron* 44 (**1988**), p. 1035

26 ELLIOTT, W. J., HROMNAK, G., FRIED, J., LANIER, G. N., *J. Chem. Ecol.* 5 (**1979**), p. 279

27 (a) SINHA, S. C., KEINAN, E., *J. Am. Chem. Soc.* 117 (**1995**), p. 3653. (b) SINHA, S. C., KEINAN, E., *Israel Journal of Chemisrty* 36 (**1996**), p. 185

28 a) REYMOND, J.-L., LERNER, R. A., *J. Am. Chem. Soc.* 114 (**1992**), p. 2257. b) REYMOND, J.-L., JAHANGIRI, G. K., STOUDT, C., LERNER, R. A., *J. Am. Chem. Soc.* 115 (**1993**), p. 3909

29 REYMOND, J.-L., JANDA, K. D., LERNER, R. A., *Angew. Chem. Int. Ed. Engl.* 30 (**1991**), p. 1711

30 SINHA, S. C., KEINAN, E., REYMOND, J.-L., *Proc. Natl. Acad. Sci. USA*, 90 (**1993**), p. 11910

31 SHABAT, D., ITZHAKY, H., REYMOND, J.-L., KEINAN, E., *Nature* 374 (**1995**), 143

32 a) KÖHLER G., MILSTEIN, C., *Nature* 256 (**1975**), p. 495. b) ENGUALL, E., *Methods Enzymol.* 70 (**1980**), p. 419

33 SINHA, S.C., BARBAS III, C. F., LERNER, R. A., *Proc. Natl. Acad. Sci.* 95 (**1998**), p. 14603

34 WAGNER, J., LERNER, R. A., BARBAS, C. F., III, *Science* 270 (**1995**), p. 1797

35 SHULMAN, H., MAKAROV, C., OGAWA, A. K., ROMESBERG, F. KEINAN. E., *J. Am. Chem. Soc.* 122 (**2000**), p. 10743

36 WIRSCHING, P., ASHLEY, J. A., LO, C.-H. L., JANDA, K. D., LERNER, R. A., *Science* 270 (**1995**), p. 1775–1782

37 BARBAS, C. F, III, HEINE, A., ZHONG, G. F., HOFFMANN, T., GRAMATIKOVA, S., BJORNESTEDT, R., LIST, B., ANDERSON, J., STURA, E. A., WILSON, E. A., LERNER, R. A., *Science* 278 (**1997**), p. 2085

38 HOFFMANN, T., ZHONG, G. F., LIST, B., SHABAT, D., LERNER R.A., BARBAS III, C. F., *J. Am. Chem. Soc.* 120 (**1998**), p. 2768

39 TANAKA, F., BARBAS, C. F. III, "Reactive innunization: a unique approach to aldolase antibodies", Chapter xxx in this book

40 ZHONG, G., LERNER, R. A., BARBAS, C. F., III, *Angew. Chem. Int. Ed. Engl.* 38 (**1999**), p. 3738

41 KARLSTROM, A., ZHONG, G., RADER, C., LARSEN, N. A., HEINE, A., FULLER, R., LIST, B., TANAKA, F., WILSON, I. A., BARBAS, C. F., III, LERNER, R. A., *Proc. Natl. Acad. Sci. USA* 97 (**2000**), p. 3878–3883

42 WONG, C. -H., MACHAJEWSKI, T. D., *Angew. Chem., Int. Ed. Engl.* 39 (**2000**), p. 1352–1374, and references cited therein

43 ZHONG, G., HOFFMANN, T., LERNER, R. A., DANISHEFSKY, S., BARBAS, C. F., 119 (**1997**) *J. Am. Chem. Soc.* p. 8131–8132

44 LIST, B., SHABAT, D., BARBAS, C. F., LERNER, R. A., *Chem. Eur. J.* 4 (**1998**), p. 881–885

45 Nicolaou, K. C., Roschanger, F., Vourloumis, D., *Angew. Chem. Int. Edn. Engl.* 37 (**1998**), p. 2014–2045

46 Zhong, G., List, B., Shabat, D., List, B., Anderson, J., Lerner, R. A., Barbas III, C. F., *Angew. Chem. Int. Ed. Engl.* 37 (**1998**), p. 2481

47 Sinha, S. C., Jian, Sun., Miller, G., Barbas III, C. F. Lerner, R. A., *Org. Lett.* 1 (**1999**), p. 1623

48 Sinha, S. C., Jian, Sun., Miller, G., Warmann, M., Lerner, R. A., *Chem. Eur.* 7 (**2001**), p. 1692

49 Sinha, S. C., Dutta, S., Jian, Sun., *Tetrahedron lett.* 41 (**2000**), p. 8243

50 Silverstein, R. M., Brownlee, R. G., Bellas, T. E., Wood, D. L., Browne, L. E., *Science* 159 (**1968**), p. 889

51 Borden, J. H., in *Comprehensive Insect Physiology Biochemistry and Pharmacology*, Vol. 9, Kerkut, G. A., Gilber, L. I. Eds., Pergamon Press, New York **1985**, p. 257

52 List, B., Shabat, D., Barbas III, C. F., Lerner, R. A., *Chemistry Eur. J.* 4 (**1998**), p. 881

53 Taniguchi, T., Ohnishi, H., Ogasawara, K., *Chem. Commun.* (**1996**), p. 1477

54 David, S., Estramareix, B., Fischer, J.-C., Therisod, M. J., Chem. Soc. *Perkin Trans 1* (**1982**), p. 2131

55 Hill, R. E., Sayer, B. G., Spenser, I. D., *J. Am. Chem. Soc.* 111 (**1989**), p. 1916

56 Shabat, D., List, B., Lerner, R.A., Barbas III, C. F., *Tatrahedron Lett.* 40 (**1999**), p. 1437

57 List, B., Shabat, D., Zhong, G., Turner, J. M., Li, A., Bui, T., Anderson, J., Lerner, R. A., Barbas III, C. F., *J. Am. Chem. Soc.* 121 (**1999**), p. 7283

58 Mori, K., Nishimura, Y., *Eur. J. Org. Chem.* 233 (**1998**), p. 3

59 Perrin, T. E., Rasmussen, L. E. L., Gunawardena, R., Rasmussen, R. A., *J. Chem. Ecol.* 22 (**1996**), p. 207

60 Krohn, K., Meyer, A., *Liebigs Ann. Chem.* (**1994**), p. 167

61 Fukuyama, T., Yang, L., Ajeck, K. L., Sachleben, R. A., *J. Am. Chem. Soc.* 112 (**1990**), p. 3712

62 Zhong, G., Hoffmann, T., Lerner, R. A., Danishefsky, S., Barbas III, C. F., *J. Am. Chem. Soc.* 119 (**1997**), p. 8131

63 Hasserodt, J., Janda, K. D., Lerner, R. A., *J. Am. Chem. Soc.* 119 (**1997**), p. 5993

64 Hasserodt, J, Janda, K. D., Lerner, R. A., *J. Am. Chem. Soc.* 122 (**2000**), p. 40

65 Lo, C-H. L., Wenthworth, P. Jr., Jung, K. W., Yoon, J., Ashley J. A., Janda, K. D., *J. Am. Chem. Soc.* 119 (**1997**), p. 10251

66 Datta, A., Wenthworth, P. Jr., Shaw, J. P., Simeonov, A., Janda, K. D., *J. Am. Chem. Soc.* 121 (**1999**), p. 10461

67 Hood E. E., Woodard S. L., Horn, M. E., *Curr. Opin. Biotech.* 13 (**2002**), p. 630–5

6
Structure and Function of Catalytic Antibodies
Nicholas A. Larsen, Ian A. Wilson

6.1
Introduction

An explosion of antibody structures has emerged in the last 15 years [1, 2]. A subset of these antibodies is catalytic, and the latest version of the Protein Data Bank (PDB) (03/2003) contains 56 such structures representing 27 different antibodies (Table 6.1). Catalytic antibodies are elicited purposely by immunization with a transition state analog. In principle, antibodies that bind such an analog may also stabilize the true transition state of a chemical reaction relative to the ground state, thereby effecting catalysis [3–7]. Remarkably, the immunoglobulin fold possesses the inherent plasticity to harness the usual combination of electrostatic, van der Waals, hydrophobic, hydrogen bond, and cation-π interactions to generate tailor-made catalysts, which, in rare instances, even rival their natural enzyme counterparts. This general plasticity has been exploited to create model systems that explore a number of common and controversial themes in classical enzymology. Thus, substrate strain, approximation, control of the microenvironment, general acid-base chemistry, induced fit, and covalent catalysis have all been systematically examined in the context of the antibody binding pocket.

The main objective of structural analysis is to develop hypotheses about key interactions in the binding pocket for binding substrate, transition state (TS), and product, and to propose plausible mechanisms for catalysis. For some antibodies, these hypotheses have been examined rigorously by biochemistry and mutagenesis (Table 6.1). Mutagenesis studies allow further quantification of the key interactions and provide an avenue for exploring the routes through the energy landscape that confer catalytic potential to the germline antibody precursor. Structural studies of non-catalytic germline precursors further illustrate how the antibodies evolve their catalytic potential by affinity maturation. Such a thorough understanding of these interactions is necessary if improvements are to be made to the antibody catalyst by either mutation or redesign of the immunizing hapten. Importantly, comparison of antibody structures has revealed common structural motifs that are elicited by different haptens during the immune response. Thus, convergent and divergent evolution has been observed directly within the immune system.

Catalytic Antibodies. Edited by Ehud Keinan
Copyright © 2005 WILEY-VCH Verlag GmbH & Co. KGaA, Weinheim
ISBN: 3-527-30688-9

6 Structure and Function of Catalytic Antibodies

Tab. 6.1: Summary of catalytic antibody structures.

Antibody	Reaction	k_{cat} (min^{-1})	K_M (mM)	$k_{cat}\,k_{uncat}^{-1}$	$K_M K_{DTSA}^{-1}$	Mutant analysis	Res. (Å)	PDB ID	Refs.
D2.3	Ester hydrol. pH = 8.3	3.6	0.28	1.3×10^5	1.1×10^5	Yes	1.85–2.10	x9 1YEC	[28]
D2.4	Ester hydrol. pH = 8.3	1.0	0.30	3.6×10^4	3.3×10^4	–	3.10	x1 1YED	[28]
D2.5	Ester hydrol. pH = 8.3	0.07	0.34	2.5×10^3	1.3×10^3	–	2.20	x1 1YEE	[28]
CNJ206	Ester hydrol. pH = 8.0	0.4	0.08	1.6×10^3	0.7×10^2	–	3.20	x1 1KNO	[26]
29G11	Ester hydrol. pH = 9.5	60	0.64	2.2×10^3	2.4×10^4	–	2.30	x1 1A0Q	[23]
17E8	Ester hydrol. pH = 9.5	220	0.46	8.3×10^3	9.0×10^2	Yes	2.50	x1 1EAP	[23, 24]
43C9	Ester hydrol. pH = 9.3	1500	0.05	2.7×10^4	6.8×10^4	Yes	2.20–2.30	x2 43CA	[27]
48G7	Ester hydrol. pH = 8.2	5.5	0.39	1.6×10^4	8.7×10^4	Yes	1.95–2.70	x4 1GAF	[130]
6D9	Ester hydrol. pH = 8.0	0.12	0.05	9.0×10^2	9.0×10^2	–	1.80	x2 1HYX	[20, 131]
7C8	Ester hydrol. pH = 8.0	0.11	0.004	7.1×10^2	1.2×10	–	2.20	x1 1CT8	[131]
15A10	Ester hydrol. pH = 8.0	2.2	0.22	2.3×10^4	2.2×10^4	Yes	2.35	x1 1NJ9	[132, 133]
13G5	Diels-Alder	0.001	3, 10$^{*)}$	7 M$^{**)}$	–	–	1.95	x1 1A3L	[91]
1E9	Diels-Alder	13	2, 30$^{*)}$	10^3 M$^{**)}$	–	–	1.90	x1 1C1E	[94]
39-A11	Diels-Alder	40	1, 7$^{*)}$	0.4 M$^{**)}$	–	Yes	2.10–2.40	x2 1A4K	[92, 93]
9D9	Retro-Diels-Alder	0.24	0.17	4.0×10^2	4.1×10^2	–	x1	[88, 134]
10F11	Retro-Diels-Alder	1.5	0.26	2.5×10^3	2.5×10^3	–	2.00–2.40	x4 1LO0	[88, 134]
33F12	Retro-aldol condensation	1.4	0.27	1.7×10^7	–	Yes	2.15	x1 1AXT	[115–117, 135]
28B4	Sulfide oxidation	490	.04, .25$^{***)}$	9 M$^{**)}$	–	Yes	1.90–2.80	x4 1KEL	[66, 68, 69]
21D8	Decarboxylation	19.2	0.4	2.0×10^4	7.5×10^3	Yes	1.61–2.10	x2 1C5C	[14, 62, 65, 136, 137]
5C8	Cyclization	1.7	0.36	ND	–	–	2.00–2.50	x3 25C8	[138]
19A4	Cationic cyclization	0.02	0.32	ND	–	–	2.70	x1 1CF8	[16, 109]
4C6	Cationic cyclization	0.02	0.23	ND	–	–	1.30–2.45	x2 1ND0	[17, 107]
1D4	Syn-elimination	0.003	0.21	ND	–	–	1.80–1.85	x2 1JGV	[100, 101]
7G12	Metal chelatase	0.41	0.15	–	–	Yes	1.60–2.60	x5 3FCT	[70, 111, 113]
AZ-28	Oxy-Cope	0.02	0.074	5.3×10^3	4.4×10^3	Yes	2.00–2.80	x4 1AXS	[80, 81]
1F7	Chorismate mutase	0.07	0.051	2.5×10^2	8.5×10^1	–	3.00	x1 1FIG	[74, 77]
4B2	Allylic isomerization	0.0016	1.3	1.5×10^3	4.3×10^2	–	1.87	x1 1F3D	[71, 72]

$^{*)}$ for diene and dienophile, respectively.
$^{**)}$ equivalent to the so-called effective molarity
$^{***)}$ for substrate and cofactor (periodate), respectively.

Most of the catalytic antibody structures found in the are complexes with their respective transition state analog, and provide detailed views of the interactions that contribute to catalysis. In nearly all cases, a high degree of electrostatic and shape complementarity is observed for the corresponding transition state analog (TSA), and, in rarer instances, putative reactive residues that initiate covalent catalysis have been proposed. Consequently, this chapter summarizes antibody-catalyzed reactions in the general context of these three broad headings and concludes with a section on future challenges in the abzyme field for structural biologists and chemists alike.

6.2
Electrostatic Complementarity

Many antibody-TSA complexes exhibit a high degree of electrostatic complementarity. Recurring binding motifs for anions (e.g., phosphate, arsenate, sulfate) have been identified, indicative of structural convergence from a variety of germline sequences [8–14]. In contrast, general motifs for binding cations (e.g., primary, secondary, tertiary amines) have not yet been identified. Eliciting an Asp/Glu carboxylate using hapten baiting strategies has met with mixed success. Frequently, cations are buried preferentially in hydrophobic pockets but surrounded by aromatic side chains, which are believed to form stabilizing cation-π interactions [15–17].

6.2.1
Anionic Binding Motifs and Ester Hydrolysis

Nearly half of the catalytic antibody structures found in the PDB correspond to various hydrolytic antibodies in complex with substrate analogs and products (Table 6.1). Thus, hydrolytic antibodies are by far the most thoroughly studied catalytic antibodies from both a structural and a biochemical vantage, and have already been reviewed extensively in the literature [18, 19]. These hydrolytic antibodies have been raised generally against either phosphonate or phosphonamidate transition state analogs (Scheme 6.1). The phosphonate moiety is tetrahedral and carries a negative charge, thereby mimicking the transition state for solvent-mediated hydrolysis that involves nucleophilic attack by a water molecule. With the exception of 6D9 [20] 7C8, and 15A10, these hydrolytic antibodies catalyze the hydrolysis of benzyl or activated 4-nitrophenol or 4-nitrobenzyl esters (Scheme 6.1), which are particularly convenient model substrates for both synthetic manipulation and sensitivity during hybridoma screening.

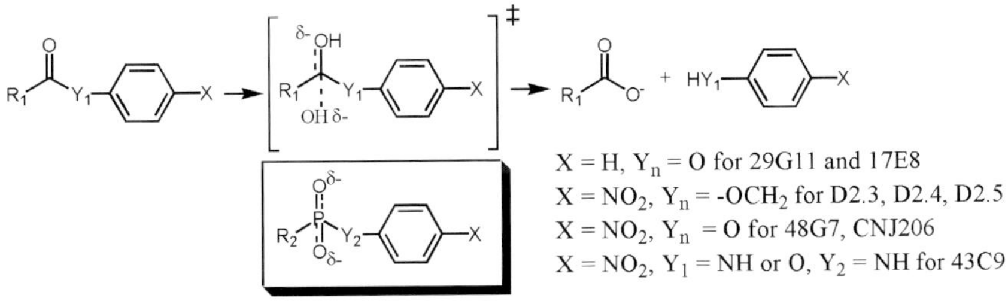

$X = H$, $Y_n = O$ for 29G11 and 17E8
$X = NO_2$, $Y_n = -OCH_2$ for D2.3, D2.4, D2.5
$X = NO_2$, $Y_n = O$ for 48G7, CNJ206
$X = NO_2$, $Y_1 = NH$ or O, $Y_2 = NH$ for 43C9

Scheme 6.1 Hydrolytic Antibodies 48G7, 17E8, 29G11, CNJ206, 43C9, D2.3, D2.4, D2.5

6.2.2
Structural Motifs

The first structural evidence that antibodies recognize antigens by electrostatic complementarity originates from the pioneering study of mAb McPC603, which binds phosphocholine [8, 9]. Further data have indicated that electronically-related phenylarsonate compounds elicit a conserved anionic binding motif from the immune repertoire [10–13]. Likewise, crystal structures of catalytic antibodies 48G7, 17E8, 29G11, CNJ206, and 43C9 reveal a recurring motif that coordinates the phosphonate moiety of the hapten (Fig. 6.1a, b) [18,19]. In general, the constellation of side-

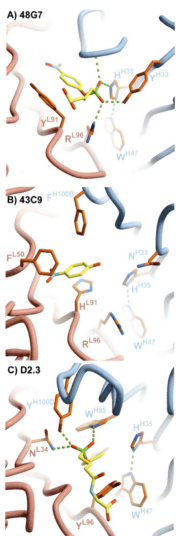

Fig. 6.1 Top view of hydrolytic antibodies. The light chain is represented in pink and the heavy chain in blue; side chains are brown and the ligand is yellow. Hydrogen bonds are represented as dashed green lines. **A** The hapten complex of 48G7 illustrates the prototypic anionic binding motif composed, for the most part, of heavy chain residues. **B** 43C9 likely has a similar motif as 48G7. The only available structure is a complex with the product p-nitrophenol. HisL91 is proposed to form an acylated intermediate at the Nδ1 atom. However, the interpretation of the kinetic and structural data is not fully convincing. **C** D2.3 has a unique anionic binding motif located on the light chain side of the antibody. The Tyr is proposed to form the key interaction responsible for oxyanion stabilization. Catalytic antibodies D2.3-5, 6D9, and 7C8 differ from this general theme. In D2.3, phosphonate binding arises from hydrogen bonds to AsnL34, TrpH95, and TyrH100D (Fig. 6.1c) [28]. Thus, the anionic binding site is located on the opposite side of the binding pocket and is composed of different residues. Finally, the phosphonate binding motifs in antibody 6D9 and 7C8 are completely different from those seen in any of the other antibodies and consist of a single, solvent-exposed, hydrogen bond donated from a single HisL27D or TyrH95, respectively [20, 29, 30]. Not surprisingly, these antibodies exhibit lower catalytic activity than any other structurally-determined hydrolytic antibodies (Table 6.1). Nevertheless, 6D9 has been recently evolved to have 20-fold higher activity with the addition of a second hydrogen bond to the phosphonate [31].

chains derived from L96 (CDR3), H33, H35 (CDR1), and the backbone amide of H96 (CDR3) constitute an anionic binding site, arising from structural convergence within the antibody binding pocket (Fig. 6.1a, b) [18, 19]. For example, antibody 48G7 utilizes ArgL96, TyrH33, HisH35, and the backbone amide of H96 for phosphonate binding (Fig. 6.1a) [21, 22]. Antibodies 17E8 and 29G11 are related to each other (92% sequence identity [23]) and use the same residues for phosphonate binding [24, 25] as 48G7, except that a hydrogen bond from the LysH93-Nz amine substitutes for the observed TyrH33 hydroxyl in 48G7. Antibody CNJ206 also similarly uses hydrogen bonds from HisH35 and the backbone amides of H96 and H97 to bind the phosphonate [26]. Although a crystal structure of 43C9 in complex with the transition state analog is not available, it is proposed that HisL91, ArgL96, AsnH33, and HisH35 likewise contribute to the anionic binding site (Fig. 6.1b) [27].

6.2.3
Mechanistic Considerations

Clearly, aryl phosphonate/arsonate/sulfonate-like haptens preferentially elicit electrostatic complementarity from the immune response and may explain the commonly observed binding motif observed in 48G7, 17E8, 29G11, CNJ206, 43C9 and other non-catalytic antibodies. For the hydrolytic antibodies, the significance of the anionic binding site is underscored by the general agreement between transition state stabilization relative to the ground state, $K_M K_{DTS}^{-1}$, and the observed rate acceleration, $k_{cat} k_{uncat}^{-1}$ [32] (Table 6.1). This agreement implies that hapten-binding interactions observed in the crystal structures correspond to the interactions found in the transition state, and that binding energy is converted into catalysis with near 100% efficiency. Thus, the simplest possible mechanism for ester hydrolysis in this family of antibodies is oxyanion stabilization within the anionic binding site. Oxyanion stabilization has a differential impact on catalysis for various natural esterase and protease enzymes. For example, biochemical studies in a bacterial cocaine esterase have demonstrated that the oxyanion hole is critical, contributing at least 5 kcal mol^{-1} to catalysis [33]. Oxyanion stabilization is estimated to contribute 2–3 kcal mol^{-1} to catalysis in papain [34] and cutinase [35], 3–4 kcal mol^{-1} in prolyloligopeptidase [36], 2–5 kcal mol^{-1} in subtilisin [37–39], and 5–7 kcal mol^{-1} in acetylcholinesterase [40] and Zn^{2+} peptidase [41].

These values are within the range ascribed to the oxyanion contribution in the hydrolytic antibodies. For example, a contribution of 5–7 kcal mol^{-1} from oxyanion stabilization was estimated for the D2.3 hydrolytic antibody [42]. Here, the oxyanion was formally assigned to TyrH100D from pH rate profile and the inactivity of a Tyr → Phe mutant (Fig. 6.1c) [42]. The alternative candidate, AsnL34, was excluded as the Asn → Gly mutant retained full activity [42]. The activity of this antibody is significantly enhanced by spiking the assay buffer with potent α-nucleophiles, such as peroxide [42]. Thus, the catalytic potential of D2.3 is far from realized, and could be increased by a carefully positioned general acid/base that could more effectively activate water for nucleophilic attack. Even with the more potent nucleophile, the activity of this antibody is apparently limited by product inhibition [42]. and conformational isomerization [43].

A largely overlooked ramification of the known anionic binding sites is the asymmetry of the coordinating residues (Fig. 6.1a–c). This asymmetry means that attack of the substrate carbonyl carbon and oxyanion stabilization has enantiofacial selectivity [44], since the phosphonate oxygens are non-equivalent (prochiral). Thus, the lack of ^{18}O incorporation in the substrates of hydrolytic antibodies should not be regarded as evidence for an acyl enzyme intermediate [45], but rather that the oxygen atoms are not racemized in the stereoselective binding pocket after hydroxide attack. This enantiofacial selectivity has been noted in other naturally-occurring hydrolytic enzymes [46, 47].

6.2.4
Mechanistic Pitfalls

There is an obvious temptation to ascribe complex mechanisms to hydrolytic antibodies to elevate them to the same status as their natural enzyme counterparts. In the case of 17E8, for example, the proximity of a serine and histidine to the phosphonate in the crystal structure led to the proposal that the mechanism involved a catalytic dyad, analogous to the familiar catalytic triad [24]. This structure-based hypothesis was further bolstered by hydroxylamine partitioning data and pH rate profiles that suggested an acyl enzyme intermediate [48]. Initially, this interpretation was criticized because the refined serine rotamer was not within hydrogen-bonding distance to the histidine in the putative dyad [18, 49], but alternative rotamers of a serine have been noted in other serine hydrolases [47, 50, 51]. Eventually, the acyl enzyme mechanism was effectively discounted by a Ser → Ala mutant, which led to twofold improvement in k_{cat} over the wild-type [52]. Likewise, the hydrolytic mechanism for 43C9 was proposed to involve an unprecedented nucleophilic attack by the Nδx atom of HisL91 on the substrate to form an acylated histidine intermediate [53]. In serine hydrolases, the nucleophilicity of histidine Nε2 atom is enhanced by a buried aspartate that H-bonds to Nδ1 stabilizing the developing positive charge on the imidazole. In the 43C9 pocket, HisL91 Nε2 H-bonds to TyrL36, but the nearest charged residue is ArgL96 (Fig. 6.1b). Hence, the available structural data [27] do not yet provide sufficient evidence for such a mechanism. Therefore, the interpretation of the kinetic, mutagenesis, and structural data [45, 53–56] should be carefully reexamined or, at the very least, regarded with skepticism until a direct observation of an acylated histidine intermediate is demonstrated in a crystal structure.

6.2.5
Structural and Functional Considerations for Hapten Design

The initial use of a phosphonate analog as the transition state for solvent-mediated hydrolysis (Scheme 6.1) provided the initial breakthrough in the catalytic antibody field [6, 7]. However, the persistent and ubiquitous use of such analogs to elicit hydrolytic antibodies has now become somewhat of a red herring. Phosphonate compounds are not transition state analog inhibitors for serine hydrolases [57]. Thus, it would be impossible to elicit a serine hydrolase-like antibody using a phosphonate analog. Nevertheless, the elicited antibodies are inevitably compared to serine hydrolases [24], or to hydrolase mutants where the entire catalytic triad has been mutated to alanine [58]. Fluorophosphonate compounds (nerve gas reagents, for example) are potent mechanistic inhibitors of serine esterases because of covalent modification of serine with phosphate and concomitant expulsion of fluoride – a good leaving group. Phosphonate and phosphonamidate compounds do not form the same modification in serine esterases, since O^- is a poor leaving-group. In contrast, phosphonate compounds are potent transition state analog inhibitors for Zn^{2+} proteases [59], which utilize a completely different mechanism from that of serine hydrolases and probably do not involve covalent acylation by the enzyme [60]. Rather, Zn^{2+} is believed to facilitate water-mediated attack of the ester/amide carbonyl carbon with arginine stabilization of the oxyanion [41, 60]. Unfortunately, no design feature in a phosphonate analog would be likely to elicit Zn^{2+} binding in an antibody combining site. Thus, the likelihood of eliciting hydrolytic antibodies with either serine hydrolase or Zn^{2+} peptidase-like features would seem unlikely without some innovative new hapten design.

6.2.6
Anionic Binding Motifs and Control of the Microenvironment in Decarboxylation

Catalytic antibody 21D8 catalyzes a classic decarboxylation reaction (Scheme 6.2) with a charge-delocalized transition state. This reaction exhibits high sensitivity to the surrounding medium, where the first-order rate constant in aqueous solution is enhanced 10^6-fold in aprotic dipolar solvents [61]. Thus, this reaction is a particularly useful model for studying medium effects and control of the microenvironment in a heterogeneous binding pocket [62, 63]. Antibody 21D8 accelerates the decarboxylation reaction by 2×10^4 (Table 6.1) [62, 64]. The hapten design incorporated two sulfonate moieties to mimic the carboxylate anion of the substrate. Not surprisingly, this antibody has an anionic binding site consisting of HisH35 and ArgL96 [14]. An additional anionic binding site was observed for the second sulfonate group on the

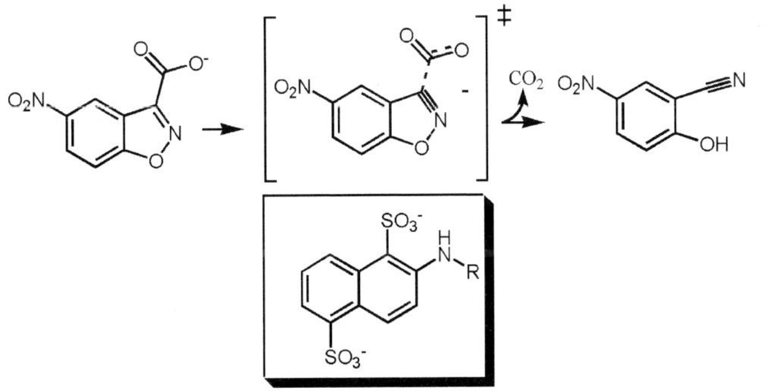

Scheme 6.2 Decarboxylation Antibody 21 D8

opposite side of the binding pocket. However, assigning which of these anionic binding sites was relevant for binding to the substrate carboxylate was hindered because of the unfortunate symmetry of the hapten. Moreover, attempts at co-crystallization with the substrate were unsuccessful [14]. However, mutagenesis studies support the hypothesis that HisH35 and ArgL96 are indeed key substrate-binding residues [65]. It is proposed that the antibody shields the substrate from solvent, but the tetrahedral hydrogen-bonding pattern of the anionic binding site is not fully optimized for binding a planar carboxylate [14, 65]. The pocket is polar with apparently non-optimal hydrogen bonds to the transition state, which may be reminiscent of a dipolar aprotic solvent. Such a model would explain, in part, how the antibody and other enzymes could control the microenvironment to facilitate catalysis.

6.2.7
Anionic Binding Motifs and the Periodate Cofactor in Sulfide Oxygenation

Catalytic antibody 28B4 catalyzes a bimolecular reaction between sodium periodate ($NaIO_4$) and a sulfide to form a sulfoxide (Scheme 6.3, Table 6.1) [66]. Oxygenation reactions in naturally-occurring enzymes normally require flavin, heme, or other cofactors as well as NADPH to regenerate the cofactor [67]. Here, periodate is exploited as an artificial chemical cofactor, readily available at high excess and obviating the need for cofactor recycling. The oxygenation reaction is believed to involve nucleophilic attack of a sulfur lone pair of electrons on a periodate oxygen through a polar transition state (Scheme 6.3). The amine in the hapten mimics the partial positive charge that develops on the sulfur, while the phosphate mimics the tetrahedral periodate. The crystal structure reveals an anionic binding site similar to that seen in the phosphocholine antibody McPC603 [8, 9] that coordinates periodate (Fig. 6.2a) [68]. However, the baiting strategy employed to stabilize the developing positive charge on the sulfur was not as effective as the postulated cation-π interaction with a nearby tyrosine does not have optimal geometry (Fig. 6.2a). Extensive mutagenesis has been performed on this antibody, and the putative germline antibody was also cloned and crystallized in complex with hapten [69]. Here, the *p*-nitrophenyl portion of the hapten adopts a completely different orientation in the binding pockets of the and affinity-matured antibodies [69]. In addition, the free and bound forms of the germline antibody structure show significant conformational differences in CDRH3

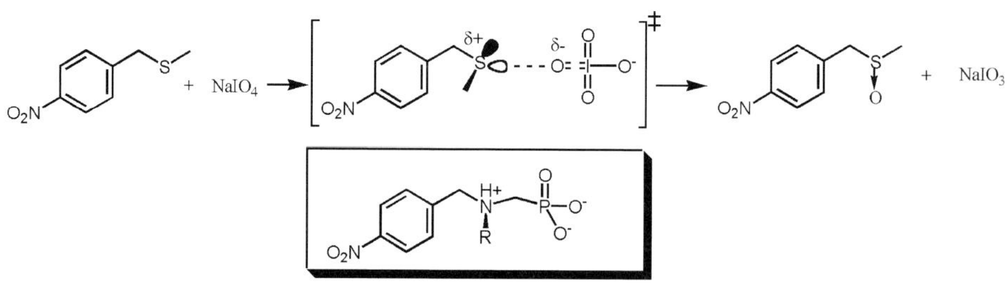

Scheme 6.3 Sulfide oxygenation Antibody 28B4

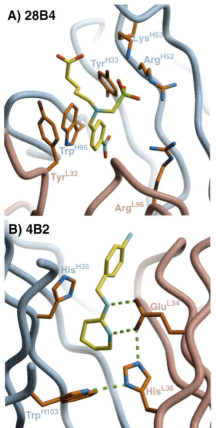

Fig. 6.2 Top and side views of sulfide oxygenation and allylic isomerization antibodies, respectively. **A** Antibody 28B4 has an anion-binding pocket that binds the cofactor periodate in a biomolecular reaction with sulfide to form sulfoxide. This pocket was elicited by a phosphate functionality in the immunizing hapten. **B** Antibody 4B2 catalyzes an isomerization reaction of a β-γ unsaturated ketone. GluL34 acts as general acid/base in the reaction by abstracting the proton from the substrate. Here, GluL34 interacts with the amidinium part of the hapten.

and CDRL1, while comparison of the free and bound mature antibody shows no significant changes on binding [69]. Thus, the germline is postulated to exhibit wider structural plasticity than the mature antibody [69]. The loss of induced fit binding during the affinity maturation process has been noted in at least two other antibodies [22, 70], but does not rule out the importance of induced fit in affinity-matured antibodies [1].

6.2.8
Cation Stabilization and Allylic Isomerization

Several antibodies have been elicited that are believed to have mechanistically important, negatively charged carboxylates in their binding pocket. In principle, such carboxylates may be elicited from a baiting strategy that incorporates positively-charged substituents in the immunizing hapten. Few structures of this type have been determined, however, and for at least one of these antibodies (1D4), the baiting strategy failed. Antibody 4B2 catalyzes an allylic rearrangement of a β–γ unsaturated ketone

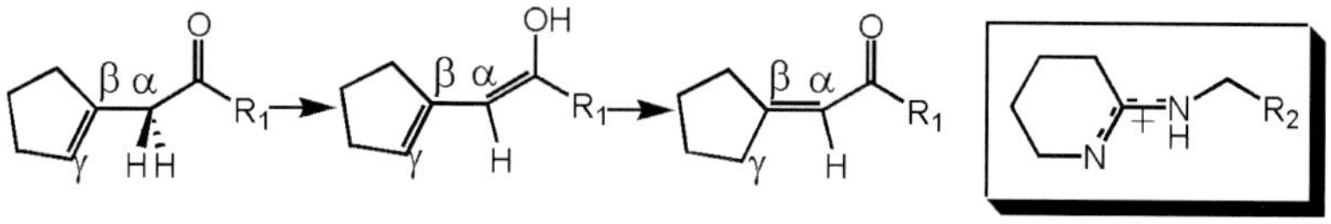

Scheme 6.4 Allylic isomerization Antibody 4B2

(Scheme 6.4, Table 6.1). Formally, the reaction would be a [1,3]sigmatropic migration of hydrogen. An amidinium group was incorporated in the hapten design to elicit a carboxylate general acid/base in the binding pocket (Scheme 6.4) [71]. The crystal structure of 4B2 shows that this baiting strategy was successful, as the two NH groups of the amidinium form H-bonds to each oxygen of the GluL34 carboxylate (Fig. 6.2b) [72]. The pH rate profile is bell shaped with optimal activity at pH 4.5, the approximate pK_a of glutamic acid. An enol intermediate was proposed, based on lack of any antibody-catalyzed ^{2}H exchange between α and γ carbons in a deuterated substrate, implying that the reaction is not concerted, but involves exchange with solvent [71]. This is a reasonable hypothesis, although it is unlikely that any antibody residue would be positioned to stabilize this transient enol intermediate [72].

6.3
Shape Complementarity and Approximation

Shape complementarity simply refers to the similarity between the contours of the binding pocket molecular surface and that of its corresponding ligand. Certainly, it is possible to conclude from most of the catalytic antibody structural studies that exquisite shape complementarity is sufficient to explain the stereoselectivity for the majority of these catalysts. Thus, when the catalysts are challenged with chiral (or prochiral) substrates, chiral products are obtained, typically in high enantiomeric excess (*ee*). In addition, every reaction requires, to some extent, approximation to bring reactive centers together to either break or make new bonds. The antibody molecule is ideally tailored for this role, as it may be programmed to bind flexible substrates in a particular conformation that would otherwise be disfavored in solution, and can bring two substrates together in bimolecular reactions to increase the effective molarity of the substrates relative to one another. In principle, approximation lowers the entropy barrier for activation and can effect catalysis. Thus, electrostatic and overall shape complementarity of the binding pocket for the transition state is a key feature for any catalyst and is one of the most important features cited for many catalytic antibody structures. However, approximation alone probably does not fully explain catalysis in all of these systems. Solvent accessibility, control of the microenvironment dielectric, and hydrogen-bonding potential are also important parameters – some reactions are accelerated simply by changing the solvent system. Indeed, not surprisingly, nonspecific interactions with bovine serum albumin (BSA) can accelerate some reactions [73]. In addition, key catalytic residues form specific interactions essential for the reaction to proceed. Catalytic residues have been elicited with mixed

6.3 Shape Complementarity and Approximation

success by TSA baiting strategies, reactive immunization, and sometimes by pure chance. Notwithstanding, approximation will be a recurrent theme throughout the remaining sections.

6.3.1
Unimolecular Rearrangements

6.3.1.1
Antibody 1F7

Antibody 1F7 catalyzes the [3,3]sigmatropic Claisen rearrangement of chorismate to prephenate (Scheme 6.5, Table 6.1) [74]. This reaction is normally catalyzed by chorismate mutase in the biosynthesis of aromatic amino acids. The enzymatic reaction proceeds through an ordered chair-like transition state ($\Delta S^{\ddagger} \approx -13$ cal mol^{-1} K^{-1}) with a rate acceleration of 3×10^6 or four orders of magnitude better than the antibody [75]. Remarkably, even with this modest activity, over-expression of 1F7 complements growth of yeast auxotrophs [76]. The original structure was determined at 3.0 Å resolution from data merged from two crystals collected at room temperature [77]. More recent (unpublished) data at 2.4 Å show some differences in the detailed interactions between the TSA and Fab [77]. However, the general placement of the hapten on the heavy chain side of the antibody is maintained and illustrates a clear and interesting difference in the orientation of the TSA in the antibody pocket as opposed to the enzyme active site [78]. For the antibody, the hapten-binding orientation is determined by the placement of the linker used for covalent attachment to the immunogen, while the enzyme has no such restriction (Scheme 6.5). The enzyme also utilizes more H-bond interactions than the antibody, which are important for stabilizing the dipolar transition state [77]. Thus, the reaction is probably being catalyzed by preferentially binding the substrate in the requisite reactive conformation, while poorly optimized hydrogen bond interactions may account for the inefficiency of the antibody catalyst. This theme recurs in which the substrate in an antibody is inserted upside down compared to that in the natural enzyme [47, 79].

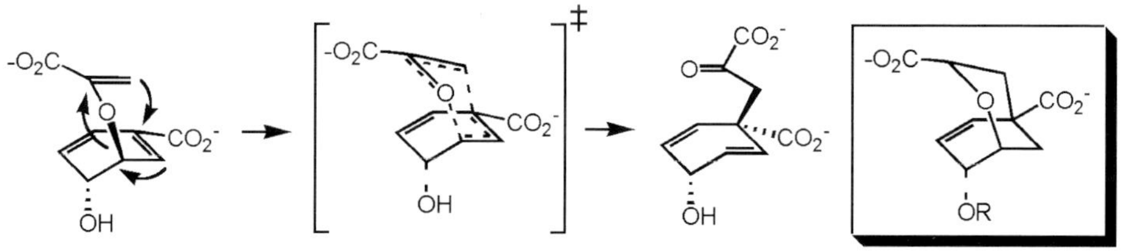

Scheme 6.5 Chorismate mutase Antibody 1F7

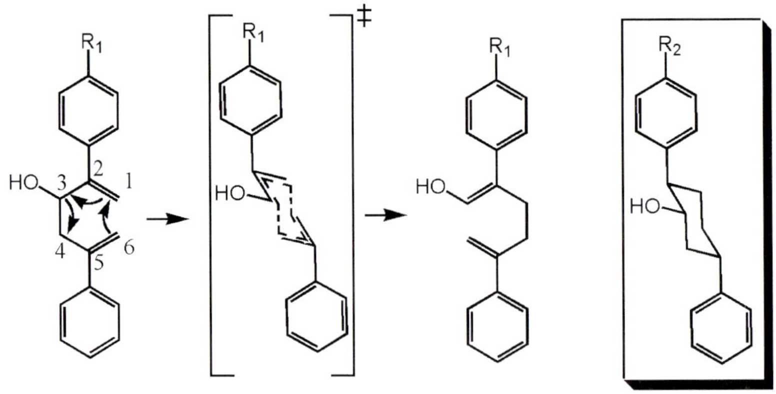

Scheme 6.6 Oxy-Cope Antibody AZ-28

6.3.1.2
Antibody AZ-28

AZ-28 catalyzes a related [3,3]sigmatropic oxy-Cope rearrangement (Scheme 6.6, Table 6.1) [80]. The chair-like transition state is highly ordered, requiring overlap of the rearranging 4π+ 2σ orbitals of the diene ($\Delta S^{\ddagger} \approx -15$ cal mol^{-1} K^{-1}). Antibodies raised against the appropriate, conformationally-restricted TSA would be expected to catalyze the reaction by acting as an "entropy trap." In addition, appropriate alignment of the 2- and 5-phenyl rings are believed to contribute stereoelectronically to the reaction through hyperconjugation with the 3-hydroxyl (Scheme 6.6) [81]. This hyperconjugation may accelerate the oxy-Cope rearrangement by an anionic substituent effect. Indeed, the crystal structure shows that packing interactions with the 2- and 5-phenyl rings constrain the cyclohexyl group in the required chair-like transition state (Fig. 6.3). However, the cyclohexyl ring of the TSA is rotated ca. 80° out of plane of the 2- and 5-phenyl rings, so that the phenyl π-orbitals do not align

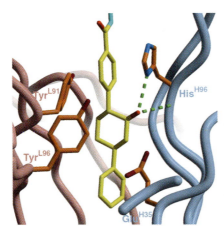

Fig. 6.3 Side view of oxy-Cope antibody AZ-28. The cyclohexyl ring of the TSA is rotated out of plane with the 2- and 5-phenyl rings because of packing interactions and hydrogen bonds to the 3-hydroxyl. The consequent lack of hyperconjugation in the rearranging π-orbitals may explain the decreased rate acceleration in the affinity mature antibody, as opposed to the germline Fab.

appreciably with the 3-hydroxyl (Fig. 6.3) [81]. The orientation of the cyclohexyl group is determined by packing interactions, as well as by hydrogen bonds to the 3-hydroxyl (Fig. 6.3). Surprisingly, biochemical analysis revealed that six mutations occurred during affinity maturation, one of which dramatically decreases the catalytic efficiency [81]. Thus, the evolution of an antibody binding pocket with increased affinity to the TSA does not always correlate with improved catalysis. The germline-encoded antibody structure was determined and revealed an induced-fit mode of binding, as opposed to a lock-and-key type binding in the mature antibody. This apparent conformational flexibility and corresponding decrease in affinity is interpreted as evidence for possible torsional flexibility in the 2-phenyl substituent [82]. This flexibility, in turn, is postulated to facilitate π-overlap and explain the improved catalysis. Studies to characterize the dynamics of the antibody catalyzed reaction would provide further evidence to support this hypothesis.

6.3.2
Bimolecular Rearrangements and the Diels-Alder Reaction

The D-A reaction is one of the most important carbon-carbon bond-forming reactions that is available for organic syntheses of complex natural products and therapeutic agents [83, 84]. Surprisingly, the D-A reaction has not been exploited widely in nature as only a handful of natural D-A enzymes are known [85]. The reaction is a pericyclic [4π+ 2π] cycloaddition of a diene and dienophile. An optimal D-A reaction occurs between an electron-rich diene (HOMO - highest occupied molecular orbital) and an electron-deficient dienophile (LUMO - lowest unoccupied molecular orbital). These abzyme-catalyzed bimolecular reactions are typically much slower and have much higher K_Ms than esterolytic antibodies (Table 6.1). The transition state for the D-A reaction is highly ordered for which $\Delta S^{\ddagger}$ is typically −30 to −40 entropy units [86]. Consequently, the recurring theme for these D-A antibodies is shape complementarity; in general, the antibodies with higher shape complementarity to the transition state are better catalysts. Thus, these antibodies may be considered as model systems for studying the effects of approximation in catalysis.

6.3.2.1
Retro-Diels-Alder Reaction

Antibody 10F11 catalyzes a unimolecular retro-D-A reaction that expels the dieneophile nitroxyl and diene anthracene (Scheme 6.7, Table 6.1). The dihedral angles between the phenyl rings in the substrate (110°) and product (180°) differ by 70° [87]. A TSA was designed with an intermediate dihedral angle of 140° [87]. Four crystal structures were determined including substrate analog (SA), transition state analog (TSA), and product analog (PA) [88]. The kinetics of the reaction suggest that binding energy is converted to catalysis with high efficiency (Table 6.1), with a clear hierarchy in the shape complementarity index between the antibody and ligands, ranked as TS > SA > PA [88]. This ranking supports the hypothesis that the antibody exerts strain on the substrate to preferentially stabilize the transition state. Antibody 9D9 catalyzes

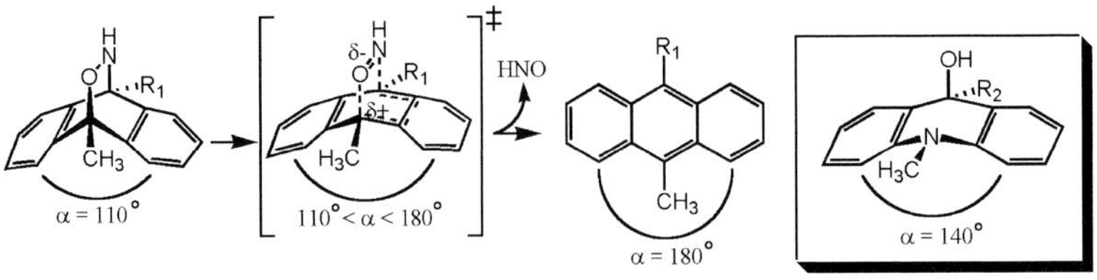

Scheme 6.7 Retro Diels-Alder Antibody 10F11

the same reaction with slower kinetics, which could result from poorer overall shape complementarity [88]. The CDRH3 loop also undergoes a slight induced-fit conformational change from binding the substrate to the transition state [88]. Finally, a water-mediated hydrogen bond was observed between the antibody backbone and the oxygen that is released from the substrate in the nitroxyl dieneophile product [88]. A second hydrogen bond is formed directly between the antibody backbone and the same oxygen of the substrate. These hydrogen bonds are postulated to act as Lewis acids that further stabilize the LUMO energy of the dieneophile and contribute to the regioselectivity of the retro-D-A reaction [88].

6.3.2.2
Disfavored *exo* Diels-Alder Reaction

For antibody 13G5, a novel hapten design was examined (Scheme 6.8, Table 6.1). Rather than using a conformationally-restricted TSA, an unrestrained ferrocene derivative was used to test whether the immune system would select a conformer that mimics one of the Diels-Alder transition states (Scheme 6.8) [89]. In principle, a panel of catalysts would be generated, each of which would selectively catalyze the formation of one of the four possible *ortho*-diastereomers. In fact, of all the antibodies examined, only one enantiomer was formed, indicative of possible structural bias in the immune response [90]. The catalyzed reaction proceeds with high regio-, diastereo-, and enantioselectivity to form the *ortho-exo* (S,S) product with ee > 95% after the background *exo* D-A reaction is subtracted [90]. The background *endo:exo* product distribution is 85:15 in PBS and 66:34 in toluene, so the *exo*-transition state

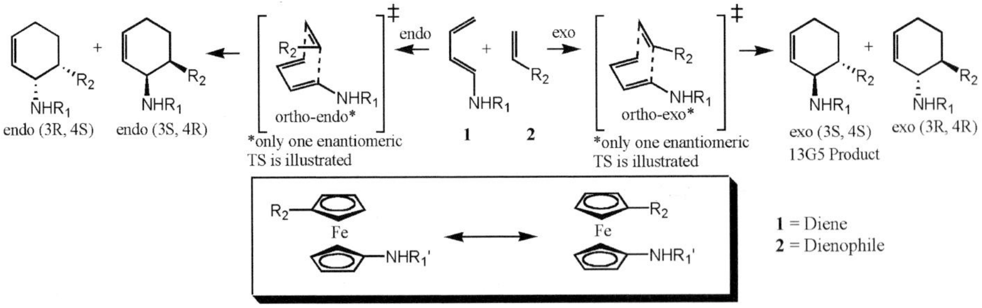

Scheme 6.8 Diels-Alder Antibody 13G5

is slightly disfavored by ca. 1 kcal mol^{-1}. The antibody rate acceleration is modest with a $k_{cat(exo)} k_{uncat(exo)}^{-1}$ of 6.9 M and a $k_{cat(exo)} k_{uncat(endo)}^{-1}$ of only 1.7 M [89, 91]. Thus, the background reaction to form unwanted *ortho-endo* product will dominate the final product distribution of the antibody-catalyzed reaction unless sufficiently high concentrations of antibody catalyst are used [90].

The crystal structure of 13G5 shows the cyclopentadienyl ferrocene deeply buried in the antibody combining site [91]. Three H-bonds are observed between the TSA and TyrL36, AspH50, and AsnL91. TyrL36 was initially hypothesized to act as a Lewis acid, activating the dienophile for nucleophilic attack, while AsnL91 and AspH50 form H-bonds to the carboxylate side chain that substitutes for the diene substrate [91]. However, the limited rate accelerations indicate that catalytic residues in the binding pocket have a modest affect, if any, on the kinetics. Rather, the observed diastereo- and enantioselectivity of the antibody could be explained simply by exclusion of solvent (e.g., note toluene product distribution) and specific van der Waals interactions and hydrogen bonds that anchor the substrates in a particular orientation that favors *ortho-exo* (S,S) product formation due to approximation [86]. The performance of this catalyst may be explained by the relatively poor shape complementarity between the antibody and the true transition state of the reaction, which differs substantially from the hapten as the distance between the cyclopentadienyl rings in the ferrocene structure is ca. 3.3 Å. But what is truly remarkable is that a specific antibody could be generated against such a flexible hapten.

6.3.2.3
Endo Diels-Alder Reactions

Antibody 39-A11 was elicited from immunization with a constrained bicyclic hapten (Scheme 6.9), and selectively accelerates formation of the kinetically-favored *endo* product [92]. The acceleration over the background or effective molarity is modest (Table 6.1). The X-ray structure of the hapten complex indicates that the antibody would bind the diene and dienophile in a reactive conformation, as preprogrammed in the hapten design strategy. The structure also suggests that the catalyst is enantioselective, although an *ee* has not been determined [93]. AsnH35 forms a hydrogen bond with the dienophile, which could make it more electron deficient and, thereby,

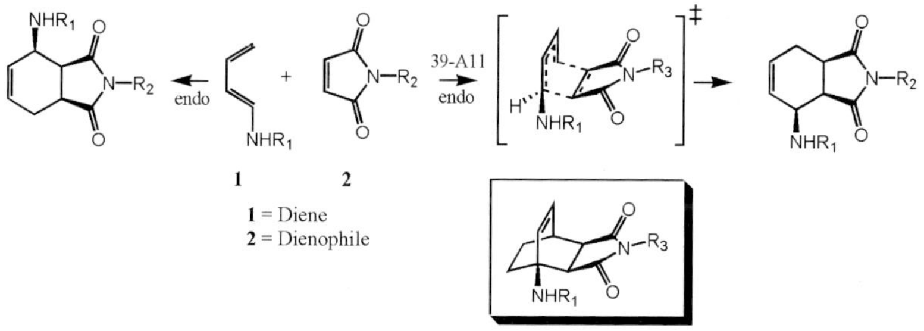

Scheme 6.9 Diels-Alder Antibody 39-A11

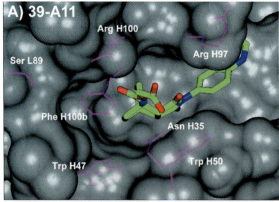

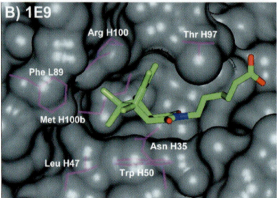

Fig. 6.4 Molecular surface representation of Diels-Alder antibodies. Antibody 39-A11 **(A)** and 1E9 **(B)** are derived from the same germline precursor, yet exhibit a large difference in activity. This difference is attributed to the poor shape complementarity of the antibody for the transition state in 39-A11 compared to 1E9. Mutations were designed in 39-A11 to improve the shape complementarity, and several improved catalysts were identified.

more reactive [93]. Interestingly, any evident shape complementarity is lacking between antibody and the *endo* transition state; the binding pocket can accommodate either the *endo* or *exo* transition states (Fig. 6.4a). Mutations were designed to improve the packing interactions around the kinetically-favored *endo* transition state, and three of six mutants led to a 5- to 10-fold improvement in k_{cat} [93]. This successful structure-based engineering is a significant and a rare accomplishment in this field. Also, only two somatic mutations were required to generate this catalytic antibody from the germline precursor. The most relevant of these two mutations is SerL91Val (CDR3), located in the binding pocket. The K_{DTSA} for the germline antibody is 380 nM compared to 130 nM in the mature 39-A11. As expected, the catalytic proficiency of the germline antibody is also slightly decreased [93].

In order to circumvent substrate or product inhibition, more sophisticated hapten design strategies incorporate extraneous chemical groups as a negative design, while still promoting high affinity interactions to the transition state (Scheme 6.10) [94]. 1E9 was originally developed to evaluate the role of proximity effects in catalysis, and is the most efficient D-A catalytic antibody yet developed (Table 6.1) with an effective molarity (EM) of 10^3 M, where the theoretical limit is 10^8 M [95]. Two structural features can explain the mechanism of the observed antibody catalysis. First is

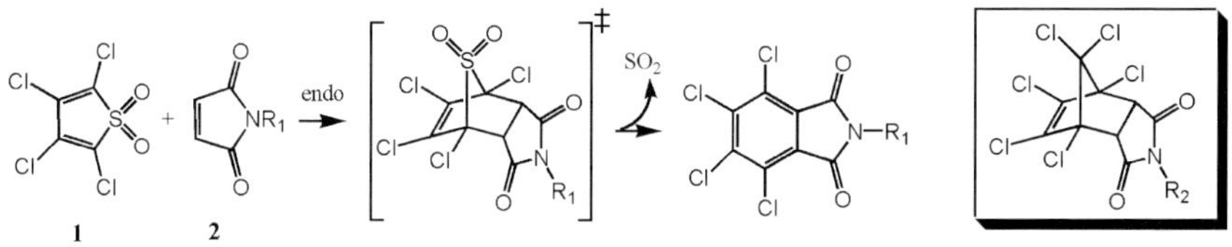

Scheme 6.10 Diels-Alder Antibody 1E9

1 = Diene
2 = Dienophile

the conserved H-bond formed between AsnH35 and the succinimide carbonyl oxygen of the hapten, which represents the carbonyl oxygen of the dienophile substrate (Scheme 6.10). Since an H-bond to the carbonyl oxygen should cause the dieneophile to be electron deficient, it is more susceptible to the attack by the diene. Second, the binding pocket features almost perfect van der Waals complementarity to the hapten, providing a clear structural demonstration of approximation in bimolecular catalysis (Fig. 6.4b) [96]. Interestingly, 1E9, *anti*-progesterone antibody DB3, and the D-A antibody 39A11 are derived from the same germline gene [93, 97–99]. Structures of these antibodies thus represent possible snapshots in the evolution of substrate binding and catalytic efficiency [94]. Subtle differences in the evolution of their binding pockets through somatic mutations have resulted in antibodies that specifically bind progesterone or catalyze a D-A reaction. The exceptional efficiency of 1E9 is hypothesized to arise from a rare somatic mutation that significantly deepens the active site, creating exquisite complementarity to the transition state [94].

6.4
Shape Complementarity and Control of the Reaction Coordinate

In many reactions, several products may be formed, but one product predominates because it is favored kinetically. The energy of activation is much lower for the favored product, which reduces the likelihood of forming the disfavored product. If the activation barrier for the reverse reaction is sufficiently large, then the reaction is essentially irreversible, or under kinetic control. Similarly, for some reactions, many possible products may result (e.g., the D-A reaction or cationic cyclization), and complex mixtures result. The antibody catalyst can steer a reaction through a complex reaction coordinate to alter the final reaction profile, often forming exclusively a single product, and, in some cases, a kinetically-disfavored one.

6.4.1
Syn-Elimination

Antibody 1D4 is a model system for studying the selective catalysis of a disfavored *syn*-elimination reaction to form a cis product (Scheme 6.11, Table 6.1) [100]. The TSA design incorporated a rigid bicyclic ring structure to constrain the two phenyl groups

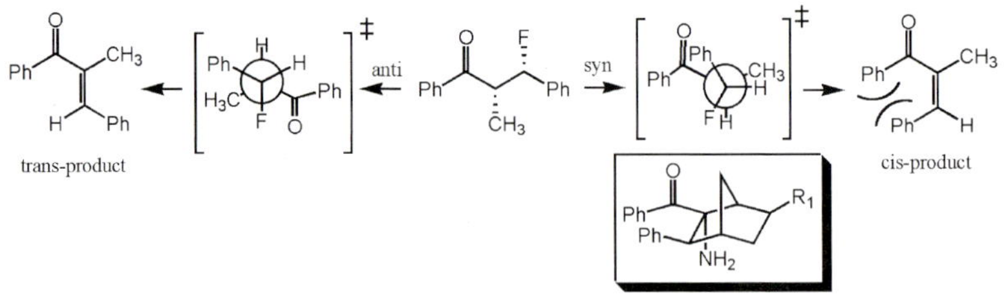

Scheme 6.11 Syn-elimination Antibody 1D4

of the substrate in the eclipsed conformation that characterizes the transition state (Scheme 6.11). In addition, the α-keto proton that is abstracted during elimination was replaced with a positively-charged amine in the hapten to elicit a complementary general base in the antibody combining site. Although the catalyzed reaction proceeds slowly (Table 6.1) [100], the background elimination reaction to the cis-product is undetectable in the absence of antibody; the competing *anti*-elimination reaction ($k = 2.5 \times 10^{-4}$ min^{-1}) dominates the background reaction, such that the product distribution is entirely trans.

Crystal structures of free and bound 1D4 show a conformational change in CDR H2, indicative of induced fit [101]. The pocket exhibits high shape complementarity to the TSA, which accounts for the selectivity for the eclipsed transition-state of the reaction (Fig. 6.5) [101]. Hence, the antibody combining site likely utilizes the majority of its binding energy toward overcoming the extreme steric barriers, and, hence, stabilizing the eclipsed transition state. The key mechanistic issue for any β-elimination process is whether the reaction is E1, E2, or E1cB [102]. Unfortunately, a crystal structure alone cannot resolve such a question. A water molecule was modeled into electron density that is near the hapten amine [101] and within

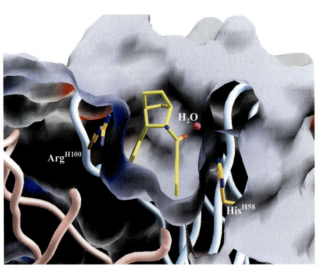

Fig. 6.5 Electrostatic potential mapped to the molecular surface of 1D4. The exquisite shape complementarity for the phenyl rings that characterize the eclipsed transition state of the reaction accounts for the selectivity of the catalyst.

hydrogen-bonding distance to a histidine. Thus, HisH58 is proposed to promote catalysis through interaction with the water [101].

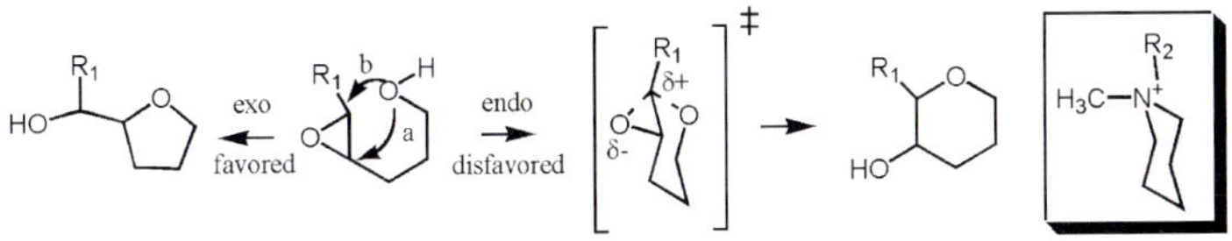

Scheme 6.12 Disfavored cyclization Antibody 5C8

6.4.2
Disfavored Ring Closure

Antibody 5C8 catalyzes the disfavored *endo*-tet cyclization reaction in violation of Baldwin's rules for ring closure (Scheme 6.12) [15]. The N-methylpiperidinium hapten was anticipated to elicit negative charge in the binding pocket that would stabilize the developing positive charge at the oxirane carbon of the substrate. Several changes were observed between the bound and unbound structures, indicative of induced fit. The largest rearrangement was observed in the backbone and side chain orientations of CDR H3 with the largest differences seen for the two tyrosine residues at the tip of the loop, which fold into the binding site, creating a solvent-inaccessible cavity. AspH95 and AspH101 bind the positive charge of the quaternary amine like a pair of tweezers (Fig. 6.6) [15]. In addition, the positive charge appears to be stabilized by a cation-π interaction with TyrL91. AspH95 and HisL89 were hypothesized to effect rudimentary general acid/base catalysis (Fig. 6.6). In the proposed mechanism, AspH95 forms an H-bond to the epoxide-oxygen, favoring the oxirane ring opening, while HisL89 forms an H-bond with the alcohol to promote nucleophilic S_N2-like attack to form the disfavored six-membered ring [15]. This stabilization of the positive charge that develops along the reaction coordinate appears to be an important

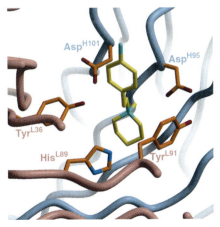

Fig. 6.6 Disfavored ring closure antibody 5C8. AspH95 is proposed to form an H-bond with the epoxide-oxygen, favoring oxirane ring opening, with HL89 forming an H-bond with the alcohol to promote SN2-like attack to form the disfavored ring. The piperidinium moiety was used as bait to elicit the desired negative charge in the binding pocket, which should stabilize the developing positive charge in the actual ring closure reaction.

factor for the rate enhancement and for directing the reaction along the otherwise disfavored pathway [15].

6.4.3
Selective Control of a Reactive Carbocation: Antibodies 4C6 and 19A4

Cationic cyclization proceeds via formation of a reactive carbocation, and may be separated into three steps: initiation, propagation, and termination. A carbocation may be formed either by ionization or by electrophilic addition to an unsaturated olefin, while cyclization is terminated by nucleophilic addition or elimination of a proton. The stereochemistry of such transformations should follow the Stork-Eschenmoser hypothesis [103, 104]. Natural enzymes catalyze polycyclization of squalene through a reactive carbocation intermediate to form a variety of cycloisoprenoids (e.g., cholesterol). For such enzymes, initiation occurs either by protonation of a double bond or by release of pyrophosphate [105]. Structural studies have shown that the enzyme active sites are typically lined with aromatic residues that stabilize the carbocation through cation-π interactions. Catalytic antibodies 4C6 and 19A4 also contain a number of aromatic residues in the binding pocket, which are postulated to play a similar role. Thus, the enzymes and antibodies exhibit similar features.

Catalytic antibody 4C6 catalyzes a model reaction that had been shown under other conditions (98% formic acid) to proceed through a carbocation [106]. This reactivity was elicited with an N-oxide hapten [107] that is structurally related to the 5C8 hapten. Structurally analogous N-oxide TSAs are known inhibitors for natural squalene cyclases, validating this design strategy [108]. The oxide was included to promote stabilization of the negative charge that develops in the sulfonic acid leaving group, while the nitrogen would elicit residues that increase the electrophilicity at C1' (Scheme 6.13, Table 6.1). The hapten also includes a silane moiety that could stabilize the carbocation via hyperconjugation [107]. In addition, conjugation with silane makes the olefin more electron rich, which facilitates attack of C1'. Carbocations are highly reactive species, yet the antibody selectively catalyzes the formation of

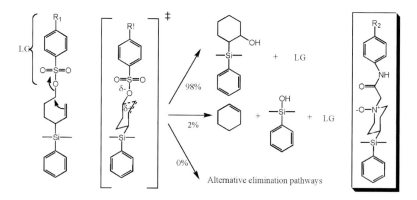

Scheme 6.13 Cationic cyclization Antibody 4C6

only one product in mild conditions where no background reaction is detectable (Scheme 6.13) [107]. The crystal structure shows several aromatic residues in the active site that could stabilize the reactive carbocation and shield it from solution [17]. In the proposed mechanism, TyrL96 and/or TyrH50 initiate carbocation formation at C1' by promoting S_N1-like departure of the acetamidobenzenesulfonic acid leaving group. Electrophilic attack of the C5'-C6' olefin by the developing C1' carbocation closes the ring and shifts the carbocation to C5', where the reaction terminates by nucleophilic addition of water [17]. This mechanism, however, does not explain why a related compound (Si exchanged for C) has a ca.100-fold reduction in k_{cat}, but similar K_M for the reaction, nor why the related substrate that lacks the C5'-C6' double bond exhibits no reaction (kinetics monitor formation of leaving group). Together, these data suggest that anchimeric assistance of the electron-rich double bond is the central feature of the mechanism [107]. Thus, the reaction could be initiated by backside attack of C1' by the electron-rich olefin, followed by termination by nucleophilic addition of water at C5'. This attack is facilitated by approximation due to the shape complementarity of the binding pocket. Additional kinetics and mutagenesis could help resolve the details of how the reaction is initiated.

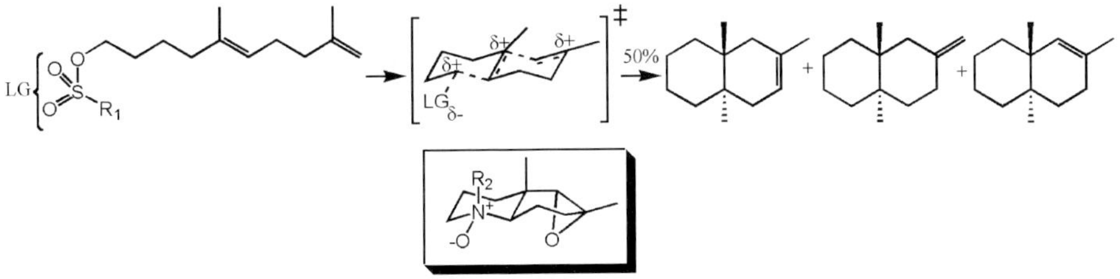

Scheme 6.14 Cationic cyclization Antibody 19A4

Catalytic antibody 19A4 catalyzes a more complicated, tandem-cationic cyclization to form bridge-methylated decalins (Scheme 6.14, Table 6.1) [109]. The hapten design was similar to that for 4C6, with the same N-oxide functionality to elicit residues that would promote initiation of the cyclization cascade by expulsion of the sulfonic acid leaving group. In contrast to 4C6, this reaction is terminated by elimination. The antibody does not possess the exquisite control over the reaction that was observed for 4C6, as three different decalins are formed accounting for only 50% of the overall products [109]. Nevertheless, the catalysis is remarkable, as no bicyclic products are formed in the uncatalyzed reaction. The crystal structure shows a hydrophobic binding pocket lined with several aromatic residues that may stabilize the carbocation [16]. In addition, the binding pocket has high shape complementarity to the productive chair-chair conformation of the substrate that is required for effective cyclization (Scheme 6.14). Here, a hydrogen bond from AsnH35a to the sulfonic acid leaving group is believed to facilitate S_N1-like departure with concerted anchimeric assistance from the C5-C6 π bond, yielding the first ring with the carbonium ion at the tertiary C5 carbon [16]. In the next step, the C9-C10 π bond undergoes electrophilic attack by the C5 carbonium to form the second ring. Although an epoxide was included in the original hapten to elicit a catalytic residue in the antibody, there is no catalytic base to direct the regiochemistry of the final termination step of the reaction [16]. Thus, the reaction results in a mixture of elimination products. Structure-based engineering would afford a unique opportunity to control the product distribution for this particular reaction.

6.5
Shape Complementarity and Substrate Strain

Many catalytic antibodies impose varying degrees of strain on the substrate to achieve reactive conformations that would otherwise be disfavored in solution (e.g., 1D4, 4C6, 19A4, 13G5, 10F11, AZ-28). The enzyme ferrochelatase, however, is a particularly attractive model system for studying substrate strain. Ferrochelatase inserts Fe^{2+} into porphyrin in heme biosynthesis. In this reaction, the enzyme is proposed to distort the porphyrin ring system to expose the pyrrole nitrogen lone pair electrons, facilitating metal ion complexation. Indeed, N-methyl protoporphyrin – a distorted porphyrin analog – is a potent inhibitor of ferrochelatase [110]. Antibodies elicited against this compound catalyze insertion of Cu^{2+} or Zn^{2+} into mesoprotoporphyrin IX, supporting the hypothesis that the hapten is a true transition state analog for the enzyme-catalyzed reaction (Scheme 6.15, Table 6.1) [111]. Thus, this reaction is a model system for studying the effects of substrate strain in catalysis [4]. In the antibody-catalyzed reaction, no saturation kinetics were observed for the metal ion, suggesting that metal binding to the antibody is not a prerequisite of catalysis, in contrast to ferrochelatase, which has a K_M of 32 μM for iron [112]. The crystal structure, however, reveals that an Asp carboxylate is directed toward the center of the protoporphyrin (Fig. 6.7) and is proposed to either guide metal complexation or act as a proton shuttle [113]. The residue is clearly important as Asp → Asn and Asp → His mutations are inactive [113]. In addition, the crystal structure shows that packing interactions between the antibody and substrate induce strain in the Michaelis complex [70]. Although accurate modeling and exact refinement of any distortions in the Michaelis complex at 2.6 Å resolution are not reasonable, the

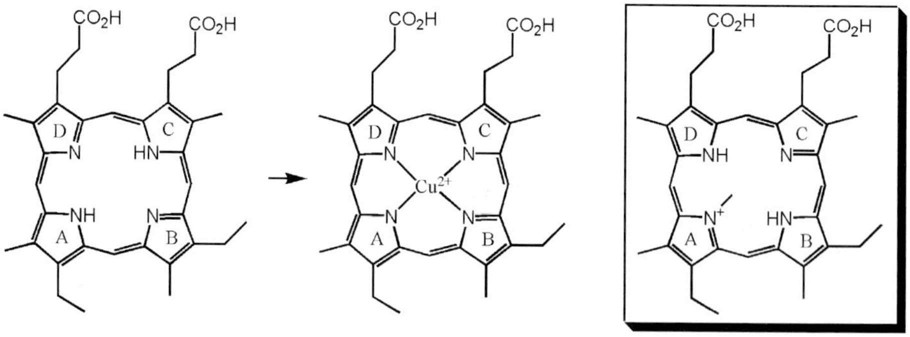

Scheme 6.15 Metal chelation antibody 7G12

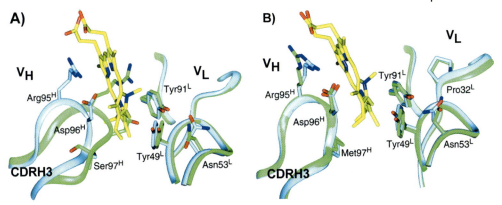

Fig. 6.7 Superposition of free (green) and TSA bound (blue) 7G12. **(A)** The germline 7G12 shows an induced-fit mode of binding, as demonstrated by the conformation change in CDRH3. **(B)** The affinity mature antibody, in contrast, has a lock-and-key mode of binding. The maturation of a preformed binding pocket is associated with higher affinity for the transition state and overall improved catalytic proficiency.

electron density does provide clear qualitative indications of deviation from ideal planarity [70]. The free and hapten-bound structures of the germline and mature antibodies were compared and nicely illustrate the evolution from an induced fit mode of binding to lock-and-key (Fig. 6.7a,b) [70]. The evolution of this lock-and-key binding results in tighter binding to the TSA (due to faster association and slower dissociation rates), and correlates with a ca. 100-fold improvement in $k_{cat}K_M^{-1}$ [70]. Together, these structural and biochemical results demonstrate the importance of substrate strain in antibody-mediated catalysis.

6.6
Reactive Amino Acids and the Possibility of Covalent Catalysis

The recently developed reactive immunization strategy [114, 115] has shown that it is possible to create antibodies that catalyze aldol reactions, one of the most important carbon-carbon bond-forming reactions in chemistry and biology. The mechanism for the natural class I aldolase enzyme involves the key formation of a Schiff base between the substrate ketone (aldol donor) and a reactive lysine, followed by condensation with an aldehyde (aldol acceptor) and subsequent hydrolysis to release product (Scheme 6.16). The binding pocket of aldolase antibody 33F12 is an elongated cleft more than 11 Å deep [116]. The only lysine residue in the active site is located at the bottom of the hydrophobic pocket (Fig. 6.8a). This environment was hypothesized to perturb the pK_a of the lysine, making it a more potent nucleophile [116]. The antibody can then use the reactive ε-amino group to form an enamine with the corresponding substrates, analogous to the natural class I aldolase. The LysH93 → Ala mutant is inactive, which confirms the importance of this residue [117]. An additional residue that may be of mechanistic interest is TyrL36, which forms a water-mediated hydro-

176 | 6 Structure and Function of Catalytic Antibodies

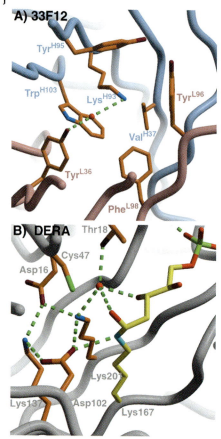

Fig. 6.8 Comparison of aldolase antibody and enzyme active sites. Red spheres represent water molecules. **A** The aldolase antibody contains a lysine buried at the bottom of hydrophobic gorge, which has been postulated to form covalent intermediates with an aldol donor (ketone) followed by condensation with an aldol acceptor (aldehyde). TyrL36 may form a water-mediated hydrogen bond with the reactive lysine ε-amino group, which may be of importance in the reaction mechanism. **B** The bacterial 2-deoxyribose-5-phosphate aldolase (DERA) likewise contains a reactive Lys167, which was observed to form a covalent intermediate with the substrate in the crystal structure. Lys167 is probably rendered nucleophilic by the sophisticated hydrogen-bonding network in the active site, which includes nearby Asp102 and Lys201. A nearby water molecule forms key stabilizing interactions to the substrate, and is believed to play a key role in the catalytic mechanism.

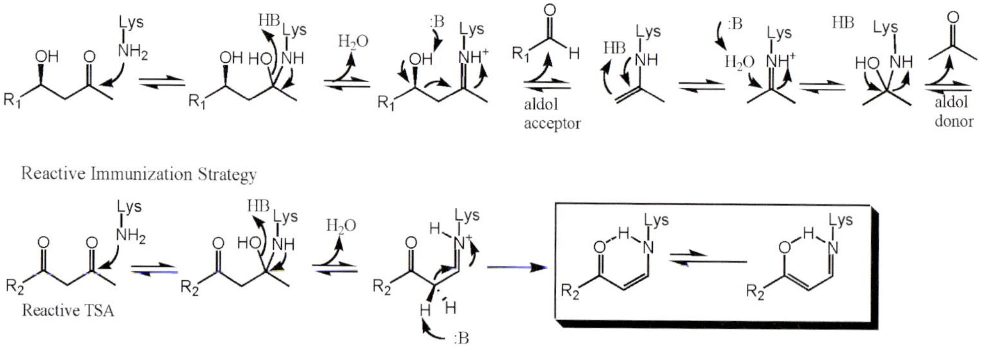

Scheme 6.16 Previously proposed retro-aldol mechanism for aldolase antibody 33F12

gen bond with the ε-amino group of lysine H93 (Fig. 6.8a). Importantly, a water molecule was postulated to play a key role shuttling protons in the mechanism of a class I aldolase [118]. Interestingly, the antibody exhibits exceptionally broad substrate specificity, implying that the binding pocket is unrefined for a particular substrate presumably due to the lack of further affinity maturation due to the proposed covalent nature of the hapten complex [116]. Antibodies 40F12, 84G3, 93F3 and others were generated with a sulfone β-diketone hapten [119]. Notably, antibodies 93F3 and 84G3 show reversed enantioselectivity compared to 33F12 [119]. Sequence analysis and our unpublished structural data [120] reveal a lysine at position L89 on the opposite side of the binding pocket, which could explain the reversed selectivity of this antibody.

The relatively unrefined binding pocket of 33F12 contrasts with that of class I aldolase enzymes, which catalyze the reversible condensation of aldehydes and ketones with high stereospecificity (121,122). The bacterial 2-deoxyribose-5-phosphate aldolase (DERA) (123,124) catalyzes the reversible aldol reaction of acetaldehyde and D-glyceraldehyde-3-phosphate to form 2-deoxyribose-5-phosphate. The ultra high resolution structure of DERA reveals a β-barrel with two lysines in the active site, one of which forms the covalent Schiff base intermediate (Fig. 6.8b) [118]. Compared to the hydrophobic antibody binding pocket, a number of charged/polar residues form a sophisticated hydrogen-bonding network within the active site, which is likely important for tuning the reactivity of the lysine and the substrate specificity [118]. The mechanism by which Lys167 is rendered nucleophilic in the enzyme is unclear – the pK_a was proposed to be depressed by a neighboring Lys201 (Fig. 6.8b) [118], but mutation of Lys201 → Leu clearly does not hinder formation of the Schiff base [118]. Surprisingly, inhibition of natural aldolases by diketones has not been fully characterized.

A lingering challenge is to determine the mode of binding of the aldolase family of antibodies to substrate-related inhibitors, as well as the diketone compound originally used as a hapten to elicit the antibody response (Scheme 6.16). Despite our best efforts, we have obtained no structural evidence demonstrating the formation of a covalent bond in the antibody binding pocket for the aldolase antibody. This contrasts with the relative ease by which we and others have obtained covalent adducts in enzyme systems [118]. Surprisingly, the Lys167Leu DERA mutant catalyzes the retro aldol reaction for its optimal substrate (3.6 min^{-1}) [118] three times faster than the antibody for its optimal substrate (1.4 min^{-1}) [125]. Clearly, catalysis of the retro-aldol condensation may occur in the absence of covalent intermediates. Alternative mechanisms for 33F12 and 38C2 were never seriously considered [116], but could involve non-covalent stabilization of an enolate or aldol donor that subsequently attacks the aldol acceptor (analogous to a class II aldolase). Such a mechanism would be more consistent with our structural observations. In any event, the mechanistic supposition that a lysine could be rendered nucleophilic in the hydrophobic environment of the antibody binding pocket led, in part, to the appreciation that secondary amines, e.g., proline, could catalyze the aldol reaction in organic and, to lesser extent, aqueous solvent [126]. Thus, structural and biochemical characterization of aldolase antibodies has led to the development of important new synthetic methodologies.

6.7
New Challenges

Classically, antibodies are effector molecules that link the recognition elements of the Fab to the destructive elements of the immune system. Recently, this paradigm has been challenged by the discovery that antibodies also possess the intrinsic capability to catalyze the formation of hydrogen peroxide from singlet oxygen and water in the absence of any other metal ions or cofactors [127, 128]. This catalytic capability would potentially link recognition and oxidative killing within the same molecule. This remarkable chemistry is likewise catalyzed by the T-cell receptor (TCR), but not β_2-microglobulin, which shares the immunoglobulin fold [128]. Very little is understood about where this reaction occurs on the antibody or TCR, and docking studies have not resolved this issue [129]. Assaying light chain dimers and camel antibodies (heavy chain only) would provide some indication of where the chemistry occurs. We have grappled with this problem indirectly by pressurizing numerous antibody crystals under a xenon atmosphere [128]. Our data reveal several hydrophobic pockets in the core of the immunoglobulin fold. Since xenon is larger than oxygen (O_2), oxygen could potentially diffuse into the same hydrophobic cavities. However, the mechanistic relevance of these hydrophobic cavities, if any, is not yet resolved.

Unfortunately, a more direct experimental approach is extremely challenging; distinguishing unambiguously between water, singlet oxygen, hydrogen peroxide, ozone, and other reactive oxygen intermediates (which are believed to be short lived) from an electron density map is not feasible, even at ultra high resolution. Moreover, discriminating between radical chemistry derived from the antibody or synchrotron radiation is non-trivial and requires numerous control experiments. However, we have observed generally that soaking antibody crystals in hydrogen peroxide and exposing them to UV light (source of singlet oxygen) has little effect on the integrity of the antibody [128], although oxidative modifications are observed on TrpLIG3 in the interfacial region between variable and constant domains [139]. Current endeavors aim to trace the origins of antibody-generated oxidants within the immunoglobulin fold.

References

1 WILSON, I. A., STANFIELD, R. L. *Curr. Op. Struct. Biol.* 4 (**1994**), p. 857–867
2 PADLAN, E. A., *Adv. Prot. Chem.* 49 (**1996**), p. 57–133
3 PAULING, L., *Am. Sci.* 36 (**1948**), p. 51–58
4 HALDANE, J. B. S., *Enzymes*, Longmans Green, London **1930**
5 JENCKS, W. P., *Catalysis in Chemistry and Enzymology*, McGraw-Hill, New York **1969**
6 TRAMONTANO, A., JANDA, K. D., LERNER, R. A., *Science* 234 (**1986**), p. 1566–1570
7 POLLACK, S. J., JACOBS, J. W., SCHULTZ, P. G., *Science* 234 (**1986**), p. 1570–1573
8 SEGAL, D. M., PADLAN, E. A., COHEN, G. H., RUDIKOFF, S., POTTER, M., DAVIES, D. R., *Proc. Natl. Acad. Sci. U.S.A.* 71 (**1974**), p. 4298–4302
9 SATOW, Y., COHEN, G. H., PADLAN, E. A., DAVIES, D. R., *J. Mol. Biol.* 190 (**1986**), p. 593–604
10 JESKE, D. J., JARVIS, J., MILSTEIN, C., CAPRA, J. D., *J. Immunol.* 133 (**1984**), p. 1090–1092
11 ROSE, D. R., STRONG, R. K., MARGOLIES, M. N., GEFTER, M. L., PETSKO, G. A., *Proc. Natl. Acad. Sci. U.S.A.* 87 (**1990**), p. 338–342
12 STRONG, R. K., CAMPBELL, R., ROSE, D. R., PETSKO, G. A., SHARON, J., MARGOLIES, M. N., *Biochemistry* 30 (**1991**), p. 3739–3748
13 STRONG, R. K., PETSKO, G. A., SHARON, J., MARGOLIES, M. N., *Biochemistry* 30 (**1991**), p. 3749–3757
14 HOTTA, K., LANGE, H., TANTILLO, D. J., HOUK, K. N., HILVERT, D., WILSON, I. A., *J. Mol. Biol.* 302 (**2000**), p. 1213–1225
15 BALDWIN, J. E., *J. Chem. Soc., Chem. Commun.*, (**1976**), p. 734–736
16 PASCHALL, C. M., HASSERODT, J., JONES, T., LERNER, R. A., JANDA, K. D., CHRISTIANSON, D. W. *Angew. Chem. Int. Ed.* 38 (**1999**), p. 1743–1747
17 ZHU, X., HEINE, A., MONNAT, F., HOUK, K. N., JANDA, K. D., WILSON, I. A. *J. Mol. Biol.* 329 (**2003**), p. 69–83
18 MACBEATH, G., AND HILVERT, D., *Chem. Biol.* 3 (**1996**), p. 433–445
19 TANTILLO, D. J., HOUK, K. N. *Chem. Biol.* 8 (**2001**), p. 535–545
20 KRISTENSEN, O., VASSYLYEV, D. G., TANAKA, F., MORIKAWA, K., FUJII, I., *J. Mol. Biol.* 281 (**1998**), p. 501–511
21 WEDEMAYER, G. J., WANG, L. H., PATTEN, P. A., SCHULTZ, P. G., STEVENS, R. C., *J. Mol. Biol.* 268 (**1997**), p. 390–400
22 WEDEMAYER, G. J., PATTEN, P. A., WANG, L. H., SCHULTZ, P. G., STEVENS, R. C., *Science* 276 (**1997**), p. 1665–1669
23 GUO, J., HUANG, W., ZHOU, G. W., FLETTERICK, R. J., SCANLAN, T. S., *Proc. Natl. Acad. Sci. U.S.A.* 92 (**1995**), p. 1694–1698
24 ZHOU, G. W., GUO, J., HUANG, W., FLETTERICK, R. J., SCANLAN, T. S., *Science* 265 (**1994**), p. 1059–1064
25 BUCHBINDER, J. L., STEPHENSON, R. C., SCANLAN, T. S., FLETTERICK, R. J., *J. Mol. Biol.* 282 (**1998**), p. 1033–1041

26 Charbonnier, J. B., Carpenter, E., Gigant, B., Golinelli-Pimpaneau, B., Eshhar, Z., Green, B. S., Knossow, M., *Proc. Natl. Acad. Sci. U.S.A.* 92 (**1995**), p. 11721–11725

27 Thayer, M. M., Olender, E. H., Arvai, A. S., Koike, C. K., Canestrelli, I. L., Stewart, J. D., Benkovic, S. J., Getzoff, E. D., Roberts, V. A., *J. Mol. Biol.* 291 (**1999**), p. 329–345

28 Charbonnier, J. B., Golinelli-Pimpaneau, B., Gigant, B., Tawfik, D. S., Chap, R., Schindler, D. G., Kim, S. H., Green, B. S., Eshhar, Z., Knossow, M., *Science* 275 (**1997**), p. 1140–1142

29 Miyashita, H., Hara, T., Tanimura, R., Fukuyama, S., Cagnon, C., Kohara, A., Fujii, I., *J. Mol. Biol.* 267 (**1997**), p. 1247–1257

30 Gigant, B., Tsumuraya, T., Fujii, I., Knossow, M., *Structure* 7 (**1999**), p. 1385–1393

31 Takahashi, N., Kakinuma, H., Liu, L., Nishi, Y., Fujii, I., *Nat. Biotech.* 19 (**2001**), p. 563–567

32 Stewart, J. D., Benkovic, S. J., *Nature* 375 (**1995**), p. 388–391

33 Turner, J. M., Larsen, N. A., Basran, A., Barbas III, C. F., Bruce, N. C., Wilson, I. A., Lerner, R. A., *Biochemistry* 41 (**2002**), p. 12297–12307

34 Menard, R., Carriere, J., Laflamme, P., Plouffe, C., Khouri, H. E., Vernet, T., *Biochemistry* 35 (**1991**), p. 8924–8928

35 Nicolas, A., Egmond, M., Verrips, C. T., de Vlieg, J., Longhi, S., Cambillau, C., Martinez, C., *Biochemistry* 35 (**1996**), p. 398–410

36 Szeltner, Z., Renner, V., Polgar, L., *Protein Sci.* 9 (**2000**), p. 353–360

37 Bryan, P., Pantoliano, M. W., Quill, S. G., Hsiao, H.-Y., Poulos, T., *Proc. Natl. Acad. Sci. U.S.A.* 83 (**1986**), p. 3743–3745

38 Carter, P., Wells, J. A., *Proteins: Struct. Funct. Genet.* 7 (**1990**), p. 335–342

39 O'Connell, T. P., Day, R. M., Torchilin, E. V., Bachovchin, W. W., Malthouse, J. G., *Biochem. J.* 326 (**1997**), p. 861–866

40 Harel, M., Quinn, D. M., Nair, H. K., Silman, I., Sussman, J. L., *J. Am. Chem. Soc.* 118 (**1996**), p. 2340–2346

41 Phillips, M. A., Fletterick, R., Rutter, W. J., *J. Biol. Chem.* 265 (**1990**), p. 20692–20698

42 Lindler, A. B., Kim, S. H., Schindler, D. G., Eshar, Z., Tawfik, D. S., *J. Mol. Biol.* 329 (**2002**), p. 559–572

43 Lindler, A. B., Eshar, Z., Tawfik, D. S., *J. Mol. Biol.* 285 (**1999**), p. 421–430

44 Tantillo, D. J., Houk, K. N., *J. Comput. Chem.* 23 (**2002**), p. 84–95

45 Janda, K. D., Ashley, J. A., Jones, T. M., McLeod, D. A., Schloeder, D. M., Weinhouse, M. I., Lerner, R. A., Gibbs, R. A., Benkovic, P. A., Hilhorst, R., Benkovic, S. J., *J. Am. Chem. Soc.* 113 (**1991**), p. 291–297

46 Derewenda, Z. S., Wei, Y., *J. Am. Chem. Soc.* 117 (**1995**), p. 2104–2105

47 Larsen, N. A., Turner, J. M., Stevens, J., Rosser, S. J., Basran, A., Lerner, R. A., Bruce, N. C., Wilson, I. A., *Nat. Struct. Biol.* 9 (**2002**), p. 17–21

48 Guo, J., Huang, W., Scanlan, T. S., *J. Am. Chem. Soc.* 116 (**1994**), p. 6062–6069

49 Haynes, M. R., Heine, A., Wilson, I. A., *Isr. J. Chem.* 36 (**1996**), p. 133–142

50 Ghosh, D., Sawicki, M., Lala, P., Erman, M., Pangborn, W., Eyzaguirre, J., Gutierrez, R., Jornvall, H., Thiel, D. J. *J. Biol. Chem.* 276 (**2001**), p. 11159–11166

51 Zhu, X., Larsen, N. A., Basran, A., Bruce, N. C., Wilson, I. A., *J. Biol. Chem.* 278 (**2003**), p. 2008–2014

52 Baca, M., Scanlan, T. S., Stephenson, R. C., Wells, J. A., *Proc. Natl. Acad. Sci. U.S.A.* 94 (**1997**), p. 10063–10068

53 Stewart, J. D., Krebs, J. F., Siuzdak, G., Berdis, A. J., Smithrud, D. B., Benkovic, S. J., *Proc. Natl. Acad. Sci. U.S.A.* 91 (**1994**), p. 7404–7409

54 Benkovic, S. J., Adams, J. A., Borders Jr, C. L., Janda, K. D., Lerner, R. A. *Science* 250 (**1990**), p. 1135–1139

55 Gibbs, R. A., Benkovic, P. A., Janda, K. D., Lerner, R. A., Benkovic, S.

J., *J. Am. Chem. Soc.* 114 (**1992**), p. 3528–3534
56 KREBS, J. F., SIUZDAK, G., DYSON, H. J., STEWART, J. D., BENKOVIC, S. J., *Biochemistry* 34 (**1995**), p. 720–723
57 SAMPSON, N. S., BARTLETT, P. A., *Biochemistry* 30 (**1991**), p. 2255
58 CARTER, P., WELLS, J. A., *Nature* 332 (**1988**), p. 564
59 BARTLETT, P. A., MARLOWE, C. K., *Biochemistry* 22 (**1983**), p. 4618–4624
60 CHRISTIANSON, D. W., LIPSCOMB, W. N., *Acc. Chem. Res.* 33 (**1989**), p. 62–69
61 KEMP, D. S., PAUL, K. G.,, *J. Am. Chem. Soc.* 97 (**1975**), p. 7305–7312
62 LEWIS, C., KRAMER, T., ROBINSON, S., HILVERT, D., *Science* 253 (**1991**), p. 1019–1022
63 TARASOW, T. M., LEWIS, C., HILVERT, D., *J. Am. Chem. Soc.* 116 (**1994**), p. 7959–7963
64 HOTTA, K., KIKUCHI, K., HILVERT, D., in *Helv. Chim. Acta* 83 (**2000**), p. 2183–2191
65 HOTTA, K., WILSON, I. A., HILVERT, D., *Biochemistry* 41 (**2002**), p. 772–779
66 HSIEH, L. C., STEPHANS, J. C., SCHULTZ, P. G., *J. Am. Chem. Soc.* 116 (**1994**), p. 2167–2168
67 HOLLAND, H. L., *Chem. Rev.* 88 (**1988**), p. 473–485
68 HSIEH-WILSON, L. C., SCHULTZ, P. G., STEVENS, R. C., *Proc. Natl. Acad. Sci. U.S.A.* 93 (**1996**), p. 5363–5367
69 YIN, J., MUNDORFF, E. C., YANG, P. L., WENDT, U., HANWAY, D., STEVENS, R. C., SCHULTZ, P. G., *Biochemistry* 40 (**2001**), p. 10764–10773
70 YIN, J., ANDRYSKI, S. E., BEUSCHER IV, A. E., STEVENS, R. C., SCHULTZ, P. G., *Proc. Natl. Acad. Sci. U.S.A.* 100 (**2003**), p. 856–861
71 GONCALVES, J., DINTINGER, T., LEBRETON, J., BLANCHARD, D., TELLIER, C., *Biochem. J.* 346 (**2000**), p. 691–698
72 GOLINELLI-PIMPANEAU, B., GONCALVES, O., DINTINGER, T., BLANCHARD, D., KNOSSOW, M., TELLIER, C., *Proc. Natl. Acad. Sci. U.S.A.* 97 (**2000**), p. 9892–9895
73 73. HOLLFELDER, F., KIRBY, A. J., TAWFIK, D. S., *Nature* 383 (**1996**), p. 60–62
74 HILVERT, D., CARPENTER, S. H., NARED, K. D., AUDITOR, M.-T. M., *Proc. Natl. Acad. Sci. U.S.A.* 85 (**1988**), p. 4953–4955
75 ANDREWS, P. R., SMITH, G. D., YOUNG, I. G., *Biochemistry* 12 (**1973**), p. 3492
76 TANG, Y., HICKS, J. B., HILVERT, D., *Proc. Natl. Acad. Sci. U.S.A.* 88 (**1991**), p. 8784–8786
77 HAYNES, M. R., STURA, E. A., HILVERT, D., WILSON, I. A., *Science* 263 (**1994**), p. 646–652
78 CHOOK, Y. M., KE, H., LIPSCOMB, W. N. *Proc. Natl. Acad. Sci. U.S.A.* 90 (**1993**), p. 8600–8603
79 LARSEN, N. A., ZHOU, B., HEINE, A., WIRSCHING, P., JANDA, K. D., WILSON, I. A., *J. Mol. Biol.* 311 (**2001**), p. 9–15
80 BRAISTED, A. C., SCHULTZ, P. G., *J. Am. Chem. Soc.* 116 (**1994**), p. 2211–2212
81 ULRICH, H. D., MUNDORFF, E., SANTARSIERO, B. D., DRIGGERS, E. M., STEVENS, R. C., SCHULTZ, P. G., *Nature* (**1997**), p. 271–275
82 MUNDORFF, E. C., HANSON, M. A., VARVAK, A., ULRICH, H. D., SCHULTZ, P. G., STEVENS, R. C. *Biochemistry* 39 (**2000**), p. 627–632
83 SAUER, J., *Angew. Chem. Int. Ed. Engl.* 5 (**1966**), p. 211–230
84 SCHMIDT, R. R., *Acc. Chem. Res.* 19 (**1986**), p. 250–259
85 POHNERT, G., *Chembiochem* 2 (**2001**), p. 873–875
86 PAGE, M. I., JENCKS, W. P., *Proc. Natl. Acad. Sci. U.S.A.* 68 (**1971**), p. 1678-1683
87 BAHR, N., GUELLER, R., REYMOND, J.-L., LERNER, R. A., *J. Am. Chem. Soc.* 118 (**1996**), p. 3550–3555
88 HUGOT, M., BENSEL, N., VOGEL, M., REYMOND, M. T., STADLER, B., REYMOND, J.-L., BAUMANN, U., *Proc. Natl. Acad. Sci. U.S.A.* 99 (**2002**), p. 9674–9678
89 YLI-KAUHALUOMA, J. T., ASHLEY, J. A., LO, C.-H., TUCKER, L., WOLFE, M. M., JANDA, K. D., *J. Am. Chem. Soc.* 117 (**1995**), p. 7041–7047
90 CANNIZZARO, C. E., ASHLEY, J. A., JANDA, K. D., HOUK, K. N., *J. Am. Chem. Soc.* 125 (**2003**), p. 2489–2506

91 Heine, A., Stura, E. A., Yli-Kauhaluoma, J. T., Gao, C., Deng, Q., Beno, B. R., Houk, K. N., Janda, K. D., Wilson, I. A., *Science* 279 (**1998**), p. 1934–1940

92 Braisted, A. C., Schultz, P. G., *J. Am. Chem. Soc.* 112 (**1990**), p. 7430–7431

93 Romesberg, F. E., Spiller, B., Schultz, P. G., Stevens, R. C., *Science* 279 (**1998**), p. 1929–1933

94 Xu, J., Deng, D., Chen, J., Houk, K. N., Bartek, J., Hilvert, D., Wilson, I. A., *Science* 286 (**1999**), p. 2345–2348

95 Page, M. I., Jencks, W. P., *Proc. Natl. Acad. Sci. U.S.A.* 68 (**1971**), p. 1678–1683

96 Hilvert, D., Hill, K. W., *Methods Enzymol.* 203 (**1991**), p. 352–369

97 Arévalo, J. H., Stura, E. A., Taussig, M. J., Wilson, I. A., *J. Mol. Biol.* 231 (**1993**), p. 103–118

98 Arévalo, J. H., Taussig, M. J., Wilson, I. A. *Nature* 365 (**1993**), p. 859–863

99 Arévalo, J. H., Hassig, C. A., Stura, E. A., Sims, M. J., Taussig, M. J., Wilson, I. A. *J. Mol. Biol.* 241 (**1994**), p. 663–690

100 Cravatt, B. F., Ashley, J. A., Janda, K. D., Boger, D. L., Lerner, R. A., *J. Am. Chem. Soc.* 116 (**1994**), p. 6013–6014

101 Larsen, N. A., Heine, A., Crane, L., Cravatt, B. F., Lerner, R. A., Wilson, I. A., *J. Mol. Biol.* 314 (**2001**), p. 93–102

102 March, J. *Advanced organic chemistry*, John Wiley & Sons, New York **1992**

103 Stork, G., Burgstahler, A. W. *J. Am. Chem. Soc.* 77 (**1955**), p. 5068

104 Eschenmoser, A., Ruzicka, L., Jeger, O., Arigoni, D. *Helv. Chim. Acta* 38 (**1955**), p. 1890

105 Lesburg, C. A., Caruthers, J. M., Paschall, C. M., Christianson, D. W., *Curr. Opin. Struct. Biol.* 8 (**1998**), p. 695–703

106 Johnson, G., *J. Am. Chem. Soc.* 86 (**1964**), p. 1959

107 Li, T., Janda, K. D., Ashley, J. A., Lerner, R. A., *Science* 264 (**1994**), p. 1289–1293

108 Abe, I., Rohmer, M., Prestwich, G. D., *Chem. Rev.* 93 (**1993**), p. 2189–2206

109 Hasserodt, J., Janda, K. D., Lerner, R. A., *J. Am. Chem. Soc.* 119 (**1997**), p. 5993–5998

110 Dailey, H. A., Fleming, J. E., *J. Biol. Chem.* 258 (**1983**), p. 11453

111 Cochran, A. G., Schultz, P. G. *Science* 249 (**1990**), p. 781–783

112 Taketani, S., Tokunaga, R., *Eur. J. Biochem.* 127 (**1982**), p. 443–447

113 Romesberg, F. E., Santarsiero, B. D., Spiller, B., Yin, J., Barnes, D., Schultz, P. G., Stevens, R. C. *Biochemistry* 37 (**1998**), p. 14404–14409

114 Wirsching, P., Ashley, J. A., Lo, C.-H. L., Janda, K. D., Lerner, R. A., *Science* 270 (**1995**), p. 1775–1782

115 Wagner, J., Lerner, R. A., Barbas III, C. F., *Science* 270 (**1995**), p. 1797–1800

116 Barbas III, C. F., Heine, A., Zhong, G., Hoffmann, T., Gramatikova, S., Bjornestedt, R., List, B., Anderson, J., Stura, E. A., Wilson, I. A., Lerner, R. A. *Science* 278 (**1997**), p. 2085–2092

117 Karlstrom, A., Zhong, G., Rader, C., Larsen, N. A., Heine, A., Fuller, R., List, B., Tanaka, F., Wilson, I. A., Barbas III, C. F., Lerner, R. A., *Proc. Natl. Acad. Sci. U.S.A.* 97 (**2000**), p. 3878–3883

118 Heine, A., DeSantis, G., Luz, J. G., Mitchell, M., Wong, C.-H., Wilson, I. A., *Science* 294 (**2001**), p. 369–374

119 Zhong, G., Lerner, R. A., Barbas, C. F., III. *Angew. Chem. Int. Ed. Engl.* submitted (**1999**)

120 Zhu, X., Wilson, I. A., personal communication (**2003**)

121 Fessner, W.-D., *Curr. Opin. Chem. Biol.* 2 (**1998**), p. 85–97

122 Chen, L., Dumas, D. P., Wong, C.-H., *J. Am. Chem. Soc.* 114 (**1992**), p. 741–748

123 Racker, E., *J. Biol. Chem.* 196 (**1952**), p. 347–365

124 Barbas III, C. F., Wang, Y.-F., Wong, C.-H., *J. Am. Chem. Soc.* 112 (**1990**), p. 2013–2014

125 Zhong, G., Shabat, D., List, B., Anderson, J., Sinha, S. C., Lerner,

R. A., Barbas III, C. F., *Angew. Chem. Int. Ed.* 37 (1998), p. 2481–2484
126 List, B., *Tetrahedron* 58 (2002), p. 5573–5590
127 Wentworth, A. D., Jones, L. H., Wentworth Jr., P., Janda, K. D., Lerner, R. A., *Proc. Natl. Acad. Sci. U.S.A.* 97 (2000), p. 10930–10935
128 Wentworth Jr., P., Jones, L. H., Wentworth, A. D., Zhu, X., Larsen, N. A., Wilson, I. A., Xu, X., Goddard, W. A., Janda, K. D., Eschenmoser, A., Lerner, R. A., *Science* 293 (2001), p. 1806–1811
129 Datta, D., Vaidehi, N., Xu, X., Goddard III, W. A., *Proc. Natl. Acad. Sci. U.S.A.* 99 (2002), p. 2636–2641
130 Patten, P. A., Gray, N. S., Yang, P. L., Marks, C. B., Wedemayer, G. J., Boniface, J. J., Stevens, R. C., Schultz, P. G., *Science* 271 (1996), p. 1086–1091.
131 Fujii, I., Tanaka, F., Miyashita, H., Tanimura, R., Kinoshita, K., *J. Am. Chem. Soc.* 117 (1995), p. 6199–6209
132 Yang, G., Chun, J., Arakawa-Uramoto, H., Wang, X., Gawinowicz, M. A., Zhao, K., Landry, D. W., *J. Am. Chem. Soc.* 118 (1996), p. 5881–5890
133 Larsen, N. A., Heine, A., de Prada, P., Redwan, R. M., Yeates, T. O., Landry, D. W., Wilson, I. A., *Acta Crystallogr.* D58 (2002), p. 2055–2059
134 Bensel, N., Bahr, N., Reymond, M. T., Schenkels, C., Reymond, J.-L., *Helv. Chim. Acta* 82 (1999), p. 44–52
135 Zhong, G., Shabat, D., List, B., Anderson, J., Sinha, S. C., Lerner, R. A., Barbas III, C. F., *Angew. Chem. Int. Ed.* 37 (1998), p. 2481–2484
136 Lewis, C., Paneth, P., O'Leary, M. H., Hilvert, D., *J. Am. Chem. Soc.* 115 (1993), p. 1410–1413
137 Tarasow, T. M., Lewis, C., Hilvert, D. *J. Am. Chem. Soc.* 116 (1994), p. 7959–7963
138 Gruber, K., Zhou, B., Houk, K. N., Lerner, R. A., Shevlin, C. G., Wilson, I. A., *Biochemistry* 38 (1999), p. 7062–7074
139 Zhu, X., Wentworth, P., Wentworth, A. D., Eschenmoser, A., Lerner, R. A., Wilson, I. A., *Proc. Natl. Acad. Sci. U.S.A.* 101 (2004), p. 2247–2252

7
Antibody Catalysis of Disfavored Chemical Reactions

Jonathan E. McDunn, Tobin J. Dickerson, Kim D. Janda

7.1
Introduction

One of the most compelling aspects of enzyme catalysis is the specificity with which enzymes process their substrates. An enzyme's specificity is doubtless the result of evolutionary pressures to carry certain substrates through a given reaction path, yielding only a subset of possible products. In many cases, enzymes catalyze reactions that would not otherwise occur. These disfavored reactions are of great interest because they reveal how nature has solved particularly difficult problems. In this way, biochemical and structural studies of the enzymes that catalyze these reactions provide chemists with insights into important mechanistic considerations as well as a starting point in the development of *de novo* catalysts for these and other disfavored processes.

While *de novo* catalyst design is often a fruitful endeavor, it is not always clear how to specifically transform one small molecule with another. Catalytic antibodies allow different features of any chemical transformation to be highlighted and invoked as haptens to probe a naïve library of proteins in the search for catalysts. Although immunization selects for binding, compounds designed to recapitulate important mechanistic features of a chemical transformation have been shown to elicit antibodies that can catalyze that transformation [1, 2]. In this way, the immune response can be thought of as a programmable combinatorial system that can be exploited to generate catalysts. This approach has been used to raise antibodies to compounds that incorporate stereoelectronic features of disfavored transformations; in many cases, resultant antibodies catalyze the desired disfavored reactions. For many of these processes there are no known natural (enzyme) or man-made (small molecule) catalysts. The inability to utilize such disfavored chemistry in total synthesis sometimes requires substantial effort to circumvent materials whose reactivity cannot be directed appropriately. Kinetic and structural characterization of antibodies which carry out disfavored transformations provides insight into the stereoelectronic factors that govern the reaction of interest, ultimately leading not only to better catalyst design, but also to a more complete understanding of complex chemical processes.

Catalytic Antibodies. Edited by Ehud Keinan
Copyright © 2005 WILEY-VCH Verlag GmbH & Co. KGaA, Weinheim
ISBN: 3-527-30688-9

One hapten design strategy that has proven particularly successful in raising catalytic antibodies for disfavored processes is the bait-and-switch [3] approach that was developed in our laboratory. This method goes beyond geometrical mimicry by incorporating electronic features of the transition state in the hapten. For several of the reactions described herein both quaternary nitrogens and N-oxide moieties were used to elicit residues within combining sites that promote the buildup of positive charge density within the substrate, thereby lowering the activation energy for the transformation.

For the purposes of this text, a disfavored chemical reaction is defined as a reaction that either has no background rate or favors a single product when a number of similar reaction pathways and their products are accessible. Furthermore, examples are discussed where the reaction proceeds by an extremely energetic mechanism. Herein are discussed in detail three examples of antibodies that catalyze disfavored chemical reactions with which we are intimately familiar. Finally, highlights of other work are presented.

7.2
Formal Violation of Baldwin's Rules for Ring Closure: the 6-*endo*-tet Ring Closure

Baldwin's rules [4, 5] summarize the stereoelectronic factors involved in intramolecular ring closure reactions and serve as a predictor for their outcome. Essentially the nucleophile must approach the olefin with a trajectory close to 109.5° (with respect to the double bond) in order to deposit electron density into the empty *p*-orbital. Although there are exceptions to these rules [6], these exceptions illustrate approximate nature of the rules and provide a means to separate geometric and electronic contributions toward the energetics of these transformations. Significantly, for most substrates the different reaction paths have different energetics.

Intramolecular cyclization reactions are typically under kinetic control, meaning that each molecule of starting material transits the reaction coordinate exactly once and that the product distribution reflects free energy differences between available transition states. Although these energetic differences are often small (a few kcal/mol), the product distribution is not easy to perturb because of similarities among the possible transition states, resulting in the inability to specifically manipulate one reaction path in the context of the others. Because of the ability of antibodies to envelop and conformationally constrain small molecules, it was thought that proper hapten design could elicit antibodies that disproportionately favor only one of the transition states available to a substrate. In the light of the estimate that antibodies typically provide up to 20 kcal/mol in binding energy to small molecules [7, 8], even a slight bias toward one transition state could lead to a substantial shift in the product distribution of such a kinetically controlled reaction.

One example of an intramolecular ring closure under kinetic control is the cyclization of an epoxy alcohol. In this regard, the first antibodies to catalyze a disfavored transformation were for the 6-*endo*-tet ring closure of epoxide 1 (Fig. 7.1) [9]. In the case of this substrate, the 5-*exo*-tet transition state is favored by approximately

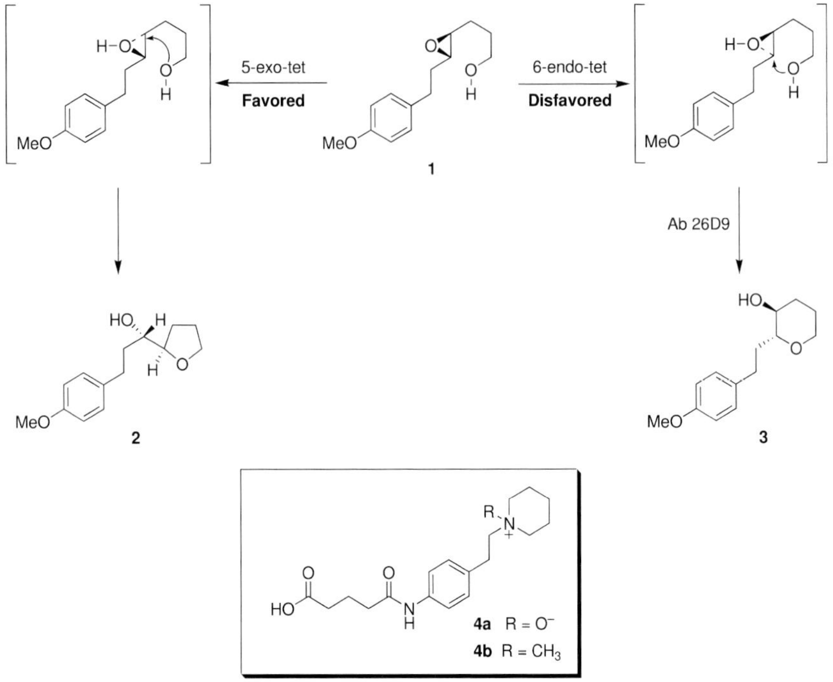

Fig. 7.1 Hapten structures (**4a,b**) and reaction schemes for both the favored and disfavored intramolecular ring closures of substrate **1**.

1.8 kcal/mol, yet, because the reaction is under kinetic control, the 5-*exo*-tet product **2** accounts for at least 96% of transformed starting material under standard, acid-catalyzed conditions. In order for tetrahydropyran **3** to form, the starting material must proceed through a six-member cyclic transition state instead of the favored, five-member cyclic transition state. In both cases, charge separation occurs concomitant with the opening of the epoxide. Hapten **4a** was designed in order to capture both the six-membered cyclic structure of the desired product as well as the polarization of the C–O bond in the desired epoxide opening as an N-oxide.

Twenty-six monoclonal antibodies were obtained that bound the hapten via ELISA and two of these (26C9 and 17F6) were regioselective catalysts for the *anti*-Baldwin ring closure. 26C9 was not only the more proficient catalyst but it also acted exclusively on the *S,S* epoxide to form the *S,R* product in high enantiomeric excess.

In order to better understand what features of the hapten were crucial in achieving 26C9's catalytic proficiency, computational analyses were performed [10]. While a simple solvent-free model (at the AM1 level) did not adequately capture the energetics of the acid-catalyzed process, more sophisticated analysis (both solvent-free MP2 and SCRF-solvent RHF/6-31G*) were correct in assigning the transition state for the 5-*exo*-tet closure the lower energy (approximately 1.8 kcal/mol). This level of modeling suggested that closure to tetrahydrofuran **2** proceeds through a transition state with only a slightly distorted angle of attack (100°) while the 6-*endo*-tet transition state has a significantly distorted angle of attack (88°). The geometry of the reaction center in the five-member transition state is much closer to tetrahedral than the highly strained six-member transition state. Thus, geometrical constraints are responsible for a significant amount of the energy difference between these two reaction paths.

Because of the higher strain necessary to achieve the 6-*endo*-tet geometry, the C–O bonds in this transition state are longer than for the 5-*exo*-tet ring closure. The longer bonds between the reaction center and both the leaving group and nucleophile result in greater cationic character developing on the carbon atom. As a result, catalysts for the disfavored reaction are likely to stabilize greater positive character on the reaction center. This model is in line with the observation and successful use of allylic substitution on the disfavored side of the epoxide to activate the 6-*endo*-tet cyclization of similar substrates [11]. The use of the N-oxide moiety in the hapten likely induced favorable electronic character within the antibody combining site that stabilizes the developing positive charge.

Based on the different product distributions observed under acid-catalyzed conditions and using antibody 26C9, the energetic stabilization of the 6-*endo*-tet transition state relative to the 5-*exo*-tet transition state was calculated to be 3.6 kcal/mol. This value is likely to be an underestimate of the relative stabilization of the 6-*endo*-tet reaction path because this disfavored product was not observed in the absence of 26C9 under the conditions that were used for the antibody-mediated reaction.

Further computational work [12] suggested that 26C9 and other catalysts for this transformation would not only poise the hydroxyl group for *endo*-tet attack of the epoxide but also should possess a microenvironment that stabilizes the partial charges that build up as the ring is closed. Protein catalysts accomplish the stabilization of charge through the placement of side chains that function as either general acids and general bases or position the π-cloud of aromatic groups to stabilize positive charge through cation-π interactions [13]. The geometry of the active site thus dictates both steric and electronic stabilization of the substrate. This computational work further suggested that the alkyl chain separating the hydroxy group from the epoxide does not play a substantial role in 26C9's catalytic proficiency.

To examine whether this was indeed the case, a ring-expanded substrate **5** was prepared (Fig. 7.2) [14]. Under acid-catalyzed conditions, the 6-*exo*-tet product **6** predominates (>98%) [15]. However, in the presence of 26C9, the 7-*endo*-tet product **7** is formed almost exclusively. As with the initial substrate, only one enantiomer of the epoxide (in this case the R,R antipode) is consumed and the S,R product is formed in greater than 70% enantiomeric excess. This result supports the model that positions the hydroxyl group for disfavored (*endo*-tet) attack on the epoxide without imposing stringent requirements on the alkyl chain separating these two moieties.

Following the proposal that the highly polarized N-oxide moiety was crucial in obtaining antibodies that catalyze the *anti*-Baldwin ring closure, mice were immunized with hapten **4b** (Fig. 7.1), replacing the N-oxide moiety with a quaternary nitrogen and thus a formal positive charge. Two additional catalytic antibodies (5C8 and 14B9) were obtained and characterized [16]. The subtle stereoelectronic differences between

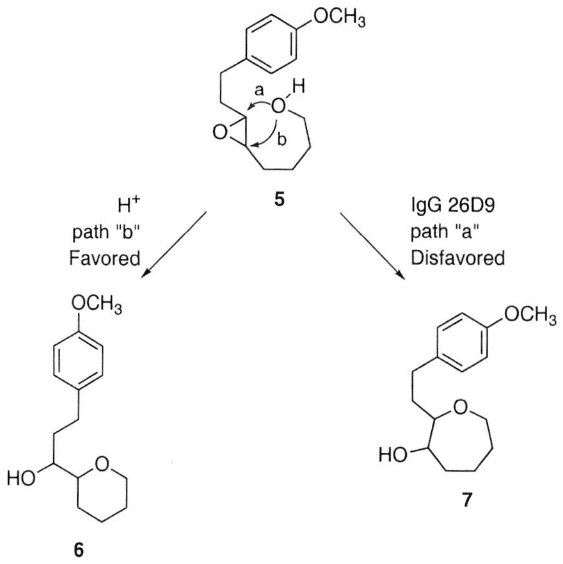

Fig. 7.2 The ring-expanded substrate (**5**) for 26C9 and its antibody-catalyzed and acid-catalyzed cyclizations.

Tab. 7.1: Kinetic parameters and product distribution for antibodies catalyzing disfavored intramolecular ring closure reactions.

Antibody	Reaction	k_{cat} (min^{-1})	K_M (µM)	% endo	% ee
5C8	1 → 3	1.7	595	70	95
26C9	1 → 3	0.9	356	70	99
26C9	5 → 7	0.9	196	> 99	78

these haptens allowed both structural data and further computational work to be used to address assumptions of how catalytic activity was dependent on hapten structure.

The most proficient antibodies raised against haptens **4a** and **4b** were compared kinetically (Table 7.1). 26C9 was found to catalyze the disfavored reaction of a broader scope of substrates as well as exert more control over both the regio- and enantioselectivities of its transformations than 5C8.

A number of mechanistic possibilities were considered, among these a single mechanism fit the biochemical and computational characterization without the need for invoking covalent catalysis or other more complex and as yet unsupported elaborations. In this mechanism, the substrate is poised in a pseudo-chair conformation where the hydroxyl oxygen attacks the reaction center opening the epoxide and inverting the configuration at the reaction center. Computational work suggested that in addition to this steric description of the reaction, electronic effects were also involved. Toward this end, computations were performed to evaluate the roles of both general acid and general base on the energetics of this reaction. A general acid would protonate the leaving epoxide while a general base would increase the nucleophilicity of the attacking alcohol. The results of these calculations were that the protonation of the epoxide would be a more significant factor [12].

Structures of the Fab of 5C8 bound to either hapten were obtained to 2.0Å resolution [16]. These structures reveal that the antibody binds these two haptens in similar conformations, with the piperidinium ring assuming the energetically-favored chair conformation with the N-oxide or N-methyl moiety in an axial position. The positive charge of the quaternary nitrogen is held in place by the tweezer-like arrangement of two aspartic acids H95 and H101. These residues are in turn rigidly positioned by hydrogen bonds to Tyr^{L91} and His^{H35} for Asp^{H95}, and Tyr^{L36} and Trp^{H103} for Asp^{H101}. The remaining interactions between the hapten and the antibody are van der Waals contacts between the piperidinium ring and mostly framework residues of the antibody. More than 90% of the solvent-accessible surface of either hapten is buried when complexed with the Fab.

Through further analysis of the structures, putative catalytic residues were assigned. While there are two aspartic acids, three histidines, and two tyrosines lining the binding cleft, their ability to act in the catalytic cycle of the antibody relies on their respective pK_a values. In the structure with hapten **4a**, Asp^{H95} is within hydrogen bonding distance of the N-oxide moiety and as a result is believed to have a perturbed pK_a and be protonated under the crystallization conditions (pH 5.5). From its distance constraint, Asp^{H95} is thought to participate as the general acid, protonating the epoxide to facilitate its opening. Docking studies with geometrically optimized transition states support this assertion, because the epoxide moiety occupies a similar position in the binding cleft for all four possible transition states. This leaves Asp^{H101} available to stabilize the developing positive charge on the reaction center.

Although the epoxide is rigidly positioned in the docked structures, the hydroxyl group is positioned quite differently, depending on whether the S,S epoxide or R,R epoxide is used. In the case of the S,S epoxide, the hydroxyl group is nearer to His^{L89} than to any other residue that could act as a general base. In contrast, the hydroxyl group of the R,R epoxide sits between Tyr^{L36} and Asp^{H101}, suggesting that these residues may act as a dyad. It is interesting to consider whether this antibody was selected based on interactions with only one enantiomer of the hapten or whether its ability to bind both, albeit differently, was key to its selection. A substantial shortcoming of these docking studies is that they do not allow for minimization of the complex as a whole, but only provide for the minimization of the small molecule structure. Without imparting flexibility to the protein, a number of improbable geometries are observed, and, as such, the docked structure may not represent a conformation that contributes significantly to the dynamics of antibody catalysis.

Although structures were not obtained for the other three antibodies which catalyze *anti*-Baldwin ring closures, sequence analysis revealed both similarities and differences that allow useful generalization. The antibody 14B9, for example, which exhibits poorer kinetics than 5C8, differs in only one residue within the entire contact surface between 5C8 and either hapten. In 5C8 the residue at H93 is an alanine, while in 14B9 H93 is occupied by a valine. The increased bulk along the interfacial surface suggests that 14B9 may be less proficient at substrate binding, which could in turn explain its poorer kinetics.

The differences between the sequences of 5C8 and 14B9 (the antibodies raised against hapten **4b**) and 17F6 and 26C9 (the antibodies raised against hapten **4a**) are more significant. Although the light chains are highly homologous (>80% identity), the heavy chains are quite different (<50% identity). Whereas the supposed general acid at H95 is an aspartic acid with a perturbed pK_a in both 5C8 and 14B9, this residue is a tyrosine in both 17F6 and 26C9. This bulkier residue is now positioned with the assistance of a serine instead of the larger histidine at position L35. Sequence analysis therefore suggests a conserved use of H95 and L35 in the catalytic mechanism of these antibodies. This result further illustrates the utility of antibodies in catalyst design by (1) using different combinations of amino acid side chains to perform the same task, (2) illustrating that complementary amino acids are selected even when they are in different CDRs, and (3) showing that subtle electronic differences in haptens may result in markedly different residues in terms of general acid/general base capacity using the bait-and-switch approach (cf. the pK_a of the Asp-His relative to the Tyr-Ser). The other residues that were identified as catalytically active from the 5C8 structures are conserved among the antibodies which catalyze the *anti*-Baldwin ring closure. Furthermore, all of the residues that form the hydrophobic pocket which accommodates the hydrocarbon portion of the haptens' piperidinium rings are conserved or highly homologous. Ala^{H93} is also mutated in both 17F6 and 26C9, but instead of a valine it is a threonine – which could hydrogen bond to the *N*-oxide oxygen.

The asserted mechanism was further corroborated by sequence analysis of a number of the antibodies which, although positive for binding the hapten (via ELISA), fail to catalyze the ring closure. While all of the sequenced non-catalytic antibodies have relevant portions of their sequences in common with the productive catalytic antibodies, none has all three of the regions that were identified to be important for catalysis: (1) stabilization of the developing positive charge on the reaction center, (2) general acid protonation of the epoxide by the residue at H95, and (3) a general base to increase the nucleophilicity of the hydroxyl group.

Finally, statistical analysis of the residue frequency from all sequenced antibodies shows that the His^{L89}, the proposed general base, is quite rare (occurring in approximately 5% of all sequenced antibodies). Glutamine is the most frequent residue at L89, occurring in nearly 60% of all sequenced antibodies [17]. The occurrence of a rare amino acid in a position believed to be crucial for catalysis strongly suggests that aspects of the hapten design selected for this residue. Furthermore, His^{L89} is absent in three of the four antibodies that bind haptens **4a** and **4b** but are not catalysts. The one antibody that does bind the haptens but is not a catalyst which has His^{L89} also contains three aspartic acids which could bind the substrate in an alternative (non-productive) conformation.

7.3
Cationic Cyclization

7.3.1
Initial Studies

In the light of the success using bait-and-switch haptens to elicit catalysts for 6-*endo*-tet ring closure, this strategy was fruitfully applied to a more demanding problem – cationic cyclization. One substantial difference between these processes is that while the *anti*-Baldwin ring closure proceeds by an S_N2-like mechanism with the development of a partial positive charge on the reaction center, cationic cyclization has more S_N1 character requiring greater cationic character if not a full positive charge to be generated and stabilized.

The burden of generating and manipulating a carbocation in the active site of an enzyme is substantial. Not only must such an active site stabilize a full point charge, but it must also be inert – unable to be alkylated by the carbocation. Additionally, to allow the cyclization to occur, the active site must also exclude solvent. If these criteria are met, one more hurdle must be overcome in order to have a useful catalyst – the stereochemistry of the cyclization process must be strictly controlled. While this demand is not so stringent in the case of a single cyclization, natural enzymes have evolved which catalyze cascade processes closing two or more rings in a tandem process. For example, 2,3-oxidosqualene cyclase stereoselectively directs the closure of a triterpenoid skeleton via cationic intermediates to yield a single product out of 128 possible stereoisomers [18].

While this accomplishment has yet to be achieved in the active site of a catalytic antibody, substantial progress has been made in generating, stabilizing, and controlling the reactivity of carbocations in antibody combining sites. This section of the chapter will review what has been done in this regard as well as point out both the successes and shortcomings of hapten design as seen from structural studies.

Cationic cyclization reactions can be thought of as occurring in three steps: initiation, propagation, and termination. The initiation step is the formation of the initial carbocation. Propagation consists of one or more ring closures with the concomitant migration of the positive charge. The termination step is where the cation is quenched either through elimination of a proton or by nucleophilic attack of the solvent. This sequence was established in the pioneering work of Stork and Eschenmoser [19, 20]. While the goals of developing catalytic antibodies for cationic cyclization processes are to mimic and ultimately expand the repertoire of product stereochemistries, initial studies [21] focused on whether an antibody combining site could be developed to stabilize a positive charge while excluding water and providing a sufficiently inert surface to avoid alkylation.

For these studies, both hapten and substrate designs were based on the compounds used in mechanistic studies of cationic cyclization reactions in organic solvent pioneered by Johnson [22]. In this work, the initial carbocation is formed by the solvolysis of a sulfonate ester. Although cyclization can occur in the absence of a template when electron-rich olefins are used, the undirected reaction results in poor

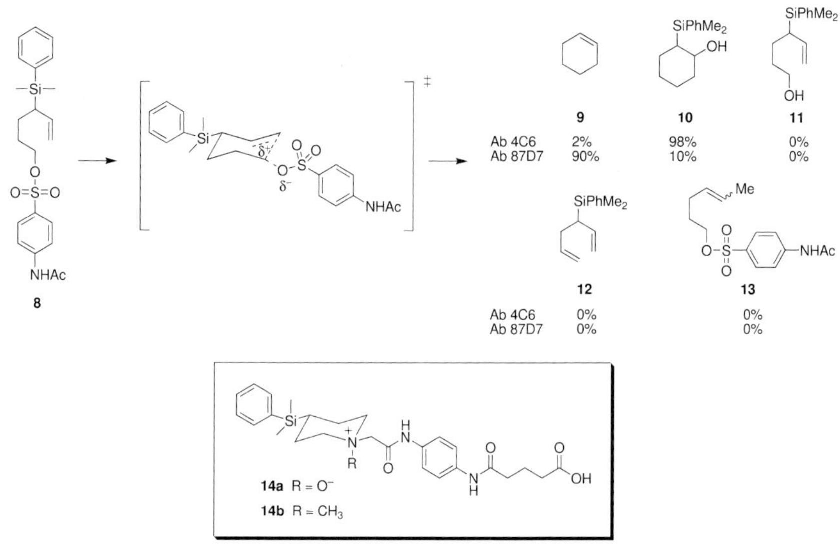

Fig. 7.3 Reaction mechanism, hapten design (**14a,b**), and product distribution for antibody-catalyzed cationic cyclization of substrate **8**.

yield and a complex mixture of products. Confinement of the substrate within an antibody combining site was expected not only to overcome entropic constraints, but also to influence the stereochemistry of the ring closure and to favor products which resemble the hapten.

Initially, hapten **14a** (Fig. 7.3) was prepared, incorporating a highly polarized N-oxide moiety to mimic the charge separation early in the departure of the leaving group. The piperidinium ring of the hapten was designed to preferentially occupy a pseudo-chair conformation. This hapten geometry was used to select an antibody with a combining site that would constrain the substrate to a similar shape. This geometry would place the sulfonate in an equatorial position so that it could be easily expelled. A silicon atom was incorporated in hapten **14a** and substrate **8** to use resonance stabilization from the C–Si bond to influence the geometry of the substrate olefin and position the π-electrons for backside attack on the desired carbocation. Following cyclization, the resultant carbocation can be quenched either by elimination or attack of another nucleophile. In addition to providing geometric constraints on the substrate, the silyl moiety in the substrate can provide mechanistic information as to the eventual fate of the cyclized product. This is because the local environment can direct how the cyclic carbocation is quenched (e.g., under formolysis conditions the silyl moiety is eliminated).

Following immunization with **14a**, twenty-nine antibodies were recovered that showed a positive ELISA signal for binding the hapten. These antibodies were further screened to determine which supported the initiating step of catalysis by monitoring release of the 4-acetamidobenzenesulfonic acid. Four of the panel of twenty-nine antibodies catalyzed the release of this material from the substrate (4C6, 16B5, 1C9,

Tab. 7.2: Kinetic parameters for the initial catalytic antibodies that performed cationic cyclizations.

Antibody	Substrate	k_{cat} (min^{-1})[1]	K_M (μM)
4C6	8	0.02	230
4C6	15	0.3	1800
4C6	17	7.8×10^{-4}	330
87D7	8	0.02	25
87D7	17	0.01	31
87D7	21a	0.013	58
87D7	21b	0.021	102

1) Measured as the rate of sulfonate release.

6H5). While this step is necessary for the formation of the carbocation, it is not sufficient. In order to determine whether cationic cyclization occurs in the combining sites of any of these antibodies, product analysis was pursued.

Antibody 4C6 was studied in more detail based on its kinetic parameters for the sulfonate ester cleavage (Table 7.2). Only two of the six possible products were observed, cyclohexene (9) (2%) and *trans*-2-dimethylphenylsilylcyclohexanol (10) (98%) (Fig. 7.3). This is significant because not only is a complex mixture of cyclized and uncyclized products (9–13) observed under solvolysis conditions, but also because one product is substantially favored within the combining site. Furthermore, it is surprising to retain the silyl moiety because it is easily expelled from the cationic intermediate. The antibody must therefore maintain strict control over the entire process such that the silyl moiety is retained in the major product. Finally, 14a is a competitive inhibitor of the reaction (Table 7.3). Taken together, these data demonstrate that antibody 4C6 provides an environment within its combining site which supports the generation and specific reactivity of a carbocation.

Tab. 7.3: Inhibition constants for compounds inhibiting 4C6 and 87D7's reaction with the designed substrate **8**.

Antibody	Compound	K_i (μM)
4C6	14a	1.0
4C6	19	200
87D7	14b	1.4

Among the other antibodies that catalyzed sulfonate ester hydrolysis, 16B5 showed single turnover kinetics. Furthermore, following reaction with the substrate, 16B5's hapten-binding activity was abrogated. These data suggest that antibody 16B5 is alkylated in the course of the reaction.

In order to gain further mechanistic insight into the cationic cyclization afforded by 4C6, a small panel of substrate analogs was synthesized and screened for ring-formation (Fig. 7.4). The fully saturated analog of the initial substrate 19 was an effective inhibitor (Table 7.3) for the antibody catalyzed process. Furthermore, 4C6 does not even hydrolyze 19, implying that the substrate olefin assists the elimination

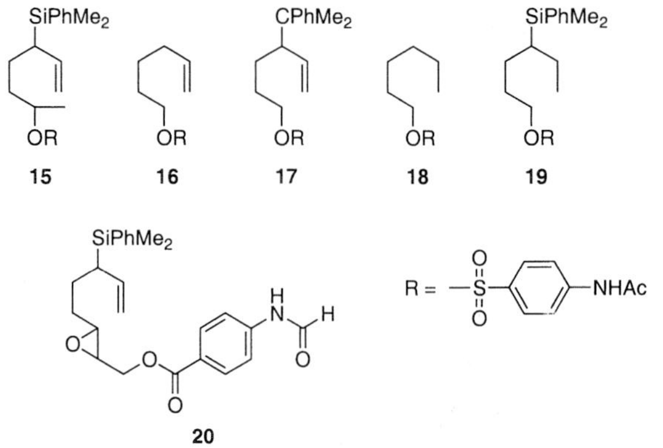

Fig. 7.4 Panel of substrate analogs to probe the mechanism of 4C6 and other catalytic antibodies which perform cationic cyclizations.

of the sulfonate ester. Compound **17** (the substrate analog with the silicon replaced by a carbon atom) is a very poor substrate for 4C6 and does not provide clean kinetics or a simple product distribution. Therefore, it is likely that the olefin in **8** is particularly electron rich – likely through resonance with the C–Si bond – and participates anchimerically to eliminate the sulfonate ester.

The secondary sulfonate ester substrate analog **15**, although processed more rapidly by 4C6 than the designed substrate **8**, does not provide a simple product distribution. Both of these effects can be explained by the increased substitution. The increased rate supports the assertion that this reaction proceeds via a carbocation because a secondary carbocation is more easily formed than a primary carbocation. The complex product distribution is interpreted to mean that 4C6 is not able to exert significant control over the ring closure of the initial carbocation. The K_m of this substrate is substantially higher than that of **8**, suggesting that the increased bulk in **15** does not allow proper recognition of this portion of the molecule. The other substrate analogs **16** and **18** that were studied were not accepted by the antibody.

Finally, in order to determine whether this antibody is capable of utilizing substrates similar to the natural substrates of terpene cyclase enzymes, compound **20** was examined. Instead of possessing the facile aryl sulfonate leaving group, the cationic cyclization of **20** would be initiated by epoxide opening. While antibodies have catalyzed epoxide openings using a variety of nucleophiles [9, 23], 4C6 was unable to process this substrate. This result was not entirely unexpected because of the structural incongruencies between the structures of substrate **20** and the hapten **14a**.

Given the success of antibody 4C6 to catalyze cationic cyclization, mice were immunized with hapten **14b**, the *N*-methyl analog of **14a** [24]. Eighteen monoclonal antibodies were recovered that were positive for hapten binding on ELISA. One of these, 87D7, was found to catalyze the transformation of **8** into the same products as 4C6 but in nearly the opposite abundance, *trans*-2-dimethylphenylsilylcyclohexanol **10** (10%) and cyclohexene **9** (90%). While the geometries of **14a** and **14b** are essentially identical, their electrostatic surfaces are not. This difference may have generated antibodies that use different strategies to bind their haptens. In the case of 87D7 it may be that recognition of the point charge on the quaternary amine may dominate antibody-hapten recognition. As a result, the combining site of 87D7 may not be as complementary to the hapten as the combining site of 4C6.

Looser recognition of the remainder of the hapten/substrate could lead to the different product distribution observed for 87D7 compared to 4C6. In order for elimination of the silyl moiety to be eliminated from the cyclic carbocation it must attain a pseudo-axial position. Yet the solution structure of hapten **14b** shows that the silyl moiety occupies an equatorial position. This conformation of the hapten is almost certainly responsible for obtaining 87D7. However it is not necessary for the solution conformation of the hapten to be strictly enforced within the antibody combining site. In the light of the proposal that 87D7 expends a substantial amount of binding energy on the quaternary nitrogen, it is possible that there is flexibility in the distal portion of the substrate. Instead of a "lock-and-key" relationship, 87D7 may recognize its substrate by induced fit [25], allowing the silyl moiety to attain a pseudo-axial position for elimination.

Given the product distribution that arose from 87D7-mediated catalysis using **8** as the substrate, further work was done to elaborate the scope of the substrates that could be transformed with this antibody [26]. At the same time, the question of the eventual fate of the carbocation in the absence of a good leaving group was examined. To address these issues, compounds **17**, **21a,b** (Fig. 7.5) were screened as substrates for 87D7, and kinetic parameters were obtained (Table 7.2). Compound **17**, the C-analog of **8**, was converted to *trans*-2-cyclohexanol (**23**) in 60% yield on incubation with 87D7. The ability to process **17** demonstrates that 87D7 can initiate the cationic cyclization using less electron-rich alkenes than 4C6.

Subtly changing the stereoelectronics of **17** by placement of an electron-donating methyl group either *cis* (**21a**) or *trans* (**21b**) on the alkene led to very interesting results. In the case of **21a**, 80% yield of a single diastereomeric exocyclic alcohol **24** was observed. The stereochemistry of **24** suggests that water adds to the cationic intermediate with facial selectivity and that the *cis*-methyl substituent occludes water from

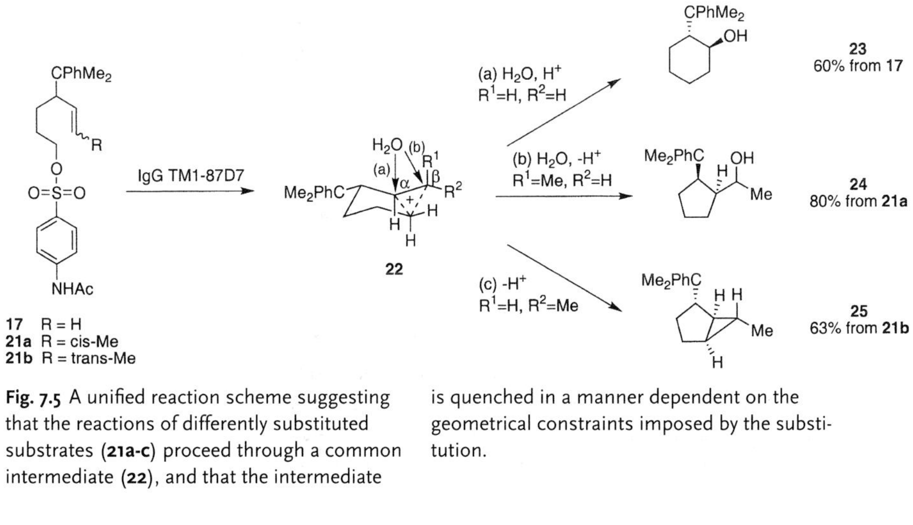

Fig. 7.5 A unified reaction scheme suggesting that the reactions of differently substituted substrates (**21a-c**) proceed through a common intermediate (**22**), and that the intermediate is quenched in a manner dependent on the geometrical constraints imposed by the substitution.

attacking that position of the alkene. When the *trans*-methyl substrate **21b** was used, the major product (63%) was bicyclo[3.1.0]hexane (**25**). Instead of either an alcohol or an unsaturated product, the *trans*-methyl substituent directed the formation of the fused bicycle **25**. This not only requires water to be excluded from the combining site, but also dictates which proton is eliminated to terminate the reaction. These products can be rationalized through the common intermediate **22** (Fig. 7.5).

Remarkably, bicyclo[3.1.0]hexane **25** is formed under the mild conditions of antibody catalysis. This product is not observed under even the harsh solvolysis conditions used by Johnson. While cationic cyclopropanation is the first step in the squalene biosynthesis cascade [27], the mechanistic details of this enzymatic process have been very difficult to study. This is because the active enzyme is intimately involved with the cellular membrane. Furthermore, small-molecule catalysts have not been able to mimic this process.

Following the remarkable formation of bicyclo[3.1.0]hexane (**25**), the methodology of antibody-mediated cationic cyclizations was extended to terpene-like electrophilic cyclization. This was accomplished by antibodies raised to hapten **32** (Fig. 7.6) [28]. In this reaction, the substrate **26** must assume a pseudo-chair conformation to properly align the olefin such that the HOMO of the π bond donates into the incipient carbocation. The concerted process involves the expulsion of the leaving group with

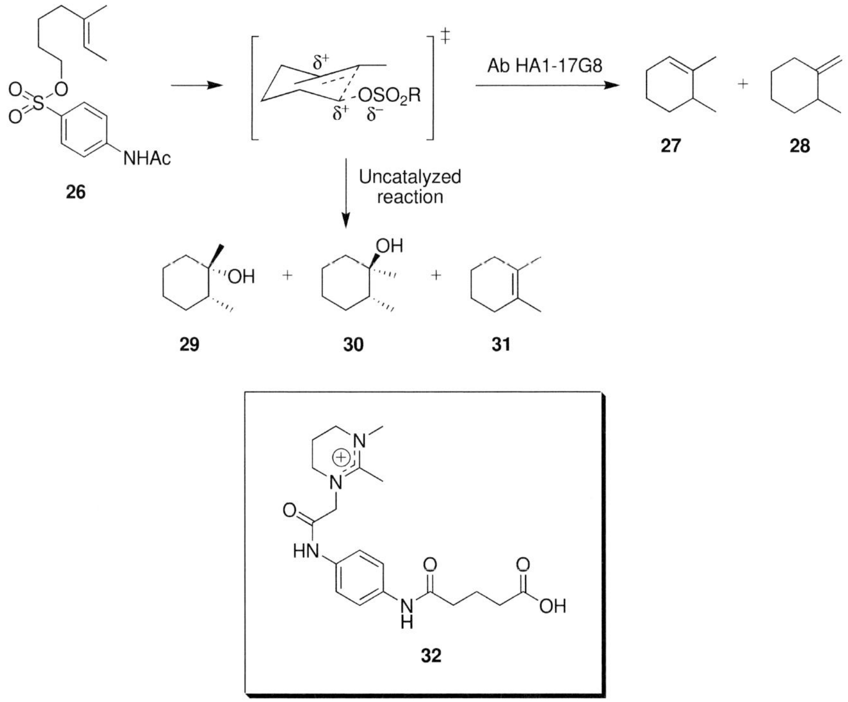

Fig. 7.6 Hapten structure (**32**), mechanism, and product distribution of antibody-catalyzed and acid-catalyzed monoterpenoid formation from substrate **26**.

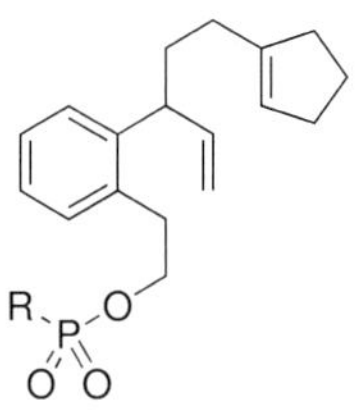

Fig. 7.7 Substrate (**33**) and hapten (**34**) structures from the first published attempt at tandem-catalyzed cationic cyclizations.

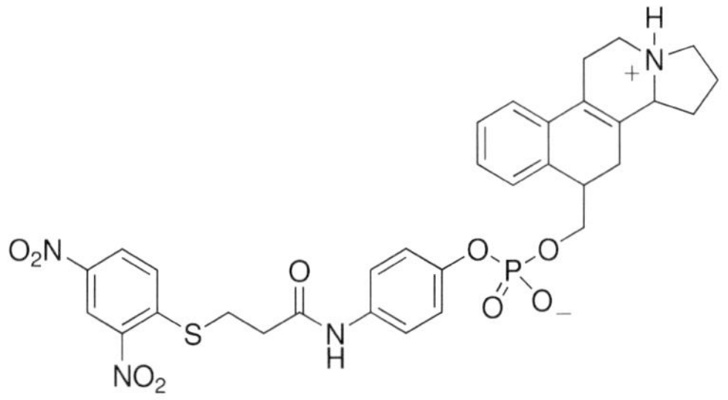

delocalization of the positive charge among several carbon atoms. Not only is the geometry of the desired transition state approximated by the hapten, but a positive charge is distributed within the ring in a manner consistent with the cationic, cyclohexyl intermediate.

Following immunization with hapten **32**, three of 24 monoclonal antibodies that bound the hapten were able to catalyze the expulsion of acetamidobenzylsulfonate from the substrate. The most active antibody, 17G8, produced only two products, **27** and **28**, (Fig. 7.8), in stark contrast to the uncatalyzed reaction, where the product mixture is more complex, containing compounds **29–31** but not **27** or **28**. The lack of any alcohol products formed in the antibody-catalyzed reaction indicates that the combining site successfully excludes solvent. The elimination products formed under antibody control are interesting because the most favored elimination product **31** is not formed. This result indicates that the antibody participates in the termination event by directing which proton will be eliminated from the cyclohexyl cation. Furthermore, the products that are formed in the antibody-mediated process possess the core structural elements found in α- and γ-irone, natural terpene components of violet derived perfumes.

7.3.2
Tandem Cationic Cyclization Mediated by Antibody

The first attempts at antibody-catalyzed cationic polyene cyclization [29] were published around the same time as the successful achievement of single cationic cyclization. Though this work did not result in antibody catalysts, the details of hapten design lend insight into the important mechanistic considerations of generating and

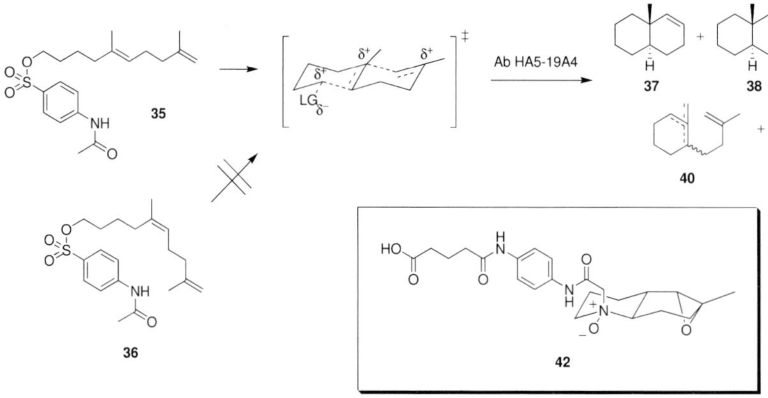

Fig. 7.8 Hapten structure (**42**), mechanism, and product distribution of antibody-catalyzed and acid-catalyzed cationic cyclization from substrate **35**. It is important to note that the antibody enables the tandem process, whereas no decalones (**37–39**) are formed in the acid-catalyzed reaction.

manipulating carbocations within antibody combining sites. Tertiary amine **34** was prepared and employed as the hapten (Fig. 7.7). This hapten was designed to raise antibodies capable of processing substrate **33** by forming the two C–C bonds that close the B and C rings of the estrogen ring system. Notably, the A ring of the estrogen skeleton is replaced with a phenyl ring in both the hapten and substrate. This was done not only to enforce the rigid planarity of this portion of the molecule but also to stabilize developing cationic character at C6, thus enhancing the sulfonate's ability to be expelled.

Hapten **34** was designed with the goal of realizing a fully concerted cascade process within the antibody combining site. Placement of the nitrogen (and thus the positive charge) at position 13 of the steroid scaffold precludes the selection of residues to stabilize cationic intermediates during the cascade process. Significantly, no part of **34** imitates the requisite cationic character at C6 of the steroid skeleton. The lack of a design element that would select for residues to stabilize the initiation step of the reaction constrains the ability to raise catalytic antibodies.

Finally, comparison of the desired reaction with mechanistic knowledge of natural terpene cyclase enzymes provides further insight into design factors for haptens to elicit catalytic antibodies for cascade cationic cyclizations. In most terpene cyclase enzymes, formation of the initial carbocation (viz. epoxide opening) and closure of ring A are concerted (often requiring participation of the olefin), while subsequent ring closures are not (due to barrierless relaxation of the substrate). This knowledge is practical inasmuch as numerous steroid structures are the result of methyl or hydride shifts that occur during the cascade process [30]. In order to faithfully mimic the details of these reactions within an antibody combining site, a hapten must therefore combine the precise placement of dipoles and point charges with conformational rigidity.

Building on all this work, the most successful antibodies to catalyze cationic cyclizations enable a tandem process to take place (Fig. 7.8) [31, 32]. An *N*-oxide moiety was again incorporated in the hapten to elicit residues to promote the initiation step in the cyclization. Three of 22 antibodies that were isolated from immunization with the diastereomeric mixture **42** could accelerate the release of the aryl sulfonate, and, among these, HA5-19A4 was the most active and studied in detail. Both the E- and Z-substrates **35** and **36** were screened, and only **35** was processed by HA-19A4. The products of this process were characterized by GC and NMR, and it was found that the alkene products **37–39** predominated (70%), although a substantial amount of the monocyclic alcohols **41** (30%) were formed. The alcohols arise from the quenching of the intermediate carbocation by water, while the olefinic products result from the elimination of a proton following the second ring closure. The product distribution suggests that there is partial solvent accessibility within the combining site (perhaps where the epoxide oxygen sits within the hapten), but that this water molecule is not poised to quench the carbocation at C9. The relative abundances of the decalone products (2:3:1) is what is expected based on the thermodynamic stability of those species. Contrary to what was observed with 17G8, HA5-19A4 does not direct which proton is eliminated.

Kinetic studies revealed that only one enantiomer of **42** was an inhibitor of the antibody-catalyzed process. Inversion at the tertiary N-oxide center is therefore sufficient to distinguish between substrate and non-substrate. Although the Stork-Eschenmoser concept requires that the leaving group should be expelled pseudo-equatorially, this hapten geometry is expected to favor axial departure. The identification of the enantiomer of the hapten that elicited HA5-19A4 coupled with the demands of the Stork-Eschenmoser proposal suggest that the linker portion of the hapten and substrate is flexible enough for a pseudo-equatorial geometry to be attained.

Compound **43** (Fig. 7.9) was found to be an inhibitor of the antibody-catalyzed process. Additionally, the aryl sulfonate of **43** is not expelled by the antibody (likely resulting from the conjugation of the alkene, which both restricts its geometry and makes it more electron-deficient). This result strengthens the argument of anchimeric assistance in antibody catalysis of cationic cyclization through the proper alignment of an electron-rich double bond in the substrate.

X-ray structural analysis of the Fab of HA5-19A4 was performed [33], and favorable comparisons were made between the antibody combining site and the active sites of numerous terpenoid cyclases. This result was somewhat surprising, because although the two evolutionarily independent families of terpenoid cyclases have numerous structural features in common [18], they are based on α-helical scaffolds, whereas antibodies have nearly pure β superstructures. From a casual inspection, it is clear that both the enzymes and antibodies that catalyze cationic cyclizations allow these processes to proceed deep within a cavity lined with hydrophobic residues. This cavity sequesters the linear polyene substrate from solvent and rigidly enforces the

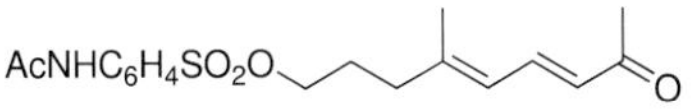

AcNHC$_6$H$_4$SO$_2$O

43

Fig. 7.9 A mechanism-based inhibitor of HA5-19A4.

desired product geometry. Furthermore, the active sites contain numerous aromatic residues, which should help stabilize the propagating, transient positive charges through cation-π interactions. Remarkably, the all-β structure of the antibody scaffold positions residues in HA5-19A4 in a manner reminiscent of the naturally-occurring terpene cyclases.

From the above discussion of haptens that were used to elicit antibodies for the cationic cyclization, it is apparent that this reaction cannot occur without the stabilization of the initial carbocation. This is uniquely achieved through the use of charged or highly polarized groups in the hapten structure in accordance with the bait-and-switch approach to antibody catalysis. The success of this technique is again seen in the structure of the antibody fold, with both acidic residues stabilizing the departing leaving group and basic and aromatic residues positioned to accept the nascent carbocation. In this way, Asn^{H35} forms a hydrogen bond to the oxygen of the N-oxide and is believed to stabilize the departing sulfonate. Three aromatic residues, Trp^{L91}, Tyr^{L96}, and Tyr^{H50}, are poised to stabilize the carbocation.

From studies of alternative substrates and inhibitors of HA5-19A4, it was shown that only electron-rich olefins are substrates. This substrate restriction suggests that the first cyclization is a concerted process which circumvents the high barrier of generating a primary carbocation. Even so, the π clouds of the aromatic residues at L91, L96, and H50 would stabilize partial positive character at C1. Immediate closure of the A ring of the *trans*-decalin results in an intermediate with a tertiary cation at C5. From the structure, there are another three aromatic residues – Tyr^{L36}, Trp^{L91}, and Tyr^{L34} – with π-cloud moments aligned to stabilize a positive charge at C5. All of these structural features are in line with what is known about how the natural cyclase enzymes such as lanosterol synthase utilize a concerted mechanism for initiation and the closure of ring A. Furthermore, because numerous side products are observed, it is likely that HA5-19A4 does not enforce concerted closure of the B ring. The disconnection between ring A and ring B closure is in accord with the cyclization pathway of tetracyclic triterpenoids. Significantly, though, there are two aromatic residues (Tyr^{L36} and Trp^{H103}) with π-clouds poised to stabilize the tertiary carbocation at C9 following closure of the B ring.

Upon closure of the B ring, the only decalins formed are the *trans*-decalins, suggesting that the substrate is conformationally constrained throughout the course of the cascade process. In accord with the ratio of the product alkene regioisomers, no catalytic base is present near any of the protons to favor one of the possible products. The simpler, prokaryotic terpene cyclases also lack a basic residue to direct the elimination of the proton which terminates the cascade process. In contrast, eukaryotic oxidosqualene cyclases have been found to possess basic residues, and thus yield a single product [34–36].

With the accomplishment of tandem cationic cyclization within antibody combining sites, two fundamental questions remain unanswered. First, can catalytic antibodies process triterpenoid substrates, and, second, can catalytic antibodies generate a carbocation from a natural leaving group (i.e., an epoxide oxygen) instead of the more facile aryl sulfonate. In order to address these questions, hapten **47** (Fig. 7.10) was synthesized [37] from lithocholic acid (**46**) and used to immunize mice and raise

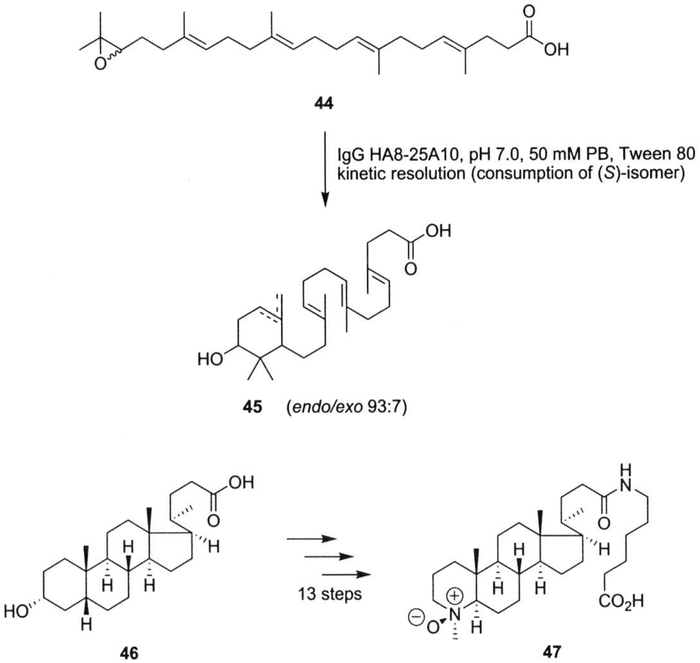

Fig. 7.10 A lithocholic acid (**46**) derived hapten (**47**) for the cationic cyclization of substrate **44** to monocycle **45**.

antibodies [38]. This hapten differs from those used previously in two important ways. With the exception of the A ring, all of the atoms throughout the tetracyclic core retain their identity, connectivity, and both relative and absolute configurations as in lithocholic acid. Also, the linker is positioned distal to the N-oxide moiety in order to direct that portion of the molecule to the deepest part of the combining site. This feature is in accord with structures of squalene-hopene synthase, which showed the epoxide moiety of the natural substrate at the deepest portion of the active site.

Out of 25 antibodies recovered, three were catalysts for the initiation of the cationic cyclization (viz., epoxide opening) and the formation of ring A of the lanosterol nucleus. The antibody 25A10 was the fastest, and on further study it was found to make perform resolution of the racemic substrate **44**. This antibody generates solely **45**, meaning that the cation is quenched by elimination of a proton, so the combining site effectively excludes water. The *endo* product **45** has the same ring system as the plant natural product Achilleol A from *Achillea odorata*. No polycyclic product was obtained, but results with cyclase enzymes show that subtle manipulation of substrate structure can drastically affect the outcome of cationic cascade reactions. For example, a non-natural diethyl squalene derivative was screened as a substrate for oxidosqualene cyclase and cyclization terminated following closure of ring A [39].

In order to assess whether the subtle manipulations of the substrate could affect ring closure, a small panel of alternative substrates was screened [40]. With all of the

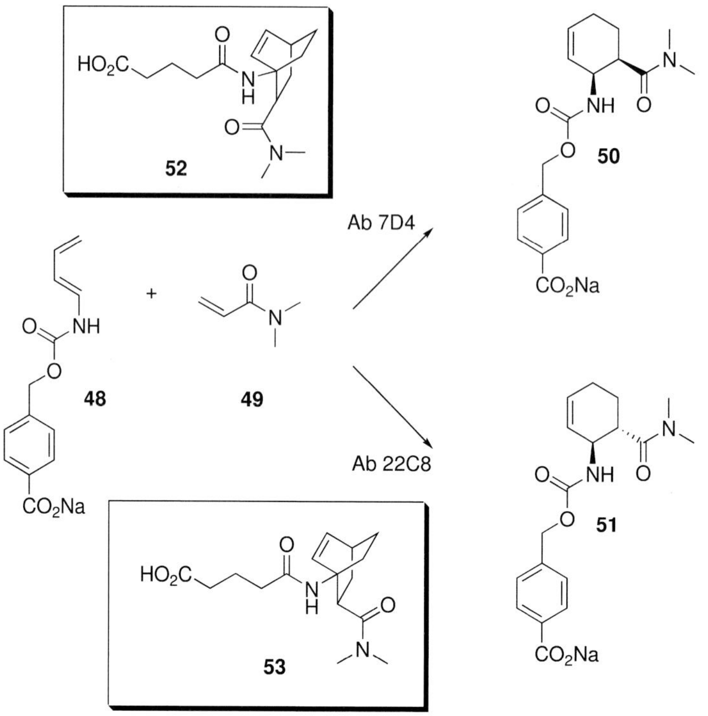

Fig. 7.11 Rational hapten design for the *endo* and *exo* Diels-Alder cycloadditions.

substrate analogs, termination followed closure of ring A. From the results with these compounds, it appears as though the methylation pattern and substitution patterns of the double bonds do not directly influence the course of 25A10 catalyzed ring closure.

This work provided partial answers to those questions set forth. It is now known that an antibody combining site can facilitate the opening of an epoxide to generate a carbocation. This work also showed that triterpenoid skeletons can be substrates for antibody cyclases. What remains to be addressed is whether catalytic antibodies can enable cascade processes with triterpenoid scaffolds and can control these reactions with a prowess rivaling that of the natural cyclase enzymes.

Recent structural work on oxidosqualene cyclase enzymes points to other obstacles that will need to be addressed before further progress can be made. For example, squalene-hopene cyclase has an active site that accommodates the linear terpenoid substrate. However, during the cyclization process the substrate's physical dimensions shrink. As rings are formed, interatomic distances are reduced, and instead of having residues positioned to precisely match the dimensions of the linear substrate, the residues within the squalene-hopene cyclase active site are poised to accommodate substrate collapse [41]. In order to address a collapsing substrate with an antibody combining site, it will be necessary to develop further hapten technology (e.g., expanding reactive immunization to substrates that undergo conformational changes on binding).

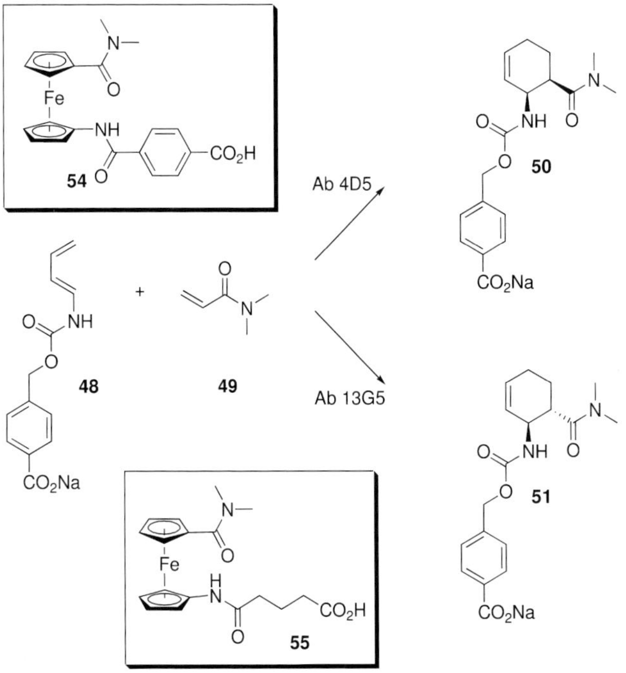

Fig. 7.12 Ferrocene-based haptens that resulted in the isolation of both *exo* and *endo* Diels-Alderase antibodies.

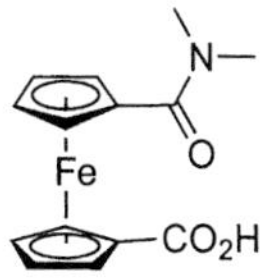

Fig. 7.13 A hapten derived inhibitor for 13G5.

7.4
exo-Diels-Alder Reactions

Diels-Alder reactions are [4+2] cycloaddition reactions in which a diene and a dienophile react to form two new C–C bonds and up to four stereocenters. If the dienophile is unsymmetrical there are two possible approach geometries. The *endo* approach positions the reference substituent of the dienophile toward the π orbitals of the diene, while the *exo* geometry orients the dienophile in the opposite manner. Although the *endo* transition state is often more sterically congested than the corresponding *exo* arrangement, *endo* addition is usually preferred. This observation, known as the Alder rule, arises from the importance of the interaction between the dienophile substituent and the π electrons of the diene [42]. In a given Diels-Alder reaction where one of the reactants is unsubstituted and the other is monosubstituted, there are as many as eight possible products. As a result, in order to produce a single one of the more disfavored products, nearly total stereochemical control of the addition reaction is required.

Although numerous Diels-Alderase antibodies have resulted from a number of different hapten designs [43–46], only a few have been obtained that catalyze the disfavored *exo* Diels-Alder cycloaddition. Two different hapten strategies resulted in the isolation of antibodies that catalyze this disfavored intermolecular reaction.

The first approach [47] exploited the differential disposition of dienophile substituents in Diels-Alder adducts between the *exo* and *endo* products. This geometrical constraint was a key factor in hapten design. By using a bicyclo[2.2.2]octene moiety to rigidly impose a boat-like conformation on the cyclohexene ring, the dienophile substituent should be directed to either the *endo* or *exo* approach accordingly. Both computational analysis (at the RHF/6-31G* level) and X-ray structure determination of the haptens revealed that the boat-shaped conformation of the cyclohexene ring adequately mimics the pericyclic transition state. Furthermore, any resultant antibodies which would bring together both diene and dienophile in a productive geometry would have a good chance of being catalysts, because there should be no product inhibition due to the greater stability of the twisted-chair conformation of the products.

Of the twenty-two antibodies that were recovered from mice immunized with hapten 52, two were found to be catalysts for the favored *endo* addition, the more proficient of these antibodies being 7D4. Twenty-five antibodies were obtained from mice immunized with hapten 53 and five antibodies were isolated that accelerate the formation of the disfavored *exo* addition, the best of these antibodies being 22C8. The kinetic parameters of these antibodies were surprisingly similar (Table 7.4), given that one catalyzes the favored addition while the other facilitates the disfavored addition. Not surprisingly, though, because of the inherent chirality of the antibody combining site, both 7D4 and 22C8 catalyze the formation of their respective adducts in greater than 98% enantiomeric excess. In comparison, the uncatalyzed reaction gives a racemic mixture of 85:15 *endo:exo* products.

The preceding work successfully employed rational hapten design to obtain antibodies for the disfavored *exo* Diels-Alder reaction using a conformationally locked bicyclo[2.2.2]octene hapten. A second strategy was pursued to obtain antibodies for the *exo*-Diels-Alder reaction that took advantage of the inter-Cp ring distance in the ferrocene molecule [48]. This distance, 3.3Å, is intermediate in the approach ge-

Tab. 7.4: Kinetic parameters and product distribution of Diels-Alderase antibodies.

Antibody	Selectivity	k_{cat} (min^{-1})	Effective molarity (M)	K_M (48)	K_M (49)	Enantio-selectivity
7D4	endo	3.44×10^{-3}	4.8	0.96 mM	1.7 mM	> 98%
22C8	exo	3.17×10^{-3}	18	0.7 mM	7.5 mM	> 98%
13G5	exo	1.20×10^{-3}	6.9	2.7 mM	10 mM	95 ± 3%
4D5	endo	3.48×10^{-3}	4.9	1.6 mM	5.9 mM	95 ± 3%

$k_{uncat}(endo) = 7.15 \times 10^{-4}$ min^{-1}
$k_{uncat}(exo) = 1.75 \times 10^{-4}$ min^{-1}

ometry for a Diels-Alder reaction. Although there is a low thermodynamic barrier to rotation of the ferrocene rings, individual antibodies were expected to be able to "freeze out" individual rotamers. Because of this ability, the population of catalytic Diels-Alderase antibodies raised against ferrocene-derived haptens could include catalysts for diastereomeric reactions from a single immunization. Furthermore, these studies would allow an investigation into the role of hapten flexibility and whether catalytic antibodies result from conformationally unrestricted molecules. Finally, the hydrophobic character of the cyclopentadienyl system should be highly immunogenic, as well as capable of inducing a hydrophobic pocket in the antibody, an important feature needed for the stabilization of the Diels-Alder transition state and improved reaction rates.

Two haptens were prepared which differed only in the linker that was used to couple them to their carrier proteins. Seven of the thirty-three antibodies recovered from mice immunized with hapten 55 were found to catalyze Diels-Alder addition (1 *endo*, 6 *exo*) and eight out of thirty-eight antibodies recovered from mice immunized with hapten 54 were found to catalyze Diels-Alder addition (7 *endo*, 1 *exo*). Although it appears as though the linker is playing a role in the selectivity of the reactions described, this is unknown.

The most proficient *exo* and *endo* Diels-Alderase antibodies were fully characterized and found to have effective molarities comparable with those antibodies obtained using the constrained bicyclo[2.2.2]octene system (Table 7.4). This result strongly supports the assertion that antibodies would freeze out specific conformations of an otherwise freely rotating molecule.

The structure of the Fab fragment of 13G5 complexed with a functionalized ferrocene molecule 56 related to hapten 55 was determined by X-ray crystallography to 1.95 Å [49]. Over 99% of the surface area of 56 was buried within the antibody combining site. In addition to 45 van der Waals contacts between the antibody and 56, there are three hydrogen bonds that act to stabilize 56 in an eclipsed conformation. One hydrogen bond is made from Tyr^{L36} to the portion of 56 equivalent of the dienophile carboxyl. This hydrogen bond acts not only to position this cyclopentadienyl ring, but also as a Lewis acid, drawing electron density away from the dienophile. The other two hydrogen bonds (from Asn^{L91} and Asp^{H50}) orient the surrogate diene by binding its carboxylate moiety. The hydrogen bonds to the diene carboxylate provide electron density into the diene system, making it more electron rich. Each of these three residues is in turn held in place by an extensive hydrogen-bonding network with amino acid side chains, the peptide backbone, and structural water molecules. While the extensive van der Waals contacts roughly shape the combining site, the precise geometry of the hydrogen-bonding network fine tunes the ferrocene ring positions and effectively freezes out the eclipsed conformer.

In addition to placing stringent geometrical constraints on 56 (and therefore the diene and dienophile), the hydrogen-bonding network also plays a crucial role in catalysis. Specifically, hydrogen bonding to the dienophile has been shown to play a role in the rate enhancement of Diels-Alder reactions. Theoretical calculations have shown that the C^+–O^- polarization of the carbonyl of the dienophile is enhanced in the transition state, and each hydrogen bond stabilizes the transition state by 1.5–2.0

kcal/mol. For example, the lowest energy transition state of the addition of methyl vinyl ketone to cyclopentadiene shows that a single water can act catalytically by providing a hydrogen bond to the carbonyl *syn* with respect to the double bond. In the structure of 13G5 with 56, this same geometry is observed for the hydrogen bond from Tyr^{L36} to dimethylacrylamide. However, because this hydrogen bond would be equivalently replaced by a water in the reaction in aqueous buffer, it most likely does not increase the rate of the reaction. Calculations reveal that the most significant hydrogen bond is the short (2.5 Å) hydrogen bond between the NH of the diene carbamate and the carboxylate of Asp^{H50}. This interaction was calculated to lower the activation energy of the cycloaddition by 5.9 kcal/mol.

Docking studies using both the liganded and unliganded structures of 13G5 found that the combining site is large enough to accommodate all four of the possible stereoisomeric transition states of diene 48 and dienophile 49. However, only the (3R,4R)-*exo* transition state docks in such a way as to have both of the catalytic hydrogen bonds as described above. The precision of the hydrogen-bonding network alone is therefore likely the source of the rate enhancement and, along with steric constraints, dictates the selectivity of 13G5.

7.5
Miscellaneous Disfavored Processes

There have been preliminary reports on a number of other disfavored processes catalyzed by antibodies. These will be described in brief, as none has received in-depth attention.

7.5.1
Syn-Elimination to a *cis* Olefin

The *syn*-elimination to a *cis* (Z) olefin was an early example of an antibody-catalyzed disfavored reaction [50]. Typically *syn* eliminations must proceed through an eclipsed conformation requiring substantial energy to overcome not only steric and torsional strain, but also to surmount orbital symmetry barriers. In most acyclic systems, antiperiplanar elimination is favored by several kcal/mol. There are a small number of acyclic substrates that undergo *syn* elimination, but these all result in the formation of the corresponding *trans* (E) olefin.

In order to obtain an antibody that catalyzes the *syn* elimination to the *cis* (Z) olefin, a conformationally locked hapten (60) was used (Fig. 7.14). This bicyclo[2.2.1]heptane displayed aryl groups in equatorial positions from neighboring carbons of the 6-membered ring. A primary amine was placed axial to one of these carbons in order to elicit a general base in the antibody combining site to promote the abstraction of the proton to facilitate formation of the olefin.

It was estimated that the best antibody for the *syn* elimination, 1D4, provides at least 5 kcal/mol of stabilization to the substrate in an eclipsed conformation. Structural analysis [51] of the Fab of 1D4 illustrated that this stabilization is accomplished by the

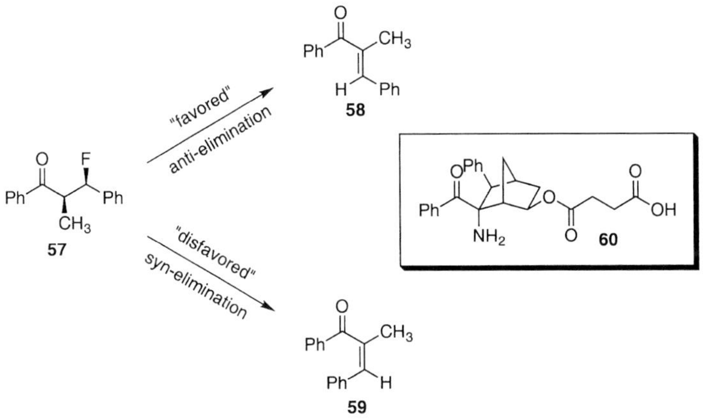

Fig. 7.14 Hapten structure (**60**), substrate, and products of antibody-catalyzed and uncatalyzed elimination.

creation of two deep hydrophobic slots for the binding of the phenyl rings. Structural work also suggests that the base responsible for initiating the elimination reaction is a hydroxide ion positioned beneath the substrate by a protonated His^{H58}. Although many details of the antibody-catalyzed *syn* elimination to the *cis* (Z) olefin are clear, the data are ambiguous with regard to the determination of the precise mechanism employed. While a strict E1 mechanism is least likely, the data do not unambiguously support either an E2 or E1cB mechanism.

7.5.2
Aryl Carbamate Hydrolysis via B$_{Ac}$2 Mechanism

An even more energetically demanding reaction for which antibodies have been elicited is the rerouting of carbamate hydrolysis [52]. Typically, carbamates are hydrolyzed via an E1cB (elimination-addition) process where the N–H group is deprotonated and the aryl alcohol is expelled, yielding the isocyanate. This intermediate is rapidly hydrolyzed to the carbamic acid, which spontaneously decarboxylates to give the amine. An alternative mechanism is available in which the addition occurs before the elimination. This is termed the B$_{Ac}$2 pathway and is typically only observed for carbamates without an N–H moiety. In this mechanism, hydroxide anion adds to the carbonyl carbon to yield a tetrahedral intermediate. The aryl alkoxide is then expelled to provide the carbamic acid, which again spontaneously decarboxylates to yield the amine (Fig. 7.15).

There are a few key features in the hapten design that resulted in catalytic antibodies for aryl carbamate hydrolysis via a B$_{Ac}$2 mechanism. First, the core element of hapten **61** is a tetrahedral phosphoryl group that mimics the geometry and anionic character of the desired intermediate. Second, the phenolic oxygen of the substrate was replaced with a benzylic methylene group in order to discourage the stabilization of anionic character at this position (a key feature of the E1cB mechanism). Finally,

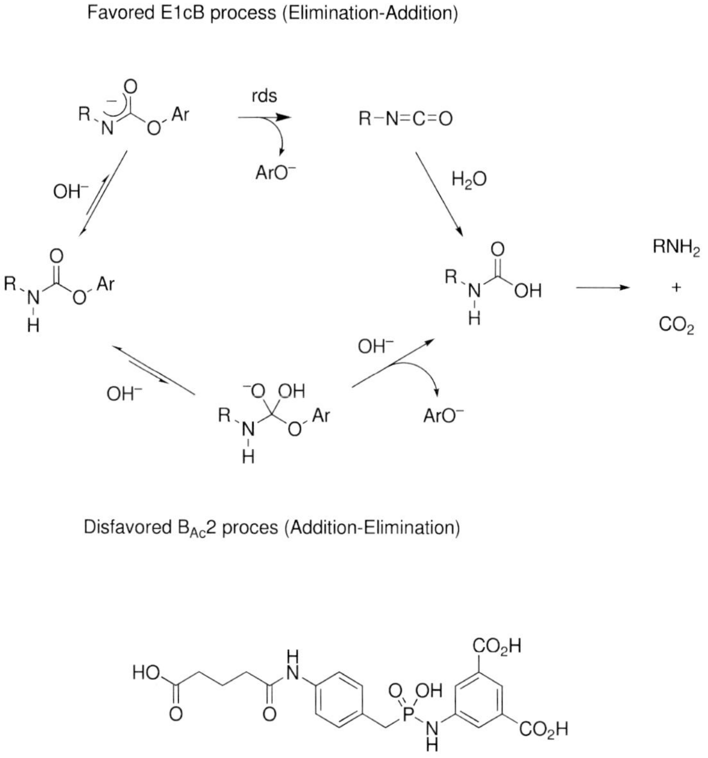

Fig. 7.15 Comparison of two mechanisms of carbamate hydrolysis, E1cB and $B_{Ac}2$, along with the hapten (**61**) used to raise antibodies to catalyze the $B_{Ac}2$ process.

the linker is positioned distal to these core structural elements, allowing the chemistry to take place in a buried pocket. The successful generation of antibodies that promote the $B_{Ac}2$ mechanism of aryl carbamate hydrolysis suggests that antibodies can provide substantially more than half of their available binding energy toward catalysis. Binding energy between antibody and small molecule has been estimated to be as high as 20 kcal/mol, and the stabilization of the $B_{Ac}2$ pathway must be at least 13 kcal/mol.

7.5.3
Transesterification

Transesterification in water is a disfavored bimolecular reaction due to the fact that water is present in vast excess over both the alcohol and the ester; thus, ester hydrolysis predominates. In order for a transesterification to effectively compete with ester hydrolysis, the substrates must be sequestered within a hydrophobic environment.

7.5 Miscellaneous Disfavored Processes

The antibody combining site provides an ideal microenvironment to facilitate this type of reaction within a greater aqueous context. As such, transesterification has been accomplished by a number of catalytic antibodies.

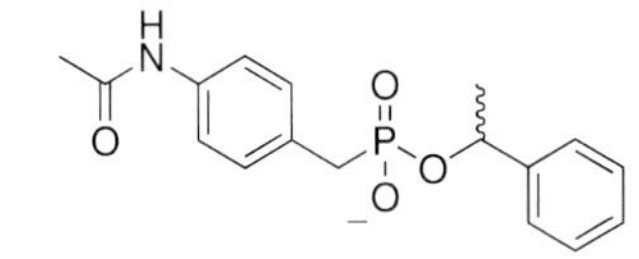

62

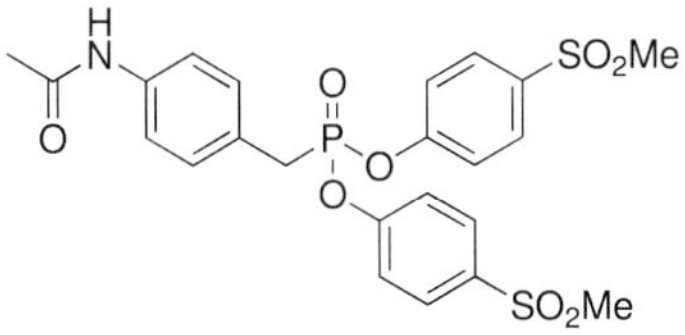

63

Fig. 7.16 Haptens designed for immunization to elicit transesterase antibodies via transition state stabilization and reactive immunization.

The first antibody-catalyzed transesterification used benzyl phosphonate hapten **62** (Fig. 7.16) for immunization, and one of the antibodies that was recovered, PCP21H3, was a catalyst for the facile transesterification of aryl esters with aryl alcohols [25]. Detailed kinetic analysis of this antibody showed that PCP21H3 uses a sophisticated mechanism in its catalytic cycle. Instead of the more straightforward random sequential addition mechanism that proceeds through an obligate ternary complex of antibody, ester and alcohol, it was shown that PCP21H3 participates covalently through an acyl-antibody intermediate. That this intermediate is crucial in catalysis was demonstrated through a number of more detailed kinetic experiments. For example, it was shown that PCP21H3 catalysis proceeds through a burst phase (indicative of the buildup of concentration of a critical intermediate) as well as substrate inhibition. Finally, the presence of the covalent acyl-antibody intermediate was demonstrated by using a radiolabeled acyl-donor and examining the retention of radioactivity by the antibody during extensive dialysis. Taken together, these data strongly suggest that PCP21H3 catalyzes transesterification using a ping-pong bi-bi mechanism [53].

In addition to the kinetic characterization that PCP21H3 proceeds through a covalent adduct, further experiments with substrate analogs demonstrated that the antibody combining site likely recognizes its substrates via induced fit. This result was rationalized as a consequence of the hapten design. While the hapten is a rigid transition-state mimic and should give rise to a binding pocket of defined geometry within the antibody combining, the actual substrate to transition state transformation is a fluid process that may require active-site remodeling for one or more of the steps of the antibody-catalyzed transesterification.

Further investigations into antibody-catalyzed transesterification revealed that one of the antibodies, SPO50C1, raised using reactive immunization for ester hydrolysis

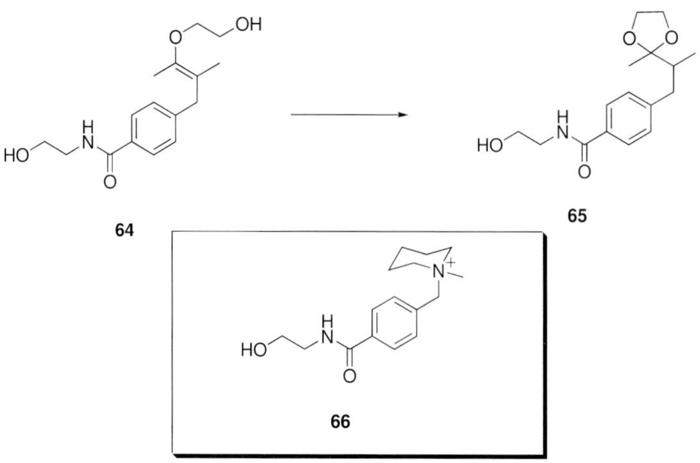

Fig. 7.17 Hapten and substrate for antibody-catalyzed ketal formation.

[54], also had transesterase activity [55]. Reactive immunization takes advantage of the energy of a chemical reaction during immunization and allows hapten **63** to form a covalent adduct with antibodies that contain facilitative residues. In this way, specific reactivity is programmed into the combining site without the need for specific shape complementarity. This key difference between the two immunization strategies is demonstrated by comparing the kinetic parameters and substrate tolerance of PCP21H3 and SPO50C1.

For antibodies which were raised against transition state analogs, the accepted substrates are limited to those which have substantial identity with the hapten. However, the effective molarity that these antibodies impose on the reaction is greater than 10^6. These parameters are effectively the reverse for SPO50C1, the transesterase antibody raised via reactive immunization. This antibody accepts a broad range of substrates including alkyl alcohols, but does not show an effective molarity greater than 4×10^3. Interestingly, antibodies raised by either process utilize a ping-pong mechanism that relies on the formation of an acyl-antibody intermediate.

7.5.4
Ketal Formation

Another example of an antibody-catalyzed reaction that is extremely disfavored in water is the transformation of an enol ether into a ketal [56]. This reaction is useful from a synthetic perspective, because the ketal is a more versatile protecting group for the carbonyl group. Antibody-catalyzed ketal formation was accomplished using a catalytic antibody originally designed with glycosidase activity, 14D9 [57]. This antibody resulted from immunization with hapten **66** and was found to possess a number of mechanistically-related reactivities including epoxide opening [23] and the hydrolysis of enol ethers [58, 59]. Detailed understanding of these mechanisms and clever substrate design resulted in antibody-catalyzed ketal formation. Substrate

64 is processed like other enol ethers except that, instead of hydrolysis, the cationic intermediate is captured by an intramolecular reaction with the pendant hydroxyl to give the ketal **65**.

7.5.5
Controlling Photoprocesses

In addition to directing the reactivity of carbocations through cationic cyclization and covalently participating in chemical transformations such as transesterification (see above), antibodies have been isolated which participate photochemically and are able to manipulate other highly reactive species including diradicals and radical anions.

For example, catalytic antibodies have been elicited that are able to cleave both thymine [60] and uracil [61] homodimers in a light-dependent manner. Pyrimidine dimerization represents the bulk of DNA damage following exposure to ultraviolet radiation. The enzyme DNA photolyase repairs these lesions by using longer wavelength light to catalyze dimer cleavage [62]. Action spectra and tryptophan fluorescence quenching experiments were performed with the antibodies that catalyze pyrimidine dimer photocleavage (Fig. 7.18), and the results support the assertion that the reaction proceeds via photosensitization through a properly positioned indole side chain. The proximity of a tryptophan to the substrate allows a single electron transfer to occur, generating the indolyl radical cation and the pyrimidine dimer radical anion. By virtue of being sequestered in the antibody combining site, this electronically excited species is shielded from solvent and directed toward scission.

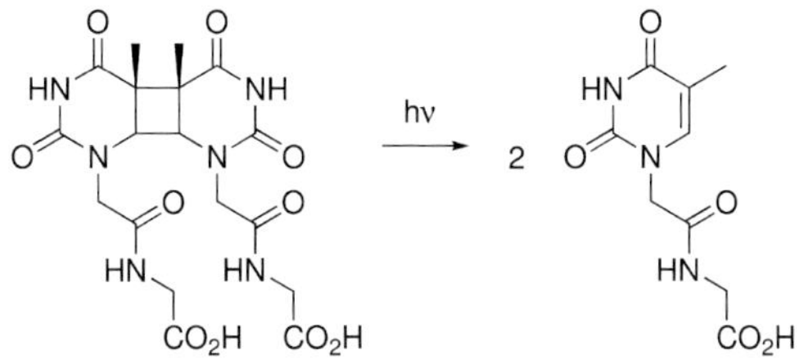

Fig. 7.18 Hapten and substrate for the antibody-catalyzed photocleavage of pyrimidine homodimers.

Antibodies for pyrimidine dimer photocleavage have kinetic parameters that compare favorably to their native enzyme counterparts. However, the antibodies require much shorter wavelengths for their activity. Because of their perceived lack of chemical potential from an evolutionary perspective, antibodies not only do not bind cofactors, but have no need to do so. As a result, antibody molecules are not able to productively absorb light at wavelengths longer than 280 nm. Enzymes that rely on light energy to accomplish their catalytic task take advantage of both flavin and folate cofactors. Cofactors allow enzymes such as DNA photolyase to carry out the same reaction using longer wavelength light (300–500 nm).

In addition to directing the reactivity of radical anions within an inert active site, antibodies have been elicited that catalyze the Norrish type II reaction – the specific recombination of a photogenerated diradical – with both ketone and α-ketoamide

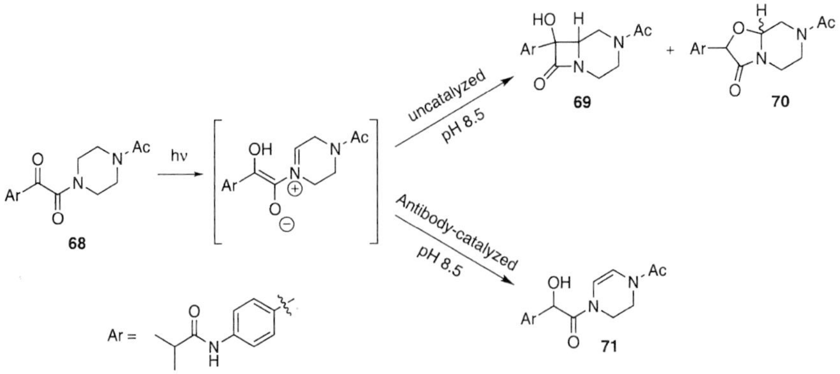

Fig. 7.19 Hapten and substrate and products generated in the uncatalyzed and antibody-catalyzed Norrish type II reaction.

substrates [63]. The desired substrate **68** (Fig. 7.19) was used as a hapten in these studies because the reaction relies on light energy to initiate the process, so the antibody catalyst is only required to direct an intermediate's reactivity. Basically, the binding energy to the ground state substrate was expected to perturb the reaction coordinate of the photogenerated diradical. Three of twenty-two monoclonal antibodies that were isolated for this study were found to quantitatively direct the reactivity of the substrate to a tetrahydropyrazine product (**71**). Under the mild conditions of antibody catalysis, not only is a mixture of products formed, but tetrahydropyrazine is not observed. A pH profile demonstrated that the product distribution is shifted in both the uncatalyzed and the antibody-catalyzed reactions. While it is known that the uncatalyzed Norrish type II rearrangement is pH sensitive, the observation of this dependence in the catalyzed reaction suggests that there is a residue participating as a general base in the antibody combining site at pH>7.5. When this residue is protonated, it is unable to assist in the processing of the substrate, leading to a complex mixture of products including substrate fragmentation.

7.6
Summary

Antibodies are capable catalysts for a broad scope of disfavored transformations, and they accomplish this feat using a combination of conformational constraints, electrostatic stabilization of high-energy intermediates with both point charges and properly directed π clouds, and exquisite control of reactivity within the active site. Remarkably, small-molecule haptens are able to convey these instructions to the immune system such that individual antibodies catalyze a single reaction path for a substrate even in the context of many similar trajectories. Although there are efficient small-molecule catalysts for a number of important transformations, catalytic antibodies provide access to difficult transformations under extremely mild reaction conditions and provide insight into how evolutionary systems perform chemical tasks with exquisite specificity.

While a substantial amount of work has already been done in this area, there are a number of challenges that remain including:

- Improving the kinetic parameters of these and other catalytic antibodies
- Increasing the scope of substrates that can be processed by a single antibody
- Increasing the complexity of the substrates that can be processed
- Elaborating the mechanisms that are employed
- Providing an upper bound on the amount of binding energy the antibody can direct toward catalysis
- Incorporating cofactors for electron transfer and photoinitiated processes.

Finally, two recent discoveries regarding the capacity of antibodies have been made that should guide new research endeavors. First, in addition to manipulating ground states of molecules, antibodies have recently been found to exert influence on molecular excited states [64]. As a result, it may be possible to expand antibody catalysis of disfavored processes into the realm of electronically excited molecules.

Second, antibodies have been found to oxidize water using singlet molecular oxygen [65]. The singlet oxygen can be from any source, exogenous photosensitization, chemically provided through the collapse of an endoperoxide, or generated by UV irradiation of the antibody through photosensitization of a tryptophan. Because of the intrinsic stereochemistry of the antibody combining site, antibodies should be able to deliver one or more of the oxidants implicated in the water oxidation pathway [66] to a substrate in a facially selective as well as stereoselective manner.

Antibody-catalyzed disfavored chemical transformations having been firmly established as an area of productive inquiry, there still remains extensive research to be done in this exciting area.

References

1 POLLACK, S. J., JACOBS, J. W., SCHULTZ, P. G., *Science* 234 (**1986**), p. 1570–1573
2 TRAMANTANO, A., JANDA, K. D., LERNER, R. A., *Science* 234 (**1986**), p. 1556–1570
3 JANDA, K. D., WEINHOUSE, M. I., DANON, T., PACELLI, K. A., SCHLOEDER, D. M., *J. Am. Chem. Soc* 113 (**1991**), p. 5427–5434
4 TENUD, L., FAROOQ, S., SEIBL, J., ESCHENMOSER, A., *Helv. Chim. Acta* 53 (**1970**), 2059
5 BALDWIN, J. E., *J. Chem. Soc., Chem. Commun.* (**1976**), p. 734–736
6 JOHNSON, C. D., *Acc. Chem. Res.* 26 (**1993**), p. 476–482
7 BENKOVIC, S. J., *Annu. Rev. Biochem.* 51 (**1992**), p. 29
8 SCHULTZ, P. G., LERNER, R. A., *Acc. Chem. Res.* 26 (**1993**), p. 391–395
9 JANDA, K. D., SHEVLIN, C. G., LERNER, R. A., *Science* 259 (**1993**), p. 490–493
10 NA, J., HOUK, K. N., SHEVLIN, C. G., JANDA, K. D., LERNER, R. A., *J. Am. Chem. Soc* 115 (**1993**), p. 8453–8454
11 NICOLAU, K. C., PRASAD, C. V. C., SOMERS, P. K., HWANG, C.-K., *J. Am. Chem. Soc* 111 (**1989**), p. 5330
12 NA, J., HOUK, K. N., *J. Am. Chem. Soc* 118 (**1996**), p. 9204–9205
13 DOUGHERTY, D. A., STAUFFER, D. A., *Science* 250 (**1990**), p. 1558–1560
14 JANDA, K. D., SHEVLIN, C. G., LERNER, R. A., *J. Am. Chem. Soc*, 117 (**1995**), p. 2659–2660
15 COXON, J. M., HARTSHORN, M. P., SWALLOW, W. H., *Aust. J. Chem.* 26 (**1973**), p. 2521
16 GRUBER, K., ZHOU, B., HOUK, K. N., LERNER, R. A., SHEVLIN, C. G., WILSON, I. A., *Biochemistry* 38 (**1999**), p. 7062–7074
17 KABAT, E. A., WU, T. T., PERRY, H. M., GOTTESMANN, K. S., FOELLER, C., *Sequences of Proteins of Immunological Interest*, 5th edn., NIH, Bethesda, Maryland **1991**
18 LESBURG, C. A., CARUTHERS, J., PASCHALL, C. M., CHRISTIANSON, D. W., *Curr. Opin. Struct. Biol.* 8 (**1998**), p. 695–703
19 STORK, G., BURGSTAHLER, A.W., *J. Am. Chem. Soc* 77 (**1955**), p. 5068–5077
20 ESCHENMOSER, A., RUZICKA, L., JEGER, O., ARIGONI, D., *Helv. Chim. Acta* 38 (**1955**), p. 1890
21 LI, T., JANDA, K. D., ASHLEY, J. A., LERNER, R. A., *Science* 264 (**1994**), p. 1289–1293
22 JOHNSON, W. S., *Acc. Chem. Res.* 1 (**1966**), p. 1–8
23 KEINAN, E., SINHA, S. C., REYMOND, J.-L., *J. Am. Chem. Soc* 115 (**1993**), p. 4893–4894
24 LI, T., HILTON, S., JANDA, K. D., *J. Am. Chem. Soc* 117 (**1995**), p. 3308–3309
25 WIRSCHING, P., ASHLEY, J. A., BENKOVIC, S. J., JANDA, K. D., LERNER, R. A., *Science* 252 (**1991**), p. 680–685
26 LI, T., JANDA, K. D., LERNER, R. A., *Nature* 379 (**1996**), p. 326–327
27 ALTMAN, L. J., KOWERSKI, R. C., LAUNGANI, D. R., *J. Am. Chem. Soc* 100 (**1978**), p. 6174–6182

28 Hasserodt, J., Janda, K. D., Lerner, R. A., *J. Am. Chem. Soc* 118 (**1996**), p. 11654–11655
29 Bell, I. M., Abell, C., Leeper, F. J., *J. Chem. Soc., Perkin* 1 (**1994**), p. 1997–2006
30 Cane, D. E., *Chem. Rev.* 90 (**1990**), p. 1089–1103
31 Hasserodt, J., Janda, K. D., Lerner, R. A., *J. Am. Chem. Soc* 119 (**1997**), p. 5993–5998
32 Hasserodt, J. Janda, K. D., *Tetrahedron* 53 (**1997**), p. 11237–11256
33 Paschall, C. M., Hasserodt, J., Jones, T., Lerner, R. A., Janda, K. D., Christianson, D. W., *Angew. Chem. Int. Ed.* 38 (**1999**), p. 1743–1747
34 Lesburg, C. A., Zhai, G., Cane, D. E., Christianson, D. W., *Science* 277 (**1997**), p. 1820–1824
35 Starks, C. M., Back, K., Cappell, J., Noel, J. P., *Science* 277 (**1997**), p. 1815–1820
36 Wendt, K. U., Poralla, K., Schultz, G. E., *Science* 277 (**1997**), p. 1811–1815
37 Hasserodt, J., Janda, K. D., Lerner, R. A., *Bioorganic and Medicinal Chemistry* 8 (**2000**), p. 995–1003
38 Hasserodt, J., Janda, K. D., Lerner, R. A., *J. Am. Chem. Soc* 122 (**1999**), p. 40–45
39 Hoshino, T., Ishibashi, E., Kaneko, K., *J. Chem. Soc., Chem. Commun.* (**1995**), p. 2401–2402
40 Kim, G. T., Wenz, M., Park, J. I., Hasserodt, J., Janda, K. D., *Bioorg. Med. Chem.* 10 (**2002**), p. 1249–1262
41 Wendt, K. U., Lenhart, A., Schulz, G. E., *J. Mol. Biol.* 286 (**1999**), p. 175–187
42 Carey, Sundberg, *Advanced Organic Chemistry*
43 Braisted, A. C., Schultz, P. G. *J. Am. Chem. Soc* 112 (**1990**), p. 7430
44 Hilvert, D., Hill, K. W., Nared, K. D., Auditor, M.-T. M., *J. Am. Chem. Soc* 111 (**1989**), p. 9261
45 Meekel, A. A. P., Resmini, M., Pandit, U. K., *J. Chem. Soc., Chem. Comm.* (**1995**), p. 571
46 Suckling, C. J., Tedford, M. C., Bence, L. M., Irvine, J. I., Stimson, W. H., *Bioorg. Med. Chem. Lett.* 2 (**1992**), p. 49

47 Gouverneur, V. E., Houk, K. N., de Pascual-Teresa, B., Beno, B., Janda, K. D., Lerner, R. A., *Science* 262 (**1993**), p. 204–208
48 Yli-Kauhaluoma, J. T., Ashley, J. A., Lo, C.-H., Tucker, L., Wolfe, M. M., Janda, K. D., *J. Am. Chem. Soc* 117 (**1995**), p. 7041–7047
49 Heine, A., Stura, E. A., Yli-Kauhaluoma, J. T., Gao, C., Deng, Q., Beno, B. R., Houk, K. N., Janda, K. D., Wilson, I. A., *Science* 279 (**1998**), p. 1934–1940
50 Cravatt, B. F., Ashley, J. A., Janda, K. D., Boger, D. L., Lerner, R. A., *J. Am. Chem. Soc* 116 (**1994**), p. 6013–6014
51 Larsen, N. A., Heine, A., Crane, L., Cravatt, B. F., Lerner, R. A., Wilson, I. A. *J. Mol. Biol.* 314 (**2001**), p. 93–102
52 Wentworth, P., Jr, Datta, A., Smith, S., Marshall, A., Partridge, L. J., Blackburn, G. M., *J. Am. Chem. Soc* 119 (**1997**), p. 2315–2316
53 Cleland, W. W. *Biochim. Biophys. Acta* 67 (**1963**), p. 104
54 Wirsching, P., Ashley, J. A., Lo, C.-H., Janda, K. D., Lerner, R. A., *Science* 270 (**1995**), p. 1775–1782
55 Lin, C.-H., Hoffman, T. Z., Xie, Y., Wirsching, P., Janda, K. D., *Chem. Commun.* (**1998**), p. 1075–1076
56 Shabat, D., Itzhaky, H., Reymond, J.-L., Keinan, E., *Nature* 374 (**1995**), p. 143–146
57 Reymond, J.-L., Janda, K. D., Lerner, R. A., *Angew. Chem. Int. Ed.* 30 (**1991**), p. 1711–1713
58 Sinha, S. C., Keinan, E., Reymond, J.-L., *Proc. Natl. Acad. Sci., USA* 90 (**1993**), p. 11910–11913
59 Reymond, J.-L., Jahangiri, G. K., Stoudt, C., Lerner, R. A., *J. Am. Chem. Soc* 115 (**1993**), p. 3909–3917
60 Cochran, A. G., Sugasawara, R., Schultz, P. G. *J. Am. Chem. Soc* 110 (**1988**), p. 7888–7890
61 Jacobsen, J. R., Cochran, A. G., Stephans, J. C., King, D. S., Schultz, P. G., *J. Am. Chem. Soc* 117 (**1995**), p. 5453–5461
62 Sancar, A. *Biochemistry* 33 (**1994**), p. 2–9

63 Taylor, M. J., Hoffman, T. Z., Yli-Kauhaluoma, J. T., Lerner, R. A., Janda, K. D., *J. Am. Chem. Soc.* 120 (**1998**), p. 12783–12790

64 Simeonov, A., Matsushita, M., Juban, E. A., Thompson, E. H., Hoffman, T. Z., Beuscher, I., A. E., Taylor, M. J., Wirsching, P., Rettig, W., McCusker, J. K., Stevens, R. C., Millar, D. P., Schultz, P. G., Lerner, R. A., Janda, K. D., *Science* 290 (**2000**), p. 307–313

65 Wentworth, A. D., Jones, L. H., Wentworth, P. J., Janda, K. D., Lerner, R. A., *Proc. Natl. Acad. Sci., USA* 97 (**2000**), p. 10930–10935

66 Wentworth, P. J., Jones, L. H., Wentworth, A. D., Zhu, X., Larsen, N. A., Wilson, I. A., Xu, X., Goddard, I., W.A., Janda, K. D., Eschenmoser, A., Lerner, R. A., *Science* 293 (**2001**), p. 1806–1811

8
Screening Methods for Catalytic Antibodies
Jean-Louis Reymond

8.1
Introduction

One of the critical steps in the process of preparing catalytic antibodies concerns the detection of catalytic activity in the antibody-containing samples. Testing for catalysis decides whether or not the efforts invested in hapten design and synthesis followed by immunization have been worth while, and this means that screening for catalysis has a strong emotional component. This emotion can be quite dangerous and must be held in check. Well-designed experimental procedures are necessary for rigorous testing, and must be capable of rapidly discounting both false positive and false negative results. I believe that anyone who has been directly involved in a catalytic antibody project understands this, as he or she has probably had to deal with screening results and the long hours, days, or weeks of wild speculation while awaiting confirmation of these results.

The general aim of this chapter is to provide a theoretical and practical perspective to help with the design of catalysis screening experiments. First we consider the governing general principles that should be taken into account for screening catalytic activity in antibodies, including the extent of the catalytic effect that one can expect with an antibody when binding energy (the usual function of an antibody) is converted into catalysis, the minimal experimental conditions that must be met to observe such catalysis, and the actual conditions of screening necessary to reveal this activity. Next we review the literature to examine which screening methods have actually been used to find catalytic antibodies and which specific problems had to be solved to realize catalysis. The practical literature highlights the use of separative techniques, in particular HPLC, as general tools for screening. Specific examples illustrate the advantages of analytical separation in terms of reliability and the opportunity for unexpected discoveries. Finally, a few selected screening methods particularly suited for high throughput are discussed. These include solid-phase-supported assays such as cat-ELISA, and the use of fluorogenic and chromogenic assays to detect catalysis with soluble substrates. In the latter case, two well-documented case studies are discussed in which the details of the screening experiment, which are often omitted in the literature, have been recorded.

Catalytic Antibodies. Edited by Ehud Keinan
Copyright © 2005 WILEY-VCH Verlag GmbH & Co. KGaA, Weinheim
ISBN: 3-527-30688-9

While striving for the most efficient, rationally designed screening protocols, one should bear in mind that one cannot take away the emotional moments of catalytic antibody discovery or project failure, but only put these moments on a more solid basis – it's really all about not fooling oneself and working efficiently.

8.2
Theoretical Consideration: What to look for during screening

8.2.1
Converting Antibody Binding to Catalysis (Quantitatively)

The genetic diversity of antibodies, which concentrates mainly in the complementarity-determining regions (CDR) that make up the antigen combining site, is given by the recombination of gene sequences during B-cell differentiation. Although this diversity may be augmented by the occurrence of random somatic mutations during antibody maturation, the total theoretical diversity available is probably not larger that 10^{12} structures. The general question regarding any chemical transformation would be whether the corresponding catalytic antibody exists and also can be found within this ensemble. Since the complete repertoire of antibodies is not accessible practically, it is necessary to preselect antibodies before testing. This is elegantly realized by immunizing an animal with a small molecule hapten which is either a stable transition state analog of the reaction or a reactive compound. One generally assumes that immunization induces the production of any antibody within the repertoire that binds this hapten, implying that the catalyst, if it exists, should be present within the subset of antibodies that will be produced by immunization.

How should one choose a reaction for antibody catalysis? The first consideration deals with binding energies. Catalytic antibodies are typically raised against small molecule haptens that are transition state analogs of chemical reactions. In the best cases one might expect that all the binding energy available from the antibody-antigen interaction will be converted into transition state stabilisation, expressed in the quantity $\Delta\Delta G^{*}$, which relates directly to the transition state dissociation constant K_{TS} (Fig. 8.1). Thus, a typical antibody-antigen complex with a dissociation constant in the nanomolar range ($K_D = 10^{-9}$ M) would translate into a dissociation constant

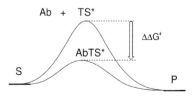

$$AbTS^{*} \underset{}{\overset{K_{TS}}{\rightleftharpoons}} Ab + TS^{*} \qquad K_{TS} = \frac{[Ab]\,[TS^{*}]}{[AbTS^{*}]}$$

Fig. 8.1 Quantitative relation between catalysis as expressed by $\Delta\Delta G^{*}$ and the dissociation constant K_{TS} for the transition state-catalysts complex.

for the reaction's transition state of $K_{TS} = 10^{-9}$ M. For a one-substrate reaction, with a typical Michaelis-Menten constant of $K_M = 10^{-3}$ M, the specific rate acceleration that can be expected should then be $k_{cat}/k_{uncat} = K_M/K_{TS} = 10^6$, corresponding to lowering the transition state energy by approximately 8.4 kcal/mol. For the antibody-catalyzed reaction to be observable, one must set a lower limit to k_{cat}, which can be for example $k_{cat} = 1$ min^{-1}. From this results a lower limit for the spontaneous reaction to be studied of $k_{uncat} = 10^{-6}$ min^{-1}. Thus one is better served by looking at relatively reactive systems for catalysis.

The second consideration for targeting an "antibody-catalyzable" reaction concerns the actual structure of the compounds to be used as haptens. Antibodies from the immune repertoire are accessed in practice either using monoclonal antibody technology or phage-display technologies starting either from an immunized animal or from artificial immunizations. In all cases, the immune system operates through binding affinity in an aqueous environment, and this works best when driven by hydrophobic forces, which are strongest with aliphatic and aromatic substituents in molecules. It is both very simple and very efficient to design haptens and the corresponding substrates (in the cases of transition state analogs of reactions) that carry aromatic groups. Indeed, aromatic groups in the substrates also facilitate kinetic analysis of the reaction products, which is most often done by UV or fluorescence.

8.2.2
Screening and the Theoretical Detection Limits

One should be well aware of the fact that there are many more antibodies that will show binding affinity to a given transition state analog than antibodies that will display actual catalytic activity for the target reaction. The impression that every single antibody binding tightly to a given TSA should also be a catalyst results from the simplification of catalysis as transition state binding along the one-dimensional reaction coordinate (Fig. 8.1). This view is correct in quantitative terms, but ignores the details and complexity of both ligand recognition and chemical catalysis. Having selected a reactive system, a suitable hapten structure, and an immunization technique, the question is therefore how to find the catalytic antibodies among the many hapten-binding antibodies.

As mentioned above, catalytic antibodies are most often set to operate with reactive systems which show a certain level of non-catalyzed, background reaction. With the reasonable expectation that the target antibodies will display substrate saturation kinetics following the Michaelis-Menten model, the ratio of catalyzed to uncatalyzed reaction increases to a maximum for the substrate concentration approaching zero (Fig. 8.2). At the same time, however, the absolute rate of the reaction tends toward zero. Generally, antibody catalysis will thus be best assayed at the lowest possible substrate concentration where product formation is still detectable. This is somewhat counterintuitive, as one would expect that the use of high substrate concentrations might be necessary to saturate substrate binding sites. The saturation of binding sites only makes sense for reactions which do not show any uncatalyzed components.

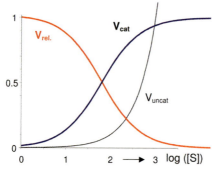

Fig. 8.2 Relationship between substrate concentration and the relative rate $V_{rel.} = V_{cat}/V_{uncat}$ in the Michaelis-Menten model, for a reaction where the uncatalyzed process is first order in substrate, displayed as a function of log(S).

The second aspect of catalyst detection sensitivity concerns the concentration of potential catalyst available in solution for testing. The observed reaction rate under optimal conditions with the substrate concentration below K_M is given by

$$V_{obs} = V_{uncat} + V_{cat} = k_{uncat} \times [S] + (k_{cat}/K_M) \times [Ab] \times [S] \quad (\text{for } [S] \ll K_M)$$
$$V_{cat}/V_{uncat} = ((k_{cat}/k_{uncat})/K_M) \times [Ab] = [Ab] / K_{TS}$$

If the detection technique is precise and sensitive, one can expect to reliably detect as little as a doubling of the reaction rate, which sets a lower limit for the catalyst concentration in solution.

detection limit: $V_{obs} = 2 \times V_{uncat}$, thus $V_{cat} = V_{uncat}$
lowest concentration of catalyst: $[Ab] = K_{TS}$

Inspection of published kinetic data for catalytic antibodies shows that many catalytic antibodies have transition state dissociation constants K_{TS} around 10^{-7} M, which corresponds to specific rate enhancements of $k_{cat}/k_{uncat} = 10^3$ [1]. In the case of antibodies having a molecular weight of approximately 70 kdal per binding site, one obtains a minimum screening concentration of 2 µg/mL = 2 mg/L to detect that level of activity. Interestingly the value of K_M does not enter into this consideration as long as it is larger that the substrate concentration used, which usually will itself be larger than the amount of catalyst present. The concentration of monoclonal antibodies in hybridoma cell culture (5–50 µg/mL) will generally be suited for screening catalysis. Recombinant expression systems, by contrast, usually do not yield such concentration for antibodies under non-optimized conditions.

8.2.3
The Proof of Selective Antibody Catalysis

When screening, it is important to establish rapidly whether an observed effect on the rate of a reaction has anything to do with the desired catalysis. Indeed the observation of a rate enhancement in a given antibody sample over the uncatalyzed reaction in an identical sample lacking the antibody does not imply that the observed effect actually stems from the combining site of the antibody, let alone from the antibody itself. Fortunately, in the case of *anti*-hapten antibodies, a tight-binding inhibitor of the hypothetical catalytic site is available in the form of the hapten used for immunization. The observed catalytic effect *must* be inhibited by the hapten, and this can be tested immediately during screening. This is a simple, necessary condition to establish the origin of catalysis. Catalysis should also be verified in terms of multiple turnovers, one of the key problems encountered with antibodies being product inhibition.

The observation of an efficient catalysis (low K_{TS}) should generally be matched by a correspondingly high affinity for the hapten (low K_i). The correlation between catalytic efficiency and hapten binding has been observed in many catalytic antibodies [2]. Claims of catalysis far in excess of observed hapten-binding affinities are not ruled out *a priori*, but have to be analyzed with caution and traced back to precise chemical catalysis effects.

The observation of a tight-binding hapten further facilitates the proof of antibody-catalysis once the antibody has been produced and purified, since it allows one to cleanly titrate the catalytic activity to the expected number of antibody combining sites. The quantitative inhibition is particularly useful in ruling out the effect of possible enzyme contaminants. Indeed, an enzyme catalyzing the reaction under study would also be expected to be inhibited quite well by the transition state analog used as hapten, but would then possibly require much less than one equivalent of hapten relative to antibody binding sites for inhibition. The danger of enzyme contaminant is usually low when studying typical "abiological" reactions, which are also the most interesting from the point of view of enzyme design. However, enzyme contamination can be very problematic when studying typical enzyme-like reactions, and practical approaches for avoiding these are dicussed in Section 8.5. The last proof of antibody catalysis is finally that the activity must be observed again when the sample is grown again and purified, either from the hybridoma cell line or in recombinant form as Fab fragment.

8.3
A Practical Perspective: Screening in the Published Literature

8.3.1
Statistical Overview

Given the large body of literature on catalytic antibodies that exists today, a straighforward manner to look at screening consists in analyzing what methods were actually used to find catalytic antibodies. There have been slightly more than one hundred publications since 1986 in which the discovery of new catalytic antibodies has been described (Table 8.1). Sixty percent of these papers have been co-authored by one of the original discoverers of catalytic antibodies, and most discoveries succeeded following the original strategy used in the very first report of raising catalytic antibodies. Ninety-five percent of the projects involved the screening of fewer than 150 antibodies, and eighty percent of the projects involved screening fewer than 50 antibodies. Furthermore, ninety-seven percent of the work published used antibodies produced by the classical hybridoma technology, with only a handful of antibodies having been

Tab. 8.1: Methods used for screening catalytic antibodies in the literature.[a]

year	luminescence	UV	HPLC	GC	other
1986	1				
1987			1		
1988		4	2		
1989		3	3		
1990		2	3		
1991		1	4	1	
1992		1	3		
1993			8		2
1994		3	11	1	2
1995		1	7		1
1996	3	2	7		2
1997		2	3		1
1998		3	2		2
1999	2	1	6	1	1
2000		2	3		1
2001	1		4		
2002			2		
TOTAL	7	25	69	3	12
%	6%	22%	59%	3%	10%

a) The number of publications is reported per year where each particular analytical method was used for screening activity. "Other" methods include radioactive counting of labeled substrates, cat-ELISA, and enzyme-coupled assays.

derived from phage-displayed antibodies. Thus, it is clear that most of the screening was done on relatively small series of antibodies, where high throughput was not a real issue. In all cases, the small numbers of antibodies were stabilized hybridoma cell lines that had been selected through the process of cell culture for their ability to bind tightly to the hapten used for immunization. The number of successfully stabilized cell lines varies enormously. Some projects have succeeded with only three monoclonal antibody-producing cell lines, while others required several hundred cell lines to achieve essentially the same result. The published literature thus indicates no correlation between these numbers and the success rates.

The methods used for screening are diverse (Table 8.1). If analyzed by publication number, i.e., different types of reactions, the bulk (60%) of the work has been carried out by HPLC, in particular reverse-phase HPLC, using a simple UV detector to analyze the compounds. This is not surprising given the fact that most antigens, and thus model substrates, were designed with strongly immunogenic aromatic groups attached to relatively polar groups such as amides or nitro groups to provide aqueous solubility.

8.3.2
HPLC Analysis

Reverse-phase HPLC analysis can be carried out on the reaction mixture containing antibody and substrate, and the use of autosamplers provides a level of automation sufficient to deal with up to one hundred samples. One of the key advantages of HPLC analysis is the opportunity to observe unexpected products. This is well illustrated in the study by the Janda group of a photochemical Norish II rearrangement of α-ketoamide 1 under the influence of an antibody against the substrate analog 2 [3]. The antibody-promoted reaction led to the formation of product which did not form under

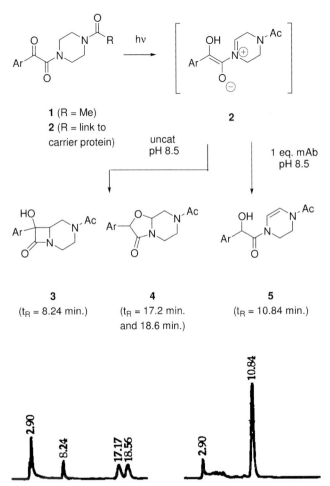

Scheme 8.1 The photochemical Norish II rearrangement of ketoamide 1 (10 µM) under broad band UV irradiation (centered on 340 nm) produces products 3 and 4 in aqueous 50 mM bicine buffer pH 8.5, 5% DMSO. The same reaction in the presence of one equivalent antibody 8C7 (anti-2) produces 5, which is identified by HPLC separation.

the reaction condition in the absence of antibody. This product was then identified as an alternative product also formed at very low pH. Its formation was interpreted in terms of conformational immobilization of the substrate within the antibody binding pocket (Scheme 8.1).

The separating power of HPLC was also crucial in the study of *endo*- and *exo*-selective pathways with Diels-Alderase catalytic antibodies to follow the formation of each of the four possible regio- and stereo-isomers that could form in the reaction of diene **6** with dienophile **7** (Scheme 8.2) [4]. In this case, and in many other case studies with catalytic antibodies, chiral phase HPLC was used to analyze the optical purity of the *exo*- and *endo*-cycloaddition products **8** and **9**.

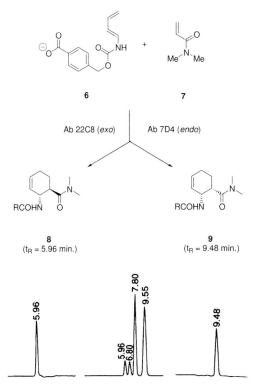

Scheme 8.2 The stereoselectivity of *exo*- (HPLC trace at lower left) and *endo*- (trace at lower right) Diels-Alderase catalytic antibodies is demonstrated by separation of reaction products on a DAICEL Chiralpak AD column (isocratic elution 70/30 (hexane +1.5% CF_3COOH)/2-propanol). The trace of the uncatalyzed reaction produces all four possible stereoisomers (middle trace). Conditions: 500 µM **6**, 500 µM **7**, 12 µM **8** is formed by incubation with 135 µM antibody 22C8 for 65 h, and 36 µM **9** is formed by incubation with 89 µM antibody 7D4 for 20 h.

HPLC was also crucial in analyzing antibody-catalyzed disfavored transformations, where the formation of isomeric products had to be detected against a background of another isomer. Examples include the formation of tetrahydropyrans, which had to be distinguished from tetrahydrofurans, from an anti-Baldwin epoxide cyclization [5], and a disfavored β-fluoride elimination leading to a Z-olefin [6]. HPLC was the only method available to easily distinguish regio-isomeric keto-alcohol products produced by the chemoselective reduction of diketones by catalytic antibodies (Scheme 8.3) [7].

In cases where substrates lacked any chromophores for HPLC-detection, detection was based on GC, as in the case of the solvolysis study of *endo*- and *exo*-norbornyl mesylate (Scheme 8.4) [8].

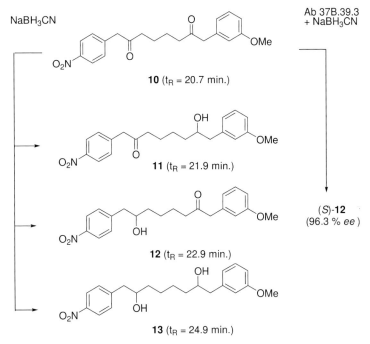

Scheme 8.3 Regioselective reduction of diketone 10 with a catalytic antibody using NaBH$_3$CN as reducing agent. Screening was carried out by RP-HPLC C$_{18}$, eluent 60–80% MeOH with 1.5% CF$_3$COOH. Antibody reactions were carried out in 50 mM MES buffer pH 5.0, 45 mM NaCl, 5% v/v methanol, 15 mM NaBH$_3$CN, 50 μM **8**. The enantioselectivity of the antibody-catalyzed reaction was determined by NMR using Mosher's esters.

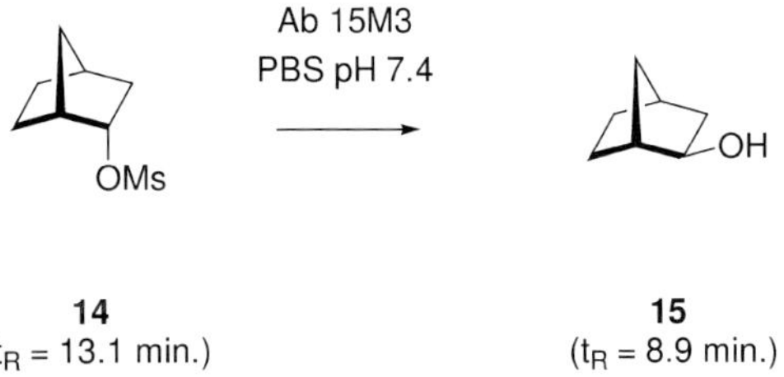

Scheme 8.4 Antibody-catalyzed solvolysis of *endo*-norbornyl mesylate **14** was followed by GC. Conditions: methyl silicone capillary column (HP-1, 25 m × 0.2 mm) T: 6 min at 60 °C, then 20 °C/min. to 200 °C. The optical purity of **15** was determined by NMR using Mosher's esters.

8.3.3
Chromogenic and Fluorogenic Reactions

The second most popular method for screening is to follow the reaction directly by spectroscopic means such as changes in UV/VIS absorbency of luminescence. Such reactions involved in many cases substrates that were directly chromogenic or fluorogenic. Interestingly, the very first report of an antibody-promoted reaction by Lerner and coworkers utilized a fluorogenic coumarin ester **16** (Scheme 8.5) [9]. This reaction, however, did not show turnover, while their next reaction with a less reactive phenyl ester (**17**) was followed by HPLC and did show turnover [10]. This

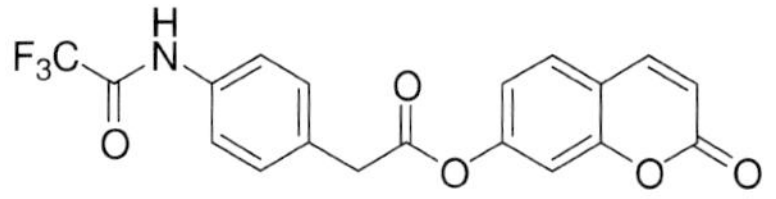

Scheme 8.5 Substrates used in the first studies of catalytic antibodies. Ester **16** is fluorogenic, carbonate **18** is chromogenic. Ester **17** was analyzed by HPLC.

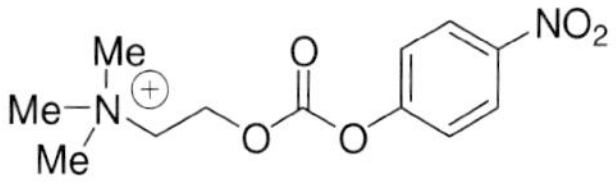

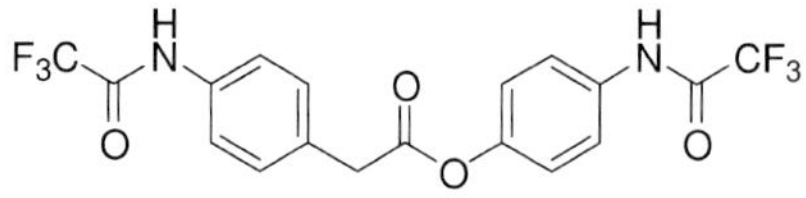

may explain in part the dominance of HPLC analysis in the work coming from this group. The second of the two founding papers, where the Schultz group reported the esterolytic activity of the *anti*-phosphocholine antibody MOPC167 [11], was based on the chromogenic nitrophenyl carbonate **18**.

A number of reactions have also been assayed in the near UV taking advantage of intrinsic changes in chromophore properties upon reaction, particularly in systems where a styrene- or benzaldehyde-type double bond is formed or modified during the reaction (Scheme 8.6). These include among others dehydrofluorination [12], dehydratation [13], imine [14] and oxime bond formation [15], and the oxidative decarboxylation of vanillin mandelic acid [16]. The chorismate mutase reaction for which two catalytic antibodies have been described can be followed by UV at 275 nm [17]. Diels-Alder reactions could be followed by following the disappearance of either a nitrosoarene as dienophile or tetrachlorothiophene dioxide as diene at 340 nm [18]. Further chromogenic and fluorogenic assays relevant to catalytic antibodies are discussed with respect to high-throughput screening in the next sections of this chapter.

8.4
High-Throughput Screening Methods

The literature of catalytic antibodies can be analyzed with respect to screening methods by considering the absolute number of catalytic antibodies ever discovered by using a particular method. However, this analysis is not reliable because initially identified catalytic antibodies have very often not been subsequently confirmed. With respect to screening methods, it is most relevant to look into the detail of the methods used by considering those that were able to quickly analyze large numbers of samples. A number of catalysis assays for high-throughput screening have been

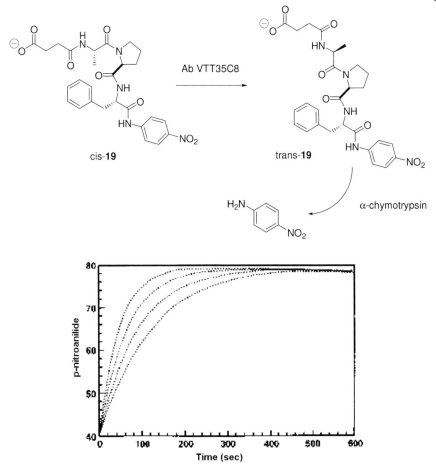

Scheme 8.6 The antibody-catalyzed cis-to-trans-prolyl isomerization is followed using the release of yellow nitroaniline (λ = 392 nm or 426 nm) after enzyme-coupled hydrolysis of the trans-prolyl peptide by α-chymotrypsin.

developed recently for the purpose of assaying enzymes and catalysts, and this subject has been reviewed recently [19]. Some of these methods were developed with particular emphasis on catalytic antibodies and are of special interest in the present context.

8.4.1
Enzyme-Coupled Assays

A number of classical enzyme assays use one or several secondary enzymes to transform product into a detectable signal. The secondary enzyme is most often a dehydrogenase, which consumes the reaction product to convert nicotinamide adenine dinucleotide (NAD) from its reduced (NADH) to its oxidized (NAD$^+$) state, or *vice* versa. Since the UV and fluorescence spectra of NADH are different from those of NAD$^+$, the reaction is readily followed spectroscopically. Several research groups have used such assays for detecting catalytic antibodies, for example for detecting the hydrolysis of vinyl esters to acetaldehyde using alcohol dehydrogenase [20], the transaminase activity of a PLP-dependent abzyme using pyruvate dehydrogenase [21], and the enantioselective hydrolysis of ethyl esters [22]. Although there has not been a very widespread use of such assays, they are very convenient for high-throughput screening.

A particularly elegant enzyme-coupled assay concerns the detection of prolyl *cis-trans* isomerase activity. The assay takes advantage of the selective chromogenic proteolytic activity of the protease α-chymotrypsin of the *trans*-prolyl-isomer of the tetrapeptide succinyl-Ala-Pro-Phe-pNA (pNA = *para*-nitroanilide) **19** [23]. The *cis-trans* prolyl isomerization reaction can also be followed by fluorescence resonance energy transfer (FRET) using a related doubly tagged peptide. Both assays were used by Janda and coworkers to find a *cis-trans*-prolyl isomerase catalytic antibody against an α-keto-amide hapten (Scheme 8.6) [24]. Schultz and coworkers reported a very similar study later using the same assay systems [25].

8.4.2
cat-ELISA

The basic tool for detecting antibodies binding to a specific antigen is the ELISA, or Enzyme Linked Immuno-Sorbent Assay [26]. In this assay, antigen-specific antibodies are detected by antigen-mediated attachment to a solid support and secondary detection with an Fc-specific enzyme-labeled secondary antibody (A in Fig. 8.3). For hapten immunization, ELISA is performed using a carrier protein, typically BSA (bovine serum albumin), different from the carrier protein used for immunization. In this manner only antibodies with binding specificity to the hapten are revealed by the assay.

A study by Tawfik et al. of nitrobenzylesterase activity stands out by reporting 2110 hits out of 17 738 screened samples [27]. The unusually high success rate was attributed to using a special strain of auto-immune mice, which were thought to be particularly well suited to producing catalytic antibodies. These experiments used an analytical method named cat-ELISA, derived from the ELISA for binding affinity, but adapted to detect catalysis, and originally reported by the same group in 1993, and which was also reported by the Hilvert group [28]. The cat-ELISA uses an antibody against the reaction product as the key reagent. A substrate-carrier protein conjugate is surface-bound in wells of a 96-well plate. The reaction is allowed to proceed in the presence of a test solution. Formation of the surface-bound product is quantitated using ELISA detection with a product specific polyclonal rabbit antibody. The presence of product indicates a possible catalyst in the test solution (B in Fig. 8.3). Similar systems applied to surface-bound substrates have been described using either biotinylated or DNA-tagged substrates [29].

A useful and very practical improvement was developed by Grassi and coworkers and involves competitive ELISA between the reaction product and a product analog

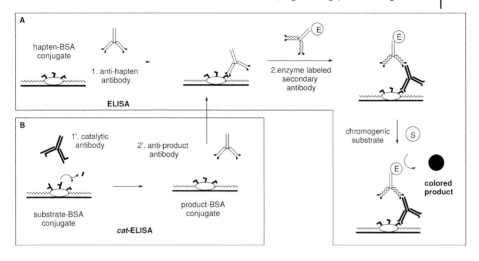

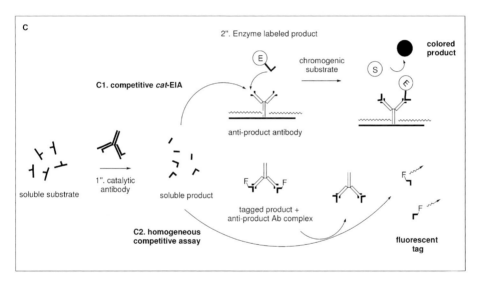

Fig. 8.3 Principle of ELISA and catalysis assays.

covalently tagged with acetylcholinesterase (AChE), which are allowed to compete for a surface-bound, product-specific polyclonal rabbit antibody (C1 in Fig. 8.3) [30]. In the absence of product, AChE-tagged product binds to the well surface. After washing unbound reagents, surface-bound AChE is revealed using a chromogenic AChE substrate. If reaction has occurred, the untagged product binds to the surface and no AChE-tagged product remains bound after washing. Thus, no coloration is observed (Fig. 8.3). The system was developed for the analysis of the oxidation chemistry of vanillic mandelic acid, and a recent application of this system concerns the measurement of optical purities in high-throughput [31].

A later version of this competitive system was also reported to analyze reaction progress instantaneously by fluorescence (C2 in Fig. 8.3) [32]. In this case, a complex of a fluorescence-quenched tagged reaction product with an anti-product antibody was used to detect the formation of product in real time. This assay allowed reaction progress to be followed, in contrast to the cat-ELISA above, which only applies for a single-point measurement of product formation.

8.4.3
Thin-Layer Chromatography

Thin-layer chromatography, which lies at the heart of synthetic chemistry, was popular in analytical chemistry and biochemistry before it was replaced by more refined analytical instruments. While HPLC is only medium-throughput, and this only given the availability of an autosampler, TLC has the ability to realize a relatively high throughput at a very low cost, and offers a very broad palette of staining reagents that allow one to detect practically any non-volatile compound. The crucial point for screening catalytic antibodies is the detection limit, since most reactions are carried out with dilute substrates (less than 1 mM). The staining or visualization method must be able to detect product formation at low conversion (less than 0.005 mM product in solution). For example, visualizing aromatic compounds under UV-light with a fluorescent plate has a detection limit around 1 mM, which is not sufficient for screening catalytic antibodies. The obvious general solution is to use specifically tagged substrates. A particularly advantageous system uses a blue fluorescent acridone which can be detected by the eye on TLC plates under UV light down to 1 picomole [33]. Acridone is completely resistant to photobleaching, in constrast to the bleach-sensitive classical dyes such as dansyl or fluorescein. For example, the activity of an enantioselective catalytic antibody 14D9 on enol ether **20** to form the (S)-aldehyde **21** was detected by indirect labeling of the aldehyde with acridone hydrazide reagent **22** to form hydrazone **23**, which can be detected on TLC plates at very low concentrations (Scheme 8.7). The method also applies to the visualization of volatile aldehydes such as butanal (hydrazone **24**).

Specific TLC-staining techniques can overcome detection problems when dealing with compounds without UV-active functional groups. Hasserodt was able to use TLC coupled with anisaldehyde staining to visualize the formation of polycyclic products not detectable otherwise (Scheme 8.8) [34]. HP-TLC was used in conjunction with an elegant pre-purification procedure to screen monoclonal antibodies for amidase activity, resulting in the isolation of one catalyst out of 1632 assays [35].

8.4.4
Chromogenic and Fluorogenic Substrates

Given the fact that catalytic antibodies are most often developed at first for model systems, the choice of a chromogenic or fluorogenic reaction can be particularly advantageous. Chromogenic substrates have been often associated with "over-reactive" systems, dealing with reactions that have a high spontaneous rate. A typical example

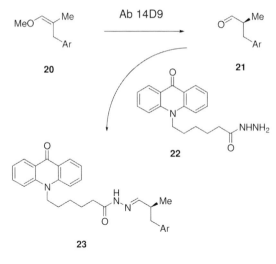

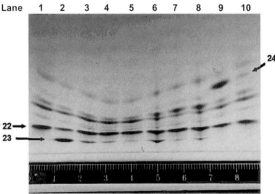

Scheme 8.7 Acridone-tagged hydrazide reagent for visualizing carbonyl products. Samples were treated with hydrazide **22** (50 µM, 3 h, 20 °C, pH 5.5) and analyzed. HP-TLC plate 10 × 10 cm (silica gel 60, 0.2 mm thickness, with pre-concentration zone). Elution: 5% v/v methanol in dichloromethane. Each lane is the analysis of 10 µL containing blank reference (lane 1); aldehyde **21** at concentration of 100 µM (lane 2), 10 µM (lane 3), 5 µM (lane 4), 2 µM (lane 5); incubation of 250 µM enol ether **20** (37 °C, pH 5.5, 15 h) with 1 mg/mL catalytic antibodies 14D9 (lane 6), 22H25 (lane 7) and 19C9 (lane 8). Butanal is detected as its hydrazone (**24**) in concentrations of 100 µM (lane 9) and 10 µM (lane 10).

is the hydrolysis of nitrophenyl esters, which has been the object of many catalytic antibody studies. In contrast to aliphatic esters or amides, in which breakdown of the tetrahedral intermediate to form the product is the decisive step, the hydrolysis of nitrophenyl esters proceeds with rate-limiting formation of a tetrahedral intermediate [36]. This mechanistic twist makes it easier to find systems that will catalyze the hydrolysis reaction. The down side concerns the possibility of simple non-specific acylation of proteins, in particular, lysine side chains, by such activated esters, which causes a high background reaction in crude samples and makes the system prone

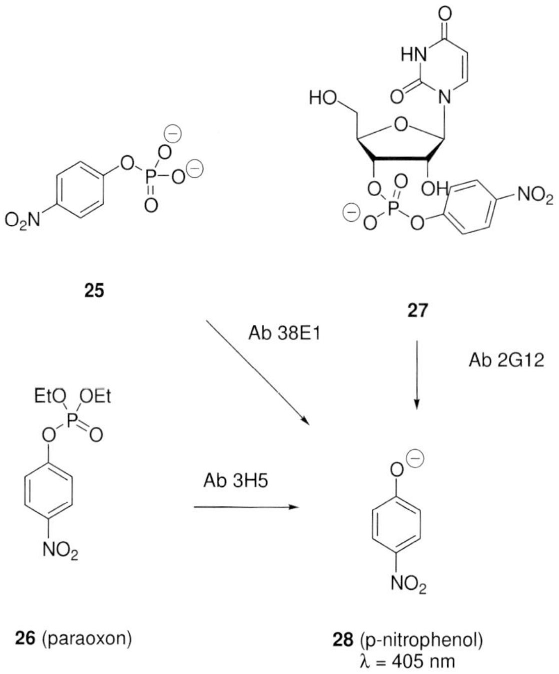

Scheme 8.8 Chromogenic phosphate derivatives hydrolyzed by catalytic antibodies.

to artifacts. Nevertheless, the high reactivity of active esters, including nitrophenyl and vinyl esters, can be put to good use for synthetic purposes by using them as acyl donor either for ester bond [20, 37] or for amide bond formation [38]. Catalytic antibodies have been described for the hydrolysis of nitrophenyl phosphate (25) [39], the insecticide paraoxon (26) [40], and for the rearrangement of uracil-3'-nitrophenyl phosphate (27) to the 2',3'-cyclic phosphate [41], all of which reactions are chromogenic (Scheme 8.8).

Activated esters such as nitrophenyl esters react non-specifically with proteins and render screening of crude extract such as cell culture supernatant practically impossible because of the high level of spontaneous hydrolysis. A similar reactivity problem is encountered with the chromogenic Kemp elimination substrate 29 (Scheme 8.9) [42], which is extremely sensitive to the combination of weak base and medium ef-

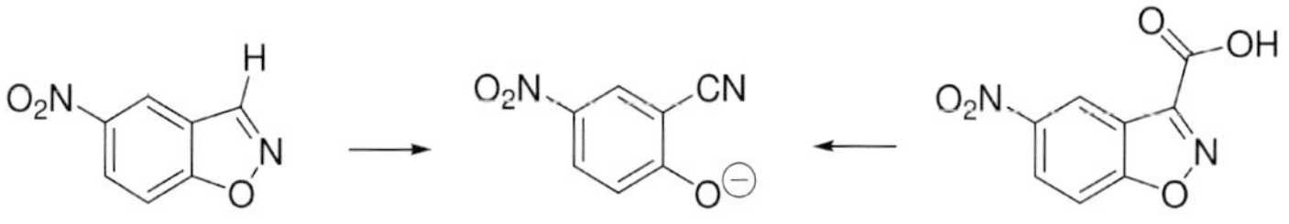

Scheme 8.9 Isoxazole substrate and their chromogenic reactions.

fects [43], and can be decomposed to its colored phenolate product **30** using either catalytic antibodies [44], bovine serum albumin [45], or more simply with natural carbons such as activated charcoal [46]. In one study Hilvert and coworkers reported finding no hits when screening the similar decarboxylation reaction of the corresponding carboxylate substrate **31** in cell culture, but succeeded in finding reasonably active decarboxylase antibodies when assaying purified antibodies [47]. Thus, it is not clear how easy it is to find catalytic antibodies for such highly reactive systems in the context of high-throughput screening.

The challenge of finding catalytic antibodies catalyzing reactions with non-activated functional groups has always been considered as the ultimate test to determine whether these designed systems can rival enzymes, particularly for the hydrolysis of non-activated ester and amide bonds. We have recently devised fluorogenic substrates that are stable enough for screening in crude cell culture extracts and are triggered by the hydrolysis of a non-activated aliphatic ester bond. Our chiral ester substrates

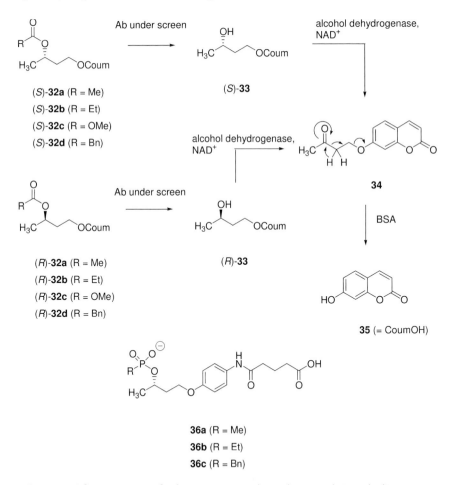

Scheme 8.10 A fluorogenic assay for detecting enantioselective lipase catalytic antibodies.

release the fluorescent product umbelliferone (**35**) by an enzyme-coupled oxidation of the primary hydrolysis product, which allows us to follow hydrolysis in real time (Scheme 8.10) [48]. We have undertaken a very broad screening program involving chiral esters and carbonates (**32a–d**) and over 400 different hybridoma generated against the corresponding phosphonate-type haptens **36a–c**. Each hybridoma culture supernatant was pre-purified and tested in microtiterplate format for esterolytic activity on each enantiomeric substrate separately, in the presence and absence of the corresponding hapten (see Section 8.5.1). While this type of assay works perfectly to identify esterolytic enzymes, we did not identify esterase-like catalytic antibodies, pointing to the fact that high-throughput screening offers no warranty for success within the numbers of candidate antibodies that can be produced by hybridoma techniques [49]. A similar assay using periodate cleavage of diol reaction products was also developed in our group to assay a very broad range of hydrolytic enzymes [50].

The key of our hydrolase screening scheme is the β-elimination triggered by bovine serum albumin (BSA), which liberates the fluorescent product umbelliferone from the carbonyl reaction product. We have devised fluorogenic substrates for other interesting organic transformations, including oxy-Cope rearrangements [51] and aldol reactions. In the latter case, a family of stereoisomeric aldol substrates, **37–39**, allowed us to rapidly determine the stereoselectivity of aldolase catalytic antibodies (Scheme 8.11) [52]. The commercially available antibody 38C6 originally reported by Wagner et al. [53] was found to react predominantly with the (S)-configured aldol **37c**, while it showed almost no reaction with the (R)-anti-anti-aldol **39f**. Two related flu-

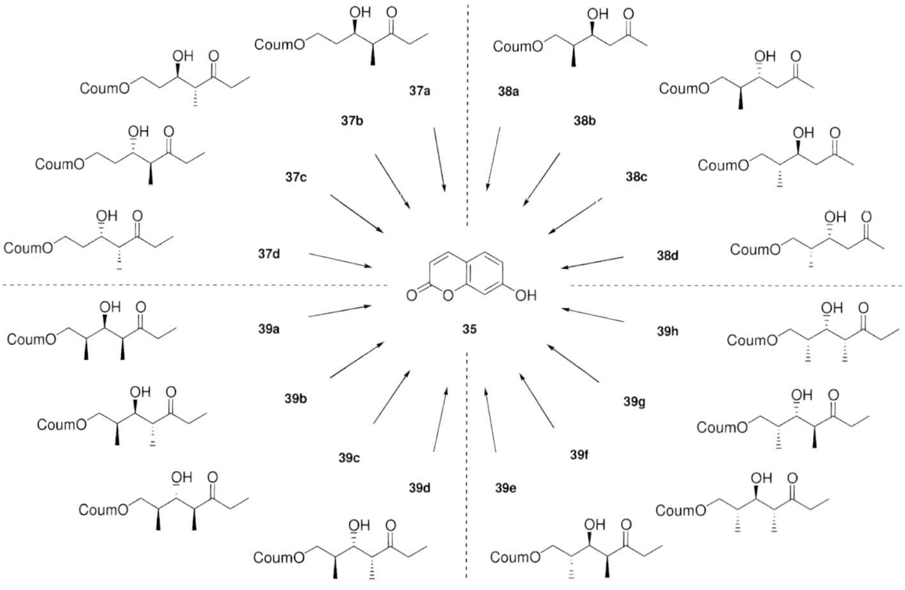

Scheme 8.11 Stereoselective fluorogenic polypropionate fragments for detecting aldolase catalytic antibodies.

orogenic retro-aldolization reactions were reported also by Barbas and coworkers to study their antibody [54]. The first releases 6-methoxy-naphthaldehyde as a fluorescent product, and the second makes use of the β-elimination reaction of the aldehyde product (also called a retro-Michael addition) described above to release resorufin as a colored product.

8.4.5
Isotopic Labeling

Landry et al. have used derivatives of cocaine bearing a titriated or ^{14}C-labeled benzoate group to selectively screen for the debenzoylation of cocaine by catalytic antibodies [55]. The samples being assayed were simply acidified and the benzoic acid released was selectively extracted with hexane, while the cocaine substrate of the methyl ester hydrolysis product remained in the aqueous phase because of the positive charge on the nitrogen. The benzoic acid formed was then quantitated in high throughput by counting radioactivity in the organic extract. The methods allowed Landry and coworkers to screen large numbers of hybridoma samples for efficient cocaine-hydrolyzing antibodies, which were subsequently shown to prevent cocaine-intoxicating effects in rats [56]. It must be mentioned that radioactive counting can in principle reveal minute amounts of product not detectable otherwise. In the cocaine study, however, the product concentration was sufficient to be detected by HPLC also, and the counting procedure was mainly useful for its high-throughput potential.

8.5
Examples with Fluorogenic Substrates and Antibodies from Hybridoma

Screening antibodies for catalysis requires an understanding of the principles and the choice of methods, as discussed in the earlier sections. The subject would not be exhausted without examining a few examples in detail. Two case studies from my laboratory are discussed to close the chapter and illuminate a few additional aspects of screening. One of these aspects concerns the issue of enzymatic impurities. Most of the early work on catalytic antibodies was done on esterolysis reactions, and there was concern that the activities observed might stem from enzyme impurities present in the samples. Indeed, fetal calf serum, which is the standard additive in hybridoma cell cultures, contains a fair amount of butyl esterase, and therefore rapidly reacts with various simple esters. Enzyme impurities can be dealt with by using high-throughput antibody purification procedures during screening, and one such procedure is described in Section 8.4.2. Nevertheless, we have also found a reproducible activity for β-galactosidase, which appears to co-purify with IgGs produced from hybridoma on protein-G columns [57].

8.5.1
Retro-Diels-Alderase Antibodies

The retro-Diels-Alder reaction of bicyclic prodrug **40** releases nitroxyl, a precursor of the regulatory molecule nitric oxide, and the anthracene product **41** (Scheme 8.12) [58]. The reaction can readily be followed by fluorescence, since the substrate is non-fluorescent, while the product **41** fluoresces strongly in the blue. This abiological retro-Diels-Alder reaction appears to be independent of pH between pH 4 and pH 10, as well as completely insensitive to buffer or cosolvent effects. At the same time, the reaction has a measurable spontaneous rate of $k_{uncat} = 10^{-5}$ s^{-1}. The reaction thus falls within the useful parameters for antibody catalysis. More importantly, the reaction is not catalyzed non-specifically by proteins such as albumins, antibodies, or enzymes, and can therefore be measured directly in cell culture.

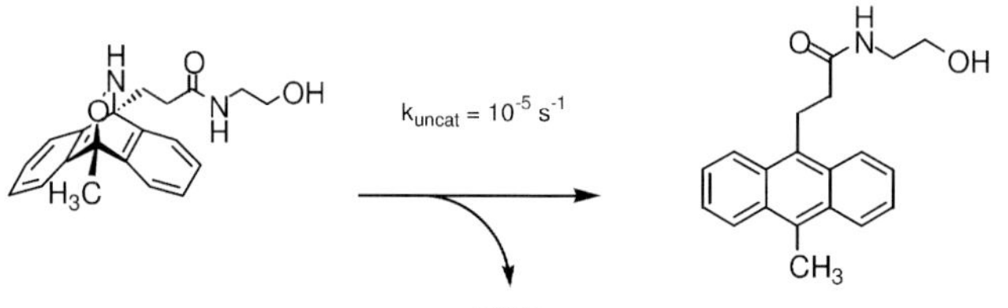

Scheme 8.12 Retro-Diels-Alder reaction catalyzed by antibodies.

Using high-throughput screening by fluorescence, we were able to screen a total of more than 14 000 hybridoma cell culture samples generated from mice immunized against transition state analogs of the reaction. Because of the ease of measurement, all samples were tested both for binding and for catalysis. There were 1143 cell culture samples showing a strong ELISA signal on hapten binding. Catalysis screening gave much lower hit rates, with only eighteen confirmed positive results on the whole series. Interetingly, there were no false positives in these series, i.e., all samples testing positive for catalysis were also positive on ELISA. Sub-cloning of the cell lines resulted eventually in eight catalytic antibodies for the reaction. These antibodies were all inhibited by stoichiometirc addition of the corresponding hapten. The rate enhancement of these antibodies were within $k_{cat}/k_{uncat} = 10^2-10^3$ [59]. These rate enhancements fall within the typical range for catalytic antibodies, with K_{TS} values ranging between 0.1 and 0.9 µM. Thus, it appeared that even weakly active antibodies with $k_{cat}/k_{uncat} = 100$ were readily detected in cell culture, establishing the high sensitivity of the detection method. Clearly, no single catalytic antibody could have escaped detection by our screening method. The most active antibodies have been studied in detail and their mechanism of action established on the basis of an X-ray crystal structure, revealing the critical role of a tryptophan side chain in triggering catalysis [60].

8.5.2
Pivalase Catalytic Antibodies

Screening directly in cell culture is feasible as long as the reaction under study is not or cannot be catalyzed by contaminant enzymes or non-specific proteins. We have recently isolated esterolytic antibodies cleaving pivaloyloxymethyl (POM) protected phenols, where this condition was not fulfilled [61]. As mentioned before, esterases may be present in cell culture samples and be responsible for the occurrence of an observed catalytic effect. We undertook immunizations against phosphonate haptens **42–44** using the fluorogenic substrate **45** as test substrate, which releases the strongly fluorescent umbelliferone (**35**) upon hydrolysis (Scheme 8.13). Although the POM-derivative **45** shows very little non-specific hydrolysis in cell culture, it is potentially a substrate for esterases. Nevertheless, we undertook a first round of immunization and screening using direct detection of activity in cell culture, doubled with a requirement for strong hapten binding as detected by ELISA and the condition of inhibition of catalysis by added hapten. This experiment led to the isolation of five monoclonal antibody cell lines showing specific, hapten-inhibited catalysis. However, purification of the antibodies showed that the activity did not bind to the protein-G antibody-affinity columns, and must have been caused by contaminant esterases. The inhibition of catalysis by the hapten was rather weak (10^{-5} M) and did not match the strong affinity of the antibodies for the hapten as detected by ELISA, providing further evidence that the catalytic activity did not originate from the antibodies.

In the light of these experiments, we decided to establish a pre-purification procedure for antibodies that we could apply during the screening of hybridoma cell culture samples. Immunoglobulins present in cell culture samples of volume 0.5 –

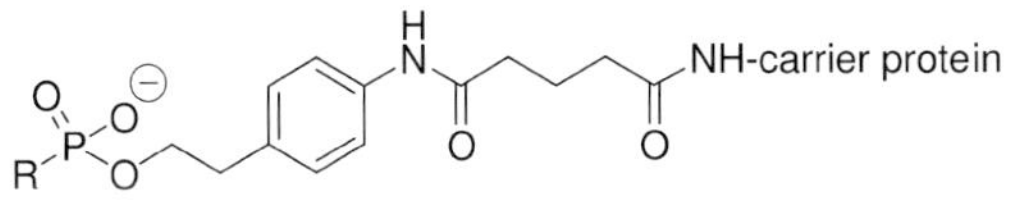

42 (R = t-Bu)
43 (R = CH$_2$Cl)
44 (R = C$_6$H$_5$)

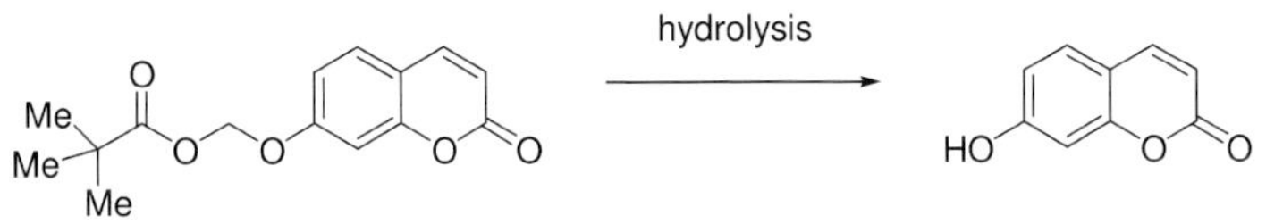

45 → hydrolysis → **35**

Scheme 8.13 Antibodies generated against haptens **42–44** were used to screen for catalysis of the hydrolysis of pivaloyloxymethyl ether (**45**), which produces the strongly fluorescent umbelliferone (**35**).

5 mL can be selectively bound by passing the sample through a very small amount of protein-G gel, which is simply placed in a Pasteur pipette (Fig. 8.3). The gel is washed with neutral buffer, and the antibody then selectively eluted using a small volume of acidic buffer. The eluted sample is neutralized and can then be assayed for specific, hapten-inhibited catalysis by comparing the rate of reaction of the fluorogenic substrate 45 in the presence or absence of the haptens.

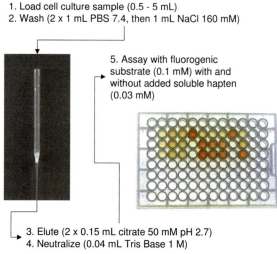

Fig. 8.4 Pre-purification of antibodies from cell culture samples on protein-G gel and activity screening.

The protein-G pre-purification procedure can be used to manually purify up to 40 samples within two hours. We used it to assay approximately 500 pre-clonal hybridoma samples generated from the different immunizations. Catalysis screening was done on hybridoma samples at a cell culture volume of approximately 5 mL, just before the first sub-cloning event. Sixteen cell lines showing selective, hapten-inhibited catalysis, were detected and cultured further. All the antibodies were successfully sub-cloned twice. Eleven of them showed confirmed, hapten-inhibited catalysis after expression and purification. Interestingly, we found that it was necessary to test for catalysis not only after the first identification of catalysis, but also after the first and second subcloning event. Indeed sub-clones from an initial catalytic hybridoma were often found positive on ELISA, but negative on catalysis, while others were positive both on ELISA and on catalysis. The "pivalase" catalytic antibodies isolated are the first antibodies capable of hydrolyzing pivalate esters, which are sterically demanding substrates. The antibodies all show rate enhancements in the range of $k_{cat}/k_{uncat} = 10^3$. In these cases, inhibition by the soluble hapten is strong and quantitative, and matches perfectly the hapten affinity as measured by ELISA.

8.6
Conclusion

Once a catalytic antibody for a given reaction has been isolated, the screening method that has been used to find it becomes irrelevant. Many catalytic antibodies have been found for reactions that are rather difficult to assay and require HPLC separation of products. Fruitful combinations of haptens and reactions, when they occur, can be detected by combining catalysis screening with ELISA as a criterion for selecting hybridoma. The implementation of this combined approach requires simply a good coordination between the immunologists and the chemists involved in the project, and succeeds best when both activities are carried out in the same laboratory.

High-throughput screening methods for catalysis offer in principle the possibility of screening much larger numbers of candidate antibodies for a given reaction. The improvement in scale, which amounted to screening approximately 10 to 50 times more antibodies for a given reaction, has not resulted in very significant improvements in catalysis. Nevertheless, there is no doubt that high-throughput screening assays are the most convenient procedures available, and the only ones that are compatible with improved biological systems where larger numbers of samples can be tested.

Acknowledgement

This work was financially supported by the University of Bern, Switzerland, and the Fonds Nationals Suisse de la Recherche Scientifique.

References

1 REYMOND, J.-L., *Top. Cur. Chem.* 200 (**1999**), p. 59–93
2 STEWART, J. D., BENKOVIC, S. J., *Nature* 375 (**1995**), p. 388–391
3 TAYLOR, M. J., HOFFMAN, T. Z., YLI-KAUHALUOMA, J. T., LERNER, R. A., JANDA, K. D., *J. Am. Chem. Soc.* 120 (**1998**), p. 12783–12790
4 GOUVERNEUR, V. E., HOUK, K. N., PASCUAL-TERESA, B. DE, BENO, B., JANDA, K. D., LERNER, R. A., *Science* 262 (**1993**), p. 203–8
5 a) JANDA, K. D., SHEVLIN, C. G., LERNER, R. A., *Science* 259 (**1993**), p. 490–494; b) JANDA, K. D., SHEVLIN, C. G., LERNER, R. A., *J. Am. Chem. Soc.* 117 (**1995**), p. 2659–60; c) GRUBER, K., ZHOU, B., HOUK, K. N., LERNER, R. A., SHEVLIN, C. G., WILSON, I. A., *Biochemistry* 38 (**1999**), p. 7062–7074
6 CRAVATT, B. F., ASHLEY, J. A., JANDA, K. D., BOGER, D. L., LERNER, R. A., *J. Am. Chem. Soc.* 116 (**1994**), p. 6013–4
7 SHIEH, L. C., YONKOVICH, S., KOCHERSPERGER, L., SCHULTZ, P. G., *Science* 260 (**1993**), p. 337–9
8 MA, L., SWEET, E. H., SCHULTZ, P. G., *J. Am. Chem. Soc.* 121 (**1999**), p. 10227
9 TRAMONTANO, A., JANDA, K. D., LERNER, R. A., *Proc. Natl. Acad. Sci. USA* 83 (**1986**), p. 6736–6740
10 TRAMONTANO, A., JANDA, K. D., LERNER, R. A., *Science* 234 (**1986**), p. 1566–70
11 POLLACK, S. J., JACOBS, J. W., SCHULTZ, P. G., *Science* 234 (**1986**), p. 1570–3
12 SHOKAT, K. M., LEUMANN, C. J., SUGASAWARA, R., SCHULTZ, P. G., *Nature* 338 (**1989**), p. 269–271
13 UNO, T., SCHULTZ, P. G., *J. Am. Chem. Soc.* 114 (**1992**), p. 6573–4
14 COCHRAN, A. G., PHAM, T., SUGASAWARA, R., SCHULTZ, P. G., *J. Am. Chem. Soc.* 113 (**1991**), p. 6670–2
15 UNO, T., GONG, B., SCHULTZ, P. G., *J. Am. Chem. Soc.* 116 (**1994**), p. 1145–5
16 TARAN, F., RENARD, P. Y., BERNARD, H., MIOSKOWSKI, C., FROBERT, Y., PRADELLES, P., GRASSI, J., *J. Am. Chem. Soc.* 120 (**1998**), p. 3332–9
17 a) JACKSON, D. Y., JACOBS, J. W., SUGASAWARA, R., REICH, S. H., BARTLETT, P. A., SCHULTZ, P. G., *J. Am. Chem. Soc.* 110 (**1988**), p. 4841–2; b) HILVERT, D., CARPENTER, S. H., NARED, K. D., AUDITOR, M. T. M., *Proc. Natl. Acad. Sci. USA* 85 (**1988**), p. 4953–5
18 a) HILVERT, D., HILL, K. W., NARED, K. D., AUDITOR, M.-T. M., *J. Am. Chem. Soc.* 111 (**1989**), p. 9261–2; b) MEEKEL, A. A. P., RESMINI, M., PANDIT, U. K., *Bioorg. Med. Chem. Lett.* 4 (**1996**), p. 1051–7
19 a) WAHLER, D., REYMOND, J.-L., *Curr. Opin. Biotechnol.* 12 (**2001**), p. 535–544; b) WAHLER, D., REYMOND, J.-L., *Curr. Opin. Chem. Biol.* 5 (**2001**), p. 152–158
20 WIRSHING, P., ASHLEY, J. A., BENKOVIC, S. J., JANDA, K. D., LERNER, R. A., *Science* 252 (**1991**), p. 680–5
21 GRAMATIKOVA, S. I., CHRISTEN, P., *J. Biol. Chem.* 271 (**1996**), p. 30583–6

22 Nakatani, T., Hiratake, J., Shinzaki, A., Umeshita, R., Suzuki, T., Nishioka, T., Nakajima, H., Oda, J., *Tetrahedron Lett.* 34 (**1993**), p. 4945–8
23 a) Garcya-Echeverrya, C., Kofron, J. L., Kuzmic, P., Kishore, V., Rich, D. H., *J. Am. Chem Soc.* 114 (**1992**), p. 2758–2759; b) Kofron, J. L., Kuzmic, P., Kishore, V., Colón-Bonilla, E., Rich, D. H., *Biochemistry* 30 (**1991**), p. 6127–6134
24 Yli-Kauhaluoma, J. T., Ashley, J. A., Lo, C.-H. L., Coakley, J., Wirshing, P., Janda, K. D., *J. Am. Chem. Soc.* 118 (**1996**), p. 5496–7
25 Ma, L., Hsieh-Wilson, L. C., Schultz, P. G., *Proc. Natl. Acad. Sci.* 95 (**1998**), p. 7251–6
26 Tijssen, P., *Practice and theory of enzyme immunoassays*, Ed. Burdon, R. H., Knippenberg, P. H., Elsevier, **1985**
27 Tawfik, D. S., Chap, R., Green, B. S., Sela, M., Eshhar, Z., *Proc. Natl. Acad. Sci. USA* 92 (**1995**), p. 2145–2149
28 a) Tawfik, D. S., Green, B. S., Chap, R., Sela M., Eshhar, Z., *Proc. Natl. Acad. Sci. USA* 90 (**1993**), p. 373; b) MacBeath, G., Hilvert, D., *J Am Chem Soc* 116 (**1994**), p. 6101
29 a) Lane, J. W., Hong, X., Schwabacher, A. W., *J. Am. Chem. Soc.* 115 (**1993**), p. 2078; b) Fenniri, H., Janda, K. D., Lerner, R. A., *Proc. Natl. Acad. Sci. USA* 92 (**1995**), p. 2278
30 a) Caruelle, D., Grassi, J., Courty, J., Croux-Muscatelli, B., Pradelles, P., Barritault, D., Caruelle, J. P., *Anal. Biochem.* 173 (**1988**), p. 328; c) Taran, F., Renard, P. Y., Créminon, C., Valleix, A., Frobert, Y., Pradelles, P., Grassi, J., Mioskowski, C., *Tetrahedron Lett.* 40 (**1999**), p. 1887–1891
31 Taran, F., Gauchet, C., Mohar, B., Meunier, S., Valleix, A., Renard, P. Y., Créminon, C., Grassi, J., Wagner, A., Mioskowski, C., *Angew. Chem* 114 (**2002**), p. 132–135
32 Geymayer, P., Bahr, N., Reymond, J.-L., *Chem. Eur. J.* 5 (**1999**), p. 1006

33 a) Reymond, J.-L., Koch, T., Schröer, J., Tierney, E., *Proc. Nat. Acad. Sci. USA* 93 (**1996**), p. 4251
34 Hasserodt, J., Janda, K. D., Lerner, R. A., *J. Am. Chem. Soc.* 122 (**2000**), p. 40–45
35 Martin, M. T., Angeles, T. S., Sugasawara, R., Aman, N. I., Napper, A. D., Darsley, M. J., Sanchez, R. I., Booth, P., Titmas, R. C., *J. Am. Chem. Soc.* 116 (**1994**), p. 6508–6512
36 Tantillo, D. J., Houk, K. N., *J. Org. Chem.* 64 (**1999**), p. 3066–3076
37 Napper, A. D., Benkovic, S. J., Tramontano, A., Lerner, R. A., *Science* 237 (**1987**), p. 1041–3
38 a) Hirschmann, R., Smith, A. B., Taylor, C. M., Benkovic, P. A., Taylor, S. D., Yager, K. M., Sprengeler, P. A., Benkovic, S. J., *Science* 265 (**1994**), p. 234–7; b) Smithrud, D. B., Benkovic, P. A., Benkovic, S. J., Taylor, C. M., Yager, K. M., Witherington, J., Philips, B. W., Sprengeler, P. A., Smith, A. B., Hirschmann, R., *J. Am. Chem. Soc.* 119 (**1997**), p. 278–282
39 Scanlan, T. S., Prudent, J. R., Schultz, P. G., *J Am Chem Soc* 113 (**1991**), p. 9397
40 Lavey, B. J., Janda, K. D., *J. Org. Chem.* 61 (**1996**), p. 7633
41 Weiner, D. P., Wiemann, T., Wolfe, M. M., Wentworth, P., Janda, K. D., *J. Am. Chem. Soc.* 119 (**1997**), p. 4088
42 Casey, M. L., Kemp, D. S., Paul, K. G., Cox, D. D., *J. Org. Chem.* 38 (**1973**), p. 2295–2301
43 a) Kemp, D. S., Casey, M. L., *J. Am. Chem. Soc.* 95 (**1973**), p. 6670–6680; b) Kemp, D. S., Cox, D. D., Paul, K. G., *J. Am. Chem. Soc.* 97 (**1975**), p. 7312–7318
44 Thorn, S. N., Daniels, R. G., Auditor, M. T., Hilvert, D., *Nature* 373 (**1995**), p. 228–30
45 a) Kikuchi, K., Thorn, S. N., Hilvert, D., *J. Am. Chem. Soc.* 118 (**1996**), p. 8184; b) Hollfelder, F., Kirby, A. J., Tawfik, D. S., *Nature* 383 (**1996**), p. 60
46 Shulman, H., Keinan, E., *Org Lett.* 2 (**2000**), p. 3747–50

47 Hotta, K., Kikuchi, K., Hilvert, D., *Helv. Chim. Acta* 83 (2000), p. 2183

48 Klein, G., Reymond, J.-L., *Helv. Chim. Acta* 82 (1999), p. 400–408

49 Klein, G., Ph. D.Thesis, University of Bern 2001

50 a) Wahler, D., Badalassi, F., Crotti, P., Reymond, J.-L., *Angew. Chem. Int. Ed.* 40 (2001), p. 4457–4460; b) Badalassi, F., Wahler, D., Klein, G., Crotti, P., Reymond, J.-L., *Angew. Chem. Int. Ed.* 39 (2000), p. 4067–4070; c) Wahler, D., Badalassi, F., Crotti, P., Reymond, J.-L., *Chem. Eur. J.* 8 (2002), p. 3211–3228

51 Jourdain, N., thesis, University of Bern 2001

52 a) Jourdain, N., Pérez Carlón, R., Reymond, J.-L., *Tetrahedron Lett.* 39 (1998), p. 9415; b) Pérez Carlón, R., Jourdain, N., Reymond, J.-L., *Chem. Eur. J.* 6 (2000), p. 4154

53 Wagner, J., Lerner, R. A., Barbas, C. F., *Science* 270 (1995), p. 1797

54 List, B., Barbas, C. F., Lerner, R. A., *Proc. Natl. Acad. Sci. USA* 95 (1998), p. 15351

55 a) Landry, D. W., Zhao, K., Yang, G. X.-Q., Glickman, M., Georgiadis, T. M., *Science* 259 (1993), p. 1899–1901; b) Yang, G., Chun, J., Arakawa-Uramoto, H., Wang, X., Gawinowicz, M. A., Zhao, K., Landry, D. W., *J. Am. Chem. Soc.* 118 (1996), p. 5881–5890

56 Mets, B., Winger, G., Cabrera, C., Seo, S., Jamdar, S., Yang, G., Zhao, K., Briscoe, R. J., Almonte, R., Woods, J. H., Landry, D. W., *Proc. Natl. Acad. Sci. USA* 95 (1998), p. 10176–10181

57 Gartenman, L., Reymond, J.-L., unpublished

58 Bahr, N., Güller, R., Reymond, J.-L., Lerner, R. A., *J. Am. Chem. Soc.* 118 (1996), p. 3550–3555

59 Bensel, N., Bahr, N., Reymond, M. T., Schenkels, C., Reymond, J.-L., *Helv. Chim. Acta* 82 (1999), p. 44–52

60 Hugot, M., Bensel, N., Vogel, M., Reymond, M. T., Stadler, B., Reymond, J.-L., Baumann, U., *Proc. Natl. Acad. Sci. USA* 99 (2002), p. 9674–8

61 Bensel, N., Reymond, M. T., Reymond, J.-L., *Chem. Eur. J.* 7 (2001), p. 4604–4612

9
In vitro Evolution of Catalytic Antibodies and Other Proteins via Combinatorial Libraries
Ron Piran and Ehud Keinan

9.1
Introduction

Natural enzymes are highly advantageous catalysts for synthetic purposes, with apparent second-order rate constants, k_{cat}/K_M, typically in the order of 10^6–10^8 M^{-1}sec^{-1}, and rate accelerations over background, k_{cat}/k_{un}, as high as 10^{17}. They also give high selectivity, including chemo-, stereo-, regio-, and enantioselectivity, minimal production of by-products, and energy-efficient operation under mild environmentally friendly conditions [1]. Yet, for many industrially important transformations [2], natural enzymes are either too unstable, too difficult to isolate, or not yet known. Therefore, there is great practical interest in the design of new enzymes with desired activities [3, 4].

This stunning array of enzyme features is a manifestation of the power of evolutionary design processes. Natural selection acting on large populations over long periods of time has generated a vast number of proteins ideally suited to their biological functions. To date, protein reengineering efforts by rational design have enjoyed only limited success, because all the key properties of an enzyme, including substrate specificity, product selectivity, stability, and activity in a non-natural environment, are the result of many amino acid residues appropriately positioned in space, all operating in concert. Thus, it is difficult to dissect these features and redesign them by point mutations. This is true even for the relatively small number of enzymes for which considerable structural and mechanistic data are available. The ineptitude of rational design highlights the benefits associated with methods of irrational design, and these take their cue from Natural processes of evolution [5].

Diversity, selection, and amplification are the three major elements of evolution. A sufficiently large library of variant proteins must be generated in order to create the space of diversity. From this space, peptides that present the desired properties are selected and amplified in order to form a biased library for the next evolutionary round. For example, the immune system, which exhibits an estimated diversity that exceeds 10^{12} different antibody molecules, can be used as a most prolific source of specific binding molecules. The magnitude of this natural library enables high-affinity antibodies to be elicited against virtually any stable molecule. Following an

Catalytic Antibodies. Edited by Ehud Keinan
Copyright © 2005 WILEY-VCH Verlag GmbH & Co. KGaA, Weinheim
ISBN: 3-527-30688-9

appropriate selection method, cells that produce an antibody with desired properties can be amplified and immortalized by the hybridoma technology.

This chapter highlights the strengths of directed evolution for the design of novel enzymes with emphasis on catalytic antibodies. Since the various techniques are characterized by significant complementarily, combined approaches may be superior to strategies that rely on a single method [6–8].

9.2
Technologies and Constructs – Basic Principles

9.2.1
Phage Display

Filamentous phage particles are flexible rods, 6 nm in diameter and approximately 1000 nm long (Fig. 9.1). They are composed mainly (87% by mass) of a protein tube that accommodates a single-stranded DNA (ssDNA) of approximately 6408 nucleotides. The wild-type virion contains 2700 copies of helically arranged molecules of the major coat protein, pVIII, which is 50 amino acids long, encoded by a single phage gene, VIII [9]. The virion is capped at one end by five copies of the minor coat protein pIII (about 200 amino acids each, encoded by gene III) and protein pVI, encoded by gene VI. The other end of the virion is capped by proteins pVII and pIX, which are encoded by genes VII and IX, respectively.

Phages are viruses that infect bacterial cells. Many of the vectors used in recombinant DNA research are phages that infect *Escherichia coli*, which is the standard recombinant DNA host. The key feature of recombinant DNA vectors, including phages, is that they can accommodate segments of foreign DNA. As a vector DNA replicates in its host, the foreign insert DNA replicates along with it. The phage infects strains of *E. coli* that display a threadlike appendage, known as the F pilus. Infection is initiated by attachment of the N-terminal domain of pIII to the tip of the pilus; then the coat proteins dissolve into the surface envelope of the cell, and the uncoated ssDNA concomitantly enters the cytoplasm. The host machinery then

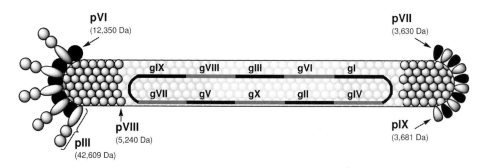

Fig. 9.1

synthesizes a complementary DNA strand, resulting in a double-stranded replicative form (RF). The RF replicates to make progeny RFs. In addition, it serves as a template for transcription of phage genes and synthesis of progeny ssDNA molecules. These progeny ssDNAs are extruded through the cell envelope, acquiring the newly synthesized coat proteins from the membrane and emerging as completed virions. Each division cycle produces several hundred virions per cell. Although progeny virions are secreted continuously without killing the host, chronically infected cells divide at a slower rate than uninfected cells. The yield of virions can exceed 0.3 mg/mL [10].

Foreign peptides have been fused to four different coat proteins, pIII, pVIII, pVI, and pIX. The first two were synthesized with an N-terminal signal peptide that is cleaved off the main polypeptide chain while it is inserted into the inner membrane of the host cell (the bacterial envelope is comprised of two membranes separated by the periplasm). Foreign peptides have been fused to the pVIII and pIII proteins by various ways that exposed them to the exterior, e.g., to the N-terminus of pVIII and to the N terminus and middle portion of pIII [10]. Thus, the phage display differs from other expression systems in that the foreign gene sequence is spliced into a foreign gene that encodes for one of the phage coat proteins, so that the foreign amino acid sequence is genetically fused to the endogenous amino acids of the coat protein to make a hybrid fusion protein. This hybrid coat protein is incorporated into the phage particles as they are released from the host cell, so that the foreign peptide or protein domain is displayed on the outer surface of the virion.

The great advantage over other strategies that use protein libraries of the phage display technology and of the other techniques that will be further discussed is that the expressed protein is physically coupled with its coding DNA sequence. The coding sequence is encapsulated within the phage envelope while its coded protein is expressed on the phage surface. This coupling enables the selected protein to be isolated and amplified and its coding gene to be sequenced, cloned, and expressed in any desired organism. Therefore, the phage display technology represents a practical tool for *in vitro* evolution.

A phage display library is a mixture of such phage clones, each carrying a different foreign DNA insert and therefore displaying a different peptide on its surface. When the phage to which a peptide is attached infects a fresh bacterial host cell, the phage multiplies to produce a huge crop of identical progeny phages displaying the same peptide. By this process, each peptide in the library replicates. If the phage DNA suffers a mutation in the peptide coding sequence, that mutation is passed on to the phage progeny and can affect the structure of the peptide. Consequently, each peptide in a phage display library has two key characteristics required for chemical evolution: mutability that enables diversity and the ability to multiply.

Because of its accessibility to solvent, a displayed peptide often behaves essentially as it would if it were not attached to the virion surface. Thus, the surface-displayed protein usually retains its intrinsic affinity and specificity. Therefore, many techniques that are usually applied to the free peptides in solution can also be applied to the phage-bound peptides. For example, in one of the earliest demonstrations of phage display, a population of phages expressing the fECO1 protein was diluted in a large access of wild-type phages [11]. This mixture was added to pre-incubated dishes with ECORI antiserum. These dishes were washed several times, and the remaining bonded phages on the dish exhibited amplified amounts of fECOI presenting phages by up to 10^3–10^4. This affinity selection has been the premier strategy of artificial selection imposed on populations of phage displayed peptides.

For *in vitro* evolution of catalytic antibodies, an immobilized transition state analog can be used for affinity screening of a large phage library that displays many different peptide structures. Specific phages that display relevant binding molecules can be captured and amplified by infecting fresh cells and culturing them to produce a large crop of progeny phages. By periodically introducing mutations into the phage population and thereby adding sequences that were not present in the initial library, the experimenter can widen the diversity of the protein library. Eventually, captured phages are cloned, so that the displayed TSA binding protein can be studied individually and its catalytic properties can be examined. The amino acid sequence of the peptide is easily obtained by determining the corresponding coding sequence in the viral DNA. One can use the phage display technique in order to manipulate the sequence and structure of Fabs and Fvs.

9.2.2
The Fab Construct

The Fab (Fragment antigen binding) comprises the full-length light chain of the antibody and half of the heavy chain. Papain cleaves the antibody molecule into three pieces (Fig. 9.2), two identical Fab fragments and one Fc (Fragment conserved). Since the Fc, which comprises half of the heavy chain, is irrelevant for the antibody binding and catalysis, there is no need to include it in libraries.

Because the Fab is made of two different peptide chains that need to assemble correctly to become functional, it is difficult to express Fab on a phage protein. In 1988 two independent efforts achieved the expression of functional Fab in *E. coli* [12, 13]. Skerra and Plückthun have cloned the Fab of an anti-phosphorylcholine

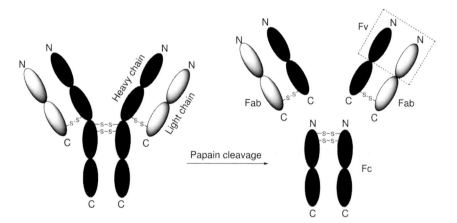

Fig. 9.2

IgA obtained from the myeloma cell line McPC603, which had a known amino acid sequence and known three-dimensional structure. The light chain and half of the heavy chain were cloned on the same expression plasmid as two separate peptide chains on the same transcription unit downstream from a Lac promoter. These peptides were cloned and expressed in *E. coli*. That work, which demonstrated that the displayed Fab had the same affinity constant towards its hapten as the whole antibody, indicated that the folding and the hetero-association of the Fab is possible and does not require any additional factor [12]. Similar results were reported by Better et al., who expressed in *E. coli* the Fab of the L6 chimeric antibody, which was elicited against the ganglioside antigen [14]. The expressed Fab was obtained with higher purity than the Fab generated by papain cleavage of the whole antibody, and both the cloned Fab and the full-length antibody were shown to bind their hapten with comparable binding constants.

In 1989, Huse et al. constructed the first bacteriophage lambda-expressed library of Fab fragments in order to discover catalytic Fabs [15]. A mouse was immunized with a KLH conjugate of *p*-nitrophenyl phosphonamidate, NPN, which is a transition state analog (TSA) of the hydrolysis reaction of amide **1** to produce *p*-nitroaniline (**2**) and arylacetic acid **3** (Fig. 9.3). NPN had already been proven to be an effective hapten for the generation of catalytic antibodies that specifically catalyzed this hydrolysis reaction [16]. The coding DNA of the polyclonal Fabs from the total spleen extract was amplified by polymerase chain reaction (PCR). Two libraries of the light chain and half of the heavy chain were co-expressed on a phage lambda vector. Many phage clones (2.5×10^7) were thus obtained, from which about 60% caused the expression of both peptide chains in their host cells. The phage-infected cells were incubated on dishes, and the resultant plaques were lifted on a nitrocellulose filter and rinsed with ^{125}I labeled NPN, and the filter was assayed for radioactivity to select the binding clones. About 10^2 of 10^6 assayed clones presented antigen-binding ability with binding affinity constants at the range of 10^7–10^8 M. One clone was analyzed by ELISA and by immunobloting, and the resultant Fab was found to be an NPN binder at the nanomolar range [15].

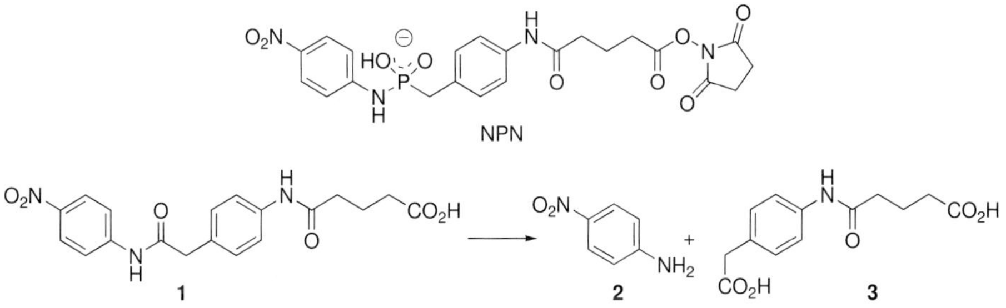

Fig. 9.3

9.2.3
The Fv Construct

Since the Fv (fraction variable), which represents half of the Fab (Fig. 9.2), is the minimal portion of the antibody molecule that is sufficient for antigen binding and catalysis, Fv molecules have become attractive objects for phage display libraries. This variable domain has several regions, some of which are more conserved than others. Analysis of the antibody coding sequences has shown two classes of variable regions within the Fv, hypervariable sequences and framework sequences [17]. Short beta-sheet peptides, which are coded by minigenes (framework), are linked together by random-order peptides that are not encoded by an original gene. The framework sequences are responsible for the correct beta-sheet folding of the variable domain of the heavy chain (V_H) and the variable domain of the light chain (V_L), and also for the inter-chain interactions that bring both domains together. Each domain contains three hypervariable sequences, known as complementarity-determining regions (CDRs). The six CDRs of both chains form the antigen-binding site.

Bird et al. have developed a single-chain Fv (scFv) by cloning the V_H together with the V_L on the same plasmid to produce a single polypeptide chain where a linker peptide tethers the C-terminus of the V_L to the N-terminus of the V_H [18]. The *scFv* was developed to circumvent problems associated with the assembly of a functional binder from two polypeptide chains that are expressed separately. The concept of using scFv for binding and catalysis was based on the notion that the binding interactions between the two peptide chains, which are donated mainly by the framework residues, are sufficient for stabilizing the conserved structure of the Fv. The first scFv molecule was constructed and expressed in *E. coli* from the V domains of 3C2, an anti-BGH (bovine growth hormone) antibody. The affinity of the scFv molecule (3C2/59) to its antigen was found to be comparable with that of the Fab that was isolated from the papain cleavage of 3C2 [18]. In another example, an anti-fluorescein scFv was created and compared with an analogous Fab that was isolated from the whole antibody, and both proteins exhibited K_a of the same order of magnitude (1.1×10^9 M^{-1} and 8.0×10^9 M^{-1}, respectively) [18]. Fuchs et al. have expressed an anti-chick lysosyme scFv molecule on the surface of *E. coli* as a fusion protein to the N-terminus of the peptidoglycan-associated protein (PAL) [19]. By using a fluorescently labeled antibody that was specific to the linker peptide of the scFv they could demonstrate that the scFv was expressed on the outer membrane of the cell. Cells expressing the scFv-PAL fusion protein were significantly more fluorescent than the wild-type bacteria.

The engineered protein, scFv, has several advantages over the cloned Fab. First, since this protein is made of a single peptide chain, its different domains do not need to reassemble in the host cell before the progeny phage is created in order to generate their tertiary structure. Accordingly, the problem of losing one of the Fab chains after the phage emerges from the host cell does not exist here. Furthermore, the assembly of the V_H and V_L domains in the scFv is a first order reaction, independent of the concentration. Finally, in comparison with the Fab, the scFv technology allows for screening of a larger library because the DNA fragments coding for scFv are considerably smaller and the sequential cloning procedures are easier.

Yet, the scFv has several drawbacks when compared with the Fab. For example, only a few scFv domains were found to contain a disulfide bond between the two chains. The cloned scFv is less stable than the Fab and tends to aggregate rather than form a heterodimer [20]. It is known that the conserved segment in the Fab molecule stabilizes the native structure, which is similar to its structure in the whole antibody. Furthermore, the peptide linker in scFv can damage its binding conformation and binding kinetics because it may obscure the antigen-binding site.

In order to solve problems of instability and aggregation, a generally modified version of Fv was developed to include a disulfide bond between the V_H and V_L chains. The first disulfide-stabilized Fv (dsFv) was cloned by Glockshuber et al. [21], who modified the Fv of the anti-phosphorylcholine antibody McPC603, to produce scFv [12]. They replaced two amino acids by cysteine, one in the V_H and one in the V_L, thus forming an inter-chain disulfide bond. This manipulation, however, requires detailed information about the three-dimensional structure of the Fv. Two different Fv fragments, each containing an intermolecular disulfide bond (L55 Tyr-Cys, H108 Tyr-Cys; L56 Gly-Cys, H106 Thr-Cys) were constructed. It was found that the intermolecular disulfide bonds in both molecules were formed in the periplasm in vivo [21]. Both mutant proteins exhibited affinity constants that were similar to that of the whole antibody. Furthermore, these mutants were much more stable than the scFv. However, this technique could not be applied to large combinatorial libraries because introduction of the disulfide bond was dependent on prior knowledge of the three-dimensional structure of the protein.

Brinkmann et al. have reported on the first application of the dsFv technology for generating combinatorial phage display libraries [22]. The disulfide bond was introduced in highly conserved framework positions away from the antigen-binding site (Kabat positions H44 and L100) [23] to allow stabilization of almost any sequence of Fv. Using the PCR technique, the V_H and V_L cDNAs were amplified, assembled, and mutated to introduce a cysteine codon at the appropriate positions. The V_L gene with a cysteine mutation at its 3' end was generated by reverse-transcriptase-PCR (rtPCR) using standard V_L5' primers. The V_L3' primer had the cysteine mutation already encoded into it. In contrast, the construction of the V_H chain was much more difficult because the required cysteine mutation is located in the middle of the gene. Therefore, the amplification of the appropriate V_H chains was achieved via several PCR steps using a set of several primers [20, 22].

The dsFv can circumvent most of the above-described problems of the scFv approach. The dsFv is much more stable than the scFv and it has no peptide linker that might disturb its activity. Yet, the dsFv library has several drawbacks in comparison with the scFv. First, the generation of dsFv in bacteria requires separate expression and secretion of mutated V_H and V_L proteins, heterodimerization, and subsequent oxidation of the disulfide bond. Second, the construction of a dsFv library is much more difficult than the construction of an scFv library. Not only does it require the cloning and assembly of V_H and V_L into the pIII fusion, but also the cysteine mutations must be placed into defined positions of both the V_H and V_L genes. Finally, the fraction of functional dsFv phage population was found to be lower than that of the analogous scFv library [22]. Since the scFv and the dsFv phage libraries have their intrinsic advantages and disadvantages, the experimenter can choose the preferred strategy, depending on the specific goals.

9.2.4
In vitro Immunization

The hybridoma technology uses the selection and amplification machinery of the natural immune system to generate highly efficient binding molecules. However, this methodology represents a single evolutionary round, because the resultant monoclonal antibodies cannot be taken back to the immune system for additional cycles of improvement and refinement. In contrast, the phage display technology allows for many cycles of evolution. Furthermore, whereas the immune system selects only for binding, when using *in vitro* methods of evolution one can screen not only for binding but also for other properties, such as catalysis.

The phage library, which imitates the natural library of B-cell lymphocytes, can undergo *in vitro* evolution to produce the desired binding molecules and catalysts. The antigen-driven selection process of the immune system can be mimicked by the ligand-driven affinity selection of phage encoding the displayed antibody. Phage displaying peptides have been selected by binding to a monoclonal antibody, and phage displaying antibodies by binding to antigen [24]. Affinity selection of the relevant phage has been achieved by several approaches, such as binding to biotinylated antigen in solution followed by capture on streptavidin-coated beads [25], binding to antigen-coated dishes or tubes [26], or binding to antigen on a column matrix [27]. Usually, release of the bound phage can be easily achieved by elution with acid or base. This strategy has resulted in enrichment factors between 20-fold and 10^6-fold for a single round of affinity selection [28]. By infecting bacteria with the eluted phage, more phage can be grown and subjected to subsequent rounds of selection. Thus, even when enrichments are low for a single round, multiple rounds of affinity selection can lead to the isolation of a rare phage along with its genetic material.

In the immune system, the improvement of binding affinity of antibodies results from somatic mutations of the V-gene pairings that occur after the initial immunization (primary response) and also from the appearance of antibodies that use V-genes that have not appeared in the primary response (repertoire shift) [29–31]. The mechanism by which the different minigenes are assembled involves replacement of different minigenes in the lymphocyte as well as linking them with a random nucleotide sequence that varies in length and composition. The number of somatically mutated B-cells arising from a given primary B-cell is much smaller than the primary library of about 10^7–10^8 B-cells in mice. In principle, since a phage display library can be larger than the natural antibody diversity, the use of phage display could lead to binding molecules with higher affinity than those produced by the immune system. The natural process of somatic mutation can be mimicked *in vitro* by scattering random mutations over the entire V-gene. For more extensive variation, artificial crossovers

(DNA shuffling) between V-genes can be done using either the PCR or somatically mutated V-genes harvested from B-cells.

The binding affinity of phage to an immobilized antigen depends not only on the affinity of each displayed antibody molecule but also on the number of antibody fragments per phage. The latter number is largely determined by the choice of either pIII or pVIII coat proteins for fusion and the use of phage or phagemid vectors. Other factors that may also contribute include possible association of antibody fragments as dimers on the phage or the proteolysis of the fusion protein. For binding to an immobilized antigen, the binding affinity will also reflect the surface density of the antigen [26]. Thus, for phage vectors, fusion of the antibody fragment to pIII should lead to not more than five copies of the fusion protein on each phage particle. With pVIII fusions, many more copies of the fusion can be displayed, typically up to 24 antibody fragments per phage. The higher binding avidity with multivalent display may help retain lower affinity phage on solid phase antigen during washing, especially for phage with fast off rates, but may hinder the discrimination between phages that have different affinities [26].

Combination of the phage display and scFv technologies allowed John McCafferty et al. to clone together an scFv that was constructed from the V_H and V_L of antibody D1.3 linked with a $(Gly_4\text{-}Ser)_3$ peptide to form a phage display library [27]. The new protein domain was able to bind its hapten, the hen egg white lysozyme, as did its ancestral antibody. The scFv-presenting clone was isolated from a mixture of phages. The scFv-presenting phage was mixed with an excess of wild-type phages in a ratio of 1 to 10^{12}. The phage population was passed over a column. Colonies derived from the eluates were analyzed by probing with an oligonucleotide that detects only the phage antibody. At least a 1000-fold enrichment of scFv-presenting phage was observed.

9.3
Screening of Libraries

The immune system can only produce binding molecules. Therefore, the traditional way of eliciting antibodies against a TSA of a given reaction affords the best binders of the TSA but not necessarily the best catalysts for this reaction. In contrast, the phage display technology can be designed for enforcing a direct evolutionary pressure for catalysis and not for binding.

Duenas and Borrebaeck [32] have created a powerful system for directed evolution by genetic manipulation of the phage pIII protein. They detached the infective particle of pIII, which includes the membrane penetration domain, N1, and the pilus-binding domain, N2 (Fig. 9.4). The Fab library was expressed on the C terminus, CT, of the remainder of pIII. In this system, a soluble fused protein comprising the antigen (hen egg lysozyme, HEL) and the missing infective particle of pIII was added to the modified phage population. The Fab-antigen interactions rendered the non-infectious Fab-displaying phage infectious. Thus, only the antigen-binding members of the phage library that possessed the essential element for becoming infectious were able to proliferate. When the soluble fusion protein, N1-N2-HEL, was added

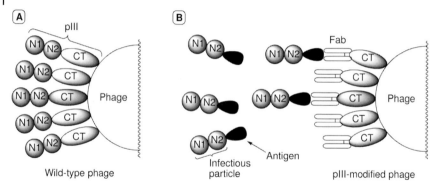

Fig. 9.4

to a mixture of three different Fab-presenting phages, only the phage clones that displayed the anti-HEL Fab could replicate. This work demonstrated that a specific phage could be isolated from a library of 10^{11} clones, affording a selection enrichment factor close to 10^{10} in only two rounds [32].

Duenas et al. have also shown that this technology can be used for the selection of different Fabs on the basis of their binding kinetics [25]. Fabs of an anti-HEL antibody that displayed high binding constant, K_a, were selected by a short-time suspension with the above-described soluble fusion protein, N1-N2-HEL. The addition of free antigen caused dissociation for the binders that had fast association-dissociation kinetics, leaving behind those with a slow dissociation rate. By using high concentrations of the free N1-N2 particles, the low affinity antibodies could be selected and amplified.

Krebber et al. have further characterized this method [33], selecting an anti-fluorescein scFv as a model. Different infectious particles were prepared by covalent coupling of fluorescein to either N1 or N1-N2 N-terminal domains (**A** and **B**, Fig. 9.5). These fusion proteins were mixed with non-infective anti-fluorescein scFv-displaying phages. Infection events were strictly dependent on the recognition of fluorescein by the displayed scFv. Up to 10^6 fluorescein-specific phage particles became infectious out of 10^{10} input phages, compared to only one case of antigen-independent infection.

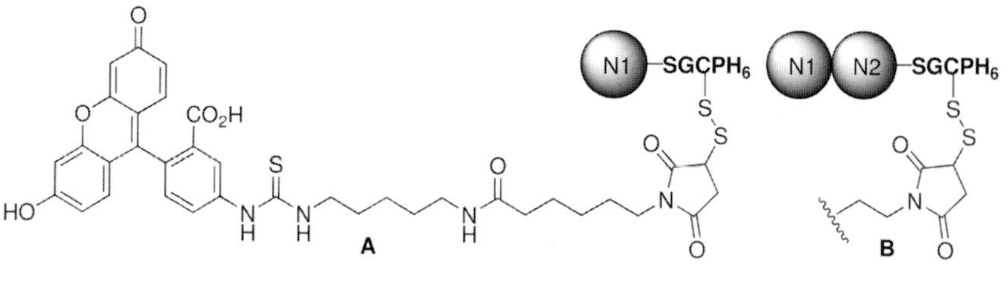

Fig. 9.5

It is difficult to create a phage display library with the protein product of cDNA expressed in the middle of pIII because the relevant cDNA contains a stop codon at the end of each gene. Consequently, insertion of cDNA into the gene of pIII would result in loss of the C-terminus domain of pIII. A cloning and expression system that allows the display of functional cDNAs or other gene products on the surface of filamentous phage has been developed by Crameri and Suter. The method is based on the high-affinity interaction of the Jun and Fos leucine zippers [34]. JUN was expressed from a lacZ promoter as a fusion protein with pIII, thereby being structurally incorporated into phage particles during infection with a helper phage (Fig. 9.6). A second lacZ promoter of the phagemid, FOS, was co-expressed as an N-terminal fusion peptide to cDNA library gene products, so that the resultant Fos-fusion proteins could become associated with the Jun-decorated phage particles. To avoid interphage scrambling of Fos-cDNA fusion products, cysteines were engineered at the N- and C-termini of each of the leucine zippers, providing a covalent bonding of the cDNA gene product to the genetic instructions required for its production. Dissociation between phage and cDNA gene products was achieved by reduction of the disulfide linkage. Phages displaying gene products covalently anchored on their surface via the modified leucine zippers could be selectively enriched 10^4-fold to 10^6-fold over nonspecific phages using antibodies. Thus, this cloning system allows rapid isolation of rare mRNA products from complex cDNA libraries by enrichment with appropriate ligands.

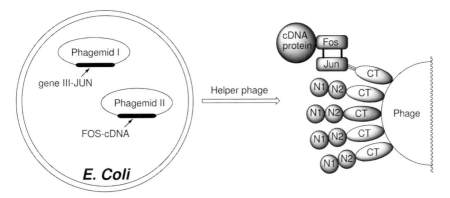

Fig. 9.6

Brunet et al. [35] significantly improved the above-described phage display technique to scan proteins for catalytic activity. They substituted domain 3 of pIII by the entire pIII and the gene encoding the Stoffel fragment of Taq DNA polymerase I was inserted at the 3' end of *fos*, the sequence coding for the leucine zipper of Fos. Phagemids were rescued using an engineered helper phage KM13, whose pIII contained a protease-cleavable site between domains 2 and 3 (Fig. 9.7) [36]. Phages that cannot express any pIII–Jun fusion displayed only cleavable pIII copies and were thereby rendered non-infective by proteolysis. Phages expressing one pIII–Jun fusion displayed one full pIII copy after proteolysis and therefore remained infective

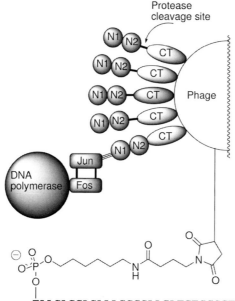

TAACACGACAAAGCGCAAGATGTGGCGT
　　　　　　　　　　　GCGTTCTACACCGAAAATAACAACAT Fig. 9.7

[36]. Thus, the protease-resistant fraction of phage became an indicator of the fraction of phage that could express the pIII–Jun fusion protein. For the *in vitro* selection of polymerase activity, model libraries composed of two types of phage particles were prepared. One type expressed a polymerase while the others did not. The substrate (a partially hybridized dsDNA) was cross-linked to the phage particles and converted to product only if the phage displayed an active enzyme in proximity to the substrate. The selection is an affinity chromatography for the reaction product, which is coupled to the phage polymerase that catalyzed the reaction. Using the system described, enrichment factors of more than 50 have been measured for *in vitro* selections of Taq DNA polymerase activity from model libraries containing phage polymerases and phages not expressing any polymerase. Following a sufficiently large number of cycles of selection and amplification, such enrichments should be sufficient for the isolation of biocatalysts and their genes from libraries of more than 10^7 proteins displayed on phage [37].

An alternative method for screening cDNA libraries was developed by Gao et al. [38, 39], who fused V_H and V_L to the N termini of pVII and pIX, respectively. Significantly, the fusion proteins interacted to form a functional Fv-binding domain on the phage surface. Both pVII and pIX are positioned, five copies each, at the same end of the phage particle that emerged first from the host cell and was required for the initiation of assembly through interaction with the packaging signal of the phage genome. Since neither protein is well studied as pIII, it was essential to know the orientation of pVII or pIX on the surface of phage to correctly display the Fv chains. For that purpose, a Flag tag was fused to the N or C terminus of pVII and pIX. It was

9.3 Screening of Libraries

found that only the Flag fused to the N terminus of pVII or pIX could be detected by ELISA, indicating that the N termini of pVII and pIX are exposed to the solvent.

Following these findings, the heavy chains of the murine antibodies 21H3, 2H6, and 92H2 were constructed as fusion proteins with pVII and the light chains were constructed as fusion proteins with pIX, and the constructs were displayed on the surface of phage particles [38]. Phage ELISA demonstrated that the Fv-displaying phages specifically bound their respective haptens. Thus, the 21H3 and 2H6 phages bound the BSA conjugate of PCP (**4**), and 92H2 phage bound the conjugate of GNC (**5**) (Fig. 9.8) [40, 41]. For the former two antibodies, the binding activity could be inhibited by free hapten PCP, and for 92H2 it could be inhibited by cocaine. Consequently, it was concluded that, when both the V_H and V_L chains were correctly displayed on the phage surface, inter-chain interaction occurred to form a functional Fv motif. In the case of the PCP21H3 phage Fv, the catalytic activity was investigated and compared with that of the previously studied whole antibody PCP21H3, which was elicited against hapten **4** and catalyzed the *trans* esterification reaction of **6** with racemic alcohol **7** in water to produce **8** [42]. A comparison of time-course curves showed that the catalytic activity with 150 nM phage Fv was roughly equal to 50 nM of the IgG. The results showed that a 10^8-fold enrichment was accomplished after only two rounds of panning.

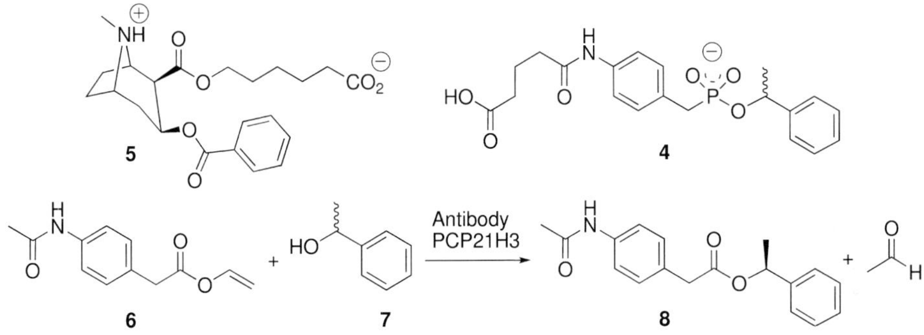

Fig. 9.8

Further work from the same group focused on an scFv library [39]. A human scFv library of 4.5×10^9 members was displayed on pIX of filamentous bacteriophage. The diversity, quality, and utility of the library were demonstrated by the selection of scFv clones against six different protein antigens. Notably, more than 90% of the selected clones showed positive binding for their respective antigens after only three rounds of panning. Kinetic analysis revealed that scFvs against staphylococcal enterotoxin B and cholera toxin B subunit had a nanomolar and subnanomolar dissociation constant, respectively, affording affinities comparable to, or exceeding, that of monoclonal antibodies obtained from regular immunization. The results suggested that the performance of pIX-display libraries could potentially exceed that of the pIII-display format.

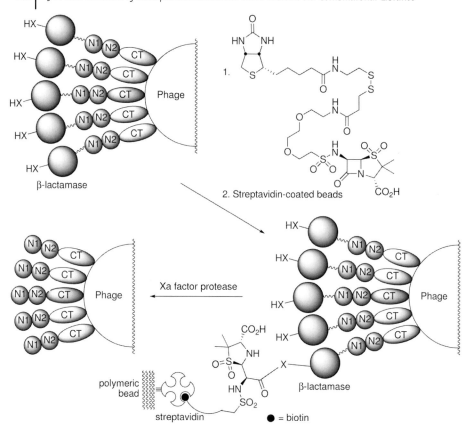

Fig. 9.9

Soumillion et al. have demonstrated that a catalytic protein presented on phage can be selected and isolated using the technique of substrate trap or suicide inhibitor. Two proteins were separately fused to the phage pIII coat protein, an active β-lactamase, and its inactive mutant [43]. The phages were incubated with a biotinylated substrate, 6-aminopenicillanic acid (Fig. 9.9). The clone that reacted with the substrate remained linked to the product and was subsequently separated from the solution by streptavidin-coated beads. Release of the phage from the beads was carried out with Xa factor, which cleaved the peptide bond that linked the enzyme and pIII. The released phage was amplified by infection of E. coli. This procedure enriched 50-fold the population of the phage displayed β-lactamase.

Janda et al. scanned semi-synthetic Fab-presenting phage libraries in an attempt to isolate a Fab that presented a cysteine residue in the CDRs. The uncoupled, free cysteine was planned to form an S-S bond with an active disulfide acceptor, which was immobilized on microtiter plates [44]. This screening technique succeeded in isolating a catalytic Fab capable of hydrolyzing a thioester bond with $K_{cat}/K_{uncat} = 30$. In more recent work, a biased Fab-phage library was screened directly for catalytic

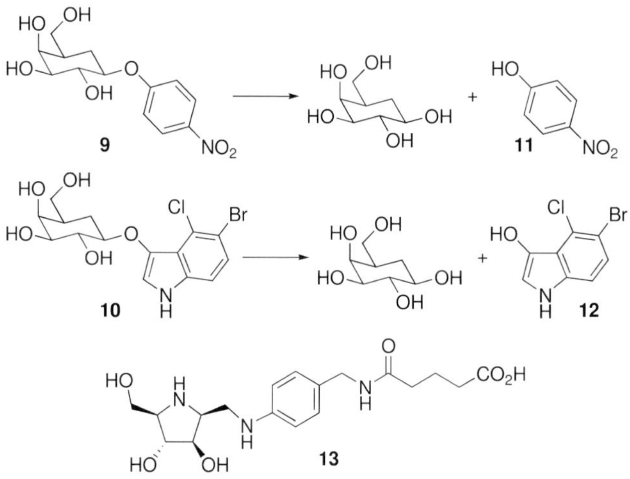

Fig. 9.10

activity [45]. The chosen reaction was the hydrolysis of galactopyranoside substrates, such as **9** and **10**, to produce **11** and **12**, respectively (Fig. 9.10). Mice were immunized with a KLH conjugate of the iminocyclitol **13**, and approximately 100 antibody clones that bound this hapten were generated by established hybridoma methodology. Using rtPCR technology, cDNA was obtained from the hybridomas and the two chains of the Fab were ligated on a plasmid in which half of the heavy chain was linked to the gene of pIII. The resultant Fab-phage library, which contained approximately 10 000 members, was screened for catalytic activity using an immobilized galactoside substrate. Upon cleavage of the glycosidic bond, a highly reactive species, a quinone methide, was formed in close proximity to the catalytic site, leading to alkylation of the catalyst and trapping it on the solid support (Fig. 9.11). Non-binders and phage that did not bind covalently were removed by washing with buffer and acid, respectively. DTT reduction of the disulfide bond in the linker released the immobilized catalyst-phage to the solution, retaining its replicative competence. These recovered phages were then amplified by infection of E. coli. The entire procedure was repeated through four consecutive cycles to maximize the selection of efficient catalytic Fab.

The DNA of gene III was isolated to allow expression of soluble Fabs in β-galactosidase-deficient XL1-Blue cells. Plating the cells onto LB agar plates containing 0.1 mM isopropyl-β-D-thiogalactopyranoside and 2% 5-bromo-4-chloro-3-indolyl-D-galactopyranoside (X-gal, **10**) confirmed that catalysis by the relevant Fab occurred also *in vivo*. The cleavage of **10** produced an indoxyl primary product that is rapidly oxidized in aqueous solution to an insoluble indigo dye, rendering the cells dark blue in color. Several such blue-colored clones were observed, and one clone termed Fab 1B was chosen for kinetic studies with substrate **9**. Indeed, Fab 1B was found to be a better catalyst (K_{cat} = 0.007 min^{-1}, K_M = 530 µM, k_{cat}/k_{un} = 7 × 10^4) than other catalytic antibodies obtained by the regular hybridoma techniques [45].

In later work, Gao et al. [46] combined the single-turnover substrate trap technique with the above-described methodology developed by Duenas and Borreback [32], by which a non-infective phage could regain infectivity upon attachment of the

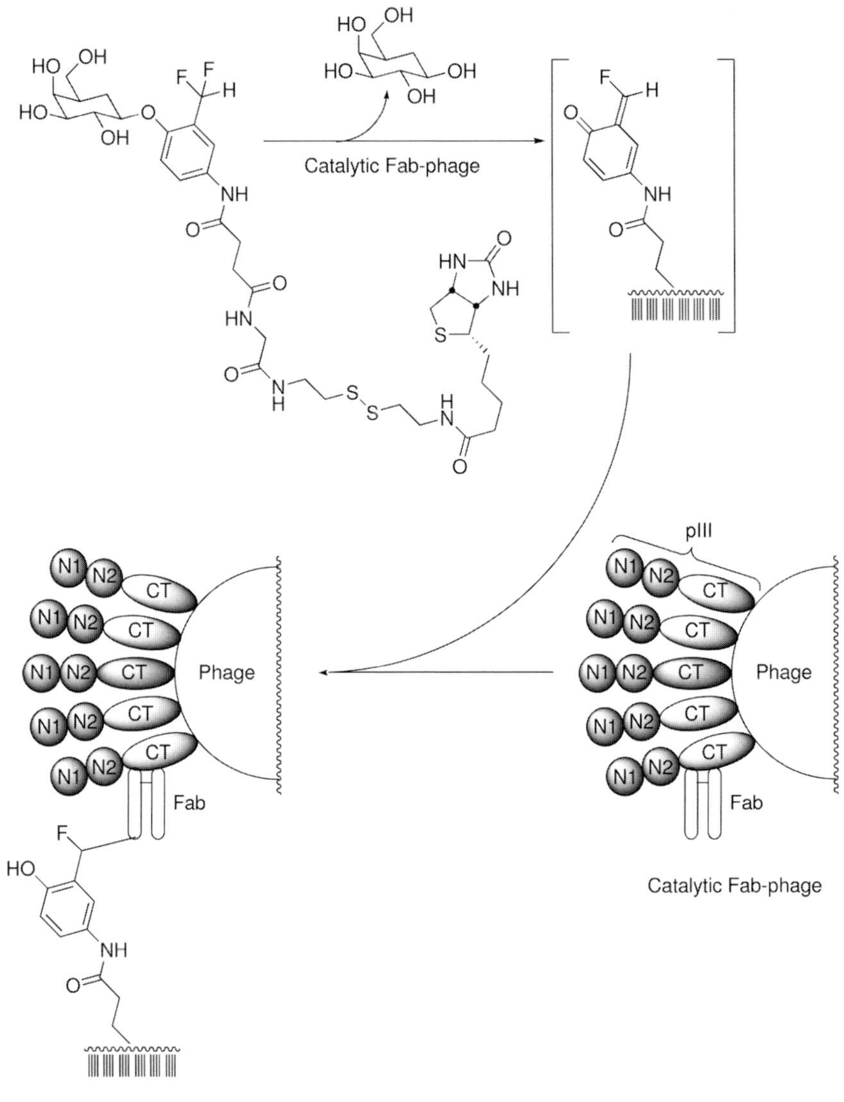

Fig. 9.11

missing N1 and N2 domains to the CT fragment of pIII (Fig. 9.12). They presented a general approach for the screening of phage display libraries for catalytic activity, using as a model the scFv of the known catalytic antibody PCP21H3. The scFv was expressed on a non-infective phage in which scFv-21H3 was mono-displayed (**14**). This scFv presented similar kinetic parameters for both *trans*-esterification and ester hydrolysis as did the whole antibody PCP21H3. A nucleophilic residue in this antibody can undergo acylation by the active ester of the substrate at pH 7.0. The resultant acyl intermediate **15** was found to be fairly stable and even isolatable at

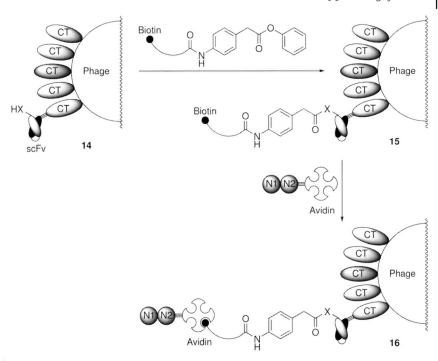

Fig. 9.12

pH 5.5. Since the trapped substrate was linked to a biotin group, it could create a linkage between the catalytic phage and an N1-N2-streptavidin fusion protein, thus rendering the phage infective and able to replicate in *E. coli* (Fig. 9.12). This paradigm links chemistry to infectivity. Two alternative linking approaches were also used instead of the avidin-biotin technology. The first used a maleimide group that was bound to the substrate and formed a covalent bond with the thiol group of an N1-N2-cysteine construct. The second technique used a substrate-bound NTA and an N1-N2-hexahistidine construct, creating a strong coordination bonding in the presence of Ni^{2+} ions [46]. As anticipated, the infectivity of the scFv-21H3-phage was restored with each of the three methods when the phages were first treated with the appropriate substrate trap. Both phage-ELISA and the DNA sequence of phagemid showed that these were indeed scFv-21H3-phage. The infectivity could be inhibited by the original phosphonate PCP hapten [40] as well as by maleimidoacetic acid in the maleimide approach, EDTA in the NTA approach, or free biotin in the biotin approach.

Several other groups followed the idea of screening for catalytic activity rather than for binding. Pedersen et al. have developed a phage that displayed both the enzyme and the substrate. The enzyme, SNase, which is known to cleave ssRNA, ssDNA and dsDNA at AT-rich regions, was fused to a single pIII domain on the phage (Fig. 9.13). The other four copies of pIII were linked to a polyanionic peptide. The latter was linked to a biotinylated ssDNA substrate via a cationic peptide,

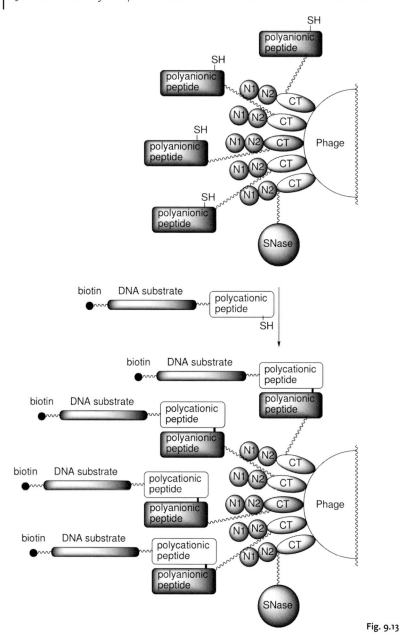

Fig. 9.13

and a disulfide bond reinforced that linkage. The entire phage was immobilized onto streptavidin-containing beads. This construct was diluted with non-catalytic constructs, where another protein was fused instead of the enzyme. Phages that expressed an active catalyst could cleave the substrate and release the phage particle to the solution, rendering it infective in *E. coli*. [47]

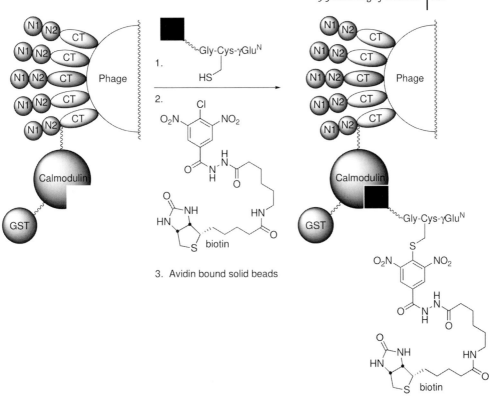

Fig. 9.14

Demartis et al. proposed a system in which an enzyme-calmodulin construct was fused to one copy pIII to allow anchoring of catalyst, reaction substrate, and product on the same phage particle (Fig. 9.14) [48]. The substrate was chemically linked to a high-affinity calmodulin-binding peptide (K_d = 2 pM). An active enzyme catalyzed arylation of the substrate thiol group by an active chloroaromatic biotinylated compound. Thus, the catalytic event linked the relevant phage particle to biotin and eventually to solid-supported avidin. Since the substrate binding to calmodulin is calcium dependent, the entrapped phages were released to solution under mild conditions upon the addition of EDTA to allow bacterial infection and amplification of the selected clones. This work demonstrated that enzymes displayed on calmodulin-tagged phage can be selected by virtue of their catalytic activity with enrichment factors greater than 50 per single round of selection.

In later work, Heinis et al. [49] implemented the above-mentioned selection scheme [48] using a small phage display library, aiming at improved catalysis with two different classes of enzymes, an endopeptidase and ligase. For example, a phage display library was prepared from a mutant of rat trypsin, His57Ala, a relatively inefficient endopeptidase, which cleaves a specific dipeptide sequence, by random mutations at positions 55, 57, 102, and 214. The enzyme library was displayed on phage in the form of fusion proteins with calmodulin and the pIII coat-protein, as shown in Fig. 9.14. The original His57Ala mutant, which exhibited the best catalytic properties, was enriched between 15- and 2000-fold per panning cycle. Similar results were obtained with the ligase BirA. In model selection experiments, phage-displaying BirA was consistently enriched between 4- and 800-fold per round of panning.

In the above-described strategies, the catalyst was employed to carry out a single turnover in a suicide-type reaction that led to a covalent bonding between the enzyme and the product. The main advantage of this approach was that the library was screened for catalysis rather than for binding. However, this screening represents a binary selection process between a catalyst and a non-catalyst but does not favor a good catalyst over a modest one. Since the substrate is either physically linked to the enzyme or the enzyme is saturated with high concentrations of soluble substrate, the values of K_M and k_{cat} are essentially meaningless for the selection process. Thus, the main benefit from using this approach is the discovery of lead proteins that require further optimization using other screening strategies.

9.4
Directed Evolution

Natural evolution has always been a role model and source of inspiration for the designers of new protein, and of new catalysts in particular. In order to evolve a desirable functional protein in real time, it is necessary to create protein diversity and apply an appropriate selective pressure. One approach to creating a diverse library of proteins involves carrying out multiple changes, often randomly, in its primary sequence of a known protein. The evolutionary pressure should give the selected clone, for example a desired catalyst, a proliferation advantage. The best-fitted protein can then be used as a parent protein for a subsequent cycle of multiple mutations, selection, and amplification.

The two major strategies of directed evolution involve (a) mutagenesis and selection and (b) DNA shuffling. The first approach mimics natural evolution as it occurs in prokaryotes, where a DNA sequence that encodes for a given protein is mutated randomly. In general, while most of the random mutations would lead to total loss of the protein function, they can rarely produce a protein with improved properties.

The alternative approach of DNA shuffling is also known as sexual evolution or recombination. Eukaryotes multiply by merging two haploid cells to form a diploid zygote from which the offspring will develop. In order to form the haploid cell, the parent cell undergoes a process called meiosis. In this process, the two equivalent chromosome sets divide twice to produce four haploid cells; recombination is a side phenomenon of this process. Recombination is a reaction between two homologous sequences of DNA. In this process, one of the two DNA strains duplicates itself into the other starting from a random point. "Upstream" of this point the sequence is identical to the original sequence, and "downstream" of this point the sequence is identical to the other one. The critical feature of recombination is that any pair of homologous sequences can undergo recombination, even inside a sequence of a gene. By shuffling the genes or their components, recombination allows favorable and unfavorable mutations to be separated and tested as individual units in a new arrangement. It provides a means of escape and spreading for favorable mutations and a means to eliminate an unfavorable mutation without bringing down the entire gene with which this mutation may have been associated. From the long perspective of evolution, a chromosome is a bird of passage, a temporary association of a particular mutation. Recombination is responsible for this flighty behavior. This is the main reason for the great diversity among eukaryotes compared to prokaryotes.

When using random mutagenesis, there is a high probability of reaching a local maximum in the protein function. Changing another individual amino acid will generally lead to loss of function of the protein. Recombination enables several beneficial mutations in the DNA sequences that were achieved separately to be combined, therefore overlapping these local barriers. J.H. Holland has demonstrated in his algorithm that the best way to achieve diversity in a DNA sequence is to combine a high recombination rate with a low mutation rate [50]. By using novel biological tools accompanied by a new scanning approach, better enzymatic properties for the catalytic antibody can be achieved.

9.4.1
Mutagenesis and Selection

Mutagenesis is a most fundamental means of driving the evolution of given protein toward any desired properties. There are only a few cases where a single point mutation has led to a significant improvement in enzymatic activity. Baldwin and Schults have demonstrated that the replacement of a single amino acid in the active site of a catalytic antibody could improve the catalytic properties by several orders of magnitude [51]. Antibody MOPC315 was elicited against the dinitroaniline derivative **17** (Fig. 9.15). The solid-state structure of this antibody revealed that a tyrosine residue at V_L34 was positioned very close to the hapten **17**. Since the imidazole group is known to catalyze ester hydrolysis, it was decided to carry out a point mutation, Tyr34His, in order to create a new hydrolase antibody. Although the binding affinity of the mutant antibody to the hapten was reduced 8-fold in comparison with the wild-type MOPC315, catalytic efficiency as ester hydrolase increased 45-fold in the hydrolysis of ester **18** to produce carboxylic acid **19**.

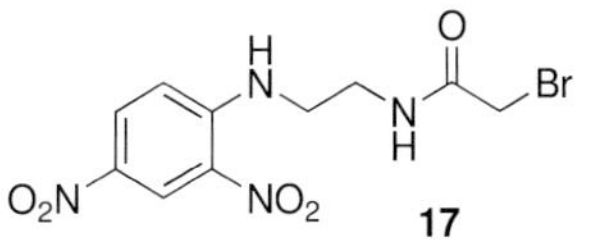

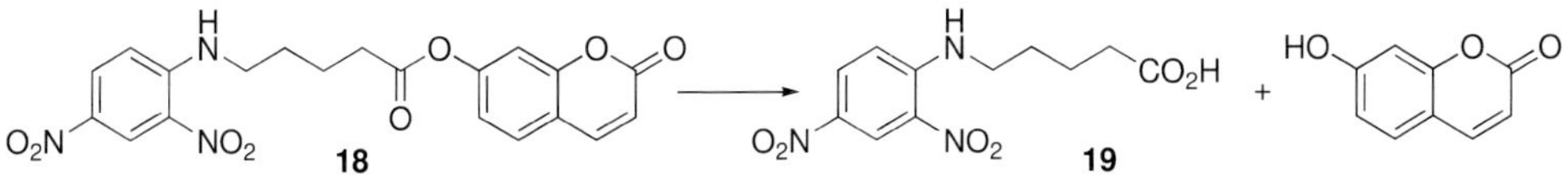

Fig. 9.15

A profound change in the catalytic activity and mechanism of 4-oxalocrotonate tautomerase (4-OT) was accomplished by a rationally designed single amino acid substitution that corresponds to a single base pair mutation [52]. The wild-type enzyme catalyzes only the tautomerization of oxalocrotonate, while the P1A mutant catalyzes two reactions – the original tautomerization reaction and the decarboxylation of oxaloacetate. Although the N-terminal amine of P1A is involved in both reactions, the results support a nucleophilic mechanism for the decarboxylase activity in contrast to the general acid/base mechanism that has been established for the tautomerase activity. These findings demonstrate that a single catalytic group in an enzyme can catalyze two reactions by two different mechanisms and support the theory that new enzymatic activity can evolve in a continuous manner. The results also suggest that P1A could be captured by either adaptive or *in vitro* evolution to optimize the new catalytic activity (Fig. 9.16).

The above work demonstrated the notion that highly evolved enzymes are optimized not only to catalyze a desired reaction but also to avoid undesired processes [53]. Mutation of active-site residues designed to decrease the optimized catalytic activity may also enhance alternative reaction pathways. Thus, even a minor change in the active-site residues could result in a dramatic change in the delicately optimized balance of their chemical reactivity. 4-OT catalyzes the isomerization of 4-oxalocrotonate to its enone tautomer using a general acid/base mechanism that involves a conserved N-terminal proline residue. The P1A mutant has been shown to catalyze this isomerization but also undergoes specific 1,4-addition to the enone product to form a stable covalent adduct. This nucleophilic reactivity of P1A is remarkable considering the complete lack of reactivity of the wild-type enzyme in this reaction.

Baca et al. have improved the catalytic efficiency of the anti-**20** antibody, 17E8, which catalyzes the hydrolysis reaction of **21** to produce carboxylic acid **22** (Fig. 9.17) [54]. From the solid-state structure of this antibody that was co-crystallized with its hapten, it was possible to identify the active-site residues that were positioned in close proximity to the hapten. Using site directed mutagenesis; these amino acids were randomized to create a phage display library of scFv. Screening for binding to hapten **20** over several rounds of panning and amplification yielded a mutant that exhibited an 8-fold increase in binding but lower catalytic activity. In contrast, a 2-fold increase in catalytic activity was achieved by a weaker binder, which had four point mutations. Rational incorporation of a fifth mutation (Tyr100$_H$Asn) at the latter variant led to a 5-fold increase in catalytic efficiency. This work has demonstrated that phage display methods can be used to optimize binding and catalytic efficiency of catalytic antibodies.

Fujii et al. have created a Fab-phage library on the basis of an anti-**23** catalytic antibody, 17E11, which catalyzes the regioselective hydrolysis of the polyacylated glucosamine **25** to produce the mono-deprotected derivative **26** (Fig. 9.18). Diversity was generated by randomizing six amino acids at the CDR3 of the heavy chain. The library was screened for binding against compound **24** over four rounds of panning and amplification. Out of the many binders obtained, 124 clones were randomly picked and their affinities to **24** were examined by ELISA. The catalytic properties of the six best binding clones were studied in comparison with the original 17E11. Two

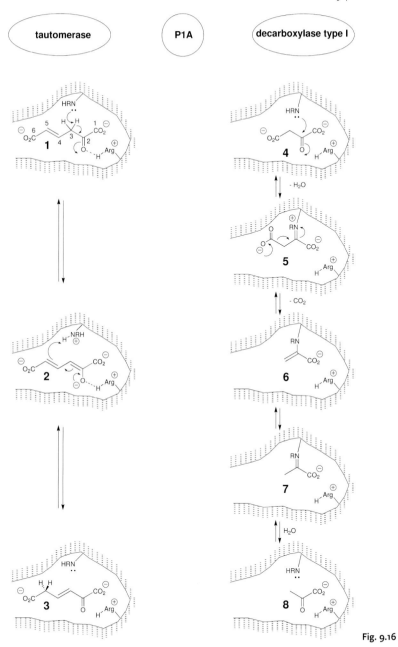

Fig. 9.16

of these were found to be better catalysts. The kinetic parameters of the best catalyst were k_{cat}/k_{un} = 2250 and K_M/K_{TSA} = 1440, while those of the parent antibody, 17E11, were 180 and 100, respectively [55].

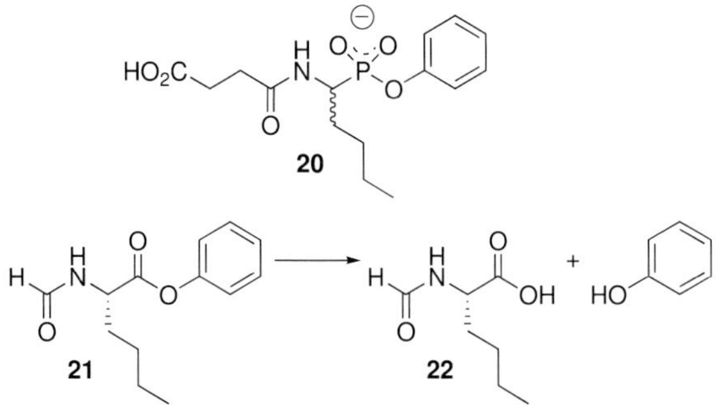

Fig. 9.17

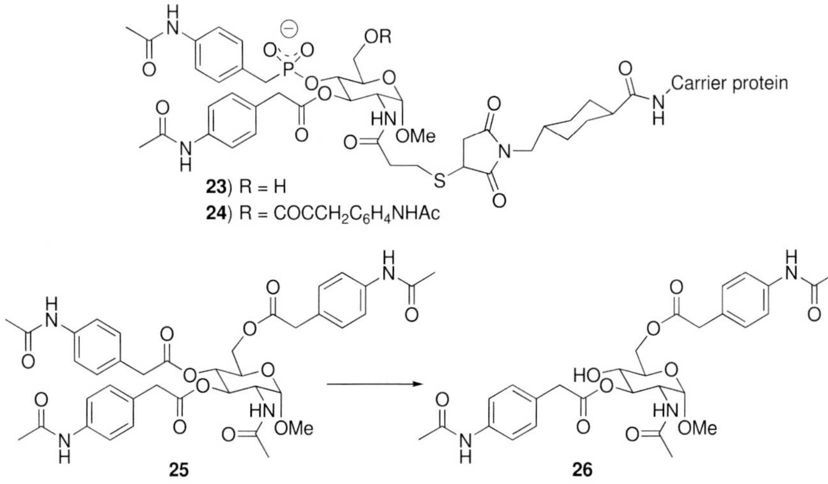

Fig. 9.18

Using a similar approach, Takahashi et al. have improved the catalytic performance of the ester-hydrolase 6D9, which was elicited against hapten **27** and was shown to catalyze the hydrolysis of chloramphenicol monoester (**29**) to produce the active drug **30** (Fig. 9.19) [56]. Based on crystallographic information, six amino acids in the active site were randomized to create a small library. The library was screened for binding against compound **28** over 8 rounds of panning and amplification. Two clones, 8Hf and 8Hg, were found to be better catalysts than the parent antibody, as shown by their kinetic parameters: catalyst; k_{cat} (min^{-1}); K_M (μM); k_{cat}/K_M; k_{cat}/k_{un}: 8Hf, 8Hg, 6D9; 2.3, 2.6, 0.13; 140, 280, 60; 16000, 9150, 2100; 16400, 18800, 940.

Atwell and Wells have used the immobilized product approach to screen a large (4×10^9 member) phage display library of mutants of subtiligase [57]. The latter is a variant of the bacterial serine protease subtilisin, BPN9, which catalyzes the ligation of activated C-terminal peptides to the N-terminus of other peptides [58]. The library was created by random mutagenesis of the 25 residues near the catalytic site, including the catalytic triad and proline that is an essential part of the oxyanion hole. In addition to fusion to the CT domain of pIII, the phage-displayed mutants were

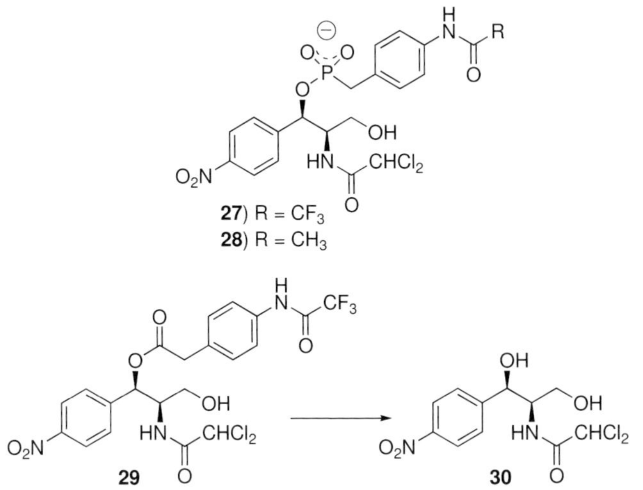

Fig. 9.19

fused to the C-terminus of a short peptide. Catalysis by this enzyme resulted in the ligation of the short peptide to a biotinylated peptide, and the resultant construct could be captured by an immobilized neutravidin. Not unexpectedly, many of the active-site residues, including the catalytic triad and oxyanion hole, were completely conserved in this selection for catalysis. Two new mutants were identified that exhibited enhanced catalytic activity. Interestingly, many other mutants were selected that exhibited increased stability and resistance to oxidation.

Cesaro-Tadic et al. combined the single-turnover screening technique with mutagenesis methods to discover new catalysts [59]. A naïve library of scFV was screened using the quinone methide technique described earlier [45]. The library was screened for phospho-esterase activity using the immobilized phosphate ester **31** as a substrate (Fig. 9.20). The catalyst was covalently trapped after a single turnover by the product. The resultant immobilized phage was washed with buffer to remove the non-binders and then with acid to remove non-covalent binders. Finally, the phage was released from the solid support by reductive treatment with DTT to allow infection and proliferation in *E. coli*. One catalyst, TT1, which was isolated after two rounds of panning, exhibited a remarkable rate enhancement, $k_{cat}/k_{un} = 2.3 \times 10^5$. This catalyst was then randomly mutated using error-prone PCR (epPCR), which introduced an average of 15 single-base-pair mutations per gene, to produce a progeny library of scFvs for additional rounds of screening using the same technique. Two such descendant catalysts, TT1.D1 and TT1.D2, have shown improved catalytic efficiency, $k_{cat}/k_{un} = 2.3 \times 10^6$ and 1.8×10^6, respectively.

9.4.2
DNA Shuffling

The DNA shuffling technique, which was developed by Stemmer [60, 61], represents an efficient strategy to improve the catalytic performance of a given enzyme. The

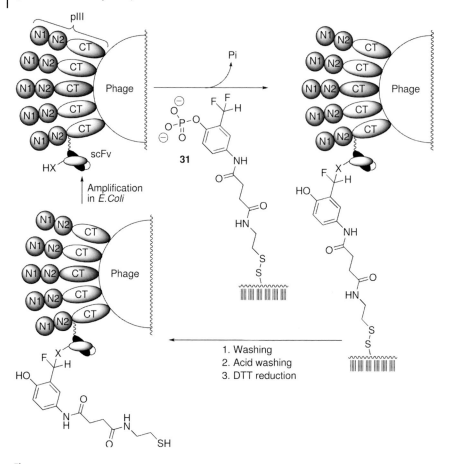

Fig. 9.20

first demonstration of this strategy was performed on a single gene, coding for TEM-1 β-lactamase, which is known to hydrolyze the antibiotic drug, cefotaxime. This gene was randomly digested with the endonuclease DnaseI and reassembled by PCR methods combined with low mutation rate (0.7%) [50]. The library of genes was cloned into an expression plasmid and transformed to *E. coli*; the bacteria were grown on a medium containing cefotaxime, leading to selection on the basis of resistance to the drug. The β-lactamase gene was isolated from the resistant colonies and subjected to another cycle of digestion and reassembly by PCR. Each round generated a better enzyme, affording better antibiotic resistance. After three rounds of DNA shuffling and screening, a 16 000-fold increase in bacterial resistance was achieved, with minimum inhibitory concentration (MIC) = 320 µg/mL as compared with MIC = 0.02 µg/mL of the wild-type TEM-1 β-lactamase. Non-essential mutations were removed by backcrossing [62], resulting in an even better mutant, MIC = 640 µg/mL.

Crameri et al. carried out DNA shuffling among four variants of the enzyme cephalosporinase, which were obtained from different bacterial species, and compared the efficiency of this process with the above-described shuffling technique of a single gene [61]. Clones originating from any of the four single-gene libraries showed up to an 8-fold increase in moxalactam resistance as compared to those expressing the wild-type genes. In contrast, the best clone originating from the gene family library showed a 540-fold increase in resistance (0.38–200 mg/mL) compared to the wild-type *Klebsiella* and *Yersinia* genes and a 270-fold increase (0.75–200 mg/mL) compared to the wild-type *Enterobacter* and *Citrobacter* genes. Thus, "family shuffling" accelerated the rate of functional enzyme improvement 34- to 68-fold in a single cycle.

A second round of shuffling was performed using the pool of colonies selected in the first round. Plating of about 5×10^4 colonies on a range of concentrations of moxalactam yielded three clones, which had a further 3.5-fold increase in moxalactam resistance over the most resistant clone from the first cycle. The reduction in the rate of improvement is expected given the limited dynamic range of the bioassay. The best clone achieved from the first cycle was found to be a chimera of the genes from *Citrobacter*, *Enterobacter* and *Klebsiella*. These two experiments highlight the importance of the screening technique. Since the evolutionary pressure used in these experiments was merely survival, the best-fitted clones could tolerate increased concentrations of the antibiotic. Obviously, for catalysts that are not survival-related, the development of other evolutionary pressure devices is required.

Crameri et al. have also tested their method for directed evolution of an antibody molecule [63]. They mimicked the immune system using DNA shuffling to generate an anti-G-CSF receptor (human granulocyte colony-stimulating factor) antibody, starting with a non-binder. An scFv was chosen on the basis of stability and expression efficiency from a naïve antibody-phage display library. This scFv was used to construct a phage display library. All six CDRs of this scFv were randomly mutated by epPCR, with the mutation frequency of each CDR residue being determined to mimic its naturally occurring variability within its V-region family. The different CDRs were shuffled to generate a phage display library containing 4×10^7 clones. Shuffling caused recombination primarily in the homologous framework sequences that separated the heterogeneous CDRs, thus creating new combinations of CDRs. The number of potential CDR combinations could therefore be as high as $(4 \times 10^7)^6$ = 4×10^{42}. The library was panned against G-CSF receptor and amplified over several rounds. Binding efficiency increased 440-fold from round 2 to round 8.

Zhao et al. have introduced the alternative technique of staggered extension process (StEP) using related genes [64]. StEP, like the above-described DNA shuffling, can be used to promote *in vitro* recombination among mutant genes. In a PCR-like reaction, mixed templates containing different mutations are primed and extended for very short periods of time to decrease polymerization rates. This step is followed by cycles of denaturation and annealing, so the partially extended fragments hybridize randomly onto templates harboring different mutations. The StEP technique is similar in efficiency to DNA shuffling, both harnessing the power of recombination in mixing different domains and in sampling large regions of sequence space. In either approach, the extent of diversity can be controlled by altering the time and temperature of the annealing steps. In this work, the StEP technique was used to combine two subtilisin E mutants, which were previously used as ancestor proteins because of their thermal stability. Five new thermostable mutants of subtilisin E were generated by epPCR, and their genes were recombined under StEP conditions. This approach led to the recovery of a mutant that exhibited a half-life at 65 °C that was 50-fold longer than that of the wild-type subtilisin E.

Jung et al. used DNA shuffling and site-directed mutagenesis to create an scFv phage display library and thereby improved the scFv stability [65]. The framework of the humanized 4D5 anti-HER2 antibody was characterized by good stability at 37 °C and by efficient folding during expression in *E. coli* when fused to pIII. Randomization of the parent scFv, 4D5Flu [66], produced a library of mutants that was screened for stability. Panning was carried out against the immobilized antigen at increased temperatures and increased concentrations of guanidinium chloride. One of the selected scFvs presented enhanced stability at 60 °C in comparison with its parent protein.

9.5
In vitro Libraries

9.5.1
Ribosomal Display

In contrast to the phage display methods, the ribosomal display technology does not involve any living organism. This technology was first developed by Mattheakis et al. [67] on the basis of the observation that mRNAs that encodes for a certain protein antigen could be isolated from the immunoprecipitated polysome (ribosome-protein-mRNA complex) [68–70]. Both ribosomal display and phage display techniques are based on the same principle of physically linking the phenotype (the polypeptide chain) with the genotype (the coding DNA sequence). In the ribosomal display technique, the DNA sequence is obtained from the mRNA by rtPCR. Then, RNA polymerase and nucleotides are added to the mixture to generate pure mRNA in large quantities. Addition of ribosomes and a mixture of aminoacyl-tRNA allows for translation of the mRNA to the appropriate protein. Since termination factors are missing, the ribosomes remain attached to both the coding mRNA and the protein product. The mRNA can then be isolated from this polysome by treatment with EDTA. The main advantage of this technique over the phage display method is that there is no loss of clones due to inefficient infectivity. Yet, the phage display technique is currently more commonly used. Mattheakis et al. used an immobilized antibody (Mab D32.39 – an anti-RQFKVV peptide antibody) to screen a large library of 10^{12} different peptides that were presented on polysomes (Fig. 9.21). A family of peptides that binds specifically to MAB D32.39 was thus isolated [67].

In 1997 two groups used ribosomal display to present scFv antibodies. Hanes and Plückthun showed that a whole protein could be refolded *in vitro* when it was linked

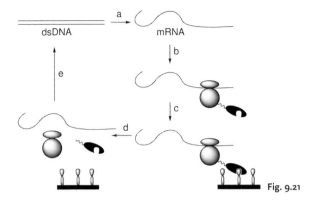

Fig. 9.21

to a ribosome, using a linker protein in its C-terminus [71]. He and Taussig showed that the ribosomal display technique could enrich an scFv from a library by a factor of 10^4–10^5 in a single cycle [72]. That experiment demonstrated that the ribosomal display technique can be effectively used for scFv display.

Jermutus et al. used the ribosomal display technique in combination with DNA shuffling to create an scFv library [73]. The initial library was created by epPCR from the fluorescein-binding antibody C12. The polysomes coding for the scFv library was first equilibrated with nanomolar concentrations of biotinylated antigen and then with a large excess of free antigen. Time periods of off-rate selections were increased in each round, from 2 h to 250 h, with DNA shuffling after each selection, and complexes still binding the biotinylated antigen were rescued by the addition of streptavidin-coated magnetic beads. The RNA coding for these proteins was purified and amplified by rtPCR and, after DNA shuffling, served as a template for the next round of directed evolution. This method led, over four rounds, to 30-fold enhancement of binding affinity. The same approach with a different screening method was used for the selection of more stable mutants. The selection pressure of DTT concentration was gradually increased over five rounds. Mutants could survive the selection process only if they folded stably in the presence of DTT and retained their antigen-binding activity. The oxidized form of the model protein in this directed evolution experiment was an anti-hag scFv. Selected mutants were found to be chemically more stable than the wild-type protein. Applying another selection pressure of increasing concentrations of urea has also led to an enhanced stability of the selected mutants [73].

9.5.2
In vitro Compartmentalization

Griffiths and Tawfik have designed an alternative system that does not involve any living organism and involves no physical connection between the coding DNA and its protein. Instead, these components are localized together with substrate and other factors in the same inversed micelle [74, 75], allowing the enzymes to operate over multiple turnovers. This technique could exert an evolutionary force to achieve optimized catalysts.

Ghadessy et al. showed that this approach can be used for directed evolution of DNA polymerases and their coding genes [76]. A library of Taq-polymerase was generated using epPCR. In order to encapsulate the specific DNA polymerase and its coding gene, the Taq-polymerase was expressed in *E. coli*. The bacteria expressing the polymerase were suspended with buffer containing flanking primers and dNTPs, and the bacteria were then segregated into aqueous compartments in oil. Thermal rupture of the bacteria released the enzyme and its coding gene from the cell to the compartment, allowing self-replication. The offspring genes were re-cloned in *E. coli* for another cycle. Among differentially active variants, the more active can be expected to produce proportionally more offspring. Two evolutionary pressures, heparin resistance and thermostability, were exerted on the polymerases. A polymerase was obtained that remained fully functional in PCR at up to 130 times the inhibitory concentration of heparin. In the thermostability experiment, an enzyme was isolated which had an 11-fold longer half-life at 97.5 °C than that of the already thermostable wild type. Interestingly, the genes themselves changed in this experiment, with preference to silent mutations of G:C to A:T, a change that promotes replication by facilitating strand separation and destabilizing secondary structures.

Griffiths and Tawfik exploited this method to evolve an improved phosphotriesterase (PTE), which exhibited a k_{cat} value 63.3 times greater than that of the wild type and a k_{cat}/K_M value that was 1.8 times greater [77]. A library of 5×10^7 PTE mutants, each with a common epitope tag, were linked to streptavidin-coated beads carrying antibodies that bind the epitope at, on average, less than one gene per bead. The beads were compartmentalized in a water-in-oil emulsion to give, on average, less than one bead per compartment. Each bead was linked to an anti-tag antibody that bound the translated proteins via their tag. The emulsion was broken and the microbeads carrying the display library were isolated. The microbe display library was then re-compartmentalized in a water-in-oil emulsion and a soluble substrate attached to caged biotin was added. The substrate was converted to product only in compartments that contained beads displaying active enzymes. The emulsion was then irradiated to release the biotin. In a compartment that contained a gene encoding for an enzyme, the product became attached to the gene via the bead. In other compartments, in which the genes did not encode for an active enzyme, the intact substrate became attached to the gene. The emulsion was broken and the beads were incubated with anti-product fluorescent antibodies. Product-coated beads could then be enriched together with the genes attached to them either by affinity purification or, after reacting with a fluorescent-labeled antibody, by flow cytometry.

9.5.3
Plasmid Display

Cull et al. constructed a library of random peptides fused to the C-terminus of the Lac repressor [78]. The DNA-binding activity of the repressor protein physically linked the peptides to the plasmid encoding them by binding to *lac*-operator sequences on the plasmid. Since the constructs were expressed in *E. coli* in access over the plasmids, upon lysis of the cells each plasmid became saturated by its own product. This linkage allowed for efficient enrichment for specific peptide ligands in the random population of peptides by affinity purification of the peptide-repressor-plasmid complexes with an immobilized receptor. After transformation of *E. coli* with recovered plasmids, the library could be amplified for additional rounds of affinity enrichment followed by isolation of the selected plasmid clones. The monoclonal antibody D32.39, an *anti* dynorphin B (a peptide sequence), was used as a model receptor to screen a random peptide library. After two rounds of enrichment, the majority of the plasmids encoded for fusion peptides that specifically bound the antibody.

9.6
In vivo Libraries

Another strategy that links the phenotype to its genotype follows the natural systems by using the expression system of an entire organism. The foreign gene is inserted into the organism and undergoes expression, and the resultant protein is displayed on the surface of this organism. Selection and isolation of a genetically modified organism from a library would provide both the functional protein and the DNA encoding it.

9.6.1
Protein Libraries Presented on an *E. coli* Surface

Francisco et al. have expressed an *anti* digoxin scFv on the surface of *E. coli* [79]. Using digoxin-FITS conjugate, the digoxin binding scFvs were fluorescently labeled. Using a fluorescence-activated cell sorter (FACS), the scFv-expressing clone was amplified 105-fold. Starting with a dilution of 1:100 000 and allowing for two rounds of growth and sorting, 79% of the cell population presented the construct.

Daugherty et al. used scFv libraries, which were displayed on the surface of *E. coli*, as a fusion protein to Lpp-OmpA in order to investigate the effects of mutation frequency on affinity maturation [80]. Libraries of scFv that binds digoxigenin were produced by epPCR with an average mutation rate between 1.7 and 22.5 base substitutions per gene. The mutant populations were analyzed by flow cytometry. At low to moderate mutation frequencies, the fraction of clones exhibiting binding to a fluorescently labeled conjugate of digoxigenin decreased significantly in comparison with the wild type. However, the most highly mutated library contained more active clones than expected. These libraries were screened for high affinity using FACS. After several rounds of enrichment, each of the three libraries yielded clones with improved affinity for the hapten.

Olsen et al. have improved the catalytic activity of the serine protease OmpT (EC3.4.21.87), which is a cell surface-displayed protease [81]. The enzymatic cleavage of the scissile bond in substrate **32a** was designed to separate the FRET quenching partner **Q** from the fluorescent moiety **F** (Fig. 9.22). The positively charged FRET

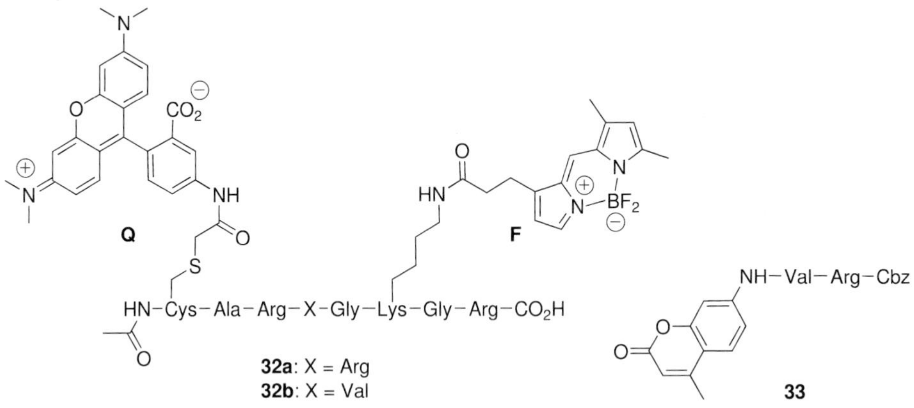

Fig. 9.22

substrate (**32a**) was designed to associate with the negatively charged cell surface of *E. coli*. Following the enzymatic cleavage of **32a**, the fluorescent product **F** was retained on the cell surface, allowing for isolation of the catalytically active clones by FACS. The FACS analysis discriminated between three different cell populations, one having no OmpT on their surface, one having active OmpT catalysts on the surface, and one that expressed an impaired OmpT variant. This strategy has created a link between the enzyme-encoding gene and the catalytic turnover. The method resulted in enrichment of *E. coli* expressing active serine protease, OmpT, from cells expressing an inactive OmpT variant by over 5000-fold in a single round. Thirty-two highly fluorescent cells were detected in the positive window of the FACS chart and were selected out of approximately 150 000 bacteria analyzed.

Since the wild-type OmpT possesses a modest ability to cleave an Arg-Val sequence, the enhancement of the cleavage rate of such substrates was chosen as the target for the directed evolution experiments. Accordingly, substrate **32b**, which contained an Arg-Val cleavage site, was used to screen libraries of OmpT random variants. The *ompT* gene was subjected to random mutagenesis by epPCR. A total of 6×10^5 transformants were obtained. The library was incubated with **32b**, and clones that exhibited fluorescence were isolated by FACS. A total of 1.9×10^6 cells were evaluated in 24 min, and 352 individual clones were recovered, corresponding to a 5400-fold enrichment in one round. Purified OmpT enzymes from three fluorescent clones were found to hydrolyze substrate **33** with apparent k_{cat}/K_M values of 1,440, 310, and 200 $s^{-1}M^{-1}$, corresponding to catalytic efficiencies that were 60-, 13-, and 9-fold better than the wild-type OmpT enzyme.

9.6.2
Protein Libraries in Yeast

Boder and Wittrup have constructed an scFv library on the membrane surface of yeast [82]. An anti-fluorescein scFv with a c-myc epitope was expressed on the cell wall of yeast by fusion to a-agglutinin, which is a two-subunit membrane glycoprotein. The scFv peptide was a C-terminal fusion to the Aga2p subunit of a-agglutinin. The fusion protein was verified by confocal fluorescence microscopy and flow cytometry and was proven to express and bind its fluorescently labeled ligand – FITS-dextran. Enrichment experiments from wild-type population have shown 600-fold enrichment. Using a bacterial mutator strain, a 5×10^5 library of descendant scFv was constructed. The pool of yeast cells that displayed the scFv were subjected to kinetic selection by competition of FITC-dextran-labeled cells with 5-aminofluorescein. Cells exhibiting the most intense FITC fluorescence were collected by flow cytometric sorting, amplified, and resorted. Cells demonstrating a substantially increased persistence time of labeling by FITC-dextran were dramatically enriched over three rounds of sorting and amplification. FITC-dextran dissociation kinetics for two individual clones selected from the scFv library differed by 2.9-fold relative to the wild-type scFv. Rate constants for the mutants were 1.9×10^{-3} s^{-1} and 2×10^{-3} s^{-1} at 23 °C, compared with 5.6×10^{-3} s^{-1} for the wild-type. Additionally, soluble fluorescein dissociation kinetics determined by spectrofluorometry demonstrated a 2.2-fold improvement for both mutants relative to wild-type, and initial equilibrium fluorescence quenching experiments suggested a similar improvement in the affinity constant of the binding reaction.

Since highly complex living organisms may respond to an evolutionary pressure in more than one way, the selected clone may not necessarily possess the desired characteristics that were defined by the experimenter. For example, instead of evolving an improved catalyst or a better binder, the cell may simply overproduce a relatively poor protein. Kieke et al. have designed a strategy to overcome this problem, performing *in vitro* evolution of an anti-T cell receptor (TCR) scFv from a library of yeast-presenting scFv [83]. The scFv was expressed as a fusion protein bound to two other protein tags: HA and c-myc (Fig. 9.23). To evaluate the binding properties of the surface-bound scFv, a soluble single-chain TCR was biotinylated, and its binding to the scFv was detected by flow cytometry. The fraction of cells that expressed active scFv was similar to that detected with anti-HA and c-myc antibodies, consistent with the expression of full-length, properly folded scFv. Moreover, two-color histograms demonstrated a tight correlation of scTCR binding with both HA and c-myc epitope display. A library was constructed from scFv-KJ16, using a bacterial mutator strain, which was predicted to introduce an average of two to three point mutations in the scFv coding sequence, yielding a library size of ca. 33×10^5.

The mutated yeast library was subjected to four successive cycles of sorting and amplification, using a double stain for either anti-c-myc or anti-HA antibody binding (FITC) and biotinylated scTCR binding (PE). Biotinylated scTCR was used at a 1:5000 dilution (~ 10 nM) that yielded just below the detectable threshold of binding by wt scFv-KJ16/yeast. Cells that exhibited phycoerythrin fluorescence more intense than antibody probe fluorescence were collected for regrowth. The rationale for this selection was based on increased antigen binding relative to the presence of displayed polypeptide fusion proteins c-myc and HA. The first two sorting rounds were performed in enrichment mode, isolating the fraction of the cell population with the highest fluorescence (~ 0.5%). The final two sorting rounds were performed

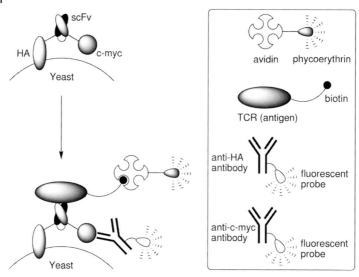

Fig. 9.23

for purity. After the fourth cycle, cells were resorted immediately and plated. Ten colonies were selected for further analysis. One clone was found to exhibit higher TCR binding levels in comparison to the original scFv.

Boder et al. have preformed directed evolution of the scFv of antibody 4-4-20, an anti-fluorescein-biotin binder [84]. Libraries of 10^5–10^7 yeast surface-displayed scFv were generated using epPCR followed by DNA shuffling. Fluorescently labeled clones exhibiting slowed antibody-hapten dissociation kinetics were identified and isolated by flow cytometry with optimal screening and sorting conditions [85]. Up to 20 improved clones were selected randomly for individual measurement of the dissociation rate constant k_{diss}, and 10 improved clones exhibiting the widest range of k_{diss} values were selected for further analysis. Modified DNA shuffling, together with further epPCR mutagenesis, then recombined the complete collection of isolated mutants. This cycle of mutagenesis and screening was repeated three times, resulting in mutant scFvs with dissociation rate constants over four orders of magnitude slower than the k_{diss} of the 4-4-20 scFv (K_d = 48 fM compared with K_d = 0.7 nM of the wild-type scFv). The half-life time for dissociation of this mutant was greater than 5 days, slower dissociation kinetics than those for the streptavidin-biotin complex. These mutants possessed the highest monovalent ligand-binding affinity yet reported for an engineered protein by over two orders of magnitude. Comparison of frequency in the Kabat database for consensus mutations vs 4-4-20 wild-type residues indicated that 9 of 10 consensus residues were changes to rare amino acids.

9.6.3
Protein Libraries in Other Organisms

Gunneriusson et al. have expressed surface-displayed scFv antibody on *S. xylosus* and *S. carnosus* [86]. The gene fragment encoding for the scFv was constructed from a mouse hybridoma producing a monoclonal antibody against human IgE. The scFv was introduced into the two expression vectors designed for surface display on *S. xylosus* and *S. carnosus*. The expressed hybrid receptors containing the functional scFv were characterized both after extraction and affinity purification and on the intact recombinant bacteria.

Rode et al. have expressed an scFv on mammalian cos-7 cell and proved that the scFvs were indeed functional using a fluorescence-labeled ligand [87]. An scFv directed against 4-ethoxy-methyl-2-phenyloxazolin-5-one (phOx) was expressed on the surface of cos-7 cells as fusion of the signal peptide of the interleukin-6 receptor and the scFv coding sequence. The cells were then suspended with (a) phOx-labeled human IgG and a second anti-human FITC-conjugated IgG, (b) phOx-labeled R-phycoerythrin, and (c) phOx-labeled ovalbumin conjugated with FLUOS. The scFv-expressing cells were marked with fluorescence.

Russel et al. have expressed an scFv molecule on the surface of a retrovirus as a fusion protein consisting of the ecotropic Moloney murine leukemia virus (MoMLV) envelope polypeptide Pr80^{env} with an anti-4-hydroxy-5-iodo-3-nitrophenacetyl caproate (NIP) scFv inserted with a spacer of 6 amino acids from the N-terminus of Pr80^{env} [88]. The fusion gene was co-expressed in ecotropic retroviral packaging cells with a retroviral plasmid carrying the neomycin-resistant gene. Retroviral particles with specific hapten-binding activities were recovered. Furthermore, the hapten-binding particles were able to transfer the *NEO* gene and the antibody-envelope fusion gene to mouse fibroblasts.

Boublik et al. have developed a system suitable for eukaryotic protein display, similar to phage display in prokaryotes [89], using the *Autographa californica* nuclear popyhedrosis virus (AcNPV) as a vector for the display of proteins on the viral surface. Analogous to the M13 pIII minor coat protein, AcNPV encodes a virion surface glycoprotein. AcNPV grows to a relatively high titer ($\sim 10^{10}$/mL). The total sequence of the viral genome, including gp64, the major baculovirus glycoprotein, is known, and the virus has a long history of safe working practice [90–92]. Prokaryotic expression of proteins places a restriction on the type of protein to which the phage display technique can be applied. Eukaryotic proteins often require complex folding and glycosylation for functional activity [93]. They are seldom expressed successfully in bacteria, in part because of the absence of the appropriate post-translational pathways when compared to their normal cellular environment. In such cases, the functional display of these mutants is possible only by expression in transfected eukaryotic cells. Boublik et al. have identified a site within gp64 that allows the display of foreign proteins on the virus surface without compromising infectivity [89].

As a model system, the gene encoding the protein glutathion-S-transferase (GST) was used to construct several fusions with the gp64 gene. Following expression in *Spodoptera frugiperda* (SF9) cells, the yield and cellular distribution of each GST-gp64 protein were assessed and examined. One fusion, in which GST was inserted between the leader peptide and the mature protein, was efficiently secreted into the cell medium. In the context of expression of the full-length gp64, the hybrid GST-gp64 was incorporated onto the virion surface. To ascertain that bigger and more complex proteins could be displayed on the viral surface fused to gp64 in a fashion similar to GST, a gp120-gp64 fusion protein was expressed. The external domain of the HIV-1 surface glycoprotein, gp120, is characterized by an intricate pattern of disulfide bonds and glycosylation. Its biological activity requires an elaborate folding and secretion pathway. The coding region of the mature form of gp120 was amplified and inserted between the signal sequence and the mature protein of gp64, and a recombinant virus, Acgp120-gp64, was produced. The new fusion protein was shown to be whole and was identified by immuno-labeling and western-blot. Mottershead et al. have produced a vector allowing for fusion of proteins to the amino-terminus of gp64 [94].

Ernst et al. have used a derivative of the above-described system to present a eukaryotic epitope library presented on the surface of insect cells [95]. They chose the membrane-associated complex hemagglutinin of the influenza A virus as a carrier molecule to provide membrane anchorage of a specific epitope library on the surface of infected insect cells. The neutralizing human monoclonal antibody (hmAb) 2F5 was identified to specifically recognize the linear, six-amino-acid (ELDKWA) epitope located in the envelope protein gp41 of HIV-1. The idea in this approach was not to alter the epitope sequence itself but to change the tertiary structure in context with the glycosylated and highly processed hemagglutinin around the epitope, resulting in variants with increased specific immunogenicity. The aim of that work was to construct a eukaryotic baculovirus surface display library and select for binders exhibiting increased binding characteristics to hmAb 2F5 using FACS.

Sf9 cells, transfected with a library of recombinant hemagglutinin constructs, were treated with hmAb 2F5 and anti-human IgG FITC-conjugate. Detected binding capacity of the infected cell-pool to hmAb 2F5 showed a quite homogenic distribution of the fluorescence signal intensity. Only a small number of individual cells were considered to contain high-affinity binders. The gate for FACS sorting, therefore, was set to include < 0.5% of the total cell population. This selected cellular fraction was enriched by re-infection of cells or alternatively directly subjected to a plaque assay to isolate individual viral clones. Analysis of this enriched fraction showed only one clone presenting an increase in binding capacity to hmAb 2F5 of about two orders of magnitude compared to the signal intensity of the initial cell pool, which was derived after transfection of the library DNA. Since only three amino acids flanking the peptide sequence XELDKWAXX were randomized, the population comprised $20^3 = 8000$ clones, representing an enrichment factor of about 8000-fold. The newly selected construct was compared to the initially characterized clones, in terms of its binding profile to hmAb 2F5. The fluorescent signal given by this clone was clearly higher than the signal of other library clones, demonstrating that screening this baculovirus expression library was successful in identifying a molecule of higher binding capacity to hmAb 2F5.

9.7
Conclusions and Outlook

The representative techniques of displayed combinatorial protein libraries that are described in this chapter do not cover the entire field of directed evolution. They demonstrate, however, the great opportunities in the employment of such techniques for the development of functional proteins, and of catalytic antibodies in particular. The various strategies outlined in this chapter harness a number of important technologies, some of which were developed only recently. These technologies include the methods of cloning and expression of Fab and scFv in different organisms, the ways of physically linking the phenotype to its genotype, the means of creating large libraries in very small volumes, methods of clone selection, and techniques of amplification of clones having the desired properties.

The importance of effective screening methods can never be overestimated, because they represent the evolutionary pressure exerted on the given library. Since in this game you always get, in the best-case scenario, what you have screened for, the quality of screening is key to success. Another key to success is the ability to create a large sequence space, including the ability to expand and mutate the library after each round using efficient mutagenesis techniques. Expansion of the library on the basis of selected clones can take the system from a local to a global optimum. Another, highly advantageous strategy is the combination of several mutation and recombination techniques. Altogether, it seems likely that highly efficient biocatalysts that rival the characteristics of natural enzymes will become readily available in the not too distant future through strategies of directed evolution.

Acknowledgement

We thank Prof. Yoram Reiter of the Technion for helpful advice and comments on this manuscript. We thank the Israel-US Binational Science Foundation, the German-Israeli Project Cooperation (DIP), and the Skaggs Institute for Chemical Biology for financial support.

References

1 RADZICKA, A., WOLFENDEN, R., *Science* 267 (**1995**), p. 90
2 SCHMID, A., DORDICK, J. S., HAUER, B., KIENER, A., WUBBOLTS, M., WITHOLT, B., *Nature* 409 (**2001**), p. 258–268
3 BOLON, D. N., VOIGT, C. A., MAYO, S. L.. *Curr. Opin. Chem. Biol.* 6 (**2002**), p. 125–9
4 ARNOLD, F. H., *Nature* 409 (**2001**), p. 253–257
5 ARNOLD F. H., *Acc. Chem. Res.* 31 (**1998**), p. 125–131
6 CEDRONE, F., MENEZ, A., QUEMENEUR, E., *Curr. Opin. Struct. Biol.* 10 (**2000**), p. 405–410
7 PETSKO, G. A., *Nature* 403 (**2000**), p. 606–607
8 ALTAMIRANO, M. M., BLACKBURN, J. M., AGUAYO, C., FERSHT, A. R., *Nature* 403 (**2000**), p. 617–622
9 MARVIN, D. A., HALE, R. D., NAVE, C., CITTERICH, M. H., *J. Mol. Biol.* 253 (**1994**), p. 260–286
10 SMITH, G. P., PETRENKO V. A., *Chem. Rev.* 97 (**1997**), p. 391–410
11 SMITH, G. P., *Science* 228 (**1985**), p. 1315–1317
12 SKERRA, A., PLÜCKTHUN, A., *Science* 240 (**1988**), p. 1038–1041
13 BETTER, M., CHANG, C. P., ROBINSON, R. R. HORWITZ, A. H., *Science* 240 (**1998**), p. 1041–1043
14 LIU, A. Y., ROBINSON, R. R., HELLSTROM, K. E., MURRAY, E. D. JR., CHANG, C. P., HELLSTROM, I., *Proc. Natl. Acad. Sci. USA* 84 (**1987**), p. 3439–3443
15 HUSE, W. H., SASTRY, L., IVERSON, S. A., KANG, A. S., ALTING-MEES, M., BURTON, D. R., BENKOVIC, S. J., LERNER, R.A., *Science* 246 (**1989**), p. 1275–1281
16 JANDA, K. D., SCHLOEDER, D., BENKOVIC, S. J., LERNER, R. A., *Science* 241 (**1988**), p. 1188–1191
17 KABAT E. A. ET AL., *Sequences of Proteins of Immunological Interest*, National Institutes of Health, Bethesda, MD **1987**, p. 1–261
18 BIRD, R. E., HARDMAN, K. D., JACOBSON, J. W., JOHNSON, S., KAUFMAN, B. M., LEE, S. M., LEE, T., POPE, S. H., RIORDAN, G. S., WHITLOW, M., *Science* 242 (**1988**), p. 423–6. Erratum: *Science* 244 (**1989**), p. 409
19 FUCHS, P., BREITLING, F., DUBEL, S., SEEHAUS, T., LITTLE, M., *Bio/technology* 9 (**1991**), p. 1369–1372
20 REITER, Y., BRINKMANN, U., LEE, B., PASTAN, I., *Nat. Biotechnol.* 14 (**1996**), p. 1239–1245
21 GLOCKSHUBER, R., MALIA, M., PFITZINGER, I., PLÜCKTHUN, A., *Biochemistry* 29 (**1990**), p. 1362–1367
22 BRINKMANN, U., CHOWDHURY, P. S., ROSCOE, D. M., PASTAN, I., *J. Immunol. Methods* 182 (**1995**), p. 41–50
23 (a) KABAT, E. A., *Sequences of Proteins of Immunological Interest*, National Institutes of Health, Washington, DC **1991**. (b) KABAT, E. A., WU, T. T. REIDMILLER, M., PERRY, H. M., GOTTESMAN, K. S., *Sequences of Proteins of Immunological Interest*, U. S. Govern-

ment Printing Office, Bethesda, MD 1991

24 MARKS, J. D., HOOGENBOOM, H. R., GRIFFITHS, A. D., WINTER, G., *J. Biol. Chem.* 267 (1992), p. 16007–16010

25 DUENAS, M., MALMBORG, A. C., CASALVILLA, R., OHLIN, M., BORREBAECK C. A. K., *Mol. Immunol.* 33 (1996), p. 279–285

26 MARKS, J. D., HOOGENBOOM, H. R., BONNERT, T. P., McCAFFERTY J., GRIFFITHS, A. D., WINTER G., *J. Mol. Biol.* 222 (1991), p. 581–597

27 McCAFFERTY, J., GRIFFITHS, A. D., WINTER, G., CHISWELL, D. J., *Nature* 348 (1990), p. 552–554

28 GARRARD, L. J., YANG, M., O'CONNELL, M. P., KELLY, R. F., HENNER, D. J., *Biotechnology* 9 (1991), p. 1373–1377

29 BEREK, C., GRIFFITHS, G. M., MILSTEIN, C., *Nature* 316 (1985), p. 412–418

30 BEREK, C., MILSTEIN, C., *Immunol. Rev.* 96 (1987), p. 23–41

31 BYE, J. M., CARTER, C., MARKS, J. D., *J. Clin. Invest.* 90 (1992), p. 2481–2490

32 DUENAS, M., BORREBAECK C. A. K., *Biotechnology* 12 (1994), p. 999–1002

33 KREBBER, C., SPADA, S., DESPLANCQ, D., KREBBER, A., GE, L., PLÜCKTHUN, A, *J. Mol. Biol* 268 (1997), p. 607–618

34 CRAMERI R., SUTER M., *Gene* 137 (1993), p. 69–75

35 BRUNET E., CHAUVIN C., CHOUMET V., JESTIN J. L. A., *Nucleic Acids Res.* 30 (2002), p. 40

36 KRISTENSEN P., WINTER G., *Fold Des.* 3 (1998), p. 321–328

37 JESTIN J. L., VOLIOTI, G., WINTER, G., *Res. Microbiol.* 152 (2001), p. 187–191

38 GAO, C., MAO, S., LO, C. H., WIRSCHING, P., LERNER, R. A, JANDA, K. D., Making, *Proc. Natl. Acad. Sci. USA* 96 1999, p. 6025–6030

39 GAO, C, MAO, S., KAUFMANN, G., WIRSCHING, P., LERNER, R. A., JANDA, K. D. A., *Proc. Natl. Acad. Sci. USA* 99 (2002), p. 12612–12616

40 JANDA, K. D, BENKOVIC, S. J, LERNER, R. A., Catalytic antibodies with lipase activity and R or S substrate selectivity, *Science* 244 (1989), p. 437–440

41 CARRERA, M. R., ASHLEY, J. A., PARSONS, L. H., WIRSCHING, P., KOOB, G. F., JANDA, K. D., *Nature* 378 (1995), p. 727–730

42 WIRSCHING, P., ASHLEY, J. A, BENKOVIC, S. J., JANDA, K. D, LERNER, R. A., An unexpectedly efficient catalytic antibody operating by ping-pong and induced fit mechanisms, *Science* 252 (1991), p. 680–685

43 SOUMILLION, P., JESPERS, L., BOUCHET, M., MARCHAND-BRYNAERT, J., WINTER, G., FASTREZ, J., *J. Mol. Biol.* 237 1994, p. 415–422

44 JANDA, K. D., LO C. H., LI T, BARBAS C. F. III, WIRSCHING P., LERNER R. A., *Proc. Natl. Acad. Sci. USA* 91 (1994), p. 2532–2536

45 JANDA, K. D., LO, L. C., LO, C. H., SIM, M. M., WANG, R., WONG, C. H., LERNER, R. A., *Science* 275 1997, p. 945–948

46 GAO, C., LIN, C. H., LO, C. H., MAO, S., WIRSCHING, P., LERNER, R. A., JANDA, K. D., *Proc. Natl. Acad. Sci. USA* 94 (1997), p. 11777–11782

47 PEDERSEN, H., HOLDER, S., SUTHERLIN, D. P., SCHWITTER, U., KING, D. S., SCHULTZ, P. G. A., *Proc. Natl. Acad. Sci. USA* 95 1998, p. 10523–10528

48 DEMARTIS, S., HUBER, A., VITI, F., LOZZI, L., GIOVANNONI, L., NERI, P., WINTER, G., NERI, D., *J. Mol. Biol.* 286 (1999), p. 617–633

49 HEINIS, C., HUBER, A., DEMARTIS, S., BERTSCHINGER, J., MELKKO, S., LOZZI, L., NERI, P., NERI D., *Protein Eng.* 14 (2001), p. 1043–1052

50 HOLLAND J. H., *Adaptation in Natural and Artificial Systems*, 2nd edn., MIT Press, Cambridge 1992

51 BALDWIN, E., SCHULTS, P. G., *Science* 245 (1989), p. 1104–1107

52 BRIK, A., D'SOUZA, L. J., KEINAN, E., GRYNSZPAN, F., DAWSON, P. E., *ChemBioChem* 3 (2002), p. 845–851

53 BRIK, A., DAWSON, P. E., KEINAN, E., *Bioorg. Med. Chem*, 10 (2002), p. 3891–3817

54 BACA, M., SCANLAN, T. S., STEPHENSON, R. C., WELLS, J. A., *Proc. Natl. Acad. Sci. USA* 94 (1997), p. 10063–10068

55 Fujii, I., Fukuyama, S., Iwabuchi, Y., Tanimura, R., *Nat. Biotechnol.* 16 **1998**, p. 463–467
56 Takahashi, N., Kakinuma, H., Liu, L., Nishi, Y., Fujii, I., *Nat. Biotechnol.* 19 (**2001**), p. 563–567
57 Atwell, S., Wells, J. A. Selection, *Proc. Natl. Acad. Sci. USA* 96 (**1999**), p. 9497–9502
58 Abrahmsen, L., Tom, J., Burnier, J., Butcher, K. A., Kossiakoff, A., Wells, J. A., *Biochemistry* 30 **1991**, p. 4151–4159
59 Cesaro-Tadic, S., Lagos, D., Honegger, A., Rickard, J. H., Partridge, L. J., Blackburn, G. M., Plückthun, A., *Nat. Biotechnol.* 21 (**2003**), p. 679–685
60 Stemmer, W. P. C., *Nature* 370 (**1994**), p. 389–391
61 Crameri, A., Raillard, S., Bermudes, E., Stemmer, W. P. C., *Nature* 391 **1998**, p. 288–291
62 Stemmer, W. P. C, *Proc. Natl. Acad. Sci. USA* 91 **1994**, p. 10747–10751
63 Crameri, A., Cwirla, S., Stemmer, W. P. C., *Nat. Med.* 2 (**1996**), p. 100–102
64 Zhao, H., Giver, L., Shao, Z., Affholter, J. A., Arnold, F. H., *Nat. Biotechnol.* 16 (**1998**), p. 258–61
65 Jung, S., Honegger, A., Plückthun, A., *J. Mol. Biol.* 294 **1999**, p. 163–80
66 Jung, S., Plückthun, A., *Protein Eng.* 10 (**1997**), p. 959–66
67 Mattheakis, L. C., Bhatt, R. R., Dower, W. J., *Proc. Natl. Acad. Sci. USA* 91 (**1994**), p. 9022–9026
68 Payvar, F., Schimke, R. T., *Eur. J. Biochem.* 101 **1979**, p. 271–282
69 Kraus, J. P., Rosenberg, L. E., *Proc. Natl. Acad. Sci. USA.* 79 **1982**, p. 4015–4019
70 Korman, A. J., Knudsen, P. J., Kaufman, J. F., Strominger, J. L., *Proc. Natl. Acad. Sci. USA* 79 (**1982**), p. 1844–1848
71 Hanes, J., Plückthun, A., *Proc. Natl. Acad. Sci. USA* 94 (**1997**), p. 4937–4942
72 He, M., Taussig, M. J., *Nucleic Acids Res.* 25 (**1997**), p. 5132–5134
73 Jermutus, L., Honegger, A., Schwesinger, F., Hanes, J., Plückthun, A., *Proc. Natl. Acad. Sci. USA* 98 (**2001**), p. 75–80
74 Tawfik, D. S., Griffiths, A. D., *Nat. Biotechnol.* 16 (**1998**), p. 652–656
75 Griffiths, A. D., Tawfik, D. S., *Curr. Opin. Biotechnol.* 11 (**2000**), p. 338–353
76 Ghadessy, F. J., Ong, J. L., Holliger, P., *Proc. Natl. Acad. Sci. USA* 98 (**2001**), p. 4552–4557
77 Griffiths, A. D., Tawfik, D. S., *EMBO* 22 **2003**, p. 24–35
78 Cull, M. G., Miller, J. F., Schatz, P. J., *Proc. Natl. Acad. Sci. USA* 89 (**1992**), p. 1865–1869
79 Francisco, J. A., Campbell, R., Iverson B. L., Georgiou G., *Proc. Natl. Acad. Sci. USA* 90 (**1993**), p. 10444–10448
80 Daugherty, P. S., Chen, G., Iverson, B. L., Georgiou, G., *Proc. Natl. Acad. Sci. USA* 97 (**2000**), p. 2029–2034
81 Olsen, M. J., Stephens, D., Griffiths, D., Daugherty, P., Georgiou, G., Iverson, B. L., *Nat. Biotechnol.* 18 (**2000**), p. 1071–1074
82 Boder, E. T., Wittrup, K. D., *Nat. Biotechnol.* 15 **1997**, 553–557
83 Kieke, M. C., Cho, B. K., Boder, E. T., Kranz, D. M., Wittrup, K. D., *Protein Eng.* 10 (**1997**), p. 1303–1310
84 Boder, E. T., Midelfort, K. S., Wittrup, K. D., *Proc. Natl. Acad. Sci. USA* 97 (**2000**), p. 10701–10705
85 Boder, E. T., Wittrup, K. D., *Biotechnol. Prog.* 14 (**1998**), p. 55–62
86 Gunneriusson, E., Samuelson, P., Uhlen, M., Nygren, P. A., Stahl, S., *J. Bacteriol.* 178 (**1996**), p. 1341–1346
87 Rode, H. J., Little, M., Fuchs P., Dorsam, H., Schooltink, H., de Ines, C., Dubel, S., Breitling, F., *BioTechniques* 21 (**1996**), p. 650–658
88 Russell, S. J., Hawkins, R. E., Winter, G., *Nucleic Acids Res.* 21 **1993**, p. 1081–10855
89 Boublik, Y., Di Bonito, P., Jones, I. M., *Biotechnol.* 13 (**1995**, p. 1079–1084
90 Whitford, M., Stewart, S., Kuzio, J., Faulkner, P., *J. Virol.* 63 (**1989**), p. 1393–1399
91 Ayres, M. D., Howard, S. C., Kuzio, J., Lopez-Ferber, M., Possee, R. D., *Virology* 202 (**1994**), p. 586–605

92 O'Reilly, D. R., Miller, L. K., Luckow, V. A., Baculovirus expression vectors: a laboratory manual, 1st edn., W. H. Freeman and Co., New York **1992**

93 Fielder, K., Simons, K., *Cell* 81 (**1995**), p. 309–312

94 Mottershead, D., van der Linden, I., von Bonsdorff, C. H., Keinanen, K., Oker-Blom, C., *Biochem. Biophys. Res. Commun.* 238 (**1997**), p. 717–722

95 Ernst, W., Grabherr, R., Wegner, D., Borth, N., Grassauer, A., Katinger, H., *Nucleic Acids Res.* 26 (**1998**), p. 1718–1723

10
Medicinal Potential of Catalytic Antibodies
Roey Amir, Doron Shabat

10.1
Introduction

Nearly two decades ago, in 1986, the first literature reports of catalytic antibodies sparked excitement and imagination among the world scientific community [1, 2]. It was realized that, by an appropriate design, a monoclonal antibody could be generated against a specific small molecule with a stable structure of the transition state analog of a chemical reaction. These antibodies had the ability to catalyze that chemical reaction similarly to enzymes. The technological and intellectual advances in this field were realized with the development of the first commercial catalytic antibody (antibody 38C2). This aldolase antibody is now being sold by Aldrich as a standard chemical reagent and was the first commercial protein offered for sale by Aldrich.

The intriguing concept of using catalytic antibodies as therapeutic agents became more appealing when it was shown that most of the amino acids in a mouse antibody molecule could be replaced with human sequences, thereby making it compatible for *in vivo* treatment in humans [3]. Furthermore, it is a molecule synthesized by a highly evolved biological system that was naturally designed for *in vivo* activity applications. Since catalytic antibodies are a kind of artificial enzymes, they can be designed to catalyze specific chemical reactions that are not catalyzed by natural enzymes. Therefore, they can be used for selective activities avoiding specific catalytic competition by natural proteins. The greatest therapeutic potential of catalytic antibodies could lie in selective prodrug activation, and this is discussed in detail below.

10.2
Prodrug Activation

10.2.1
Catalytic Antibodies Designed for Carbamate/Ester Hydrolysis

Prolonged administration of effective concentrations of chemotherapeutic agents is usually not possible because of the observed dose-limiting systemic toxicities and

---

Catalytic Antibodies. Edited by Ehud Keinan
Copyright © 2005 WILEY-VCH Verlag GmbH & Co. KGaA, Weinheim
ISBN: 3-527-30688-9

strong side effects involving non-malignant tissues. Thus, new strategies to target cytotoxic agents specifically to sites of metastatic or solid cancer are required. One such targeting approach involves selective enzymatic activation of a non-toxic prodrug to a toxic drug. The enzyme is either directed to the tumor by a targeting device (i.e. a specific monoclonal antibody) or selectively secreted by the tumor cells. This strategy has been exploited for prodrug activation by a catalytic antibody [4, 5].

In principle, a catalytic antibody can be designed to change a prodrug into an active drug. Several groups have reported prodrug activation by antibody catalysis. All of these except one used phosphonate hapten as a tetrahedral transition state analog for the hydrolytic reaction, which releases the free drug. One catalytic antibody, 38C2, was generated through a different approach and is comprehensively discussed below.

The first example of prodrug activation by a catalytic antibody was reported in 1993 by Fujii et al. [6]. They found that antibody 6D9 elicited against hapten 3 (Fig. 10.1) is able to catalyze the hydrolysis of ester prodrug 1 to chloroamphenicol 2 with a rate enhancement of 1.8×10^3. They then demonstrated the possibility of prodrug therapy by achieving inhibition of growth of *Bacillus subtilis* cells, which were inhibited by prodrug 1 when combined with antibody 6D9.

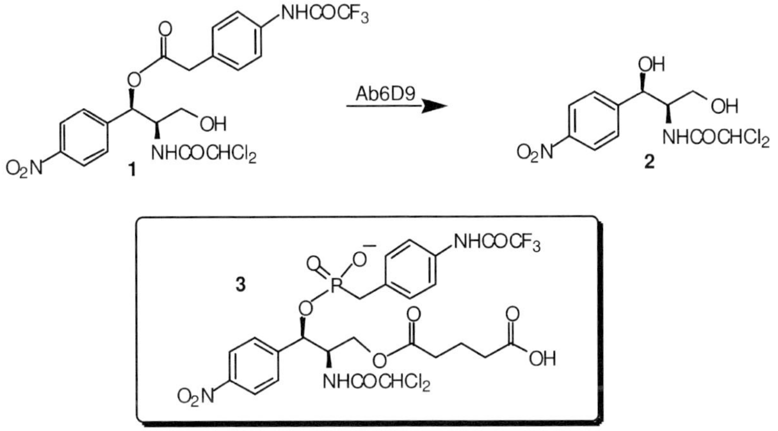

Fig. 10.1 Monoclonal antibody 6D9 elicited against hapten 3 catalyzes the hydrolysis of ester 1 prodrug to chloroamphenicol 2.

One year later, in 1994, Schultz et al. [7] utilized phosphonate 6 to elicit antibody 49.AG.659.12, which catalyzes the hydrolysis of ester prodrug 4 to release 5-fluorodeoxyuridine 5, a known anticancer drug that inhibits thymidylate synthetase. The rate enhancement of the catalysis reaction was 968-fold over the uncatalyzed reaction. The free drug completely inhibits the growth of *Escherichia coli* HB101 at a concentration of 20 µM, whereas prodrug 4 does not affect the growth of bacteria at a concentration of 400 µM. Incubation of prodrug 4 (400 µM) with antibody 49.AG.659.12 (20 µM) led to complete inhibition of the *E. coli* HB101 growth, showing the effective activation of the prodrug in this system.

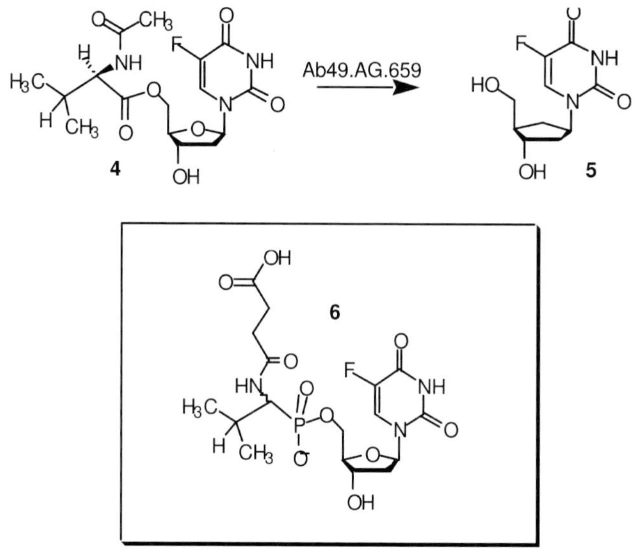

Fig. 10.2 Monoclonal antibody 49.AG.659.12 elicited against phosphonate hapten **6** catalyzes the hydrolysis of ester **4** prodrug to 5-fluorodeoxyuridine **5**.

The next progress in prodrug activation by antibody catalysis was reported by Blackburn et al. in 1996 [8]. They used the nitrogen mustard anticancer drug **8** to generate a carbamate prodrug **7**, which could be hydrolyzed to form the parent drug and glutamic acid. Phosphoamidate **9** was used as a stable transition state analog for the purpose of immunization. One monoclonal antibody, EA11-D7, was found to efficiently catalyze the hydrolysis of prodrug **7** and was selected for growth inhibition studies. Incubation of the antibody (1 µM) and prodrug **7** (100 µM) led to significant in vitro cell-kill of human colorectal tumor cell line (LoVo), whereas incubation of prodrug **7** alone results in no toxicity.

An interesting concept for prodrug activation by antibody catalysis was reported by Taylor et al. in 2001 [9]. They generated antibodies against a phosphonate transition state analog that can catalyze the hydrolysis reaction of the drug-masking linker. In this approach, the drug can remain outside of the antibody's catalytic pocket, and the prodrug is activated through a carbamate hydrolysis followed by spontaneous 1,6 elimination of methide species and subsequent release of the free drug (Fig. 10.4). However, the obtained catalytic antibody ST51 could not catalyze the cleavage reaction in a physiological pH, and only at pH 9.0 could a 5000-fold rate enhancement be observed in their model system, in which L-tryptophan was used to simulate a drug.

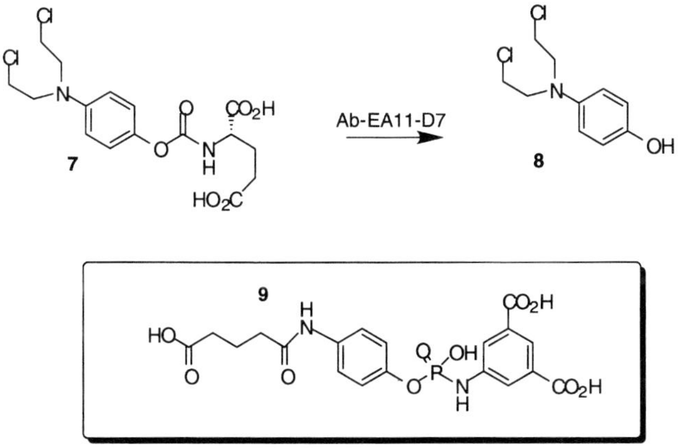

Fig. 10.3 Monoclonal antibody EA11-D7 elicited against phosphonamidate hapten **9** catalyzes the hydrolysis of carbamate **7** prodrug to release drug **8**.

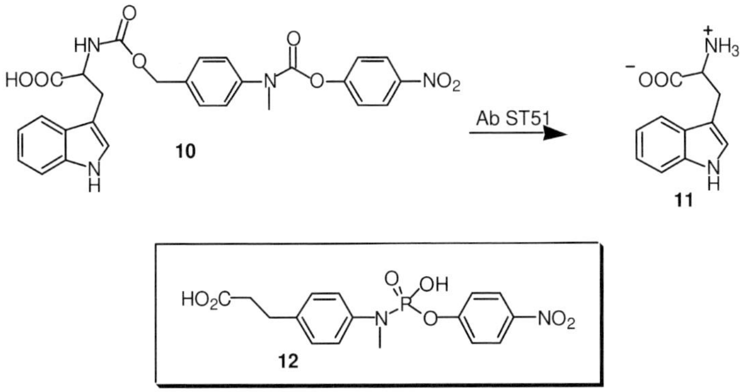

Fig. 10.4 Monoclonal antibody ST51 elicited against phosphonamidate hapten **12** catalyzes the activation of tripartite prodrug **10** through a tandem carbamate hydrolysis- 1,6-elimination reaction to release the model drug **11**.

10.2.2
Catalytic Antibody Designed for *retro*-Diels-Alder Reaction

Reymond and Lerner et al. [10] reported in 1996 that catalytic antibody 9D9 (elicited against hapten **15**) catalyzes the release of an anthracene derivative (**14**) and a nitroxyl from prodrug **13** with multiple turnovers. The biological relevance of the reaction has been demonstrated by the production of physiological concentrations of nitric oxide in the presence of enzyme superoxide dismutase. Releasing nitroxyl by antibody catalysis at a controlled rate under physiological conditions also offers a unique opportunity to study its direct biological effect. Most importantly, both the rate and location of nitroxyl release from prodrug **13** can be controlled by the localization of

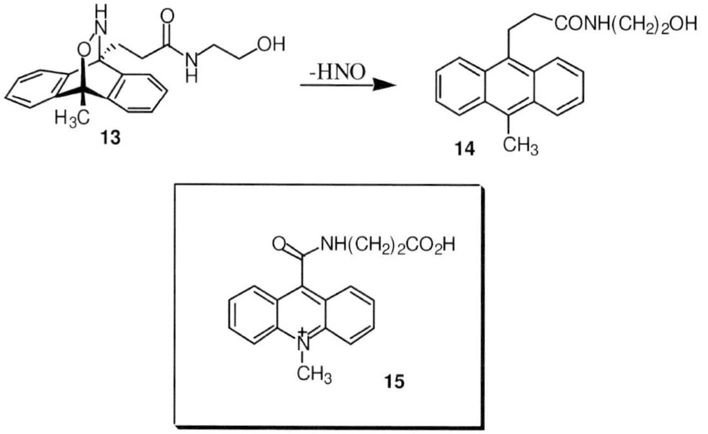

Fig. 10.5 Monoclonal antibody 9D9 elicited against hapten **15** catalyzes the *retro*-Diels-Alder reaction of prodrug **13** to release nitroxyl and anthracene derivative **14**.

the catalytic antibody. Preliminary studies have shown that either **13** or anthracene **14** has no measurable cytotoxicity at the concentration used for nitric oxide production, which suggest that the antibody might be a useful source of nitric oxide *in vivo*.

10.2.3
Catalytic Antibody 38C2 (*retro*-Aldol-*retro*-Michael Cleavage Reaction)

Barbas, Lerner and coworkers achieved a major breakthrough with the development of a new immunization concept. Instead of immunizing against transition state analogs, they immunize with a compound that is highly reactive in order to create a chemical reaction during the binding of the antigen to the antibody. The same reaction becomes part of the mechanism of the catalytic event. In other words, the antibodies are elicited against a chemical reaction instead of a transition state analog. This strategy was termed reactive immunization [11, 12].

1,3-Diketone **16** was used as a trap for an amino lysine residue in an antibody active site (Fig. 10.6). Two antibodies, 38C2 and 33F12, which contained the desired lysine, were found to mimic type I aldolases very efficiently.

Antibody 38C2 has a major advantage regarding the approach of prodrug activation. Since it has the ability to accept a broad variety of substrates, the antibody may potentially activate any prodrug [13]. A general prodrug-masking chemistry was developed and was designed to take advantage of the broad scope and mechanism of catalytic antibody 38C2. The drug-masking/activation concept was based on a sequential *retro*-aldol-*retro*-Michael reaction catalyzed by antibody 38C2 (Fig. 10.7) [14].

This reaction sequence is not known to be catalyzed by any other enzyme and has a very low background, i.e., the reaction is very slow in the absence of the catalyst.

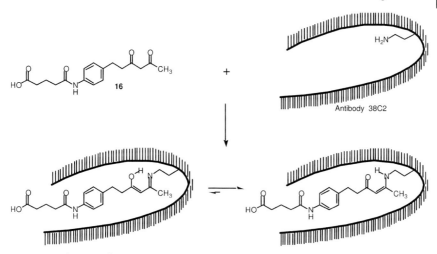

Fig. 10.6 Mechanism of trapping the essential ε-amino group of a lysine residue in the antibody's binding pocket by using the 1,3-diketone hapten **16**.

Fig. 10.7 Prodrug activation via a tandem *retro*-aldol-*retro*-Michael reaction. X stands for heteroatoms N, O, or S.

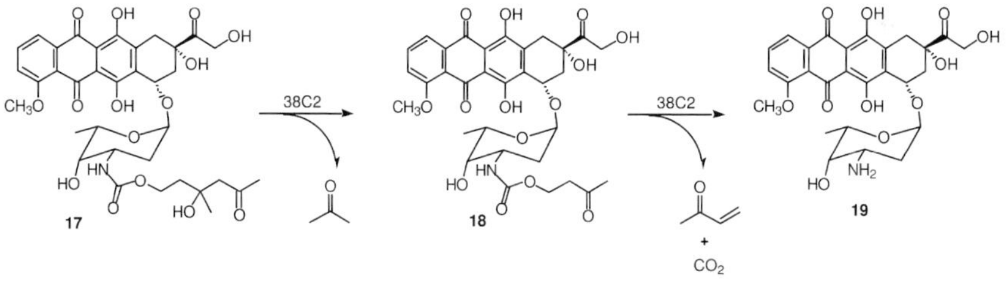

Fig. 10.8 Antibody 38C2 catalyzes the activation of prodoxorubicin prodrug **17** via the tandem *retro*-aldol-*retro*-Michael reaction to release doxorubicin **19** (through intermediate **18**).

This chemistry was first demonstrated in 1999 with the anticancer drugs doxorubicin and camptothecin [14]. The doxorubicin prodrug **17** (Fig. 10.8) was constructed, by masking the amine functionality with the *retro*-aldol-*retro*-Michael linker, whereas the camptothecin prodrug **20** (Fig. 10.8) has an additional extension linker that is self-cyclized after cleavage of the *retro*-aldol-*retro*-Michael trigger by the catalytic antibody. The rate enhancement of the doxorubicin prodrug **17** activation reaction was > 10^5-fold better than the uncatalyzed reaction.

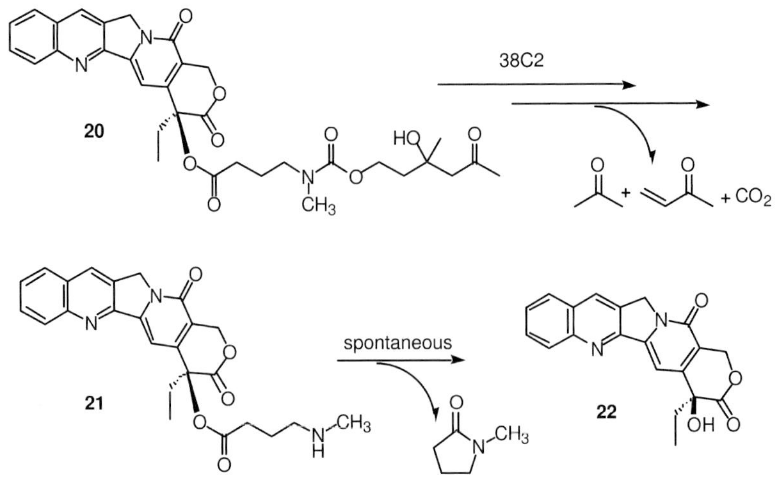

Fig. 10.9 Antibody 38C2 catalyzes the activation of prodrug **20** via the tandem *retro*-aldol-*retro*-Michael reaction and a spontaneous cyclization to release free camptothecin **22**.

It was shown that weakly or nontoxic concentrations of the corresponding prodrug **17** can be activated by therapeutically relevant concentrations of antibody 38C2 to kill colon and prostate cancer cell lines (Fig. 10.10). To further test the therapeutic relevance of this model system, it was shown that antibody 38C2 remained catalytically active over weeks after intravenous injection into mice. Based on these findings, it is possible that the system described here has the potential to become a key tool in selective chemotherapy.

The development of strategies that provide selective chemotherapy presents significant multidisciplinary challenges. Selective chemotherapy might, in the case of cancer, be based on the enzymatic activation of a prodrug at the tumor site. The enzymatic activity must be directed to the site with a targeting molecule, usually an antibody, that recognizes a cell surface molecule selectively expressed at the tumor site. Since a single molecule of enzyme catalyzes the activation of many molecules of prodrug, a localized and high drug concentration may be maintained at the tumor site. This concept of antibody-directed enzyme prodrug therapy, which was termed ADEPT [15, 16], holds promise as a general and selective chemotherapeutic strategy if several specific criteria can be met.

A number of antigens that are expressed on the surface of tumor cells or in their supporting vasculature have been shown to be effective targets for antibody-mediated cancer therapy. Thus, for the most part, the targeting antibody component of this strategy is not limiting. By contrast, the requirements of the enzyme component and complementary prodrug chemistries for ADEPT are difficult to achieve. First of all, selective prodrug activation requires the catalysis of a reaction that must not be accomplished by endogenous enzymes in the blood or normal tissue of the patient. Enzymes of non-human origin that meet these needs are, however, likely to be highly

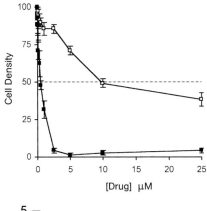
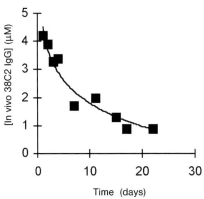

Fig. 10.10 (*Top*) Growth inhibition of LIM1215 human colon carcinoma cells *in vitro* by doxorubicin (■) and prodoxorubicin (□). (*Bottom*) *In vivo* activity of antibody 38C2. Mice were injected with 100 mL of 15 mg/mL 38C2 IgG in PBS on day 0. The concentration of 38C2 IgG in mouse sera was studied as a function of time after the injection. Activity was calculated based on the antibody 38C2- catalyzed conversion of the fluorogenic aldol sensor methodol into fluorescent 6-methoxy-2-naphthaldehyde. Typical data derived from one mouse are shown. The initial 38C2 IgG concentration on day 0 can be estimated to be about 6 mM based on a blood volume of 1.5 mL and 1.5 mg of injected antibody. Catalysis was not detectable in sera from mice injected with 100 mL of 15 mg/ml control antibody in PBS.

immunogenic, a fact that makes repeated administration impossible. Finally, the chemistry used to convert a drug into a prodrug should be versatile enough to allow for the modification of many drug classes while not interfering with the operation of the enzyme, so that a single enzyme could be used for the activation of a multiplicity of prodrugs.

The limitations of the ADEPT complex described above encourage scientists to suggest that the enzyme component for ADEPT might be replaced by a catalytic antibody. The potential of catalytic antibodies for ADEPT is indeed compelling. The catalysis of reactions that are not catalyzed by human enzymes and minimal immunogenicity through antibody humanization are feasible. Combining these features, the ADEPT conjugate translates into a bispecific antibody consisting of a targeting arm and a catalytic arm.

The general concept of bispecific antibodies is illustrated in Fig. 10.11. Two parent antibodies are combined in a manner such that each antibody contributes one light chain and one heavy chain. The bispecific construct contains two different binding regions. One originates from the targeting antibody and can bind specifically to antigens, which are expressed specifically on tumor cells, and the other originates from the catalytic antibody and is used to activate a prodrug.

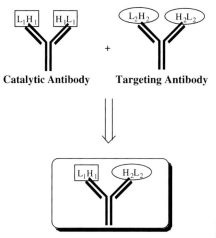

Fig. 10.11 Generation of bispecific antibody hybrid originating from a corresponding catalytic and a targeting antibody.

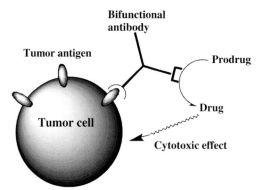

Fig. 10.12 Schematic illustration of targeting chemotherapy selectively to a tumor cell. The prodrug is transformed to the active drug by the catalytic arm of the bifunctional antibody. The other arm binds to an antigen on a tumor cell.

Fig. 10.12 illustrates how such a bispecific antibody could be used for selective chemotherapy. This possible future treatment can consist of two steps. First, a dose of the bispecific antibody can be administered and the targeting arm can locate and attach to specific antigens on the tumor cell surface. Excess of the antibody can be cleared after a limited time from the patient's blood, thereby preventing non-specific prodrug activation. In the second step, several doses of prodrug can be administered with suitable time gaps. The prodrug can reach the tumor site through the blood circulation and can be activated by the catalytic arm in very close proximity to the tumor cell. The damage to non-cancer cells should therefore be minimized, and the free drug can specifically target cancer cells. An animation movie that illustrates selective prodrug targeting and activation by bispecific antibody can be found at the following web site (http://www.scripps.edu/mb/barbas/multimedia.html).

In order to test this approach, Barbas, Lerner and coworkers performed and reported in 2001 the first *in vivo* study of prodrug activation by antibody catalysis [17]. A new etoposide (VP16) prodrug (23) was designed and synthesized for the purpose

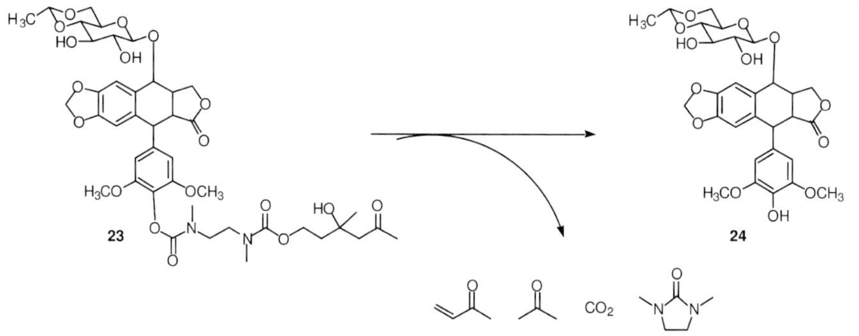

Fig. 10.13 Antibody 38C2 catalyzes the activation of prodrug **23** via the tandem *retro*-aldol-*retro*-Michael reaction and a spontaneous cyclization to release free etoposide **24**.

of this study. The phenol functionality was masked by the *retro*-aldol-*retro*-Michael trigger, which was attached to N,N-dimethyl-ethylenediamine, a self-immolative extension. The prodrug activation was achieved after cleavage of the trigger followed by spontaneous cyclization to release the free etoposide drug **24**, as shown in Fig. 10.13

To evaluate the efficacy of antibody 38C2-mediated etoposide prodrug activation, the activity of the prodrug was evaluated *in vitro* in the presence or absence of 1 µM antibody 38C2 with cultured murine NXS2 neuroblastoma cells. Interestingly, the prodrug revealed a clearly reduced (> 100-fold) potential for growth inhibition (Fig. 10.14-I). Activation of the prodrug with antibody 38C2 resulted in growth profiles similar to the unmodified etoposide control.

The efficacy of localized catalytic antibody-mediated etoposide prodrug activation against syngeneic murine NXS2 neuroblastoma was evaluated on established primary tumors (Fig. 10.14-II). Primary tumors were established in the lateral flank of animals, and studies were initiated when an average tumor size of 100 mm^3 was obtained. Each experimental group consisted of eight animals. Catalytic antibody 38C2 was delivered locally by intratumoral injection. Prodrug or etoposide was delivered systemically by i.p. injection at three time points. Two major findings were obtained in this experiment. First, an increased antitumor efficacy of locally activated etoposide prodrug over systemically applied unmodified etoposide was observed at the maximum tolerated dose of etoposide. In fact, a dramatic 75% reduction in s.c. tumor growth was observed only in the group of mice receiving both intratumoral injections of catalytic antibody 38C2 and systemic treatments with etoposide prodrug. This is in contrast to control groups receiving each agent as monotherapy or injections with PBS. Because all animals that received both catalytic antibody 38C2 and prodrug survived the 24-day experiment, they were treated at day 25 with a second cycle of catalytic antibody 38C2 and prodrug therapy. After this, three of eight mice revealed the complete absence of a primary tumor, underlining the efficacy of the principle of catalytic antibody-mediated prodrug activation *in vivo*. Furthermore, etoposide prodrug demonstrated dramatically reduced toxicity *in vivo* as compared with unmodified etoposide itself. In fact, no toxicity was observed at the dosage ad-

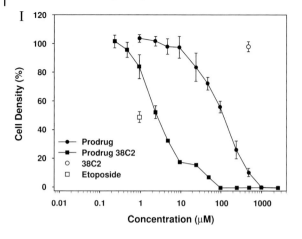

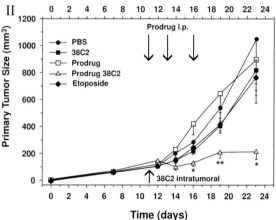

Fig. 10.14 (I) Growth inhibition activity of etoposide prodrug **23** in the presence and absence of catalytic antibody 38C2. The growth response of NXS2 neuroblastoma cells to a 72 h incubation with increasing concentrations of etoposide prodrug **23** in the presence and absence of 1 mM 38C2 catalytic antibody was analyzed by using a standard lactate dehydrogenase release assay. A serial dilution of etoposide and wells containing 1 mM 38C2 only are shown as controls. **(II)** Effect of catalytic antibody 38C2-mediated etoposide prodrug **23** activation on the growth of primary neuroblastoma tumors. Treatment of mice bearing established s.c. primary neuroblastoma tumors induced by s.c. injection with 2×10^6 NXS2 neuroblastoma cells was initiated 11 days after tumor cell inoculation. The treatment consisted of intratumoral injection of 38C2 (0.5 mg) followed by three i.p. etoposide prodrug **23** (total dose 1250 mg/kg) injections. Arrows indicate days of treatment with each reagent, respectively. One control group received unmodified etoposide (total dose 40 mg/kg).

ministered to the animals (1,250 mg/kg), as defined by the absence of any decrease in body weight in contrast to a 20% weight loss observed in mice treated with the maximal tolerated dose of etoposide (40 mg/kg)).

The previously described study has shown that prodrug activation by antibody catalysis is feasible *in vivo*. Next, it was necessary to attach targeting devices to the catalytic antibody in order to direct it selectively to the tumor site. Shabat and coworkers have recently report the first targeting device conjugated to a catalytic antibody. They have used a water-soluble synthetic polymer, N-(2-hydroxypropyl)methacrylamide (HPMA), which is biocompatible, non-immunogenic and non-toxic [18]. Moreover, its *in vivo* body distribution is well characterized [19] and is known to accumulate selectively at tumor sites because of the enhanced permeability and retention (EPR) effect [20]. This effect occurs because of the difference between the vasculature physiology of solid tumors and that of normal tissues. Compared with the regular ordered vasculature of normal tissues, blood vessels in tumors are often highly abnormal. The growth of the tumor creates a constant need for the continuous supply of new blood vessels. This process, termed angiogenesis, often results in the construction of vessels with leaky walls, which allows enhanced permeability of macromolecules within the tumor. In addition, poor lymphatic drainage at the tumor site promotes accumulation of large molecules.

Similarly to the previous described ADEPT system, polymer molecules can be used as passive targeting devices instead of specific monoclonal antibodies. Polymer-directed enzyme prodrug therapy (PDEPT) is a two-step antitumor approach in which both the prodrug and the enzyme are targeted to the tumor site with a polymer molecule [21, 22] (Fig. 10.15). In the first step, a polymer-prodrug conjugate is administered and trapped in tumor tissues through the EPR effect. The excess of the conjugate is cleared out from the blood in a relatively short time. In the second step, a polymer-enzyme conjugate is injected. The polymer molecule carries the enzyme to the tumor site, where it releases the drug from the polymer.

Step I. Administration of polymer - drug

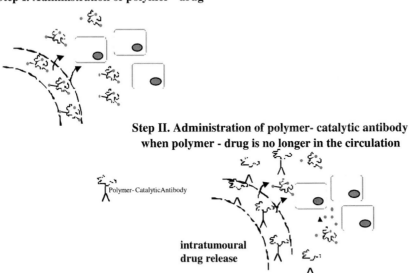

Step II. Administration of polymer- catalytic antibody when polymer - drug is no longer in the circulation

Polymer- CatalyticAntibody

intratumoural drug release

Fig. 10.15 Schematic representation of PDEPT (polymer directed enzyme prodrug therapy).

296 | 10 Medicinal Potential of Catalytic Antibodies

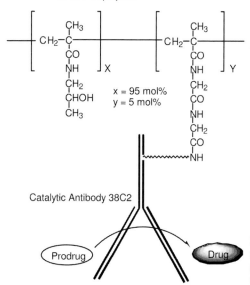

Fig. 10.16 Schematic representation of antibody 38C2-HPMA-copolymer conjugate, designed from selective prodrug activation by the PDEPT approach.

Shabat and coworkers [23] prepared a conjugate of catalytic antibody with the targeting moiety HPMA copolymer based on amide bond formation between an external lysine of the antibody and an active ester of the HPMA copolymer (Fig. 10.16). The conjugation yield was very high and the antibody retained most of its catalytic activity in the conjugate. Furthermore, they showed that the antibody-polymer conjugate can activate an etoposide prodrug *in vitro* and consequently inhibit proliferation of two different cancer cell lines (Molt3-T-cell leukemia and NXS2-neurtoblastoma).

One more step toward selective drug delivery by applying catalytic antibodies was recently reported by Shabat and coworkers [24]. A new concept that combines a tumor targeting device, a prodrug, and a prodrug activation trigger in a single entity was presented. They designed a generic module chemical adaptor that is based on three chemical functionalities, as shown in Fig. 10.17. The first functionality is attached to an active drug and, thereby, masks it to yield a prodrug. The second is linked to a targeting moiety which is responsible for guiding the prodrug to the tumor site, and the third is attached to an enzyme substrate. When the corresponding enzyme cleaves the substrate, it triggers a spontaneous reaction that releases the active drug

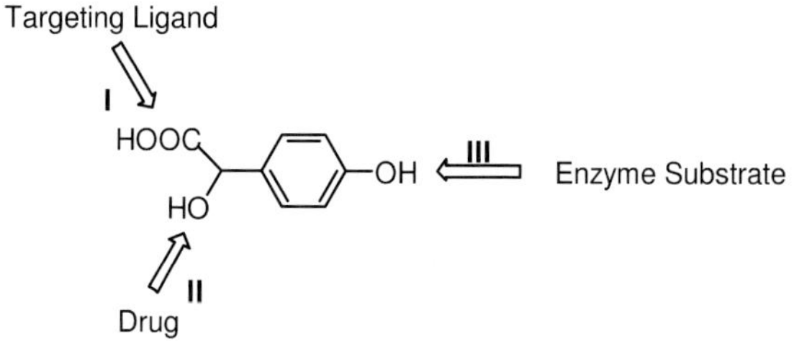

Fig. 10.17 4-Hydroxy-mandelic acid, the central core of the chemical adaptor system.

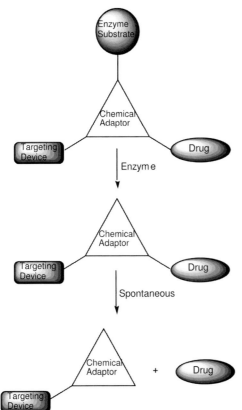

Fig. 10.18 General design of the chemical adaptor system. Cleavage of the enzyme substrate generates an intermediate that spontaneously rearranges to release the drug from the targeting device.

from the targeting moiety. As a result, prodrug activation will preferentially occur at the tumor site.

The central core of the chemical adaptor is based on 4-hydroxy-mandelic acid (Fig. 10.18), which has three functional groups suitable for linkage. Group **I** is a carboxylic acid that is conjugated to a targeting moiety via an amide bond. The drug is linked through the benzyl alcohol group **II**, and the enzyme substrate is attached via the phenol group **III** by a carbamate bond. Fig. 10.19 illustrates the release mechanism of the drug. Cleaving the enzyme substrate from compound **25** generates a free amine group (compound **26**) that spontaneously cyclizes to form a dimethyl urea derivative and phenol **27**. The latter undergoes spontaneous rearrangement to give a quinone methide intermediate (trapped by water to give benzyl alcohol **28**) that releases the drug in the form of carbamic-acid, which then decarboxylates spontaneously, yielding the active drug.

The design of the generic module enables one to potentially link any targeting device to a variety of drugs and to release them with any enzyme by using the corresponding substrate as a trigger. As proof of the concept, a pilot system was designed for which catalytic antibody 38C2 was chosen as the cleaving enzyme. As a targeting device, they chose a polymer molecule and etoposide as the drug. Next, they tested

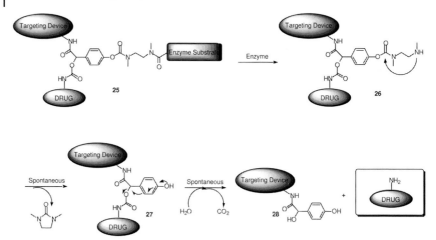

Fig. 10.19 General mechanism of drug release starting with specific enzymatic cleavage.

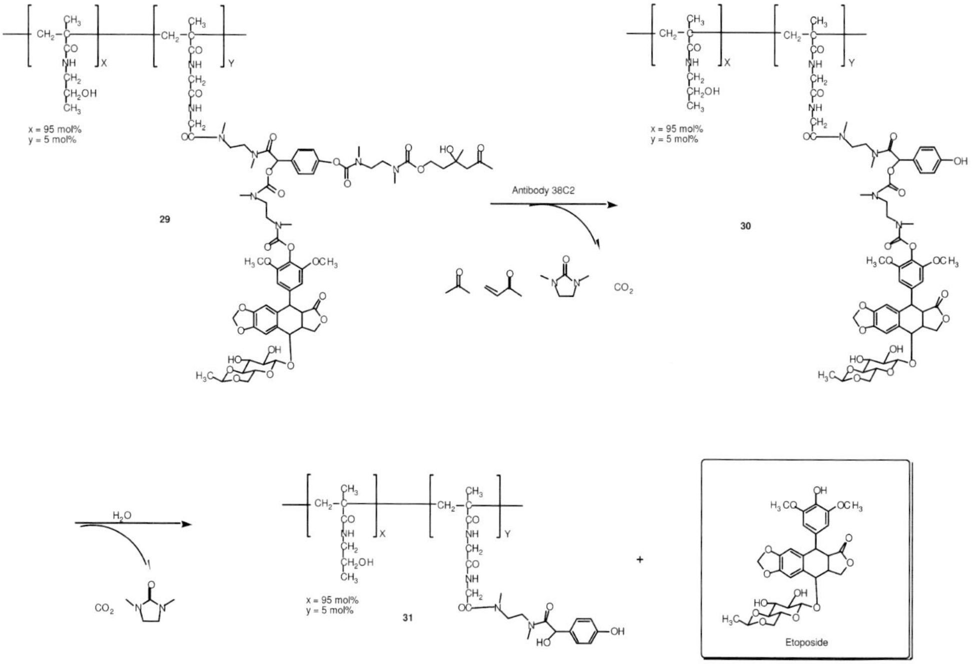

Fig. 10.20 Mechanism of etoposide drug release from the HPMA-copolymer, using catalytic antibody 38C2 as the triggering enzyme.

whether the etoposide drug could be released from complex **29** by the catalytic activity of antibody 38C2. According to the design, the drug should be spontaneously released after the generation of phenol **30**, as illustrated in Fig. 10.20. Complex **29**

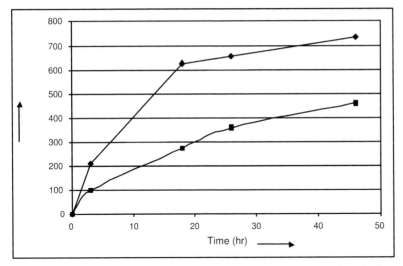

Fig. 10.21 Determination of etoposide release from the HPMA-copolymer vs drug formation from a known etoposide prodrug by catalytic antibody 38C2. ◆–◆: Complex **29** (500 µM) with antibody 38C2 (66 µM) in PBS-7.4. ■–■: Etoposide prodrug (500 µM) with antibody 38C2 (66 µM) in PBS (pH 7.4). Reactions were incubated at 37 °C for the indicated time. The drug release concentration was monitored by HPLC analysis.

was incubated with catalytic antibody 38C2 in PBS (pH 7.4) at 37 °C and monitored the appearance of etoposide using an HPLC assay. As a positive control, a previously described etoposide prodrug, **23**, which is activated by antibody 38C2, was used. As Fig. 10.20 shows, etoposide was released by the catalytic activity of antibody 38C2 to form compound **31** and the free drug. The rate of drug release was similar to the activation rate of the known etoposide prodrug. No spontaneous etoposide release was observed in the absence of the antibody.

Olefsky, Lerner and Barbas et al. [25] have developed a methodology of prodrug delivery by using a modified insulin species whose biological activity potentially can be regulated *in vivo*. Native insulin was derivatized with *retro*-aldol-*retro*-Michael linker that can be selectively removed by the catalytic antibody 38C2 under physiological conditions (Fig. 10.22). The derivatized organoinsulin (insulin-D) was defective with respect to receptor binding and stimulation of glucose transport. The affinity of

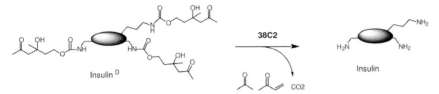

Fig. 10.22 Antibody 38C2 catalyzes the transformation of insulin D to free insulin via the tandem *retro*-aldol-*retro*-Michael reaction.

insulin-D for the insulin receptor was reduced by 90% in binding studies using intact cells. The ability of insulin-D to stimulate glucose transport was reduced by 96% in 3T3-L1 adipocytes and by 55% in conscious rats. Incubation of insulin-D with the catalytic antibody 38C2 cleaved all of the aldol-terminated modifications, restoring native insulin. Treatment of insulin-D with antibody 38C2 also restored insulin-D's receptor binding and glucose transport-stimulating activities *in vitro*, as well as its ability to lower glucose levels in animals *in vivo*. They proposed that these results are the foundation for an *in vivo* regulated system of insulin activation using the prohormone insulin-D and catalytic antibody 38C2 with potential therapeutic application.

10.3
Cocaine Inactivation

Drug inactivation is an obvious potential medicinal application for catalytic antibodies. Since these catalysts can be specifically tailored against any molecule, they could be designed to hydrolyze a toxic drug to form a non-toxic compound. This concept has been explored by two groups that elicited antibodies against a phosphonate transition state analog (compound 34) of cocaine hydrolysis reaction. Landry et al. [26] reported in 1993 the generation of catalytic antibodies that can hydrolyze cocaine 32 (Fig. 10.23). They showed that rats which were treated with this antibody did not suffer from cocaine's reinforcing and other toxic effects. Janda et al. [27] reported in 2001 the generation of a new set of catalytic antibodies for cocaine hydrolysis, using a related hapten for immunization purpose. Their antibodies had similar activity to those reported by Landry.

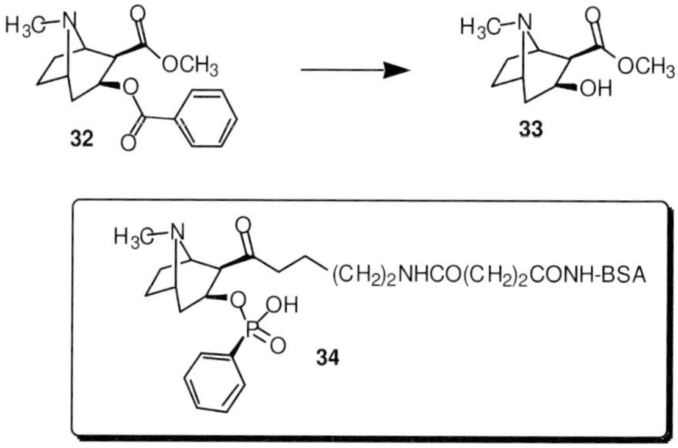

Fig. 10.23 Monoclonal antibody 3B9 elicited against phosphonate hapten 34 catalyzes the hydrolysis of cocaine 32 to ecgonine methyl ester 33.

10.4
Conclusions

Catalytic antibodies have been proven to be useful synthetic tools in enantioselective synthesis and also have played an important role in performing organic reactions with unprotected chemical functionalities. However, the most promising use for them is probably in the medicinal field. These molecules were generated by the immune system and adjusted for *in vivo* activity by nature. Most of the research studies performed so far with catalytic antibodies have dealt with the application of prodrug activation. The concept of a nontoxic prodrug which is transformed into a toxic drug selectively at the tumor site is a promising approach for chemotherapy treatment. Catalytic antibodies have proved themselves to be efficient reagents for the concept of prodrug activation. However, in order to make this approach practical, the catalytic antibody needs to be targeted to the tumor site. The first step toward targeting was when an HPMA-copolymer was conjugated with catalytic antibody 38C2. The concept of targeting an antibody molecule to a tumor with polymers may be achieved, similarly to the targeting of enzymes in the PDEPT approach. Drug inactivation, demonstrated by the prevention of cocaine's reinforcing, is another intriguing application of catalytic antibodies. Finally, it should be noted that in a relatively young research field such as this, a new breakthrough can appear at any time, sparking off new and exciting progress toward a practical medicinal application.

References

1 Tramontano, Janda, K. D., Lerner, R. A., *Science* 234 (**1986**), p. 1566–70
2 Pollack, S. J., Jacobs, J. W., Schultz, P. G., *Science* 234 (**1986**), p. 1570–3
3 Gussow, D., Seemann, G., *Methods Enzymol.* 203 (**1991**), p. 99–121
4 Blackburn, G. M., Datta, A., Partridge, L. J., *Pure Appl. Chem.* 68 (**1996**), p. 2009–2016
5 Jones, L. H., Wentworth, P. Jr., *Mini-Reviews in Medicinal Chemistry* 1 (**2001**), p. 125–132
6 Miyashita, H., Karaki, Y., Kikuchi, M., Fujii, I., *Proc. Natl. Acad. Sci. USA* 90 (**1993**), p. 5337–40
7 Campbell, D. A., Gong, B., Kochersperger, L. M., Yonkovich, S., Gallop, M. A., Schultz, P. G., *J. Am. Chem. Soc.* 116 (**1994**), p. 2165–2166
8 Wentworth, P., Datta, A., Blakey, D., Boyle, T., Partridge, L. J., Blackburn, G. M., *Proc. Natl. Acad. Sci.* 93 (**1996**), p. 799–803
9 Dinaut, A. N., Taylor, S. D., *Chem. Commun.* (**2001**), p. 1386–1387
10 Bahr, N., Gueller, R., Reymond, J.-L., Lerner, R. A., *J. Am. Chem. Soc.* 118 (**1996**), p. 3550–3555
11 Wagner, J., Lerner, R. A., Barbas, C. F. III., *Science* 270 (**1995**), p. 1797–800
12 Wirsching, P., Ashley, J. A., Lo, C. H., Janda, K. D., Lerner, R. A., *Science* 270 (**1995**), p. 1775–82
13 Hoffmann, T., Zhong, G., List, B., Shabat, D., Anderson, J., Gramatikova, S., Lerner, R. A., Barbas, C. F. III., *J. Am. Chem. Soc.* 120 (**1998**), p. 2768–2779
14 Shabat, D., Rader, C., List, B., Lerner, R. A., Barbas, C. F. III, *Proc. Natl. Acad. Sci. USA.* (**1999**), p. 6925–6930
15 Bagshawe, K. D., *Biochem. Soc. Trans.* 18 (**1990**), p. 750–752
16 Bagshawe, K. D., Sharma, S. K., Springer, C. J., Rogers, T., *Ann. Oncol.* 5 (**1994**), p. 879–891
17 Shabat, D., Lode, H. N., Pertl, U., Reisfeld, R. A., Rader, C., Lerner, R. A., Barbas, C. F. 3rd, *Proc. Natl. Acad. Sci. USA* 98 (**2001**), p. 7528–33
18 Rihova, B., Bilej, M., Vetvicka, V., Ulbrich, K., Strohalm, J., Kopecek, J., Duncan, R., *Biomaterials* 10 (**1989**), p. 335–342
19 Seymour, L. W., Ulbrich, K., Strohalm, J., Kopecek, J., Duncan, R., *Biochem. Pharmacol.* 39 (**1990**), p. 1125–1131
20 Maeda, H., Wu, J., Sawa, T., Matsumura, Y., Hori, K., *J. Controlled Release* 65 (**2000**), p. 271–284
21 Satchi, R., Connors, T. A., Duncan, R., *Br J Cancer* 85 (**2001**), p. 1070–6
22 Duncan, R., Gac-Breton, S., Keane, R., Musila, R., Sat, Y. N., Satchi, R., Searle, F., *J. Controlled Release* 74 (**2001**), p. 135–146
23 Satchi-Fainaro, R., Wrasidlo, W., Lode, H. N., Shabat, D., *Bioorg. Med. Chem.* 10 (**2002**), p. 3023–9
24 Gopin, A., Pessah, N., Shamis, M., Rader, C., Shabat, D., *Angew. Chem.* 74, (**2003**), p. 327–332

25 WORRALL, D. S., MCDUNN, J. E., LIST, B., REICHART, D., HEVENER, A., GUSTAFSON, T., BARBAS, C. F. 3RD, LERNER, R. A., OLEFSKY, J. M., *Proc. Natl. Acad. Sci. USA* 98 (2001), p. 13514–13518

26 LANDRY, D. W., ZHAO, K., YANG, G. X., GLICKMAN, M., GEORGIADIS, T. M., *Science* 259 (1993), p. 1899–1901

27 MATSUSHITA, M., HOFFMAN, T. Z., ASHLEY, J. A., ZHOU, B., WIRSCHING, P., JANDA, K. D., *Bioorg. Med. Chem. Lett.* 11 (2001), p. 87–90

11
Reactive Immunization:
A Unique Approach to Aldolase Antibodies
Fujie Tanaka, Carlos F. Barbas, III

11.1
Introduction

Natural selection of enzyme catalysis occurs as a consequence of improved function that results in increased fitness of the organism during the course of evolution. By contrast, selection of antibody catalysts has, until recently, been based on binding to transition state analogs of the reactions or charged compounds designed using information from the reaction coordinate of a given chemical transformation. If the selection criteria operating within the immune system could be switched from simple binding to catalytic function, efficient catalytic antibodies should be generated. Addressing to this type of selection, a strategy termed reactive immunization has been developed. Reactive immunization provides a means to select antibody catalysts *in vivo* on the basis of their ability to carry out a chemical reaction. In this approach, a designed reactive immunogen is used for immunization, and chemical reactions such as the formation of a covalent bond occur in the binding pocket of the antibodies during their induction. The chemical reactivity and mechanism integrated into the antibody by the covalent trap with the reactive immunogen are used in the catalytic reaction with substrate molecules. The reactive immunization strategy has provided for the preparation of the most highly proficient catalytic antibodies described to date. A unique feature of antibody catalysts prepared in this way is their unusually broad substrate scope.

In this chapter, we describe aldolase antibodies generated by reactive immunization, their features and applications, *in vitro* evolution of these aldolase antibodies, as well as catalytic antibodies selected by using reactive compounds *in vitro*.

Catalytic Antibodies. Edited by Ehud Keinan
Copyright © 2005 WILEY-VCH Verlag GmbH & Co. KGaA, Weinheim
ISBN: 3-527-30688-9

11.2
Generation of Aldolase Antibodies by Reactive Immunization

11.2.1
Development of the Concept of Reactive Immunization

Traditionally, catalytic antibodies have been obtained by immunization with chemically inert transition state analogs (TSAs). This approach was based on Linus Pauling's notion that enzymes provide rate acceleration for chemical reactions by binding transition states [1, 2]. William Jencks restated Pauling's insight concerning catalytic antibodies more precisely, realizing that his insight could be used not only to explain catalysis but also as a roadmap to create new catalysts [3]: "If complementarity between the active site and the transition state contributes significantly to enzymatic catalysis, it should be possible to synthesize an enzyme by constructing such an active site. One way to do this is to prepare an antibody to a haptenic group which resembles the transition state of a given reaction. The combining sites of such antibodies should be complementary to the transition state and should cause an acceleration by forcing bound substrates to resemble the transition state." For example, phosphonate **1** can be considered a mimic of the transition state for the hydrolysis of ester **2**. Indeed, antibodies obtained from immunization with TSA **1** catalyze the hydrolysis of ester **2** to give the acid **3** (Scheme 11.1) [4–6].

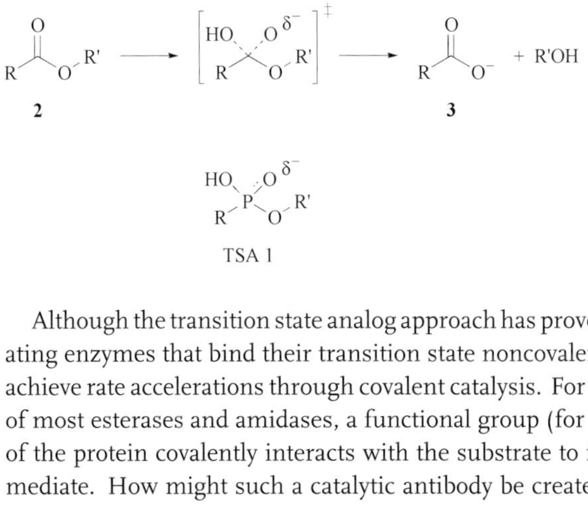

Scheme 11.1

Although the transition state analog approach has proven to be a robust route to creating enzymes that bind their transition state noncovalently, many natural enzymes achieve rate accelerations through covalent catalysis. For example, in the mechanism of most esterases and amidases, a functional group (for example, a serine hydroxyl) of the protein covalently interacts with the substrate to form a protein-bound intermediate. How might such a catalytic antibody be created when the transition state analog approach "teaches" noncovalent interactions? It has been noted in particular cases that immunization with a transition state analog has provided catalysts that use a covalent mechanism. Covalent catalysis in these cases is fortuitous and was not an element of the experimenter's design. We were intrigued with the challenge of directing a covalent mechanism. With this as our guide, our initial studies attempted to capitalize on the findings achieved with transition state analogs. In 1992, with the aid of postdoctoral fellow Willi Amberg, a phosphonate diester hapten was synthesized with the aim of trapping a nucleophilic serine in the active site of an esterase antibody during the course of immunization or phage selections from synthetic antibody libraries. This early attempt, while unsuccessful, was later validated in 1995 with the use of a phosphonate diester prepared by more active leaving groups [7]. Unfettered by the early failure, we embarked on creating aldolase antibodies. One of us, CFB, had considered aldolase enzymes to be fascinating from a synthetic perspective since his days as a graduate student in Chi-Huey Wong's laboratory where he worked with aldolases. Given the ubiquity of the aldol reaction in nature and in traditional organic synthesis, the quest to create aldolase antibodies presented itself as one of the most formidable challenges in the field of catalytic antibodies from a mechanistic perspective and potentially the most rewarding from a synthetic perspective. Nature's most fundamental carbon-carbon bond-forming enzymes, class I aldolases, use covalent catalysis in an elaborate mechanistic dance of enzyme-bound intermediates. Class I aldolases utilize the ε-amino group of a Lys in their active site to form a Schiff base with one of their substrates that becomes the aldol donor. Schiff base formation lowers the activation energy for proton abstraction from the Cα atom and subsequent enamine formation. The enamine, a nascent carbon nucleophile, is then poised to react with an aldehyde substrate, the aldol acceptor, to form a new C–C bond. The Schiff base is then hydrolyzed and the product is released (Scheme 11.2) [8, 9]. With nature as a model, we set out to develop antibody catalysts that mimic the covalent mechanism of natural class I aldolases.

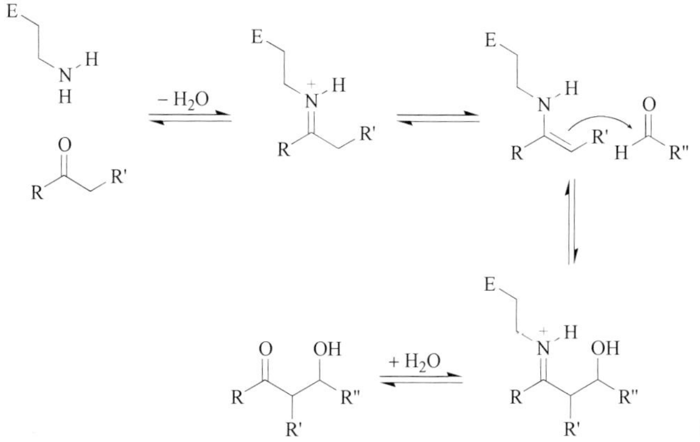

Scheme 11.2

To address this sort of reaction we needed to develop a new approach to catalytic antibodies or enzymes in general that differed significantly from the non-covalent approach of Pauling and Jencks. We sought a method that would allow for the programming of detailed aspects of chemical reaction mechanisms, down to the level of the chemical identity of a residue to be used in the catalytic reaction. Our proposed solution to this problem was to use an antigen that would be reactive enough to form a stable covalent conjugate with an antibody that possessed a reactive amino group. The hapten we chose was 1,3-diketone **4** [10]. Historically, Frank Westheimer introduced 1,3-diketones in the mid 1960s as mechanistic probes of the enzyme acetoacetate decarboxylase [11]. In these studies he demonstrated that 1,3-diketones could trap the catalytic lysine of acetoacetate decarboxylase.

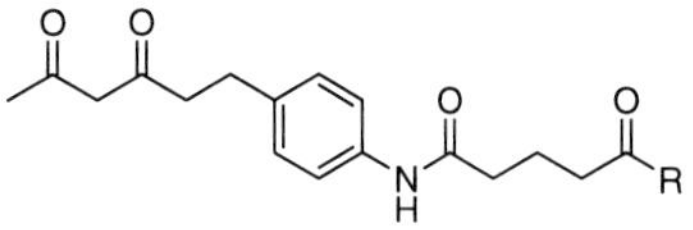

4 (R = OH or NH-carrier protein)

With this foundation, we anticipated that if **4** encountered a reactive amino group in the correct chemical microenvironment, it would form an enaminone. In contrast to regular enamines, this enamine (or enaminone or vinylogous amide) would be a stable species because of conjugation and hydrogen bonding. Thus, unlike a transition state analog, this hapten would be covalently complementary over the course of a reaction with a low pKa amine with the protein adapting itself to the variety of structures on this reaction coordinate. The formation of the enaminone, which significantly could be monitored by UV spectroscopy, would indicate that parts of the catalytic machinery, beyond the reactive amine, that were necessary for aldol catalysis were present (Scheme 11.3). These would include: 1) catalysis of carbinolamine formation, 2) dehydration, and 3) deprotonation of the imminium intermediate. The carbon-carbon bond-forming step may further be catalyzed by an acid that protonates the second carbonyl group in the hapten:

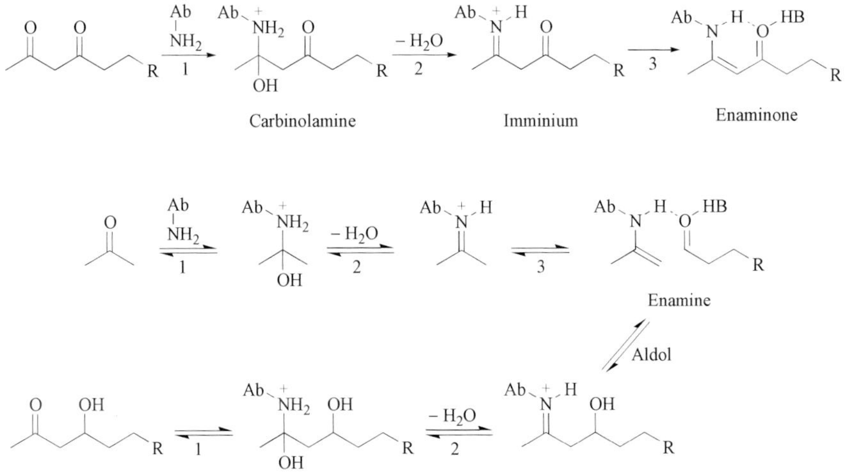

Scheme 11.3

The molecular steps involved in trapping the requisite Lys residue are essentially the same chemical steps involved in activating a substrate ketone to an enamine. This mechanistic mimicry provides for the selection of an active site that can adapt to the chemical and steric changes that occur over this reaction coordinate. Further, the diketone structure provides for appropriate binding sites for the two substrates of the intermolecular reaction, facilitating crossing of the entropic barrier intrinsic to this bimolecular reaction. The driving force for the reaction of 1,3-diketone hapten with the antibody is the formation of a stable covalent conjugated enaminone between hapten **4** and the ε-amino group of Lys. The symmetry of the aldol mechanism would then allow for catalysis of the hydrolysis of the resulting hydroxyimminium compound to the aldol.

With this as our design premise, mice were immunized with 1,3-diketone hapten **4** [10]. Antibodies prepared from hybridomas derived following immunization with the 1,3-diketone were screened for their ability to form the stable enaminone, which has an absorption maximum $\lambda_{max} = 316$ nm with hapten **4** (R=OH). Two antibodies, 38C2 and 33F12, out of 20 reacted covalently with the 1,3-diketone and demonstrated the enaminone absorption band by ultraviolet spectroscopy. These two antibodies catalyzed the addition of acetone to aldehyde **5** to give β-hydroxyketone **6** (Scheme 11.4). Both antibodies catalyzed the diastereoselective addition of acetone to the *si*-face of **5** regardless of the stereochemistry at C2 of this aldehyde. These antibodies also catalyzed the *retro*-aldol reaction of **6**. The remaining antibodies were unable to catalyze the aldol and *retro*-aldol reactions, indicating that only those that formed the critical enaminone intermediate were active. The catalytic activity of 38C2 and 33F12 was completely inhibited when either hapten **4** (R=OH) or 2,4-pentanedione was added prior to the catalytic assay. For the aldol reaction of acetone and aldehyde **5** at pH 7.5, aldolase antibodies 38C2 and 33F12 showed a $(k_{cat}/K_m)/k_{uncat} \approx 10^9$. The efficiency

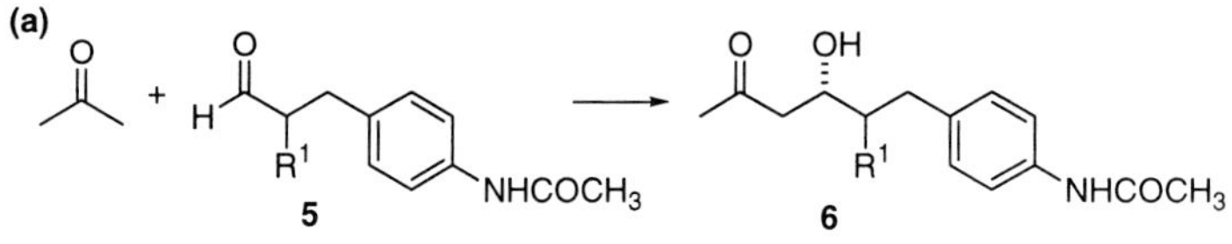

Name of aldolase Ab	Substrate R¹	k_{cat} (min⁻¹)	K_m (μM)
38C2	H	6.7×10^{-3}	17
38C2	CH₃	4.0×10^{-2}	48
33F12	CH₃	8.3×10^{-2}	125

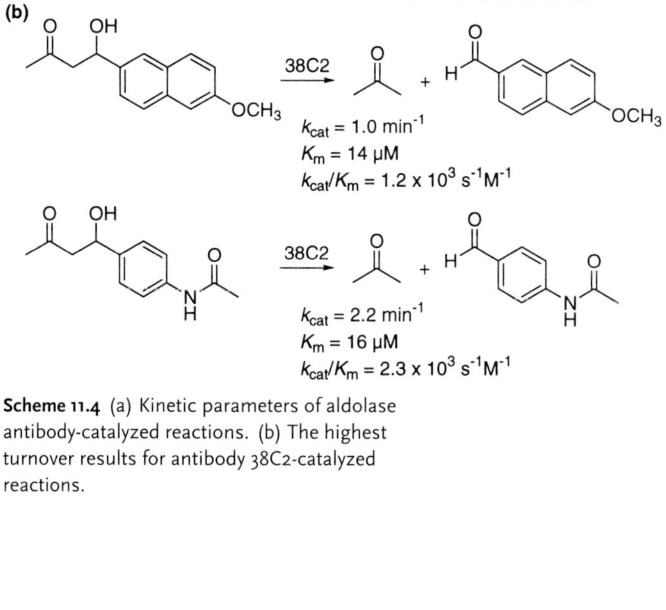

Scheme 11.4 (a) Kinetic parameters of aldolase antibody-catalyzed reactions. (b) The highest turnover results for antibody 38C2-catalyzed reactions.

of catalysis is due in a large part to an entropic advantage in the antibody-catalyzed reaction, which is reflected as a high effective molarity, $k_{cat}/k_{uncat} > 10^5$ M. With the best substrates identified to date for the *retro*-aldol reaction [12–14], the catalytic proficiency $(k_{cat}/K_m)/k_{uncat}$ is ca. 6×10^{10}. The catalytic efficiency (k_{cat}/K_m) of these aldolase antibodies with these substrates is only ca. 20 to 40-fold lower than that of the most studied aldolase enzyme, fructose-1,6-bisphosphate aldolase a central enzyme in glycolysis, 4.9×10^4 s^{-1}M^{-1} [9].

Later, other haptens possessing the 1,3-diketone functionality were used for immunization by the Keinan group and antibody 24H6, which also catalyzes aldol reactions via an enamine mechanism was obtained [15].

11.2.2
Combining Reactive Immunization with Transition State Analog for Aldolase Antibodies

Although 1,3-diketone hapten 4 provided for the generation of the efficient aldolase antibodies 38C2 and 33F12, the hapten does not address the tetrahedral geometry of the rate-determining transition state of the C–C bond forming step. With this, we set out to address this shortcoming by recruiting transition state analog features and thus the teachings of Pauling and Jencks into the already successful diketone structure. We designed and synthesized hapten 7 to contain features common to the transition state analog approach that has been successful for many reactions and the 1,3-diketone functionality key to the reactive immunization strategy (Scheme 11.5) [16]. The tetrahedral geometry of sulfone moiety in hapten 7 mimics the tetrahedral transition state of the C–C bond forming step and therefore should facilitate nucleophilic attack of the enamine intermediate on the acceptor aldehyde substrate.

Hapten 7-carrier protein conjugate was used for the immunization. Nine antibodies out of 17 reacted with 2,4-pentanedione to form a stable enaminone, a significant increase of catalysts as compared with simple diketone immunization, and an unusually high percentage of catalysts as compared to other systems. These nine antibodies were catalytic and their catalytic activity was inhibited by the addition of 2,4-pentanedione. These results are consistent with a covalent catalytic mechanism in which a reactive amine is programmed in these antibodies. Of practical interest is that immunization with hapten 7 generated the two families of catalysts that operate on opposite optical isomers. Antibodies 93F3 and 84G3 generated with hapten 7 exhibit antipodal reactivity to antibodies 38C2 and 33F12 generated with hapten 4, and antibodies 40F12 and 42F1 generated with hapten 7 exhibit the same enantiopreferentially to antibodies 38C2 and 33F12. When kinetic resolution of aldol (±)-8 was preformed with antibody 93F3, (S)-8 (> 99% ee) was obtained at 52% conversion. In the aldol reaction between acetone and aldehyde 9, antibody 93F3 gave (R)-8. Antibody 38C2 provides (R)-8 in the kinetic resolution of (±)-8 and (S)-8 in the aldol reaction between acetone and 9. The catalytic proficiency of antibodies 93F3 and 84G3 exceeds that of antibody 38C2 in most cases. Antibodies 93F3 and 84G3 catalyzed the *retro*-aldol reaction of (R)-10 with k_{cat}s 1.2 sec^{-1} and 1.4 sec^{-1}, respectively. The rate enhancement k_{cat}/k_{uncat} of the best 84G3-catalyzed reaction is 2.3×10^8, approximately 1000-fold higher than that reported for any other catalytic antibodies

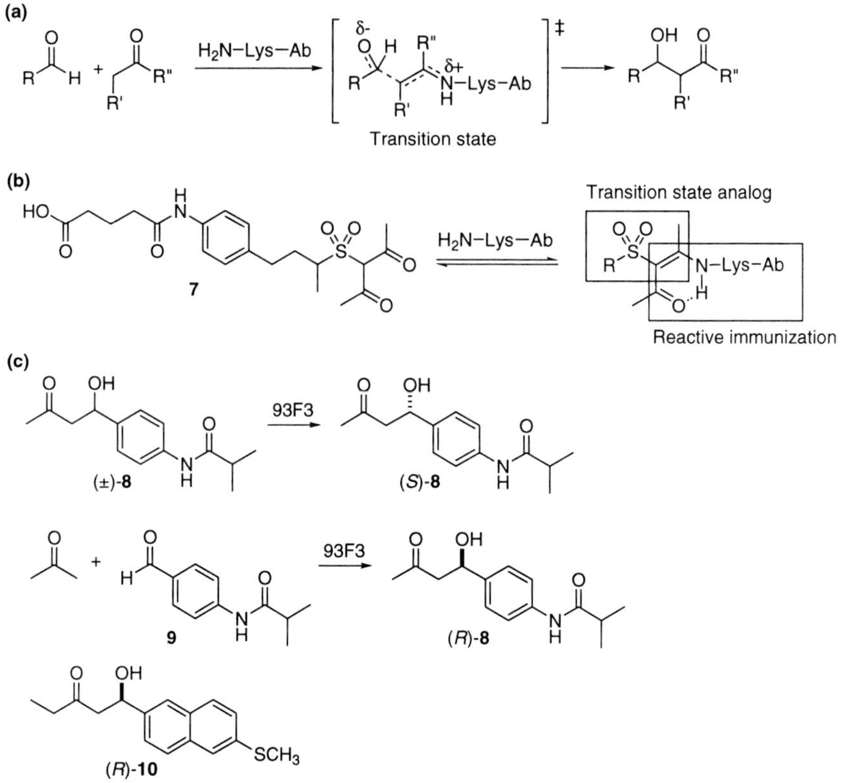

Scheme 11.5 (a) Mechanism of the antibody-catalyzed aldol reaction. (b) Reactive immunization with **7** for the generation of aldolase antibodies. (c) Example of aldolase antibody 93F3-catalyzed reactions.

[17]. In this case the catalytic efficiency (k_{cat}/K_m) of 84G3, 3.3×10^5 s^{-1}M^{-1}, is slightly greater than that of the most studied aldolase enzyme, fructose-1,6-biphosphate aldolase [9]. As such, 84G3 and 93F3 are currently the most efficient man-made protein catalysts ever prepared.

11.3
Structural Insight into Aldolase Antibodies

Antibodies 38C2 and 33F12 are highly homologous with respect to sequence. The X-ray crystal structure of the Fab fragment of 33F12 was determined in collaboration with the Wilson laboratory (Fig. 11.1) [18]. The structure showed that the entrance of the antigen binding site of 33F12 is a narrow elongated cleft. The binding pocket is more than 11 Å deep and is comparable to combining sites of antibodies raised against other small haptenic molecules [19–22]. At the bottom of the pocket, the catalytic lysine, LysH93, was found within a hydrophobic environment. LysH93 is common to both antibodies 38C2 and 33F12 and had been identified prior to the solution of the structure in site-directed mutagenesis experiments. Within this pocket, only one charged residue was found to be within an 8Å radius of the nitrogen of the ε-amino group of LysH93. No salt bridges or hydrogen bonds can be formed for LysH93 in 33F12; thus it resides in a rather hydrophobic reaction vessel that is the binding pocket. This microenvironment serves to perturb the pK_a of LysH93, which allows it to exist in its uncharged form, facilitating its function as a strong nucleophile. The first step in the aldol reaction is nucleophilic attack of the ε-amino group of lysine on a carbonyl group. For the ε-amino group to be nucleophilic, it must be in its uncharged form. However, the reader should note that the pK_a of the ε-amino group of "normal" lysine in aqueous solution is 10.7 [23]. Because both natural and antibody aldolases depend on a nonprotonated lysine as a nucleophile and operate with maximal activity at neutral pH where the ε-amino group of lysine normally would be protonated, the pK_a of this group must be perturbed. The pK_a of the active-site lysine was determined to be 5.5 and 6.0 for antibodies 33F12 and 38C2, respectively, by the pH dependence of enaminone formation with 3-methyl-2,4-pentanedione [18]. Thus, the active site of the aldolase antibodies serves to modulate the acidity of this amine group as much as 10^5-fold.

Notably, with the exception of a single residue, the residues in van der Waals contact with LysH93 are conserved in both antibodies, and all other residues are encoded in the germline gene segments used by these antibodies. This conservation suggests that LysH93 appeared early in the process of antibody evolution in a germline antibody. The insertion of this residue into this hydrophobic microenvironment resulted in chemical reactivity that was efficient enough to be selectable. Once this covalent process appeared, the binding pocket did not further evolve toward high specificity as would be indicated by the selection of somatic variants of the germline sequence. The broad substrate specificity, (see Section 11.4) of aldolase antibodies prepared by reactive immunization is likely to be the result of the special ontogeny of antibodies induced by immunogens that form covalent bonds within the binding pocket during the induction. This broad specificity was unprecedented in the catalytic antibody area, standing in contrast to what is observed with catalysts prepared using transition-state analogs. Antibody catalysts generated from the transition-state analog approach typically exhibit highly complementary binding pockets that restrict the substrate specificity of the catalyst like many natural enzymes [24–28]. The antibody catalysts are typically very specific and only catalyze the reaction of a single substrate (or substrate combination). This is because, in the normal case of immunization, a series of somatic mutations usually leads to an increase in binding specificity toward the inducing antigen. In reactive immunization, however, there is no meaningful biological selection toward refinement of the active site once the covalent chemistry becomes fixed. Chemical reactivity then trumps conformational complementarity in binding.

The antipodal reactivity of aldolase antibodies can be correlated to the amino acid sequences of the catalysts. Aldolase antibodies 40F12 and 42F1 generated with hapten 7, which exhibit the same enantiopreference as antibodies 38C2 and 33F12 in the catalyzed reactions, also share a high homology with 38C2 and 33F12 in their amino

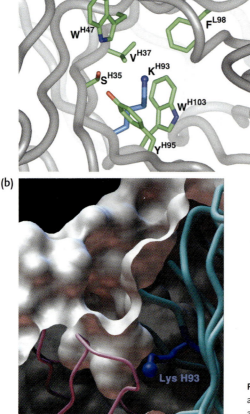

Fig. 11.1 X-ray structure of aldolase antibody 33F12. (a) Top view of the active site. (b) A slice through the molecular surface.

acid sequences and possess an essential lysine catalytic residue at the same position (H93) as 38C2 and 33F12 [29]. On the other hand, aldolase antibodies 93F3 and 84G3 generated with hapten 7, which exhibit antipodal reactivity to antibodies 38C2 and 33F12, place their reactive lysine in a position different from that observed in antibodies 38C2, 33F12, 40F12, and 42F1. Antibodies 93F3 and 84G3 possess an essential catalytic lysine residue at L89, as identified by site-directed mutagenesis and later in structural studies (Fig. 11.2).

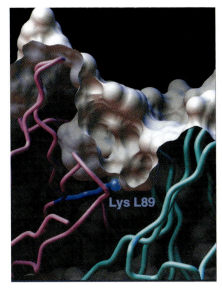

Fig. 11.2 X-ray structure of aldolase antibody 93F3.

11.4
Aldolase Antibody-catalyzed Reactions

11.4.1
Broad Scope and High Enantioselectivity

All aldolase antibodies generated by the immunization with diketones 4 and 7 are very broad in scope, accepting a wide variety of substrates. For example, aldolase antibodies 38C2 and 33F12 are capable of accelerating more than 100 different aldol reactions including aldehyde-aldehyde, ketone-aldehyde, and ketone-ketone reactions (Fig. 11.3) [10, 12–14, 18, 29–31]. For cross-aldol reactions, various ketones are accepted as donors, such as oaliphatic open chain (for example, acetone to pentanone), aliphatic cyclic (cyclopentanone to cycloheptanone), functionalized open chain (hydroxyacetone, dihydroxyacetone, fluoroacetone), and functionalyzed cyclic (2-hydroxycyclohexanone) ketones. The active-site lysine residue of the catalyst is able to convert these ketones into the corresponding enamines, which are key intermediates that are able to attack both aldehydes and ketones. As with the nucleophilic component, the antibodies also accept different kinds of aldehyde substrates as electrophiles, such as pentanal, 4-acetamidobenzaldehyde, or 2,4-hexadienal. The antibodies also catalyze self-aldol condensations of acetone or cyclopentanone provided that no acceptor aldehyde is present for a cross-aldol reaction. The antibodies also catalyze *retro*-aldol reactions [13, 14].

In addition to broad-scope substrate specificity, the antibody-catalyzed aldol and *retro*-aldol reactions are highly enantioselective in most cases. The absolute stereochemistry of the product aldol provided under antibody catalysis is controlled be the antibody and varies with the antibody used. Antibodies 38C2 and 33F12 generated

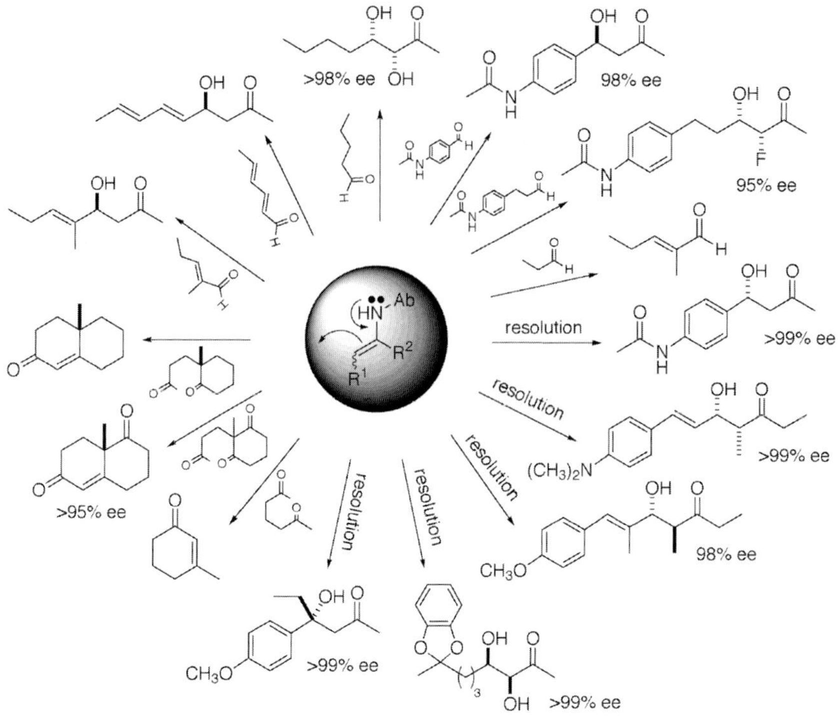

Fig. 11.3 Aldol reactions and kinetic resolutions catalyzed by aldolase antibody 38C2.

with diketone **4** and antibodies 40F12 and 42F1 generated with diketone **7** provide for the synthesis of the same aldol enantiomers. Antibodies 93F3 and 84G3 generated with diketone **7** belong to the antipodal family of catalysts and provide for the synthesis of the antipodal aldols as compared to the aldols provided under 38C2-, 33F12-, 40F12-, and 42F1-catalysis. The rules governing the enantioselectivity of these families of aldolases are simple and general (see Sections 11.4.2–11.4.4).

11.4.2
Aldol Reactions Catalyzed by Aldolase Antibodies 38C2 and 33F12

Aldolase antibodies 38C2 and 33F12 catalyze aldol reactions of a variety of substrates and these aldolase antibody-catalyzed reactions are frequently highly enantioselective (up to > 99% *ee*) [10, 12, 30, 31]. For example, antibody 38C2 catalyzes the aldol reaction of acetone and aldehyde **9** and provide (*S*)-**8** with an excellent enantioselectivity (> 99% *ee*) (Scheme 11.6). Aldols **11–15** can also be prepared with 38C2. The rules for the enantioselectivity of 38C2-catalyzed aldol reactions are simple and general, although this selectivity is not directly programmed in the diketone hapten used for the immunization. Asymmetric induction is a consequence of the asymmetry of the active site that directs the attack of the enamine intermediate while stereochemically

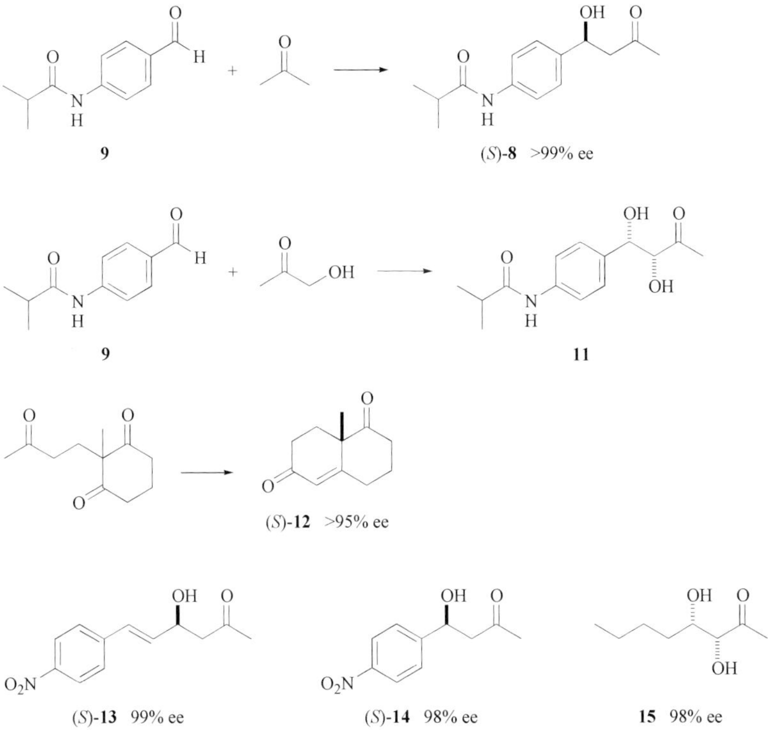

Scheme 11.6 38C2-catalyzed aldol reactions.

fixing the face of the acceptor aldehyde. With acetone as the aldol donor substrate, a new stereogenic center is formed by attack of the *si*-face of the aldehyde, as shown in the formation of **8**, **13**, and **14**. With hydroxyacetone as the donor substrate, attack occurs on the *re*-face of the aldehyde, as shown in the formation of **11** and **15**. The antibody-catalyzed reactions of hydroxyacetone as acceptor provide only the single regioisomer.

11.4.3
Antibody 38C2 and 33F12-catalyzed *retro*-Aldol Reactions and Their Application to Kinetic Resolutions

The antibodies also catalyze *retro*-aldol reactions of secondary [13] and tertiary aldols [14]. In the *retro*-aldol reactions, antibodies 38C2 and 33F12 operate on the product enantiomer of the aldol reaction [13]. Therefore, kinetic resolutions via *retro*-aldol reactions provide enantiomers of the opposite optical configuration as compared with the forward aldol reaction (Scheme 11.7). The antibodies accept a variety of substrates in *retro*-aldol reactions, and the recovered aldols are typically highly enantiomerically enriched [13, 14]. The degree of optical resolution can be readily modified experimentally. Significantly, the very mild nature of this approach allows for the resolution not

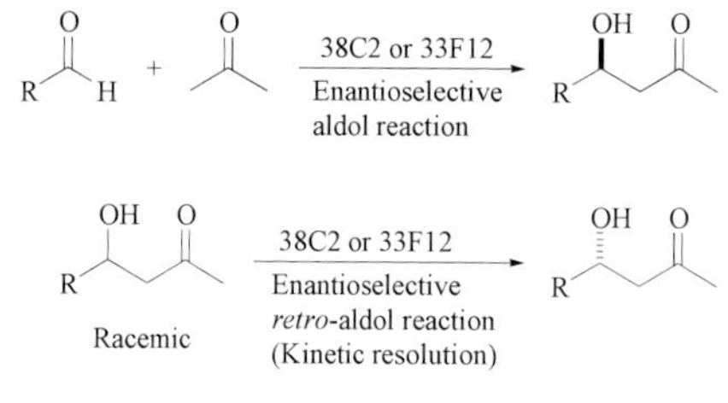

Scheme 11.7 Aldol and *retro*-aldol reactions in antibodies 38C2- and 33F12-catalyzed reactions.

only of keto aldols but also of aldehyde aldols (Scheme 11.8). For example, (*R*)-**8**, and (*R*)-**16**–**20** are obtained by the 38C2-catalyzed kinetic resolution of the racemic aldols. In the 38C2-catalyzed forward aldol reaction, (*S*)-**17** was obtained in 58% ee, while (*R*)-**18** was obtained in > 99% *ee* in the 38C2-catalyzed kinetic resolution after 67% conversion. A variety of enantiomerically enriched tertiary aldols can be prepared by the aldolase antibody-catalyzed kinetic resolutions; general access to optically active tertiary aldols is not available by traditional synthetic methods.

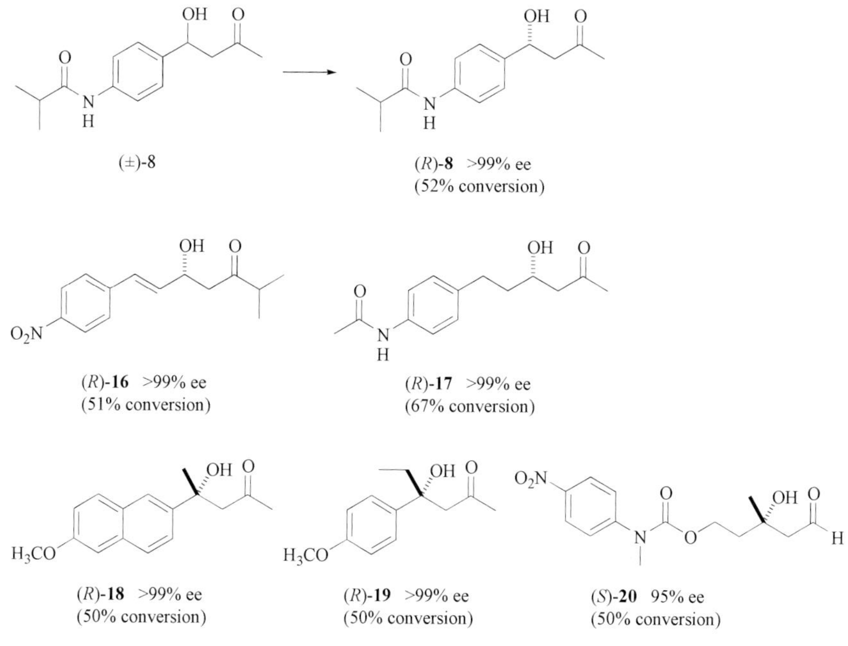

Scheme 11.8 Antibody 38C2-catalyzed kinetic resolutions.

11.4.4
Aldol and *retro*-Aldol Reactions Catalyzed by Aldolase Antibodies 93F3 and 84G3

Aldolase antibodies 93F3 and 84G3 also possess promiscuous active sites and catalyze reactions with a wide variety of substrates [16, 32, 33]. These antibodies provide an enantioselectivity opposite to that of selectivity to the 38C2 and 33F12. For example,

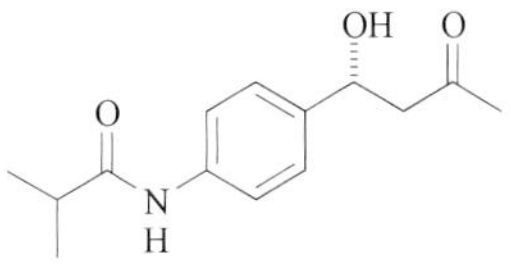

Scheme 11.9 Antibody 93F3- or 84G3-catalyzed aldol reaction products.

(R)-8 >99% ee by 93F3

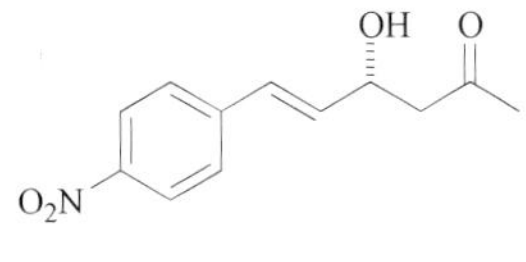

(R)-13 98% ee by 93F3

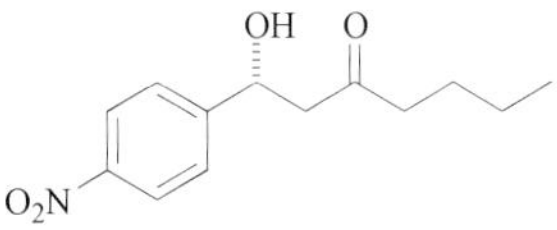

(R)-21 95% ee (dr 99:1) by 84G3

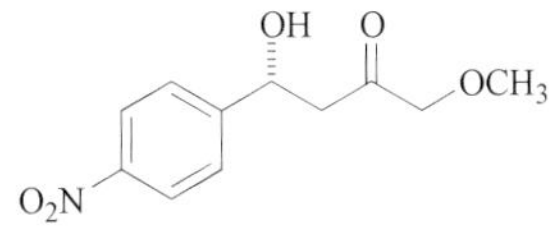

(R)-22 95% ee (dr 100:1) by 84G3

antibody 93F3- and 84G3-catalyzed aldol reactions provide (R)-**8**, (R)-**13**, (R)-**21**, and (R)-**22** (Scheme 11.9). When unsymmetrical ketones are used in antibody 84G3-catalyzed cross aldol reactions with aldehydes, the reactions occur exclusively at the less substituted carbon atom of the ketones, regardless of the presence of heteroatoms in the ketones: aldol products (R)-**21** and (R)-**22** are regioselectively obtained. This is a distinguishing feature, since antibodies 38C2 and 33F12 provide the corresponding regioisomeric mixtures and since the background reaction favors the formation of the other regioisomers. In contrast to antibodies 38C2 and 33F12, aldolase antibodies 93F3 and 84G3 are very poor catalysts for aldol reactions involving hydroxyacetone as a donor with aldehyde acceptors.

When the kinetic resolution of aldol (±)-**8** was preformed with antibody 93F3, (S)-**8** (> 99% ee) was obtained at 52% conversion, while antibody 38C2 provides (R)-**8** via kinetic resolution (Scheme 11.10). The catalytic proficiency of antibodies 93F3 and 84G3 exceeds that of antibody 38C2 in most cases, especially in *retro*-aldol reactions of 3-keto-5-hydroxy-type substrates, for example, **23**. In contrast to antibodies 38C2 and 33F12, antibodies 93F3 and 84G3 cannot process *retro*-aldol reactions of tertiary aldols.

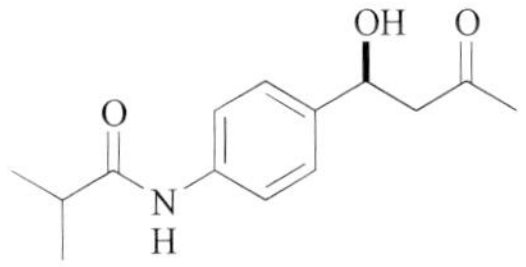

(S)-8 >99% ee
(52% conversion)

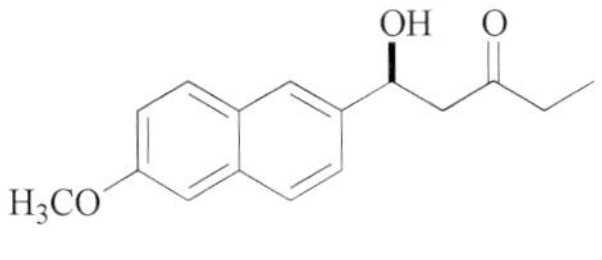

(S)-23 >99% ee
(50% conversion)

Scheme 11.10 Antibody 93F3- or 84G3-catalyzed *retro*-aldol reaction products.

Tab. 11.1: Preparative scale kinetic resolutions in biphasic system.

product	antibody	time	recovery	ee
(R)-24	38C2 (255 mg, 0.025 mol%)	88 h	1.55 g (49%)	>97%
(R)-18	38C2 (18 mg, 0.12 mol%)	193 h	22 mg (44%)	99%
(S)-23	84G3 (500 mg, 0.0086 mol%)	65 h	10 g (50%)	>99%

11.4.5
Preparative scale kinetic resolutions using aldolase antibodies in a biphasic aqueous/organic solvent system

Antibody-catalyzed reactions are typically performed in aqueous buffer, since catalytic antibodies function optimally in an aqueous environment. On the other hand, many organic molecules of interests exhibit low solubility in water. For the transformation of such molecules by aldolase antibodies, biphasic aqueous/organic solvent systems are useful, especially for large-scale reactions (Table 11.1) [34].

11.4.6
Aldolase Antibody-catalyzed Reactions in Natural Product Syntheses

Aldol reactions and kinetic resolutions using aldolase antibodies provide an efficient route to highly enantiomerically pure aldols. These processes were used for the total synthesis of cytotoxic natural products, epothilone A (**25**) and C (**26**), which exhibit a taxol-like mode of action, functioning through stabilization of cellular microtubules (Scheme 11.11) [35]. The structural moieties (+)-*syn*-**27** and (–)-**28** were prepared by antibody 38C2-catalyzed reactions and converted into the intermediates **29** and **30**, respectively. The key compound (–)-**28** (98% ee) was also obtained by kinetic resolution using antibodies 93F3 and 84G3 [36]. The chiral precursors for the syntheses of other epothilones have also been prepared by aldolase antibody-catalyzed reactions [36, 37].

Other natural product syntheses have capitalized on the 38C2-catalyzed addition of hydroxyacetone to install 1,2-*syn*-diol functionalities. In this respect, aldolase antibodies were used for the synthesis of brevicomins **31** and **32** (Scheme 11.12) [38], 1-deoxy-L-xylulose (**33**) (Scheme 11.13) [39], and (+)-frontalin [14].

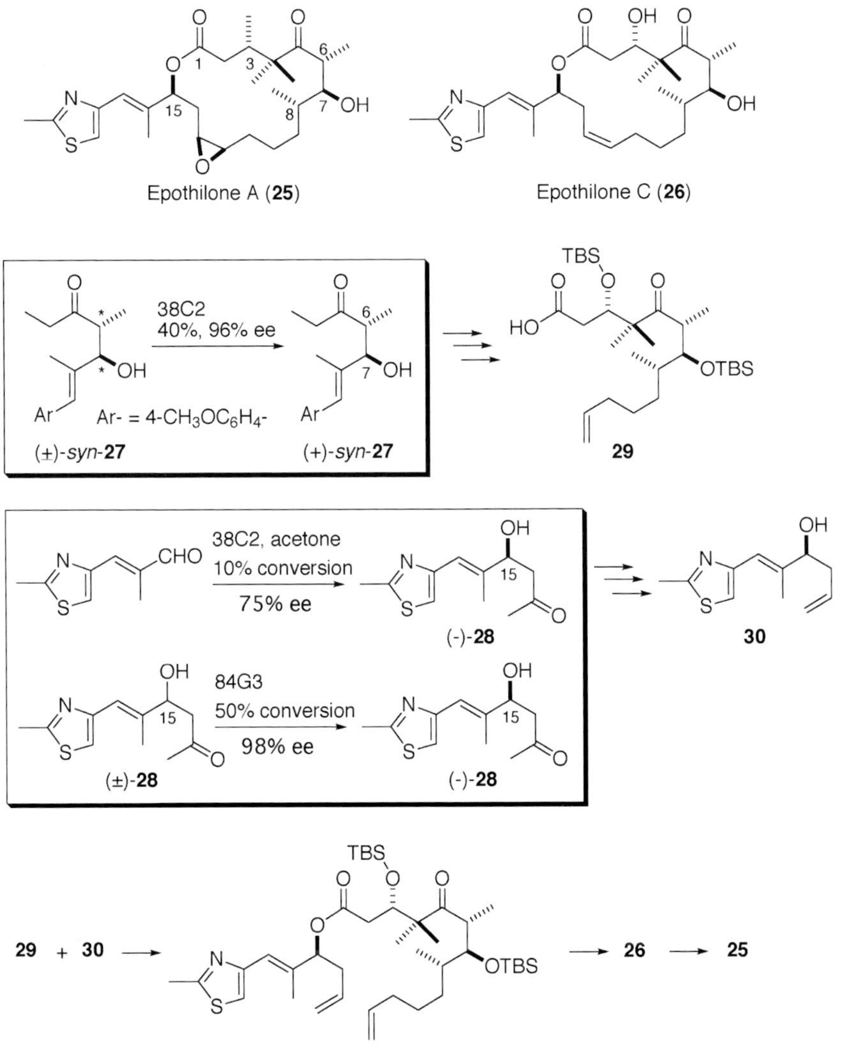

Scheme 11.11 Syntheses of epothilones using aldolase antibody-catalyzed reactions.

11.4.7
Aldolase Antibodies in Reactions Involving Imine and Enamine Mechanisms and Exploitation of the Nucleophilic Lysine ε-Amino Group

Since aldolase antibodies operate via an enamine mechanism, they also catalyze not only aldol and *retro*-aldol reactions but also other reactions that are processed via an enamine mechanism. For example, as described in the prodrug activation reactions, antibody 38C2 catalyzes β-elimination (*retro*-Michael) reactions (see Sections 11.4.8 and 11.4.9). Antibodies 38C2 and 33F12 also catalyze decarboxylation of β-keto acids

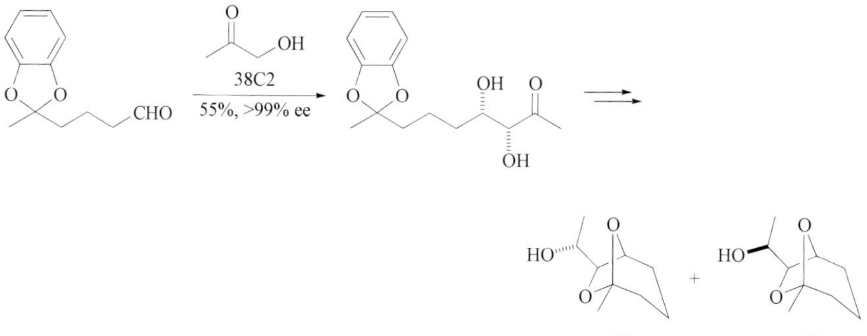

Scheme 11.12 Syntheses of brevicomins using aldolase antibody-catalyzed reactions.

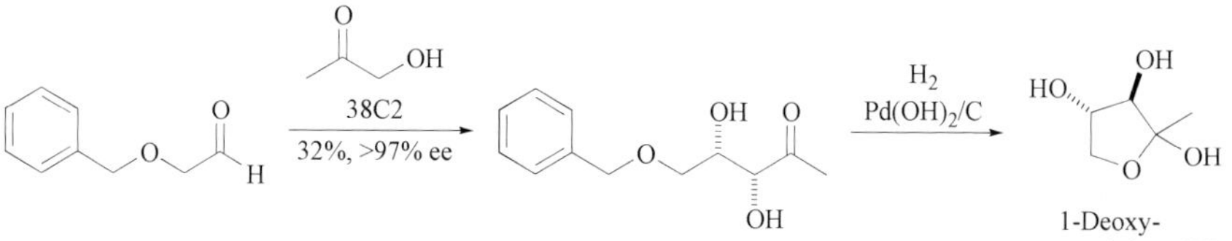

Scheme 11.13 Synthesis of 1-deoxy-L-xylulose using aldolase antibody-catalyzed reaction.

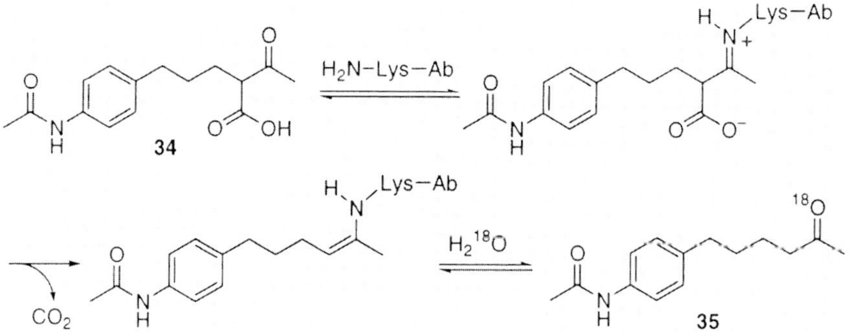

Scheme 11.14 Antibody 38C2-catalyzed decarboxylation.

(Scheme 11.14) [40], allylic rearrangement of steroids [41], and deuterium-exchange reactions [42].

In the antibody 38C2-catalyzed decarboxylation of **34**, incorporation of ^{18}O into the product **35** was observed in the presence of $H_2^{18}O$, consistent with decarboxylation proceeding via an enamine intermediate [40].

Antibody 38C2 catalyzes deuterium-exchange reactions at the α-position of a variety of ketones and aldehydes [42]. Since aldehydes bearing a longer alkyl chain (≥ valeraldehyde) are not the substrates for the 38C2-catalyzed self-aldol reactions, deuterium-exchanged aldehydes are accumulated in the presence of 38C2 in D_2O.

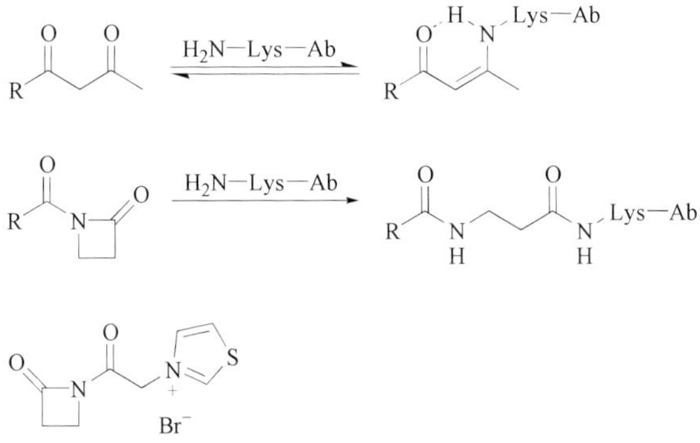

Scheme 11.15 Cofactor introduction at a unique active-site lysine of aldolase antibodies.

Aldolase antibodies can be covalently modified with cofactor derivatives at the active-site lysine residue for the catalyzing cofactor-dependent reactions [43, 44]. A variety of 1-acyl-β-lactam derivatives form the stable amide linkage to active-site lysine, while 1,3-diketones bind in a reversible manner (Scheme 11.15) [43]. Antibody 38C2 was modified with lactam **36**, and the modified antibody catalyzed thiazolium-dependent decarboxylation of PhCOCOOH. Antibody 38C2 was also covalently modified with the succinic anhydride derivative bearing bis-imidazole functionality that chelates Cu(II), and the modified antibody was used for the Cu(II)-dependent ester hydrolysis [44]. The effect of metal-cofactors on antibody 38C2-catalyzed aldol reactions has also been reported [45].

11.4.8
Concise Catalytic Assays for Aldolase Antibody-catalyzed Reactions

Spectroscopic or visible detection of antibody-catalyzed reactions enhances rapid characterization of catalysts on a small scale. Such detection systems are also useful for the high-throughput screening for new aldolase antibody catalysts in their development and in evolution of aldolase antibody *in vitro*. Examples of the substrates for spectroscopic or visible detection of the antibody-catalyzed reactions and their detectable methods are shown in Scheme 11.16. Substrate **37** was the first UV/VIS-active aldol substrate designed for following the *retro*-aldol reaction. *retro*-Aldolization of **37** results in the liberation of the yellow product **38** [13]. The establishment of this approach served as the basis for the development of fluorescent versions when fluorescent aldehydes and ketones were later identified [46]. Substrates **39**, **18**, and **40** liberate the fluorescent products **41**, **42**, and **43**, respectively, by the aldolase antibody-catalyzed *retro*-aldol reactions [46]. Thus the progress of the reactions with these fluorogenic substrates can be followed by fluorescence. A variety of other substrates have also been designed to facilitate the screening and characterization of aldolase antibodies; for example substrate **44** generates umbelliferone (**45**) by the *retro*-aldol-*retro*-Michael reaction [47, 48] while substrate **46** generates 2-naphthol derivative **47** by the *retro*-aldol-*retro*-Michael reaction, which forms a visible colored azo dye with diazonium salts [49]. Maleimide derivative **48** has recently been developed for the detection

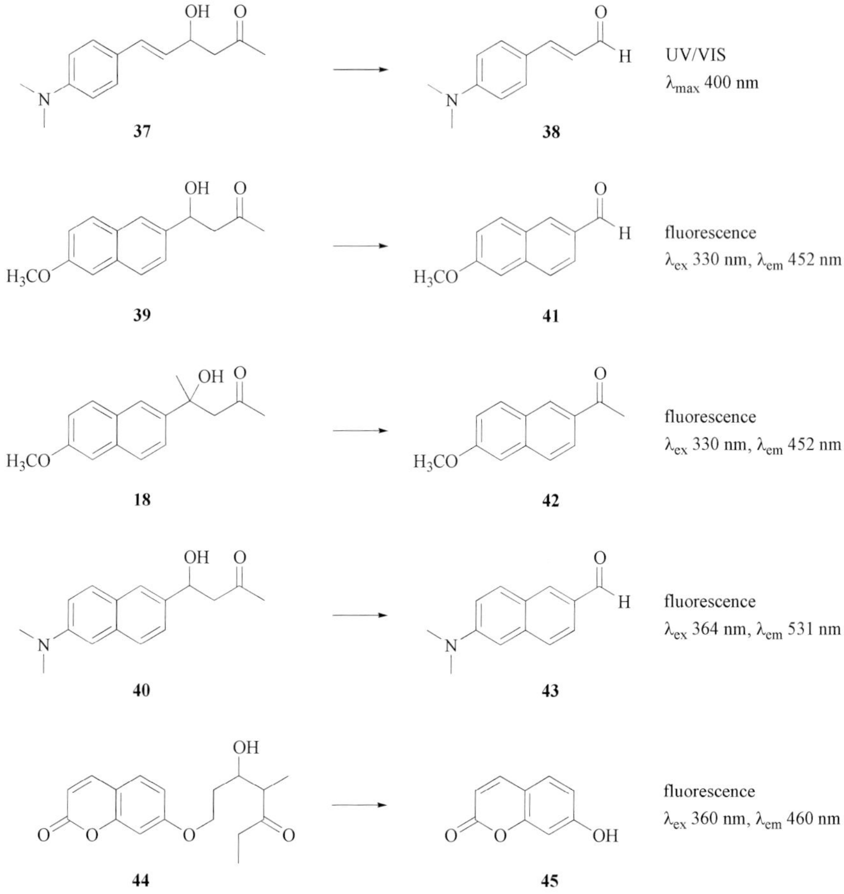

Scheme 11.16 Substrates for fluorescent and visible detection of aldolase antibody-catalyzed reactions.

of carbon-carbon bond formation catalyzed by aldolase antibodies [50]. It should be noted that all the other systems monitor carbon-carbon bond cleavage, not formation. The reaction of **48** with acetone provides **49**, which posesses a markedly increased fluorescence.

11.4.9
Prodrug Activation by an Aldolase Antibody

Aldolase antibody 38C2 has also been developed as a chemotherapeutic tool. In this approach we designed a unique prodrug masking chemistry that complemented the catalytic activity of the antibody, allowing it to work as a specific tool to mediate prodrug activation (Scheme 11.17) [51, 52]. The overall goal of this project is to develop a specific chemotherapeutic approach to cancer. We found that antibody 38C2

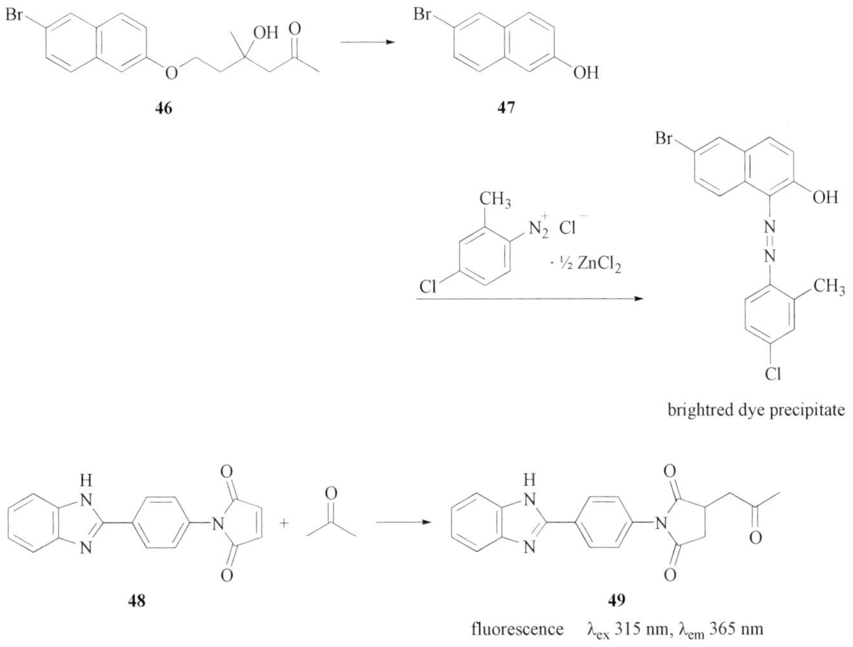

Scheme 11.16 (continued)

catalyzed the unmasking of doxorubicin prodrug **50**, which incorporates a trigger portion designed to be released by sequential *retro*-aldol-*retro*-Michael reactions [51]. Since this reaction cascade is not catalyzed by any known natural enzymes, masking of the anti-cancer drug at a key position can be used to substantially reduce its toxicity. Ideally then the toxicity of the drug can be selectively released at the site of the cancer, limiting undesirable peripheral toxicity that plagues current chemotherapeutic regimes. We found that the combination of prodrug **50** and antibody 38C2 strongly inhibited the growth of cancer cell lines, whereas the same concentration of **50** alone was far less potent. Camptothecin- and etoposide-prodrugs **51** and **52**, designed using the same principles, are also activated by 38C2 and can be applied in chemotherapeutic strategies. These studies led to the first application of catalytic antibodies in a disease model *in vivo* where we have demonstrated that our aldolase antibody-prodrug system can produce a profound therapeutic effect in animal models of cancer [52].

The prodrug activation strategy can also be applied in the context of protein activation. In this case, native insulin modified with aldol-terminated linkers at the primary amines crippled the biological activity of insulin. The modified insulin was defective with respect to receptor binding and stimulation of glucose transport. Antibody 38C2 cleaved the linker on insulin and restored insulin activity in an animal model of diabetes [53]. This approach might ultimately lead to therapeutic time-released versions of insulin or other protein drugs.

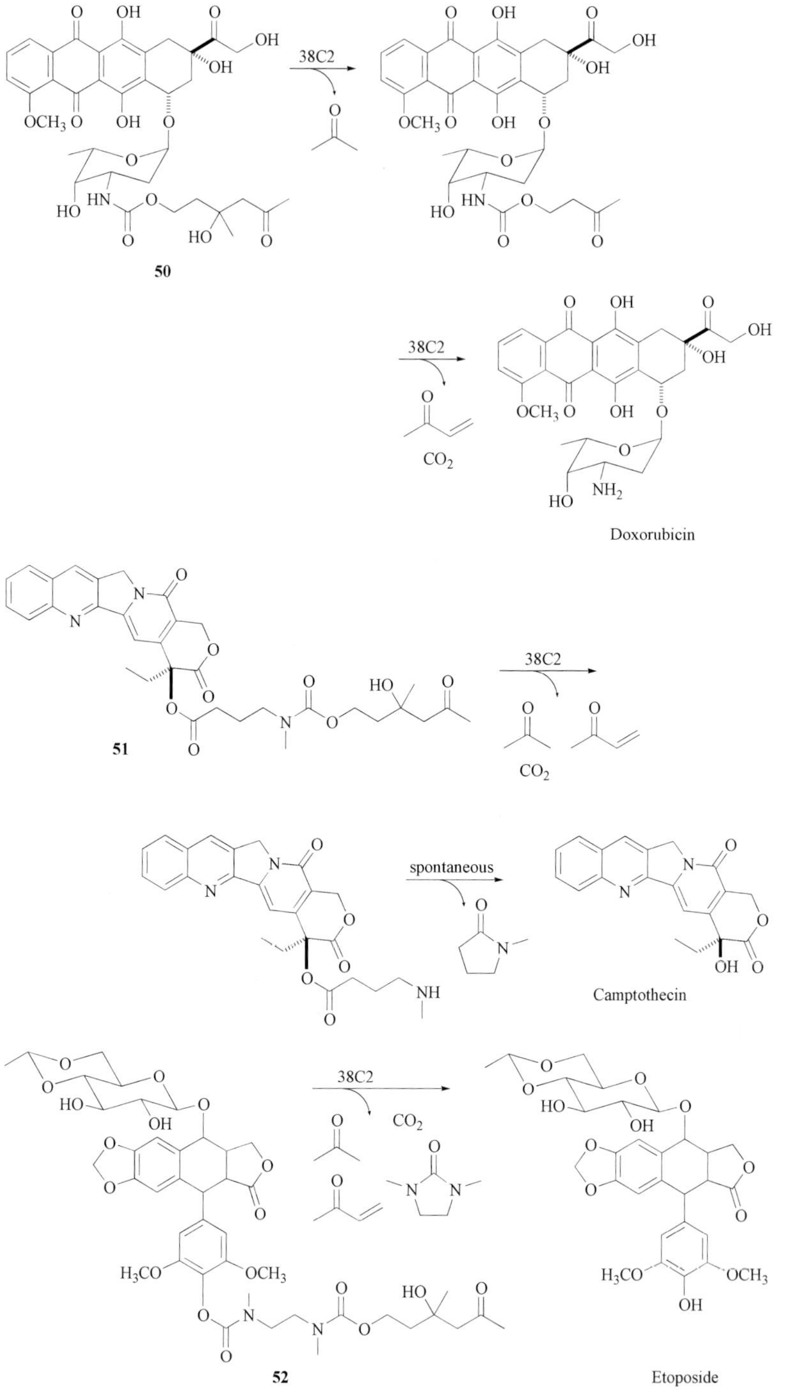

Scheme 11.17 Prodrug activation via a tandem *retro*-aldol-*retro*-Michael reaction catalyzed by antibody 38C2.

The most recent medical application of aldolase antibodies has capitalized on the unique reactivity of the active-site lysine residue. The approach is based on using the catalytic antibody as a device to equip small synthetic molecules with both effector function and the long serum half-life of a generic antibody molecule. As a prototype, we developed a targeting device that is based on the formation of a covalent bond of defined stoichiometry between a 1,3-diketone derivative of an integrin $α_vβ_3$ and $α_vβ_5$ targeting RGD peptidomimetic and the reactive lysine of aldolase antibody 38C2. The resulting complex was shown to (i) spontaneously assemble *in vitro* and *in vivo*, (ii) selectively retarget antibody 38C2 to the surface of cells expressing integrins $α_vβ_3$ and $α_vβ_5$, (iii) dramatically increase the circulatory half-life of the RGD peptidomimetic, (iv) effectively reduce tumor growth in animal models of human Kaposi's sarcoma and colon cancer. This immunotherapeutic has the potential to target a variety of human cancers, acting on both the vasculature that supports the tumor's growth as well as the tumor cells themselves. Further, by use of a generic antibody molecule that forms a covalent bond with a 1,3-diketone functionality, essentially any compound can be turned into an immunotherapeutic agent thereby not only increasing the diversity space that can be accessed but also multiplying the therapeutic effect [54]. As such, this approach promises a new class of pharmaceuticals that combines the merits of traditional small-molecule drugs and with the distinct advantages of antibodies.

11.5
Evolution of Aldolase Antibodies *in vitro*

11.5.1
Selection of Aldolase Antibodies with Diketones using Phage Display

Phage-displayed antibody libraries have been explored as an efficient strategy for the identification of monoclonal antibodies [55–59]. Since antibody phage libraries can be made from both immune and non-immune sources [60–63], chemical selection with reactive compounds using antibody phage libraries *in vitro* provides access to catalytic antibodies that are not limited by animal sources or immune responses.

Although aldolase antibodies are broad in scope, the efficiency with which any given aldol is processed can vary significantly. To access aldolase antibodies with altered substrate specificity and turnover, a strategy based on screening-designed phage libraries using different diketone derivatives has been developed. In this approach, libraries are prepared by recombining the catalytic machinery of aldolase antibodies with a naive V gene repertoire [64]. *In vitro* selection systems the provide for the use of multiple haptens without animal reimmunization, allowing the experimenter to combine insights gained by the study of existing catalytic antibodies with the diversity of the immune repertoire.

This strategy was used for preparing catalysts that would efficiently process cyclohexanone-aldols **53**, since *retro*-aldol reactions of **63** are slower than those involving acetone-aldols **54** (Scheme 11.18). The phage libraries were selected by panning against **55**- and **4**-bovine serum albumin (BSA). Fab 28 obtained from this selection

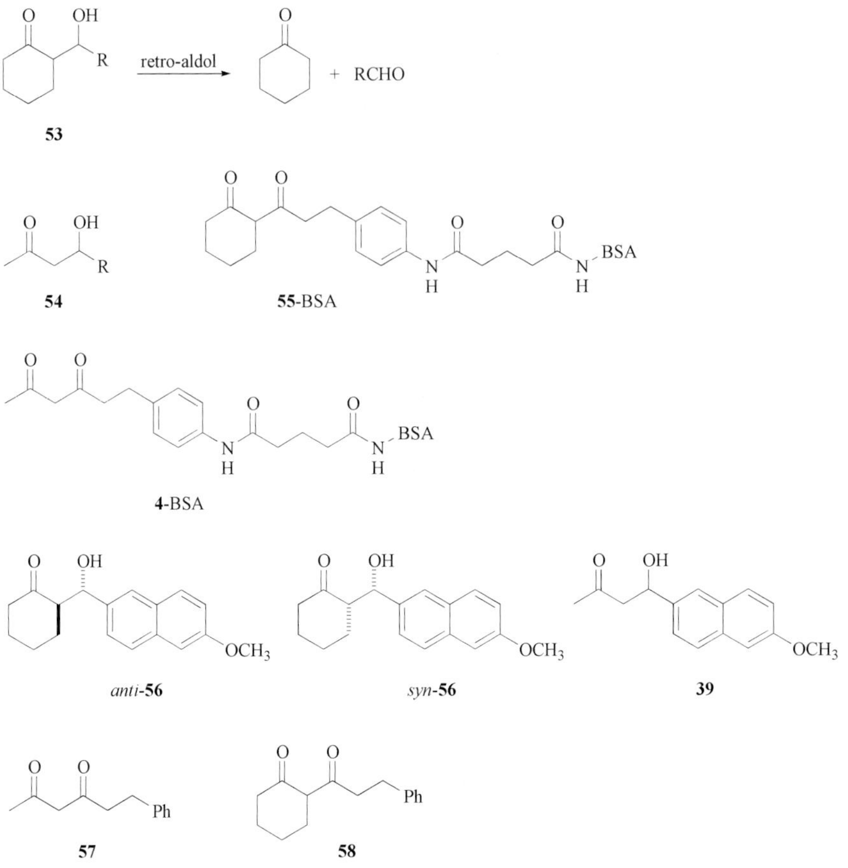

Scheme 11.18

catalyzed the *retro*-aldol reactions of *anti*-56, *syn*-56, and 39. The k_{cat} values of Fab 28 were superior to those of parental antibodies for cyclohexanone-aldols *anti*-56 and *syn*-56 by approximately 3- to 10-fold. In addition, Fab 28 catalyzed acetone-aldol 39 with a similar k_{cat} value to that of antibody 33F12. The stereochemistries of the preferred substrate enantiomers of Fab 28 are the same as those of the parental antibodies. Based on the design of the library, Fab 28 retained specific sequence elements of the parental antibodies and the essential LysH93 of the catalytic mechanism. The remaining primary sequence of Fab 28 is not related to the parental antibodies. Since a naive V gene library was generated using human bone marrow cDNA in this reconstruction, Fab 28 is a humanized aldolase antibody. This strategy is useful for providing humanized antibodies for the catalytic antibody-mediated prodrug activation described above.

To evaluate the library strategy and selection strategy with the diketones, the parental aldolase antibodies and the family of *in vitro* selected aldolase antibodies obtained with Fab 28 were analyzed. The *in vitro* selected aldolase antibodies also

11.5 Evolution of Aldolase Antibodies in vitro

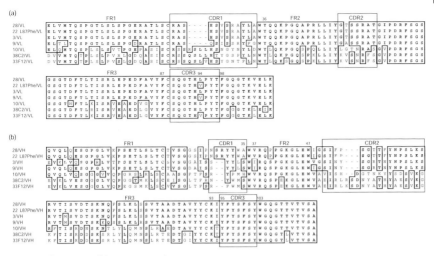

Fig. 11.4 Alignment of the amino acid sequences of aldolase antibodies (a) VL segments. (b) VH segments. CDRs are indicated.

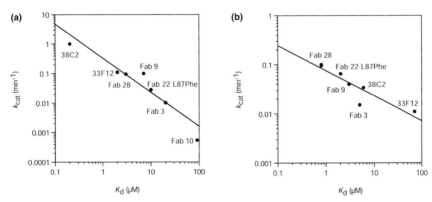

Fig. 11.5 (a) Plot of k_{cat} of retro-aldol reaction of **39** versus K_d of diketone **57**. (b) Plot of k_{cat} of retro-aldol reaction of anti-**56** versus K_d of diketone **58**.

formed the enaminone with the diketones and had sequence elements of the parental antibodies as expected based on the libraries' design as described for Fab 28 (Fig. 11.4) [65]. Correlation was observed between the k_{cat} of the antibody-catalyzed retro-aldol reaction and the apparent K_d of the corresponding diketones (i.e., the reactivity to the diketones) determined by competitive enzyme-linked immunosorbent assay within the family of aldolase antibodies [65]. A stronger binding (lower K_d value) to acetone-diketone **57**, i.e., a higher reactivity with acetone-diketone **57**, correlated with a higher k_{cat} value for the reaction of acetone-aldol **39**, and a stronger binding (lower K_d value) to cyclehexanone-diketone **58**, i.e., a higher reactivity with cyclohexanone-diketone

58, correlated with a higher k_{cat} value for the reaction of cyclohexanone-aldol *anti*-56 (Fig. 11.5) [65]. Selection using a structure-altered diketone provided catalytic antibodies that have altered substrate specificity as directed by the structure of the selecting diketone. The library strategy and selection strategy with compounds designed for catalysis are efficient for the evolution of aldolase antibodies utilizing an enamine-based reaction mechanism. 1,3-Diketones have also been used for phage selections of small peptides that catalyze aldol and *retro*-aldol reactions via an enamine mechanism [66, 67]. Selection with 1,3-diketones appears to be a general and effective route to catalysts that operate via an enamine mechanism.

11.5.2
Selection Systems for Aldolase Antibodies

Although *in vitro* selection strategies using 1,3-diketones can be used to access aldolase antibodies, direct selection of catalysis of both *retro*-aldol and *retro*-Michael reactions could result in improved catalysts for prodrug activation. For this, a visual detection system of antibody-catalyzed *retro*-aldol-*retro*-Michael reactions using substrate 46 (Scheme 11.16) has been developed to facilitate catalyst screening [49].

Genetic selection is also useful for the selection of catalysts. Antibody-catalyzed reaction of substrate 59, a prodrug, generates *p*-aminobenzoic acid (PABA), an essential metabolite for *E. coli*. When an *E. coli* strain that cannot synthesize PABA because of a genetic defect is provided with the gene for an aldolase antibody, the expression of the aldolase antibody can provide the *E. coli* with the ability to metabolize 59 and survive. Such a genetic selection has been shown in preliminary experiments to provide a growth advantage to a strain expressing aldolase antibody 38C2 (Scheme 11.19) [68]. Ideally, in the future, such a system would be used to rapidly evolve aldolase antibodies based on turn-over efficiency as well as substrate specificity.

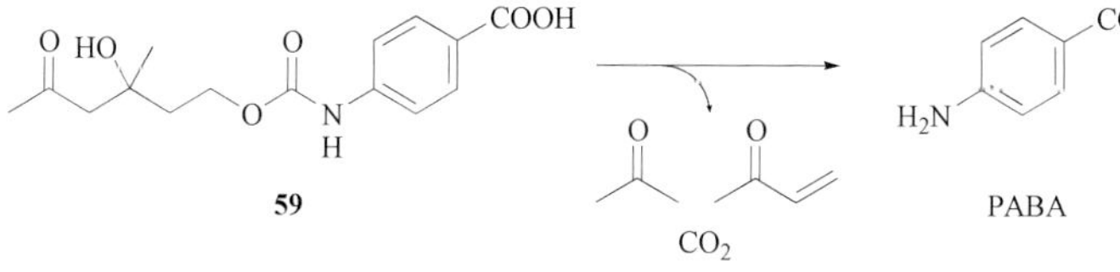

Scheme 11.19

11.6
Other Catalytic Antibodies Selected with Reactive Compounds *in vitro*

Since mechanism-based inhibitors react to form covalent adducts with the enzymes that process them along a defined reaction pathway, they should allow for the direct selection of catalytic antibodies that utilize particular features of a designed mechanism. A mechanism-based inhibitor of β-lactamase 60 was used to obtain β-lactam hydrolytic antibodies (Scheme 11.20a) [69]. The inhibitor 60-carrier protein conjugate was used for the immunization, and single-chain antibody (scFv) libraries were

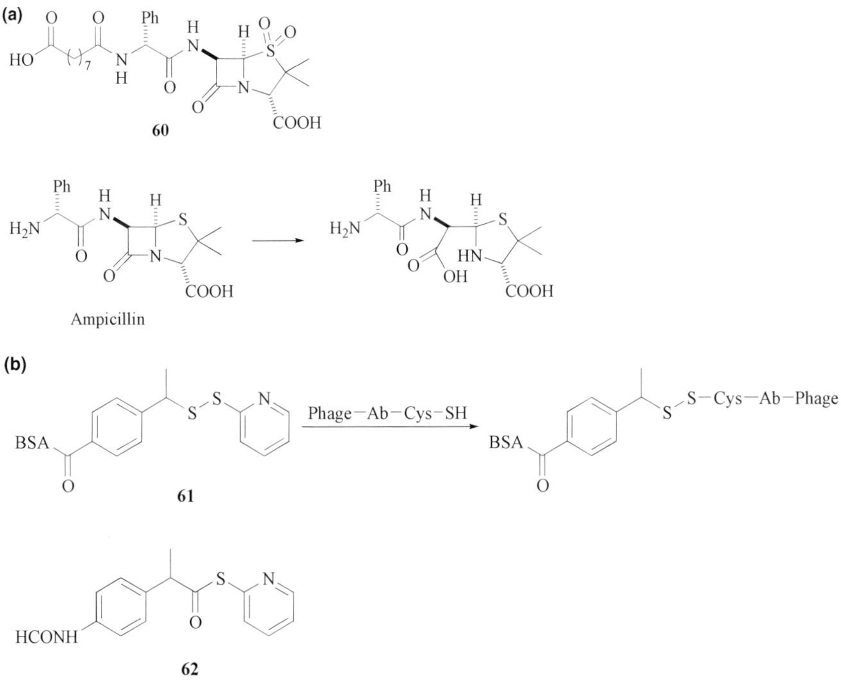

11.6 Other Catalytic Antibodies Selected with Reactive Compounds in vitro

Scheme 11.20 Selection of catalytic antibodies with reactive compounds.

prepared using the spleen of an immunized mouse. Selection of covalently binding antibodies against the inhibitor **60** was performed using the phage display system. Acidic washing conditions were used during the binding selection to remove non-covalently bound phage. The selected scFv FT6 catalyzed the ampicillin hydrolysis with a rate acceleration (k_{cat}/k_{uncat}) of 5200.

To select antibodies possessing a reactive cysteine residue within the active site, semisynthetic Fab libraries were panned with pyridyl disulfide **61**, which undergoes disulfide interchange (Scheme 11.20b) [70]. The selected Fab 32-7 has a sulfur nucleophile and forms a covalent intermediate with **62**. Antibodies that possesses glycosidase activity have also been obtained by a chemical selection (Scheme 11.20c) [71]. Transition state analog **63**-carrier protein conjugate was used for immunization, and hybridoma cells were prepared. Approximately 100 hybridoma clones that bound to **63** were used to prepare a combinatorial Fab library. The Fab phage libraries were selected with mechanism-based inhibitor **64**. A selected antibody Fab 1B catalyzed the hydrolysis of **65** with a rate acceleration (k_{cat}/k_{uncat}) of 7×10^4.

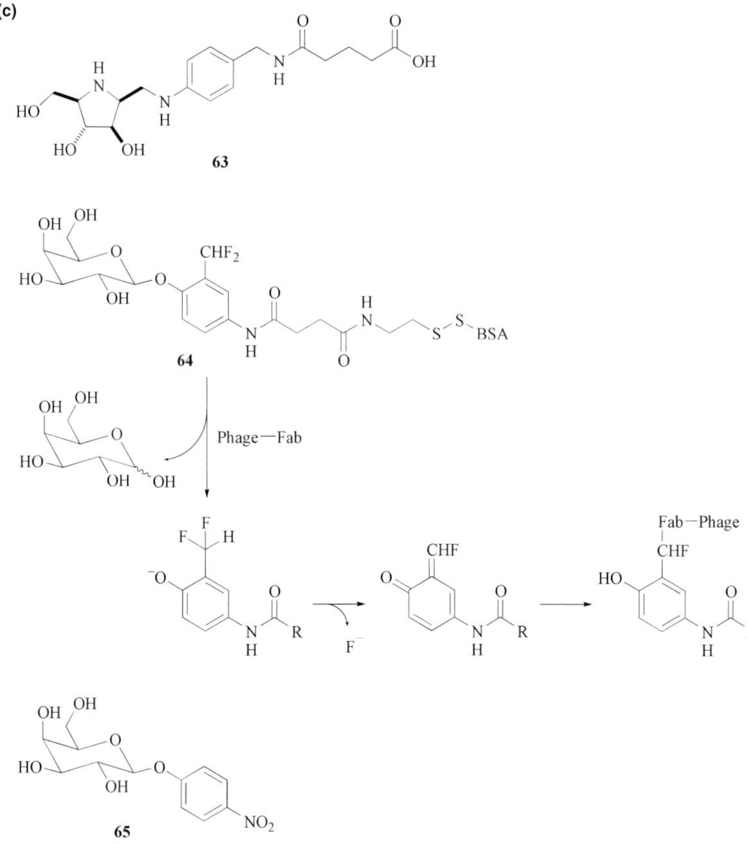

Scheme 11.20 (continued)

11.7
Summary

The development of the concept of reactive immunization and its application to the creation of aldolase antibodies *in vivo* has been discussed. The *in vitro* application of this strategy can be achieved using phage display selections with reactive compounds. Reactive immunization *in vivo* and reactive selection *in vitro* utilize designed reactive compounds that covalently react with antibodies during their induction and selection in such a way as to effectively program a reaction mechanism into the selected protein. Compounds designed for reactive immunization are not simple affinity labels but require chemical processing to ultimately trap the antibody catalyst. The process of reactive immunization then selects for a defined chemical reactivity and a reaction mechanism that can be multi-step. The direct consequence of this strategy is the development of efficient catalysts that operate via experimenter-defined mechanisms. This approach is not limited to antibodies but can be applied to other proteins, peptide catalysts, and nucleic acids. Some features of reactive immunization, particularly access to covalent reaction mechanisms, can be realized using *in vitro* selection protocols with designed compounds. Reactive immunization has provided highly proficient aldolase antibodies that had not been accessible by traditional immunization with transition state analogs. Broad scope, enhanced catalytic activity, and defined chemical mechanism are three features that distinguish antibodies derived from reactive immunization from those obtained by immunization with transition state analogs [72]. Antibodies 38C2 and 84G3 are commercially available from Aldrich.

11.8 Acknowledgments

We would like to thank the many wonderful scientists that have participated with us in the development of science presented in this chapter, in particular Richard Lerner. We would also like to thank the U.S. National Institute of Health and the Skaggs Institute for Chemical Biology for their financial support of these studies.

References

1 PAULING, L., Molecular architecture and biological reactions, *Chem. Eng. News* 24 (**1946**), p. 1375
2 PAULING, L., *Silliman Lecture*, Yale University Press, New Haven **1947**
3 JENCKS, W. P., *Catalysis in Chemistry and Enzymology*, McGraw-Hill, New York **1969**; Dover edition, New York **1987**, p. 288
4 JANDA, K. D., BENKOVIC, S. J., LERNER, R. A., Catalytic antibodies with lipase activity and R or S substrate selectivity, *Science* 244 (**1989**), p. 437
5 POLLACK, S. J., HSIUN, P., SCHULTZ, P. G., Stereospecific hydrolysis of alkyl esters by antibodies, *J. Am. Chem. Soc.* 111, (**1989**) p. 5961
6 TRAMONTANO, A., JANDA, K. D., LERNER, R. A., Antibody catalysis approaching the activity of enzymes, *J. Am. Chem. Soc.* 110 (**1988**), p. 2282
7 WIRSHING, P., ASHLEY, J. A., LO, C.-H. L., JANDA, K. D., LERNER, R. A., Reactive immunization, *Science* 270, (**1995**), p. 1775
8 LAI, C. Y., NAKAI, N., CHANG, D., Amino acid sequence of rabbit muscle aldolase and the structure of the active center, *Science* 183 (**1974**), p. 1204
9 MORRIS, A. J., TOLAN, D. R., Lysine-146 of rabbit muscle aldolase is essential for cleavage and condensation of the C3-C4 bond of fructose 1,6-bis(phosphate), *Biochemistry* 33 (**1994**), p. 12291
10 WAGNER, J., LERNER, R. A., BARBAS III, C. F. Efficient aldolase catalytic antibodies that use the enamine mechanism of natural enzymes, *Science* 270 (**1995**), p. 1797
11 TAGAKI, W., GUTHRIE, J. P., WESTHEIMER, F. H., Acetoacetate decarboxylase. Reaction with acetopyruvate, *Biochemistry* 7 (**1968**), p. 905
12 HOFFMANN, T., ZHONG, G., LIST, B., SHABAT, D., ANDERSON, J., GRAMATIKOVA, S., LERNER, R. A., BARBAS III, C. F., Aldolase antibodies of remarkable scope, *J. Am. Chem. Soc.* 120 (**1998**), p. 2768
13 ZHONG, G., SHABAT, D., LIST, B., ANDERSON, J., SINHA, R. A., LERNER, R. A., BARBAS III, C. F., Catalytic enantioselective *retro*-aldol reactions: Kinetic resolution of β-hydroxyketones with aldolase antibodies, *Angew. Chem., Int. Ed.* 37 (**1998**), p. 2481
14 LIST, B., SHABAT, D., ZHONG, G., TURNER, J. M., LI, A., BUI, T., ANDERSON, J., LERNER, R. A., BARBAS III, C. F., A catalytic enantioselective route to hydroxy-substituted quaternary carbon centers: Resolution of tertiary aldols with a catalytic antibody, *J. Am. Chem. Soc.* 121 (**1999**), p. 7283
15 DHULMAN, H., MAKAROV, C., OGAWA, A. K., ROMESBERG, F., KEINAN, E., Chemically reactive immunogens lead to functional convergence of the immune response, *J. Am. Chem. Soc.* 122 (**2000**), p. 10743
16 ZHONG, G., LERNER, R. A., BARBAS III, C. F., Broadening the aldolase catalytic antibody repertoire by combining reactive immunization and transition state theory: New enantios-

electivities, *Angew. Chem. Int. Ed.* 38 (1999), p. 3738

17 BLACKBURN, G. M., GARCON, A., Catalytic antibodies, in KELLY, D. R. Ed., Biotechnology Vol. 8b, Biotechnology Biotransformations II Wiley-VCH, Weinheim 2000

18 BARBAS III, C, F., HEINE, A., ZHONG, G., HOFFMANN, T., GRAMATIKOVA, S., BJORNESTSDT, R., LIST, B., ANDERSON, J., STURA, E. A., WILSON, I. A., LERNER, R. A., Immune versus natural selection: antibody aldolases with enzymic rates but broader scope, *Science* 278 (1997), p. 2085

19 DAVIS, D. R., PADLAN, E. A., SHERIFF, S., Antibody-antigen complexes, *Annu. Rev. Biochem.* 59 (1990), p. 439

20 AREVALO, J. H., TAUSSING, M. J., WILSON, I. A., Molecular basis of cross reactivity and the limits of antibody-antigen complementarity, *Nature* 365 (1993), p. 859

21 HSIEH-WILSON, L. C., SCHULTA, P. G., STEVENS, R., Insights into antibody catalysis: Structure of an oxygenation catalyst at 1.9-Å resolution, *Proc. Natl. Acad. Sci. U.S.A.* 93 (1996), p. 5363

22 WEDEMAYER, G. J., WANG, L. H., PATTEN, P. A., SCHULTZ, P. G., STEVENS, R., Crystal structures of the free and liganded form of an estrolytic catalytic antibody, *J. Mol. Biol.* 268 (1997), p. 390

23 LIDE, D. R. ED., CRC Handbook of Chemistry and Physics 83rd Edition, CRC Press, New York 2002, p. 7-1

24 FUJII, I., TANAKA, F., MIYASHITA, H., TANIMURA, R., KINOSHITA, K., Correlation between antigen-combining-site structures and functions within a panel of catalytic antibodies generated against a single transition state analog, *J. Am. Chem. Soc.* 117 (1995), p. 6199

25 TANAKA, F., KINOSHITA, K., TANIMURA, R. FUJII, I., Relaxing substrate specificity in antibody-catalyzed reactions: Enantioselective hydrolysis of N-Cbz-amino acid esters, *J. Am. Chem. Soc.* 118 (1996), p. 2332

26 WEDEMAYER, G. J., PATTERN, P. A., WANG, L. H., SCHULTZ, P. G., STEVENS, R. C., Structural insights into the evolution of an antibody combining site, *Science* 276 (1997) p. 1665

27 YIN, J., MUNDORFF, E. C., YANG, P. L., WENDT, K. U., HANWAY, D., STEVENS, R. C., SCHULTZ, P. G., A comparative analysis of the immunological evolution of antibody 28B4, *Biochemistry* 40 (2001), p. 10764

28 TANAKA F., Catalytic antibodies as designer proteases and esterases, *Chem. Rev.* 102 (2002), 4885

29 KARLSTROM, A., ZHONG, G., RADER, C., LARSEN, N. A., HEINE, A., FULLER, R., LIST, B., TANAKA, F., WILSON, I. A., BARBAS III, C. F., LERNER, R. A., Using antibody catalysis to study the outcome of multiple evolutionary trials of a chemical task, *Proc. Natl. Acad. Sci. U.S.A.* 97 (2000), p. 3878

30 ZHONG, G., HOFFMANN, T., LERNER, R. A., DANISHEFSKY, S., BARBAS III, C. F., Antibody-catalyzed enantioselective Robinson annulation, *J. Am. Chem. Soc.* 119 (1997), p. 8131

31 LIST, B., LERNER, R. A., BARBAS III, C. F., Enantioselective aldol cyclohydrations catalyzed by antibody 38C2, *Org. Lett.* 1 (1999), p. 59

32 MAGGIOTTI, V., RESMINI, M., AND GOUVERNEUR, V., Unprecedented regiocontrol using an aldolase I antibody, *Angew, Chem. Int. Ed.* 41 (2002), p. 1012

33 MAGGIOTTI, V., WONG, J.-B., RAZET, R., COWLEY, A. R., GOUVERNEUR, V., Asymmetric synthesis of aldol products derived from unsymmetrical ketones: assignment of the absolute configuration of antibody aldol products, *Tetrahedron: Asymmetry* 13 (2002), p. 1789

34 TURNER, J. M., BUI, T., LERNER, R. A., BARBAS III, C. F., LIST, B., An efficient benchtop system for multigram-scale kinetic resolutions using aldolase antibodies, *Chem. Eur. J.* 6 (2000), p. 2772

35 SINHA, S., BARBAS III, C. F., LERNER, R. A., The antibody catalysis route to the total synthesis of epothilones, *Proc. Natl. Acad. Sci. U.S.A.* 95 (1998), p. 14603

36 SINHA, S., SUN, J., MILLER, G., BARBAS III, C. F., LERNER, R. A., Sets

of aldolase antibodies with antipodal reactivities. Formal synthesis of epothilone E by large-scale antibody-catalyzed resolution of thiazole aldol, *Org. Lett.* 1 **(1999)**, p. 1623

37 SHINHA, S, C.; SUN, J., MILLER, G. P., WARTMANN, M., LERNER, R. A., Catalytic antibody route to the naturally occurring epothilones: Total synthesis of epothilones A-F, *Chem. Eur. J.* 7 **(2001)**, p. 1691

38 LIST, B., SHABAT, D., BARBAS III, C. F., LERNER, R .A., Enantioselective total synthesis of some Brevicomins using aldolase antibody 38C2, *Chem. Eur. J.* 4 **(1998)**, p. 881

39 SHABAT, D., LIST, B., LERNER, R. A., BARBAS III, C. F., A short enantioselective synthesis of 1-deoxy-L-xylulose by antibody catalysis, *Tetrahedron Lett.* 40 **(1999)**, p. 1437

40 BJORNESTEDT, R., ZHONG, G., LERNER, R.A., BARBAS III, C. F., Copying nature's mechanism for the decarboxylation of β-keto acids into catalytic antibodies by reactive immunization, *J. Am. Chem. Soc.* 118 **(1996)**, p. 11720

41 LIN, C.-H., HOFFMAN, T. Z., WIRSHING, P., BARBAS III, C. F., JANDA, K. D., LERNER, R. A., On roads not taken in the evolution of protein catalysts: antibody steroid isomerases that use an enamine mechanism, *Proc. Natl. Acad. Sci.* 94 **(1997)**, p. 11773

42 SHULMAN, A., SITRY, D., SHULMAN, H., KEINAN, E., Highly efficient antibody-catalyzed deuteration of carbonyl compounds, *Chem. Eur. J.* 8, **(2002)**, p. 229

43 TANAKA, F., LERNER, R. A., BARBAS III, C. F., Thiazolium-dependent catalytic antibodies produced using a covalent modification strategy, *Chem. Commun.* **(1999)**, p. 1383

44 NICHOLAS, K. M., WENTWORTH, P. JR., HARWIG, C. W., WENTWORTH, A. D., SHAFTON, A., JANDA, K. D., A cofactor approach to copper-dependent catalytic antibodies, *Proc. Natl. Acad. Sci. U.S.A.* 99 **(2002)**, p. 2648

45 FINN, M. G., LERNER, R. A., BARBAS III, C. F., Cofactor-induced refinement of catalytic antibody activity: a metal-specific allosteric effect, *J. Am. Chem. Soc.* 120 **(1998)**, p. 2963

46 LIST, B., BARBAS III, C. F., LERNER, R. A., Aldol sensors for the rapid generation of tunable fluorescence by antibody catalysis, *Preoc. Natl. Sci. Acad. U.S.A.* 95 **(1998)**, p. 15351

47 JOURDAIN, N., CARLON, R. P., REYMOND, J.-L., A stereoselective fluorogenic assay for aldolases: detection of an anti-selective aldolase catalytic antibody, *Tetrahedron Lett.* 39 **(1998)**, p. 9415

48 CARLON, R. P., JOURDAIN, N., REYMOND, J.-L., Fluorogenic polypropionate fragments for detecting stereoselective aldolases, *Chem. Eur. J.* 6 **(2000)**, p. 4154

49 TANAKA, F., KERWIN, L., KUBITZ, D., LERNER, R. A., BARBAS III, C. F., Visualizing antibody-catalyzed *retro*-aldol-*retro*-Michael reactions, *Bioorg. Med. Chem. Lett.* 11 **(2001)**, p. 2983

50 TANAKA, F., THAYUMANAVAN, R., AND BARBAS III, C. F., Fluorescent detection of carbon-carbon bond formation, *J. Am. Chem. Soc.* **(2003)**, submitted

51 SHABAT, D.. RADER, C., LIST, B., LERNER, R. A., BARBAS III, C. F., Multiple event activation of a generic prodrug trigger by antibody catalysis, *Proc. Natl. Acad. Sci. U.S.A.* 96 **(1999)**, p. 6925

52 SHABAT, D., LODE, H. N., PERTL, U., REISFELD, R. A., RADER, C., LERNER, R. A., BARBAS III, C. F., *In vitro* activity in a catalytic antibody-prodrug system: Antibody catalyzed etoposide prodrug activation for selective chemotherapy, *Proc. Natl. Acad. Sci. U.S.A.* 98 **(2001)**, p. 7528

53 WORRALL, D. S., MCDUNN, J. E., LIST, B., REICHART, D., HEVENER, A., GUSTAFSON, T., BARBAS III, C. F., LERNER, R. A., OLEFSKY, J. M., Synthesis of an organoinsulin molecule that can be activated by antibody catalysis, *Proc. Natl. Acad. Sci. U.S.A.* 98 **(2001)**, p. 13514

54 RADER, C., SINHA, S., POPKOV, M., LERNER, R. A., BARBAS III, C., F., Chemically programmed monoclonal antibodies for cancer therapy: adaptor immunotherapy based on a cova-

lent antibody catalyst, *Proc. Natl. Sci. Acad. U.S.A.* 100 (**2003**), p. 5396

55 BARBAS III, C. F., KANG, A. S., LERNER, R. A., BENKOVIC, S. J., Assembly of combinatorial antibody libraries on phage surfaces: the gene III site, *Proc. Natl. Acad. Sci. U.S.A.* 88 (**1991**), p. 7978

56 BARBAS III, C. F., BURTON, D. R., Selection and evolution of high-affinity human anti-viral antibodies, *Trends in Biotechnology* 14 (**1996**), p. 230

57 RADER, C., BARBAS III, C. F., Phage display of combinatorial antibody libraries, *Curr. Opin. Biotechnol.* 8 (**1997**), p. 503

58 BARBAS III, C. F., RADER, C., SEGAL, D., LIST, B., TURNER, J. M., From catalytic asymmetric synthesis to the transcriptional regulation of genes: *in vivo* and *in vitro* evolution of proteins, *Advances in Protein Chemistry* 55 (**2000**), p. 317

59 BARBAS III, C. F., BURTON, D. R., SCOTT, J. K., SILVERMAN, G. J. EDS., *Phage Display: A Laboratory Manual*, Cold Spring Harbor Laboratory Press, Cold Spring Harbor, New York **2001**

60 BURTON, D. R., BARBAS III, C. F., Human antibodies from combinatorial libraries, *Adv. Immunol.* 57 (**1994**), p. 191

61 BARBAS III, C. F., Synthethic human antibodies, *Nat. Med.* 1 (**1995**), p. 837

62 RADER, C., CHERESH, D. A., BARBAS III, C. F., A phage display approach for rapid antibody humanization: Designed combinatorial V gene libraries, *Proc. Natl. Acad. Sci. U.S.A.* 95 (**1998**), p. 8910

63 ANDRIS-WIDHOPF, J., STEINBERGER, P., BARBAS III, C. F., Bacteriophage display of combinatorial antibody libraries. *Encycl. of Life Sci.* (**2001**), www.els.net

64 TANAKA, F., LERNER, R. A., BARBAS III, C. F., Reconstructing aldolase antibodies to alter their substrate specificity and turnover, *J. Am. Chem. Soc.* 122 (**2000**), p. 4835

65 TANAKA, F., FULLER, R., SHIM, H., LERNER, R. A., BARBAS III, C. F., Evolution of aldolase antibodies *in vitro*: correlation of catalytic activity and reaction-based selection, *J. Mol. Biol.* 335 (**2004**), p. 1007

66 TANAKA, F., BARBAS III, C. F. Phage display selection of peptides possessing aldolase activity, *Chem. Commun.*, p. 769 (**2001**)

67 TANAKA, F., BARBAS III, C. F. A modular assembly strategy for improving the substrate specificity of small catalytic peptides, *J. Am. Chem. Soc.* 124 (**2002**), p. 3510

68 GILDERSLEEVE, J., JANES, J., ULRICH, H., YANG, P., TURNER, J., BARBAS, C., SCHULTZ, P. G., Development of a genetic selection for catalytic antibodies, *Bioorg. Med. Chem. Lett.* 12 (**2002**), p. 1691 and 2789

69 TANAKA, F., ALMER, H., LERNER, R. A., BARBAS III, C. F., Catalytic single-chain antibodies possessing β-lactamase activity selected from a phage displayed combinatorial library using a mechanism-based inhibitor, *Tetrahedron Lett.* 40 (**1999**), p. 8063

70 JANDA, K. D., LO C.-H. L., LI. T., BARBAS III, C. F., WIRSCHING, P., LERNER, R. A., Direct selection for a catalytic mechanism from combinatorial antibody libraries, *Proc. Natl. Acad. Sci. U.S.A.* 91 (**1994**), p. 2532

71 JANDA, K. D., LO, L.-C., LO, C.-H., SIM, M.-M., WANG, R., WONG, C.-H., LERNER, R. A., Chemical selection for catalysis in combinatorial antibody libraries, *Science* 275 (**1997**), p. 945

72 TANAKA, F., BARBAS III, C. F., Reactive immunization: a unique approach to catalytic antibodies, *J. Immunol. Methods* 269 (**2002**), p. 67

12
The Antibody-Catalyzed Water Oxidation Pathway
Cindy Takeuchi, Paul Wentworth Jr.

A central concept within immunology is that antibodies are the key molecular link between recognition and destruction of antigens/pathogens [1]. The antibody catalysis field introduced by Lerner [2, 3] and Schultz [4], which is being celebrated in this volume, has demonstrated that the antibody molecule is capable of performing sophisticated chemistry, but there was no compelling evidence that antibodies use this catalytic potential in their normal immune function [5]. Recently, however, we have uncovered a reaction that all antibodies can catalyze, regardless of source or antigenic specificity. This reaction involves the oxidation of water by singlet oxygen ($^1\Delta_g$) via a pathway, which is postulated to include trioxygen species such as dihydrogen trioxide (H_2O_3) and ozone (O_3), to the ultimate product, hydrogen peroxide (H_2O_2) [6–10]. A significant amount of theoretical and experimental evidence has been generated that supports this postulated pathway, which we have termed the "water-oxidation pathway". We have shown that oxidants generated by antibodies can kill bacteria [8], and maybe generated by activated human neutrophils [8, 10]. This chapter details this research from its initial discovery to our present understanding.

At an early stage we were alerted to the key involvement of the low-energy singlet state of oxygen, $^1O_2^*$, as the substrate for this reaction. Preliminary kinetic studies revealed that the rate of formation of H_2O_2 by antibodies is increased in deuterated water (D_2O), reduced in sodium azide (NaN_3) and proportional to the UV absorbance profile of the protein [6]. Subsequent studies revealed that the $^1O_2^*$ could be generated by either direct UV irradiation of the antibody molecule, by visible light and a $^3O_2$ sensitizer such as hematoporphyrin IX, or by thermal decomposition of endoperoxides, and in each case would trigger the antibody-catalyzed formation of H_2O_2 (Fig. 12.1).

The efficiency of H_2O_2 formation by antibodies, upon long-term UV irradiation, is unparalleled by non-immunoglobulin proteins. Typically, other proteins display a short burst of H_2O_2 production followed by a rapid quenching in rate as photooxidation occurs. Antibodies typically exhibit linear formation of H_2O_2 for up to 40 mole equivalents of H_2O_2 before the rate declines. It appears that H_2O_2 reversibly inhibits its own formation (apparent $IC_{50} \approx 225\ \mu M$), and antibodies can resume photo-production of H_2O_2 at the same initial rate if catalase is added. However, if a cycle is established that involves photo-irradiation of an immunoglobulin solution in

Catalytic Antibodies. Edited by Ehud Keinan
Copyright © 2005 WILEY-VCH Verlag GmbH & Co. KGaA, Weinheim
ISBN: 3-527-30688-9

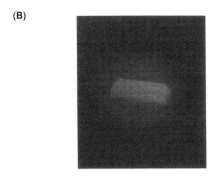

Fig. 12.1 **(A)** Antibody-catalyzed water oxidation by $^1O_2^*$. The singlet oxygen can be generated either by thermal decomposition of endoperoxides, irradiation of antibody samples with UV light, or irradiation of a sensitizer with visible light. The initial read-out for hydrogen peroxide was a horseradish peroxidase (HRP)-based assay using the conversion of resorufin to the fluorescent dye Amplex™ Red. **(B)** Fluorescent micrograph of a single crystal of murine antibody 1D4 Fab fragment after UV irradiation and H_2O_2 detection with the Amplex™ red reagent.

aqueous buffer followed by removal of the generated H_2O_2 by catalase and further photo-irradiation, > 500 equivalents of H_2O_2 can be generated by each antibody molecule.

The photo-production of > 500 equivalents of H_2O_2 from $^1O_2^*$, per antibody molecule, creates a critical electron inventory problem. Preliminary studies to identify the electron source ruled out a metal-mediated redox process, amino acid oxidation, and chloride ion. At this point our accumulated observations pointed to the involvement of a non-obvious electron source that does not deactivate the protein catalyst and that could account for the high turnover numbers, hence that is quasi-unlimited.

Isotopic labeling experiments were then undertaken to determine the oxygen source for the hydrogen peroxide. The content of $^{16}O/^{18}O$ in H_2O_2 formed by antibodies in either $H_2^{18}O$ and with $^{16}O_2$ or *vice versa* was determined by trapping experiments with triscarboxyethylphosphine. These studies revealed that water is oxidized by $^1O_2^*$, demonstrated by the incorporation of oxygen from water into H_2O_2 upon UV irradiation of antibodies. Under saturating $^{16}O_2$ conditions in a solution of $H_2^{18}O$ phosphate buffer, the relative abundance of the $^{16}O/^{18}O$ ratio observed in the MS of the phosphine oxide after irradiation of sheep polyIgG was (2.2 ± 0.2):1. When the converse experiment was performed with an ^{18}O-enriched molecular oxygen mixture in $H_2^{16}O$ phosphate buffer, the reverse ratio, 1:(2.0 ± 0.2), was observed. These ratios exhibited good reproducibility and were equivalent for all antibodies studied.

Thus, the isotope incorporation experiments suggested that a molecule of water, in the presence of an antibody, may hypothetically add as a nucleophile to $^1O_2^*$ and form dihydrogentrioxide (H_2O_3) as an intermediate on a pathway that ultimately leads to H_2O_2. Thus, water, in becoming oxidized to H_2O_2, fulfills the role of the electron source.

Having postulated such a reaction course, we sought thermodynamic information on such a pathway [Eqs. (1A)–(1C)]. It was clear that the thermodynamics regulating the balance between reactants and products for the oxidation of H_2O by $^1O_2^*$, which has an enthalpy of reaction given by ΔH_r = +28.1 kcal/mol [Eq. (1A)], demands a stoichiometry in which more than one molecule of $^1O_2^*$ participates per molecule of oxidized water during its conversion into two molecules of H_2O_2. Qualitative chemical reasoning on hypothetical mechanistic pathways, together with thermodynamic considerations, leads to a likely overall stoichiometry in accordance with either Eq. (1B) or Eq. (1C) (heats of formation are reported in kcal/mol):

$$^1O_2^* + 2H_2O \rightarrow 2H_2O_2 \quad \Delta H_r^\circ = +28.1 \quad (1A)$$
$$2\,^1O_2^* + 2H_2O \rightarrow 2H_2O_2 + {}^3O_2 \quad \Delta H_r^\circ = +5.6 \quad (1B)$$
$$3\,^1O_2^* + 2H_2O \rightarrow 2H_2O_2 + 2\,^3O_2 \quad \Delta H_r^\circ = -16.9 \quad (1C)$$

A fundamental assumption for the antibody-mediated process is the addition of a water molecule to a molecule of $^1O_2^*$ to form an H_2O_3 intermediate on the pathway to H_2O_2. Dihydrogen trioxide has received considerable attention in the chemical literature [11–17]. Plesnicar has shown that H_2O_3, reductively generated from ozone, decomposes into H_2O and $^1O_2^*$ in a process catalyzed by a water molecule [14]. Applying the principle of microscopic reversibility, we anticipated that one or more molecules of water should also catalyze the reverse reaction. Using quantum chemical methods, Xu et al. [18] have calculated the reaction enthalpies (ΔH_r°) and activation enthalpies (ΔH_a°) for the formation of H_2O_3 from $^1O_2^*$ and water [Eqs. (2A)–(2C); values are reported as kcal/mol).

While the direct reaction of $^1O_2^*$ with H_2O has an activation barrier too high for it to play an important role in the observed process (> 60 kcal/mol) [Eq. (2A)], involvement of a second water molecule as a catalyst provides an energetically more feasible route [Eq. (2B)]. A concerted process with two water molecules decreases the reaction barrier to 31.2 kcal/mol, and involvement of a third water decreases it further to 12.4 kcal/mol [Eq. (2C)].

$$H_2O + {}^1O_2^* \rightarrow H_2O_3 \quad \Delta H_r^\circ = +16.0;\ \Delta H_a^\circ = +64.7 \quad (2A)$$
$$2H_2O + {}^1O_2^* \rightarrow [H_2O_3\ \ H_2O] \quad \Delta H_r^\circ = +4.7;\ \Delta H_a^\circ = +31.2 \quad (2B)$$
$$3H_2O + {}^1O_2^* \rightarrow [H_2O_3\ \ 2H_2O] \quad \Delta H_r^\circ = -5.4;\ \Delta H_a^\circ = +12.0 \quad (2C)$$

In addition, H_2O_3 forms stabilizing hydrogen-bonded complexes with water molecules, such that the equilibrium constant for the reaction increases as more water molecules are involved. Therefore, the antibody's function as a catalyst may involve providing a molecular environment that stabilizes this intermediate relative to its reversible formation and/or accelerates the consumption of the intermediate by channeling its conversion into H_2O_2. An essential feature of such an environment must consist of a constellation of organized water molecules at an active site conditioned by an antibody-specific surrounding.

$$2t\text{-}H_2O_3 \rightarrow 2H_2O_2 + {}^1O_2^* \quad \Delta H_a° = +57.8;\ \Delta H_r° = -1.9 \quad (3A)$$
$$2c\text{-}H_2O_3 \rightarrow H_2O_2\text{-}O_3 \quad\quad \Delta H_a° = +12.0;\ \Delta H_r° = -10.0 \quad (3B)$$

While direct conversion of two resulting H_2O_3 molecules into two H_2O_2 molecules involves an intermediate too high in energy ($\Delta H_a° \approx 57.8$ kcal/mol) to be viable in the antibody-catalyzed process [Eq. (3A)], there are a number of possible pathways for the conversion of H_2O_3 into H_2O_2. An alternative route with a low barrier involves the disproportionation of an HOOOH dimer [Eq. (3B)], which gives a linear HOO-HOOO intermediate that allows for subsequent rearrangement to an active $H_2O_2\text{-}O_3$ complex. Subsequent formation of H_2O_4 from this complex with the aid of another catalytic water molecule precedes decomposition into H_2O_2 and ${}^3O_2$. The 2.2:1 ${}^{16}O/{}^{18}O$ incorporation ratio coincides exactly with the value predicted for mechanisms in which two molecules of ${}^1O_2^*$ and two molecules of H_2O are transformed into two molecules of H_2O_2 and one molecule of ${}^3O_2$. Thus, our experimental evidence, supported by quantum chemical analysis, leads to strong support for the oxidation of water via H_2O_3.

Host defense against infection requires the integrated use of several systems in innate and adaptive immunity. Mechanisms of innate immunity include phagocytosis by macrophages and neutrophils, through which engulfment and digestion of microorganisms occurs. This and other preexisting mechanisms act as a first line of host defense, but lack the element of specific recognition and the ability to provide protective memory associated with adaptive immunity. In contrast, clonal expansion of lymphocytes induces the generation of antigen-specific effector cells and receptors in response to an infection that has overwhelmed innate mechanisms. Acting synergistically, humoral and functions lead to the production of protective memory cells to increase defenses against reinfection by the same organism. Immunoglobulins are the recognition molecules produced and secreted by differentiated B lymphocytes, which specifically recognize and bind extracellular pathogens and toxins, selectively targeting the offending species for their removal. The traditional view of the antibody component of this system contends that antibodies do not carry out any sophisticated chemistry, but rather mark antigens for destruction by remote effector systems [1]. This position assumes that separate entities must be the executors of recognition and killing events. For example, antibody binding to bacteria in the bloodstream can initiate binding and cleavage of a series of complement proteins, ultimately yielding an attack complex that is capable of damaging membranes and recruiting phagocytic cells to the site of antigen-antibody union. Opsonizaton of organisms by antibody can also activate cellular effectors such that the membrane-bound NAD(P)H oxidase of phagocytes initiates a chemical cascade that reduces molecular oxygen to reactive oxygen species, thereby killing phagocytosed organisms [19, 20]. The discovery of the water-oxidation pathway raised an important question, i.e., whether antibodies could also behave as bactericidal agents when presented with a chemical or biological source of $^1O_2^*$. Our preliminary bactericidal studies focused on two strains of the gram-negative bacteria *Escherichia coli*, the common laboratory serotype XL1-blue, and an enteroinvasive strain O-112a,c, which can infect malnourished and immuno-compromised individuals. Given the known microbicidal action of $^1O_2^*$ itself, a $^1O_2^*$ generating system was chosen such that it would not, on its own, kill *E. coli*. Conditions including white light irradiation of hematoporphyrin IX had negligible bactericidal activity against the two serotypes. In a typical experiment, bacteria from a log phase culture were washed and resuspended in PBS. After cooling to 4 °C in glass vials, the photosensitizer and antibody were added and the samples exposed to white light for 1 h. Viability was determined by recovery of colony-forming units (CFUs) on agar plates. Subject to these conditions, whether the antibodies used are antigen-specific or non-specific, over 99% killing of bacteria is observed (Fig. 12.2).

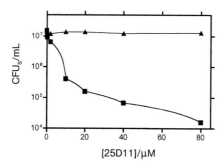

Fig. 12.2 Graph showing antibody-dependent microbicidal action of murine monoclonal antibody 25D11.

This activity directly reflects differences in antibody concentration as well as irradiation time and sensitizer concentration (hematoporphyrin IX). The observations support the vital role of both $^1O_2^*$ and the water-oxidation pathway in the antibody-mediated process. The bactericidal potential of antibodies appears to be general. Investigation of 12 murine monoclonals of various isotypes and one rabbit polyclonal IgG that are specific for *E. coli* cell-surface antigens shows each to be bactericidal. Given that the activation of the water-oxidation pathway is independent of antibody-antigen union and requires only $^1O_2^*$, it follows that of 10 non-specific murine monoclonal antibodies, one non-specific sheep and one horse polyclonal IgG studied, which have no specificity for *E. coli* cell-surface antigens, all generate bactericidal agents.

The morphology of the killed bacteria was examined by electron microscopy. Gold-labeled secondary antibodies were used to establish a correlation between morphological damage and sites to which cell wall-specific antibodies were bound. The formation of holes in the bacterial cell wall is in fact seen at the sites of antibody binding. The process appears to be a gradual one, as evidenced by the range of bacterial morphologies sampled. Distinct stages in the bactericidal pathway may also be interpreted, in which increasing cell wall and plasma membrane permeability is assumed to result from oxidative damage. A slight disruption between the cell wall and cytoplasm is first observed; this becomes more distinct, eventually separating the cell wall from the cytoplasmic contents of the bacterium. Given that the bacterium is under an internal pressure of ca. 30 atmospheres, weakening of the membrane by any mechanism leads to catastrophic rupture of the cell well and plasma membrane. In this regard, it is interesting that the observed morphologies induced by antibody-mediated killing are similar to those seen when bacteria are destroyed by phagocytosis.

Given that H_2O_2 is ultimately produced by the antibody-catalyzed oxidation pathway [6–9], by logical progression one might presume that it acts as the killing agent. This hypothesis was bolstered by the finding that catalase affords complete protection against the bactericidal activity of non-specific antibodies. However, experiments on two E. coli cell lines reveal that the toxicity of H_2O_2 is insufficient to account for the potent bactericidal activity derived from the antibodies. The amount of H_2O_2 generated by non-specific antibodies, 35 ± 5 µM, lies 1–4 orders of magnitude below that required to kill 50% of tested strains. The implications of the experiments with catalase were thus reevaluated. Hydrogen peroxide could feasibly react with another chemical species generated by the antibody to produce the bactericidal molecule(s), and thus, by destroying H_2O_2, catalase prevents their formation. Alternatively, bactericidal species that are formed on the way to H_2O_2 may also be substrates for catalase.

Reasoning suggested that an oxidant more powerful than H_2O_2 was produced by antibodies during the water-oxidation pathway. Subsequent analysis identified an ozone (O_3)-like intermediate, and all data to date support this as the sought oxidant. Under aqueous conditions, ozone is quite long lived ($t_{1/2} = 66$ s) and thus can be detected by sensitive chemical probes such as indigo carmine (**1**) [21, 22]. In the presence of chemically generated ozone, **1** is oxidized to the cyclic α-ketoamide **2** with bleaching of the characteristic absorbance of **1** ($\lambda_{max} = 610$ nm, $\varepsilon = 20\,000$ $LM^{-1}cm^{-1}$) (Scheme 12.1).

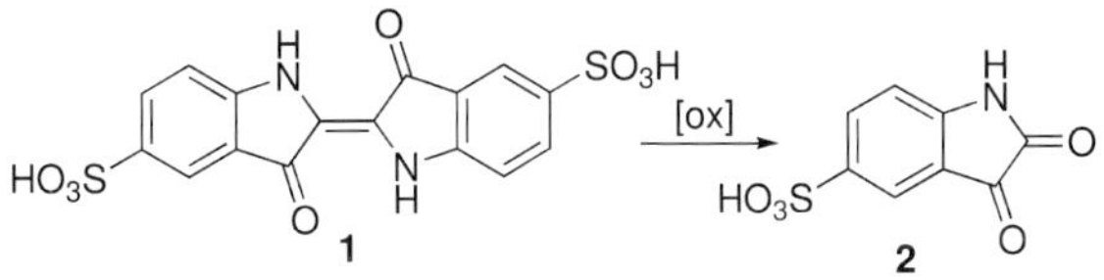

Scheme 12.1 Oxidation of indigo carmine (**1**) into isatin sulfonic acid (**2**).

When a solution of **1** in PBS is irradiated with UV light, no bleaching is observed. However, the same experiment carried out in the presence of a polyclonal or monoclonal antibody gives catalase-insensitive bleaching of **1**, with initial rates of antibody-mediated conversion of **1** into **2** that are linear. Electrospray mass-spectrometry and HPLC analyses confirmed formation of product **2**. While indigo carmine is a sensitive probe for ozone it is not completely selective. Singlet oxygen, $^1O_2^*$, among other oxidants, is also found to oxidatively cleave **1** to **2** under aqueous conditions. Differential reactivity with water, however, allows discrimination between the two. Singlet oxygen adds across the central double bond to form a dioxetane that is cleaved to form the two amide oxygens of **2** without incorporation from water. In contrast, the carbonyl-oxide intermediate generated during the O_3-mediated oxidation is trapped by water molecules, leading to incorporation of oxygen from solvent into the amide carbonyl of the α-ketoamide **2**.

The observations with the indigo carmine probe **1** strongly implicate ozone in the antibody-mediated oxidation of water. This claim was further strengthened with experiments with 3- and 4-vinyl-benzoic acid (**3** and **4** respectively) (Fig. 12.3).

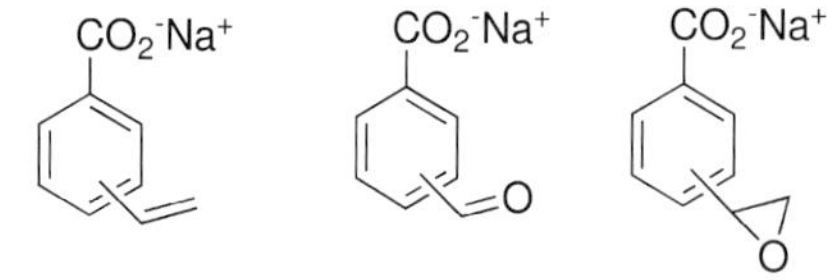

Fig. 12.3 3- and 4-Vinyl benzoic acid (**3, 4**) and oxidation products (**5a–6b**) generated by ozonolysis.

3-vinyl **3**   3-carboxy **5a**   3-oxiranyl **6a**
4-vinyl **4**   4-carboxy **5b**   4-oxiranyl **6b**

Irradiation of solutions of **3** or **4** with UV light, in the presence of either a murine monoclonal antibody or a sheep polyclonal IgG, leads to the formation of 3-carboxy-benzaldehyde (**5a**) and 3-oxiranyl benzoic acid (**6a**) (ratio 15:1, 1.5% conversion of **5**) or 4-carboxybenzaldehyde **5b** and 4-oxiranyl-benzoic acid **6b** (ratio of 10:1, 2% conversion relative to **6**). These products would all be expected to arise from reaction of **3** and **4** with ozone. By contrast, $^1O_2^*$ generated by HPIX and visible light, does not cause any detectable oxidation of either **5** or **6** under these conditions.

Ozone itself is known to be highly bactericidal. Moreover, the formation of ozone from the trioxygen species H_2O_3 by either disproportionation, or further oxidation by $^1O_2^*$ is expected to proceed via, or be accompanied by, the formation of the hydrotrioxy radical ($HO_3\bullet$), a species deserving special consideration in the context of radicals that display the chemical signature of the notorious hydroxyl radical (HO•) [23]. Since the end product of the antibody-catalyzed water oxidation pathway is H_2O_2, its interaction with ozone could also act as a source of the very same radical.

Ozone is known to decompose in the presence of H_2O_2 [24] to form oxidants that include species with the signature of HO• [25, 26]. Therefore we instigated a search for evidence that the hydroxy radical or a "masked" form of this radical was generated when the water-oxidation pathway was activated. Among our large collection of X-ray structures of antibodies, murine monoclonal antibody 4C6 [27] turned out to be of special interest in the context of the search for the radicals generated during the antibody-mediated water-oxidation pathway, because it was discovered to bind one molecule of the widely used hydroxyl radical probe, benzoic acid, within the antibody combining site. Interestingly, it was also noted that the X-ray crystal structure of Fab 4C6 contains a unique water channel that links the interfacial region of the constant and variable domains (where we have speculated that the chemistry of the water-oxidation pathway [see below] occurs [7]) to the combining site that contains the bound benzoic acid. It was considered that the antibody-bound benzoic acid, being so sequestered in this "molecular reaction chamber", may be expected to react *in situ* with short-lived reactive species that, because of their low diffusion radius and high reactivity, might otherwise not be detected by the use of external probes in solution. When sodium benzoate (2 mM) and antibody 4C6 (20 µM) in PBS is irradiated with UV light (312 nm, 0.8 µW cm^{-2}), regioselective hydroxylation of benzoic acid occurs to yield exclusively *para*-hydroxybenzoic acid. No *ortho*- or *meta*-isomers were detected. The observed initial rate of formation of the *para*-isomer is ca. 0.8 µM/min, and

Fab 4C6 generates > 10 mole equivalent (10% of substrate) of *para*-hydroxybenzoic acid without a significant decrease in this initial rate. The addition of catalase (13 U/mL) during the irradiation of 4C6 does not affect the hydroxylation reaction. The product conversion is 10% (relative to the starting concentration of benzoic acid) after 3 h of irradiation. By contrast, no hydroxylation of sodium benzoate occurs either by irradiation in PBS, or when exposed to $^1O_2^*$, generated by irradiation of HPIX (40 µM) with visible light (2.7 mW cm^{-2}). Similarly, no hydroxylation is observed when benzoic acid is irradiated with UV light (312 nm, 0.8 µW cm^{-2}) in the presence of H_2O_2 (1 mM). Significantly, when this same experiment is repeated with an antibody that does not bind benzoic acid, murine IgG 6G6 [28], no hydroxylation of benzoate is observed. The combined experimental observations that catalase does not affect the 4C6-catalyzed hydroxylation process and that hydroxylation by non-specific antibodies, such as 6G6, is not detectable under our assay conditions, suggest that the hydroxylating species is being generated within the antibody fold and is sequestered and/or very short-lived. The reaction of HO• radical with benzoic acid in aqueous medium has been intensively investigated and is known to lead to an isomeric mixture of hydroxy-isomers whose constitution is dependent on the method of formation of the radical species, the pH of the media, and the species that oxidizes the intermediary cyclohexadienyl radical. For example, Klein et al. [29] observed that at pH 7, with O_2 as the oxidant, the observed *o:m:p*: ratio was 1:0.6:0.5. Therefore, the fact that the hydroxylation reaction of benzoic acid with 4C6 is highly regioselective is consistent with a process that occurs when the benzoic acid is bound within the antibody-combining site.

While water oxidation generates such powerful oxidants as H_2O_2, O_3, H_2O_3, and HO_3•, the antibody structure itself seems remarkably inert toward these molecules. SDS-PAGE analysis of antibody samples revealed that UV irradiation for 8 h does not significantly fragment or agglomerate the antibody. Also, the overlaid native and H_2O_2-treated structures of the murine Fab 4C6 are superimposable at the level of side-chain positions, reinforcing the evidence of stability of the antibody fold to H_2O_2. Beside antibodies, the only other protein known to generate H_2O_2 catalytically is the αβ T cell receptor (αβ TCR), which shares a similar arrangement of its immunoglobulin fold domains with antibodies. However, possession of this structural motif does not necessarily confer an H_2O_2-generating ability on proteins. For example $β_2$-microglobulin, although a member of the immunoglobulin superfamily, does not generate H_2O_2.

Given that the ability to mediate this reaction is conserved in all antibodies and the αβ TCR, X-ray structural studies were instigated to locate a common reaction site within these immunoglobulin fold proteins. Potential loci were expected to have molecular oxygen (either $^1O_2^*$ or $^3O_2$ with a potential sensitizing residue in proximity, preferably Trp) co-localized with water, and the predicted transition states and intermediates along the pathway to be stabilized were expected to be either within the site or in close proximity. Xenon gas was used as a heavy atom tracer to locate cavities that may be accessible to O_2 within the murine monoclonal antibody 4C6. Three xenon sites (Xe1, Xe2, and Xe3) were identified, and all occupy hydrophobic cavities, as observed in other Xe-binding sites in proteins. Upon overlaying the refined native and Xe-derivatized structures, little discernible change is seen in the protein backbone or side-chain conformation or in the location of bound water molecules. The Xe1 site is conserved in all studied antibodies and in the αβ TCR. Xe1 is located within a highly conserved region between the β-sheets of the variable domain of the light chain (V_L), 7 Å from an invariant Trp. This site is critical to the immunoglobulin fold composition of the V_L, approximately 5 Å from the outer molecular surface. Xe2 sits at the base of the antigen-binding pocket, directly above several highly conserved residues that form the structurally conserved interface between the heavy and light chains of an antibody. The residues in the interface between the heavy and light chain of the variable region ($V_L V_H$) are primarily hydrophobic and include conserved aromatic side chains, such as Trp^{H109}.

The contacting side chains for Xe1 in Fab 4C6 are Ala^{L19}, Ile^{L21}, Leu^{L73}, and Ile^{L75}, which are highly conserved aliphatic side chains in all antibodies; additionally, only slight structural variation was observed in this region in all antibodies surveyed. Notably, several other highly conserved and invariant residues are in the immediate vicinity of this xenon site, including Trp^{L35}, Phe^{L62}, Tyr^{L86}, Leu^{L104}, and the disulfide-bridge between Cys^{L23} and Cys^{L88}. Trp^{L35} stacks against the disulfide-bridge and is only 7 Å from the xenon atom. In this structural context, Trp^{L35} may be a putative molecular oxygen sensitizer, since it is the closest Trp to Xe1. Comparison with the 2C αβ TCR structure and all available TCR sequences shows that this Xe1 hydrophobic pocket is also highly conserved in TCRs. Thus, at least one site was identified that is both accessible to molecular oxygen and is in a conserved region (V_L) in close proximity to an invariant Trp; an equivalent conserved site is also possible in the fold of V_H. Analysis of the sequence and structure around these sites shows that they are highly conserved in both antibodies and TCRs, thus providing a possible understanding of why the Ig fold in antibodies and the TCR can be involved in this unusual chemistry.

Our postulate that highly reactive intermediates are generated by the antibody-catalyzed water oxidation process brings with it a reasonable question as to whether the actual antibody molecule might be modified at or near the sites where they are generated. An X-ray structural analysis of the 4C6 Fab crystallized after activation of the water oxidation pathway by UV irradiation provided a unique insight into this issue [9]. The analysis revealed that, remarkably, only one residue, Trp^{L163}, within the antibody molecule was modified. Trp^{L163} is located in the constant region of the 4C6 Fab light chain, where its side chain protrudes into the interfacial region of the constant and variable domains. It is only the 4 position of the indole nucleus that is hydroxylated, and the hydroxyl group appears to form a hydrogen bond to a proximate water molecule. While Trp^{L163} is the most solvent-accessible tryptophan residue of the 4C6 Fab (with a solvent-accessible area of 113 $Å^2$, 1.4 Å probe radius), it is unlikely that this would be the sole reason for it being uniquely oxidized, because no oxidation of Trp^{H97}, which possesses a nearly identical solvent-accessible area (100 $Å^2$), is observed. A more plausible explanation for the regioselective oxidation is that Trp^{L163} is located in close proximity to where the short-lived intermediates are generated. Under aerobic conditions in solution, it is known that "hydroxyl radical-mediated" hydroxylation of tryptophan leads to a mixture of 4-, 5-, 6- and 7-hydroxytryptophan in addition to N-formylkynurenine (NFK) [30].

While these studies decisively depict the capacity of antibodies to mediate this oxidation process through trioxygen intermediates, they do not expose the relevance of the reaction, if any, in an existing biological context. With this in mind, it was important to investigate whether a biological source of $^1O_2^*$ could be utilized by antibodies. This activity would then broaden the scope of the reaction beyond conditions that require photochemical activation and put the newly discovered oxidation reactions into the perspective of cellularly generated oxidants.

Neutrophils (PMNs) are the most abundant leucocytes in the bloodstream. Their function is the killing of bacteria and fungi by the triggering of an oxidative burst that comprises a set of enzymatic and chemical reactions, ultimately leading to the formation of hypohalous acid, $^1O_2^*$, and hydroxyl radical (HO$^•$) [19, 20]. The first step in this cascade, the reduction of dioxygen, is initiated by the enzyme oxidase. This oxidase is a complex enzyme composed of 5 components: gp91^{phox} (a heavily glycosylated 56 kDa protein that contains the electron-carrying components of the oxidase), p67^{phox}, p47^{phox}, and p22^{phox} (proteins named according to their approximate molecular weights), and rac2 (a low-molecular weight GTPase). In the resting cell, p47^{phox} and p67^{phox} form a complex in the cytosol (which also contains p40^{phox}, a protein whose effect on oxidase activity is unclear), while gp91^{phox} and p22^{phox} are in the membrane. When the neutrophil is activated by antibody-coated bacteria, p47^{phox} is phosphorylated on particular serines and moves to the membrane to assemble the active oxidase, carrying with it its cargo of p67^{phox} and the enigmatic p40^{phox}. Rac 2, also necessary for oxidase activity, picks up a GTP and moves into the oxidase assembly. The NAD(P)H oxidase then produces superoxide anion ($O_2^{•-}$) [Eq. (4)]:

$$NAD(P)H + 2^3O_2 \rightarrow NAD(P)^+ + H^+ + 2O_2^{•-} \qquad (4)$$

The reactivity of $O_2^{•-}$ with cellular components is relatively low, although it is known to oxidize iron-sulfur clusters. Its function, it seems, is to give rise to a large variety of more powerful reactive oxygen species (ROS). The key reactions of $O_2^{•-}$ within the phagosome are either protonation to its conjugate acid, the hydroperoxyl radical, (HO$_2^•$, pK_a = 4.9), followed by a fast (k_{bi} = 1.0 × 10^8 M^{-1} s^{-1}) dismutation into H_2O_2 and $^3O_2$ [31] or to function as a substrate for superoxide dismutase (SOD)-catalyzed bimolecular dismutation into H_2O_2 and $^3O_2$ [32, 33].

The H_2O_2 is then used to oxidize Cl$^-$ to hypochlorite (OCl$^-$), a reaction catalyzed by myeloperoxidase. Hypochlorite is itself highly bactericidal, but is also used to form chloramines, some of which are even more microbicidal than OCl$^-$ (e.g. NH$_2$Cl). An alternative fate of H_2O_2 is conversion into the hydroxyl radical (HO$^•$) via Fenton chemistry. Given that H_2O_2 and HOCl are present in such high concentrations within the phagosome, there exists the prospect of a chemical production of high levels of $^1O_2^*$ [Eq. (5)]. Recent evidence points to such a reaction taking place, as evidenced by the detection of $^1O_2^*$ within neutrophils [34, 35].

$$H_2O_2 + HOCl \rightarrow {}^1O_2^* + HCl + H_2O \qquad (5)$$

Additionally, PMNs express, on their outer surface, multiple antibody Fc receptors, which are associated with antibody molecules whether in a resting or an activated state. Thus, these cells, which are central to host defense, uniquely juxtapose a biological source of $^1O_2{}^*$ with the antibody catalysts capable of processing it. In this regard, the ability of activated human neutrophils to generate the same oxidant that is produced by the water oxidation pathway was explored. The results contribute to the possible physiological relevance of the newly discovered catalysis.

Following treatment with 4-β-phorbol 12-myristate 13-acetate, a protein kinase C activator, human neutrophils generate an oxidant that oxidatively cleaves indigo carmine (1) and generates isatin sulfonic acid (2). In ^{18}O water, isotope incorporation into the amide carbonyl of 2 occurs during the oxidation of 1 by products from the oxidative cascade of neutrophils. This ^{18}O incorporation result parallels the observation of both ozone and antibody-mediated oxidative cleavage of 1. Hypochlorous acid (HOCl) is a powerful oxidant known to be produced by neutrophils. It oxidizes 1, but does not oxidatively cleave the double bond. Interestingly, almost 50% of the oxidation of 1 does yield isatin sulfonic acid, revealing the significant concentration of the ozone-like oxidant present during the oxidative burst. The primary implication is that antibody-coated activated human neutrophils produce an oxidant with a chemical signature similar to ozone, if not ozone itself.

However, the lack of absolute specificity of indigocarmine oxidation by ROS has meant that we are seeking more specific probes to strengthen the evidence that the IgG-mediated pathway has biological relevance. Recently, we have shown that products derived from ozonation of cholesterol are present in atherosclerotic artery tissue [36], but we still continue searching for a more direct chemical probe for ozone.

A proper analysis of the role of this newly discovered chemical potential of the humoral immune system in defense must address questions concerning how it fits into the global organization of this system. Also, one must question why there should be partial redundancy in terms of production of toxic products that have long been thought to be produced solely by cells and not isolated proteins. Analysis of these issues begins by consideration of the availability of the key substrate $^1O_2{}^*$. The employment of $^1O_2{}^*$ by immunoglobulins is significant not only because of the energetic demands of the reaction (see below), but also because $^1O_2{}^*$ is produced during a variety of physiological events, including reperfusion and neutrophil activation during phagocytosis, where the presence of antibodies could have an effect on its fate.

A system that functions by killing presents the ultimate challenge to the host in that, by definition, toxic materials must be produced to accomplish the task while minimizing self damage. It is well known in immunology that protection comes at a price, and much collateral damage is done during an infection or when the system inappropriately recognizes itself in autoimmunity. In this sense, one wonders what the adverse consequences of this newly discovered potential of antibodies to make toxic materials might be. In terms of global organization of the immune system, generation of the substrate $^1O_2{}^*$ is driven by adverse events such as phagocytosis or reperfusion, and, unless these events occur, the antibody-catalyzed water oxidation pathway is silent. This is similar to all the other effector systems, whether they are in the form of pro-enzymes in the complement cascade or the inducible redox enzyme system in the membrane of phagocytes. In terms of redundancy, this is not unusual for important pathways among which protection from infection must be counted. Adverse events may occur because production of $^1O_2^*$ at sites where antibody is also present may be relatively common during episodic ischemic events, high oxygen demand, or inappropriate macrophage activation. On the other hand, antibodies may actually defend against $^1O_2^*$ by converting it to hydrogen peroxide, which, in turn, can be deactivated by catalase.

While a true evolutionary account remains unsubstantiated, it is possible that the immune system began with a single protein with killing capacity and acquired diversity and recognition components later. Thus, the ability of certain other proteins to perform this process, although usually at lower rates, offers the prospect that antibodies could be evolved by coupling this specific property of some proteins with a diversity-generated targeting device. Alternatively, antibodies may function in defending a host organism against $^1O_2^*$. This postulate would require the further processing of H_2O_2 into water and $^3O_2$ by catalase. Given that catalase activity can be found in archaebacteria and was likely derived early in the phylogenetic tree, the question can be posed as to whether the structural element responsible for the catalytic destruction of $^1O_2^*$ is equally ancient and considerably precedes what is known today as antibodies. Singlet oxygen may have even played a decisive role in the initiation of the evolution of the immunoglobulin fold. Thus, it makes sense to search among ancient aerobic organisms for proteins that can accomplish similar chemistry. From an evolutionary perspective, the key issue is that this ability seems to be conserved in all antibodies.

References

1 Burton, D. R., *TIBS* 15 (1990) p. 64
2 Tramontano,, A., Janda, K. D., Lerner, R. A., *Proc. Natl. Acad. Sci. U.S.A.* 83 (1986), p. 6736
3 Tramontano, A., Janda, K. D., Lerner, R. A., *Science* 234 (1986), p. 1566
4 Pollack, S. J., Jacobs, J. W., Schultz, P. G., *Science* 234 (1986), p. 1570
5 Wentworth Jr., P., *Science* 296 (2002), p. 2247
6 Wentworth, A. D. Jones, L. H. Wentworth, P. J. Janda, K. D. Lerner, R. A., *Proc. Natl. Acad. Sci. U.S.A.* 97 (2000), p. 10930
7 Wentworth Jr. P., et al., *Science* 293 (2001), p. 1806
8 Wentworth Jr., P., et al., *Science* 298 (2002), p. 2195
9 Wentworth Jr., P., et al., *Proc. Natl. Acad. Sci. U.S.A.* 100 (2003), p. 1490
10 Babior, B. M., Takeuchi, C., Ruedi, J., Guitierrez, A., Wentworth Jr., P., *Proc. Natl. Acad. Sci. U.S.A.* 100 (2003), p. 3920
11 Plesnicar, B., Cerkovnik, J., Tekavec, T., Koller, J., *Chem. Eur. J.* 6 (2000), p. 809
12 Cerkovnik, J., Plesnicar, B., *J. Am. Chem. Soc.* 115 (1993), p. 12169
13 Cacace, F., dePetris, G., Pepi, F., Troiani, A., *Science* 285 (1999), p. 81
14 Koller, J., Plesnicar, B., *J. Am. Chem. Soc.* 118 (1996), p. 2470
15 Plesnicar, B., Cerkovnik, J., Tuttle, T., Kraka, E., Cremer, D., *J. Am. Chem. Soc.* 124 (Sep 25, 2002), p. 11260
16 Kraka, E., Cremer, D., Koller, J., Plesnicar, B., *J. Am. Chem. Soc.* 124 (Jul 17, 2002), p. 8462
17 Cerkovnik, J., Erzen, E., Koller, J., Plesnicar, B., *J. Am. Chem. Soc.* 124 (Jan 23, 2002), p. 404
18 Xu, X., Muller, P. R., Goddard III, W. A., *Proc. Natl. Acad. Sci. U.S.A.* 99 (2002), p. 3376
19 Klebanoff, S., *Encyclopedia of Immunology* 3 (1998), p. 1713
20 Klebanoff, S. in *Inflammation: Basic Principles and Clinical Correlates*, Lippincott Williams & Wilkins, Philadelphia 1999
21 Takeuchi, K., Takeuchi, I., *Anal. Chem.* 61 (1989), p. 619
22 Takeuchi, K., Kutsuna, S., Ibusuki, T., *Anal. Chim. Acta* 230 (1990), p. 183
23 Datta, D. Nagarajan, V. Xu, X. Goddard III, W. A., *Proc. Natl. Acad. Sci. U.S.A.* 99 (2002), p. 2636
24 Weiss, J., *Trans. Faraday Soc.* 31 (1935), p. 668
25 Hoigne, J., Bader, H., *Water Res.* 10 (1976), p. 381
26 Bray, W. C., *J. Am. Chem. Soc.* 60 (1938), p. 82
27 Li, T., Janda, K. D., Ashley, J. A., Lerner, R. A., *Science* 264 (1994), p. 1289
28 Lo, C.-H. L., et al., *J. Am. Chem. Soc.* 119 (1997), p. 10251
29 Klein, G. W., Bhatia, K., Madhavan, V., Schuler, R. H., *J. Phys. Chem.* 79 (1975), p. 1767

30 Maskos, Z., Rush, J. D., Koppenol, W. H., *Arch. Biochem. Biophys.* 296 (1992), p. 514
31 Bielski, B. H. J., Allen, A. O., *J. Phys. Chem.* 81 (1977), p. 1048
32 Hassan, H. M., Fridovich, I., *J. Bacteriol.* 129 (1977), p. 1574
33 McCord, J. M., Fridovich, I., *J. Biol. Chem.* 243 (1968), p. 5753
34 Steinbeck, M. J., Khan, A. U., Karnovsky, M. J., *J. Biol. Chem.* 267 (1992), p. 13425
35 Steinbeck, M. J., Khan, A. U., Karnovsky, M. J., *J. Biol. Chem.* 268 (1993), p. 15649
36 Wentworth Jr., P., Nieva, S., Takeuchi, C., Galve, R., Wentworth, A. D., Dilley, R. B., Delaria, G. A., Saven, A., Babior, B. M., Janda, K. D., Eschenmoser, A., Lerner, R. A., *Science* 302 (2003), p. 1053

13
Photoenzymes and Photoabzymes
Sigal Saphier, Ron Piran, Ehud Keinan

13.1
Introduction

Life on earth depends on light-induced chemistry. The most obvious example is photosynthesis in plants, where solar radiation is used to produce carbohydrates and oxygen from carbon dioxide and water. The chemistry of vision represents another major type of light-induced transformations. Nevertheless, when searching for a single protein that exploits light to catalyze a specific chemical reaction, one finds that only three such enzymes have been reported. These include protochlorophylide reductase and two enzymes, DNA photolyase and [6-4] photoproduct lyase, which catalyze the fragmentation of thymine dimers as part of the DNA repair mechanism [1, 2].

A thorough understanding of the origin of the catalysis and the mechanism of action in the very few known naturally occurring photocatalysts could help in the design of new catalysts. Therefore, in this chapter we discuss some mechanistic details of the photoenzymes.

In addition, the catalysis of photochemical reactions offers a unique opportunity of studying the conformational effects conferred by these catalysts on the reaction pathway. Biocatalysis [3] seems a most attractive and useful strategy by which mechanistic manifolds could be restrained and a reactive intermediate such as the excited species in photochemical reactions could be channeled into a single product. In some of the following cases, antibodies were elicited against the substrate or product, and not against a transition state analog (TSA). Since a photochemical reaction proceeds through a light-induced, highly activated intermediate, stabilization of the transition state in the classical sense of lowering the activation energy barrier is irrelevant. Other concepts of enzyme catalysis such as "entropic trap" and restriction of the conformational space or "negative catalysis" are more appropriate in understanding the activity of photoenzymes and the design of new photocatalytic antibodies.

Recently, it was also found that antibodies have the intrinsic ability to catalyze the generation of hydrogen peroxide from singlet molecular oxygen and water [4], with the possibility that ozone is also formed during this pathway of water oxidation [5]. Singlet molecular oxygen was formed from molecular oxygen by UV irradiation, al-

---

Catalytic Antibodies. Edited by Ehud Keinan
Copyright © 2005 WILEY-VCH Verlag GmbH & Co. KGaA, Weinheim
ISBN: 3-527-30688-9

though a chemical source of singlet oxygen with an antibody in the dark also led to the same result of hydrogen peroxide formation. This reaction seems to be general for all antibodies and is not linked to specific substrate binding and catalysis of a photochemical step within the binding site. This discovery has great importance for the field of immunology and for the understanding of antibody function in nature. Nevertheless, since another chapter in this book covers this issue [6], it is not discussed here.

Photocontrol of enzyme activity is a different yet related field of interest [7]. Light has very important regulatory functions such as circadian rhythms, plant flowering, seed germination, phototropism, and photomorphogenesis. The mechanisms of these photobiological processes are, in large part, still not well understood. When the light-absorbing chromophore is an inhibitor or an active-site gating molecule, the enzyme activities can also be regulated photochemically. Systems that regulate activity by light have appeared in the literature [8], including a photoinduced release of a light-active antigen from an antibody [9]. These systems are beyond the scope of this review, which focuses only on systems in which a specific chemical transformation is observed.

13.2
Photoenzymes

13.2.1
Protochlorophyllide Reductase

This enzyme is involved in the transformation of yellow, dark-grown plants to green plants when exposed to light. In addition to its central biosynthetic role, the greening reaction serves as a light-activated trigger to initiate the development of the etioplast into the chloroplast and thereby to control plant flowering [10]. The light-dependent, enzyme-catalyzed step is the reduction of the C_{17}-C_{18} double bond in protochlorophyllide to produce chlorophyllide (1). Protochlorophyllide reductase is a membrane-bound protein that utilizes dihydronicotinamide adenine dinucleotide phosphate (NADPH) and flavin as cofactors.

Although the catalytic mechanism of protochlorophyllide reductase remains largely unresolved, studies with cofactor analogs have revealed that NADPH is absolutely essential for the photoreduction [11]. By using 4R and 4S tritiated NADPH, it was found that the hydride is transferred from the pro-S face of the nicotinamide ring to the C_{17} position of the protochlorophyllide molecule [12, 13]. In addition, it has been proposed that the conserved Tyr189 donates a proton to the C_{18} position (Fig. 13.1), whereas a closely positioned Lys residue lowers the apparent pK_a of the phenolic group to facilitate the proton transfer [14]. The enzyme activity was found to be highly sensitive to small structural changes in the substrate, such as replacement of the central Mg atom [15] or changes in the macrocyclic ring [16].

The catalytic mechanism involves two additional steps, which do not require light. The first of these dark reactions involves the conversion of the initial photoproduct,

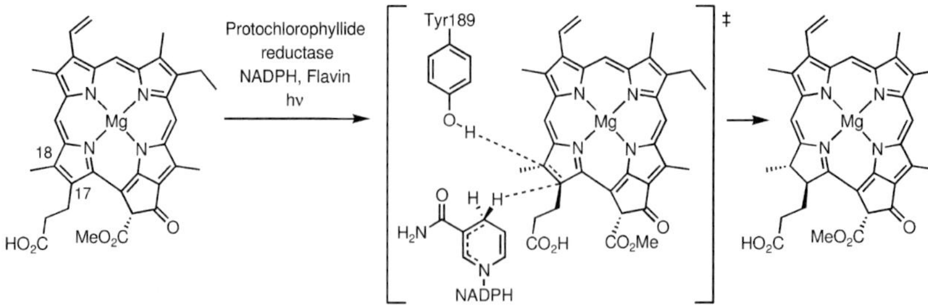

Fig. 13.1 Reduction of the C17-C18 double bond of protochlorophyllide to produce chlorophyllide.

a non-fluorescent radical species, into a new intermediate that has an absorbance maximum at 681 nm and a fluorescence peak at 684 nm. During the second dark step, this species gradually blue shifts to yield the product, chlorophyllide. Temperature dependence studies of these two processes revealed that they could only occur close to or above the "glass transition" temperature of proteins, suggesting that domain movements and/or reorganization of the protein is required for these stages of the catalytic mechanism [17].

13.2.2
DNA Photolyase

The best-known photoenzymes are the DNA photoreactivating enzymes. DNA damage caused by UV light (200–400 nm) includes the formation of more then a dozen photoproducts, such as various pyrimidine dimers, which are mutagenic and carcinogenic, and can be lethal [18]. The two most abundant photoproducts are dimers of adjacent thymine bases (1, Fig. 13.2). Compounds 2 and 5 consist of 70–80% and 20–30%, respectively, of the total UV photoproducts [19]. The two enzymes that catalyze fragmentation of thymine dimers, as part of the DNA repair mechanism, are DNA photolyase and [6-4] photoproduct lyase. DNA photolyase directly reverses the pyrimidine dimer, 2, back to two pyrimidine nucleotides by using the energy of near-UV and visible light (300–500 nm).

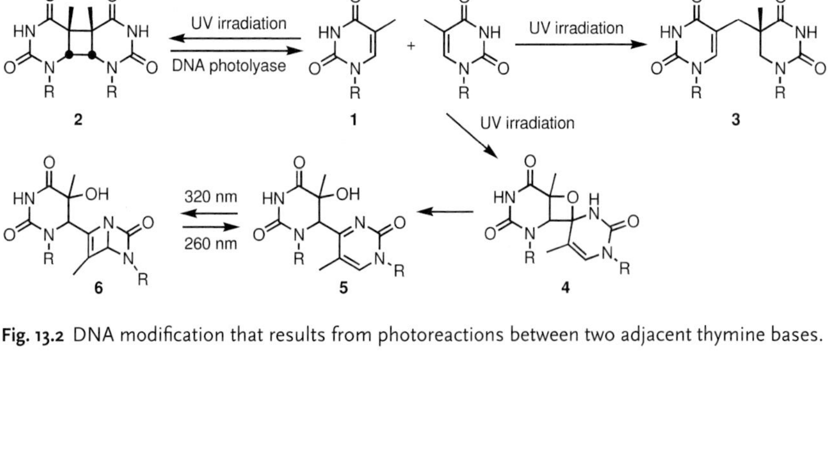

Fig. 13.2 DNA modification that results from photoreactions between two adjacent thymine bases.

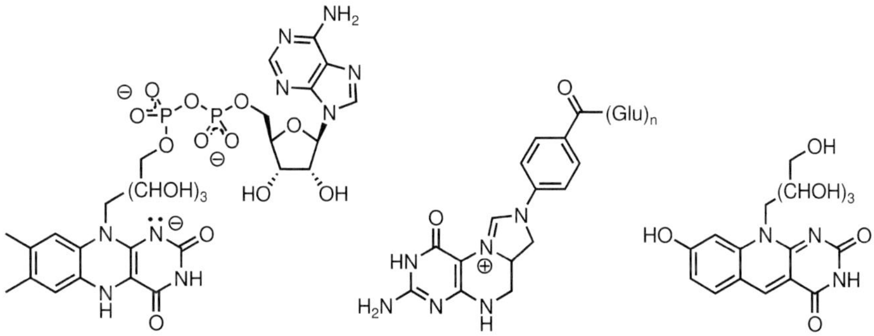

Fig. 13.3 Cofactors of the photolyase/cryptochrome family. All family members contain FAD.

All photolyases contain FAD and either MTHF or 8-HDF as a second chromophore (Fig. 13.3). Flavin is the most abundant naturally occurring cofactor, with at least 151 enzymes dependent on FAD and/or FMN [20]. FAD is the most common form of flavin found in enzymes. Flavin can be reduced and oxidized by one- or two-electron transfer reactions. Consequently, in electron transfer reactions FAD functions as a redox switch between NADH, which can transfer only two electrons, and heme, which can transfer only one electron [21]. The active form of flavin in photolyase is a two-electron-reduced form [22].

The photo-antenna in most photolyases is pterin, which appears as 5,10-methenyl-tetrahydropteroylpolyglutamate (methenyltetrahydrofolate, MTHF). The second chromophore of the deazaflavin class enzymes is 8-hydroxy-7,8-didemethyl-5-deazariboflavin (8-HDF). Upon excitation, this component becomes a strong one-electron reductant that can reduce flavoproteins. Since the second chromophore has a higher extinction coefficient than FADH$^-$ and an absorption maximum at a longer wavelength than that of the two-electron-reduced flavin, this chromophore may increase the rate of repair by 10- to 100-fold under limiting light, depending on the wavelength.

Although the kinetics of photolyase catalysis follows the Michaelis-Menten model, the transition from the enzyme-substrate complex to the enzyme-product complex is absolutely light dependent [23].

Upon binding of the modified DNA to the enzyme, the pyrimidine dimer is flipped out of the double helix into the active-site cavity, which is hydrophobic on one side and polar on the other. This asymmetry fits well the asymmetric polarity of the pyrimidine dimer, since the cyclobutane ring is hydrophobic and the opposite edges, which are lined up with oxygens and nitrogens atoms, is hydrophilic. The folate (or 8-DHF) absorbs a near-UV/blue-light photon and transfers the excitation energy via dipole-dipole interaction to the flavin cofactor. The latter transfers an electron to the pyrimidine dimer, rendering the 5-5 and 6-6 bonds of the cyclobutane ring in violation of the Hückel rules, thus splitting the pyrimidine dimer to form two pyrimidine monomers. An electron is concomitantly transferred back to the newly produced, uncharged FADH to regenerate the FADH$^-$ form [2].

Cyclobutane pyrimidine dimers are formed by a [2+2] cycloaddition reaction, which is a classic example of a photochemical reaction that is allowed by the Woodward-Hoffmann rules of orbital symmetry conservation. The same rules apply for the splitting of the photodimer by a [2+2] cycloreversion with far ultraviolet light (240 nm), which occurs with a quantum yield of near unity. However, photolyase splits dimers with near- UV/blue photons (300–500 nm) that have an energy of 250–300 kJ/mol, far less than the 500 kJ/mol needed to excite a pyrimidine dimer [24]. Consequently, the photolyase-mediated splitting of the cyclobutane dimer cannot occur via the above-mentioned concerted cycloreversion. Alternatively, the splitting occurs by a stepwise mechanism where the enzyme promotes an electron transfer to the dimer. The resultant negatively charged dimer is not a photochemically excited species, and the splitting is a thermal reaction. Hence, it must follow the rules of the conservation of orbital symmetry for thermal reactions. Nevertheless, it can be shown that conversion of a cyclobutane radical anion into ethene + ethane radical anions is also forbidden by the rules of orbital symmetry conservation [24]. As a consequence, photolyase does not mediate photocycloreversion of the cyclobutane ring by converting a symmetry-forbidden reaction into a symmetry-allowed reaction. The remaining mechanistic option is that the enzyme lowers the splitting activation energy to approximately 0.45 eV (~10 Kcal/mol) [25], allowing a formally "symmetry-forbidden" reaction to proceed very efficiently at ambient temperatures.

On the basis of studies with various photolyases and with model systems, a 5-step mechanistic model for the *E. coli* photolyase has been proposed (Fig. 13.4). In step 1, a blue light photon (350–450 nm) is absorbed by MTHF. In step 2, the resultant $^1(MTHF)^*$ transfers energy to $FADH^-$. In step 3, $^1(FADH^-)^*$ transfers an electron to

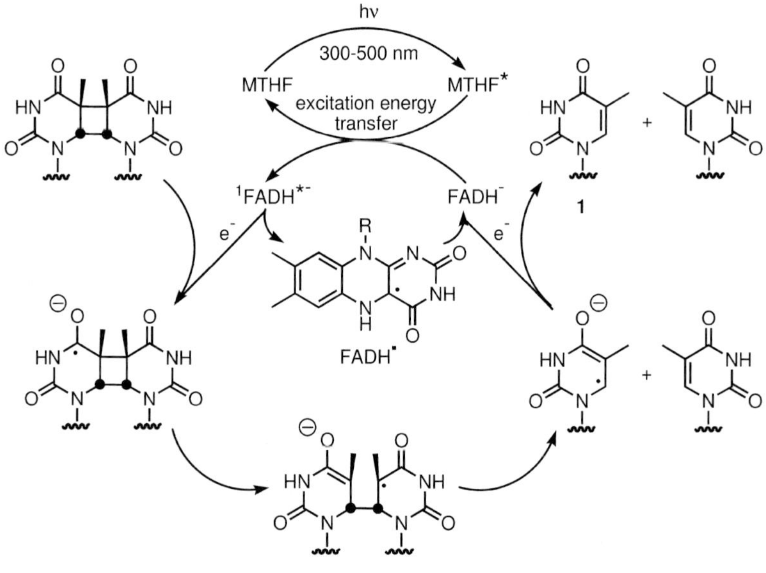

Fig. 13.4 Catalytic mechanism of DNA photolyase.

a pyrimidine dimer. In step 4, the 5-5 and 6-6 bonds are cleaved by an asynchronous, concerted mechanism. Finally, step 5 involves back electron transfer from either the 5-5 cleaved dimer radical anion [24] or the pyrimidine anion radical that results from the 5-5 and 6-6 cleavage restores the FADH$^-$ from neutral FADH [26].

Remarkably, it has been shown for the photolyase of *E. coli* that direct excitation of one specific tryptophan residue with far UV irradiation leads to the splitting of the cyclobutane ring with high quantum yield, independent of other chromophores. This Trp residue, which is a highly conserved in all photolyases and is one out of 15 such residues in the *E. coli* protein, is positioned at the DNA binding site of the enzyme. Although calculations have predicted that the real contribution of the Trp residue to photoreversal of the Pyr-Pyr dimer by sunlight could be very small, the authors suggest that it may have conferred some selective advantage in early life when higher flux of short wavelength UV reached the earth surface. Nevertheless, the ability of the photoreactivating enzyme to function without any cofactor is fundamental for the application of catalytic antibodies in photochemical reactions [1].

It is noteworthy that nearly all species have a potent and general repair system that operates via an excision repair mechanism, which eliminates the pyrimidine dimers and other mutations and thus ensures survival even after relatively high doses of UV [27, 28]. In animals that express photolyase, the enzyme appears to be uniformly expressed in all tissues, whether the tissues are exposed to light or not. The excision repair system is a multisubunit, ATP-dependent machinery that restricts single-stranded DNA, bracketing the lesion and thus excising the damage. In general, lesions that grossly distort the duplex structure are more efficiently recognized and removed by the excision nuclease system [28]. In contrast to [6-4] photoproducts, cyclobutane pyrimidine dimers cause relatively modest perturbations in the duplex structure and are therefore removed relatively slowly by the excision nuclease. Photolyase increases the helical deformity, probably by flipping out the pyrimidine dimer. This deformation accelerates the rate-limiting damage recognition step of the excision nuclease and hence the overall rate of excision [29]. Interestingly, in addition to recognizing pyrimidine dimer, photolyase binds other DNA lesions which are caused by various chemicals and which cannot be repaired by photoreactivation. Thus, it is conceivable that photolyase expressed in internal organs participates in the recognition and removal of certain lesions produced by internal and external chemical genotoxicants.

13.2.3
[6-4] Photoproduct Lyase

The pyrimidine excited singlet state leads to formation of the [6-4] photoproduct, which is the less common product [30]. Cyclobutane dimers can be cleaved to produce the original monomers by simply breaking the C5-C5 and C6-C6 σ-bonds, either by direct excitation or by photolyase. However, the breaking of the C5-OH and C6-C4 bonds of [6-4] photoproducts by any means would not result in repair. Instead, it would generate two damaged bases. Therefore the reversal of [6-4] photoproducts by photolyase seemed unlikely. Indeed, classical photolyase does not repair this

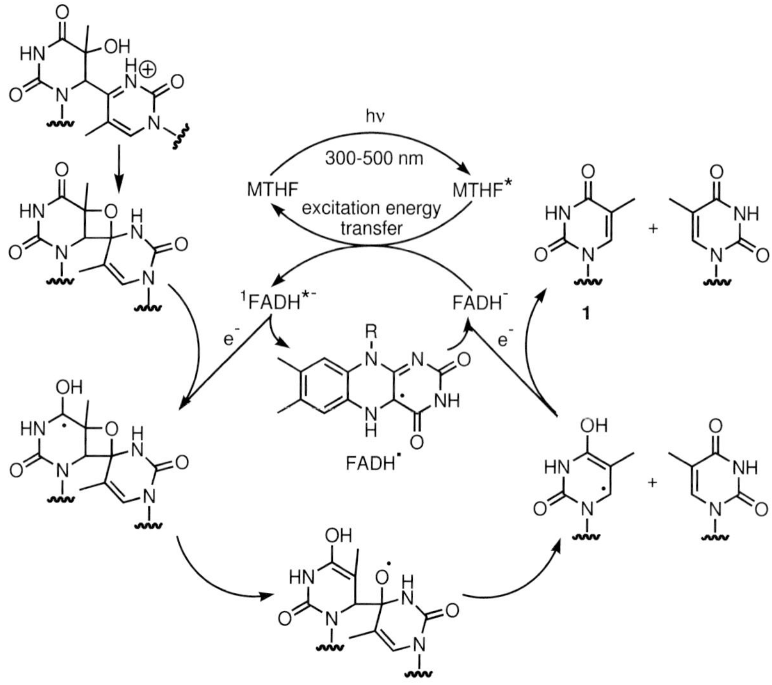

Fig. 13.5 Catalytic mechanism of [6-4] photolyase.

mutation [31]. Furthermore, the [6-4] photoproduct and even its Dewar isomer are removed very efficiently by the excision repair system, which was considered to solely operate in the removal of [6-4] photoproducts in all organisms. Therefore, it was quite surprising to discover photolyase, which repairs the [6-4] photoproduct, in *Drosophila* [32] *xenopus*, rattlesnake, and *Arabidopsis* [19, 33].

The discovery of [6-4] photolyase and the subsequent identification of structural and cofactor similarities to the classical photolyase has suggested that its reaction mechanism could be similar to that of cyclobutane photolyase [19, 34]. [6-4] Photolyase binds specifically to the damaged sites and causes conformational change, which is known as base flipping [35], allowing for the photochemical reaction (Fig. 13.5).

Although classical and [6-4] photolyases share similar structures, the same chromophores, and the same basic reaction mechanism, there are significant differences between the two classes of enzymes. For example, while cyclobutane photolyases repair the photodimers with a uniformly high quantum yield (0.7–0.98), the quantum yield of [6-4] photolyases is much lower, being in the range 0.05–0.10 [19, 36].

13.2.4
General Considerations

Several thousands of different enzymatic activities have evolved over 3.5 billion years of evolution. Only three of these have been identified as requiring light. Even for these, alternative light-independent strategies have evolved such as the excision repair mechanism for DNA dimers [27, 28]. Thus, while it is possible that some additional photoenzymes remain to be discovered, it is clear that evolution has selected against catalysis with photoenzymes and found alternative routes in multicellular organisms.

A few suggestions were made to explain this negative selection for photoenzymes in living systems: (a) photoenzyme-dependent organisms can only function in sunlight; (b) in multicellular organisms the fraction of cells able to absorb light energy is less than that in single-cell organisms, and this limitation necessitated the replacement of photoenzymes with light-independent enzymes; (c) electron transfer can occur over relatively long distances, and photoenzymes may be more difficult to regulate by allosteric control; (d) the generation of excessively reactive intermediates at the photoenzyme active site may shorten the catalytic lifetime of the enzyme [10]. The fact that the three existing photoenzymes are stable and function well suggests that the lack of natural photoenzymes is the result of negative evolution and not due to inherent inefficiency.

13.3
Photocatalytic Antibodies

Although very few antibodies that catalyze photochemical reactions have been reported to date, there are strong indications that photocatalytic antibodies can become attractive tools in organic synthesis. The current status of immunization techniques and hybridoma technology allows for sophisticated design of the antibody binding site, including the positioning of catalytic groups and cofactors. Furthermore, antibodies are generally known to be quite robust under UV radiation [4] and they are capable of handling reactive intermediates, including free radicals, carbonium ions [37] and reactive oxygen species [38]. The latter property is of particular significance

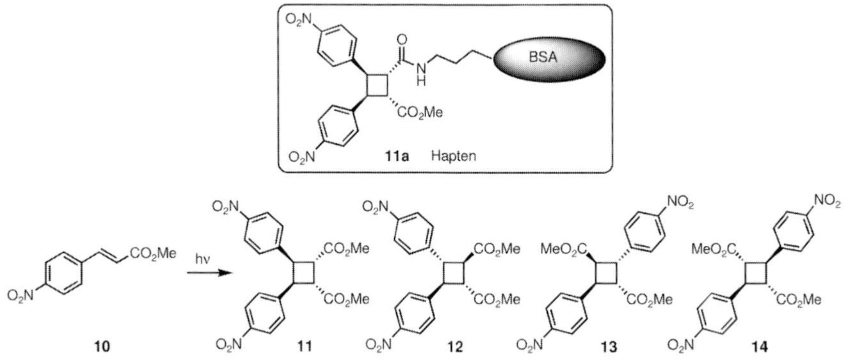

Fig. 13.6 Selective photodimerization of p-nitrocinnamate in the presence of antibodies that were raised against one of the isomeric photoproducts, **8**.

for photocatalysis because reactive intermediates are often formed in photochemical reactions.

The first attempt to generate a photocatalytic antibody, which was one of the very first attempts to create any catalytic antibody, targeted the photodimerization of methyl *p*-nitrocinnamate (**10**) [39]. In solution, irradiation of **10** produces a mixture of four strereoisomeric dimers, **11–14** (Fig. 13.6). Antibodies were raised against the BSA conjugate of one of them, **11a**. The rationale behind this hapten design was that antibodies raised against a derivative of **11** would preferentially bind and orientate pairs of the monomeric reactant molecule, **10**, in the appropriate geometry necessary for the formation of photodimer **11**. Thus, the antibody binding site was designed to function as an entropic trap. In the presence of the antisera of two rabbits, which contained specific polyclonal anti-**11** antibodies, irradiation of **10** at 350 nm led to a product mixture that was enriched in **11** in comparison with the background reaction mixture. In addition, the initial consumption rate of **10** was appreciably enhanced in the presence of the antibodies. Unfortunately, because of severe product inhibition, no catalytic turnover could be observed. This pioneering work also demonstrated the ability of antibodies to simultaneously bind two molecules, a fact that was later utilized in the design of catalytic antibodies for other bimolecular reactions, such as the Diels-Alder transformation [40].

An additional strategy relevant to catalysis of a photochemical reaction is the elicitation of a catalytic protein side chain in the binding site. This can be done by the design of an antigen with structural features complementary to that of the desired side chain. At about the same time that the above-described work was published, this second strategy was applied to mimic the action of DNA photolyase by generating antibodies that catalyze the photochemical cleavage of a thymine cyclobutane dimer [41]. This work was followed by a more detailed mechanistic investigation of the cleavage of uracil cyclobutane dimer [42]. The reasoning was that antibodies generated against the polarized π system of a pyrimidine dimer might contain a complementary tryptophan residue in the combining site. Thus, antibodies were elicited against photodimer **15** (Fig. 13.7), and one of these antibodies catalyzed the cleavage of **15** under irradiation at 300 nm. The kinetic behavior of this antibody-catalyzed reaction was consistent with the Michaelis-Menten model. For example, reaction with substrate **16** exhibited enhancement of rate ($k_{cat}/k_{uncat} = 380$) and Michaelis constant (K_M(**16**) = 280 µM).

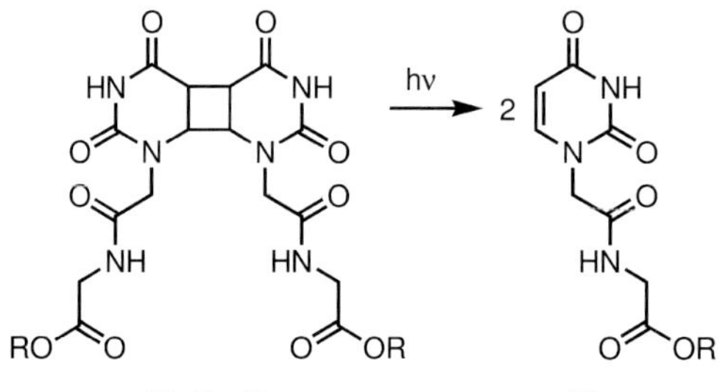

15: R = H
16: R = Me

Fig. 13.7 Uracil dimer cleavage reactions catalyzed by anti-**12** antibody, UD4C3.5.

The rate of the antibody-catalyzed cleavage reaction was found to depend on the irradiation wavelength, and this dependence followed the absorbance spectrum of indole. In addition, fluorescence-quenching experiments with increased amounts of 15 resulted in a 40% decrease in antibody fluorescence. These results suggested that a photoexcited tryptophan residue in the antibody could cause transfer of an electron to the dimer followed by a cascade of reactions leading to the fragmentation of the dimer to produce two molecules of monomer 17.

Scatchard analysis of the fluorescence-quench titration of this antibody with 15 afforded a dissociation constant of K_D = 54 nM. In agreement with this observation, the measured K_M, which was lower then 8 µM, also indicated tight binding. In contrast, a similar titration experiment with substrate 16 exhibited only very modest fluorescence quenching.

Other photocatalytic antibodies were obtained by immunization with the substrate [43]. Thus, α-ketoamide 18 was linked to a carrier protein via the acetate group (Fig. 13.8). In solution, irradiation of 18 effected cyclization via zwitterionic intermediates to produce the β-lactam derivative 20 along with the oxazolidinone 21 [44]. An antibody catalyst, 8C7, increased product selectivity by precluding formation of any cyclic product and enhancing the intramolecular hydrogen transfer reaction to give tetrahydropyrazine, 22, with k_{cat} = 1.4 ×10^{-3} min^{-1} at 280 nm and with an enantiomeric excess of 78%. The same reaction occurs in solution without the antibody only at low pH, but in this case the product was racemic. Although tight binding of the substrate to the antibody prevented a full Michaelis-Menten analysis to give the K_M value, an equilibrium binding constant, K_d = 7.6 nM, was measured for an amine-tethered substrate.

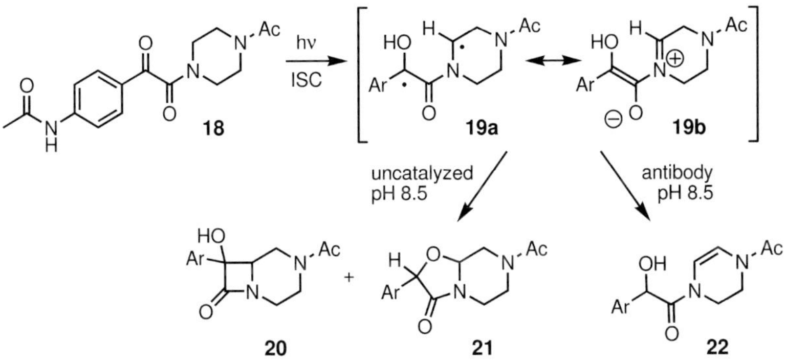

Fig. 13.8 The products of the uncatalyzed and antibody-catalyzed reactions with anti-18 antibodies.

An optimal reaction rate at 310 nm, which is the absorption maximum of the substrate, was found for the uncatalyzed reaction. In contrast, maximal rates in the antibody-catalyzed reaction were observed at the 280–290 nm region, which corresponds to the absorption maximum of the protein indole. Although this phenomenon could reflect sensitization by a tryptophan indole, the authors have suggested that the change in optimal wavelength could also reflect a shift in the absorption maximum of the substrate upon binding to the antibody. Nevertheless, since the spectrum of the antibody-substrate complex in this region is dominated by the antibody absorbance, it was difficult to prefer one of these two rationales.

It was suggested that in this case chemical catalysis occurred by an active-site residue acting as a general base. Catalysis by antibody 8C7 exhibited a pH-dependent product distribution. Moreover, the dependence of V_{max} of the production of **22** on the pH was found to be bell-shaped with a maximum at pH 7.5. This profile suggested a change of reaction mechanism when the pH changed. At low pH the reaction led to fragmentation products rather than proton transfer, yielding multiple products, some of which were identified. A model consistent with these observations was proposed with a general base in the antibody combining site performing proton abstraction to form **22**. Protonation of this basic amino acid residue at low pH is expected to result in preferred fragmentation rather than proton abstraction. On the basis of these mechanistic considerations and the X-ray crystal structure of the substrate, it was concluded that in order to produce a single product in this reaction there must be a delicate interplay of conformational control and chemical catalysis of active-site residues.

Antibody-catalyzed Yang cyclization was demonstrated in our laboratories (Fig. 13.9) [45]. Generally, the Norrish type II photochemical reaction involves abstraction of a γ-hydrogen by an excited carbonyl oxygen, e.g., **23**, to produce a 1,4-diradical intermediate, **I** [46]. The latter can undergo three alternative reactions: (a) reverse hydrogen transfer to regenerate the ground state of **23**; (b) C-C bond cleavage to form an alkene, **24**, and an enol that tautomerizes to the carbonyl compound, **25**; and (c) radical recombination (Yang cyclization [47]) to produce cyclobutanols **26** and **27**. Usually, (b) is the dominant route, while (c) represents a minor side reaction. Intense mechanistic studies over the past three decades have made the Norrish type II reaction one of the most well-understood photochemical processes [46].

Antibodies were elicited against the *cis* and *trans* oxetanes **28** and **29** and were tested for catalysis by irradiating their solution in the presence of **23** (using either a xenon or a mercury UV lamp). Three antibodies, 12B4, 20F10 and 21H9, catalyzed the formation of the *cis*-cyclobutanol **26**. No reaction was observed in the dark with or without antibodies [45].

The opportunity of achieving enantioselective cyclization is of particular interest because it is generally difficult to achieve stereochemical control over reactions that involve radical intermediates. Interestingly, all three antibodies produced the same enantiomer of **26**, the absolute configuration of which is yet unknown (antibody, *ee*): 20F10, 96%; 12B4, 80%; 21H9, 78%. The most enantioselective antibody, 20F10, also yielded the highest cyclization/fragmentation ratio (70:30) at 312 nm. At low antibody concentrations the background photochemical cleavage reaction becomes dominant. This limitation precluded measurements of the kinetic parameters (k_{cat} and K_M) under the Michaelis-Menten approximation, where low catalyst concentrations relative to substrate loads are normally used. To circumvent this problem an alternative model was used employing excess catalyst with constant substrate concentration [45]. Under these conditions rate constants were obtained from a series of pseudo-first-order kinetic data. The results were modeled according to the general format of Michaelis-Menten as described by Klotz [48].

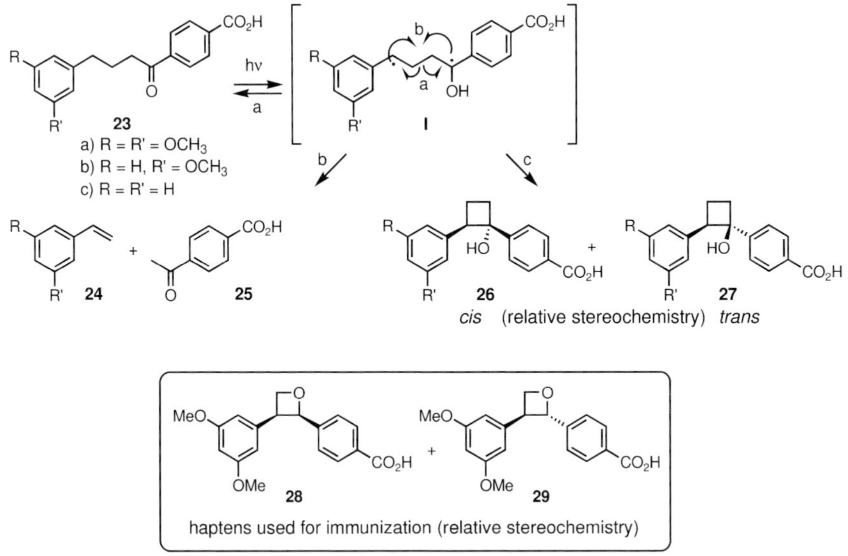

Fig. 13.9 Alternative pathways in the Norrish type II reaction with substrate 23.

Expectedly, the rate constant is highly dependent on the flux of light. For example, when using a 150W xenon lamp the reaction was complete within 10 s. This high k_{cat} made it technically difficult to obtain accurate results. Therefore, a weaker (75W) xenon lamp was used to slow down the reaction over a time span of 20 min. Under these conditions, the values of $k_{cat} = 0.008 \pm 0.001$ min^{-1} and $K_M = 58 \pm 13$ µM were obtained. Obviously, much higher values of k_{cat} could be obtained with more intense UV lamps. Noticeably, no cyclization products could be detected in the absence of the antibody. To evaluate the antibody catalytic efficiency over multiple turnovers, the antibody was recycled several times by dialysis and its activity was examined. The photochemical reaction was performed in each cycle with equal concentrations of antibody and substrate. No reduction in activity was observed after five cycles, indicating that the protein active site was remarkably stable under the photochemical conditions. Total product formation amounted to more than three times the initial concentration of the antibody.

The dependence of the cyclization efficiency on the irradiation wavelength within the range of 245–320 nm was examined. The relative rates followed quite faithfully the antibody absorbtion spectrum, with a red shift of approximately 10 nm. This points to a possible active-site tryptophan acting as a sensitizer that transfers the light energy to the substrate. By contrast, the fragmentation efficiency was found to follow the absorption spectrum of the substrate ($\lambda_{max} = 250$ nm). Consequently, antibody catalysis probably involved a dual effect: stabilization of the productive conformation for cyclization, and light harvesting followed be transferring energy to the substrate.

In long-chain ketones undergoing Norrish type II reactions, the γ-hydrogen abstraction occurs with very high regioselectivety. Approximately 5% δ-hydrogen abstraction may compete, and direct β-hydrogen abstraction is very rare [49]. This

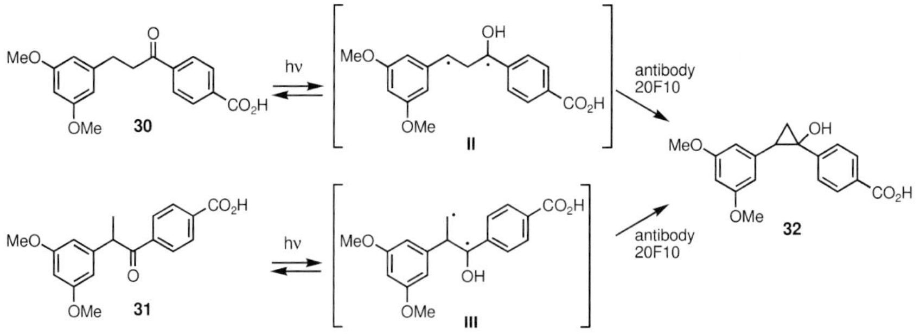

Fig. 13.10 Proposed formation of one cyclopropanol photoproduct, **32**, in the Norrish cyclization of either **30** or **31**.

specificity is a good demonstration of conformational equilibrium determining reactivity. The strong preference for a six-atom transition state makes the Norrish type II reaction ideal for monitoring basic structural effects on rate constants for hydrogen abstraction. Since a six-membered transition state ring conformation leading to cyclobutanol is highly preferred, the first prerequisite for photochemical formation of a three-membered ring is the absence of a γ-hydrogen [50]. Only when favored by exceptional stereoelectronic factors has β-hydrogen abstraction ever been seen [51]. Cyclopropanol is also formed from several β-amino-ketones because the electron-donating amine functionality activates the C-H bond for β-abstraction [52].

The possibility of antibody-catalyzed photochemical formation of cyclopropanol derivatives via β-hydrogen abstraction was examined with substrates **30** and **31**, which lack γ-hydrogens (Fig. 13.10) [53]. Neither substrate has any special stereoelectronic features that encourage β-hydrogen abstraction in solution. No evidence for such an event has been observed in the absence of an antibody. When substrates **30** and **31** were irradiated with an Hg(Ar) lamp in PBS solution only highly polar photoproducts were observed, including 4-hydroxy benzoic acid, 4-formyl benzoic acid, and other products of the Norrish type I fragmentation reaction.

We envisioned that conformational constraints on the substrates within the antibody-binding site could bring the β-hydrogen into close proximity with the excited carbonyl and thus enable the abstraction of that hydrogen, leading to the formation of a cyclopropanolic product, **32**. Indeed, antibody 20F10 was found to catalyze the formation of a new product from either **30** or **31**, and that product exhibited HPLC characteristics that were comparable with those of the above-mentioned cis-cyclobutanol **26** [53]. Although the structure of the new product has not yet been confirmed either spectroscopically or by independent synthesis, co-injection of both reaction mixtures to a reverse-phase HPLC column suggested that this product was likely to be the cyclopropanol **32**.

A 3D model of the 20F10 antibody was prepared by fitting crystal structures of homologous heavy and light chains of structurally known antibodies to overlap the peptide chains of 20F10, and the new sequence was corrected and some amino acids mutated according to the 20F10 sequence [54]. The final structure was obtained after energy minimization using the AMBER 7.0 program [55]. Docking simulations were done for haptens 28 and 29 (Fig. 13.11), substrate 23 (Fig. 13.12) and transition

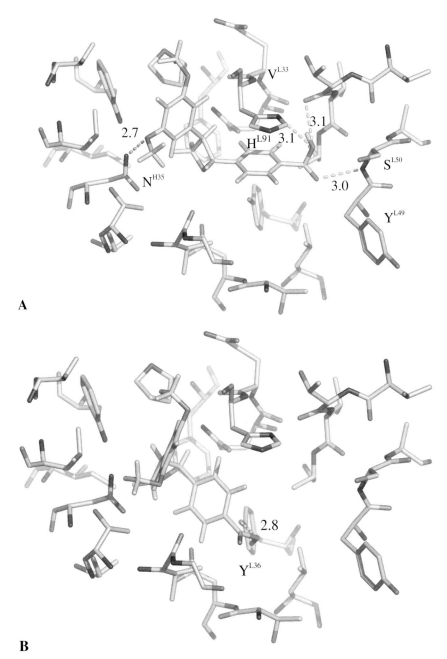

Fig. 13.11 Trans, A, and cis, B, haptens docked in antibody 20F10 combining state.

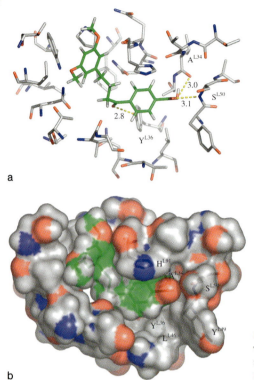

Fig. 13.12 Flexible substrate, **23**, docked within antibody 20F10 binding pocket shown in stick-type (**a**) and surface representation (**b**).

state (Fig. 13.13). All single-bond rotation freedoms, including those between phenyl groups and their substituents were allowed in the docking experiments. The antibody was chosen to be rigid. The binding free energies predicted by docking experiments were used to rank the binding affinities between the antibody and various species.

The antibody 20F10 homology model was found to bind the *trans* hapten, **28**, more favorably than the *cis* hapten, **29**, with the computed binding free energies being −8.1 and −6.9 kcal/mol, respectively (Fig. 13.11). The carboxylic acid of **28** forms three hydrogen bonds with Ser L50, Val L33 backbone and His L91 at about 3.0 Å (Fig. 13.11A). One methoxy group forms a hydrogen bond with Asn H35 at 2.7 Å. There is also one hydrogen bond between the oxetane oxygen and Tyr L36. His L91 also stacks with one phenyl ring in the hapten. The *cis* hapten, **29**, has a different binding mode in that there is only one hydrogen bond between the hapten and the antibody and no π stacking. The phenyl ring on the right side is away from the cleft formed by His L91 and Tyr L36 (Fig. 13.11B). These results correlate well with the observed 10-fold preference for binding of the *trans* hapten by antibody 20F10.

Substrate docking with all single-bond flexibilities generated only one dominant binding mode with the most favorable binding free energy at −8.5 kcal/mol (Fig. 13.12). The carboxylic acid in substrate **23** forms two hydrogen bonds with Ser L50 and Ala L34 amide at about 3.0 Å. A third hydrogen bond is formed between carbonyl group and Tyr L36 at 2.8 Å. His L91 along with Ala L34, Tyr L36, Leu L46, and Ser L50

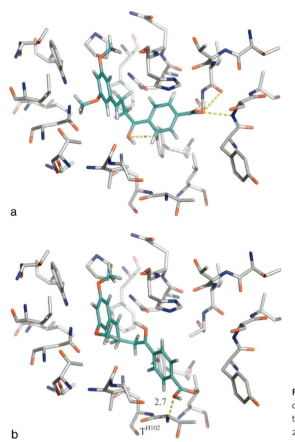

Fig. 13.13 Transition states of cyclization (**a**) and fragmentation (**b**) docked in antibody 20F10 binding pocket.

define a relatively narrow corridor and shallow concave site at the right hand side of the binding pocket and push the phenyl group into the middle.

The binding free energies of cyclization and fragmentation transition states were obtained. The cyclization transition state (Fig. 13.13a) adopts a nearly identical binding mode as the substrate and trans hapten. One phenyl group resides between His L91 and Tyr L36 while the other one sticks out of the binding pocket. Three hydrogen bonds with Ser L50, Ala L34, and Tyr L36 are formed in a similar fashion to that in the substrate binding. The extraordinary resemblance of the cyclization transition state to the substrate and trans hapten suggests a minimal conformational requirement in the antibody-catalyzed cyclization process. The fragmentation transition state binds in a completely different manner from those of both the substrate and the cyclization transition states (Fig. 13.13b).

Another photochemical reaction that was found to be catalyzed by antibody 20F10 is the photocyclization of substituted *cis*-stilbenes, such as **33**, to the corresponding substituted dihydrophenanthrenes, **34**, that are easily oxidized to phenanthrenes **35** (Fig. 13.14) [56].

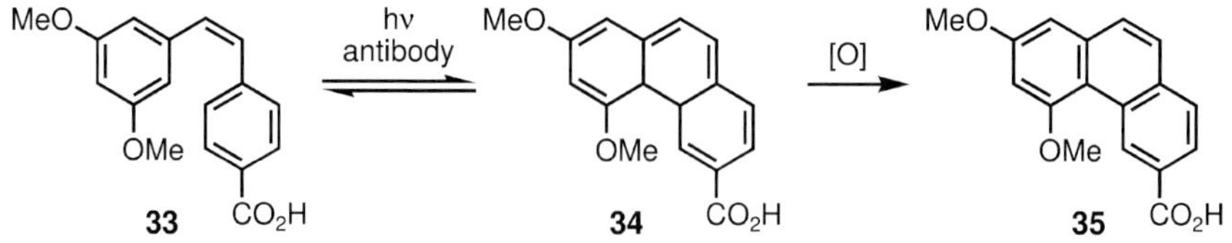

Fig. 13.14 General, antibody-catalyzed cyclization of *cis*-stilbene, **33**, to phenanthrene, **35**.

13.4
Conclusions

Biocatalysis is an attractive and useful strategy by which mechanistic manifolds can be restrained and reactive intermediates such as excited species in photochemical reactions can be channeled into a single product. The lesson we learn from the very few known natural photoenzymes is that catalysis originates mainly from entropic stabilization of a productive conformer and from the involvement of a photoactive group. Such groups may be available in the form of either a cofactor or an amino acid residue. The few reported examples of photocatalytic antibodies have demonstrated that this approach can be utilized successfully for the design of novel photocatalysts and that this opportunity should be considered seriously, particularly because natural evolution has decided against the development of photocatalytic enzymes.

13.5
Acknowledgement

Dr. Yunfeng Hu of the Scripps Research Institute is acknowledged for helpful comments on this manuscript. We thank the Israel-US Binational Science Foundation, the German-Israeli Project Cooperation (DIP), and the Skaggs Institute for Chemical Biology for financial support.

References

1 Kim, S. T., Li, Y. F., Sancar, A., *Proc. Natl. Acad. Sci. USA* 89 (1992), p. 900
2 Sancar, A., *Chem. Rev.* 103 (2003), p. 2203–2237
3 (a) Retey, J., *Angew. Chem., Int. Ed. Engl.* 29 (1990), p. 355–361. (b) Walsh, C., *Enzymatic Reaction Mechanisms*, W. H. Freeman and Company, New York **1979**
4 Wentworth, A. D., Jones, L. H., Wentworth, P. Jr., Janda, K. D., Lerner, R. A., *Proc. Natl. Acad. Sci.* 97 (2000), p. 10930–10935
5 Wentworth, P. Jr., McDunn, J. E., Wentworth, A. D., Takeuchi, C., Nieva, J., Jones, T., Bautista, C., Ruedi, J. M., Gutierrez, A., Janda, K. D., Babior, B. M., Eschenmoser, A., Lerner, R. A., *Science* 298 (2002), p. 2143–2414
6 Wentworth. P., Jr., chapter in this book
7 a) Willner, I., *Angew. Chem. Int. Ed. Engl.* 108 (1996), p. 419–439. b) Wilner, I., *Acc. Chem. Res.* 30 (1997), p. 347–356
8 Tai, L. A., Hwang, K. C., *Photochem Photobiol*, 73(4) (2001), p. 439–446 and references cited therein.
9 a) Harada, M., Sisido, M., Hirose, J., Nakanishi, M., *FEBS* 286 (1991), p. 6–8; b) Blonder, R., Levi, S., Tao, G., Ben-Dov, I., Willner, I., *J. Am. Chem. Soc.* 119 (1997), p. 10467–10478
10 Begley, T. P., *Acc. Chem. Res.* 27 (1994), p. 394

11 a) Griffiths, W. T., *Biochem. J.* 174 (1978), p. 681. b) Heyes, D. J., Martin, G. E., Reid, R. J., Hunter, C. N., Wilks, H. M., *FEBS Lett.* 483 (2000), p. 47–51. c) Oliver, R. P., Griffiths, W. T., *Biochem. J.* 195 (1981), p. 93–101
12 Valera, V., Fung, M., Wessler, A. N., Richards, W. R., *Biochem. Biophys. Res. Commun.* 148 (1987), p. 515–520
13 Begley, T. P., Young, H., *J. Am. Chem. Soc.* 111 (1989), p. 3095–3096
14 Wilks, H. M., Timko, M. P. *Proc. Natl. Acad. Sci. U.S.A.* 1995, 92, 724-728.
15 Griffiths, W. T., *Biochem. J.* 186 (1980), p. 267–278
16 Klement, H., Helfrich, M., Oster, U., Schoch, S., Rudiger, W., *Eur. J. Biochem.* 265 (1999), p. 862–874
17 Heyes, D. J., Ruban, A. V., Hunter, C. N., *Biochemistry* (2003), p. 523–528
18 Kim, S. T., Sancar, A., *Photochem. Photobiol.* 57 (1993), p. 895–904
19 Kim, S. T., Malhotra, K., Smith, C. A., Taylor, J. S., Sancar, A., *J. Biol. Chem.* 269 (1994), p. 8535–8540
20 Ames, B. N., Elson-Schwab, I., Silver, E. A., *Am. J. Clin. Nutr.* 75 (2002), p. 616–658
21 Walsh, C. T., *Acc. Chem. Res.* 19 (1986), p. 216–221
22 Payne, G., Heelis, P. F., Rohrs, B. R., Sancar, A., *Biochemistry.* 26 (1987), p. 7121–7127
23 Rupert, C. S., *J. Gen. Physiol.* 43 (1960), p. 573–595
24 Heelis, P. F., Hartman, R. F., Rose, S. D., *Chem. Soc. Rev.* 24 (1995), p. 289–297

25 Langenbacher, T., Zhao, X., Bieser, G., Heelis, P. F., Sancar, A., Michel-Beyerle, M. E., *J. Am. Chem. Soc.* 119 (**1997**), p. 10532–10536

26 Sancar, A., *Biochemistry* 33 (**1994**), p. 2–9

27 Sancar, A., Sancar, G. B., *Annu. Rev. Biochem.* 57 (**1987**), p. 29–67

28 a) Sancar, A., *Annu. Rev. Biochem.* 65 (**1996**), p. 43–81

29 Sancar, A., Franklin, K. A., Sancar, G. B., *Proc. Natl. Acad. Sci. U.S.A.* 81 (**1984**), p. 7397–7401

30 Lamola, A. A., *Photochem. Photobiol.* 8 (**1968**), p. 601–616

31 Brash, D. E., Franklin, W. A., Sancar, G. B., Sancar, A., Haseltine, W. A., *J. Biol. Chem.* 260 (**1985**), p. 11438–11441

32 Todo, T., Takemori, H., Ryo, H., Ihara, M., Matsunaga, T., Nikaido, O., Sato, K., Nomura, T., *Nature* 361 (**1993**), p. 371–374

33 Nakajima, S., Sugiyama, M., Iwai, S., Hitomi, K., Otoshi, E., Kim, S. T., Jiang, C. Z., Todo, T., Britt, A. B., Yamamoto, K., *Nucleic Acids Res.* 26 (**1998**), p. 638–644

34 a) Zhao, X., Liu, J., Hsu, D. S., Zhao, S., Taylor, J. S., Sancar, A., *J. Biol. Chem.* 272 (**1997**), p. 32580–32590. b) Sancar, A., *Science* 272 (**1996**), p. 48–49. c) Todo, T., Ryo H., Yamamoto, K., Toh, H., Inui, T., Ayaki, H., Nomura, T., Ikenaga, M., *Science* 272 (**1996**), p. 109–112

35 Zhao, X., Liu, J., Hsu, D. S., Zhao, S., Taylor, J. S., Sancar, A., *J. Biol. Chem.* 272 (**1997**), p. 32580–32590

36 Hitomi, K., Kim, S. T., Iwai, S., Harima, N., Otoshi, E., Ikenaga, M., Todo, T., *J. Biol. Chem.* 272 (**1997**), p. 32591–32598

37 Hasserodt, J., Janda, K. D., Lerner, R. A., *J. Am. Chem. Soc.* 122 (**2000**), p. 40–45

38 a) Wentworth, A. D., Jones, L. H., Wentworth, P. Jr., Janda, K. D., Lerner, R. A., *PNAS* 97 (**2000**), p. 10930–10935 b) Wentworth, P. Jr., McDunn, J. E., Wentworth, A. D., Takeuchi, C., Nieva, J. Jones, T., Bautista, C., Ruedi, J. M., Gutierrez, A., Janda, K.D., Babior, B.M., Eschenmoser, A., Lerner, R.A., *Science* 298 (**2002**), p. 2195–2199

39 Balan, A., Doctor, B. P., Green, B. S., Torten, M., Ziffer, H., *J. Chem. Soc. Chem. Commun.* (**1988**) p. 106–108

40 Chen, J., Deng, Q., Wang, R., Houk, K., Hilvert, D., *Chembiochem.* 1 (**2000**), p. 255–261

41 Cochran, A. G., Sugasawara, R., Schultz, P. G., *J. Am. Chem. Soc.* 110 (**1988**), p. 7888–7890

42 Jacobsen, J. R., Cochran, A. G., Stephans, J. C.; King, D. S., Schultz, P. G., *J. Am. Chem. Soc.* 117 (**1995**), p. 5453–5461

43 Taylor, M. J., Hoffman, T.Z., Yli-Kauhaluoma, J.T., Lerner, R. A., Janda, K. D., *J. Am. Chem. Soc.* 120 (**1998**), p. 12783–12790

44 Aoyama, H., Sakamoto, M., Kuwabara, K., Yoshida, K., Omote, Y., *J. Am. Chem. Soc.* 105 (**1983**), p. 1958–1964

45 Saphier S., Sinha S. C., Keinan E., *Angew Chem Int Ed Engl.* 42(12) (**2003**), p. 1378–1381

46 Wagner, P. J. in *CRC handbook of organic photochemistry and photobiology*, CRC Press, New York **1995**, p. 449

47 Yang, N. C., Yang, D. H., *J. Am. Chem. Soc.* 80 (**1958**), p. 2913–2914

48 Suh, J., Scarpa, I. S., Klotz, I. M., *J. Am. Chem. Soc.* 98 (**1976**), p. 7060–7064

49 P. Wagner, B.-S. Park, *Org. Photochem.* 11 (**1991**), p. 227

50 Henning, H.-G. in "CRC handbook of organic photochemistry and photobiology", CRC Press, New York **1995**, p. 484.

51 Ariel, S., Askari, S. H., Scheffer, J. R., Trotter, J., *Tett. Lett.* 27 (**1986**), p. 783–786

52 Abdul-Baki, A. Rotter, F., Schrauth, T., Roth, H. J., *Arcg. Pharm (Weinheim, Ger.)* 311 (**1978**), p. 341

53 S., Saphier, Vebenov, D., Keinan, E., Unpublished results

54 Saphier, S., Hu, Y., Sinha, S.C., Houk, K.N., Keinan, E., Unpublished results

55 Case, D. A., Pearlman, D. A., Caldwell, J. W., Cheatham T. E. III, Wang, J., Ross, W. S., Simmerling, C. L., Darden, T. A., Merz, K. M., Stanton, R. V., Cheng, A. L., Vincent, J. J., Crowley, M., Tsui, V., Gohlke, H., Radmer, R. J., Duan, Y., Pitera, J., Massova, I., Seibel, G. L., Singh, U. C., Weiner, P. K., Kollman, P. A., AMBER 7 (2002), University of California, San Francisco

56 . Vebenov, D., Saphier, S., Keinan, E., Unpublished results

14
Selectivity with Catalytic Antibodies – What can be achieved?
Veronique Gouverneur

14.1
Introduction

The field of catalytic antibodies has undergone a rapid maturation process since its creation in 1986. Since the initial "proof of concept", many transformations, including difficult and unfavorable chemical reactions, have been catalyzed by antibodies. Practically, the applicability of antibody catalysis in organic synthesis relies on the existence of catalysts that are both highly efficient and selective. Regarding efficiency, current levels of activity might be adequate for some selected laboratory applications, but, generally speaking, higher efficiencies would certainly be beneficial. In particular, enhanced proficiencies will certainly be necessary if one wants to generate antibodies for transformations that are energetically more demanding, such as site-specific protease, glycosidase, or nuclease. Interestingly, antibodies are very selective catalysts for a wide range of reactions, achieving selectivities that are sometimes difficult to obtain with natural enzymes or chemical catalysts. Selectivity may be manifested in the transformation of a mixture of two or more compounds (e.g., isomers or stereoisomers), and this is defined as substrate selectivity, or between groups or faces in the same molecule in reactions giving more than one product from a single substrate (product selectivity). In antibody-catalyzed transformations, both types of selectivity are common. As most synthetically important molecules contain more than one functional group, and most functional groups can react in more than one way, organic chemists often have to predict which functional group will react, where it will react, and how it will react. These questions are all related to selectivity. Selectivity falls into three main categories: chemoselectivity (which functional group will react), regioselectivity (where it will react) and stereoselectivity (how it will react). Because the selectivities observed with antibody-catalyzed transformations generally reflect the structure of the hapten and can rival those of natural enzymes, transformations that cannot be achieved efficiently and selectively via more traditional chemical methods are possible and are still the subject of much current research. The programmable nature of the antibody's binding site allows both activity and selectivity to be specified through appropriate design of the hapten, with a fraction of the binding energy serving the purpose of controlling selectivity. In this chapter, we aim at giving

Catalytic Antibodies. Edited by Ehud Keinan
Copyright © 2005 WILEY-VCH Verlag GmbH & Co. KGaA, Weinheim
ISBN: 3-527-30688-9

a compact overview of which types and what level of selectivity can be achieved for the most important antibody-catalyzed transformations [1].

14.2
Acyl Transfer Reactions: Ester Hydrolysis, Transacylations, and Amide Hydrolysis

Hydrolytic antibodies with tailored specificities could significantly extend the scope of naturally occurring esterase, lipase, or amidase enzymes, and therefore it is not surprising that many examples of catalytic antibodies exhibiting esterase activity have been reported in the literature. In terms of selectivity, these esterase-like antibodies could be easily exploited for processes such as the kinetic resolution of alcohols, carboxylic acids, and amino acids, or the enantioselective desymmetrization of *meso* compounds. The first esterase antibodies have been raised against phosphonate haptens selected as transition state analogs of the rate-determining step of the hydrolysis [2]. Other haptens have also been used, such as phosphates [3], phosphonamidates [4], phosphorodithioates [5], sulfones [6], fluorinated ketones [7], or secondary alcohols [8]. In terms of efficiency, rate accelerations of up to 10^6-fold are usually observed and therefore do not yet approach those of natural enzymes, which can be as high as 10^{10} over background. This could reflect the imperfect mimicry of the rate-determining transition states by the haptenic analogs. Houk et al. have recently found that, for the hydrolysis of *p*-nitrophenyl acetate, aryl phosphonate haptens are better analogs of the tetrahedral intermediate itself and the "elimination-step" transition state than they are of the rate-limiting "addition-step" transition state. In addition to a haptenic strategy based on the stabilization of the transition state by hydrogen binding of the oxyanion, alternative strategies have been explored in an attempt to improve activities [9]. Such new hapten strategies include bait-and-switch [10], heterologous immunization [11], and more recently reactive immunization [12]. The key feature of this latter strategy is the use of a hapten that can react with antibody binding site residues at the B-cell level of the immune response; in other words, the immune system is challenged by a mechanism-based inhibitor. Finally, a hapten that addressed more the electrostatic than the geometric features of the reaction has also been used successfully to yield esterase antibodies, therefore highlighting the importance of charge rather than shape complementarity as a design element of hydrolytic antibodies [13]. One interesting feature of most esterase-like antibodies generated to date is their ability to display very high levels of selectivity. For hydrolytic reactions, selectivity is usually manifested in the transformation of a mixture of two or more compounds (usually enantiomers or diastereomers), a type of selectivity that we have defined as substrate selectivity. Numerous esterase antibodies have allowed the kinetic resolution of chiral racemic alcohols, α-substituted carboxylic acids, and amino acids. Racemic haptens are generally used, because their synthesis is usually more expeditious and they can potentially generate in a single immunization process antibody catalysts displaying opposite enantioselectivities. Lerner et al. performed this experiment in 1989 and generated eleven catalytic antibodies against the racemic phosphonate 1 [14]. Nine of these antibodies were enantiospecific for the hydrolysis of the ester (R)-2, whereas the other two were stereospecific for the hydrolysis of the antipodal ester (S)-2. The best two antibodies, 2H6 (R specific) and 21H3 (S specific), were fully characterized, and it was found that the hydrolysis of the preferred stereoisomer was catalyzed at a rate > 50 times that of the antipode. It is interesting to note that, out of the eleven antibodies, the ratio of S specific / R specific antibodies was not even close to 1/1 (Scheme 14.1).

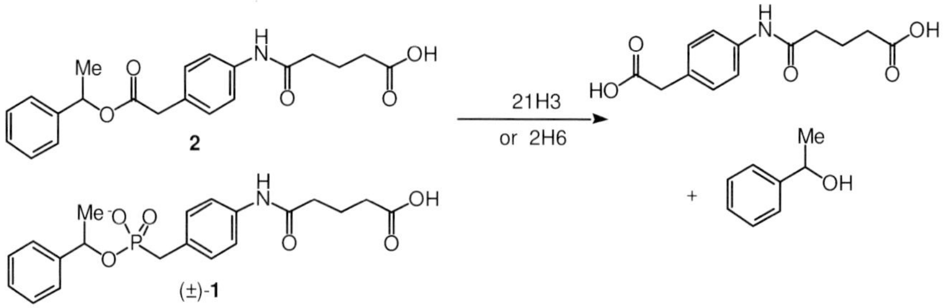

Antibody	K_M (10^{-6}M)	k_{cat}(min^{-1})	k_{cat}/k_{uncat}	specificity
21H3	394	0.09	1619	(S)-2
2H6	3994	4.6	82733	(R)-2

Scheme 14.1

P. Schultz isolated eighteen catalytic antibodies elicited against hapten 3, a tripeptide transition state analog for the hydrolysis of esters 4 under basic conditions (Scheme 14.2) [15].

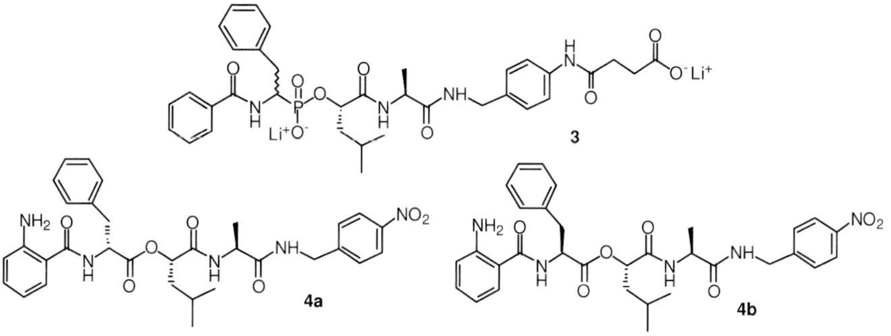

Antibody	K_M (10^{-6} M)	k_{cat}(min^{-1}x10^3)	k_{cat}/k_{OH^-}	k 4a/k 4b
6E4D5	4.4	2.3	33	<0.005
3E10D8	6.0	7.0	100	0.025
3E9F2D10	4.5	5.8	83	0.018
2H12E4	14.8	18.7	267	<0.005
2B5B11	6.2	9.8	140	<0.005

Scheme 14.2

14.2 Acyl Transfer Reactions: Ester Hydrolysis, Transacylations, and Amide Hydrolysis

The hapten 3 was synthesized and used as a roughly equimolar mixture of two diastereomers. All 18 antibodies raised against 3 were selective for the D-phenylalanine-containing ester 4a, suggesting therefore that haptens containing non-coded D-amino acids might be more immunogenic than those containing L-amino acids. Kinetic parameters for selected antibodies reveal that these antibodies show a high degree of substrate specificity.

In some cases, the use of a racemic hapten does not always allow the preparation of antibodies that can sufficiently differentiate between two enantiomers to provide enantiopure product at acceptable levels of conversion. Janda's group reported that amongst the five most active antibodies raised against the racemic "reactive" hapten 5, the best antibody 5A9 produces the desired (S)-naproxen acid 6 with a 90% ee [16]. A detailed study revealed that the poorer substrate enantiomer (R)-7 binds more tightly and inhibits the antibody-catalyzed hydrolysis of the (S)-7 enantiomer, thus contributing to the observed reduction of enantiomeric excess. The authors reasoned that this unwanted recognition of the R enantiomer could be avoided by using a homochiral hapten.

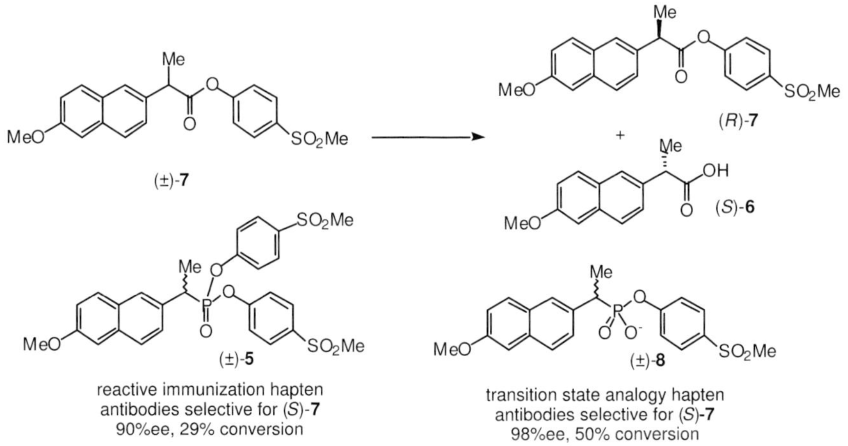

Scheme 14.3

The same group also reported that three antibodies obtained from the racemic "inert" hapten 8 perform a useful kinetic resolution of (±)-7 in 98% enantiomeric excess at 50% conversion [17]. Unfortunately, product inhibition was a significant problem for these antibodies. This study comparing these two hapten strategies (reactive immunization vs transition state analogy) suggests that the corresponding antibodies, although elicited for the same reaction and the same substrate, exhibit quite different catalytic behavior and stereopreference (Scheme 14.3).

Another hydrolytic antibody that proved to be highly stereoselective was generated by Scanlan et al. [18]. Antibody 17E8 is an esterase that catalyzes the hydrolysis of N-acyl amino acid phenyl esters. This antibody was raised against the norleucine phenyl phosphonate analog (±)-9 and catalyzed the enantioselective hydrolysis of

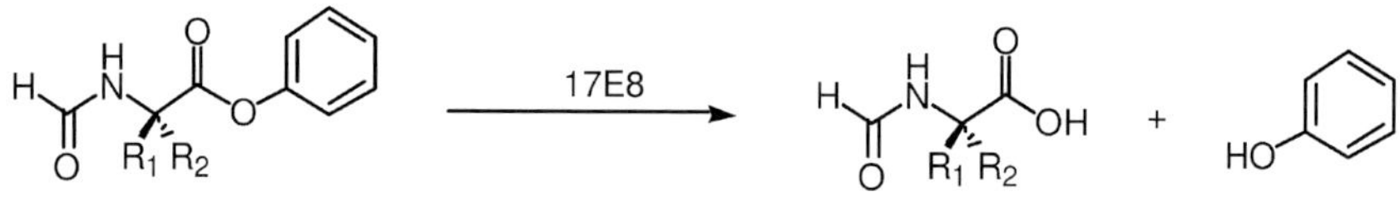

Scheme 14.4

10 R$_1$ = nBu, R$_2$ = H
11 R$_1$ = Me, R$_2$ = H

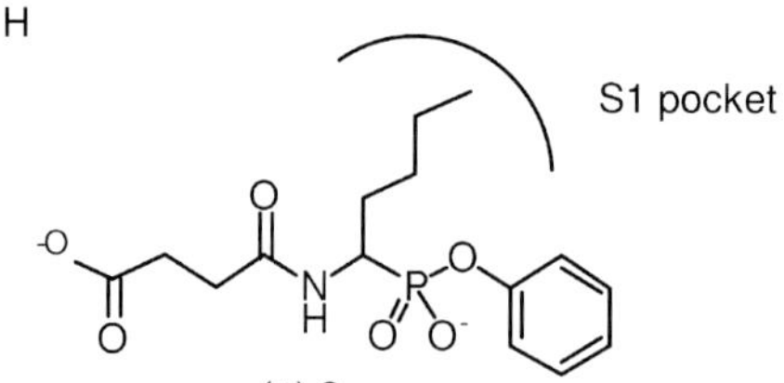

Substrate	Cα configuration	k_{cat} (s^{-1})	K_M (10^{-3}M)
10	S	2.1	0.1
10	R	-	-
10	racemic	2.0	0.18
11	S	0.9	11
11	R	-	-
11	racemic	0.8	9

unactivated esters derived from N-formyl-L-norleucine (**10**) and L-alanine (**11**), with rate accelerations up to 10^4 over background (Scheme 14.4).

The antibody is specific for enantiomers that possess the natural S configuration (L) at the alpha carbon. The X-ray structure, in addition to site-directed mutagenesis experiments, allowed the authors to revise their initial hypothesis as an attempt to explain the origin of catalysis for this antibody [19]. In addition to shedding light on the mechanism, it was confirmed that the binding site of antibody 17E8 is complementary to the L-enantiomer of phosphonate but not to the D-enantiomer. A more detailed analysis revealed that enantioselectivity is controlled by hydrophobic interactions (remote from the reaction center) between the substrate P1 side chain and the enzyme S1 pocket. Interestingly, it was found that the magnitude of these hydrophobic interactions is greater for 17E8 than for chymotrypsin (Scheme 14.4) [20].

In some cases, the use of a racemic hapten such as compounds **5**, **8** or **9** does not allow the preparation of antibodies displaying opposite enantiopreference. However, one could imagine synthesizing the two enantiomerically pure haptens and carrying two immunization processes rather than one to ensure the production of antibodies with high and predictable stereopreference for either one or the other enantiomer of a racemic substrate. The group of Kitazume performed this experiment for the synthesis of enantiopure fluorinated materials [21]. An antibody elicited against the homochiral hapten **R-12** catalyzed the stereospecific hydrolysis of the racemic ester

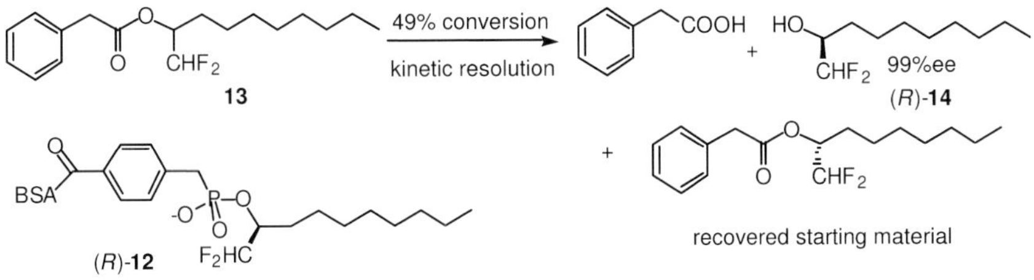

14.2 Acyl Transfer Reactions: Ester Hydrolysis, Transacylations, and Amide Hydrolysis

Scheme 14.5

(±)-**13**. This transformation allowed the isolation of R-1,1-difluoro-2-decanol (**14**) in 99% ee (at 49% conversion), the S enantiomer being prepared from the recovered ester (Scheme 14.5).

In addition, an antibody induced from hapten S-**12** catalyzed the same hydrolytic reaction but with a preference for the S enantiomer of **13** in more than 98% ee after 48.5% conversion. For this experiment, the R enantiomer was obtained from the recovered non-hydrolyzed ester (not illustrated).

In a subsequent study, the same group attempted to separate a diastereomeric mixture of substrates by antibody catalysis [22]. They considered the possibility to separate the four stereoisomers (RR', SS', RS', SR') of 4-benzyloxy-3-fluoro-3-methylbutan-2-ol (**15**) through the antibody-mediated hydrolysis of a diastereomeric mixture of the corresponding phenacetyl esters (**16**). Catalytic antibodies raised separately against the corresponding four stereochemically related haptens **17** were produced. This approach was successful and allowed the efficient separation of the substrate mixture of the four stereoisomers, with conversion close to 25% for each isomer (Scheme 14.6).

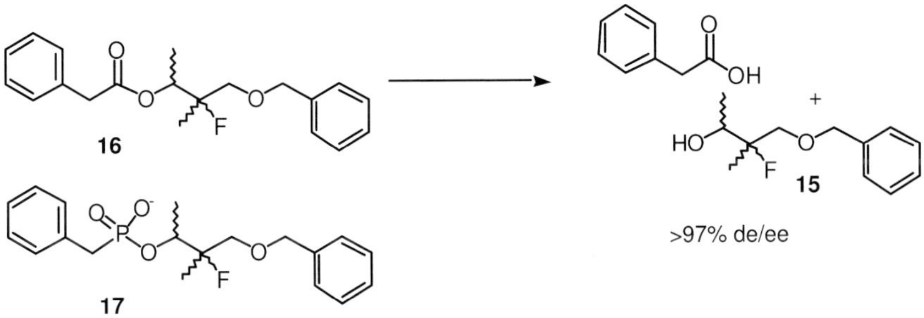

Product alcohol	k_{cat}(min^{-1})	K_M (10^{-6} M)	ee/de (%)
(2R,3R)-**15**	0.88	390	99.0
(2S,3S)-**15**	0.91	400	98.5
(2R,3S)-**15**	0.94	410	98.5
(2S,3R)-**15**	0.86	380	98.0

Scheme 14.6

The authors also found that, in addition to the separation of a 1,2-diastereomeric mixture, a 1,3 double stereoselection is also feasible with antibodies elicited against the corresponding stereochemically related haptens (not illustrated).

With these few selected examples in mind, it seems that a very high level of specificity of the antibody for their substrate is usually observed, and this may be an advantage for medical or sensor applications. However, if the antibody is to be used in synthesis, a broad and predictable substrate specificity might be preferred, avoiding the need to develop new catalytic antibodies for each new transformation and substrate. There have been a number of attempts to endow esterase catalytic antibodies with broad substrate specificity, whilst retaining high selectivity. Fujii et al. have generated esterase antibodies that discriminate between identical functional groups, as first exemplified by the chemo- and stereoselective deprotection of a variety of acylated carbohydrates [23]. Phosphonate hapten **18** was prepared, as it was anticipated that the two large hydrophobic groups might induce a deep chiral pocket capable of recognizing both the chemo- and the stereochemistry at C-3 and C-4. The point of attachment of the linker, in addition to the combined size of the two esters being enough to occupy the antigen-combining site, should address the problem of broad substrate specificity.

Consistent with this design, antibody 17E11, induced in response to hapten **18**, exhibited high selectivity at C-4 in addition to good tolerance to substitutions elsewhere on the sugar ring. The antibody is also stereoselective, as 17E11 hydrolyzes the glucopyranoside derivative **19** but not the corresponding galactopyranoside **20**. In the presence of 20% mol antibody, the reaction with substrate **19** is fast enough to avoid acyl migration from C-3 to C-4 (Scheme 14.7).

The same group subsequently demonstrated that a similar strategy could be applied to generate antibodies displaying broad substrate specificity for the enantioselective hydrolysis of N-Cbz-amino acid esters (Scheme 14.8) [24].

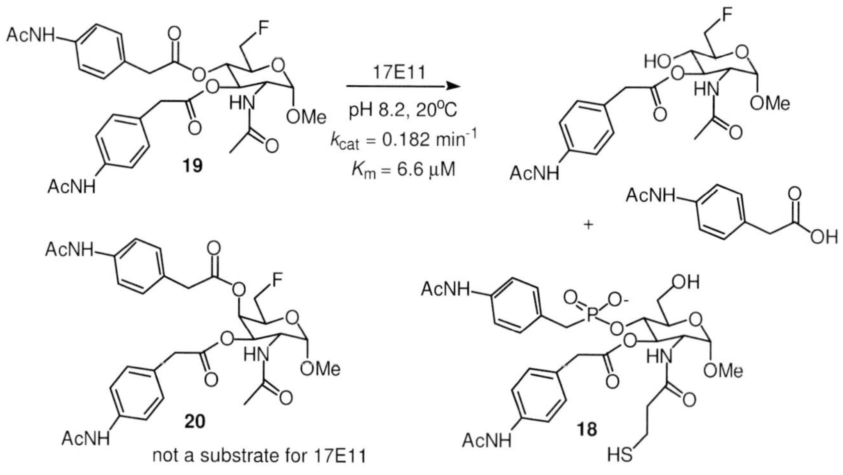

Scheme 14.7

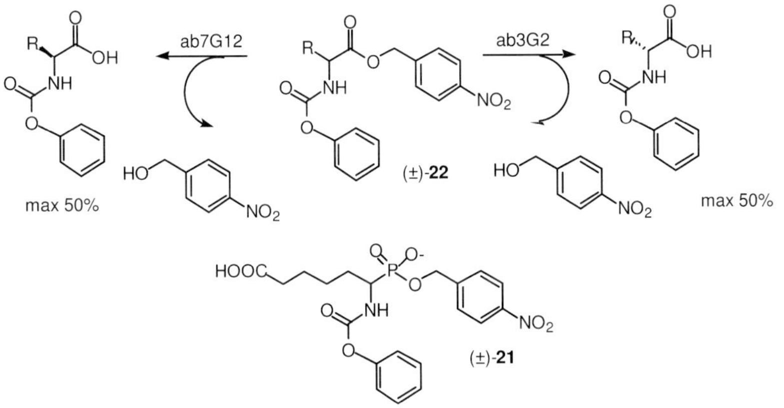

Scheme 14.8

	antibody 7G12				antibody 3G2		
ratio of velocities L-22:D-22	k_{cat} (min^{-1})	K_M (μM)	Substrate (R group)	K_M (μM)	k_{cat} min^{-1}		ratio of velocities L-22:D-22
88:12	2.8x10^{-2}	13	CH$_3$	16	2.1x10^{-2}		5:95
98:2	3.7x10^{-2}	23	(CH$_3$)$_2$CHCH$_2$	21	1.0x10^{-2}		5:95
98:2	2.8x10^{-2}	36	CH$_3$(CH$_2$)$_3$	41	8.6x10^{-2}		4:96
95:5	2.4x10^{-2}	4.9	CH$_3$SCH$_2$CH$_2$	5.5	3.0x10^{-2}		6:94
98:2	1.2x10^{-1}	64	Ph CH$_2$	27	6.2x10^{-1}		7:93

The racemic phosphonate **21** was prepared with the linker attached to the stereogenic center in order to accommodate as many structural modifications as possible for the R substituent of the amino esters. This hapten induced two separate classes of catalytic antibodies that hydrolyze either the L- or the D- isomers of Cbz-amino acid esters **22**. Antibodies 3G2 and 7G12 were found to accelerate the hydrolysis of D- and L-amino esters respectively. The preparative scale kinetic resolution of selected substrates **22** yielded the chiral acids with enantiomeric excesses up to 94% ee.

The concept of relaxed specificity will also be most useful when applied to protecting group chemistry. The group of Janda et al. developed esterase-like antibodies for the cleavage of alcohol ester of 4-nitrophenylacetyl protecting groups [25]. The basis of their approach was that antibody epitope recognition should be directed toward only the key elements contained within the 4-nitrophenylacetyl group and not the entire haptenic molecule. This strategy was reasonably successful, as exemplified by the different subtrates accepted by the catalyst. An interesting result is the failure to observe catalysis with any of the 3-nitrophenylacetates, therefore highlighting antibody protecting group selectivity (Scheme 14.9).

A similar strategy was adopted by the group of Fujii et al., who generated antibody 7B9 against the p-nitrobenzyl phosphonate **23** [26]. They found that the antibody catalyzed the hydrolysis of p-nitrobenzylmonoesters of nonsubstituted, and β- or γ-substituted glutaric acids (**24**) with almost identical K_M and k_{cat} values. The antibody also displayed substrate tolerance towards the α-substituents and accepted the p-

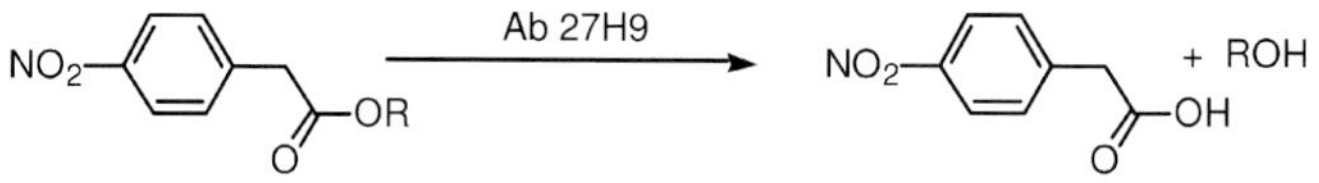

Scheme 14.9

R = ethyl, benzyl, allyl, 2-chloroethyl, isobutyl, *m*-nitrobenzyl, neopentyl, CH(CF$_3$)$_2$
k_{cat} ranging from 0.0038 min^{-1} to 0.47 min^{-1}
not recognised: substrates with R = (CH$_2$)$_2$-*t*-Bu, isopropyl, cyclohexyl

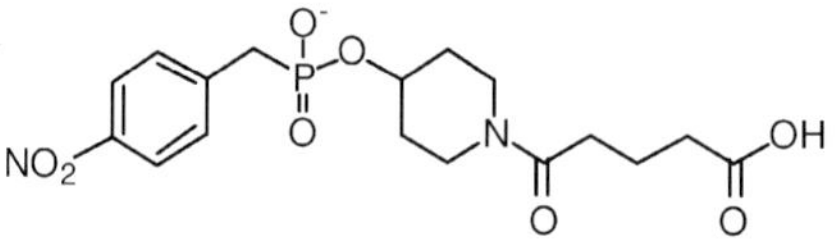

nitrobenzyl esters **25** of Leu, Norleu, and Phe. A model of the binding site constructed after cloning and sequencing the antibody revealed a relatively shallow pocket of the antigen-combining site that can accommodate the *p*-nitrobenzyl group with the α-, β- or γ- substituents of the substrates being outside the binding site. This is consistent with the broad substrate specificity of this antibody. Interestingly, antibody 7B9 hydrolyzed both enantiomers (L- and D-) with almost the same values for the specificity constants k_{cat}/K_M, indicating that the antibody binding site stabilized the two enantiomeric transition states to a similar extent (Scheme 14.10).

Lipase-like antibodies that are able to differentiate between two enantiotopic groups of a *meso* substrate are valuable catalysts [27]. The group of Lerner was the first to challenge a catalytic antibody with the enantioselective perturbation of a *meso* diester compound [28]. Out of the 33 antibodies raised against the homochiral phospho-

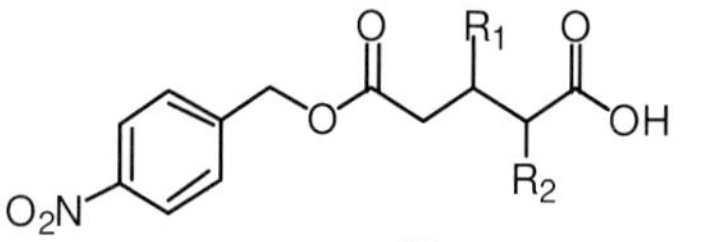

Substrate	R$_1$	R$_2$	R$_3$	K_M	k_{cat}	k_{cat}/K_M
24	Me	H		456	5.08x10^{-2}	758
24	H	H		346	2.01x10^{-2}	1020
L-25			(CH$_3$)$_2$CHCH$_2$	22.4	1.67x10^{-3}	74.6
D-25			(CH$_3$)$_2$CHCH$_2$	10.6	4.68x10^{-4}	44.2
L-25			PhCH$_2$	4.7	1.83x10^{-3}	389
D-25			PhCH$_2$	1.9	4.00x10^{-3}	211

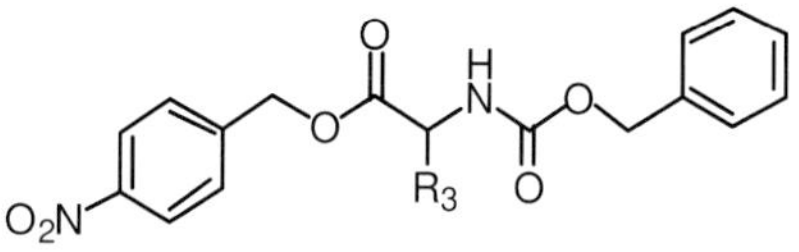

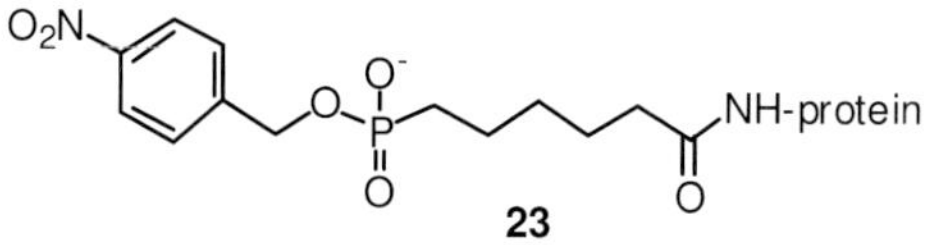

Scheme 14.10

14.2 Acyl Transfer Reactions: Ester Hydrolysis, Transacylations, and Amide Hydrolysis

nate hapten **26**, one antibody 37E8 was found to be catalytic. This catalyst allowed the enantioselective desymmetrization of the *meso* diester **27** to afford the (1R,4S)-hydroxy-2-cyclopentenyl acetate **28** with an uncorrected enantiomeric excess of 84%. This enantiomeric excess appears to be limited only by the catalytic activity of antibody 37E8 rather than by an inherent lack of enantiotopic group differentiation. The product is a direct precursor of prostaglandin prostaglandin $F_{2\alpha}$ (Scheme 14.11).

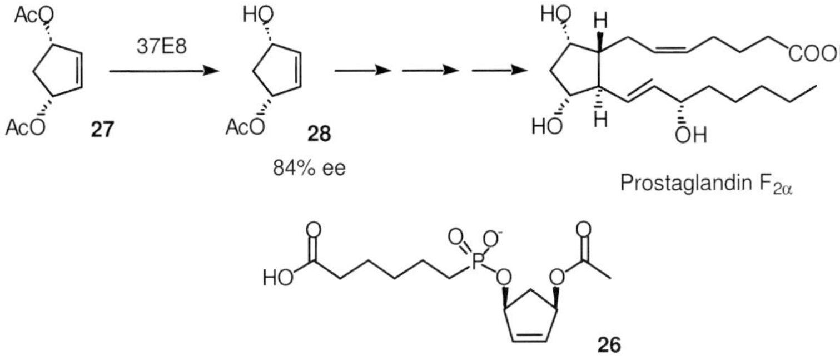

Scheme 14.11

In addition to hydrolytic processes, transacylation reactions have also been successfully catalyzed by antibodies. With enzymes, these transacylation reactions are often performed in media that contain high concentrations of organic solvents to minimize competing hydrolysis. Mechanistically, antibodies such as 21H3 (see Scheme 14.1) that catalyze the hydrolysis of esters, apparently through a covalent intermediate, might also accelerate transacylation reactions in the presence of alcohols, amines, or thiols. Preliminary studies revealed that 21H3 catalyzed the acyl transfer in water from enol ester **29** and various analogous aromatic alcohols (Scheme 14.12) [29].

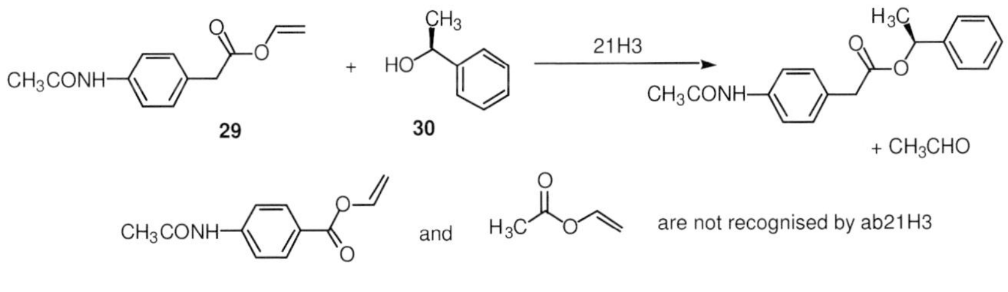

Scheme 14.12

Comparison of the kinetic data for the lipase-catalyzed reactions and 21H3-catalyzed reactions with compound **29** revealed that 21H3 in aqueous solution is as effective as lipase SAM II in dichloromethane. In all cases, the stereochemistry of acceptable alcohol substrate is the same as that of compound **30**, and the antibody is very selective for the enol ester **29**.

In 1992, Schultz et al. reported the generation of an antibody, raised against the phosphonate diester, **31** capable of selectively acylating the 3′-hydroxyl group of

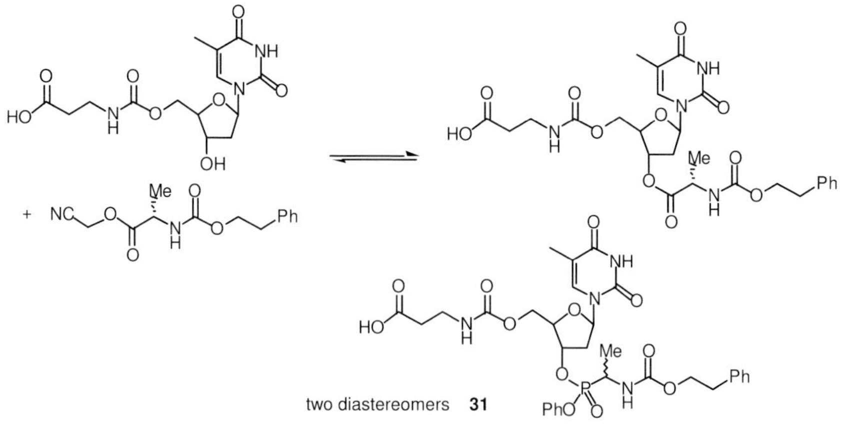

Scheme 14.13

thymidine [30]. The hapten which contained elements of both the acyl donor and the acceptor was used as a mixture of diastereomers (Scheme 14.13).

Out of the eight antibodies found to be catalytic, four of them accelerate the reaction of L-alanine ester reagent, three were selective for the D isomer, and one of them was not selective. The origin of stereoselectivity was not through preferential binding of one of the enantiomers, as suggested by the similar K_M values, but it seems that the antibody discriminates instead between the diastereomeric transition states for the transesterification process. The differential binding affinity of the antibody to the phosphonate diester relative to the substrates appears to account for a large fraction of the catalytic activity, therefore contrasting with the mechanism of antibody 21H3, based on the formation of an acyl intermediate. Unfortunately, product inhibition was a major drawback for this bimolecular process.

Another synthetically very useful antibody 16G3 has been obtained from hapten 32 [31]. This catalyst accelerated the coupling of p-nitrophenyl esters of N-acetyl-valine, -leucine and -phenylalanine with compound 33 to form the corresponding dipeptides (Scheme 14.14).

All possible stereoisomeric combinations of the ester and the amide substrate 33 were coupled at comparable rates. To account for the lack of selectivity for this coupling reaction, it was suggested that this might be the result of the small size of the substrates compared to that of the hapten. This might be a general strategy to raise antibodies displaying broader stereospecificities. It is also possible that the large hapten whose structure did not match directly with the product or the leaving group might account for the relatively low product inhibition for this antibody. Finally, it is noteworthy that this catalyst was capable of multiple turnovers at rates that exceeded the rate of spontaneous ester hydrolysis.

An intramolecular transacylation was successfully catalyzed by antibody 24B11 raised against the racemic cyclic phosphonate ester 34 [32]. The rate acceleration was modest (k_{cat}/k_{uncat} = 167), but gratifyingly it was found that the reaction ceased at 50% of the initial ester concentration, suggesting that the antibody catalyzed the cyclization

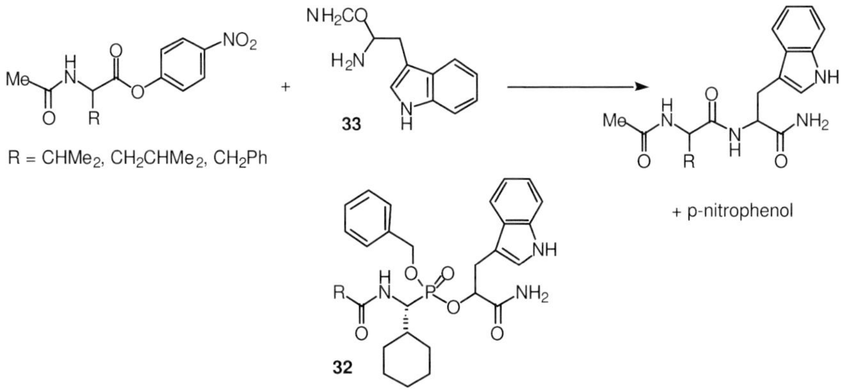

14.2 Acyl Transfer Reactions: Ester Hydrolysis, Transacylations, and Amide Hydrolysis

Scheme 14.14

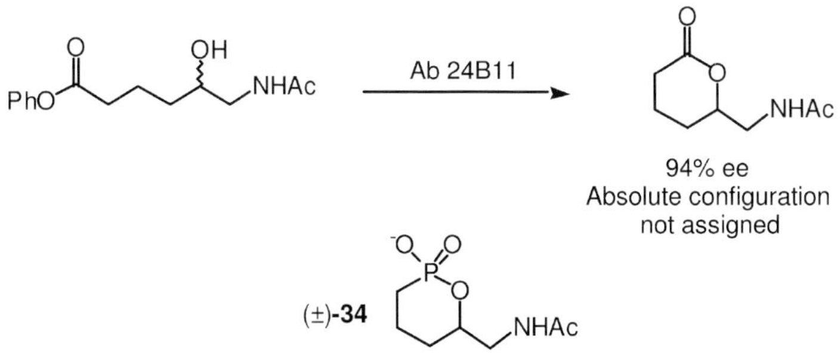

Scheme 14.15

of a single enantiomer. The corrected enantiomeric excess for this transformation was found to be 94% (Scheme 14.15).

In contrast to esterase antibodies or catalytic antibodies for transacylation processes, the generation of amidase antibodies represents an important challenge, as the uncatalyzed hydrolysis of the amide bond is a relatively kinetically inert reaction [33]. However, much effort has been invested in this area of research, as sequence-specific proteolytic antibodies would have enormous therapeutic or industrial applications. Numerous attempts to produce amidase antibodies have met with only limited success because of the intrinsic stability of the amide bond along with the necessity to activate the amine as a better leaving group prior to departure [34]. Amongst all the amidase-like antibodies reported in the literature, the most efficient catalyst produced to date is antibody 43C9 raised against the phosphonamidate transition state analog **35** [35]. This antibody accelerates the hydrolysis of a *p*-nitroanilide by a factor of 2.5×10^5 over the background rate in addition to catalyzing the hydrolysis of a series of aromatic esters. Mechanistic studies have revealed that the catalytic mechanism involves two key residues, His-L91, which acts as a nucleophile to form the acyl antibody intermediate and Arg-l96, which stabilizes the anionic tetrahedral moieties. Only achiral substrates were the object of this report (Scheme 14.16).

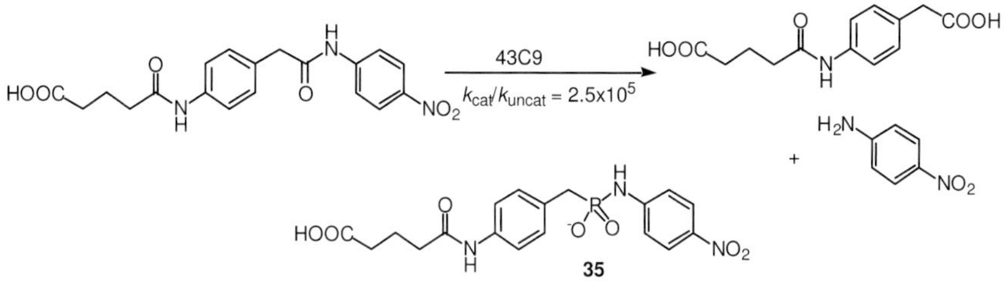

Scheme 14.16

The phosphinate hapten **36** was used successfully by the group of Martin et al. to generate antibody 13D11, which is able to catalyze, albeit with a modest rate-acceleration (k_{cat}/k_{uncat} = 132), the hydrolysis of the primary amide **37** without cofactor assistance [36]. The stereoselectivity of this reaction was found to be opposite to known naturally occurring primary amide-hydrolyzing enzymes, suggesting that the catalytic activity associated with antibody 13D11 was not due to a contaminating protease, a hypothesis supported by additional control experiments. Indeed, only the R-amide **37**, which sterically corresponds to a D-amino-acid, was hydrolyzed. The S isomer of **37** and a shorter acylated amide were not hydrolyzed (Scheme 14.17).

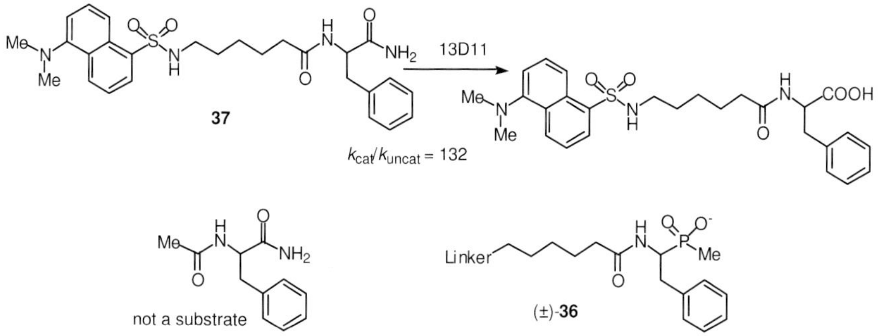

Scheme 14.17

Another phosphinate hapten (**38**) was used successfully in raising antibodies 14A8 and 39F3, two catalysts that accelerate the rearrangement of the asparaginyl-glycine peptide bond of compound **39**, an example of substrate-assisted amide cleavage involved in the denaturation and degradation of proteins *in vivo* [37]. This rearrangement proceeds via an unsymmetrical succinimide intermediate that can lead to either the corresponding aspartyl glycinamide or to isoaspartyl glycinamide, according to which hydrolytic pathway operates. The hapten possessing both a phosphinate group and a secondary alcohol can potentially generate antibodies that could catalyze both hydrolytic pathways. It was found that several antibodies raised against hapten **38** accelerate the rate-determining step of succinimide formation. In terms of product selectivity, a mixture of products **41** and **42** is more often obtained in the presence of

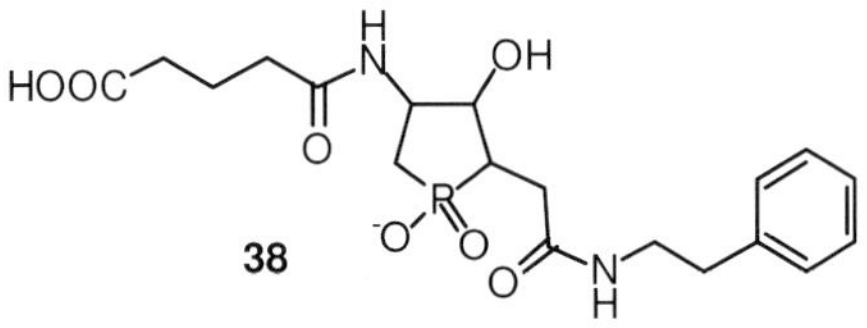

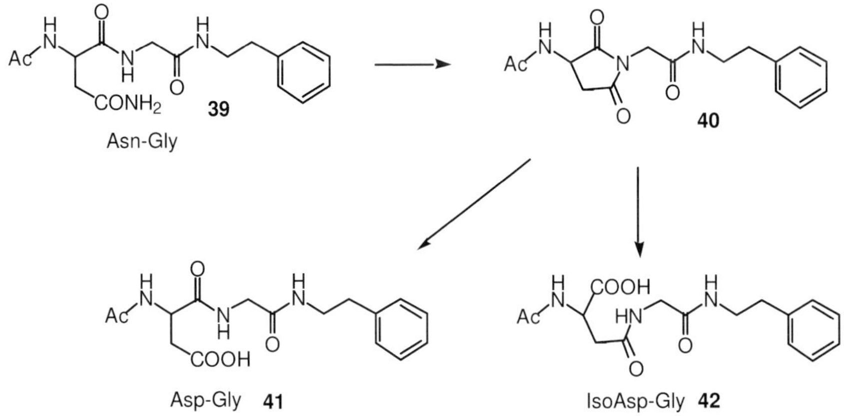

Scheme 14.18

these catalytic antibodies, suggesting that these proteins catalyze the two hydrolytic pathways simultaneously with different degrees of efficiency (Scheme 14.18).

Other examples of antibodies endowed with amidase activities have been reported, but the low activity of most amidase-like antibodies reported so far has precluded a detailed investigation of selectivity issues for this important class of reaction [38].

14.3
Glycosyl and Phosphoryl Group Transfer

The development of catalytic antibodies that selectively cleave or form glycosidic bonds (glycosidase or glycosyltransferase-like antibodies) would contribute enormously to our ability to chemically manipulate the structures of carbohydrates [39]. Early efforts in the production of glycosidase-like antibodies (cleavage of glycosidic bond) focused on cyclic acetals as substrates and antibodies that were generated to positively charged or conformationally restricted analogs of the oxocarbenium intermediate. The resulting antibodies accelerated the hydrolysis reaction from 100- to 1000-fold [40]. More recently, a few other haptens were employed such as five-membered ring iminocyclitols for the hydrolysis of aryl glucopyranosides and galactopyranosides or compounds designed to mimic the half-chair conformation of the oxocarbenium intermediate and an appropriately disposed leaving group [41]. Antibody ST-8B1 raised against hapten **43** catalyzed the hydrolysis of substrates (1R,

4R)-44 and (1R, 4S)-44 but not the other stereoisomers [42]. Comparison of the kinetic constants reveals that (1R, 4R) is the substrate most efficiently catalyzed by ST-8B1. This result suggests that the antibody also discriminated between the equatorial 4-hydroxy group and the axial stereoisomer, as expected from the stereochemistry of the hapten. Thus, antibody ST-8B1 has high stereospecificity with respect to both C-1 and C-4 stereogenic centers of the substrates, indicating that this antibody may have been generated from (1S, 4R, 6S)-43 (Scheme 14.19).

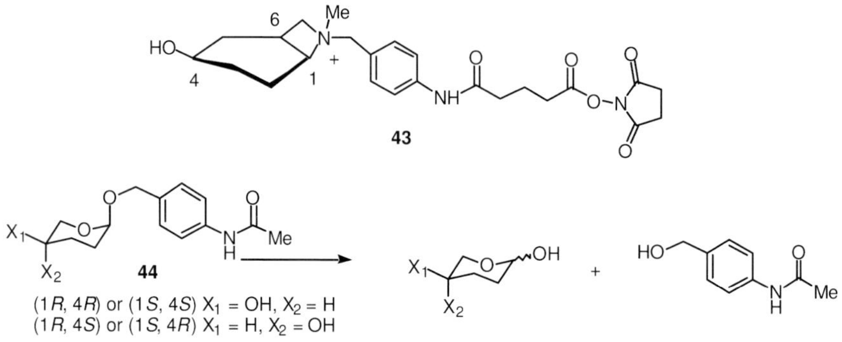

Scheme 14.19

More recently, Lerner et al. successfully employed a mechanism-based inhibitor to select antibodies displayed on filamentous phage (library constructed from genes of monoclonal antibodies raised *in vivo* against an iminocyclitol inhibitor of β-galactosidase) which might promote glycosidic bond cleavage covalently. However, no detailed study was carried out on the stereoselectivity of these catalysts (not illustrated) [43].

Finally, a glycosidase antibody Ab 4f4f elicited against the chair-like transition state analog 45 by *in vitro* immunization was produced by the group of Schultz [44]. The pH dependence of the reaction and chemical modification studies suggested the presence of an active-site Asp or Glu residue that may function as a general acid. Substrate specificity studies suggested that, in general, ab4f4f could bind the β-anomers with higher affinity than the corresponding α-anomers, suggesting that the N-benzyl group of the hapten may better overlap the β-nitrophenol isomer of the pyranoside ring. Neither anomer of p-nitrophenyl D-mannopyranoside showed detectable hydrolysis, suggesting that equatorial substitution at C_2 hydroxyl group is required. Other equatorial substituents are tolerated, and the fact that D-galactopyranosides are good substrates indicates that the relative configuration of the C4 hydroxyl group is not critical.

Similarly to the hydrolysis of amides, the hydrolytic cleavage of phosphate esters is an energetically demanding transformation [45]. This reaction proceeds through a pentacoordinate transition state that is difficult to mimic. In addition, significant differences in the precise mechanisms of phosphate ester cleavage vary significantly between monoester, diester and triester derivatives. Each of these targets was already considered for antibody catalysis using haptens such as phosphonate, O-

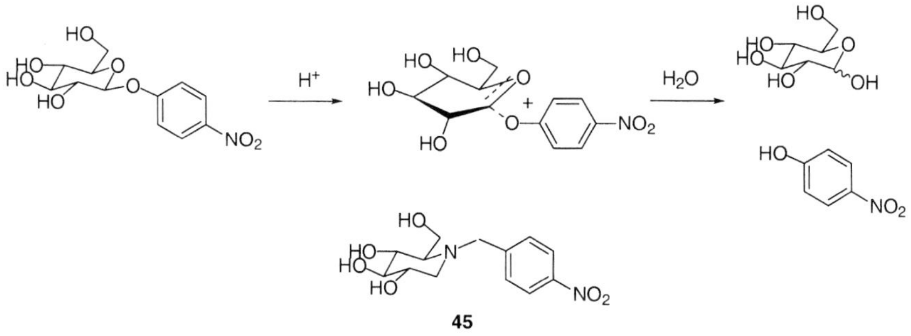

Scheme 14.20

phosphorylated hydroxylamine derivatives, pentacoordinated oxorhenium chelates, or amine oxides [46]. Only catalytic antibodies displaying low activity were obtained, and no data are therefore available on the selectivity of these antibodies.

In 1995, Lerner and coworkers used the organophosphonate **46** as a reactive immunogen. Following spontaneous hydrolysis *in vivo*, it becomes a stable monoester transition state analog **47** which in turn could challenge the immune system [12]. Reactive immunization with compound **46** generated 11 monoclonal antibodies able to hydrolyze the phosphonate **48**. The most efficient antibody AbSPO49H4 is highly stereoselective and allowed the resolution of this substrate. The same antibody also catalyzed the hydrolysis of esters, suggesting that this antibody has indeed undergone reactive immunization (Scheme 14.21).

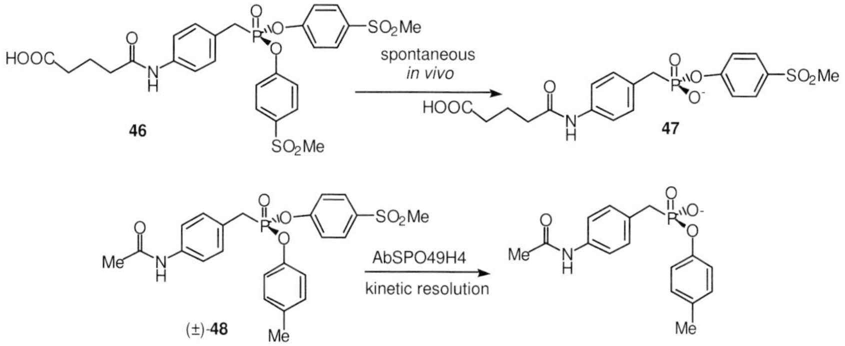

Scheme 14.21

14.4 Pericyclic Reactions

Given the importance of pericyclic reactions in organic synthesis, it is not surprising that much effort has gone into developing selective catalytic antibodies for this class of transformations. So far, numerous pericyclic reactions have been successfully catalyzed by antibodies, and most of them have no enzymatic precedent. These include sigmatropic rearrangements such as Claisen and oxy-Cope rearrangements, pericyclic *syn* elimination of selenoxides or amine oxides to give the corresponding alkenes, Diels-Alder as well as 1,3-dipolar cycloadditions.

14.4.1
Sigmatropic Rearrangement

In asymmetric synthesis, the important characteristics of the Claisen and related [3,3]-sigmatropic rearrangements can be traced back to the highly ordered chair transition state and the preference of the various groups to adopt axial or equatorial positions. Chemically, the development of an enantioselective catalytic Claisen rearrangement remains an important, yet elusive goal in chemical synthesis, but recently a few promising catalysts have been described in the literature [47]. As this class of reactions is highly sensitive to and proximity effects but not typically subjected to general acid-base catalysis, these transformations are attractive targets for antibody catalysis. Bartlett and Johnson synthesized a potent inhibitor for the enzyme chorismate mutase, an enzyme that catalyzes the conversion of chorismate into prephenate, a rare example of natural pericyclic reaction in cellular metabolism [48]. This inhibitor, which could be regarded as a transition state analog for this transformation, was converted into two very similar haptens of general structure **49** by the groups of Schultz [49] and Hilvert [50] respectively. Both groups successfully produced two catalytic antibodies, 11F1-2E11 and 1F7, with rate enhancements of 10^4 and 250. The two catalysts displayed good stereoselectivity in that they catalyzed the rearrangement of the (–)-isomer of chorismic acid only. Indeed, for both catalysts, enantioselectivities greater than 40:1 were observed. It is noteworthy that even though both haptens were used in racemic form, antibodies that could catalyze the rearrangement of (+)-chorismic acid were not found. The crystal structure of antibody 1F7 bound to the transition state analog has been solved and revealed preferential binding of the (–)-chorismate enantiomer in its pseudodiaxial conformation as required for this rearrangement to proceed (Scheme 14.22) [51].

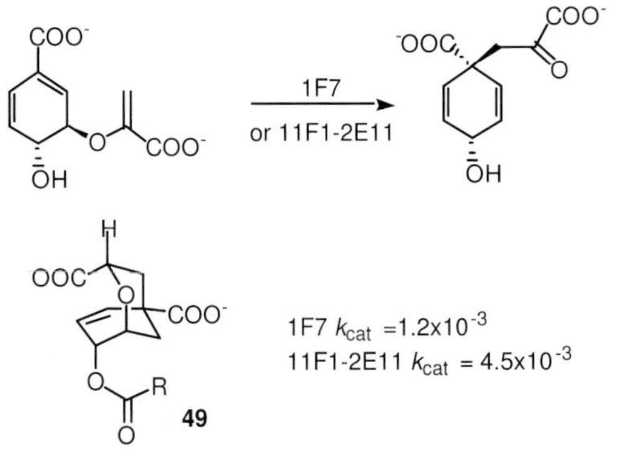

1F7 k_{cat} = 1.2×10^{-3}
11F1-2E11 k_{cat} = 4.5×10^{-3}

Scheme 14.22

14.4 Pericyclic Reactions

A catalytic antibody has also been generated for an oxy-Cope [3,3] sigmatropic rearrangement [52]. The neutral cyclohexanol 50, mimicking the highly ordered chair-like transition state of this reaction, has been used as the hapten and has allowed the generation of several catalytic antibodies, the most active of which accelerated this rearrangement 5300-fold relative to the uncatalyzed process. *In situ* trapping of the aldehyde group of the rearranged product in the form of an oxime was necessary to prevent both chemical modification of the catalyst and product inhibition. The antibody was found to prefer the *S* alcohol 51 (kinetic resolution, substrate selectivity), the other enantiomer being converted only at the background rate. It was suggested that better rate accelerations could be achieved if one used a cationic hapten that might induce antibodies, including a complementary carboxylate residue that could be part of a general base-catalyzed anionic oxy-Cope rearrangement (Scheme 14.23).

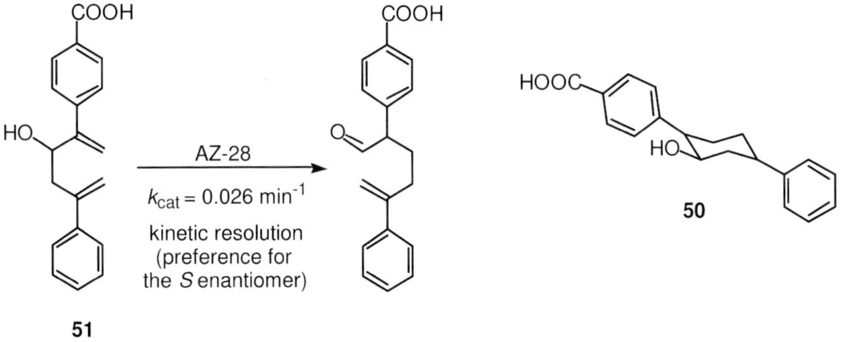

AZ-28
$k_{cat} = 0.026$ min^{-1}
kinetic resolution (preference for the *S* enantiomer)

51

50

Scheme 14.23

14.4.2
Pericyclic Eliminations

Conformational constraints imposed through the hapten combined with medium effect have been successfully explored as a strategy to control the chemical reactivity and selectivity of an antibody-catalyzed selenoxide elimination [53]. For this reaction, proceeding via a planar pericyclic transition state less polar than the initial state, the neutral proline derivatives 52 and 53 were used as haptens. Two stereoisomers were prepared in order to assess how the relative disposition of the carboxylate and aryl group dictates the orientation of the corresponding substituents in the transition state and therefore in the product. Antibody SZ-*trans*-28F8 raised against the *trans*-hapten 53 showed no selectivity with the racemic substrate 54, but antibodies SZ-*cis*-39C11 and SZ-*cis*-42F7 convert only 50% of the racemic substrate 54 into the product. In addition, although the formation of the *E* olefin is favored in the uncatalyzed reaction, SZ-*cis*-39C11 affords a mixture of two diastereomeric olefins in a 45:55 ratio of *cis* and *trans* alkene. As in the Kemp elimination, it is worth noting that bovine serum albumin is also capable of catalyzing the selenoxide elimination of these substrates,

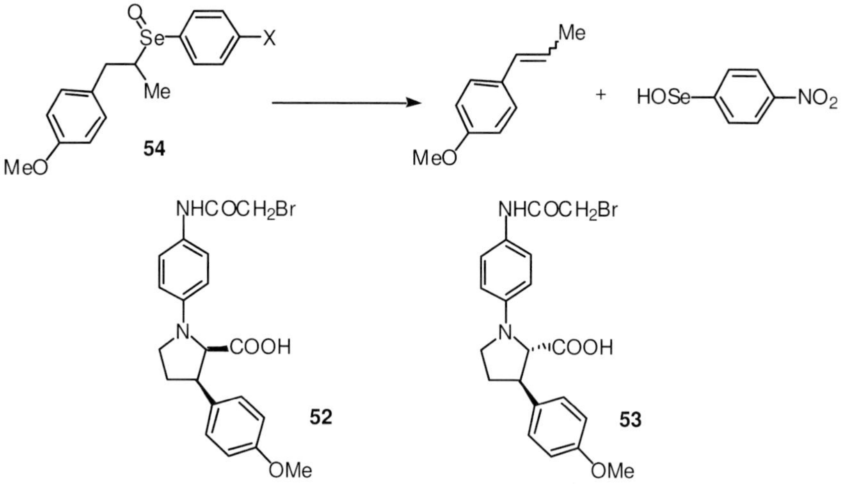

Scheme 14.24

but it does not discriminate between substrates or exhibit any product selectivity (Scheme 14.24).

14.4.3
Cycloaddition

Antibody-catalyzed pericyclic cycloadditions are of particular interest, as the evidence for the existence of natural enzymes for Diels-Alder reactions has only recently been obtained. Hilvert [54] and Schultz [55] were the first to generate catalytic antibodies for the Diels-Alder reaction. They both addressed the problem of product inhibition for this bimolecular process using a bicyclic hapten and relying either on a chemical or a conformational change of the product such that turnover can be achieved. The crucial problem of producing Diels-Alderases that could control the stereochemical outcome of a cycloaddition process between a monosubstituted diene and a dienophile that can potentially yield to eight possible isomers was addressed by Lerner's group [56]. A pair of antibodies was generated where one catalyzed the formation of the favored *endo* isomer and the other the disfavored *exo* Diels-Alder product. The hapten strategy (compound 55) features an appropriately substituted boat-shaped cyclohexene ring as a mimic for the transition state for either the *endo* or the *exo* approach of the dienophile. In the design, the stereochemical relationship between the two amido substitutents on the bicyclic ring was essential in order to control diastereoselectivity. Antibodies 7D4 and 22C8 which accelerated the formation of the *endo* and *exo* adducts 56 respectively in greater than 98% enantiomeric excess, were isolated. In comparison, under similar conditions, the uncatalyzed reaction gives a racemic mixture of 85:15 *endo:exo* products (Scheme 14.25).

Similar results for the same reaction were obtained with antibodies 4D5 and 13G5 raised against the single hapten 57 featuring a conformationally unrestricted fer-

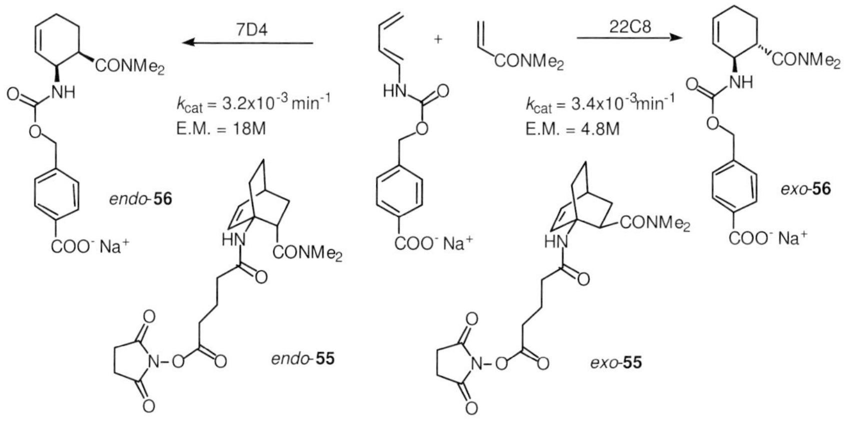

14.4 Pericyclic Reactions | 389

Scheme 14.25

rocene [57]. The success of the approach relies on the ability of the immune system to select a conformer that resembled either the *endo* or *exo* Diels-Alder transition state. The X-ray structure of the Fab fragment of 13G5 in complex with a ferrocene molecule related to the original hapten revealed that a complex network of hydrogen bonding is believed to be responsible for both the modest rate acceleration and the high level of stereoselectivity [58]. Indeed, close examination of the antibody 13G5 binding site and docking of the putative transition state structures indicated that the (3R, 4R)-*exo* transition state isomer is preferentially bound. Tyr L36 probably acts as a Lewis acid activating the dienophile, and AsnL91 and AspH50 form hydrogen bonds to the carboxylate of the diene, both activating and orienting it for the *exo* approach (Scheme 14.26).

In a study carried out by the group of Pandit, an antibody-catalyzed Diels-Alder reaction in which an arylnitroso reacts as the dienophile was generated against the

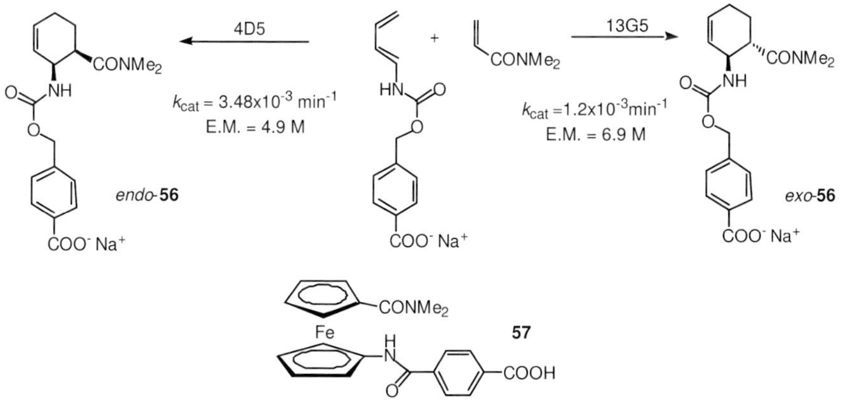

Scheme 14.26

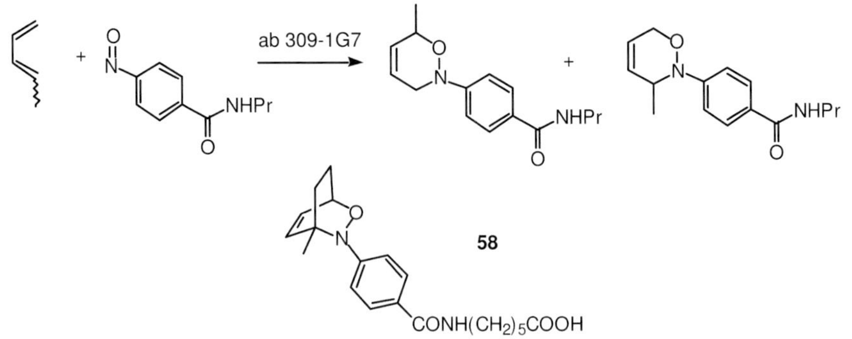

Scheme 14.27

bicyclic [2.2.2.] hapten **58** with the purpose of controlling both the regio- and enantioselectivity of the reaction [59].

The uncatalyzed reaction of the *cis* diene leads to a mixture of racemic regioisomers in a nearly equimolar ratio, 48:52. In the presence of antibody 309-1G7, the reaction was significantly accelerated (effective molarity = 2618 M), but the product distribution was only slightly modified, to a ratio of 32:68. Interestingly, in the presence of a stoichiometric amount of antibody, forcing therefore the entire reaction to take place in the antibody active site, only one regioisomer was obtained, with an enantiomeric excess of 82% (Scheme 14.27).

A mixture of polyclonal antibodies was raised against the *endo*-like hapten **59** and was shown to be a modest catalyst for the hetero-Diels-Alder reaction between diene **60** and ethyl glyoxylate [60]. The polyclonal mixture accelerates the formation of the more favored *endo* adduct **61**. Although the *exo*-like hapten **59** was synthesized, further studies using this hapten were not reported (Scheme 14.28).

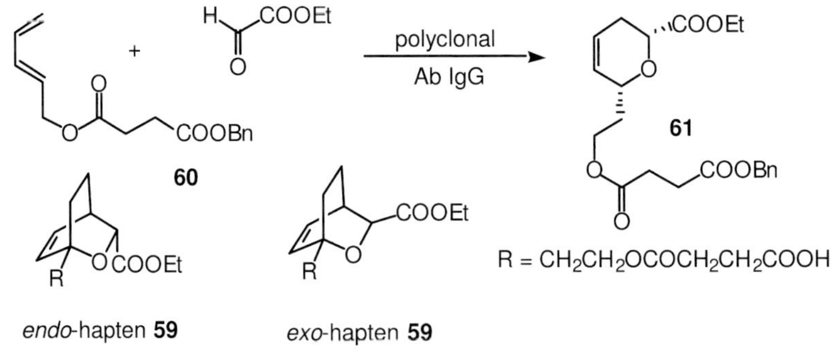

Scheme 14.28

The same group recently reported an example of an antibody-catalyzed aza Diels-Alder reaction [61]. Immunization of rabbits with the bicyclic hapten **62** designed to mimic the *exo* transition state of an aza Diels-Alder reaction provided polyclonal antibodies. One of the polyclonal antibody mixtures, Aza-BSA-3, could catalyze the

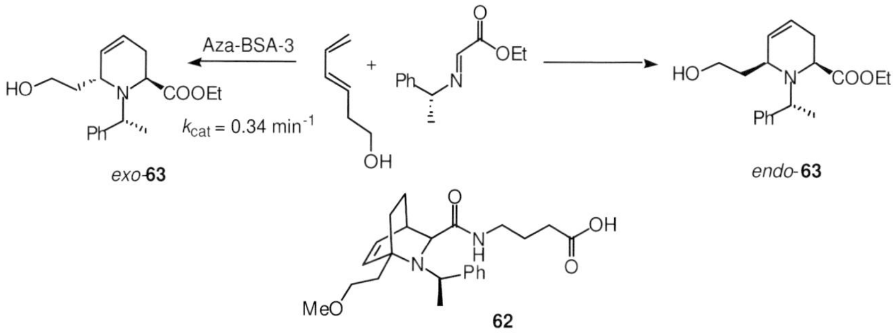

Scheme 14.29

cycloaddition with preferential formation of the *exo* adduct **63** as programmed in the hapten. The rate constant was modest ($k_{cat} = 0.34$ min^{-1}) (Scheme 14.29).

Janda and coworkers have reported the preparation of antibodies capable of catalyzing the 1,3-dipolar cycloaddition of the benzonitrile N-oxide **64** and N,N-dimethylacrylamide. The hapten **65** including flexible benzyl carbamate substituents around the central planar core was designed to potentially yield regioselective catalysts for the generation of either the favored 5- or the disfavored 4-substituted products **66** or **67** [62].

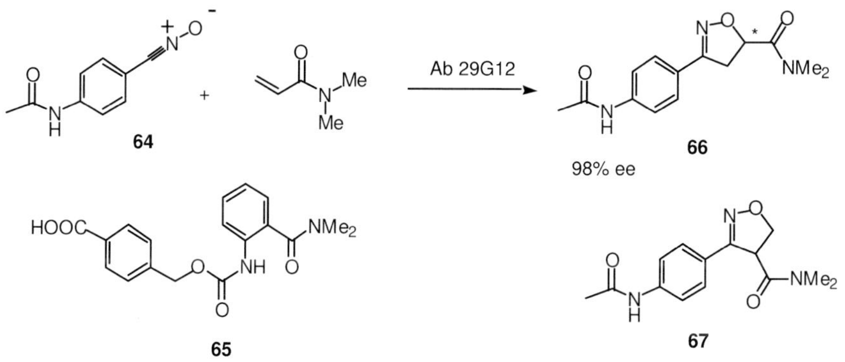

Scheme 14.30

The most efficient catalytic antibody, 29G12, catalyzes the formation of the favored 5-substituted adduct **66** in 98% *ee*, with no evidence of product inhibition. It should be emphasized that, in this case, the hapten **65** contains no chirality in the vicinity of the stereogenic center in the corresponding product. The degree of regio- and enantioselectivity of antibody 29G12 further demonstrates the ability of the immune system to elicit a chiral environment capable of exquisite stabilization of the enantiomeric transition state leading to adduct *R*-**66** only, and this without programmation within the hapten structure.

14.5
Aldol Reactions

The aldol reaction is one of the most powerful methods for the construction of carbon-carbon bonds. The value of the reaction relies upon our ability to generate catalysts that not only accelerate the rate of the reaction but also control the relative and absolute configurations of the newly formed stereogenic centers [63]. In recent years, the search for asymmetric catalytic methods for direct aldol reactions that do not require preactivation of the starting materials has become a major effort in synthetic organic chemistry [64]. To date, two type I aldolase catalytic antibodies are commercially available (Aldrich Chemical Co.). They have the ability to match the efficiency of the natural aldolase while accepting a more diverse range of substrates.

Initial progress in this area was made by Reymond and Lerner, who demonstrated that the generation of antibodies for a multistep process such as the aldol reaction is feasible. They discovered that antibody 78H6 raised against the piperidinium-based hapten 68 promoted the intramolecular aldol condensation of the keto-aldehyde 69 to give the substituted 2-benzyl-3-hydroxy-cyclohexanone 70 as the primary product, which subsequently eliminated water to give the corresponding unsaturated ketone 71 [65]. In the presence of the antibody, all four possible stereoisomers of the aldol product 70 are formed in equal amounts, but the elimination sequence occurred with complete *anti*-selectivity to give a single enantiomer of the intermediate 70 (Scheme 14.31).

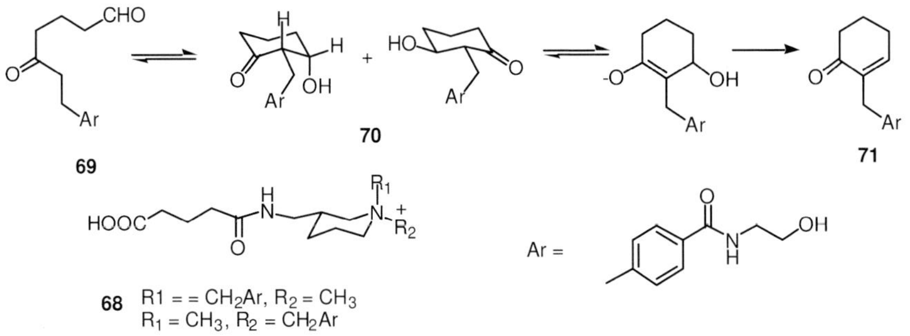

Scheme 14.31

Reymond and Chen subsequently investigated bimolecular aldol processes and found that antibody 72D4, raised against the same piperidinium hapten 68, displays aldolase-like activity in the presence of the primary amine 72 [66]. The exogenously added amine serves as a cofactor for the antibody, mimicking the role of the essential lysine residue present in the active site of type I aldolase. The antibody catalyzed the stereoselective addition of acetone to the *si* face of the aldehyde 73 to give (4S, 5S)-74 (> 95% de) from S-73 and (4S, 5R)-74 from R-73 (65% de). When the racemic aldehyde is used, these products are formed in a 1:2.8 ratio. The aldol reaction is a reversible process, and therefore aldolase antibodies are potentially able to catalyze both the forward aldolization and *retro*-aldolization processes and might be therefore

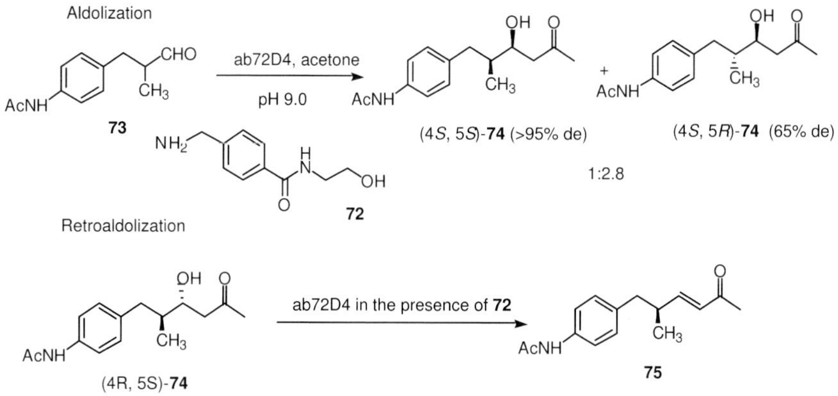

Scheme 14.32

useful for the kinetic resolution of aldol products. Reymond has shown that antibody 72D4 accelerated the *retro* aldol reactions of (4S, 5S)-74 in the presence of the amine 72, but elimination of water to form the unsaturated ketone 75 competes with the *retro* aldol process (Scheme 14.32).

A major breakthrough in aldolase antibody catalysis came with the report of Lerner and Barbas III, who generated aldolase antibodies against the β-diketone hapten 76, which serves as a chemical trap to imprint the dependent type I aldolase mechanism in the active site [67]. This concept of "reactive immunization" [12] allowed the formation of two aldolase antibodies 38C2 and 33F12 that were found to have remarkable scope. The structure of the hapten and the suggested mechanism for the selection process of antibodies 38C2 and 33F12 during immunization is presented below (Scheme 14.33).

The X-ray structure of 33F12 has been solved and revealed the presence of the Schiff base-forming lysine residue buried in a hydrophobic pocket at the base of the binding site, which accounts for its lower pK_a value of 5.5 [68]. Both aldolase antibodies 38C2 and 33F12 catalyze highly stereoselective inter- as well as intramolecular aldol

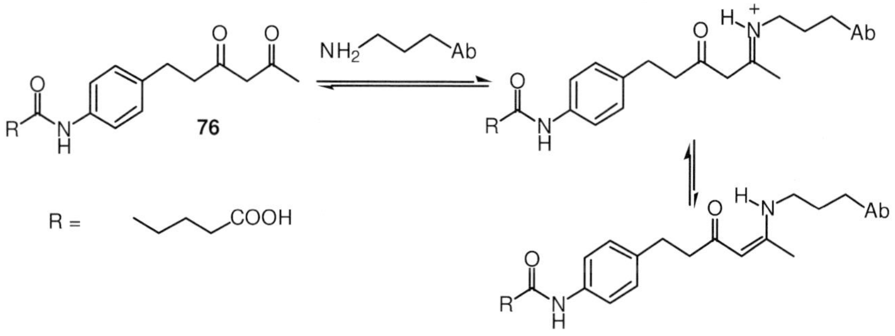

Scheme 14.33

reactions [69]. In particular, the Wieland-Miescher ketone has been prepared with an enantiomeric excess greater than 95% [70].

As a general rule, acetone adds to aldehydes with *si*-facial selectivity, whereas with hydroxyacetone a reversal of enantioface selectivity results in addition to the *re*-face. The aldol products are usually obtained in *ee*s up to > 99%. The donor, the acceptor, and the antibody catalyst all exert an effect on the enantioselectivity of the reaction. The highest enantioselectivities were observed for reactions where conjugated aldehydes served as acceptors with acetone as the donor. Lower enantioselectivities were usually obtained with aldol acceptors containing an sp^3 center at the α position. However, in this latter case, enantioselectivity might be increased by addition of steric bulk at this carbon center. The lowest enantiomeric excesses were obtained for the reaction of acetone with 4-(4′-acetamidophenyl)butyraldehyde as the acceptor (20% *ee* and 3% *ee* with antibodies 38C2 and 33F12 respectively). The use of hydroxyacetone as an aldol donor instead of acetone provides the corresponding aldol products with better enantiomeric excesses (77% and 70% with antibodies 38C2 and 33F12 respectively). The product is formed as a single *syn*-diastereoisomer and regioisomer, with the reaction taking place exclusively on the carbon bearing the hydroxy group (Scheme 14.34).

Interestingly, in the presence of 38C2, the regioselectivity of fluoroacetone is opposite to that observed with the natural aldolase DERA, thus providing a complementary approach even though the antibody reaction is not highly selective [71]. The major product of the reaction of fluoroacetone with aldehyde **77** is the *syn*-α-fluoro-β-hydroxyketone (**78**) formed in 95% *ee*. The *anti*-isomer **78** is formed with an isolated chemical yield of 21% and an *ee* of 34%. The regioisomer product **79** was also detected (7% isolated yield and 97% *ee*) (Scheme 14.35).

Product		ee		Product		ee	
		38C2	33F12			38C2	33F12
		> 99 %	> 99 %			de> 95 %	de> 95 %
		99 %	98 %			77% (de>99%)	70% (de >99%)
		58 %	69 %			> 98%	89%
		20 %	3 %			> 95 %	> 95 %

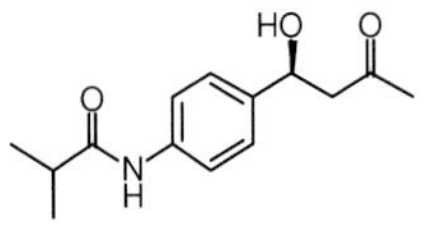

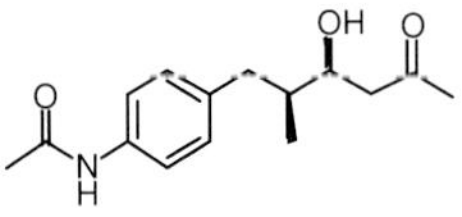

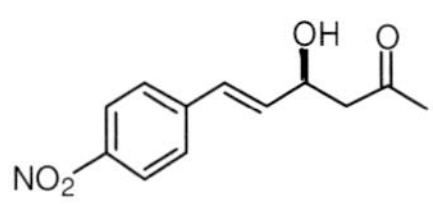

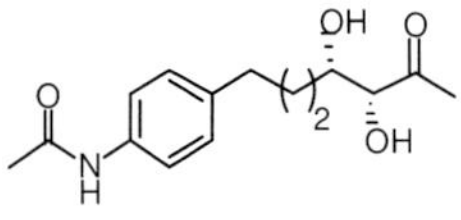

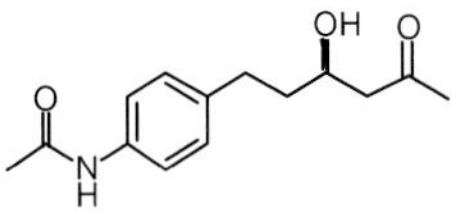

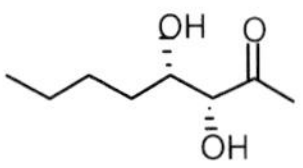

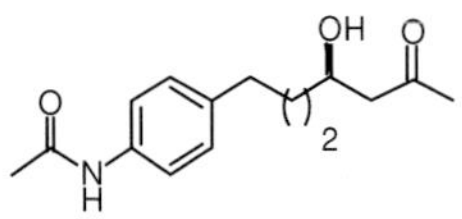

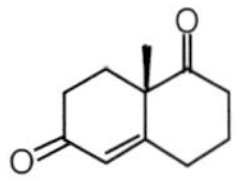

Scheme 14.34

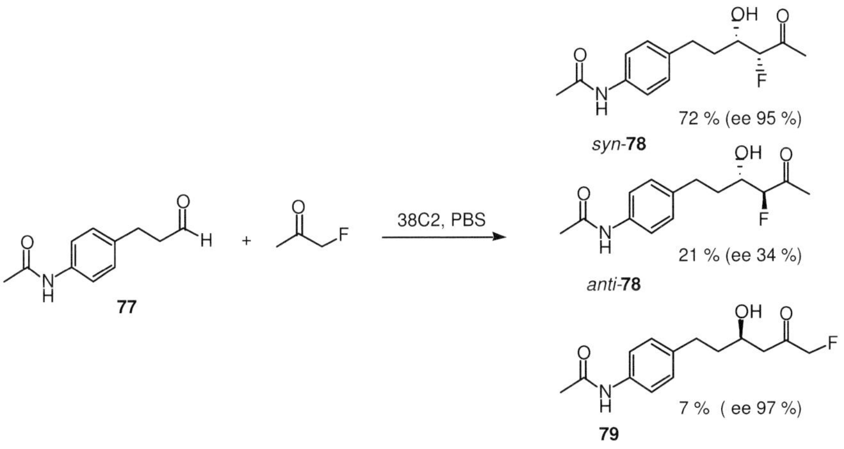

Scheme 14.35

The reversible nature of the aldol reaction allows the possibility to use these aldolase antibodies for both a forward aldol and *retro*-aldolization process. Therefore, a single antibody could then be used for the preparation of both aldol enantiomers (Scheme 14.36) [72].

This was exemplified by the enantioselective preparation of a series of compounds including β-hydroxyketones, which could be obtained at approximately 50% conversion with enantiomeric excesses up to 99%. Antibodies 38C2 and 33F12 were also successfully used in the enantioselective *retro*-aldol reaction of tertiary aldols containing structurally diverse heteroatom-substituted quaternary carbon centers [73]. These compounds represent challenging targets and were successfully prepared with enantiomeric excesses values typically greater than 95%. The catalytic proficiency, $(k_{cat}/K_M)/k_{uncat}$, of the antibody for these reactions was on the order of 10^{10} M^{-1}. The utility of this methodology for natural product syntheses has been demonstrated with the preparation of (+)-frontalin, the side chain of Saframycin H, and key precursors for the synthesis of mevalonolactone.

A rare example of an aldolase antibody-catalyzed enantiogroup-differentiating process was reported by Lerner and Barbas III [74]. Indeed, it was shown that antibody 38C2 catalyzed the enantioselective aldol cyclodehydration of 4-substituted-2,6-heptanediones **80** to give the enantiomerically enriched corresponding cyclohexenones **81** with moderate enantiomeric excesses up to 62% (Scheme 14.37).

In order to further increase the repertoire and the efficiency of the aldol reaction, a second generation β-diketosulfone **82** was used as hapten with the purpose of combining both reactive immunization and transition state theory [75]. The overall trend toward increased efficiency of the resulting antibodies (84G3, 93F3) is consistent with the idea that the inclusion of transition state analogy into the hapten design is beneficial. Interestingly, the *ee* values obtained with 84G3 and 93F3 are similar to those obtained with 38C2 and 33F12, but the enantioselectivity is reversed (Scheme 14.38).

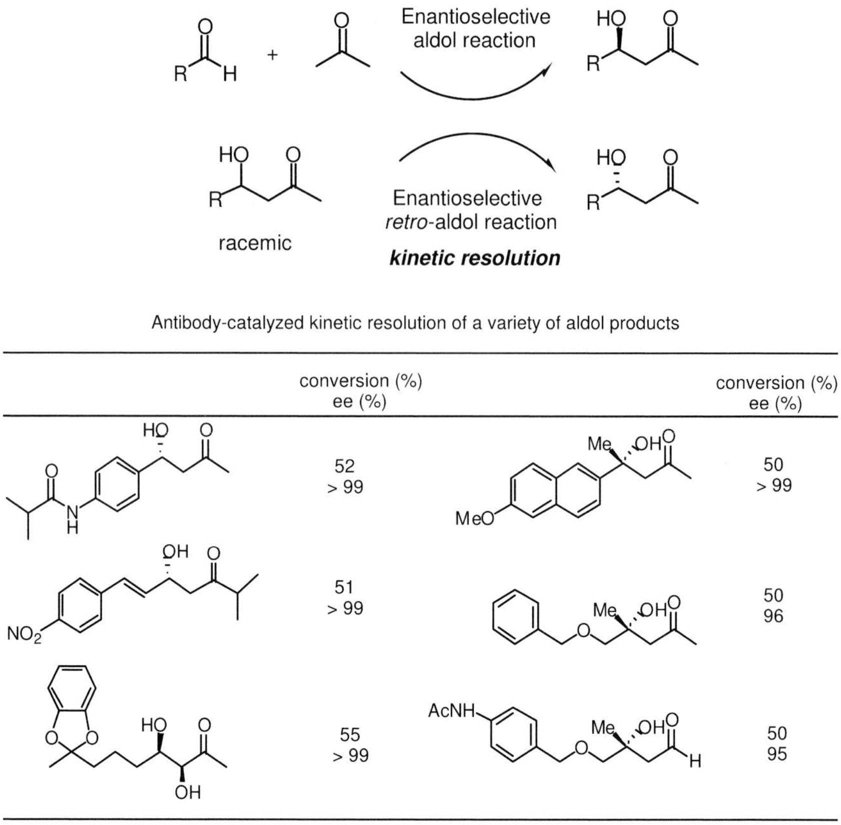

Scheme 14.36

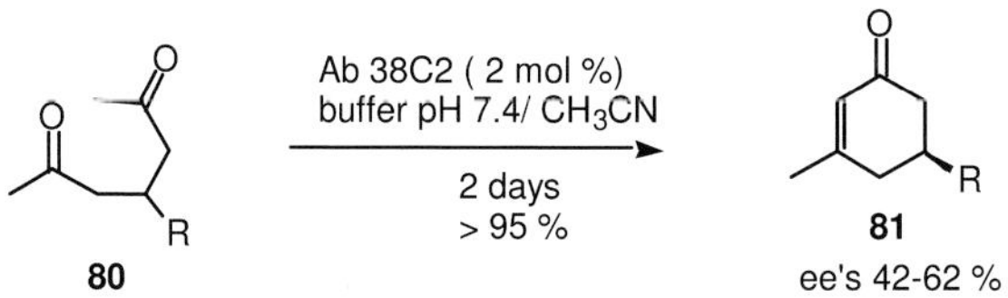

with R = Ph, n-C_5H_{11}, $pAcNHC_6H_4CH_2CH_2$

Scheme 14.37

In addition, antibody 93F3 exhibited diastereoselectivities that differ from that obtained with ab38C2, as exemplified by the addition of 3-pentanone to aldehyde **77**. Antibody 93F3 provides **83** with 90% *de* (*syn*-isomer) and 90% *ee* while ab38C2 provides the *anti*-isomer with 62% *de* and 59% *ee* (Scheme 14.39).

Gouverneur et al. recently reported their findings that antibody 84G3 is an effective asymmetric catalyst for the regio- and enantiocontrolled formation of a series of disfavored aldol products, therefore controlling reactivity in a unique way [76]. Indeed, in the presence of unmodified unsymmetrical methyl ketones, the antibody-

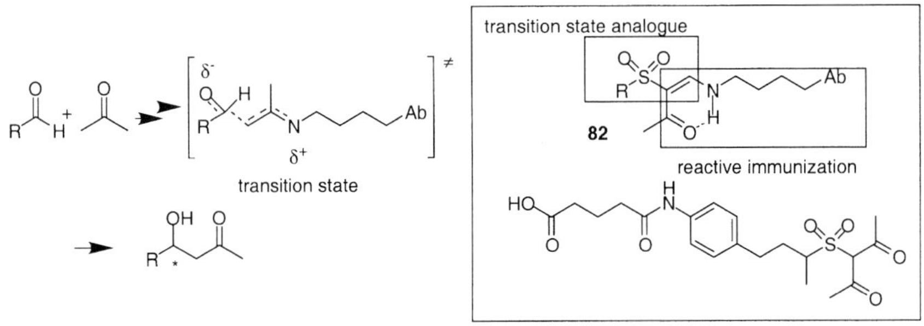

Scheme 14.38

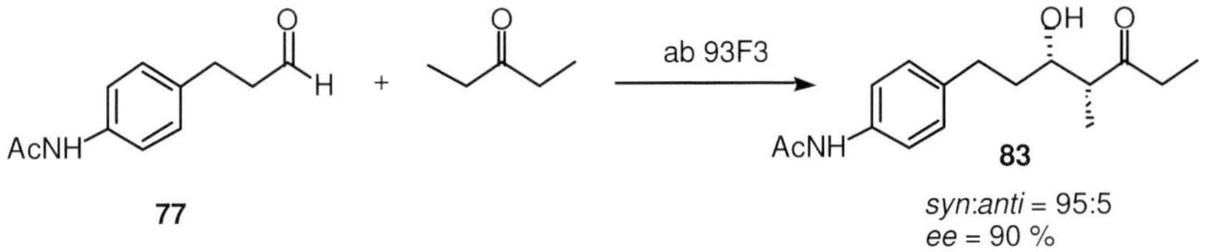

Scheme 14.39

catalyzed reactions occur exclusively at the less substituted carbon, independently of the presence of heteroatoms in the donor ketones. This reactivity contrasts with the results obtained for the uncatalyzed reactions carried out under the same conditions (PBS, pH=7.4). Indeed, the uncatalyzed reactions occur preferentially at the more substituted carbon, resulting in a mixture of *syn* and *anti* stereoisomers. The other regioisomer is either not observed or is formed only as a minor product. Antibody 84G3 is unique in the sense that this antibody is capable of reversing the regioselectivity for these reactions. The typical enantioselectivities of these ab84G3-catalyzed aldol or *retro*-aldol reactions are typically greater than 94%. This new reactivity highlights the scope of the reactive immunization strategy developed by Lerner and Barbas III for catalyst design and contrast with the regioselectivity obtained in the presence of antibody 38C2. With this latter antibody, the regioselectivity of the aldol reactions in the presence of unsymmetrical methyl ketones as donors occur typically in favor of the methylene group (Scheme 14.40).

The sense of the regioselectivity observed with 38C2 has been confirmed by the group of Keinan et al., who demonstrated that for the antibody-catalyzed deuteration of hexan-2-one, the discrimination in favor of the methylene group over the methyl group is of the order of 6 (k_{CH_2}/k_{CH_3} = 5.8) [77].

All the examples presented in this section are related to antibodies mimicking or possessing an aldolase type I activity. To date, no example of aldolase type II antibodies has been reported in the literature. The only report of an antibody-catalyzed *retro*-aldol reaction not involving enamine chemistry was published by the P. G. Schultz group [78]. They demonstrated that Ab 29C5.1 raised against the phosphinate hapten 84 catalyzed the Henry type *retro*-aldol reaction of 85. The stereoselectivity of this

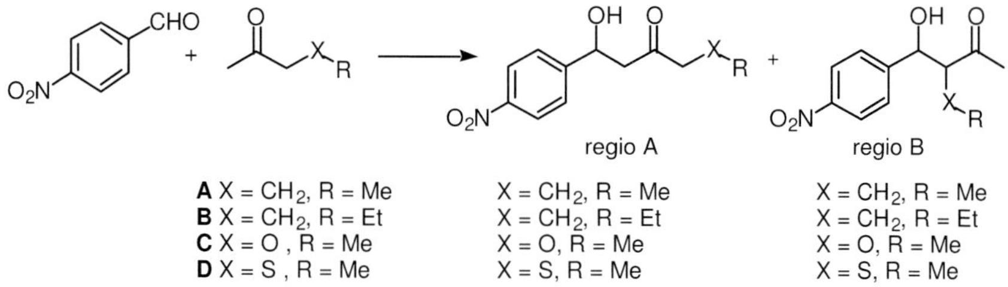

14 Selectivity with Catalytic Antibodies – What can be achieved?

Scheme 14.40

	A	X = CH$_2$, R = Me
	B	X = CH$_2$, R = Et
	C	X = O, R = Me
	D	X = S, R = Me

regio A: X = CH$_2$, R = Me; X = CH$_2$, R = Et; X = O, R = Me; X = S, R = Me

regio B: X = CH$_2$, R = Me; X = CH$_2$, R = Et; X = O, R = Me; X = S, R = Me

Ketone	Condition	[%] conversion[a]	[%] Ratio[a] regio A/ regio B[b]
A	PBS, pH=7.4, rt, 3h	0	0/0
A	PBS, 9%mol Ab84G3, pH=7.4, rt, 3h	88	100/0[c]
B	PBS, pH=7.4, rt, 26h	1	0/100
B	PBS, 9%mol Ab84G3, pH=7.4, rt, 26h	66	99/1[c]
C	PBS, pH=7.4, rt, 15h	54	4/96
C	PBS, pH=7.4, 0°C, 41h	0	0/0
C	PBS, 25%mol Ab84G3, pH=7.4, 0°C, 41h	76	100/0[c]
D	PBS, pH=7.4, rt, 3h	49	0/100
D	PBS, pH=7.4, 0°C, 1h40	9	10/90
D	PBS, 25%mol Ab84G3, pH=7.4, 0°C, 1h40	58	98/2[c]

[a]assigned by HPLC; [b]mixture of syn and anti isomers; [c]uncorrected ratio

antibody is modest. The *syn* diastereomer was found to be a better substrate than the *anti* diastereomer for antibody catalysis. Unfortunately, no evidence of enantioselectivity was observed, as verified by monitoring the relative consumption of the enantiomers of *syn*-85 by chiral HPLC (Scheme 14.41).

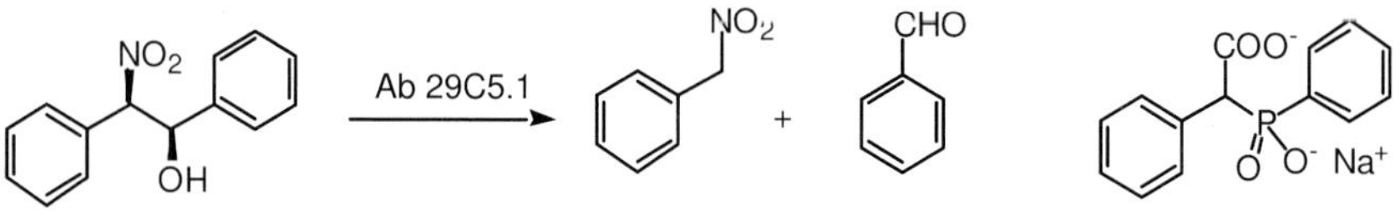

Scheme 14.41

syn-85 → Ab 29C5.1 → products + 84

14.6 Cyclization

We have seen previously that antibody-catalyzed intramolecular aldol reactions can lead to cyclic products such as the Wieland-Miescher ketone, a key compound in the synthesis of steroids. Another synthetically useful cyclization process is the intramolecular ring opening of epoxides, as this transformation could lead to the formation of oxygen-containing heterocycles such as tetrahydropyrans, tetrahydrofurans, or tetrahydrooxepanes. The groups of Lerner and Janda generated two antibodies, 26D9 and 5C8, raised against haptens **86** and **87** respectively, that catalyzed the regioselective *endo* ring closure of epoxyalcohol **88** to form the tetrahydropyran **89** exclusively [79]. No product arising from an *exo* addition was observed in the presence of catalyst 26D9. This process further exemplifies the antibody-mediated formation of products along a disfavored reaction pathway. Calculations confirmed that the catalyst favors an energetically more demanding pathway, as the 6-*endo* transition state is 3.6 kcal/mol less favorable than the 5-*exo* transition state [80]. Antibody 26D9 is stereospecific, transforming only the (S,S)-enantiomer of substrate **88** to (2R,3S)-**89**. Antibody 5C8 was found to transform both antipodes of **88** but with different product distribution. The (S,S)-enantiomer of **88** was converted by 5C8 to the disfavored tetrahydropyran, whereas (R,R)-**88** yields the tetrahydrofuran **90** as major product that is also the product of the uncatalyzed process. Therefore, the antibody 26D9-catalyzed reaction is more useful synthetically, as this catalyst allows the kinetic resolution of racemic **88**. This particular transformation has been automated for large-scale production under biphasic conditions (Scheme 14.42) [81].

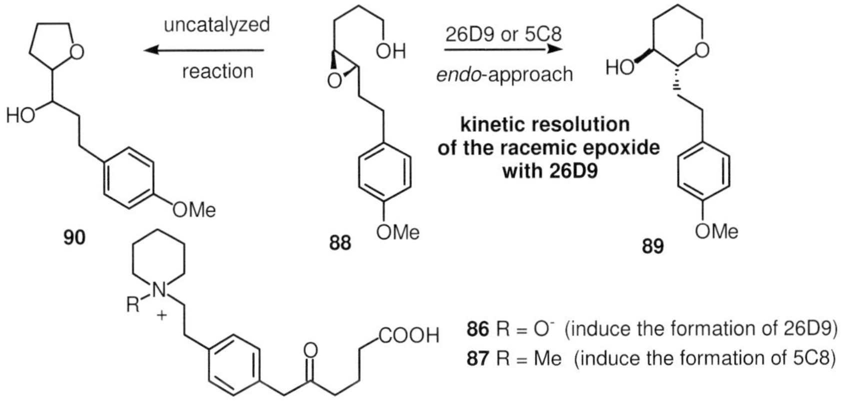

Scheme 14.42

This study was extended to the preparation of hydroxyoxepane [82]. In the presence of the antibody 26D9 the substrate **91** undergoes the 7-*endo*-tet process in preference to the 6-*exo*-process with almost complete regiocontrol. Product **92** is formed with an enantiomeric excess of 78%. The absolute configuration of the product is consistent with a process that proceeds by inversion at the reacting center (Scheme 14.43).

The *exo*- and *endo*-lactonizations of the α-trifluoromethyl-γ,δ-unsaturated acid **93** are two other examples of cyclizations successfully catalyzed by an antibody [83]. The group of Kizamune elicited antibodies against the tetrahydrofuran **94** and the tetrahydropyran **95**, which catalyzed the regiospecific formation of the γ- or δ-lactones **96** and **97** respectively. The antibody-dependent *exo*-cyclization was found to proceed with high diastereoselectivity (> 96%) (Scheme 14.44).

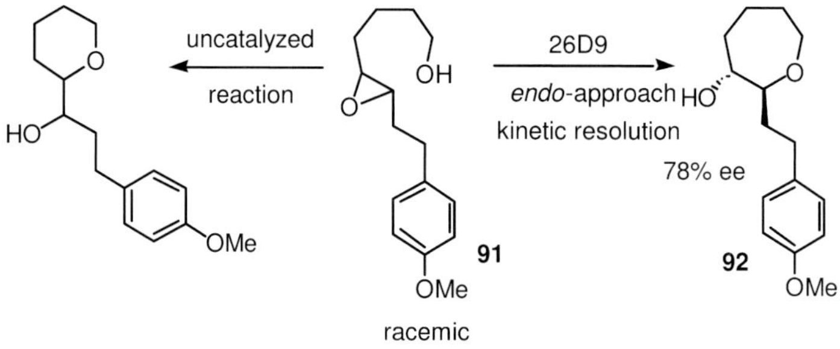

Scheme 14.43

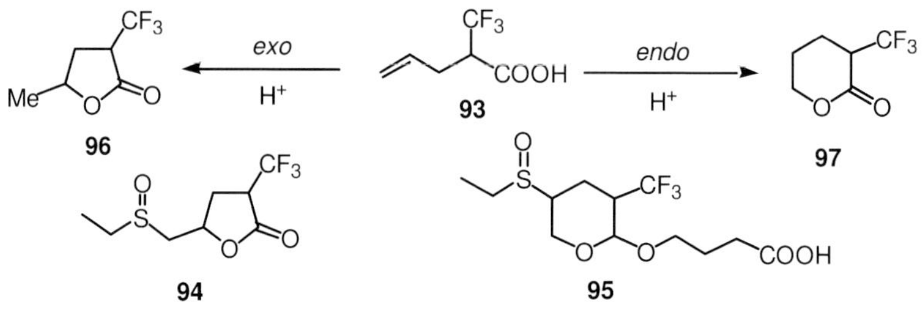

Scheme 14.44

Electrophilic ring closure is a key transformation involved in the biosynthesis of many natural terpenoid products. For example, the conversion of 2,3-oxidosqualene to lanosterol is one of the most complex processes controlled by an enzyme, as it involves the simultaneous construction of four new carbon-carbon bonds and six stereocenters [84]. Recently, much effort has been devoted toward the generation of antibodies that could catalyze such a demanding process. In 1994 and 1996, catalytic antibodies were elicited against two different haptens that catalyze the simplest form of this cyclization reaction, the formation of a single cyclohexane ring [85, 86]. In both cases, an arylsulfonate was chosen as the leaving group, possibly for contribution to substrate recognition by the antibody and also presumably for ease of assay. Lerner and Janda generated first antibody 4C6 using the N-oxide compound 98 as the hapten. This antibody was able to catalyze the almost exclusive formation of trans-cyclohexanol 99 from 100 (Scheme 14.45).

Subsequently, they prepared antibody TM1-87D7 elicited against hapten 101 including an ammonium group instead of an N-oxide [87]. Interestingly, for this antibody, subtle changes in substrate structure led to remarkable changes in the product outcome. Substrate 102, possessing a structure very similar to that of compound 100, led to the trans-cyclohexanol 103. The trans olefin 104 was exclusively transformed into the bicyclic compound 105, a product not observed in the uncatalyzed reaction. The production of this strained product is of particular interest, as its formation via a chemically catalyzed cationic cyclization process is unprecedented. Changing to the cis-olefinic substrate 106 yielded the five-membered ring 107. The enantiomeric excesses of these products were not reported. The rate acceleration observed with TM1-87D7 was within an order of magnitude of those of natural enzymes catalyzing similar processes (Scheme 14.46).

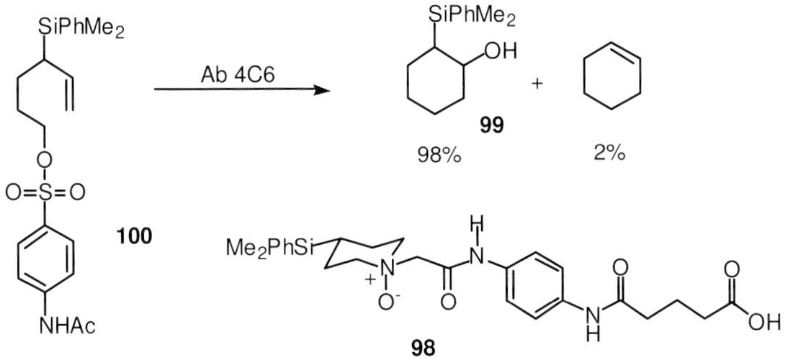

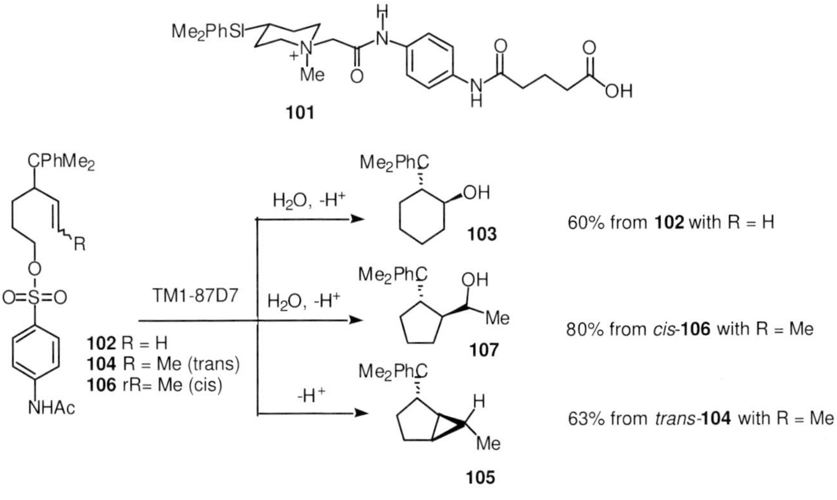

Scheme 14.45

Scheme 14.46

In a subsequent report, antibody HA5-19A4 was described as a powerful catalyst for the formation of the first bridge-methylated decalins via a tandem cationic cyclization process [88]. The antibody was raised against hapten **108** used as a mixture of diastereomers differing only in the configuration at the tertiary N-oxide center and co-immunized as racemates. The hapten was designed to generate a catalyst that will enforce the appropriate reactive conformation of the substrate and stabilize any developing cationic centers during the cyclization cascade. Hapten **108** has the advantage of having little similarity to the possible products, therefore minimizing product inhibition. In addition, the presence of the epoxide within its structure might lead to an enhanced immune response by covalent interaction and function therefore as a reactive immunogen. Substrate **109** was found to be a cyclization substrate for antibody HA5-19A4. The products **110**, **111**, and **112** were formed by the antibody in a ratio of 2:3:1, which is in accord with the expected stabilities of these regioisomers. This suggests that the antibody active site does not provide tight regiocontrol on the deprotonation process as part of the termination. The regioisomeric olefins 110 and 111 were formed with an enantiomeric excess around 50%. Compound 112 was formed with an enantiomeric excess of 80% (Scheme 14.47).

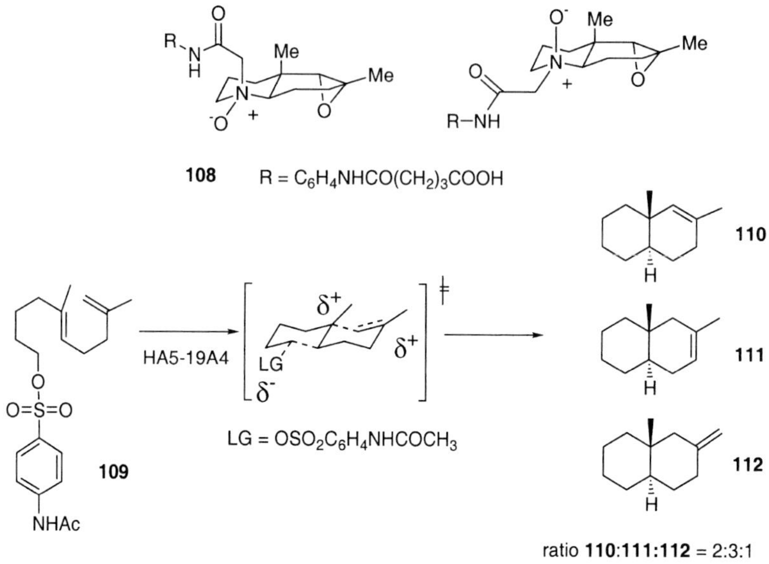

Scheme 14.47

In 2000, Hasserodt et al. used a 4-aza-steroid N-oxide to generate three antibodies that catalyzed the cationic cyclization of an oxidosqualene derivative to form the A ring of the lanosterol skeleton. Therefore, these catalysts are able to exert control over the initiation process of this multistep transformation. This result opened up the possibility of the antibody formation of steroidal carbon frameworks (not illustrated) [89].

14.6.1
Reduction and Oxidation

Although an impressive level of selectivity could be achieved with naturally occurring redox enzymes, natural biocatalysts might require added cofactors for their *in vitro* functioning, and in many cases cofactor recycling complicates their use [90]. Therefore, alternative biocatalysts that could employ inexpensive oxidants or reductants are of particular interest, especially if these catalysts achieve a high degree of selectivity. Initial progress toward antibody-catalyzed stereoselective reductions was made by Arada and coworkers [91]. They reasoned that an *anti*-dansyl monoclonal antibody possessing a highly hydrophobic subsite might serve as a kind of chiral auxiliary for the stereoselective reduction of 5-(dansylamino) levulinic acid in the presence of

NaBH$_4$. Incubation of the starting material with the antibody (molar ratio 2:1) in the presence of NaBH$_4$ at room temperature for three days afforded, after acidic work-up, the expected reduced and cyclized product **113** from **114** with an enantiomeric excess of 36%. This preliminary approach suffers from moderate enantiomeric excess for an antibody-promoted reduction that is stoichiometric instead of catalytic (Scheme 14.48).

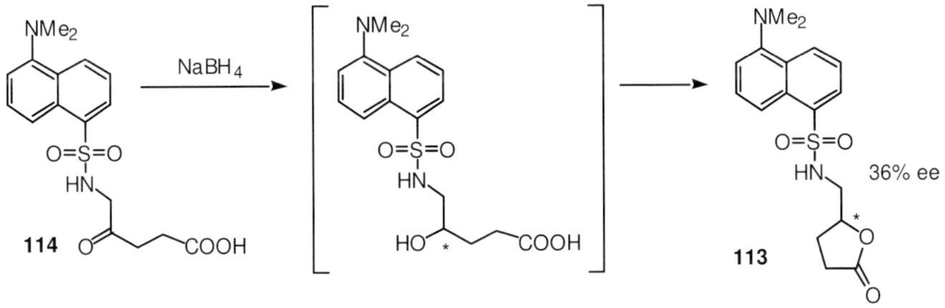

Scheme 14.48

Schultz et al. have used the phosphonate hapten **115** to generate antibody A5, capable of accelerating the NaBH$_3$CN-dependent reduction of the ketoamide **116** at pH 5.0 with greater than 25 turnovers and no apparent change in V$_{max}$, suggesting that the reducing agent does not inactivate the antibody at a significant rate [92]. The reduced product **117** was obtained with 99% diastereoselectivity, indicating that the antibody active site efficiently discriminated the diastereomeric transition states for carbonyl reduction. Interestingly, the uncatalyzed reaction provides as a major product the opposite diastereoisomer with a *de* of 56% (Scheme 14.49).

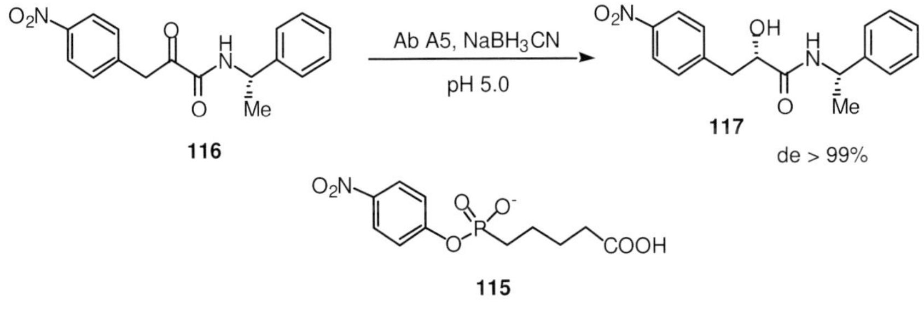

Scheme 14.49

For compounds possessing more than one functional group sensitive to reduction, it is important to design catalysts that can achieve simultaneously a high degree of chemo- and stereoselectivity. The remarkable specificity of an antibody catalyst has been explored to accomplish a highly selective ketone reduction not attainable by current chemical methods without recourse to protective-group chemistry. Antibody 37B39.3 was raised against the amine oxide hapten **118** by the group of Schultz and was found to catalyze the reduction of the diketone **119** to the hydroxyketone **120** with >75:1

chemoselectivity [93]. This chemoselectivity for one of two nearly equivalent ketone groups is remarkable as the two ketones within the substrate are distinguishable only by methoxy and nitro groups several atoms away from the reacting center. In addition, this antibody-catalyzed reaction is highly stereoselective, affording the hydroxyketone **120** in very high enantiomeric excess (96.3%). In this case, it is the amine oxide hapten **118** that served to induce a chiral environment capable of discriminating the enantiotopic faces of the prochiral carbonyl group and of stabilizing the tetrahedral transition state resulting from hydride attack. This simple approach may find general applicability to perform highly selective functional group transformations regardless of chemical environment and substrate complexity (Scheme 14.50).

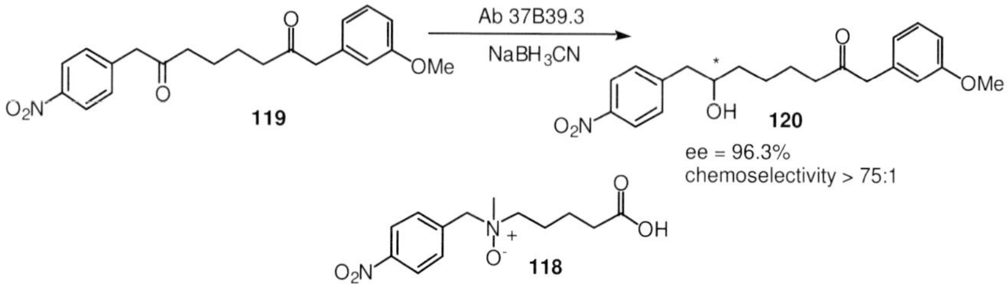

Scheme 14.50

Given their chemical and biological importance, oxygenation reactions have also long been targets for antibody catalysis. The preparation of enantiopure epoxides from non-functionalized alkenes with the assistance of a biocatalyst has a special synthetic relevance as this transformation is difficult to achieve by other means [94]. J.-L. Reymond et al. have reported that peroxycarboximidic acid generated *in situ* from acetonitrile and hydrogen peroxide is compatible with catalytic antibodies in physiological buffers and can be used as a reagent for antibody-catalyzed enantioselective epoxidations of non-functionalized alkenes [95]. Antibody 20B11 raised against the piperidinium hapten **121** is a catalyst for the epoxidation of several alkenes such as **122**. The enantiomeric excesses for these reactions are modest but could reach 98% for selected transformations. Screening of a library of 46 monoclonal antibodies against **121** with the milder reagent formamide-hydrogen peroxide gave three additional catalytic antibodies that display complementary enantioselectivities (Scheme 14.51) [96].

In 1994, it was reported that sodium periodate is an oxidant compatible with monoclonal antibodies and could be used for the antibody-catalyzed oxidation of sulfides to form the corresponding sulfoxides [97]. Eight antibodies raised against hapten **123** were found to be catalytic. Hapten **123** contains both a protonated amine at physiological pH expected to stabilize the incipient positive charge on sulfur present in the transition state and a phosphonic acid group aimed to provide a binding site for the periodate ion (Scheme 14.52).

Antibody 28B4.2 is particularly efficient, with a turnover number (k_{cat} = 8.2 s^{-1}) that compares favorably with some naturally occurring monooxygenase enzymes, but selectivity is low, as reflected by an enantiomeric excess of only 16% for the oxidation of the sulfide **124**. This low enantioselectivity could be explained by a

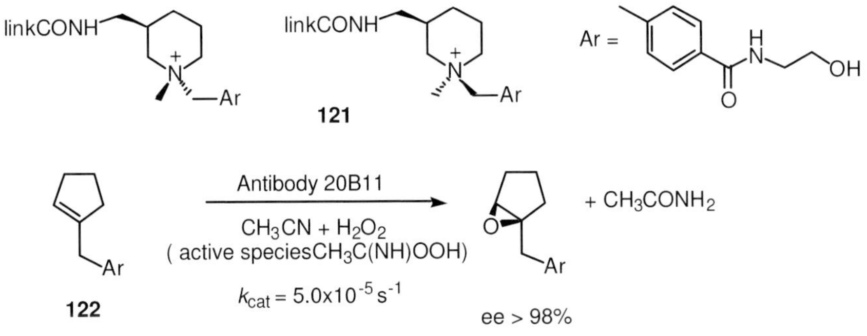

Scheme 14.51

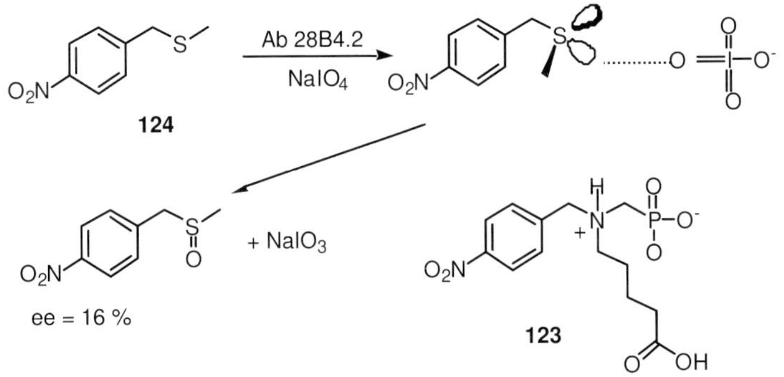

Scheme 14.52

limited complementary between the active site and the linker region of the hapten, as revealed by the structural work on this antibody [98].

14.7 Additions and Eliminations

The enantioselective protonation of a prochiral enol ether is a challenging process as exemplified by numerous examples reported in the literature [99]. In addition to chemical systems, biological catalysts could be very attractive candidates for this process. Enantioselectivities between 41% and 96% *ee* for enol protonation were reported for the yeast esterase-catalyzed hydrolysis of 1-acetoxycycloalkenes [100]. In 1991, an antibody was prepared that allowed the enantioselective protonation of an enolate with an *ee* of 42% [101]. In 1992, Reymond, Lerner, and Janda demonstrated that is possible to achieve a much higher level of enantioselectivity for the hydrolysis of enol ethers under acidic conditions (Scheme 14.53) [102].

The versatile antibody 14D9 raised against hapten **125** showed a good activity for the cleavage of enol ethers such as **126** and **127**, and the corresponding aldehyde was

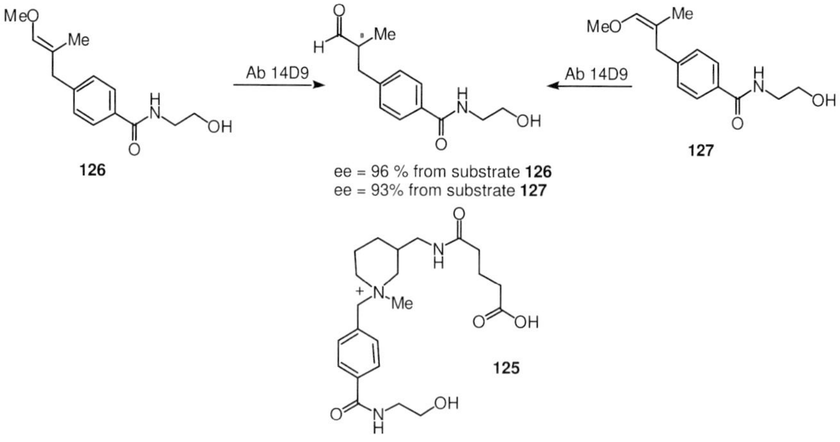

14 Selectivity with Catalytic Antibodies – What can be achieved?

Scheme 14.53

obtained with enantiomeric excesses of 96% and 93% respectively. Catalysis occurs via stabilization of the intermediate oxocarbonium ion by an ionizable protein side chain. This antibody-catalyzed transformation was undertaken on a gram scale and used as the key step of the synthesis of the pheromone multastriatin [103]. Further studies on this enantioselective antibody-catalyzed protonation revealed that the observed selectivity originated from the discrimination of the competing enantiomeric transition state (diastereomeric within the active site), since the antibody binds the two enantiomeric products with similar affinity.

Reversal of chemoselectivity is a particularly interesting challenge for catalytic antibodies as it involves modification of the intrinsic order of reactivity in a given series of functional groups. Sinha and Keinan addressed this problem with the same antibody 14D9 [104]. They reported two examples of reversal of chemoselectivity, including the activation of enol ether in the presence of ketal and the inversion of the order of reactivity in structurally similar ketals. In 1995, they also demonstrated that antibody 14D9 catalyzed the conversion of the enol ether **128** to the cyclic ketal **129** in water with a chemical yield of 12% [105].

This result is remarkable because, although the ketal is the normal product in organic solvent, it is never observed in water because the highly reactive oxocarbenium intermediate reacted immediately with water to give the ketone **130** as the end product (Scheme 14.54).

In addition to the elimination processes described above and proceeding through a pericyclic transition state, a few other examples of elimination have been reported in the literature including an energetically demanding *syn*-process [106]. A positively charged ammonium-based hapten, **131**, was prepared to elicit an antibody possessing within its active site a complementary negatively charged carboxylate capable of serving as a base to trigger the elimination of HF from substrate **132**. Antibody 43D4-3D12 was identified as a catalyst for this elimination in a pH-dependent manner and with a rate acceleration (compared to the acetate-catalyzed process) in the range of 10^5.

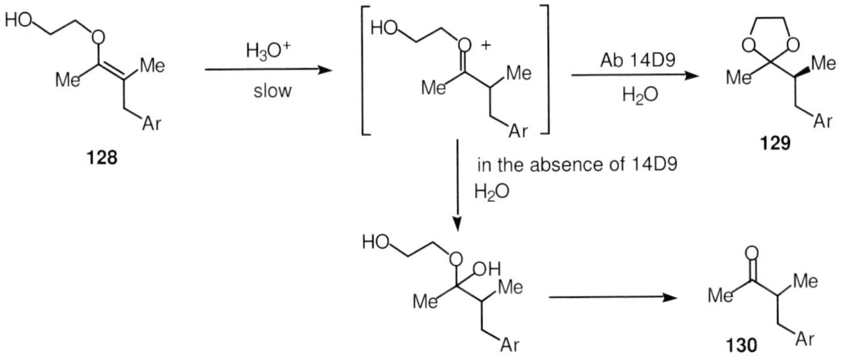

Scheme 14.54

Surprisingly, the antibody displayed little stereoselectivity, abstracting both the pro-R and pro-S proton, but the *anti*-elimination geometry was favored over the *syn*-process, suggesting important conformational flexibility of the substrate within the antibody active site (Scheme 14.55).

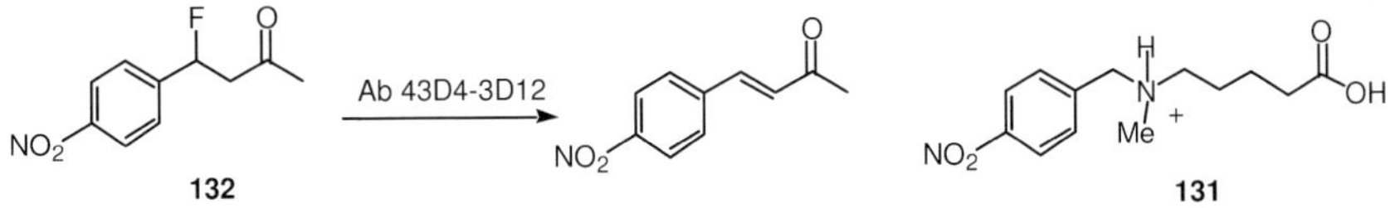

Scheme 14.55

For an acyclic system, it is generally accepted that antiperiplanar elimination is greatly favored over *syn* elimination [107]. Moreover, a *syn* elimination yielding a *cis* olefin is an even more difficult transformation. Hapten **133** was designed and synthesized to generate antibodies that could catalyze the *syn* elimination of substrate **134** to yield the *cis* olefin **135**. In the uncatalyzed reaction, this same substrate undergoes an *anti* elimination to give exclusively the *trans* olefin **136**. The bicyclic ring of the hapten ensured a presentation of the phenyl and the benzoyl groups in the requested eclipsed arrangement for a *syn* elimination to proceed. As in the previous example, an ammonium group was selected to induce a catalytic base within the active site of the antibody. Antibody 1D4 elicited against hapten **133** converts substrate **134** exclusively to the Z-isomer **135** via the programmed *syn* pathway with a $k_{cat} = 3 \times 10^{-3}$ min^{-1}. The inability of 1D4 to convert the diastereomeric substrate **137** demonstrates the strong preference for the eclipsed conformation of the substrate within the active site. Preliminary estimates of the energy difference between the *syn* and the *anti* pathways indicate an up to 5 kcal mol^{-1} gap, which suggests exciting opportunities to use antibodies as catalysts for numerous other disfavored processes (Scheme 14.56).

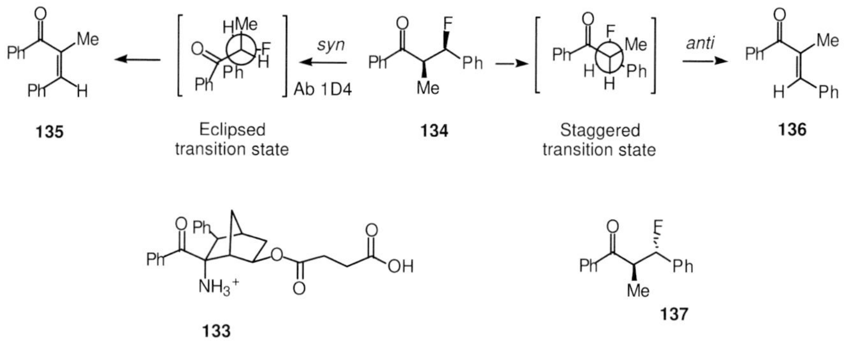

408 | 14 Selectivity with Catalytic Antibodies – What can be achieved?

Scheme 14.56

14.8
Conclusion

Amongst the selective processes successfully catalyzed by antibodies, numerous kinetic resolutions were conducted efficiently in the presence of esterase and aldolase antibodies, but no dynamic kinetic resolution (DKR) process has been reported [108]. Surprisingly, only a few examples of antibody-mediated enantioselective desymmetrization of *meso* compounds and enantiogroup-differentiating reactions have been reported despite the synthetic potential behind this type of stereoselective transformations [109]. Asymmetric carbon-carbon-forming transformations have been widely studied, many processes having no enzymatic equivalent. Difficult asymmetric transformations involving C-C bond formation have been achieved with excellent levels of selectivity, such as numerous direct aldol reactions, Diels-Alder or 1,3-dipolar cycloadditions. However, there are still many reactions of interest to the synthetic chemist that have been insufficiently explored, such as asymmetric Michael addition [110] and asymmetric reduction or oxidation. From a practical point of view, the impact of the Sharpless oxidation [111] and dihydroxylation [112], asymmetric hydrogenation [113] and Jacobsen's epoxidation [114] have been tremendous in recent years, and the generation of efficient antibodies for similar transformations still remains a great challenge.

Nevertheless, catalytic antibodies are clearly well suited for asymmetric catalysis or selective transformations. The examples described herein demonstrate that antibodies are clearly able to discriminate between closely related configurational or stereochemical isomers. In addition, we have seen that antibodies can easily achieve preferential stabilization of a single enantiomeric or diastereomeric transition state. Therefore, two reaction pathways that in the absence of a chiral reagent would be equal in free energy (enantiomeric) become unequal in free energy (diastereomeric) in the presence of the antibody catalyst and therefore lead to a stereoselective reaction. The programmable nature of the antibody binding site is a highly attractive feature of these proteins, as selectivity can, if necessary, be programmed within the hapten. This is especially true for the control of chemo-, regio-, and diastereoselectivity, with a hapten acting as an inert transition state analog. The enantioselectivity of the reaction is more often not programmed within the hapten, even though it has been suggested that the more rigid or immunogenic elements in the linker may improve antibody-hapten complementarity, thus affording a more enantioselective catalyst. Both chiral and achiral haptens have been prepared to elicit antibodies that catalyze enantioselective additions. This observation demonstrates the ability of the immune system to elicit a chiral environment capable of exquisite stabilization of a single enantiomeric transition state without programmation within the hapten structure. More demanding processes such as reversal of chemo- regio-, or diastereoselectivity have also been successfully catalyzed with antibodies using haptens designed for the purpose. The high level of stereoselectivity obtained with catalytic antibodies allows the chemist to use them as custom catalysts for synthetic schemes, thereby arriving at synthetic routes that are otherwise difficult or even inaccessible. Recently, this feature of antibody catalysis has been well exemplified in the literature by numerous total syntheses of natural products such as α-multistriatin [103], epothilones [115] or brevicomins [116].

It seems clear that future advances in this field will rely on detailed mechanistic studies, not only for a true understanding of asymmetric induction steps but also for the improvement of catalytic efficiency [117].

References

1 For a selection of reviews on a similar topic, see for example: (a) Keinan, E., Sinha, S. C., Shabat, D., Itzhaky, H., Reymond, J.-L., "Asymmetric Organic Synthesis with Catalytic Antibodies", *Acta Chem. Scand.* 50 (**1996**), p. 679–687; (b) Hasserodt, J., "Organic Synthesis Supported by Antibody Catalysis", *Synlett* 12 (**1999**), p. 2007–2012; (c) Hilvert, D., "Stereoselective Reactions with Catalytic Antibodies", *Top. Stereochem.* 22 (**1999**), p. 83–135

2 For a selection of reviews covering hydrolytic antibodies, see for example: Stevenson, J. D., Thomas, N. R., "Catalytic antibodies and other biomimetic catalysts", *Nat. Prod. Rep.* 17 (**2000**), p. 535–577; Thomas, N. R., "Catalytic Antibodies – Reaching Adolescence?", *Nat. Prod. Rep.* 17 (**1996**), p. 479–511; Blackburn, G. M., Garcon, A., "Catalytic Antibodies", *Biotechnology* (**2000**), p. 403–490.

3 Pollack, S. J., Jacobs, J. W., Schultz, P. G., "Selective chemical catalysis by an antibody", *Science* 234 (**1986**), p. 1570

4 Janda, K. D., Schloeder, D., Benkovic, S. J., Lerner, R. A., "Induction of an antibody that catalyzes the hydrolysis of an amide bond", *Science* 241 (**1988**), p. 1188–1191

5 Brummer, O., Wentworth P. Jr., Weiner, D. P., Janda, K. D., "Phosphorodithioates: synthesis and evaluation of new haptens for the generation of antibody acyl transferases", *Tetrahedron Lett.* 40 (**1999**), p. 7307–7310

6 Benedetti, F., Berti, F., Colombatti, A., Ebert, C., Linda, P., Tonizzo, F., "*Anti*-sulfonamide antibodies catalyze the hydrolysis of a heterocyclic amide", *J. Chem. Soc. Chem. Commun.* (**1996**), p. 1417–1418

7 Kitazume, T., Tsukamoto, T., Yoshimura, K., "A catalytic antibody elicited by a hapten of tetrahedral carbon-type', *J. Chem. Soc. Chem. Commun.* (**1994**), p. 1355

8 Shokat, K. M., Ko, K. M., Scanlan, T. S., Kochesperger, L., Yonkovich, S., Thaisrivongs, S., Schultz, P. G, "Catalytic antibodies: a new class of transition-state analogs to produce hydrolytic antibodies", *Angew. Chem. Int. Ed. Engl.* 29 (**1990**), p. 1296

9 Tantillo, D. J., Houk, K. N., "Canonical binding arrays as molecular recognition elements in the immune system: tetrahedral anions and the ester hydrolysis transition state", *Chem. Biol.* 8(6) (**2001**), p. 535–545; Tantillo, D. J., Houk, K. N., "Fidelity in Hapten Design: How Analogous Are Phosphonate Haptens to the Transition States of Alkaline Hydrolyzes of Aryl Esters", *J. Org. Chem.* 64 (**1999**), p. 3066–3076

10 Janda, K. D., Weinhouse, M. I., Schloeder, D. M., Lerner, R. A., Benkovic, S. J., "Bait and switch strategy for obtaining catalytic antibodies with acyl-transfer capabilities", *J. Am. Chem. Soc.* 112 (**1990**),

p. 1274–1275; Janda, K. D., Weinhouse, M. I., Danon T., Pacelli, K. A., Schloeder, D. M., "Antibody bait and switch catalysis: a survey of antigens capable of inducing antibodies with acyl-transfer properties", *J. Am. Chem. Soc.* 113 (**1991**), p. 5427–5434

11 Saga, H., Ersoy, O., Williams, S. F., Tsumuraya, T., Margolies, M. N., Sinskey, A. J., Masamune, S., "Catalytic antibodies generated via heterologous immunization", *J. Am. Chem. Soc.* 116 (**1994**), p. 6025–6026; Tsumuraya, T., Saga, H., Meguro, S., Tsunakawa, A., Masamune, S., "Catalytic Antibodies Generated via Homologous and Heterologous Immunization", *J. Am. Chem. Soc.* 117 (**1995**), p. 11390–11396

12 Wirsching, P., Ashley, J. A., Lo, C.-H. L., Janda, K. D., Lerner, R. A., "Reactive immunization", *Science* 270 (**1995**), p. 1775–1782

13 Grynszpan, F., Keinan, E., "Use of antibodies to dissect the components of a catalytic event. The cyclopropenone hapten", *Chem. Commun.* 8 (**1998**), p. 865–866

14 Janda, K. D., Benkovic, S. J., Lerner, R. A., "Catalytic antibodies with lipase activity and R or S substrate selectivity", *Science* 244 (**1989**), p. 437–440

15 Pollack, S. J., Hsiun, P., Schultz, P. G., "Stereospecific Hydrolysis of Alkyl Esters by Antibodies", *J. Am. Chem. Soc.* 111 (**1989**), p. 5961–5962

16 Lo, C-H. L., Wentworth, P., Jung, K. W., Yoon, J., Ashley, J. A., Janda, K. D., "Reactive Immunization Strategy Generates Antibodies with High Catalytic Proficiencies", *J. Am. Chem. Soc.* 119 (**1997**), p. 10251–10252

17 Datta, A., Wentworth, P., Shaw, J. P., Simeonov, A., Janda, K. D., "Catalytically Distinct Antibodies Prepared by the Reactive Immunization versus Transition State Analog Hapten Manifolds", *J. Am. Chem. Soc.* 121 (**1999**), p. 10461–10467

18 Guo, J., Huang, W., Scanlan, T. S, "Kinetic and Mechanistic Characterization of an Efficient Hydrolytic Antibody: Evidence for the Formation of an Acyl Intermediate", *J. Am. Chem. Soc.* 116 (**1994**), p. 6062–6069

19 Zhou, G. W., Guo, J., Huang, W., Fletterick, R. J., Scanlan, T. S., "Crystal structure of a catalytic antibody with a serine protease active site", *Science* 265 (**1994**), p. 1059–1064

20 Wade, H., Scanlan, T. S., "Pi-Si Interactions Control the Enantioselectivity and Hydrolytic Activity of the Norleucine Phenylesterase Catalytic Antibody 17E8", *J. Am. Chem. Soc.* 118 (**1996**), p. 6510–6511

21 Kitazume, T., Lin, J. T., Takeda, M., Yamazaki, T., "Stereoselective Synthesis of Fluorinated Materials Catalyzed by an Antibody", *J. Am. Chem. Soc.* 113 (**1991**), p. 2123–2126

22 Kitazume, T., Lin, J. T., Yamamoto, T., Yamazaki, T., "Antibody-Catalyzed Double Stereoselection in Fluorinated Materials", *J. Am. Chem. Soc.* 137 (**1991**), p. 8573–8575

23 Iwabuchi, Y., Miyashita, H., Tanimura, R., Kinoshita, K., Kikuchi, M., Fujii, I., "Regio- and Stereoselective Deprotection of Acylated Carbohydrates via Catalytic Antibodies", *J. Am. Chem. Soc.* 116 (**1994**), p. 771–772

24 Tanaka, F., Kinoshita, K., Tanimura, R., Fujii, I., "Relaxing Substrate Specificity in Antibody-Catalyzed Reactions: Enantioselective Hydrolysis of N-Cbz-Amino Acid Esters", *J. Am. Chem. Soc.* 118 (**1996**), p. 2332–2339

25 Li, T., Hilton, S., Janda, K. D., "The Potential Application of Catalytic Antibodies to Protecting Group Removal: Catalytic Antibodies with Broad Substrate Tolerance", *J. Am. Chem. Soc.* 117 (**1995**), p. 2123–2127

26 Kurihara, S., Tsumuraya, T., Susuki, K., Kuroda, M., Liu, L., Takaoka, Y., Fujii, I., "Antibody-Catalyzed Removal of the *p*-Nitrobenzyl Ester Protecting Group: The Molecular Basis of Broad Substrate Specificity", *Chem. Eur. J.* 6(9) (**2000**), p. 1656–1662

27 For several examples, see Wong, C-H., Whitesides, G. M., "Enzymes in Synthetic Organic Chemistry", *Tetrahedron Organic Chemistry Series Volume 12* (**1994**), p. 41–130

28 Shoji, I., Weinhouse, M. I., Janda, K. D., Lerner, R. A., Danishefsky, S. J., "Asymmetric induction via a catalytic antibody", *J. Am. Chem. Soc.* 113 (**1990**), p. 7763–7764

29 Fernholz, E., Schloeder, D., Liu, K.-C., Bradshaw, C. W., Huang, H., Janda, K., Lerner, R. A., Wong, C.-H., "Specificity of Antibody-Catalyzed Transesterifications Using Enol Ethers: A Comparison with Lipase Reactions", *J. Org. Chem.* 57 (**1992**), p. 4756–4761

30 Jacobsen, J. R., Prudent, J. R., Kochersperger, L., Yonkovich, S., Schultz, P. G., "An efficient antibody-catalyzed aminacylation reaction", *Science* 256 (**1992**), p. 365–367

31 Hirschmann, R., Smith III, A. B., Taylor, C. M., Benkovic, P. A., Taylor, C. M., Yager, K. M., Witherington, J., Philips, B. W., Sprengler, P. A., Benkovic, S. J., "Peptide synthesis catalyzed by an antibody containing a binding site for variable amino acids", *Science* 265 (**1994**), p. 234–237

32 Napper, A. D., Benkovic, S. J., Tramontano, A., Lerner, R. A., "A stereospecific cyclization catalyzed by an antibody", *Science* 237 (**1987**), p. 1041–1043.

33 Bryant, R. A. R., Hansen, D. E.,, "Direct Measurement of the Uncatalyzed Rate of Hydrolysis of a Peptide Bond", *J. Am. Chem. Soc.* 118 (**1996**), p. 5498–5499

34 Wiberg, K. B. In, *"The Amide Linkage: Selected Structural Aspects in Chemistry, Biochemistry, and Material Science"*, Greenberg, A., Breneman, C. M., Liebman, J. F., Eds.; John Wiley & Sons, Inc, New York **2000**, p. 35–45

35 Stewart, J. D., Krebs, J. F., Siuzdak, G., Berdis, A. J., Smithrud, D. B., Benkovic, S. J., "Dissection of an Antibody-Catalyzed Reaction", *Proc. Natl. Acad. Sci. USA* 91 (**1994**), p. 7404–7409

36 Martin, M. T., Angeles, T. S., Sugasawara, R., Aman, N. I., Napper, A. D., Darsley, M. J., Sanchez, R. I., Booth, P., Titmas, R. C., "Antibody-Catalyzed Hydrolysis of an Unsubstituted Amide", *J. Am. Chem. Soc.* 116 (**1994**), p. 6508–6512

37 Gibbs, R. A., Taylor, S., Benkovic, S. J., "Antibody-catalyzed rearrangement of the peptide bond", *Science* 258 (**1992**), p. 803–805

38 Ersoy, O., Fleck, R., Sinskey, A., Masamune, S., "Antibody catalyzed cleavage of an amide bond using an external nucleophilic cofactor", *J. Am. Chem. Soc.* 120 (**1998**), p. 817–818 and references therein

39 Dwek, R. A., Edge, C. J., Harvey, D. J., Wormald, M. R., Parekh, R. B., "Analysis of glycoprotein-associated oligosaccharides", *Annu. Rev. Biochem.* 62 (**1993**), p. 65–100

40 Reymond, J.-L., Janda, K. D., Lerner, R. A., "Antibody-catalyzed hydrolysis of glycosidic compounds", *Angew. Chem. Int. Ed. Engl.* 30 (**1991**), p. 1711–1713; Yu, J., Hsieh, L., Kochersperger, L., Yonkovich, S., Stephans, J. C., Gallop, M. A., Schultz, P. G., "On the path to antibody glycosidases", *Angew. Chem. Int. Ed. Engl.* 33 (**1994**), p. 339–341

41 Schramm, V. L., Horenstein, B. A., Kline, P. C., "Transition state analysis and inhibitor design for enzymatic reactions", *J. Biol. Chem.* 269 (**1994**), p. 18259–18262; Kuroki, R., Weaver, L. H., Matthews, B. W., "A covalent enzyme-substrate intermediate with saccharide distortion in a mutant T4 lysozyme", *Science* 262 (**1993**), p. 2030–2033

42 Suga, H., Tanimoto, N., Sinskey, A. J., Masamune, S., "Glycosidase Antibodies Induced to a Half-Chair Transition-State Analog", *J. Am. Chem. Soc.* 116 (**1994**), p. 11197–11198

43 Janda, K. D., Lo, L.-C., Lo, C.-H. L., Sim, M.-N., Wang, R., Wong, C.-H., Lerner, R. A., "Chemical selection for catalysis in combinatorial antibody libraries", *Science* 275 (**1997**), p. 945–948

44 Yu, J., Choi, S. Y., Moon, K-D., Chung, H. H., Youn, H.J, Jeong, H. J., Park, H., Schultz P. G., "A glycosidase antibody elicited against a chair-like transition state analog by in

vitro immunization", *Proc. Natl. Acad. Sci. USA* 95 (1998), p. 2880–2884
45 CHIN, J., "Developing artificial hydrolytic metalloenzymes by a unified mechanistic approach", *Acc. Chem. Res* 24 (1991), p. 145–152
46 SCANLAN, T. S., PRUDENT, J. R., SCHULTZ, P. G., "Antibody-catalyzed hydrolysis of phosphate monoesters", *J. Am. Chem. Soc.* 113 (1991), p. 9397–9398; LAVEY, B. J., JANDA, K. D., "Catalytic Antibody mediated Hydrolysis of Paraoxon", *J. Org. Chem.* 61 (1996), p. 7633–7636; ROSENBLUM, J. S., LO, S.-C., LI, T., JANDA, K. D., LERNER, R. A., "Antibody-catalyzed Phosphate Triester Hydrolysis", *Angew. Chem. Int. Ed. Engl.* 34 (1995), p. 2275–2277; WEINER, D. P., WIEMANN, T., WOLFE, M. M., WENTWORTH, P. J., JANDA, K. D., "A Pentacoordinate Oxorhenium (V) Metallochelate Elicits Antibody Catalysts for Phosphodiester Cleavage", *J. Am. Chem. Soc.* 117 (1997), p. 4088–4089
47 See for example: (a), ABRAHAM, L., CZERWONKA, R., HIERSMANN, M., "The Catalytic Enantioselective Claisen Rearrangement of an Allyl Vinyl Ether", *Angew. Chem. Int. Ed.* 40, (2001), p. 4700–4703; (b), YOON, T. P., MACMILLAN, D. W. C., " Enantioselective Claisen Rearrangements: Development of a First Generation Asymmetric Acyl-Claisen Reaction", *J. Am. Chem. Soc.* 123 (2001), p. 2911–2912
48 BARTLETT, P. A., NAKAGAWA, Y., JOHNSON, C. R., REICH, S. H., LUIS, A., "Chorismate Mutase Inhibitor: Synthesis and Evaluation of Some Potential Transition-State Analogs", *J. Org. Chem.* 53 (1988), p. 3195–3210
49 JACKSON, D. Y., JACOBS, J. W., SUGASAWARA, R., REICH, S. H., BARTLETT, P. A., SCHULTZ, P. G., "An antibody-catalyzed Claisen rearrangement", *J. Am. Chem. Soc.* 110 (1988), p. 4841–4842; D. Y. JACKSON, K. D., LIANG, M. N., BARTLETT, P. A., SCHULTZ P. G., "Activation Parameters and Stereochemistry of an Antibody-Catalyzed Claisen Rearrangement", *Angew. Chem. Int. Ed. Engl.* 31 (1992), p. 182–183
50 HILVERT, D., CARPENTER, S. H., NARED, K. D., AUDITOR, M.-T. M., "Catalysis of concerted reactions by antibody: the Claisen rearrangement", *Proc. Natl. Acad. Sci. USA* 85 (1988), p. 4953–4955.; HILVERT, D., NARED, K. D., "Stereospecific Claisen rearrangement catalyzed by an antibody", *J. Am. Chem. Soc.* 110 (1988), p. 5593–5594
51 HAYNES, M. R., STURA, E. A., HILVERT, D., WILSON, I. A., "Routes to catalysis: structure of a catalytic antibody and comparison with its natural counterpart", *Science* 263 (1994), p. 646–652; TANG, Y., HICKS, J. B., HILVERT D., "In vivo catalysis of a metabolically essential reaction by an antibody", *Proc. Natl. Acad. Sci. USA* 88 (1991), p. 8784–8786
52 BRAISTED, A. C., SCHULTZ, P. G., "An Antibody-Catalyzed Oxy-Cope Rearrangement", *J. Am. Chem. Soc.* 116 (1994), p. 2211–2212; ULRICH, H. D., DRIGGERS, E. M., SCHULTZ P. G., "Antibody catalysis of pericyclic reactions", *Acta Chim. Scand.* 50 (1996), p. 328–332; ULLRICH, H. D., MUNDORFF, E., SANTARSIERO, B. D., DRIGGERS, E. M., STEVENS, R. C., SCHULTZ, P. G., "The interplay between binding energy and catalysis in the evolution of a catalytic antibody", *Nature* 389 (1997), p. 271–275
53 ZHOU, Z. S., JIANG, N., HILVERT, D., "An Antibody-Catalyzed Selenoxide Elimination", *J. Am. Chem. Soc.* 119 (1997), p. 3623–3624
54 HILVERT, D., HILL, K. W., NARED, K. D., AUDITOR M.-T. M., "Antibody catalysis of the Diels-Alder reaction", *J. Am. Chem. Soc.* 111 (1989), p. 9261–9262
55 BRAISTED, A. C., SCHULTZ, P. G., "An Antibody-catalyzed bimolecular Diels-Alder reaction", *J. Am. Chem. Soc.* 112 (1990), p. 7430–7431
56 GOUVERNEUR, V. E., HOUK, K. N., DE PASCUAL-TERESA, B., BENO, B., JANDA, K. D., LERNER, R. A., "Control of the exo and endo Pathways of the Diels-alder Reaction by Antibody Catalysis", *Science* 262 (1993), p. 204–208

57 Yli-Kauhaluoma, J. T., Ashley, J. A., Lo, C.-H., Tucker, L., Wolfe, M. M., Janda, K. D., "Anti-Metallocene Antibodies: A New Approach to Enantioselective Catalysis of the Diels-Alder Reaction", *J. Am. Chem. Soc.* 117 (1995), p. 7041

58 Heine, A., Stura, E. A., Yli-Kauhaluoma, J. T., Gao, C., Deng, Q., Beno, B. R., Houk, K. N., Janda, K. D., Wilson, I. A., "An antibody exo Diels-Alderase inhibitor complex at 1.95 angstrom resolution", *Science* 279 (1998), p. 1934

59 Meekel, A. A. P., Resmini, M., Pandit, U. K., "Regioselectivity and Enantioselectivity in an Antibody Catalyzed Hetero Diels-Alder Reaction", *Bioorg. Med. Chem.* 4 (1996), p. 1051–1057; Meekel, A. A. P., Resmini, M., Pandit, U. K.,, "First example of an antibody-catalyzed hetero Diels-Alder reaction", *J. Chem. Soc., Chem. Commun.* 571 (1995); Resmini, M., Meekel, A. A. P., Pandit, U. K., "Catalytic antibodies: regio- and enantioselectivity in a hetero Diels-Alder reaction", *Pure Appl.Chem.* 68 (1996), p. 2025–2028

60 Hu, Y.-J., Ji, Y.-Y., Wu, Y.-L., Yang, B. H., Yeh, M., "Polyclonal Catalytic Antibody for hetero-Cycloaddition of Hepta-1,3-diene with Ethyl Glyoxylate An Approach to the Synthesis of 2-Nonulosonic Acid Analogs", *Bioorg. Med. Chem. Lett* 7 (1997), p. 1601–1606

61 Shi, Z.-D., Yang, B.-H., Wu, Y.-L., Pan, Y. J., Ji, Y. Y., Yeh, M., "First Example of an Antibody-Catalyzed Aza Diels-Alder Reaction", *Bioorg. Med. Chem. Lett.* 12 (2002), p. 2321–2324

62 Toker, J. D., Wentworth, P., Hu, Y., Houk, K. N., Janda, K. D., "Antibody-Catalysis of a Biomolecular Asymmetric 1,3-Dipolar Cycloaddition Reaction", *J. Am. Chem. Soc.* 122 (2000), p. 3244–3245

63 Machajewski, T. D., Wong, C-H., "The Catalytic Asymmetric Aldol Reaction", *Angew. Chem. Int. Ed.* 39 (2000), p. 1352–1374

64 Yamada, Y. M. A., Yoshikawa, N., Sasai, H., Shibasaki, M., "Direct catalytic asymmetric aldol reactions of aldehydes with unmodified ketones", *Angew. Chem. Int. Ed. Engl.* 36 (1997), p. 1871–1873

65 Koch, T., Reymond, J.-L., Lerner, R. A., "Antibody catalysis of multistep reactions: an aldol addition followed by a disfavoured elimination", *J. Am. Chem. Soc.* 117 (1995), p. 9383–9387

66 Reymond, J.-L., Chen, Y., "Catalytic, Enantioselective Aldol reaction with an Artificial Aldolase Assembled from a Primary Amine and an Antibody", *J. Org. Chem.* 60 (1995), p. 6970–6979; Reymond, J.-L., *Angew. Chem. Int. Ed.* 34 (1995), p. 2285–2287

67 Wagner, J., Lerner, R. A., Barbas III C. F., "Efficient aldolase catalytic antibodies that use the enamine mechanism of natural enzymes", *Science* 270 (1995), p. 1797–1800

68 Barbas III, C. F., Heine, A., Zhong, G., Hoffman, T., Gramatikova, S., Bjornestedt, R., List, B., Anderson, J., Stura, E. A., Wilson, I. A., Lerner, R. A., "Immune versus natural selection: antibody aldolases with enzymic rates but broader scope", *Science* 278 (1997), p. 2085

69 Hoffman, T., Zhong, G., List, B., Shabat, D., Anderson, J., Gramatikova, S., Lerner, R. A., Barbas III, C. F., "Aldolase Antibodies of Remarkable Scope", *J. Am. Chem. Soc.* 120 (1998), p. 2768–2779

70 Zhong, G., Hoffman, T., Lerner, R. A., Danishefsky, S., Barbas III, C. F., "Antibody-Catalyzed Enantioselective Robinson Annulation", *J. Am. Chem. Soc.* 119 (1997), p. 8131–8132

71 Chen, L., Dumas, D. P., Wong, C.-H., "Deoxyribose 5-phosphate aldolase as a catalyst in asymmetric aldol condensation", *J. Am. Chem. Soc.* 114 (1992), 741

72 Zhong, G., Shabat, D., List, B., Anderson, J., Sinha, S. C., Lerner, R. A., Barbas III, C. F., "Catalytic Enantioselective Retro-Aldol Reactions: Kinetic Resolution of β-Hydroxyketones with Aldolase Antibodies", *Angew. Chem. Int. Ed.* 37 (1998), p. 2481–2484

73 LIST, B., SHABAT, D., ZHONG, G., TURNER, J. M., LI, A., BUI, T., ANDERSON, J., LERNER, R. A., BARBAS III, C. F., "A Catalytic Enantioselective Route to Hydroxy-Substituted Quaternary Carbon Centers: Resolution of Tertiary Aldols with a Catalytic Antibody", *J. Am. Chem. Soc.* 121 (**1999**), p. 7283–7291

74 LIST, B., LERNER, R. A., BARBAS III, C. F., "Enantioselective Aldol Cyclodehydrations Catalyzed by Antibody 38C2", *Org. Lett.* 1 (**1999**), p. 59–61

75 ZHONG, G., LERNER, R. A., BARBAS III, C. F., "Broadening the Aldolase Catalytic Antibody Repertoire by Combining Reactive Immunization and Transition State Theory: New Enantio- and Diastereoselectivities", *Angew. Chem. Int. Ed.* 38 (**1999**), p. 3738–3741

76 MAGGIOTTI, V., RESMINI, M., GOUVERNEUR, V., "Unprecedented Regiocontrol Using an Aldolase I Antibody", *Angew. Chem. Int. Ed.* 41 (**2002**), p. 1012–1014

77 SHULMA, A., SITRY, D., SHULMAN, H., KEINAN, E., "Highly Efficient Antibody-Catalyzed Deuteration of Carbonyl Compounds", *Chem. Eur. J.* 8 (**2002**), p. 229–239

78 FLANAGAN, M. E., JACOBSEN, J. R., SWEET, E., SCHULTZ, P. G., "Antibody-Catalyzed Retro-Aldol Reaction", *J. Am. Chem. Soc.* 118 (**1996**), p. 6078–6079

79 JANDA, K. D., SHEVLIN, C. G., LERNER, R. A., "Antibody catalysis of a disfavoured chemical transformation", *Science* 259 (**1993**), p. 490–493

80 NA, J., HOUK, K. N., SHEVLIN, C. G., JANDA, K. D., LERNER, R. A., "The Energetic Advantage of 5-Exo Versus 6-Endo Epoxide Openings: A Preference Overwhelmed by Antibody Catalysis", *J. Am. Chem. Soc.* 115 (**1993**), p. 8543

81 SHEVLIN, C. G., HILTON, S., JANDA, K. D.,, "Automation of antibody catalysis: a practical methodology for the use of catalytic antibodies in organic synthesis", *Bioorg. Med. Chem. Lett.* 4(2) (**1994**), p. 297–302

82 JANDA, K. D., SHEVLIN, C. G., LERNER, R. A., "Oxepane Synthesis Along a disfavored Pathway: The Rerouting of a Chemical Reaction Using a Catalytic Antibody", *J. Am. Chem. Soc.* 117 (**1995**), p. 2659–2660

83 KITAZUME, T., TAKEDA M., "A cyclization reaction catalyzed by antibodies", *J. Chem. Soc., Chem. Commun.* (**1995**), p. 39–40

84 ABE, I., ROHMER, M., PRESTWICH, G. D., "Enzymatic cyclization of squalene and oxidosqualene to sterols and triterpenes", *Chem. Rev.* 93 (**1993**), p. 2189–2206; PALE-GROSDEMANGE, C., FEIL, C., ROHMER, M., PORALLA, K., "Occurrence of Cationic Intermediates and Deficient Control during the Enzymatic Cyclization of Squalene to Hopanoids", *Angew. Chem. Int. Ed.* 37 (**1998**), p. 2237–2240

85 LI, T., JANDA, K. D., ASHLEY, J. A., LERNER, R. A., "Antibody catalyzed cationic cyclization", *Science* 264 (**1994**), 1289–1293

86 LI, T., JANDA, K. D., LERNER, R. A., "Cationic cyclopropanation by antibody catalysis", *Nature* 379 (**1996**), p. 326–327

87 LI, S. HILTON, K. D. JANDA, "Remarkable Ability of Different Antibody Catalysts To Control and Diversify the Product Outcome of Cationic Cyclization Reactions", *J. Am. Chem. Soc.* 117 (**1995**), p. 3308–3309

88 (a) HASSERODT, J., JANDA, K. D., LERNER, R. A., "Formation of Bridge-Methylated Decalins by Antibody-Catalyzed Tandem Cationic Cyclization", *J. Am. Chem. Soc.* 119 (**1997**), p. 5994–5998; (b), PASCHALL, C. M., HASSERODT, J., JONES, T., LERNER, R. A., JANDA, K. D., CHRISTIANSON, D. W., "Convergence of Catalytic Antibody and Terpene Cyclase Mechanisms: Polyene Cyclization Directed by Carbocation-π Interactions", *Angew. Chem. Int. Ed.* 38 (**1999**), p. 1743–1747

89 HASSERODT, J., JANDA, J. D., LERNER, R., "Antibodies Mimic of Natural Oxidosqualene-Cyclase Action in Steroid Ring A Formation", *J. Am. Chem. Soc.* 122 (**2000**), p. 40–45

90 JAKOBY, W. B., ZIEGLER, D. M., "The enzymes of detoxification", *J. Biol.*

Chem. 265 **(1990)**, p. 20715–20718; HOLLAND, H. L., "Chiral sulfoxidation by biotransformation of organic sulfides", *Chem. Rev.* 88 **(1988)**, p. 473–485

91 KIM, J. I., NAGANO, T., HIGUCHI, T., HIROBE, M., SHIMADA, I., ARADA, Y., "Conformation and stereoselective reduction of hapten side chains in the antibody combining site", *J. Am. Chem. Soc.* 113 **(1991)**, p. 9392–9394

92 NAKAYAMA, G. R., SCHULTZ, P. G., "Stereospecific antibody-catalyzed reduction of an alpha-keto amide", *J. Am. Chem. Soc.* 114 **(1992)**, p. 780–781

93 L. C. HSIEH, S. YONKOVICH, L. KOCHERSPERGER, P. G. SCHULTZ, "Controlling chemical reactivity with antibodies", *Science* 260 **(1993)**, p. 337–339

94 JACOBSEN, E. N., FINNEY, N. S., "Synthetic and biological catalysts in chemical synthesis: how to assess practical utility?", *Chem. Biol.* 1 **(1994)**, p. 85–90; ALLAIN, E. J., HAGER, L. P., DENG, L., JACOBSEN, E. N., "Highly enantioselective epoxidation of disubstituted alkenes with hydrogen peroxide catalyzed by chloroperoxidase", *J. Am. Chem. Soc.* 115 **(1993)**, p. 4415

95 KOCH, A., REYMOND, J.-L., LERNER, R. A., "Antibody-Catalyzed Activation of Unfunctionalized Olefins for Highly Enantioselective Asymmetric Epoxidation", *J. Am. Chem. Soc.* 116 **(1994)**, p. 803–804

96 CHEN, Y., REYMOND, J.-L., "Enantioselective epoxidation with a library of catalytic antibodies", *Synthesis* 6 **(2001)**, p. 934–936

97 HSIEH, L. C., STEPHANS, J. C., SCHULTZ, P. G., "An Efficient Antibody-Catalyzed Oxygenation Reaction", *J. Am. Chem. Soc.* 116 **(1994)**, p. 2167–2168

98 HSIEH-WILSON, L. C., SCHULTZ, P. G., STEVENS, R. C., "Insights into antibody catalysis: Structure of an oxygenation catalyst at 1.9-A resolution", *Proc. Natl. Acad. Sci. USA* 93 **(1996)**, p. 5363–5367

99 EAMES, J., WEERASOORIYA N. "Recent advances into the enantioselective protonation of prostereogenic enol derivatives", *Tetrahedron: Asymmetry* 12 **(2001)**, p. 1–24

100 MATSUMOTO, K., TSUTSUMI, S., IHORI, T., OHTA, H., "Enzyme-mediated enantioface-differentiating hydrolysis of alpha-substituted cycloalkanone enol esters", *J. Am. Chem. Soc.* 112 **(1990)**, p. 9614

101 FUJII, I., LERNER, R. A., JANDA, K. D., "Enantiofacial protonation by catalytic antibodies", *J. Am. Chem. Soc.* 113 **(1991)**, p. 8528–8529

102 REYMOND, J-L., JANDA, K. D., LERNER, R. A., "Highly Enantioselective Protonation Catalyzed by an Antibody", *J. Am. Chem. Soc.* 114 **(1992)**, p. 2257–2258

103 SINHA, S. C., KEINAN, E., "α-Multistriatin: the first total synthesis of a natural product via antibody catalysis", *Isr. J. Chem.* 36 **(1996)**, p. 185–193; SINHA, S. C., KEINAN, E., "Catalytic Antibodies in Organic Synthesis. Asymmetric Synthesis of (–)-α-Multistriatin", *J. Am. Chem. Soc.* 117 **(1995)**, p. 3653–3654

104 SINHA, S. C., KEINAN, E., REYMOND, J.-L., "Antibody-catalyzed reversal of chemoselectivity", *Proc. Natl. Acd. Sci. USA* 90 **(1993)**, p. 11910–11913

105 SHABAT, D., SHULMAN, H., ITZHAKY, H., REYMOND, J-L., KEINAN, "Enantioselectivity vs. kinetic resolution in antibody catalysis: formation of the (S) product despite preferential binding of the (R) intermediate", *Chem. Commun.* 8 **(1998)**, p. 1759–1760

106 BARTSCH, R. A., ZAVADA, J., "Stereochemical and base species dichotomies in olefin-forming E2 eliminations", *Chem. Rev.* 80, **(1980)**, p. 453–494

107 CRAVATT, B. F., ASHLEY, J. A., JANDA, K. D., BOGER, D. L., LERNER, R. A., "Crossing Extreme Mechanistic Barriers by Antibody Catalysis: *Syn* Elimination to a Cis Olefin", *J. Am. Chem. Soc.* 116 **(1994)**, p. 6013–6014

108 BRIOT, A., BUJARD, M., GOUVERNEUR, V., MIOSKOWSKI, C., "Design and Synthesis of Haptens for the Catalytic

Antibody-Promoted Dynamic Kinetic Resolution of Oxazolin-5-ones", *Eur. J. Org. Chem.* (**2002**), p. 139–144

109 VEDEJS, E., CHEN, X., "Kinetic Resolution of Secondary Alcohols. Enantioseelctive Acylation Mediated by a Chiral (Dimethylamino)pyridine Derivative", *J. Am. Chem. Soc.* 118 (**1996**), p. 1809–1810; WALLACE, J. S., BALDWIN, B. W., MORROW, C. J., "Separation of remote diol and triol stereoisomers by enzyme-catalyzed esterification in organic media or hydrolysis in aqueous media", *J. Org. Chem.* 57 (**1992**), p. 5231–5239

110 COOK, C. E., ALLEN, D. A., MILLER, D. B., WHISNANT, C. C., "Antibody-Catalyzed Michael Reaction of Cyanide with an α,β-Unsaturated Ketone", *J. Am. Chem. Soc.* 117 (**1995**), p. 7269–7270

111 SHARPLESS, K. B., VERHOEVEN, T. R., "Metal-catalyzed highly selective oxygenations of olefins and acetylenes with tert-butyl hydroperoxide. Practical considerations and mechanisms", *Aldrichimica Acta* 12 (**1979**), 63–74; HOVEYDA, A. H., EVANS, D. A., FU, G. C., "Substrate-directable chemical reactions", *Chem. Rev.* 93 (**1993**), p. 1307–1370

112 JOHNSON, R. A., SHARPLESS, K. B., In *Catalytic Asymmetric Synthesis*; OJIMA, I., Ed.; VCA, Weinheim **1993**, p. 227–272

113 TAKAYA, H., OHTA, T., NOYORI, R. ref 112, p. 1–39

114 JACOBSEN, E. N., ZHANG, W., MUCI, A. R., ECKER, J. R., DENG, LI., "Highly enantioselective epoxidation catalysts derived from 1,2-aminocyclohexane", *J. Am. Chem. Soc.* 113 (**1991**), p. 7063

115 (a) SINHA, S. C., SUN, J., MILLER, G. P., WARTMANN, M., LERNER, R. A., "Catalytic antibody route to the naturally occurring epothilones: total synthesis of epothilones A-F", *Chem. Eur. J.* 7 (**2001**), p. 1691–1702; (b) SINHA, S. C., BARBAS III, C. F., LERNER, R. A., "The antibody catalysis route to the total synthesis of epothilones", *Proc. Natl. Acad. Sci* 25 (**1998**), p. 14603–14608; (c) SINHA, S.C, SUN, J., MILLER, G. P., WARTMANN, M., LERNER, R. A., "Catalytic antibody route to the naturally occurring epothilones: total synthesis of epothilones A-F", *Chemistry* 7 (**2001**), p. 1691–1702; (d) SINHA, S. C., SUN, J., WARTMAN, M., LERNER, R. A., "Synthesis of Epothilones Analogs by Antibody-Catalyzed Resolution of Thiazole Aldol Synthons on a Multigram Scale. Biological Consequences of C-13 Alkylation of Epothilones", *ChemBioChem* 2 (**2001**), p. 656–665

116 LIST, B., SHABAT, D., BARBAS III, C. F., LERNER, R. A., "Enantioselective Total Synthesis of Some Brevicomins Using Aldolase Antibody 38C2", *Chem. Eur. J.* 4 (**1998**), p. 881–885

117 HILVERT, D., "Critical analysis of Antibody Catalysis", *Annu. Rev. Biochem.* 69 (**2000**), p. 751–793

15
Catalytic Antibodies as Mechanistic and Structural Models of Hydrolytic Enzymes

Ariel B. Lindner, Zelig Eshhar, Dan S. Tawfik

Abbreviations

TS – transition state; TSA – TS analog; OAS – oxyanion stabilization; OAH – oxyanion hole; AChE – acetylcholinesterase; D-Abs – D-Antibodies (esterolytic antibodies D2.3, D2.4, and D2.5).

15.1
Introduction

During the past 16 years, knowledge gained from studies of natural hydrolases and the mechanism of hydrolytic reactions led to a variety of strategies for generating antibodies that catalyze hydrolytic reactions. In particular, ester hydrolysis became the most intensely studied antibody-catalyzed reaction. Esterolytic antibodies provided the first 'proof of principle' of antibody-based catalysis, spearheading the wide spectrum of antibody catalysts detailed in the chapters of this book. Although esterolytic antibodies do not, in general, exhibit enzyme-like rates, knowledge may be extracted from the many mechanistic and structural studies of hydrolytic antibodies concerning the mechanism of action of hydrolytic enzymes. Comparing hydrolytic antibodies to hydrolytic enzymes is the underlying theme of this chapter.

Specific-base-(hydroxyl)-catalyzed and general-base-catalyzed is the mechanism by which most natural esterases act (Fig. 15.1). The reaction proceeds via relatively low activation barriers (the *addition* TS being higher than the *elimination* TS [1]), and a well-defined singular tetrahedral, oxyanionic intermediate (I_1, Fig. 15.1a). Inherent in hydroxyl-catalyzed ester hydrolysis is the linear of the hydrolytic rates. In implementing these basic mechanisms, which are also seen in solution, enzymes apply complex mechanisms, combining a variety of catalytic forces, as exemplified by three major families of hydrolytic enzymes: seryl/cysteyl hydrolases (Fig. 15.1b), aspartyl hydrolases (Fig. 15.1c), and zinc hydrolases (Fig. 15.1d).

In essence, two major catalytic forces are utilized by these enzymes [2]:

(i) A nucleophilic attack is mediated by a precisely positioned nucleophile, either an endogenous general base (seryl alkoxide (Fig. 15.1b, step I)) or an explicitly bound

Catalytic Antibodies. Edited by Ehud Keinan
Copyright © 2005 WILEY-VCH Verlag GmbH & Co. KGaA, Weinheim
ISBN: 3-527-30688-9

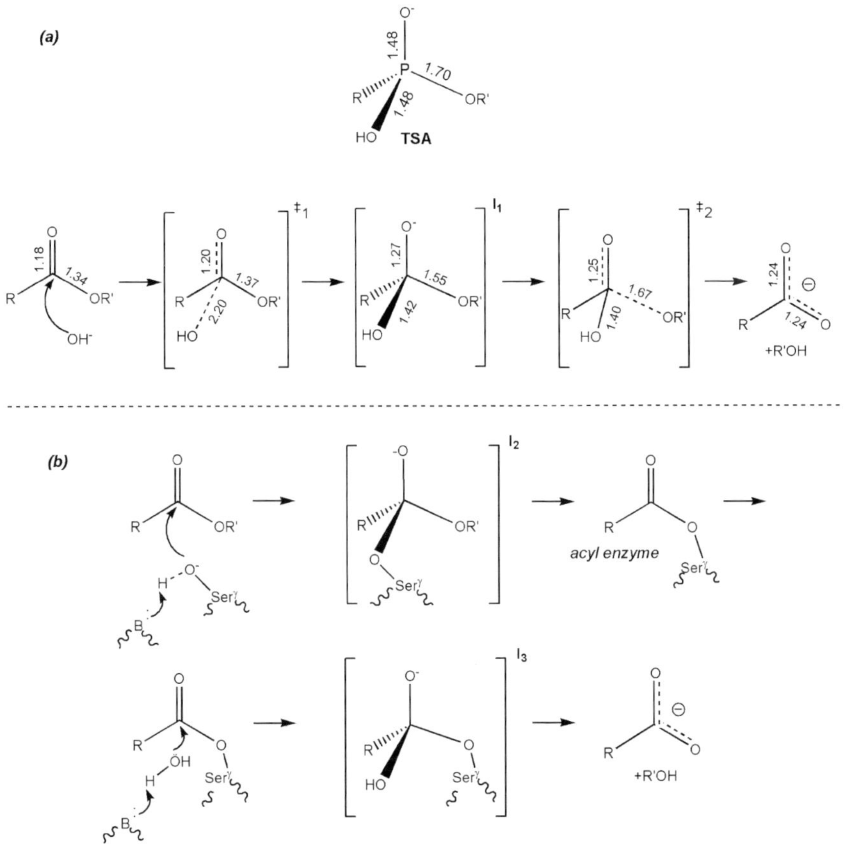

Fig. 15.1 Mechanisms of ester hydrolysis.

(a) Hydroxide-catalyzed ester hydrolysis. The rate-limiting step consists of an *exogenous* hydroxide attack on the ester's carbonyl, forming, via the *addition TS1*, a tetrahedral oxyanionic intermediate (I_1). The latter collapses, via the *elimination TS2*, to form the carboxylate and alcohol products. Top formula: a representative phosphonate TSA, where, for the D-antibodies, R=glutaryl glycine and R' = p-nitrobenzyl [24]. The numbers correspond to bond lengths derived from ab initio calculations [1].

(b) The endogenous nucleophile mechanism of *Serine hydrolases*. The nucleophilic attack is carried out by the active site's seryl alkoxide, activated by an active-site general-base residue (B:) to form a tetrahedral oxyanionic intermediate (I_2). This is followed by a rate-limiting deacylation step mediated by a hydroxide ion generated *in situ* by the base-catalyzed proton elimination of an active-site water molecule, releasing the products via a second intermediate (I_3) [2].

specific base (a hydroxyl generated via general-base-catalyzed deprotonation of an active-site water molecule; Fig. 15.1a, step II, 1c and 1d), to yield a tetrahedral, oxyanionic intermediate.

(ii) Oxyanionic stabilization (OAS) of the tetrahedral, oxyanionic intermediate and the TSs leading to its formation and breakdown. The negative charge developed

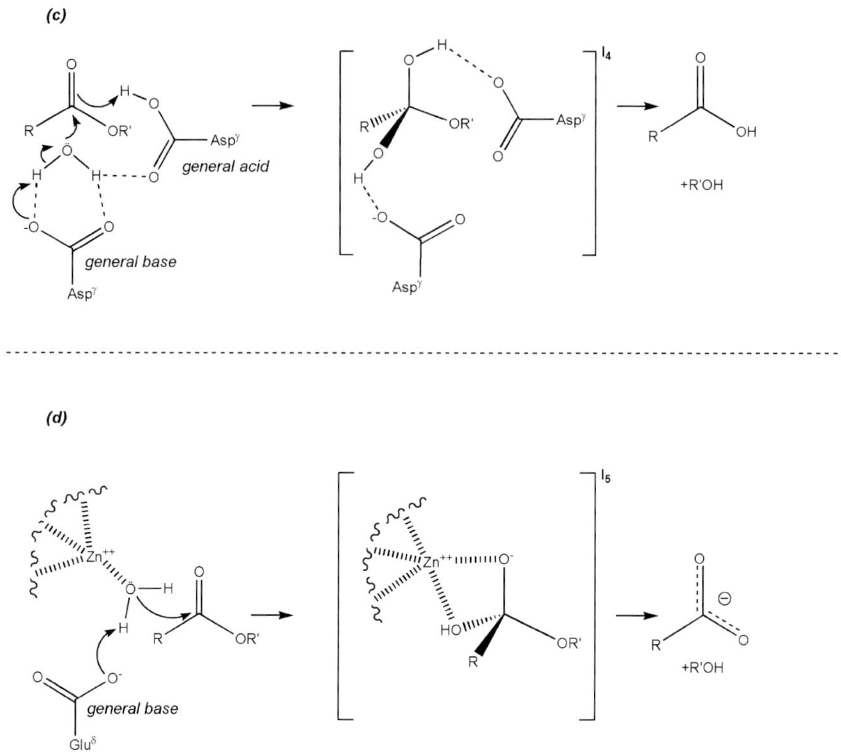

Fig. 15.1 Mechanisms or ester hydrolysis (*cont.*)
(c) The general base – general acid ("push-pull") mechanism of *Aspartic hydrolases*. An active-site bound water molecule is activated by a general-base carboxylate residue, whereas the oxyanion tetrahedral intermediate (I_4) is stabilized by a proton donation from a second aspartyl residue. Residues maintaining the carboxylates' protonation level [135] are not shown for clarity. An alternative "symmetric" model can be found in [136].

(d) The metal-mediated mechanism of *Zinc hydrolases*. The nucleophilic attack is carried out by a zinc-bound water molecule, activated by a general-base Glu residue. The formed oxyanionic intermediate (I_5) is stabilized by its coordination to the metal ion. Though a similar five-coordinated zinc intermediate has recently been observed [137], a four-coordinated intermediate was previously suggested [2, 138].

on the carbonyl oxygen is stabilized by hydrogen bonding (Fig. 15.1b, c) or proton transfer (general-acid catalysis), or by a positive charge (Fig. 15.1d).

(iii) Protonation of the alkoxide leaving group by general-acid catalysis.

The relative positioning of various active-site residues involved in the above mechanisms provides the recognition and stabilization of the tetrahedral transition state (TS). It is therefore expected that, although described as separate forces or contributions, their effect on catalysis is coupled. This coupling is indeed encoded in the enzymes' active-site structure – for example, the seryl residue (Fig. 15.3), the aspartyl residues (positioned by a shared a hydrogen bond; Fig. 15.1b) and the zinc ion

(Fig. 15.1c) all provide both the endogenous nucleophile as well as a hydrogen bond to the TS oxyanion.

In addition to mechanism above, enzymes take advantage of a combination of other forces, including medium (solvation/desolvation effects), substrate (ground-state) destabilization, remote recognition, as well as co-ordinated conformational changes [2, 3]. Thus, the remarkable catalytic efficiency of natural enzymes, at times limited only by the rate of diffusion, is the outcome of many forces acting in concert. The separation of these factors and the assessment of their independent contributions to catalysis is far from trivial, as exemplified by various mutagenesis studies (e.g., see [4, 5]). Catalytic antibodies, raised to mimic single facets of the complex enzymatic regimes, provide a unique opportunity to examine and quantify enzymatic forces and mechanisms of action individually.

15.2
Chapter Overview

The first and most widely applicable strategy for generating esterolytic antibodies is by immunizing with analogs of the tetrahedral, oxyanionic intermediates of ester hydrolysis (Fig. 15.1a), which are generally referred to as transition state analogs (TSAs). Analysis of the structure, mechanism and catalytic efficiency of these antibodies provides the opportunity to compare and assess oxyanion stabilization (OAS) as the primary source of catalysis in both esterolytic antibodies and natural, hydrolytic enzymes (Section 15.2.1). Surprisingly, several antibodies directed towards OAS (e.g., 43C9 [6]) have incorporated, haphazardly, an endogenous nucleophile as part of their mechanism of action. These antibodies, together with studies attempting to directly obtain antibodies with an endogenous nucleophile (e.g., reactive immunization [7] and chemical modification strategies [8, 9]), form the basis for our discussion on nucleophilic reactivity (Section 15.2.2). This is accompanied by an analysis of "chemical rescue" experiments, where strong exogenous nucleophiles supplement the initial OAS activity of various antibodies [10–13] (Section 15.2.3). Several studies, attempting at the generation of esterolytic antibodies with general-acid/base mechanism either by a "bait and switch" strategy (Section 15.2.4) or metal incorporation (Section 15.2.5), are then qualitatively assessed. Finally, the role of conformational changes, suggested in a number of catalytic antibodies [14–20], is described *vis-a-vis* known contributions of conformational changes to natural enzymes (Section 15.3).

15.2.1
Catalysis by Oxyanion Stabilization (OAS)

The most commonly used TSA haptens for the generation of esterolytic antibodies are phosphonates, although other phosphate derivatives such as phosphinates, phosphonoamidates, and phosphothioates have also been used. This indirect approach for the generation of biocatalysts was first formulated by Jencks [21]. Immunization with these analogs generates antibodies that catalyze ester hydrolysis via preferential binding of the reaction's tetrahedral, oxyanionic TS compared to the planar, uncharged ester substrate (Fig. 15.1a). This approach relies on optimal representation of the reaction's TSs by the TSA. It has been shown, for example, that substrate analogs completely fail to generate esterolytic antibodies [22]. Thus, as discussed below, TSA design is a primary issue.

15.2.1.1
Fidelity of TSA design

Phosphate derivatives that are potent inhibitors of hydrolases were shown to exhibit TSA characteristics, albeit with some important discrepancies ([23] and references therein). *Ab initio* calculations suggest that these TSAs better present the *elimination* TS_2, *vis-a-vis* the *addition* TS_1, the higher and thus the rate-limiting barrier of the two TSs [1] (Fig. 15.1a). The TSA P-O bonds overestimate the corresponding lengths of the TS, apart from their being significantly shorter (1.48 Å) than the 2.2 Å distance in the *addition* TS between the incoming hydroxyl nucleophile and the ester's carbonyl carbon (Fig. 15.1a). The symmetrical charge partition between the phosphonates' oxygens overestimates the charge formed in the actual TS and of the incoming hydroxyl nucleophile. Both of the phosphonates' oxygens are expected to bait antibodies with H-donor residues. This is crucial to stabilizing the developing oxyanionic charge, yet "baiting" a basic residue would be preferable in the position representing the nucleophilic attack. The latter may act as a general base to activate the incoming nucleophile or as an endogenous nucleophile. Nonetheless, despite the above-described infidelities, phosphonate TSAs (and related derivatives) are fairly close mimics of the TSs of ester hydrolysis and hence elicit antibodies in which OAS is the primary, or often the sole, source of catalysis.

15.2.1.2
Esterolytic Antibodies Based Solely on Oxyanion Stabilization

Our comparison of OAS in esterolytic antibodies with natural enzymes is based on a number of antibodies in which OAS was shown to be the only source of the catalytic activity (group I antibodies). This group includes antibodies D2.3, D2.4 and D2.5 [10, 24–26], 48G7 [27, 28], 29G11 [29, 30], and 6D9 [31–33] (Group I; Fig. 15.2), which show most if not all of the following characteristics:

(i) The observed rate acceleration correlates well with the differential affinity toward the TSA vs the substrate: $K_{TSA}/K_S \geq k_{cat}/k_{uncat}$ (Table 15.1)

(ii) A linear phase is observed in the pH-rate profile

(iii) Crystal structure indicating active-site residues that comprise the oxyanion hole

(iv) Mutation of OAS residues eliminates catalysis (Table 15.2).

Other antibodies (Group II; Fig. 15.2) for which structural data is available, CNJ206 [14, 34], 17E8 [29, 30, 35], and 7C8 [11] were all raised against similar or identical

Tab. 15.1: Oxyanion stabilization by hydrolytic antibodies.

Antibody	k_{cat}/k_{uncat} × 10^3	$\Delta\Delta G^{\ddagger}_{cat\text{-}uncat}$ Kcal/mol	$\Delta\Delta G^{\circ}_{TSA\text{-}S}$ Kcal/mol	Catalytic OAH residues[a]	Secondary OAH residues[b]
Group I					
D2.3	110	7.0	7.0	Tyr100dH, Asn34L	Trp95H, Tyr96L-H_2O
D2.4	36	6.3	6.3	Tyr100eH, Asn34L	Trp95H, Tyr96L-H_2O
D2.5	1.9	4.5	4.3	Tyr100eH, Ser34L-H_2O	Trp95H, Tyr96L-H_2O
48G7	16	5.8	6.3	His35H, (Tyr96H)	Arg96L, Tyr33H
29G11	2.2	4.6	6.0	Lys93H, (Tyr96H)	His35H, Tyr96L
6D9	0.9	4.1	4.1	His27d	–
Group II					
43C9	27	6.1	8.1	His35H, AsnH33	His91L, Arg96L
17E8	13	5.7	3.8	Lys93H, (Tyr96H)	His35H, Tyr96L
CNJ206	1.6	4.4	3.4	(Asp96H), (Tyr97H)	(Asp96H), (Tyr97H)
7C8	0.7	3.9	1.5	Tyr95H	–

a) Residues in hydrogen-bond distance from the phosphonate's oxyanion represnting the ester substrate carbonyl (Fig. 15.1a), as-signed by comparison of the antibodies' phosphonate- and substrate analog-bound structures (D2.3, D2.4 and D2.5) [128], mutagenesis analysis (see Table 15.2), pH-dependency analysis (D-antibodies [10], CNJ206 [129], 7C8 [11]), and docking (48G7, 29G11, and 17E8 [39]).

b) Residues in hydrogen-bond distance from the phosphonate's oxyanion representing the incoming hydroxide nucleophile (Fig. 15.1a) (see text for assignment).

Tab. 15.2: Oxyanion hole mutants of hydrolytic antibodies.

Antibody	OAH residue mutant	k_{cat}/k_{uncat} wt = 1.0
D2.3/4[b]	AsnL34 → Gly	1.1
	TyrH100d → Phe	n.d.
	TyrH100d → Gly	n.d.
	TyrH100d → Lys	n.d.
	TyrH100d → Ser	0.08
48G7[a]	His35H → Glu	0.04
	His35H → Gln	0.6
6D9	His27d → Ala	n.d.

a) Mutants expressed as recombinant Fab fragments [28].
b) Mutants expressed as scFv chimera of D2.4 V_L/D2.3 V_H [10].
n.d. – catalytic activity not detected

TSAs to those of Group I, and exhibit $K_{TSA}/K_S \leq k_{cat}/k_{uncat}$. This suggests that forces additional to or other than OAS take part (Sections 15.2.2 and 15.2.4). Nonetheless, their catalytic efficiencies may be taken as an upper limit to the contribution of OAS to their activity. Antibody 43C9, where an endogenous nucleophile was shown to be involved [6], is not considered, as the two mechanisms that affect its catalysis (OAS and nucleophilic catalysis) are difficult to separate (Section 15.2.2).

Detailed comparisons of the hydrolytic antibodies' crystal structures in the presence of the respective TSAs are available in several reviews [36–38]. A common motif ("canonical binding array") of interactions with the TSA oxyanionic core was revealed, identifying the residues contributing to the binding of the two oxyanions (Table 15.1, Fig. 15.2). Convergent evolution is evident by the sequence similarities as well as structural homology. Antibodies CNJ206 and 17E8 exhibit only 41–52% differences in the primary sequence of the heavy- and light-variable regions, yet share 8 out of 10 contact residues with the hapten. Antibody 48G7 shares the same light chain with CNJ206 and has 7 of 10 of the heavy chain's CDR residues in common with CNJ206 (for the detailed analysis see [36–38]). Another example of convergence is seen in the structures of D2.3 and D2.4, sharing the same germline genes with 16 positions difference in their variable regions. In particular, the CDR3H loop of D2.4 differs in 4 positions and includes an amino acid insertion compared to D2.3, as a result of different D-J segments' junction. Consequently, their CDR3H loops adopt distinctly different structures yet place the critical OAH residue Tyr100d (D2.3) and Tyr100e (D2.4) at the exact same position with respect to the TSA's oxyanion [26]. The TSAs of Group I antibodies share two epitopes: an aryl or benzyl leaving group, and a phosphonate monoester (Fig. 15.2). While aromatic moieties are well known as immunogenic epitopes (hence their common use in TSA haptens), Tantillo and Houk suggested that the determining epitope may be the bidentate "transition state epitope" oxyanionic moiety. This is supported by the identification of similar motif in crystal structures of arsonates, sulfonates and DNA binding antibodies. Interestingly, the origin of many of the residues in contact with the TSA phosphonates can be tracked down to the germline genes (*e.g.* TyrH33 of 48G7, HisH35 of CNJ206 and 48G7, His27d of 6D9). Exceptions do occur though, as in the structures of 6D9 and 7C8 raised against a secondary *p*-nitrophenyl phosphonate. In contrast to other Group I TSAs, in antibodies 6D9 and 7C8, the linker via which the TSA hapten is linked to a carrier protein, is located on the leaving group and not on the phosphonate (Fig. 15.2). The active site of 6D9 and 7C8 is shallower, the phosphonate is less buried than in other Group I antibodies, and the active-site residues in contact with the TSA are fewer and different [11].

Despite the evident convergence between Group I's active sites, the immune system provided some variations within the interactions forming the oxyanion hole (Table 15.1). The distinction between the position of the oxyanion hole and that of the entering hydroxy nucleophile was determined by the respective structure in the presence of substrate analogs (as in the D-antibodies), by pH profiles pointing to key hydrogen-bond donor residues (D-Abs, CNJ206), and by mutagenesis (D-Abs, 48G7, 17E8), and was complemented by docking *in silico* (Table 15.1, Fig. 15.2) [39].

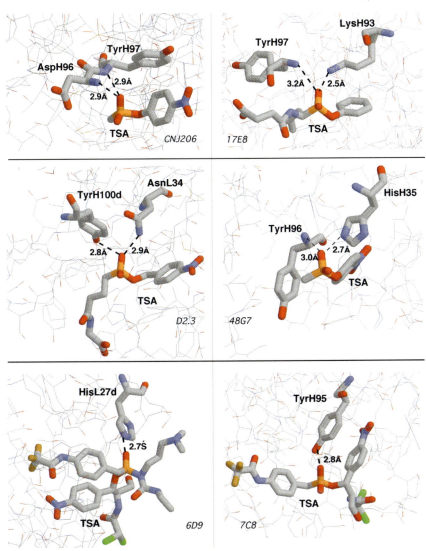

Fig. 15.2 The *"oxyanion hole"* of esterolytic antibodies. Oxyanion-stabilizing residues are depicted from the TSA complexes of antibodies CNJ206 (1KNO in PDB), 17E8 (1EAP), D2.3 (1YEC), 48G7 (1AJ7), 6D9 (1HYX), and 7C8 (1CT8), viewing from the oxyanion representing the incoming hydroxyl nucleophile. CPK color codes: gray: carbon, red: oxygen, blue: nitrogen, orange: phosphate, and green: chlorine.

15.2.1.3
Antibody Oxyanion Holes

The D-antibodies (D2.3 and D2.4), the most efficient antibodies of Group I, are taken here as a model for examining the contribution of active-site residues acting OAH.

Two such residues (Tyr100dH and Asn34L) were suggested to form the crystal structures of the D-antibodies in complex with the TSA and a substrate analog (Fig. 15.2). In the crystal structure of D2.3 with an amide substrate analog (1YEF in PDB), 3.8 Å is the distance between the aspargine's δ-amide proton donor and the scissile bond carbonyl-oxygen (compared to the 2.6 Å distance from the oxygen of TyrH100d), suggesting that Asn34L interaction may be specific to the transition state. In addition, D2.5, lacking Asn34L, has 20–50 fold lower catalytic efficiency than D2.3 and D2.4 (Table 15.1). However, mutating Asn34L to Glycine did not affect the rate of catalysis (Table 15.2), leaving Tyr100d the primary functional OAH residue. This was further supported by several Tyr100d mutants (Table 15.2) and pH profiles of binding and catalysis [10]. The pH-binding profiles revealed a 1.6–1.9 units difference in pK_a between the substrate-antibody and the TSA-antibody complexes. Thus, the negative charge marking the difference between the substrate and the TSA (and hence the actual TSs of the reaction) dramatically strengthens the tyrosyl-oxyanion H-bond, thus rendering it the key residue in the D-Abs oxyanion hole [10]. The specific free energies, 6.3 and 7 Kcal/mol for the complexation of D2.4 and D2.3 with the TSA $\Delta G°_{TSA} - \Delta G°_S$) are identical to the reduction in the activation energy barrier derived for the D-Abs from their rate acceleration ($\Delta G^{\ddagger}_{cat} - \Delta G^{\ddagger}_{uncat}$) (Table 15.1). This tight correlation suggests that the D-Abs active sites convert binding energy into catalysis with close to 100% efficiency, corresponding to the contribution of the tyrosyl OAH.

Judging from Tables 15.1 and 15.2, tyrosine is the most efficient single residue for OAS. Further, a hierarchy may be derived where Tyr (contributing up to 7.0 Kcal/mol) > His (4.1 Kcal/mol, assumed from 6D9 single His OAH and 48G7 His35HGlu mutant) ≥ Lys (~ 3 Kcal/mol, 29G11 Lys-main chain amide OAH combination is weaker than 48G7) > backbone amide (2.2 Kcal/mol, judged by comparison of CNJ206 to wt48G7; 48G7 HisH35Glu mutant's activity, which is derived from Tyr97H backbone amide, is approximately half that of CNJ206 with two backbone amides).

Though highly speculative, this order of activities may be justified by the oxyanion pK_as found for serine hydrolase substrates, ranging between 7 and 10 [40, 41]. It is expected that hydrogen bonds between donors and acceptors with matching pK_a would form the strongest hydrogen bonds [42]. Tyrosine's hydroxyl ($pK_a \approx 10$) may afford the closest match to ester's TS pK_a. It is expected that this pK_a would be considerably lower than the pK_a of the TSs, leading, for example, to amide hydrolysis. The former is often stabilized in nature by OAH comprised of amide bonds that have a pK_a much higher than the hydroxyl of tyrosine (see below) [5, 43]. Matching pK_a values may also explain the significantly stronger interaction of the D-Abs' tyrosine (~ 7.0 Kcal/mol) compared to the average binding energy of a charged group via a single hydrogen bond (3.5–4.5 Kcal/mol) [44].

15.2.1.4
Oxyanion holes – Antibodies vs. Enzymes

The versatility of OAS solutions adopted by the immune system is comparable to, or perhaps even higher than, that observed in natural hydrolytic enzymes. Excluding metal hydrolases, which are not easily mimicked by antibodies (Section 15.2.5), the overall majority of enzymes use a combination of two (e.g., chymotrypsin, trypsin) or three (e.g., AChE, Fig. 15.3) backbone amides to form their OAHs. Their contribution to the catalytic activity cannot be addressed directly by mutation analysis or separated according to the individual contribution of each backbone NH. OAS could only be assessed by attributing the residual activity of the triad mutants to OAS [e.g., trypsin [45], acetylcholinesterase (AChE) [46]; Table 15.3]. A minority of enzymes were found to utilize a combination of a backbone NH groups as H-donor together with a side-chain NH group [e.g., Asn in subtilisin (Fig. 15.3), Gln, or Ser (Table 15.3)]. Interestingly, the recently characterized prolyl oligopeptidase-like family [43, 47, 48] and the structurally similar cocE esterase (Fig. 15.3) [49] utilize a tyrosine side chain (Table 15.3) together with the backbone NH of the seryl nucleophile. Indeed, the efficient OAS by catalytic antibodies such as the D-antibodies may indicate that tyrosine could be as efficient as or even more efficient than NH groups. The contribution of the OAH side chain to the overall catalytic rates of these enzymes was assessed by mutation of the putative side chain (Table 15.3). Such studies are complicated by mutation-induced structural changes of the active site as clearly seen in the cocE Tyr44Phe mutant's structure. As a result of a loss of H-bonding to the mutated tyrosine, three Trp residues change their position, and the Phe ring is tilted by 20° compared to the wild-type tyrosyl ring [5]. In addition, given the average bacterial translation error rate of 5×10^{-4} per amino acid [50], wild-type revertants may significantly affect these single-mutants' analyses.

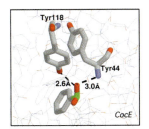
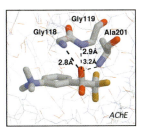
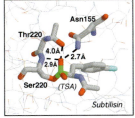

Fig. 15.3 The 'oxyanion hole' of hydrolytic enzymes. Oxyanion-stabilizing residues are depicted from the following inhibitor-enzyme crystal structures: acetylcholinesterase (AChE) with m(N,N,N-trimethylammonio)-2,2,2- trifluoroacetophenone (TMTFA) (1AMN in PDF), cocaine esterase I (cocE) with phenyl boronic acid (1JU3), and subtilisin with D-p-chlorophenyl-1-acetamido- boronic acid (1AVT).

Histidine-containing OAH (as in antibody 6D9; Fig. 15.2) have not yet been found in nature. The lower efficiency of His in OAS is seen in hydrolytic antibodies (see above). The latter may be due to the relatively low pK_a of the His side chain or the more demanding desolvation of the charged His residue compared to the more hydrophobic tyrosine, aspargine, or serine. Interestingly, several structural studies (NMR of trypsin-leupeptin complex [51]; X-ray of proteaseA-chymostatin [52]) identified the oxygen carbonyl of the aldehyde inhibitors pointing toward the active-site's His, suggesting that the latter has comparable affinity to the oxyanion with respect to the original OAH two backbone NH groups. This observation is in agreement with the OAS efficiency scale postulated from the catalytic antibodies' data.

Tab. 15.3: Contribution of oxyanion stabilization to enzymes, mutagenesis studies.

Enzyme	OAH	Mutation	OAS[a] $\Delta\Delta G^{\ddagger}_{cat-uncat}$ Kcal/mol	Ref.
Subtilisin		Asn155Gly	3	[4]
	Asn155,	Asn155Ala	4.2	[130]
	Thr220[c],	Thr220Ala	1.8	[130]
	(Ser221)	Asn155Ala Thr220Ala	6.0	[130]
		Ser221Ala[b]	4.8	[4]
CarboxypeptidaseA	Arg127	Arg127Ala[a]	6.0	[131]
Trypsin	(Gly193) (Ser195)	Ser195Ala[c]	4.2	[45]
Gln19	Gln19Ala (Cys25) Ser42	3	[132]	
(Ser42)	Ser42Ala (Gln121)	3.6	[133]	
Prolyloligopeptidase	Tyr473 (Asn555)	Tyr473Phe	3.7	[43, 47]
Oligo-peptidaseB	Tyr452 (?)[e]	Tyr452Phe	3.6	[48]
CocE	Tyr44 (Tyr114)	Tyr44Phe	> 4.4[d]	[5]
AChE	(Gly118) (Gly119) (Ala201)	–[f]	5–7	[46]

a) Contribution to oxyanion stabilization derived from OAH-residue mutation's effect on catalytic rates or:
b) Contribution to oxyanion stabilization derived from residual activity after mutation of the enzyme's endogenous nucleophile.
c) Though Thr220 is only within 4.0 Å of the oxyanion, dynamic simulations suggest this residue's participation in subtilisin's OAS.
d) Limited by detection of mutant's activity. Significant structural perturbation of mutant's active site
e) As crystal structure is unavailable, the identity of an assumed second OAH residue (analogous to the homologous Asn555 of prolyloligopeptidase (see separate entry) is unknown.
f) Estimated by the authors based on AChE-TSA crystal structure and accumulative data.

The estimated values for the individual residue's contribution to OAS are generally lower for enzymes (Table 15.3) than for the D-antibodies and most other esterolytic antibodies (Tables 15.1 and 15.2). It may be that the enzymes' mutagenic studies underestimate the OAH side chain's contribution, as the remaining H-bond by the backbone NH at the OAH is strengthened in the absence of its side-chain partner, hence compensating for its absence. Such compensation may explain the mild effect of 48G7 His35HGlu mutation (40-fold reduction in catalytic rate), compared to 6D9 His27dAla mutation. The latter lacks a backbone NH, which is present in 48G7's OAH (Table 15.2).

The only natural OAH suggested to have a similar contribution to catalysis as that of the antibodies described above is of AChE, which has three hydrogen donors (Fig. 15.3) (compared to 1–2 in most antibodies). Did enzyme evolution miss out on an efficient OAS strategy that is easily implemented by antibodies? Few plausible answers come to mind. The OAH may not have evolved independently in nature, but paralleled (or followed) the appearance of a catalytic triad that takes care of nucleophilic and proton-transfer catalysis (Fig. 15.1b). Once the latter was established, together with other mechanisms such as substrate destabilization [46] [52], the hydrolytic activity may have been sufficiently high (diffusion-limited at times [46]) to make further improvements in OAS unnecessary. Alternatively, it was recently suggested that the charged imidazole of the subtilisin triad's histidine has a major role of lowering the pK_a of the TS oxyanion, thereby diminishing the OAH binding energy [40]. In such a scenario, the major role of the OAH is substrate alignment rather than TS stabilization as was initially suggested [53]. It remains to be seen whether these results would extend to other natural enzymes.

In conclusion, hydrolytic antibodies validate the fundamental role of OAS in enzyme catalysis. They do so with similar or perhaps better efficiency than enzymes. OAS measured for antibodies could be thus used as an upper limit to the contribution of OAS in natural hydrolytic enzymes, providing a reference value for the contribution of their other mechanistic features (*see below*).

15.2.1.5
Antibody Affinity and Rate Acceleration Limitation

Despite the limitations of TSA design, our analysis suggests that many antibodies do convert TSA binding energies to catalysis with nearly 100% efficiency (Table 15.1). This may suggest a facile route to increased catalytic efficiency via antibodies with higher affinity. However, it appears that antibodies with affinities higher than 1 nM may not be easily obtained.

Data collected on the affinity of a large number of antibodies [54] suggest an antibody of 0.1–1 nM [55]. This is derived from the upper range for the association rate ($< 5 \times 10^6$ M^{-1}s^{-1}) and a dissociation rate (k_{off}) that is sufficient to trigger B-cell activation (10^{-3}–10^{-4} s^{-1}). These studies suggest (see [54, 55] and references therein) that 0.1–1nM affinities are sufficient for mounting an efficient immunological response, arguing that the immune system is unlikely to produce higher affinity antibodies. By inference, with a low-average K_M ($\approx K_S$) in the mM range [56–58] a "catalytic ceiling" of $K_{TS}/K_S = k_{cat}/k_{uncat} = 10^6$–$10^7$ may be derived for catalytic antibodies raised against TSA haptens even if these happen to perfectly mimic the reaction TSs. Thus, the turnover number (k_{cat}) of antibody-mediated hydrolysis of esters commonly targeted by catalytic antibodies (k_{uncat} 10^{-6}–10^{-5} s^{-1}) is expected to be well under 10s^{-1} and the specificity constant (k_{cat}/K_M) $\leq 10^4$ M^{-1}s^{-1}. Under the immunological constraints

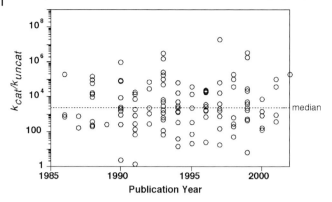

Fig. 15.4 Rate enhancements exhibited by catalytic antibodies: a compilation of rate accelerations exhibited by all catalytic antibodies [139]. The theoretical threshold is set by the affinity ceiling of antibodies as described in the text. The only antibody to surpass the theoretical threshold dictated by the immune system is 38C2 [140], achieved by *reactive immunization* (see text for details).

limiting K_{TSA}, a lower K_M (or higher substrate affinity) would result in a lower rate acceleration, without lowering the specificity constant. Indeed, a survey of the published catalytic antibodies conforms with the above theoretical limitation (Fig. 15.4). Moreover, in some instances, not all of the TSA binding energy is directed towards TS stabilization, but rather to non-contributing epitopes of the TSA hapten. For example, antibody 48G7 was shown to bind its TSA with 30 000-fold higher affinity compared to its germline counterpart, resulting in only ~80-fold improvement of k_{cat}/K_M [16]. Although independent values of k_{cat} and K_M were not reported, it is likely that a significant part of the increase in binding energy upon maturation was directed to both the substrate and the TSA, and therefore led to no increase in rate acceleration. Similarly, attempts to improve the rate of esterolytic antibodies by directed evolution aimed toward higher TSA affinities have shown that major increases in affinity may result in minor rate improvements [59, 60]. Improving the rates of esterolytic antibodies is not a purely academic challenge. The limitations in rate hinder at least two of the most promising biomedical applications of hydrolytic antibodies – cocaine detoxification [61] and pro-drug therapy [62]. Catalytic forces other than OAS may help in closing the gap of ~100-fold reactivity [5, 62] required for their successful implementation. These are discussed below.

15.2.1.6
Nucleophilic Catalysis

The vast majority of esterolytic antibodies make use of a solution (exogenous) hydroxide ion to generate the tetrahedral, oxyanionic intermediate that leads to ester hydrolysis. This is not particularly effective, as at or close to neutral pH, hydroxide ion concentrations are very low. The enzymatic solutions to this problem are basically two: (i) nucleophilic catalysis by an active-site residue (an endogenous nucleophile) via an acyl-enzyme intermediate; (ii) a general-base residue and a bound water molecule leading to the *in situ* generation of an appropriately positioned hydroxide ion.

Inspired by natural hydrolytic enzymes that utilize a covalent catalytic mechanism (Fig. 15.1b), and with the goal of catalyzing the more demanding reaction of amide hydrolysis, several attempts were made to introduce an endogenous nucleophile into antibodies' combining sites. But, in fact, the more effective cases of endogenous nucleophilic catalysis seem to be in antibodies in which the endogenous nucleophile was incorporated haphazardly. Antibodies in which endogenous nucleophilic catalysis plays a role are discussed in Section 15.2.2 below. In the same vein, attempts were made to generate antibodies in which the nucleophilic hydroxide is generated *in situ* by general-base catalysis (discussed in Section 15.2.4 below). Alternatively, the efficiency of the exogenous nucleophilic attack can be enhanced by applying nucleophiles more potent than hydroxide ("chemical rescue"; Section 15.2.3).

15.2.2
Endogenous Nucleophiles in Hydrolytic Antibodies

15.2.2.1
Reactive immunization (RI)

Several attempts were made to introduce endogenous nucleophiles into antibody combining sites, either by chemical modification [9, 63] or by immunization. The latter was aimed at "reactive immunization", whereby the labile phosphonate diester **1** (Scheme 15.1), a known covalent inhibitor of hydrolases, was used as a hapten [7, 64].

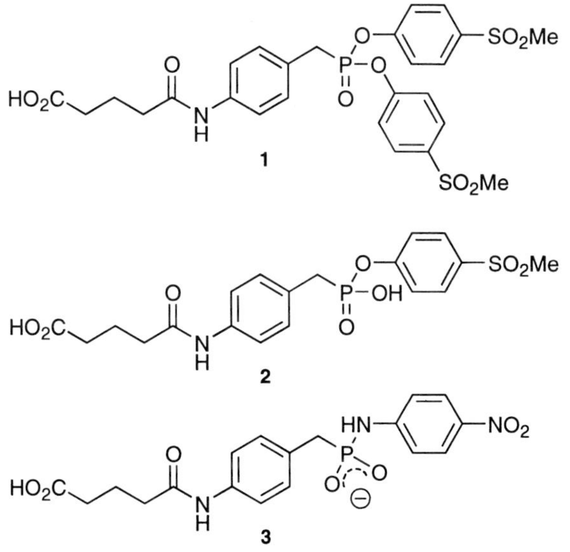

Scheme 15.1

During the immunization process, an RI hapten should covalently trap antibodies with a nucleophilic residue, thereby forming a highly stable complex that is selectable by the immune system. There is, however, a major caveat to this methodology. Hydrolysis of the activated *p*-methylsulfonate ester substrate via nucleophilic catalysis will result in an that is significantly more stable than the original substrate. For effective catalysis to occur, an would be necessary to activate the hydrolysis of the acyl-enzyme intermediate (Fig. 15.1b). Such a residue is counter-selected in this system, as it would also hydrolyze the covalent phosphonyl hapten-antibody bond, thus eliminating the selectable advantage of the reactive immunization. Moreover, hapten 1 is labile (estimated half-life mice's serum is 24 h), resulting in a mixture of haptens, 1 and 2, (the bi- and mono-phosphonate esters, respectively). Unexpectedly, the affinity to the "baiting" hapten 1 is rather low (in the micromolar range) compared to the high nanomolar affinity of antibodies raised against hapten 2. The kinetic parameters of *anti*-1 antibodies , suggest that OAS (represented by hapten 2) does not account for the RI-derived antibodies' catalytic efficiencies (e.g., for antibody 15G12, $K_{TSA(2)}/K_M$ = 2; k_{cat}/k_{uncat} = 6.6 × 10^5). However, there is no other evidence to support nucleophilic catalysis in the hydrolytic antibodies raised against RI hapten 1 (e.g., burst kinetics, trapping of the covalent intermediate, non-competitive inhibition by hapten 1). Some of the RI antibodies were reported to catalyze the dephosphonylation of their cognate hapten 1, albeit with a low turnover (< 3), because of product inhibition [7] (such single turnover rate enhancement by ester-hydrolytic antibodies, raised against the ground state hapten, was reported as early as 1980 [65]). Schowen has recently suggested that TS flexibility, where the classical bipyramidal dephosphonylation TS may adopt a more tetrahedral-like conformation, renders it available for stabilization by the *anti*-phosphonate antibodies [66]. It should be mentioned that, although RI achieved a limited success with ester hydrolysis, it has proven highly effective for reactions such as aldol condensation, *retro*-aldol and Michael addition reactions that proceed via a covalent Schiff-base enzyme-substrate intermediate [67, 68].

15.2.2.2
Serendipity and 43C9 Antibody

It has been suggested that several hydrolytic antibodies raised against phosphonate TSAs include an endogenous nucleophile as part of their mechanism of action. Unfortunately, in most cases, only partial evidence is available. In general, proving a nucleophilic mechanism is not a trivial matter, and several independent pieces of evidence need to provided, including the direct identification of an acyl-enzyme intermediate, before a definite statement can be made [2]. A "ping pong" mechanism by antibody PCP21H3 (k_{cat}/k_{uncat} not determined) in a transesterification reaction of a vinyl ester has been suggested. The antibody exhibits burst kinetics for the hydrolysis of a related *p*-nitrophenyl ester [17]. Product inhibition, common to many catalytic antibodies, was detected as well, yet this does not outweigh the evidence for the covalent mechanism. A similar burst in product formation interpolated to a 1:1 molar ratio of product to antibody was reported for antibody 6F11 (k_{cat}/k_{uncat} = 3.4 × 10^3) [69]. Chemical modification of tyrosyl residues resulted in complete activity loss.

Antibody 20G9 (k_{cat}/k_{uncat} = 500) exhibits "burst kinetics", and this led to the suggestion of nucleophilic catalysis [70–72]. However, the kinetic analysis is complicated by strong product inhibition, and no evidence apart from the "burst" is available. Incubation of the antibody with N-cbz-glycine -O-phenyl ester resulted in multiple acylation of the antibody and a significant decrease in its activity, yet acylation of an active-site residue such as lysine that is not directly involved in catalysis may account for this result [73]. In addition, the reported acid-limb pH profile fitting a tyrosyl pK_a (9.6), as well as the loss of activity when chemically modifying tyrosyl residues with tetranitromethane (TNM), could well be explained by this residue's participation in OAS [10, 73, 74].

Antibody 7C8 (k_{cat}/k_{uncat} = 3.4 × 10^3) also exhibits a pH profile with an acid-limb with a pK_a ≈ 9.0. The rate is also insensitive to the addition of a strong nucleophile (hydroxylamine; see also Section 15.2.3 below). These led to the suggestion of nucleophilic catalysis by an active-site tyrosyl hydroxyl [11]. However, D2.3 and D2.4 antibodies also manifest the above characteristics [10], yet their active site tyrosine (H100d) was shown to serve as an OAH rather than a nucleophile. Similarly, 7C8's crystal structure reveals a sole residue that may serve as an OAH, Tyr95H. Another clear demonstration of the difficulties associated with proving a nucleophilic mechanism is antibody 17E8. Its crystal structure and hydroxylamine partitioning experiments suggested the existence of a Ser-His dyad, reminiscent of natural serine hydrolases [13, 35]. However, direct evidence of an acyl-enzyme intermediate was not provided, and the proposed mechanism involved the disruption of a hydrogen bond between His35H and TrpH47 that is conserved in all known antibody structures [36]. Indeed, nucleophilic catalysis was ruled out when a variant of 17E8 in which the Ser, assumed to act as nucleophile, was replaced by Gly exhibited a similar rate enhancement [30]. Most significantly, analysis of the antibodies described above suggests that even if an is indeed part of their mechanism, the resulting rate enhancements are 2–3 orders of magnitude lower than is achieved with antibodies raised against the same haptens and that act by OAS only (see above).

Antibody 43C9, raised against the phosphonoamidate 3 (Scheme 15.1) [75, 76] stands out from all the above examples, not only because of its high reactivity (it catalyzes the hydrolysis of both a p-nitrophenyl ester and a p-nitroanilide and, although its rate enhancement is within the range of other known esterolytic antibodies, its specificity constant is so far unrivalled (k_{cat}/K_M = 2.8 × 10^7 M^{-1}s^{-1})), but also because of the unambiguous assignment of a nucleophilic mechanism.

In the case of antibody 43C9, there is ample evidence to prove a nucleophilic mechanism. The pH-rate profile suggests a change in the rate-limiting step: while at pH > 9 product release is limiting, at more acidic pH an acylation step determines the rate [77]. This was demonstrated by pre-steady-state kinetics, where a burst of antibody-equivalent product release kinetics was observed [77]. Finally, the effect of p-substituents of the phenolic leaving group on the rate of catalysis [78] and the identification by electrospray mass spectroscopy of an acyl-enzyme intermediate (with HisL91 as the acylated residue) unambiguously support the proposed mechanism [6]. A detailed catalytic mechanism based on 43C9 crystal structure was put forward

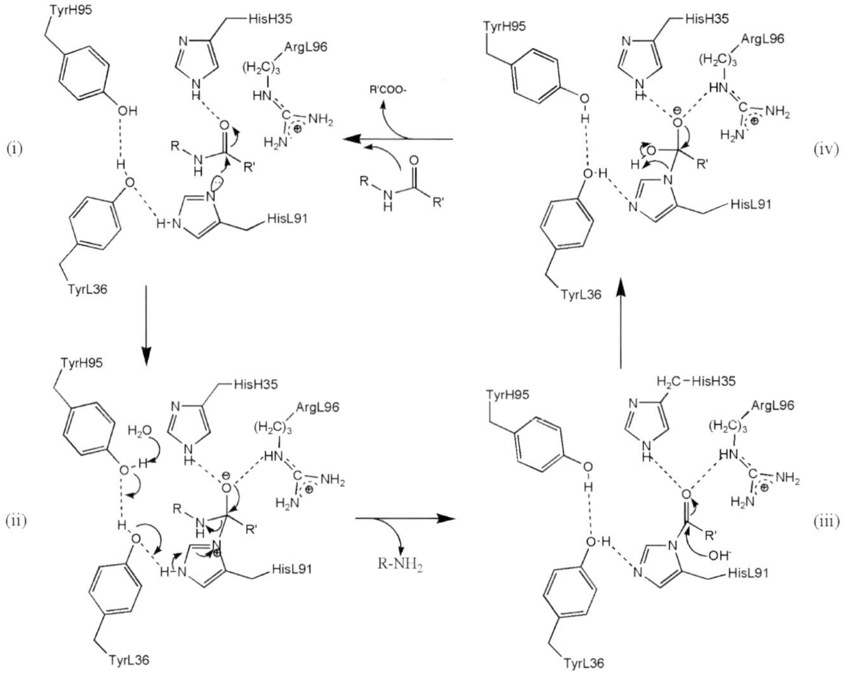

Fig. 15.5 The mechanism of catalysis by antibody 43C9 [79]). **(i)** A nucleophilic attack by an endogenous nucleophile - HisL91 side chain. The nucleophilic imidazole is maintained in the correct tautomer by a network of H-bonds with TyrL36 and TyrH95 side chains. **(ii)** The resulting oxyanionic intermediate is stabilized by H-bonding to His35H and ArgL96. Bond rearrangement, coupled with a proton transfer cascade, results in the acyl intermediate and the release of the aniline leaving group. The aniline release is mediated by protonation by a water molecule (not shown). **(iii)** An exogenous nucleophilic attack by a hydroxide ion on the acyl-enzyme intermediate. **(iv)** Bond rearrangement results in the release of the carboxylate product and the free antibody (see hapten 3 structure for R and R').

where a nucleophilic attack, carried out by HisL91, is accompanied by OAS via the side chains of HisH35 and ArgL96 (Fig. 15.5) [79].

The serendipitous selection of a His side chain as the nucleophilic catalyst of 43C9 is quite interesting. Not only is the His's imidazole side chain a better nucleophile toward activated esters than the seryl alcohol moiety, but also its pK_a is optimal for catalysis and the resulting acyl-enzyme intermediate is much more labile than an ester intermediate. Thus, His makes an ideal nucleophilic catalyst. Nevertheless, His was never recorded as a nucleophile in natural hydrolytic enzymes. One possible explanation is that with non-activated ester substrates, imidazole prefers acting as a general base rather than a nucleophile, hinting at nature's preference for other nucleophiles as alcohols and thiols [80]. The OAH of 43C9 is equivalent to other known antibodies (Table 15.1, Fig. 15.2), suggesting a comparable contribution to its rate enhancement. Thus, the nucleophilic mechanism does not seem to contribute much to the rate acceleration of ester hydrolysis. This can be explained by the absence of a general base residue (Fig. 15.1b) that would catalyze the rate-limiting de-acylation step, thus replacing an activated *p*-nitrophenyl ester with a less active acyl-imidazolium moiety. However, nucleophilic catalysis may contribute significantly to anilide hydrolysis, where formation of the intermediate, rather than its breakdown, is rate-determining. The nucleophilic mechanism may also release 43C9 from having a k_{cat}/K_M limited by low affinity to the ground state substrate, as prescribed for efficient catalytic antibodies utilizing solely the energy gained from preferential TS *vis-a-vis* substrate binding, yielding ca. 2 orders of magnitude higher specificity constant, in parity with natural enzymes' values. Quantifying the individual contribution of the nucleophilic attack is further complicated by the fact that a HisL91Gln mutation of 43C9 results in total loss of hydrolytic activity, yet retains wild-type levels of affinity to its hapten and substrates [81]. Thus, in agreement with natural enzymes, even in a somewhat simplified model, different catalytic mechanisms seem to be coupled.

Importantly, the nucleophilic mechanism endows 43C9 with the ability to hydrolyze the more stable *p*-nitroanilide substrate, unattainable by the antibodies using OAS as their only source of reactivity. Unlike natural amidases, 43C9 does not catalyze the hydrolysis of non-activated amides. The rate-limiting step in their hydrolysis involves the protonation of the amine-leaving group via a general acid residue (as in natural amidases, Fig. 15.1b), not found in 43C9 active site.

15.2.3
Exogenous Nucleophiles and Chemical Rescue in Hydrolytic Antibodies

The nucleophilic attack of a solution's (exogenous) hydroxide anion seems to constitute the rate-limiting step of all hydrolytic antibodies. In most cases, the hydroxide reacts directly with the ester substrate and, in rarer cases, with an acyl-antibody intermediate. This step therefore merits careful examination. One of the most useful analytical tools is the replacement of hydroxide by alternative nucleophiles. Negatively charged, tetrahedral intermediates, homologous to the hydroxyl-mediated catalytic mechanism (Fig. 15.1a) can be formed by the attack on the ester's carbonyl by other nucleophiles (Fig. 15.6). Thus, it may be expected that small nucleophiles, stronger than the hydroxyl anion, may be active in the antibodies' active sites, resulting in higher catalytic rates. A similar strategy was previously used to "chemically rescue" hydrolytic enzymes where the nucleophilic residue was mutated. The mutants' catalytic activity can often be restored to rates only 100- to 300-fold lower than the intact enzyme [82–84].

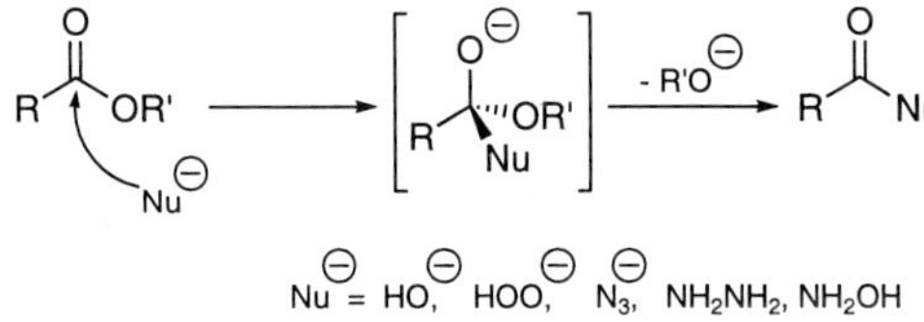

Fig. 15.6 The tetrahedral intermediate formed by various exogenous nucleophiles. $Nu^- = HO^-, HOO^-, N_3^-, NH_2NH_2, NH_2OH$

One example of effective "chemical rescue" of hydrolytic antibodies is with the peroxide anion, a ~200-fold stronger nucleophile than hydroxide, that was found to enhance the turnover rates of D2.3 and D2.4 antibodies to a similar extent (Fig. 15.7a,

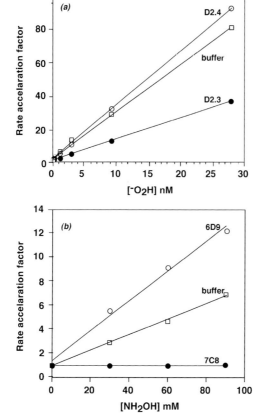

Fig. 15.7 The effect of exogenous nucleophiles on the rate of antibody-mediated ester hydrolysis. Rate acceleration factors are the ratios of net initial rates observed in the presence of the nucleophile at a given concentration and in its absence. (a) Effect of peroxide anion on the hydrolysis of p-nitrophenyl ester by D2.3 and D2.4 [10]. (b) Effect of hydroxylamine on chloramphenicol ester hydroysis (data derived from [11]).

Tab. 15.4: Nucleophile effect on D-Abs' catalysis.

Nucleophile[a] $(\times 10^6\ M^{-1})$[b]	pK_a	Buffer $(\times 10^6\ M^{-1})$[b]	Antibody D2.3 $(\times 10^6\ M^{-1})$[b]	Antibody D2.4
$^-$OH	15.4	13	12	14
$^-$O$_2$H	11.6	2800	1270	3200
NH$_2$NH$_2$	8.1	0.24	0	0.008
NH$_2$OH	6.0	0.45	0	0.006
N$_3^-$	4.0	1×10^{-3}	n.d.	0.01×10^{-3}

a) A linear correlation links the nucleophilicity of most nucleophiles to their pK_as. However, the reactivity of the α-nucleophiles cited above (i.e. excluding OH$^-$) is far higher than predicted from their pK_a (e.g. 10^3-fold for hydroxylamine).

b) Results are given in "molar reactivity" – the rate acceleration factor at 1 M nucleophile as depicted in [10].

Table 15.4) [10]. The concentration-dependent rate acceleration exhibits a rate ceiling at a k_{cat} that is 500- to 2000-fold higher than that of the antibodies' hydroxide-mediated catalysis. The rate ceiling is due to a change in the rate-limiting step observed at high peroxide concentrations – a conformational change becomes rate limiting instead of the nucleophilic attack. The observed rate enhancement of the D-antibodies in the presence of peroxide compares favorably with the estimated contribution of the deacylation rate-limiting step of various natural hydrolases (e.g., ca. 200-fold for cocaine esterase cocE [10, 49]. The D-Abs' maximal rate of 6 s^{-1} is also comparable to many natural esterolytic enzymes, such as cocE [5], thioesterases [85], and carboxylesterases [86], and is within the lower range of typical rates of (1–200 s^{-1}). This suggests that a potent exogenous nucleophile can match, in the framework of efficient OAS antibodies, the role of the enzyme's general-base deacylation mechanism (e.g., the His-Asp dyad in Fig. 15.1b). The second-order rate for peroxide anion reactivity with the D-antibodies was calculated to be 6×10^5 $M^{-1}s^{-1}$. This is significantly lower than the directly measured second-order rate for water-mediated deacylation of chymotrypsin ($< 4 M^{-1}s^{-1}$) and compares favorably with its deacylation by an exogenous amine nucleophile (tyrosinamide: 1.7×10^4 $M^{-1}s^{-1}$), a direct result of the enzyme's specificity for the leaving group [87]. Importantly though, the D-Abs' rate accelerations (k_{cat}/k_{uncat}) are relatively unchanged when peroxide replaces hydroxide, suggesting that the D-Abs are equally complementary to the respective TSs of both hydroxide and peroxide. The D-Abs exhibit absolute preference toward small, negatively charged nucleophiles (i.e., hydroxylate and peroxylate) for the displacement of the ester's alcohol moiety, whilst other nucleophiles (e.g., hydroxylamine) do not affect the rate (Table 15.4). We believe that this specificity is a result of (i) the selection, by the negatively charged phosphonate anion, of antibodies with a cluster of positively charged residues, creating a channel to the antibody active site, and (ii) the snug fit of the antibody to its hapten and therefore to the TS, preventing the entry of bulky nucleophiles such as azide [10]. These results suggest that it may be possible to optimize the electrostatic properties of the channel through which the nucleophile approaches the active site, perhaps by engineering positively charged residues in and around this channel without interfering with the properties of the active-site core [88], thereby increasing the rate of catalysis.

The rate of hydrolysis by antibody 6D9 (raised against hapten 4, Scheme 15.2) is also increased by the presence of the α-nucleophile hydroxylamine. Unlike the D-abs, both the rate (k_{cat}, or k_{cat}/K_M) and the rate acceleration (k_{cat}/k_{uncat}) increase, albeit to a minor extent (up to ~2 fold at 100 mM hydroxylamine) (Fig. 15.7b) [11]. The source of recognition of hydroxylamine by 6D9 was not addressed, yet the Lys32L residue [11] is located in a position suitable for general-base activation of the nucleophile. Alternatively, it may act as an H-bond donor to orient the nucleophile toward attacking the chloramphenicol ester's carbonyl. It would thus be interesting to study the effect of Lys32L mutants on the hydroxylamine-mediated reaction, as well as attempting the hydrolysis of the more demanding amide hydrolysis in this system.

Another reported case of "chemical rescue" concerns antibody 14-10. Using the heterologous immunization approach, where mice were challenged with both haptens 5 and 6 (Scheme 15.2), Masamune and coworkers attempted the generation

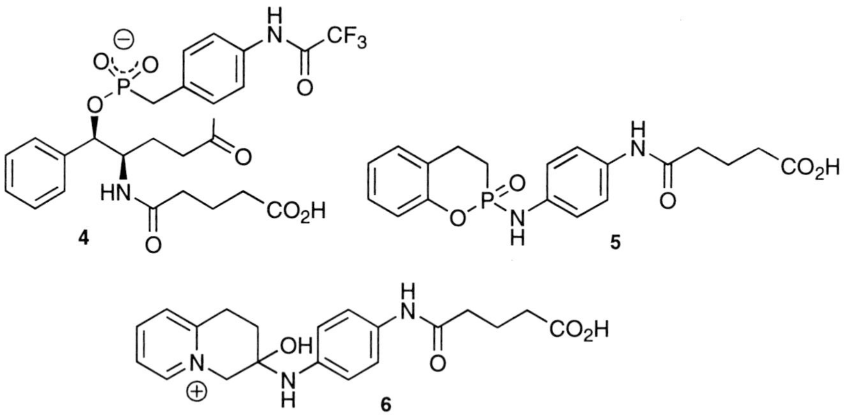

Scheme 15.2

of a *p*-nitroanilide hydrolyzing antibody that makes use of OAS, general-base catalysis (by "baiting" with a positively-charged amine hapten) and an exogenous phenol nucleophile. The best antibody achieved, 14-10, catalyzed both the lactonization of substrate **7** (Scheme 15.3) via an intramolecular phenol nucleophilic attack ($k_{cat} = 1.5 \times 10^{-2}$ min^{-1}, $K_M = 1.27$ mM [89]) and the intermolecular attack of added phenol nucleophile to substrate **8** to produce phenyl propionate, **9** ($k_{cat} = 1.3 \times 10^{-4}$ min^{-1}, $K_{M(2)} = 0.37$ mM, $K_{M(phenol)} = 0.14$ mM [12]).

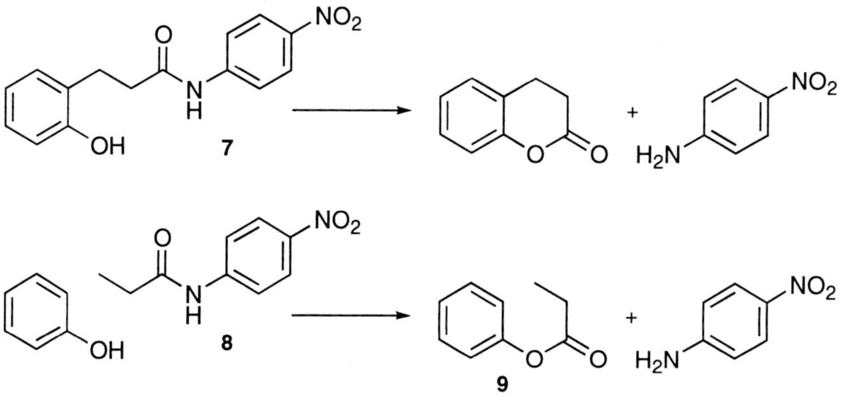

Scheme 15.3

In both cases the phenol does not seem to participate in the background buffer-catalyzed reactions. Thus, the antibody catalyzes kinetically "disfavored" reactions, resulting in $k_{cat}/k_{uncat} = 2 \times 10^4$ and 630 for the intramolecular and intermolecular reactions, respectively (the uncatalyzed rates serving as an upper limit of the background rates). Interestingly, 14-10 does not hydrolyze the ester product **9**. This could be only partly explained by the lack of a negatively charged TSA to elicit an OAH, as similar, uncharged haptens (e.g., phosphono-amidates [62], and β-amino alcohols

[90]) were shown to elicit efficient hydrolytic antibodies. Nonetheless, as the buffer-mediated hydrolysis rate of 9 proceeds twice as fast as the antibody-mediated acyl transfer, the net reaction can be formally considered as the activated amide's hydrolysis. Future studies may reveal whether the phenol nucleophilicity is indeed enhanced by a general-base residue, as implied by 14-10's pH optimum (8.0 compared with the phenol's solution pK_a of 10).

15.2.4
General Acid/Base Mechanisms in Hydrolytic Antibodies

The most notable example of hydrolytic enzymes that make use of general acid and base catalysis is aspartic peptidases. This family of hydrolases utilizes the concerted action of two carboxylate residues (Fig. 15.1c). The first carboxylate serves as a general base to align and activate the water nucleophile, and subsequently as a general acid to protonate the leaving group. The second serves as the OAH and may be regarded as a general acid, donating a proton to the formed carboxylate. In order to mimic such mechanisms with catalytic antibodies, several "bait and switch" haptens were devised with the aim of "baiting" oppositely charged residues in the antibody's active site that may serve in a general acid/base mechanism. An example of this strategy (haptens 10-12) and the kinetic rates of the resulting antibodies is summarized in Table 15.5.

Tab. 15.5: Kinetic parameters of "bait and switch"-raised antibodies.

Antibody	Hapten	k_{cat} [a] (min^{-1}) pH = 7.8	K_M (mM)	k_{cat}/K_M (M^{-1} s^{-1})	k_{cat}/k_{uncat} [b]	Ref.
27A6	10	0.004	0.25	0.7	1.4×10^5	[134]
30C8	11	0.005	1.1	0.07	1.2×10^5	[90]
2D5	12	3.5	0.5	120	1.3×10^5	[69]

a) Values for antibodies 30C8 and 27A6 were derived from pH-rate profiles.
b) Although the same substrate and reaction are involved, a ~1000-fold difference exists between the uncatalyzed rates reported of one group [134] and those of the other [69].

Haptens 10–12 (Scheme 15.4) were designed to bait different types of residues in the elicited antibodies' binding sites. The positively-charged hapten 11 was designed to bait a general-base residue (e.g., a negatively-charged carboxylate), whereas hapten 10 aims at baiting a general-acid residue that would facilitate the rate-determining protonation of the amine-leaving group of substrate 14. Nonetheless, the catalytic activity of the best antibodies raised against haptens 10 and 11 is remarkably similar (Table 15.5), suggesting that even if "baiting" did occur, it does not effect to a significant extent the antibodies' catalytic efficiency. Moreover, whilst the antibodies raised against haptens 10–12 all catalyze the hydrolysis of the ester substrate, 13, with similar

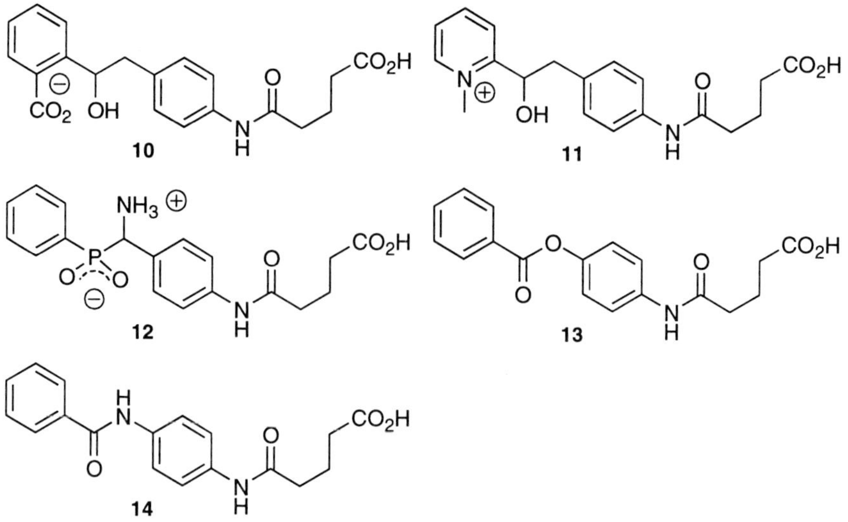

Scheme 15.4

kinetic parameters, the amide substrate **14** is hydrolyzed by none of these antibodies, not even by those antibodies raised against hapten **12**. The effectiveness of the "bait and switch" strategy in raising hydrolytic antibodies is further questioned by the fact that hapten **12** was designed to bait a general-base catalyst while maintaining the phosphonate tetrahedral negatively charged TSA core for the generation of OAH residues. On the other hand, in haptens **10** and **11**, a planar, uncharged carbon represents the tetrahedral TS. The catalytic rate of the best *anti*-**12** antibody, 2D5, is also in agreement with rates of the above-described OAS antibodies, suggesting that even if an active-site general base has been generated, its contribution to catalysis is not very significant. By comparing the catalytic rates of antibodies 30C6 and 27A6 to 2D5, one could conclude that the tetrahedral phosphonate is a better mimic of the TSs of ester hydrolysis than the β-alcohol moiety, or, in other words, the phosphonate TSA "baits" a ~500-fold more efficient OAH. Other attempts to utilise the "bait and switch" strategy yielded similar rate enhancements to the examples described above, with k_{cat}/k_{uncat} values in the range 10^2–10^4 (see, for example [91–93]). Future studies of antibodies raised by the "bait and switch" strategy may shed more light on their mechanism and may allow an improvement in their rates. Notably, as is the case with reactive immunization, the "bait and switch" approach was applied to non-hydrolytic reactions with greater success. This is evident, for example, in the Glu residue acting as a general base in antibody 4B2's mediated catalysis of an allylic rearrangement reaction (PDB code 1F3D [94]). In contrast, the crystal structure of antibody 1D4, raised against a charged β-amino phosphate, does not show the expected positioning of a catalyzing general-base residue in its active site [95].

Finally, it is encouraging to note that a structural model of the chemiluminescent *anti*-phosphonate 7F11 antibody ($k_{cat}/k_{uncat} = 1.2 \times 10^3$, $k_{cat}/K_M = 5.9 \times 10^5$ M^{-1}s^{-1} [96]) indicates the presence of two aspartic acid residues in positions resembling the natural aspartyl protease. Based on this model, the authors suggested that the 7F11 mechanism is similar to the natural counterpart, but that its much inferior catalytic efficiency may be due to the imperfect positioning of the apartyl residues. Antibody 7F11 also suffers from severe product inhibition, enabling an estimated turnover of ca. 5 cycles [97]. Nevertheless, if its active-site structure does prove similar to its natural counterpart, it may serve as a starting point for the generation of more potent hydrolases.

15.2.5
Metal-Activated Catalytic Antibodies

Several studies have attempted to raise catalytic antibodies with metal ion-mediated hydrolysis, following the *zinc-protease* mechanism (Fig. 15.1d). Many enzyme models based on zinc have been created that exhibit high rate enhancements [98], suggesting that the generation of metallo-antibodies with enzymatic activities is a promising avenue. However, this approach has thus far been explored with limited success. Lerner and coworkers reckoned that the "bait and switch" hapten **11** may elicit a carboxylate residue in the antibodies' active site that would serve as a metal-binding ligand. Furthermore, the pyridinium's methyl group will induce a cavity for the metal ion. Indeed, in the presence of Zn^+, antibody 84A3 catalyzes with high specificity the hydrolysis of a pyridinium ester (k_{cat}/k_{uncat} = 1200 [99]), the structure of which is related to the immunizing hapten. No catalysis was observed in the absence of the metal ion or with other, structurally similar ester substrates. The antibody suffers from severe substrate inhibition, probably due to non-productive binding of the substrate, probably in a mode not involving a zinc ion, as judged by the affinity difference to the substrate in the presence or absence of Zn^+ (3.5 µM and 840 µM, respectively). In an attempt to overcome this hurdle, the mercury(II)-phosphorodithioate unsaturated complex **15** (Scheme 15.5) was used to elicit Hg-dependent hydrolysis of ester **16** [100].

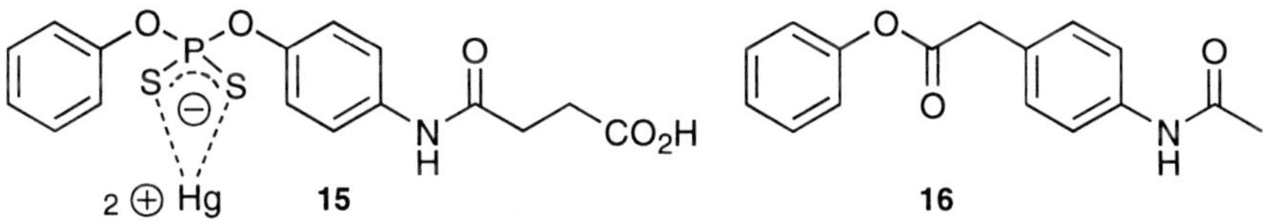

Scheme 15.5

The best *anti*-**15** antibody, 38G2, exhibited Michaelis-Menten kinetics (k_{cat}/k_{uncat} = 300, $K_{M(Hg)}$ = 87 µM and $K_{M(16)}$ = 345 µM), multiple turnovers and absolute dependency on Hg(II) for catalysis. While far from enzyme-like reactivity, these studies widen the spectra of possible mechanisms that could be tailored by antibody mimics.

15.2.6
Substrate Destabilization

Remote interactions, away from the ester or amide carbonyl reaction core, play an important role in many natural enzymes. Such interactions may contribute to catalysis via substrate destabilization (for a recent review see [101] and references therein). For example, Sussman and coworkers postulated that as much as 5×10^5 (corresponding to TS stabilization of 8 kcal/mol) of AchE's spectacular rate enhancement (10^{13}) is derived from orienting the substrate to a highly unfavorable conformation, perfectly aligning the substrate for the nucleophilic attack to take place. This is achieved by strong binding to the choline ammonium charged moiety [46]. Several examples of the identification and possible role of substrate destabilization by hydrolytic antibodies are given below.

The docking of the activated anilide substrate in the crystal structure of 43C9 suggests that the substrate's amide bond is twisted by ~140° out of planarity, thus destabilizing the bond's resonance and facilitating its hydrolysis [79]. Similarly, 6D9 antibody's structure suggests that the antibody locks its ester substrate in the *sin* unfavorable conformation, resulting in substrate destabilization. In both the above examples, a quantitative assessment of the contribution of substrate destabilization to the rate acceleration is not available.

A 5-fold increase in the catalytic rate accompanied by a smaller effect of higher K_M (2-fold) was seen between the catalysis of the glutaryl-glycine nitrophenylester substrate and a caproyl-elongated substrate by antibody D2.3, suggesting a remote effect of the substrate's linker moiety. Crystal structure analysis revealed different interactions with the two linkers that may account for the kinetic differences [102]. When the linker terminal glycine was replaced by either the D- or L-Ala enantiomers, a ~20-fold difference in D2.3's specificity constant (k_{cat}/K_M) was observed in favor of the L-Ala enantiomer [103] (for the enantioselectivity potential of the esterolytic active sites of various antibodies see [39]).

Another example is set by antibody 17E8 catalyzing the hydrolysis of *n*-formylnorleucine (nLeu) phenyl ester. The specificity factors (k_{cat}/K_M) observed with a series of substrates where the nLeu side chain was replaced by different linear and branched alkyl side chains extend over 2 orders of magnitude. But rate enhancements higher than those with the original substrate (namely with a Leu side chain) were not obtained. These results suggested that 17E8 interactions with the side chain are used to stabilise *both* the TS and the ground state. However, the replacement of nLeu side chain with a more hydrophilic ethyl-cysteyl side chain resulted in a 20-fold increase in k_{cat} (43 vs 2.1 s^{-1}) accompanied by a similar decrease in ground state binding (K_M = 4 vs 0.18 mM with nLeu). This rendered the new substrate as efficient as the original substrate ($k_{cat}/K_M \approx 1.2 \times 10^4$ $M^{-1}s^{-1}$) and suggested that substrate destabilization may play a key role in the mechanism of 17E8 with the ethyl-cysteyl substrate [104].

15.3
The Role of Conformational Changes in Catalytic Antibodies

Structural dynamics are part of the nature of proteins, though their identification on the molecular level and their mechanistic roles are still a subject of much research. In enzymes, the identification of conformational changes is generally elusive, as they are typically much faster than the chemical step and product release [2]. Nonetheless, structural fluctuations were observed in serine proteases when comparing their free and boronic acid inhibitor-complexed structures. Such plasticity may be an integral part of the catalytic cycle, mediating, amongst other steps, substrate binding, TS formation, stabilization, and product release (for a recent example see [105, 106]). Several hydrolytic enzymes were also shown to utilize an *induced-fit* mechanism to tune their specificity, either to a broad (e.g., alpha lytic protease) or extremely narrow (e.g., factor D) substrate spectrum (see [101] and reference therein). In addition, a recent kinetic study suggests that conformational changes control the aperture of AChE's active site and are coupled to its catalytic activity [107, 108].

There had been several speculations regarding the role and effect that conformational changes may have on catalysis by antibodies. Padlan, for example, speculated that the of antibodies may hinder their catalytic efficiency [109], whilst Benkovic and Stewart suggested that antibodies elicited against TSA might be limited by a "lock-and-key" rigidity and will lack the conformational isomerism that characterizes enzymatic catalytic cycles [20]. In a more general context, the issue of conformational isomerism in antibodies has been a subject of much research and debate. In the past, kinetic (e.g., [110, 111]) and structural (e.g., [112] and references therein) studies provided strong evidence of the dominance of structural changes in antibodies. Conformational changes can be generally described by the *induced-fit* model (the ligand binding a low-affinity form of the antibody and inducing a subsequent conformational change that leads to a high affinity complex) or by *pre-equilibrium* between two different conformational species of the same antibody, only one of which binds the ligand with high affinity. The two models can be combined in one scheme (Fig. 15.8). While kinetic evidence supports the existence of both mechanisms, structural evidence has been put forward thus far for induced fit only. However, a recent study combining pre-steady-state kinetics in solution and X-ray crystallography provides conclusive evidence for pre-equilibrium in an antibody [113]. The identification of conformational changes in catalytic antibodies and their possible role in catalysis are discussed in this section.

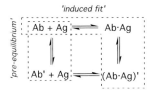

Fig. 15.8 Schematic representation of the two mechanisms of conformational changes linked to antibody-ligand complex formation. Ab and Ab' represent two antibody isomers exhibiting low and high affinity, respectively.

15.3.1
Conformational changes in catalytic antibodies

The advent of catalytic antibodies has brought a renewed interest in antibody structure and evolution. In particular, structures of catalytic antibodies underlined the understanding of their mechanism at the molecular level (over 50 published structures of catalytic antibodies in PDB as of end of 2002). As is the case with many ordinary antibodies, conformational changes are observed with catalytic antibodies as well – in several catalytic antibodies, the structure of the TSA-bound antibody differs from the uncomplexed structure. But, in the absence of kinetic studies in solution, the interpretation of the crystallographic data regarding the mechanism of these conformational changes is ambiguous. Further complications from the comparison of bound vs unbound antibody crystal structures arise from the effect of packing forces (e.g., the antibodies' arrangement in the crystal in antibody 1D4 [95], CDR1 packed inside the active site of antibody 43C9, blocking the diffusion of the hapten [79]), and the buffer used for crystallization (e.g., antibody D2.3, see below and [15]). A recurrent structurally observed conformational change is a relative movement between the heavy- and light-chains' variable regions [114], as exemplified by antibody CNJ206 where a 7° rotation and a ~1 Å translation change were found between the TSA unbound and bound structures. In addition, a distinct isomerization was noted in the active site, where a dramatic change of the entire H3 CDR loop – from an open state to a closed state – is observed. This changes the binding site completely – from a shallow form in uncomplexed antibody to a deep and narrow cavity in the TSA-bound state. This is assumed to be a TSA-induced conformational change, but in fact the TSA-bound conformation may be pre-existing, as neither the oxyanion hole nor the part of the active site that binds the phenolic leaving group is formed in the free form [14]. *Induced-fit* movement of CDR-H3 residues (H99-H102) was also suggested to enhance surface complementarity to the TSA vs the substrate and product of the catalyzed *retro*-Diels-Alder reaction, thus implying that better rate enhancement is gained by the conformational changes (antibody 10F11, [19]). Crystal structures of antibody 1D4, catalyzing a *syn* elimination reaction, indicated conformational changes limited to CDR-H2. The conformational change increases the contacts with the TSA (notably residue Arg58), and may contribute to catalysis by constraining the substrate to an elimination-favorable orientation [95].

The free and TSA-complexed structures of several catalytic antibodies were solved, alongside the structures the putative germline antibodies from which these catalytic antibodies diverged [27, 115–117]. A general trend has been suggested, whereby the germline antibodies exhibit a different free and bound structure, whilst the mature antibodies exhibit the same structure. The observed isomerizations between the free and bound antibodies seem to occur mainly at CDR3H residues. In antibody 48G7's precursor, tyrosines H98 and H99 shift by up to 5Å upon TSA binding together, with a 4.5°4 change in the V-regions' orientation, resulting in better packing of the TSA's *p*-nitrophenyl ring by TyrH99 [27, 117]. The hapten of antibody AZ-28, catalyzing an oxy-Cope reaction, induces a rotation of CDR-H3 away from the binding site to yield better π–π stacking with the hapten's phenyl ring as well as enhanced orbital overlap to the reaction's TS [115]. In the precursor of the oxidating, periodate-dependent antibody 28B4, better packing of the TSA phenyl ring is also achieved by movement of residues H95–H99 [116]. As judged by crystallography, the matured antibodies related to these germline antibodies exhibit the same structure when free and bound to the TSA. Interestingly, in antibodies 48G7 and 28B4, the TSA's *p*-nitrophenyl ring adopts a different conformation in the mature vs the germline-complexed structures, locking them in a different binding pocket. This is surprising, as, in the case of the mature 48G7, the germline's pocket remains unchanged.

The above studies [118] promoted the hypothesis that the germline antibodies exhibit high conformational diversity, and that, during affinity maturation, the hapten binding conformation becomes dominant [110]. However, structural differences between the bound and the free state are commonly observed in mature antibodies ([112] and references therein), and at least one study has shown similar structures of the mature Diels-Alderase antibody 39-A11 and its putative germline precursor in both bound and unbound states [119]. The crystallographic studies were not accompanied by kinetics in solution that indicate the existence of a pre-equilibrium between different unbound conformations, in either the mature or the germline antibodies. Indeed, a kinetic study of mature antibodies that showed no structural differences between their free and TSA-bound states clearly indicated a pre-equilibrium between two antibody conformers, one exhibiting low affinity toward the TSA and another high affinity and catalytic activity. The slow, minute-scale isomerization between the two conformers resulted in hysteretic, "memory-like" behavior [15]. Moreover, it has also been shown that the crystallization buffer shifts the equilibrium toward the active conformer, resulting in both the free and TSA-bound antibodies yielding the same structure [15].

Also of interest are the mutations that lead to affinity maturation and their effect on TSA binding and rate of catalysis. In antibodies 48G7 and AZ-28 the mutations occur away from the active site. The importance of remote residues on catalytic efficiency was also noted in mutants of antibody 17E8 selected by phage display for better TSA binding. A double mutant was achieved, with ten-fold improvement of k_{cat}/K_M [120]. In antibody 28B4, however, three of the somatic mutations (nine in total), are in direct contact with the hapten [116]. In all three antibodies, somatic mutations increased the TSA-binding affinity (up to 40 000-fold in 48G7 [117]). Interestingly, the catalytic activity of the mature 48G7 was increased compared to its germline counterpart (see Section 15.2.1), whereas that of antibody AZ-28 was diminished, hinting at the role of structural flexibility in this antibody's catalytic activity (the catalytic activity of 28B4 germline was not reported). Structural flexibility may also explain the increase in the catalytic activity of the ArgH100aGln mutant of antibody 43C9 (a CDR-H3 mutation remote from the active site), which reduces product inhibition [81].

15.3.2
The contribution of structural isomerization to catalysis

The structural studies described above do not indicate a direct role for conformational changes in antibody catalysis, nor do they provide a measure of their contribution to

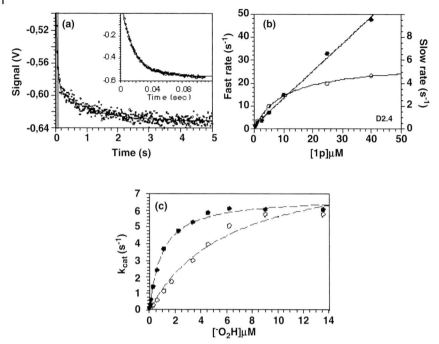

Fig. 15.9 Induced fit exhibited by the D-antibodies. (a) Two-step kinetics reflected by the stopped-flow trace of fluorescence quenching of antibody D2.4 in the presence of TSA. The fast phase (shaded) is enlarged in the inset. Each phase was fitted to a single exponential rate equation. (b) The effect of TSA concentration on the slow bimolecular binding step and on the fast isomerization step [15]. (c) D2.4 (full circles) and D2.3 (empty circles) catalytic rates dependency on the peroxide anion exogenous nucleophile's concentration [10].

TS stabilization and rate of catalysis. In the first quantitative study of conformational changes in catalytic antibodies, two independent kinetic measurements (one related to TSA binding and another directly to catalysis) performed with the D-Abs (D2.3, D2.4, D2.5) were used to reveal and quantify the contribution of conformational changes to catalysis [10, 15]. Pre-steady-state TSA-binding kinetics of D2.3, D2.4 and D2.5 antibodies, based on the change of intrinsic tryptophan fluorescence upon hapten binding, indicated the formation of a low micromolar affinity primary complex followed by a conformational change toward the final nanomolar affinity complex (Fig. 15.9a,b and Table 15.6 [15]).

The rate of isomerization (6 s^{-1}) revealed by fluorescence quenching measurements with both the **1b** and **1p** TSAs is significantly higher than the rate of hydrolysis of the D-Abs (0.05–0.22 s^{-1}) and is therefore invisible in the catalytic, product-release assay. However by increasing the turnover rate by means of a strongly reactive nucleophile (i.e., peroxide anion, see Section 15.2.3 above), a "rate ceiling" equal to the isomerization rate was observed (Fig. 15.9c) [10]. The isomerization provides 25 to 140fold higher affinity to the TSA (Table 15.6) and a concomitant increase to the D-Abs catalytic rates. This is yet another manifestation of the true representation of

Tab. 15.6: Kinetic parameters of D-Abs' binding to p-nitro-phenyl (1p)- and benzyl (1b)-phosphonate TSAs.

$$Ab + TSA \underset{k_{fast-}}{\overset{k_{fast+}}{\rightleftharpoons}} [Ab \cdot TSA] \underset{k_{slow-}}{\overset{k_{slow+}}{\rightleftharpoons}} [Ab \cdot TSA]' \underset{k_{slowest-}}{\overset{k_{slowest+}}{\rightleftharpoons}} [Ab \cdot TSA]''$$

Ab	TSA	k_{fast+}[a] $\mu M^{-1}s^{-1}$	k_{fast-} s^{-1}	K_{fast+}[b] μM	k_{slow+} s^{-1}	k_{slow-} s^{-1}	K_{slow}	$k_{slowest+}$ s^{-1}	$k_{slowest-}$ s^{-1}	$K_{slowest}$	K_D[c] nM
D2.3	1b	4×10^5	0.24 ± 0.02	0.6	5 ± 1	0.2 ± 0.1	0.04	0.013 ± 0.002	$0.002 \pm 3 \times 10^4$	0.15	3 ± 1
	1p	3×10^5	0.20 ± 0.02	0.7	6 ± 1	0.10 ± 0.05	0.02	$0.001 \pm 3 \times 10^4$	$0.0023 \pm 3 \times 10^4$	2.3	25 ± 3
D2.4	1b	2×10^6	0.15 ± 0.01	0.1	1 ± 0.1	0.01 ± 0.003	0.01	–	–	–	1 ± 0.4
	1p	5×10^5	0.3 ± 0.03	0.6	6 ± 1	0.04 ± 0.01	0.007	–	–	–	4 ± 1
D2.5	1b	5×10^5	0.20 ± 0.03	0.4	7 ± 2	0.2 ± 0.1	0.03	–	–	–	10 ± 2
	1p	4×10^5	0.20 ± 0.02	0.5	4 ± 1	0.08 ± 0.03	0.02	–	–	–	10 ± 2

a) Because of high error in its determination k_{fast+} was back-calculated from the overall dissociation constant (K_d) after measuring all the other kinetic parameters.
b) Calculated from K_d.
c) Measured by fluorescence equilibrium titration.

the actual TS by the TS analog. The "induced-fit" isomerization contributes not only to rate enhancement but also to specificity. A second, minute-scale, isomerization step is unique to antibody D2.3 [15]. This very slow conformational change *increases* the affinity to the TSA 1b antigen ~7 fold yet *decreases* its affinity to TSA 1p by ~2 fold (Table 15.6). These results may suggest that the immune system may select high-affinity and high-specificity antibodies that still exhibit a considerable degree of conformational change, and that affinity maturation does not necessarily involve a rigid "lock-and-key" binding mechanism [27, 37].

15.4 Conclusions

The examples given in this review emphasize the ability of catalytic antibodies to mimic multiple aspects of enzyme catalysis. In addition, several other enzyme characteristics are faithfully mimicked by hydrolytic antibodies but are beyond the scope of this review. Hydrolytic antibodies can exhibit both a fine substrate specificity that rivals natural enzymes (e.g., 2H6 and 21H3 [121], the D-Abs [25] and 12F12 [93] antibodies), as well as very broad specificity that may be induced by bulky TSA linker moieties (e.g., [122, 123]). In addition, high enantioselectivity was reported for several ester hydrolytic antibodies (e.g., [124, 125]). Metal (Pd)-mediated allostery was also described [126].

Rivaling the rates of enzyme catalysis, by catalytic antibodies or any other enzyme mimic, is a daunting challenge. Antibodies, like all other enzyme mimics, are still much inferior to natural enzymes in their catalytic efficiency. Yet, despite the limited rates achievable by hydrolytic antibodies, their generation and the associated research effort have provided unique insights into enzyme catalysis and antibody diversity, evolution and structure. The clear message from 16 years of research into hydrolytic antibodies is that it is possible to mimic *individual* forces and contributions to enzymic catalysis with enzyme-like efficiencies. However, it is the *combination* of the different catalytic forces and contributions described here that may yield antibodies with efficiencies rivaling natural enzymes. Moreover, a significant enhancement in rate may require the incorporation of several catalytic elements and a parallel improvement of the antibody's dynamics. These complex structural and mechanistic elements are not likely to be induced by designing better haptens or engineering existing antibodies. Thus, the most general and effective way of improving the catalytic potency of antibodies would be to select esterolytic antibodies *directly* for higher rates and turnover ([127] and references therein). Such selections can affect all the factors concerned, including those linked directly with the active site (e.g., inducing endogenous active-site nucleophiles) and those peripheral to it (e.g., improving the approach of exogenous nucleophiles or the rate of conformational isomerism). The existing hydrolytic antibodies described above that exhibit low to medium rate accelerations and effective turnover may serve as the ideal starting point for such an evolutionary process.

References

1 Tantillo, D. J., Houk, K. N., *J. Org. Chem.* 64 (**1999**), p. 3066–3076
2 Fersht, A. R., *Enzyme structure and mechanism*, W.H. Freeman, New-York **1985**
3 Hammes, G. G., *Biochemistry* 41 (**2002**), p. 8221–8
4 Carter, P., Wells, J. A. *Proteins* 7 (**1990**), p. 335–342
5 Turner, J. M., Larsen, N. A., Basran, A., Barbas, C. F. 3rd, Bruce, N. C., Wilson, I. A., Lerner, R. A., *Biochemistry* 41 (**2002**), p. 12297–307
6 Krebs, J. F. Siuzdak, G. Dyson, H. J. Stewart, J. D., Benkovic, S. J., *Biochemistry* 34 (**1995**), p. 720–723
7 Wirsching, P., Ashley, J. A., Lo, C. H., Janda, K. D., Lerner, R. A., *Science* 270 (**1995**), p. 1775–82
8 Baldwin, E., Schultz, P. G., *Science* 245 (**1989**), p. 1104–7
9 Pollack, S. J., Nakayama, G. R., Schultz, P. G., *Science* 242 (**1988**), p. 1038–40
10 Lindner, A. B., Kim, S. H., Schindler, D. G., Eshhar, Z., Tawfik, D. S., *J. Mol. Biol.* 320 (**2002**), p. 559–72
11 Gigant, B., Tsumuraya, T., Fujii, I., Knossow, M., *Struct. Fold Des.* 7 (**1999**), p. 1385–93
12 Ersoy, O. Fleck, R. Sinskey, A. Masamune, S., *J. Am. Chem. Soc.* 120 (**1998**), p. 817–8
13 Guo, J. Huang, W. Scanlan, T. S., *J. Am. Chem. Soc.* 116 (**1994**), p. 6062–6069
14 Charbonnier, J. B., Carpenter, E., Gigant, B., Golinelli-Pimpaneau, B., Eshhar, Z., Green, B. S., Knossow, M., *Proc. Natl. Acad. Sci. U S A* 92 (**1995**), p. 11721–5
15 Lindner, A. B., Eshhar, Z., Tawfik, D. S., *J. Mol. Biol.* 285 (**1999**), p. 421–430
16 Wedemayer, G. J., Patten, P. A., Wang, L. H., Schultz, P. G., Stevens, R. C., *Science* 276 (**1997**), p. 1665–9
17 Wirsching, P., Ashley, J. A., Benkovic, S. J., Janda, K. D., Lerner, R. A., *Science* 252 (**1991**), p. 680–685
18 Mundorff, E. C., Hanson, M. A., Varvak, A., Ulrich, H., Schultz, P. G., Stevens, R. C., *Biochemistry* 39 (**2000**), p. 627–32
19 Hugot, M., Bensel, N., Vogel, M., Reymond, M. T., Stadler, B., Reymond, J. L., Baumann, U., *Proc. Natl. Acad. Sci. U S A* 99 (**2002**), p. 9674–8
20 Stewart, J. D., Benkovic, S. J., *Nature* 375 (**1995**), p. 388–91
21 Jencks, W. P., *Catalysis in chemistry and enzymology*, McGraw-Hill Book Company, New-York **1969**
22 Tawfik, D. S., Chap, R., Green, B. S., Sela, M., Eshhar, Z., *Proc. Natl. Acad. Sci. U S A* 92 (**1995**), p. 2145–9
23 Mader, M. M., Bartlett, P. A., *Chem. Rev.* 97 (**1997**), p. 1281–1302
24 Tawfik, D. S., Green, B. S., Chap, R., Sela, M., Eshhar, Z., *Proc. Natl. Acad. Sci. U S A* 90 (**1993**), p. 373–7

25 Tawfik, D. S., Lindner, A. B., Chap, R., Eshhar, Z., Green, B. S., *Eur. J., Biochem.* 244 (**1997**), p. 619–26
26 Charbonnier, J. B., Golinelli, P. B., Gigant, B., Tawfik, D. S., Chap, R., Schindler, D. G. Kim, S. H., Green, B. S., Eshhar, Z., Knossow, M., *Science* 275 (**1997**), p. 1140–1142
27 Wedemayer, G. J., Wang, L. H., Patten, P. A., Schultz, P. G., Stevens, R. C., *J Mol. Biol.* 268 (**1997**), p. 390–400
28 Patten, P. A., Gray, N. S., Yang, P. L., Marks, C. B., Wedemayer, G. J., Boniface, J. J. Stevens, R. C., Schultz, P. G., *Science* 271 (**1996**), p. 1086–91
29 Buchbinder, J. L., Stephenson, R. C., Scanlan, T. S., Fletterick, R. J., *J. Mol. Biol.* 282 (**1998**), p. 1033–41
30 Guo, J., Huang, W., Zhou, G. W., Fletterick, R. J., Scanlan, T. S., *Proc. Natl. Acad. Sci. U S A* 92 (**1995**), p. 1694–8
31 Miyashita, H., Karaki, Y., Kikuchi, M., Fujii, I., *Proc. Natl. Acad. Sci. U S A* 90 (**1993**), p. 5337–40
32 Miyashita, H., Hara, T., Tanimura, R., Fukuyama, S., Cagnon, C., Kohara, A., Fujii, I., *J. Mol. Biol.* 267 (**1997**), p. 1247–57
33 O., Kristensen, Vassylyev, D. G., Tanaka, F., Morikawa, K., Fujii, I., *J. Mol. Biol.* 281 (**1998**), p. 501–11
34 R., Zemel, Schindler, D. G., Tawfik, D. S., Eshhar, Z., Green, B. S., *Mol. Immunol.* 31 (**1994**), p. 127–37
35 Zhou, G. W., Guo, J., Huang, W., Fletterick, R. J., Scanlan, T. S., *Science* 265 (**1994**), p. 1059–64
36 MacBeath, G., Hilvert, D., *Chem. Biol.* 3 (**1996**), p. 433–45
37 Tantillo, D. J., Houk, K. N., *Chem. Biol.* 8 (**2001**), p. 535–45
38 Golinelli-Pimpaneau, B., *J. Immunol. Methods* 269 (**2002**), p. 157–71
39 Tantillo, D. J., Houk, K. N., *J. Comput. Chem.* 23 (**2002**), p. 84–95
40 O'Connell, T., Day, P. R. M., Torchilin, E. V., Bachovchin, W. W., Malthouse, J. G., *BioChem. J.* 326 (**1997**), p. 861–6

41 O'Sullivan, D. B., O'Connell, T. P., Mahon, M. M., Koenig, A., Milne, J. J., Fitzpatrick, T. B., Malthouse, J. P., *Biochemistry*, (**1999**) 38 p. 6187–94
42 Gerlt, J. A., Gassman, P. G., *Biochemistry* 32 (**1993**), p. 11943–52
43 Szeltner, Z., Renner, V., Polgar, L., *Protein Sci.* 9 (**2000**), p. 353–360
44 Fersht, A. R., Shi, J. P., Knill-Jones, J., Lowe, D. M., Wilkinson, A. J., Blow, D. M. Brick, P., Carter, P., Waye, M. M., Winter, G., *Nature* 314 (**1985**), p. 235–8
45 Corey, D. R., Craik, C. S., *J. Am. Chem. Soc.* 114 (**1992**), p. 1784–90
46 Harel, M., Quinn, D. M., Nair, H. K., Silman, I., Sussman, J. L., *J. Am. Chem. Soc.* 118 (**1996**), p. 2340–2346
47 Szeltner, Z., Rea, D., Renner, V., Fulop, V., Polgar, L., *J. Biol. Chem.* 277 (**2002**), p. 42613–22
48 Juhasz, T., Szeltner, Z., Renner, V., Polgar, L., *Biochemistry* 41 (**2002**), p. 4096–4106
49 Larsen, N. A., Turner, J. M., Stevens, J., Rosser, S. J., Basran, A., Lerner, R. A., Bruce, N. C., Wilson, I. A., *Nat. Struct. Biol.* 9 (**2002**), p. 17–21
50 Parker, J., *MicroBiol. Rev.* 53 (**1989**), p. 273–98
51 Ortiz, C., Tellier, C., Williams, H., Stolowich, N. J., Scott, A. I., *Biochemistry* 30 (**1991**), p. 10026–34
52 Delbaere, L. T., Brayer, G. D., *J. Mol. Biol.* 183 (**1985**), p. 89–103
53 Henderson, R., *J. Mol. Biol.* 54 (**1970**), p. 341–54
54 Chappey, O., Debray, M., Niel, E., Scherrmann, J. M., *J. Immunol. Methods* 172 (**1994**), p. 219–25
55 Foote, J., Eisen, H. N., *Proc. Natl. Acad. Sci. USA* 92 (**1995**), p. 1254–6
56 Thomas, N. R., *Appl. BioChem. Biotechnol.* 47 (**1994**), p. 345–72
57 Thomas, N. R., *Nat. Prod. Rep.* 13 (**1996**), p. 479–511
58 Stevenson, J. D., Thomas, N. R., *Nat. Prod. Rep.* 17 (**2000**), p. 535–77
59 Baca, M., Scanlan, T. S., Stephenson, R. C., Wells, J. A., *Proc.*

Natl. Acad. Sci. U S A 94 (**1997**), p. 10063–8
60 Takahashi, N., Kakinuma, H., Liu, L., Nishi, Y., Fujii, I., *Nat. Biotechnol.* 19 (**2001**), p. 563–7
61 Deng, S. X., de Prada, P., Landry, D. W., *J. Immunol Methods* 269 (**2002**), p. 299–310
62 Wentworth, P., Datta, A., Blakey, D., Boyle, T., Partridge, L. J., Blackburn, G., *Proc. Natl. Acad. Sci. USA* 93 (**1996**), p. 799–803
63 Pollack, S. J., Jacobs, J. W., Schultz, P. G., *Science* 234 (**1986**), p. 1570–3
64 Datta, A., Wentworth Jr., P., Shaw, J. P., Simeonov, A., Janda, K. D., *J. Am. Chem. Soc* 121 (**1999**), p. 10461–7
65 Kohen, F., Kim, J. B., Lindner, H. R., Eshhar, Z., Green, B., *FEBS Lett* 111 (**1980**), p. 427–31
66 Schowen, R. L., *J. Immunol. Methods* 269 (**2002**), p. 59–65
67 Wagner, J., Lerner, R. A., Barbas III, C. F., *Science* 270 (**1995**), p. 1797–800
68 Zhong, G. F., Lerner, R. A., Barbas, C. F., *Angew. Chem. Int. Ed.* 38 (**1999**), p. 3738–3741
69 Tsumuraya, T., Takazawa, N., Tsunakawa, A., Fleck, R., Masamune, S., *Chem. Eur. J.* 7 (**2001**), p. 3748–55
70 Angeles, T. S., Smith, R. G., Darsley, M. J., Sugasawara, R., Sanchez, R. I., Kenten, J. Schultz, P. G., Martin, M. T., *Biochemistry* 32 (**1993**), p. 12128–35
71 Angeles, T. S., Martin, M. T., *BioChem. Biophys. Res. Commun.* 197 (**1993**), p. 696–701
72 Martin, M. T., Napper, A. D., Schultz, P. G., Rees, A. R., *Biochemistry,* (**1991**), 30, p. 9757–61
73 Gigant, B., Charbonnier, J. B., Golinelli-Pimpaneau, B., Zemel, R. R., Eshhar, Z., Green, B. S., Knossow, M., *Eur J., BioChem.* 246 (**1997**), p. 471–6
74 Tawfik, D. S., Chap, R., Eshhar, Z., Green, B. S., *Protein Eng.* 7 (**1994**), p. 431–4
75 Stewart, J. D., Krebs, J. F., Siuzdak, G., Berdis, A. J., Smithrud, D. B., Benkovic, S. J., *Proc. Natl. Acad. Sci. USA* 91 (**1994**), p. 7404–9
76 Janda, K. D., Schloeder, D., Benkovic, S. J., Lerner, R. A., *Science* 241 (**1988**), p. 1188–91
77 Benkovic, S. J., Adams, J. A., Borders, Jr., C. L., Janda, K. D., Lerner, R. A., *Science* 250 (**1990**), p. 1135–9
78 Gibbs, R. A., Benkovic, P. A., Janda, K. D., Lerner, R. A., Benkovic, S. J., *J. Am. Chem. Soc* 114 (**1992**)
79 Thayer, M. M., Olender, E. H., Arvai, A. S., Koike, C. K., Canestrelli, I. L., Stewart, J. D., Benkovic, S. J., Getzoff, E. D., Roberts, V. A., *J. Mol. Biol.* 291 (**1999**), p. 329–45
80 Jencks, W. P., *Catalysis in chemistry and enzymology,* McGraw-Hill Book Company, New-York **1969**, p. 67–72
81 Stewart, J. D., Roberts, V. A., Thomas, N. R., Getzoff, E. D., Benkovic, S. J., *Biochemistry* 33 (**1994**), p. 1994–2003
82 Kimball, A. S., Lee, J., Jayaram, M., Tullius, T. D., *Biochemistry* 32 (**1993**), p. 4698–4701
83 Viladot, J. L., de Ramon, E., Durany, O., Planas, A., *Biochemistry* 37 (**1998**), p. 11332–11342
84 Admiraal, S. J., Schneider, B., Meyer, P., Janin, J., Veron, M., Deville-Bonne, D., Herschlag, D., *Biochemistry* 38 (**1999**), p. 4701–11
85 Lee, Y. L., Chen, J. C., Shaw, J. F., *BioChem. Biophys. Res. Commun.* 231 (**1997**), p. 452–6
86 Stoops, J. K., Horgan, D. J., Runnegar, M. T., De Jersey, J., Webb, E. C., Zerner, B., *Biochemistry* 8 (**1969**), p. 2026–2033
87 Fastrez, J., Fersht, A. R., *Biochemistry* 12 (**1973**), p. 2025–2034
88 Selzer, T., Albeck, S., Schreiber, G., *Nat. Struct. Biol.* 7 (**2000**), p. 537–541
89 Ersoy, O., Fleck, R., Sinskey, A., Masamune, S., *J. Am. Chem. Soc.* 118 (**1996**), p. 13077–13078
90 Janda, K. D., Weinhouse, M. I., Schloeder, D., Lerner, R. A., *J. Am. Chem. Soc.* 112 (**1990**), p. 1274–1275
91 Suga, H., Ersoy, O., Tsumuraya, T., Lee, J., Sinskey, A. J., Masamune, S., *J. Am. Chem. Soc.* 116 (**1994**), p. 487–494
92 Grynszpan, F., Keinan, E., p. –, A., *Chem. Commun.* 21 (**1998**)

93 Iwabuchi, Y., Kurihara, S., Oda, M., Fujii, I., *Tetrahedron Lett.* 40 (**1999**), p. 5341–5344
94 Golinelli-Pimpaneau, B., Goncalves, O., Dintinger, T., Blanchard, D., Knossow, M., Tellier, C., *Proc. Natl. Acad. Sci. USA* 97 (**2000**), p. 9892–5
95 Larsen, N. A., Heine, A., Crane, L., Cravatt, B. F., Lerner, R. A., Wilson, I. A., *J. Mol. Biol.* 314 (**2001**), p. 93–102
96 Cross, S. S., Brady, K., Stevenson, J. D., Sackin, J. R., Kenward, N., Dietel, A., Thomas, N. R., *J. Immunol. Methods* 269 (**2002**), p. 173–95
97 Stevenson, J. D., Dietel, A., Thomas, N. R., *Chem. Commun.* 1999 (**1999**), p. 2105–2106
98 Williams, N. H., Takasaki, B., Wall, M., Chin, J., *Acc. Chem. Res.* 32 (**1999**), p. 485–493
99 Wade, W. S., Ashley, J. A., Jahangiri, G. K., McElhany, G., Janda, K. D., Lerner, R. A., *J. Am Chem. Soc* 115 (**1993**), p. 4906–4907
100 Brummer, O., Hoffman, T. Z., Janda, K. D., *Bioorg. Med. Chem.* 9 (**2001**), p. 2253–7
101 Hedstrom, L., *Chem. Rev.* 102 (**2002**), p. 4501–24
102 Gigant, B., Charbonnier, J. B., Eshhar, Z., Green, B. S., Knossow, M., *J. Mol. Biol.* 284 (**1998**), p. 741–50
103 D'Souza, L. J., Gigant, B., Knossow, M., Green, B. S., *J. Am. Chem. Soc.* 124 (**2002**), p. 2114–5
104 Wade, H., Scanlan, T. S., *J. Am. Chem. Soc.*, (**1999**), 121, p. 11935–11941
105 Finer-Moore, J. S., Santi, D. V., Stroud, R. M., *Biochemistry*, (**2003**), 42, p. 248–56
106 Stroud, R. M., Finer-Moore, J. S., *Biochemistry*, (**2003**), 42, p. 239–47
107 Shi, J., Radic, Z., Taylor, P., *J. Biol. Chem.*, (**2002**), 277, p. 43301–8
108 Shi, J., Boyd, A. E., Radic, Z., Taylor, P., *J. Biol. Chem.* 276 (**2001**), p. 42196–204
109 Padlan, E. A., *Mol. Immunol.* 31 (**1994**), p. 169–217
110 Foote, J., Milstein, C., *Proc. Natl. Acad. Sci. USA* 91 (**1994**), p. 10370–4
111 Lancet, D., Pecht, I., *Proc. Natl. Acad. Sci. USA* 73 (**1976**), p. 3549–53
112 Wilson, I. A., Stanfield, R. L., *Curr. Opin. Struct. Biol.* 4 (**1994**), p. 857–867
113 James, L. C., Roversi, P., Tawfik, D. S., *Science* 299 (**2003**), p. 1362–7
114 Pellequer, J. L., Chen, S., Roberts, V. A., Tainer, J. A., Getzoff, E. D., *J. Mol. Recognit.* 12 (**1999**), p. 267–75
115 Ulrich, H. D., Mundorff, E., Santarsiero, B. D., Driggers, E. M., Stevens, R. C., Schultz, P. G., *Nature* 389 (**1997**), p. 271–5
116 Yin, J., Mundorff, E. C., Yang, P. L., Wendt, K. U., Hanway, D., Stevens, R. C., Schultz, P. G., *Biochemistry* 40 (**2001**), p. 10764–73
117 Yang, P. L., Schultz, P. G., *J. Mol. Biol.* 294 (**1999**), p. 1191–201
118 Schultz, P. G., Lerner, R. A., *Nature* 418 (**2002**), p. 485
119 Romesberg, F. E., Spiller, B., Schultz, P. G., Stevens, R. C., *Science* 279 (**1998**), p. 1929–33
120 Arkin, M. R., Wells, J. A., *J. Mol. Biol.* 284 (**1998**), p. 1083–94
121 Janda, K. D., Benkovic, S. J., Lerner, R. A., *Science* 244 (**1989**), p. 437–40
122 Tanaka, F., Kinoshita, K., Tanimura, R., Fujii, I., *J. Am. Chem. Soc.* 118 (**1996**), p. 2332–2339
123 Kurihara, S., Tsumuraya, T., Suzuki, K., Kuroda, M., Liu, L., Takaoka, Y., Fujii, I., *Chem. Eur. J.* 6 (**2000**), p. 1656–62
124 Shi, Z. D., Yang, B. H., Zhao, J. J., Wu, Y. L., Ji, Y. Y., Yeh, M., *Bioorg. Med. Chem.* 10 (**2002**), p. 2171–5
125 Wade, H., Scanlan, T. S., *J. Am. Chem. Soc.* 118 (**1996**), p. 6510–6511
126 Finn, M. G., Lerner, R. A., Barbas, C. F., *J. Am. Chem. Soc.* 120 (**1998**), p. 2963–2964
127 Griffiths, A. D., Tawfik, D. S., *Curr. Opin. Biotechnol.* 11 (**2000**), p. 338–53
128 Gigant, B., Charbonnier, J. B., Eshhiar, Z., Green, B. S., Knossow, M., *Proc. Natl. Acad. Sci., USA* 94 (**1997**), p. 7857–7861
129 Charbonnier, J. B., Golinelli-Pimpaneau, B., Gigant, B., Green, B. S., Knossow, M., *Isr. J., Chem.* 36 (**1996**), p. 171–6

130 BRAXTON, S., WELLS, J. A., *J. Biol. Chem.* 266 (1991), p. 11797–800
131 PHILLIPS, M. A., FLETTERICK, R., RUTTER, W. J., *J. Biol. Chem.* 265 (1990), p. 20692–8
132 MENARD, R., STORER, A. C., *Biol. Chem. Hoppe-Seyler*, 373 (1992), p. 393–400
133 NICOLAS, A., EGMOND, M., VERRIPS, C. T., DEVLIEG, J., LONGHI, S., CAMBILLAU, C., MARTINEZ, C., *Biochemistry* 35 (1996), p. 398–410
134 JANDA, K. D., WEINHOUSE, M. I., DANON, T., PACELLI, K. A., SCHLOEDER, D., *J. Am. Chem. Soc.* 113 (1991), p. 5427–5434
135 ANDREEVA, N. S., RUMSH, L. D., *Protein Sci.* 10 (2001), p. 2439–50
136 NORTHROP, D. B., *Acc. Chem. Res.* 34 (2001), p. 790–7
137 KLEIFELD, O., FRENKEL, A., MARTIN, J. M. L., SAGI, I., *Nat. Struct. Biol.* 10 (2003), p. 98–103
138 PENNER-HAHN, J. E., *Nat. Struct. Biol.* 10 (2003), p. 75–77
139 LINDNER, A. B., Ph.D. Thesis, Feinberg 2002, graduate school, Weizmann Institute of Science, Rehovot, Israel
140 HOFFMANN, T., ZHONG, G. F., LIST, B., SHABAT, D., ANDERSON, J., GRAMATIKOVA, S., LERNER, R. A., BARBAS, C. F., *J. Am. Chem. Soc.* 120 (1998), p. 2768–2779

16
Transition State Analogs –
Archetype Antigens for Catalytic Antibody Generation
Anita D. Wentworth, Paul Wentworth, Jr., G. Michael Blackburn

Modern explanations of the catalytic power of enzymes date from Haldane's [1] classic volume on enzymatic activity and comments made by Pauling [2, 3] in the 1940s. Pauling clearly outlined the theory that enzymes achieve catalysis because of their complementarity to the transition state for the reaction being catalyzed [2]. This concept is now seen as a logical extension of the then contemporary "transition state theory", developed to explain chemical catalysis [4]. The theory rests on two fundamental assumptions: (a) that the reactant(s) in a chemical reaction are in equilibrium with an "activated complex", and (b) that the rate of reaction is governed solely by the decomposition of this "activated complex". The term "transition state" (TS‡) is applied to the structure represented by the saddle point on the potential energy surface of the reaction between reactant(s) and product(s). More correctly, however, the TS‡ corresponds to any of the vibrational states of the "activated complex" near the saddle point that links reactant(s) and product(s). Thus, the basic proposition was that the rate of a reaction is related to the difference in Gibbs free energy (ΔG_o) between the ground state of the reactant(s) and the TS‡ for that reaction. To manifest catalysis, therefore, either the energy of the transition state has to be lowered (transition state stabilization) or the energy of the substrate has to be elevated (substrate destabilization). Pauling applied this concept to enzyme catalysis by stating that an enzyme preferentially binds to and thereby stabilizes the transition state for a reaction relative to the ground state of substrate(s) (Fig. 16.1). This has become a classical theory in enzymology and is still used to explain the way in which such biocatalysts are able to enhance specific processes with rate accelerations of up to 10^{17} over background [5, 6].

The relationship between transition state stabilization and enzymatic catalysis is effectively illustrated by a thermodynamic cycle (Fig. 16.2). This formalism was first proposed by Kurz [7] and was later developed by both Wolfenden [8, 9] and Lienhardt [10] to account for the classification of certain enzyme inhibitors as transition state analogs. The designations, E, S, and P relate to enzyme, substrate (in the ground state), and product respectively. The thermodynamic cycle relates the dissociation constant K_S and the hypothetical dissociation constant K_{TS} to the pseudo-equilibrium

Catalytic Antibodies. Edited by Ehud Keinan
Copyright © 2005 WILEY-VCH Verlag GmbH & Co. KGaA, Weinheim
ISBN: 3-527-30688-9

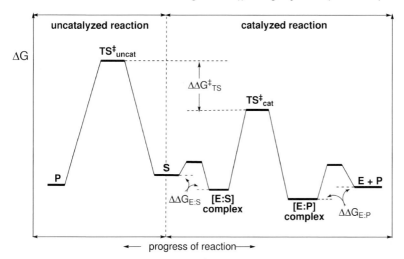

Fig. 16.1 Catalysis is achieved by lowering the free energy of activation for a process, i.e. the catalyst must bind more strongly to the transition state ($TS^\ddagger$) of the reaction than to either reactants or products. Thus: $\Delta\Delta G^\ddagger_{TS} \gg \Delta\Delta G_{E:S}$ and $\Delta\Delta G_{E:P}$. E – a theoretical enzyme, S – substrate, P – product.

constants $K_{un}^\ddagger$ and $K_{cat}^\ddagger$ [Eq. (1)]. From transition state theory, there is a clear relationship between the rates of the catalyzed and uncatalyzed reactions, k_{cat} and k_{uncat} respectively, and K_S and K_{TS} [Eqs. (2 and 3)]. If we assume that the Michaelis constant K_m and the inhibition constant of a transition state analog K_i correspond to the reversible dissociation constants for the enzyme-substrate complex K_S and the enzyme-transition state complex, K_{TS} respectively, this leads to the key equivalence in Eq. (4). Simply put, the anticipated rate enhancement of a biocatalyst, k_{cat}/k_{uncat}, should be directly proportional to the ratio of the dissociation constants of substrate and "transition state" K_m/K_i for a given reaction.

While transition states have been discussed in terms of their free energies, there have been relatively few attempts to describe their structure at atomic resolution for most catalyzed reactions. Transition states are high-energy species with lifetimes on the order of 10^{-14} s, often involving incompletely formed bonds, and this makes their specification very difficult. Zewail [11, 12] has pioneered the study of transition states of chemical reactions using laser femtochemistry, for which he received the Nobel prize for chemistry in 1999, but the majority of our knowledge of the charge and geometry of these species comes from quantum chemical calculations [13].

Intermediates along the reaction co-ordinate also have a very short lifetime, though some of their structures have been studied under stabilizing conditions, and their existence and general nature can often be established using spectroscopic techniques or trapping experiments. The Hammond postulate predicts that if a high-energy intermediate occurs along a reaction pathway, it will resemble the transition state nearest to it in energy [14]. Conversely, if the transition state is flanked by two such intermediates, the one of higher energy will provide a closer approximation to the

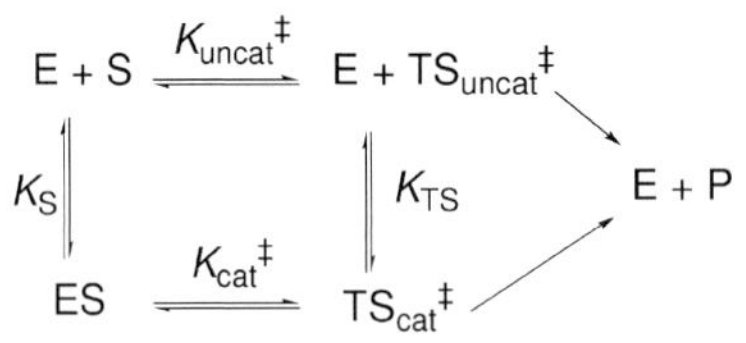

$$E + S \xrightleftharpoons{K_{uncat}^{\ddagger}} E + TS_{uncat}^{\ddagger}$$

$$K_S \updownarrow \qquad K_{TS} \updownarrow \qquad \searrow E + P$$

$$ES \xrightleftharpoons{K_{cat}^{\ddagger}} TS_{cat}^{\ddagger}$$

a) From the thermodynamic cycle

$$\frac{K_{cat}^{\ddagger}}{K_{uncat}^{\ddagger}} = \frac{K_S}{K_{TS}} \qquad (1)$$

b) From transition-state theory

$$\frac{k_{cat}}{k_{uncat}} = \frac{K_{cat}^{\ddagger}}{K_{uncat}^{\ddagger}} \qquad (2)$$

therefore (3)

$$\frac{k_{cat}}{k_{uncat}} = \frac{K_S}{K_{TS}}$$

c) Assumptions

$$K_S = K_m$$
$$K_{TS} = K_i$$

therefore
$$\boxed{\frac{k_{cat}}{k_{uncat}} = \frac{K_m}{K_i}} \qquad (4)$$

Fig. 16.2 Thermodynamic box illustrating the relationship between ground-state and transition-state binding by an enzyme (E) with a single substrate (S) in a reaction to give product (P). k_{cat} – rate constant for the enzyme-catalyzed reaction, k_{uncat} – rate constant of the uncatalyzed reaction.

transition state structure. This assumption provides a strong basis for the use of mimics of unstable reaction intermediates as transition state analogs (TSAs) [15, 16].

Enzymes and antibodies are distinct families of proteins that have evolved pockets to bind small molecules specifically and reversibly. However, the fundamental difference in their biological roles has meant that enzymes have evolved to bind transition states and antibodies have evolved to bind ground states. Pauling apparently did not bring ideas about antibodies, catalytic or otherwise, into his concept of enzyme catalysis, so it fell to Jencks [17], in his seminal 1969 treatise on catalysis, to disclose the opportunity for synthesis of an enzyme by the use of antibodies that had been programmed by appropriate exploitation of the immune repertoire.

"One way to do this [i.e. synthesize an enzyme] is to prepare an antibody to a haptenic group which resembles the transition state of a given reaction."

Following Jencks's *gedanken* experiment, early attempts to demonstrate that catalytic activity could be invoked within enriched inhomogeneous (i.e. polyclonal) antibody preparations met with only limited success [18, 19]. Ultimately, the practical achievement of the "antibody-enzyme" analogy was delayed until it was possible to isolate, purify and characterize single-sequence proteins, monoclonal antibodies, from the vast immune repertoire of $> 10^8$ immonoglobulins [20, 21]. This problem was resolved by Köhler and Milstein's [22,23] hybridoma technology, for which they were rewarded the Nobel prize for medicine in 1984, that made it possible both to screen rapidly the "complete" immune repertoire and to produce large amounts of one specific monoclonal antibody *in vitro*.

Lerner [24, 25] and Schultz [26] published the first reports showing that antibodies could be programmed to be catalysts via the use of stable analogs of a reaction's TS‡ as haptens. Since that time, the majority of all published work on catalytic antibodies supports the notion that this original guided methodology proposed by Jencks, i.e. the design of stable (TSAs) for use as haptens to induce the generation of antibody catalysts, has served as the bedrock of this research [27–32].

Most studies have been directed at acyl transfer processes, particularly hydrolytic reactions, perhaps because of the broad knowledge of the nature of the mechanisms for such reactions and the wide experience of deploying phosphoryl species as stable mimics of unstable tetrahedral intermediates. Thus, the course of the hydrolysis of aryl or alcyl esters, thioesters and amides is accepted to proceed through a high-energy tetrahedral intermediate (TI‡) (1) (Fig. 16.3a) [33]. The faster of such reactions

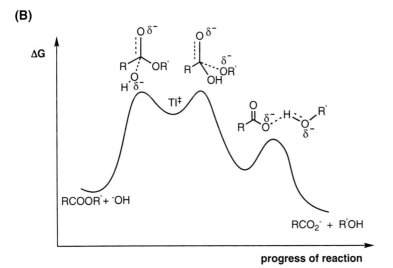

Fig. 16.3 (A) A generic equation showing the two-step hydrolytic process of an acyl species progressing through the anionic tetrahedral intermediate (1) to products. (B) Schematic representation of base-catalyzed hydrolysis of an alkyl or aryl ester, progressing through TI‡. The theorized transition structures both precede and follow the formation of the TI‡.

generally involve good leaving groups, and the addition of the nucleophile is the rate-determining step (Fig. 16.3B). It should be noted that in certain cases, e.g. with some alkyl amides (X=NH, R'=alkyl), the rate determining step can become proton-assisted expulsion of the poor tearing group. However, this broad conclusion from much detailed kinetic analysis has been endorsed by computation for the hydrolysis of methyl acetate [34]. This places the energy for product formation from an anionic TI$^-$ some 7.6 kcal mol^{-1} lower than that for its reversion to reactants. Thus, for the generation of antibodies for the hydrolysis of aryl esters, alkyl esters, carbonates, and activated anilides, the design of the TSA hapten has focused on facilitating nucleophilic attack, and with considerable success. From more than 80 examples of hydrolytic antibodies that have been reported, almost 50 casas involve acyl group transfer to water [35]. The tetrahedral intermediates used for this purpose initially deployed phosphorus (V) systems, relying on the strong polarization of the P=O bond (arguably more accurately represented as P$^+$–O$^-$). A wealth of mechanistic information has been generated regarding acyl transfer reactions and the role of phosphonic acids and their derivatives as TSAs [15, 36, 37]. The range has included many of the possible species containing an ionized P–OH group (Fig. 16.4). One particularly good feature of such systems is that the length of the P–O$^-$ bond (1.521 Å) is intermediate between that of the C–O$^-$ bond calculated for a TI$^-$ (0.2–0.3 Å shorter) and that estimated for the C–O bond breaking in the transition state (some 0.6 Å longer) [34]. Other tetrahedral systems used to induce amidase or esterase activity have included sulfonamides [38] and sulfones [39], secondary alcohols [40], and α-fluoroketone hydrates [41].

In-depth analysis of catalytic antibodies at atomic resolution is dealt with elsewhere in this book. However, a cursory overview of the combining sites of several antibody

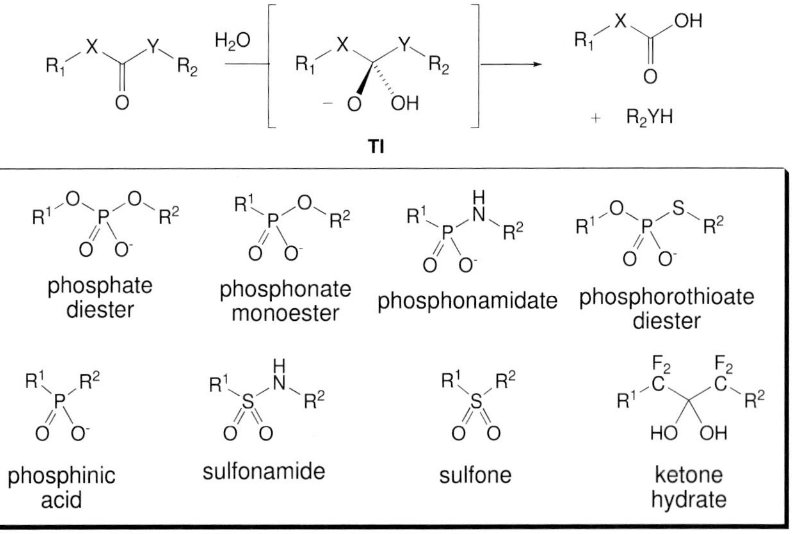

Fig. 16.4 A generic hydrolysis reaction (acyl transfer to water). The stereoelectronic features of the TI have been mimicked by a range of phosphorus(V)-containing molecules and other species.

structures elicited to such TSAs is illuminating at this point. A number of X-ray analyses of antibody acyl transferases, elicited to phosphonic acid derivatives, have revealed a common structural feature, the oxyanion hole [42]. This feature is usually formed by a basic residue that provides a coulombic interaction with the phosphorus (V) core of the TSA hapten (Figures 16.5 and 16.6) [28, 42, 43]. In general, these antibodies possess fairly rudimentary catalytic mechanisms with the primary source of the observed rate acceleration being supplied by complementarity to the rate-determining TS‡. A number of examples are discussed below.

Antibody 48G7, elicited to the TSA hapten, p-nitrophenyl 4-carboxybutylphosphonate (3), catalyzes the hydrolysis of the p-nitrophenyl ester 4 (Fig. 16.5A) [44]. The 48G7 Fab-3 complex exemplifies the oxyanion feature, with electrostatic stabilization from a proximate Arg^{L96} residue and hydrogen bonds from His^{H35} and Tyr^{H33} serving to stabilize the polarized phosphoryl bonds of 3 [45, 46]. Furthermore, main-chain amide bonds, from Tyr^{L91} and Tyr^{H100}, provide hydrogen bond stabilization to the TSA (not shown).

Antibody D2.3, elicited to the p-nitrophenyl 4-carboxamidobutylphosphonate hapten 5, catalyzes the hydrolysis of 4-nitrobenzyl ester 6 (Fig. 16.5B) [47]. The crystal structure of the D2.3 Fab complexed with 5 reveals an oxyanion hole formed by Tyr^{H100d} and Asn^{L34}; both residues are within hydrogen bond distance of one of the phosphoryl oxygen bonds of 5. Additional stabilization of the second phosphoryl oxygen is supplied by Trp^{H95} and a water molecule.

Antibodies 17E8 and 29G11, elicited against the phosphonate TSA 7, catalyze the stereoselective hydrolysis of phenyl esters of n-formyl norleucine (8) with a rate enhancement, k_{cat}/k_{uncat}, of > 8300 and 2200 respectively (Fig. 16.5C) [48]. The binding affinity of 29G11 for hapten 7 is 19-fold greater (K_i = 27 nM) than that exhibited by 17E8 (K_i = 500 nM). These two antibodies are highly illustrative of divergent evolution within the immune repertoire to the same hapten. The two antibodies differ in their sequence by eight substitutions in the variable region of the heavy chain, the light chains being identical. Relative to 17E8, 29G11 has four substitutions in the hapten combining site: Ala^{33}Val, Ser^{95}Gly, Ser^{99}Arg and Tyr^{100A}Asn.

A pH rate profile of the 29G11 antibody-catalyzed reaction implicated two residues, one with a pK_a of 8.1, which must be deprotonated, and another with a pK_a of 10.2, which must be protonated during ester hydrolysis [49]. The 17E8 antibody exhibited a similar pH profile with two residues, one with a pK_a of 9.1 and one with a pK_a of 10.0 being revealed.

The X-ray structure of the 17E8 Fab-7 complex, determined to 2.5 Å resolution, reveals the expected oxyanion hole formed by coulombic stabilization of both the pro-R phosphoryl oxygen by Arg^{L96} and His^{H35} and the pro-S oxygen by Lys^{H97} (Fig. 16.5C) [49]. In addition, a main-chain amide bond of Tyr^{H96} makes a hydrogen bond to the bridging oxygen of 7 (not shown). The X-ray structure (recorded at pH 7.5) of 17E8 revealed a potential serine-histidine catalytic dyad with Ser^{H99} and His^{H35}. However, the side-chain of Ser^{H99} was facing away from the phosphorus center, being H-bonded to Tyr^{H101}. It was theorized that, as the pH increases, the H-bond between Ser^{H99} and Tyr^{H101} is destroyed, allowing the side-chain of Ser^{H99} to rotate around the C$_\alpha$–C$_\beta$ bond and facilitating attack on the carbonyl of 8. However, recent side-

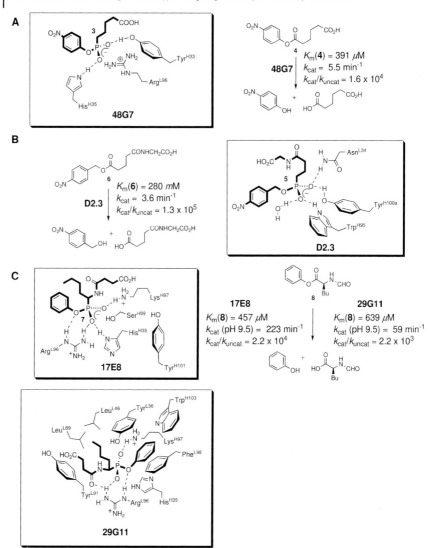

Fig. 16.5 A Antibody 48G7, elicited against the phosphonate TSA **3**, catalyzes the hydrolysis of aryl ester **4** (kinetic parameters shown). The 48G7 Fab-**3** complex is shown (boxed) with the key contacts that comprise the oxyanion hole. **B** Antibody D2.3, elicited against TSA **5**, catalyzes the hydrolysis of benzyl ester **6** (kinetic parameters shown). The D2.3 Fab-**5** complex is shown (boxed) with the key contacts that comprise the oxyanion hole. **C** Antibodies 17E8 and 29G11, elicited against the phosphonate TSA **7**, catalyze the hydrolysis of aryl ester **8** (kinetic parameters shown). The 17E8 Fab-**7** complex (boxed) shows the residues that comprise the oxyanion hole and offers an insight into how Ser^{H99} was initially proposed to act as a nucleophile in the antibody-catalyzed process.

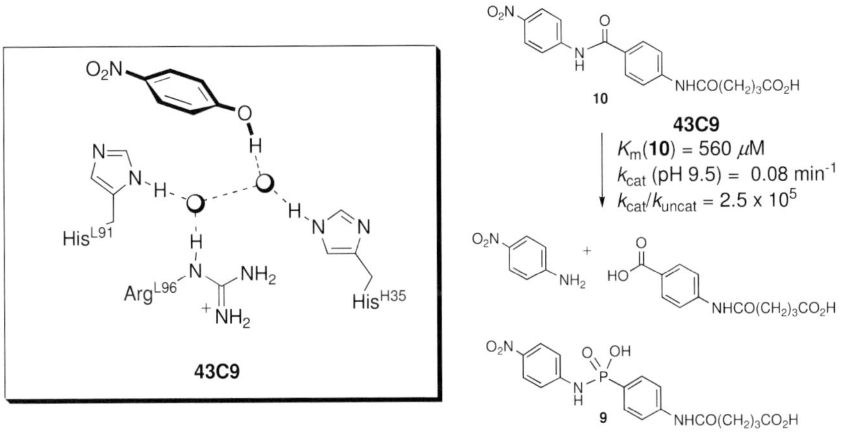

Fig. 16.6 Antibody 43C9, elicited to phosphonamidate hapten **9**, catalyzes the hydrolysis of p-nitroanilide substrates such as **10** (kinetic parameters shown). It is so far the only known antibody that has unequivocally been assigned a nucleophilic component to its catalytic acyl transfer mechanism. The scFv of 43C9, co-crystallized with p-nitrophenol (boxed) reveals that His^{L91} is in an ideal position for nucleophilic attack on the anilide's carbonyl carbon.

directed mutation studies have cast doubt on the role of Ser^{H99} as a nucleophile [50]. A Ser^{H99}Ala mutant possesses a twofold *increase* in k_{cat}/K_m relative to its parent clone. Thus the overall mechanism has been revised to suggest that 17E8 catalyzes ester hydrolysis by either general base-assisted catalysis or direct hydroxide attack.

The X-ray crystallographic structure of the Fab 29G11-7 complex was determined at 2.2 Å resolution (Fig. 16.5C) [51]. Hapten recognition is mediated by a combination of positively charged, aromatic and hydrophobic side-chains. Arg^{L96} forms hydrogen bonds with the amide carbonyl oxygen atom, the pro-*R* phosphonate oxygen atom, and the bridging oxygen atom of **7**. Charged hydrogen bonds are formed between the protonated terminal NH$_2$ group of Lys^{H97} and the pro-*S* phosphonate oxygen atom of **7**. The norleucine side-chain of the hapten is surrounded by hydrophobic residues: Tyr^{L36}, Leu^{L46}, Leu^{L89}, and Tyr^{L91}. Recognition of the phenyl ring of **7** is mediated by van der Waals contacts with the aromatic side-chains of Phe^{L98}, Tyr^{L36}, Trp^{H47} and Trp^{H103}.

Among the growing family of antibodies elicited to TSAs that catalyze acyl transfer processes, 43C9 is rare in its ability to catalyze the difficult reaction of selective anilide [52] as well as ester bond hydrolysis [53, 54]. The 43C9 antibody was elicited to the *p*-nitrophenylphosphonamidate hapten **9** and catalyzes the hydrolysis of the *p*-nitroanilide **10** (Fig. 16.6). Pre- and steady-state reaction kinetics [55] combined with a Hammett correlation of the effects of *para* substituents on key rate constants [56] implicated a neutral nitrogen nucleophile donated by 43C9 in forming an acyl-antibody intermediate, which was further confirmed by electrospray mass spectrometry [57]. A computational model of the 43C9 Fv fragment had been constructed to assist in the determination of the structural basis underlying its kinetic mechanism [58, 59]. The model implicated Arg^{L96} in transition state stabilization and His^{L91} as the neutral nitrogen nucleophile. Both hypotheses were supported by site-directed mutagenesis

[59, 60]. The X-ray crystallographic structure of the 43C9 scFv with and without p-nitrophenol, a product of ester hydrolysis, was determined to be 2.2 and 2.3 Å respectively (Fig. 16.6) [61]. While the results would have been more significant with hapten 9 bound, antigen-binding sites of two adjacent scFvs interacted during crystal packing, preventing antigen binding. However, the crystal structure fully supported the computer model. His^{L91} is centered at the bottom of the antigen-binding site with the imidazole ring in perfect range for nucleophilic attack. His^{L91}, Arg^{L96}, and the bound p-nitrophenol are linked into a hydrogen-bonding network by two well-ordered water molecules. These water molecules are speculated to occupy the positions of the phosphonamidate oxygen atoms of 9, which in turn mimics the transition state of the reaction. This network also includes His^{H35}, suggesting a previously overlooked role for this residue in transition state stabilization.

In 1991, Jacobs [62] analyzed 18 examples of antibody catalysis of acyl transfer reactions as a test of the Pauling concept, i.e. delivering catalysis by $TS^{\ddagger}$ stabilization. The range of examples included the hydrolysis of both aryl and alkyl esters as well as aryl carbonates. In some cases more than one reaction was catalyzed by the same antibody, and in others the same reaction was catalyzed by different antibodies.

Much earlier, Wolfenden [63] and Thompson [64] established a criterion for enzyme inhibitors working as TSAs. They proposed that such activity should be reflected by a linear relationship between the inhibition constant for the enzyme K_i and its inverse second-order rate constant, K_m/k_{cat}, for pairs of inhibitors and substrates that differ in structure only at the TSA/substrate locus. This assumption is a clear extension of Eq. (4). This principle has been well validated, *inter alia*, for phosphonate inhibitors of thermolysin [37] and pepsin [65]. In order to apply such a criterion to a range of catalytic antibodies, Jacobs assumed firstly that the spontaneous hydrolysis reaction proceeds via the same $TS^{\ddagger}$ as that for the antibody-mediated reaction and secondly that all corrective factors due to medium effects are constant. By treating the hydrolysis reactions as pseudo-first-order processes, one can derive the simple relationship with approximations of K_{TS} and K_S to provide a mathematical statement in terms of K_i, K_m, k_{cat}, and k_{uncat} [Eq. (4), Fig. 16.7A] [8, 62, 66, 67]. A log-log plot using K_i, K_m, k_{cat}, and k_{uncat} data from 18 separate cases of antibody catalysis exhibited a linear correlation over four orders of magnitude and with a gradient of 0.86 (Fig. 16.7a) [62]. Considering the assumptions made, this value is sufficiently close to unity to suggest that the antibodies do stabilize the transition state for their respective reactions. However, even the highest k_{cat}/k_{uncat} value of 10^6 in this series [68] compares poorly with enhancement ratios seen for enzyme catalysts [10].

A second use of this type of analysis has been presented by Stewart and Benkovic [69] (Fig. 16.7B). They showed that the observed rate accelerations for some 60 antibody-catalyzed processes can be predicted with some degree of accuracy from the ratio of equilibrium binding constants for the reaction substrate, K_m, and for the TSA, K_i, used to raise the antibody. In particular, this analysis supports a rationalization of product selectivity shown by many antibody catalysts for kinetically-disfavored reactions [70–73] and predictions of the extent of rate accelerations that may ultimately be achieved by catalytic antibodies. This analysis also underlines some differences between the mechanism of catalysis by enzymes and that by antibody catalysts [69].

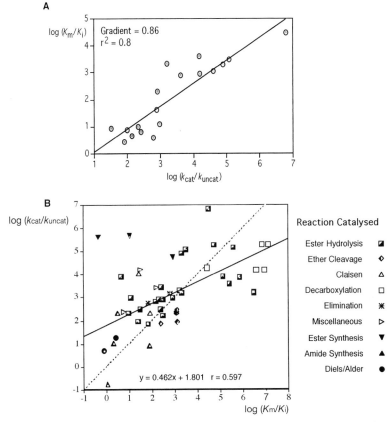

Fig. 16.7 A Jacobs's [62] correlation between the enhancement ratio (k_{cat}/k_{uncat}) and the ratio of the equilibrium dissociation constants for the substrate and TSA (K_m/K_i). The slope is an unweighted linear regression analysis ($r^2 = 0.80$).
B The Stewart-Benkovic [69] plot of rate enhancement versus relative binding of substrate and TSA for 60 abzyme-catalyzed reactions. The theoretical unit slope (—) diverges from the linear regression slope (—) for these data (for which the equation is shown).

The fact that many values of log K_m/K_i fall below the curve in Fig. 16.7 may suggest that interactions between the antibody and the substrate are largely passive in terms of potential catalytic benefit. Of course, even slight changes in hapten/inhibitor orientation in the antibody combining site can undermine the anticipated correlation, as has been observed for closely related phosphonate inhibitors of carboxypeptidase A [74]. Perhaps what this correlation most clearly exposes, however, is a limitation in the design of haptens. What is interesting to note, from the data plotted in Fig. 16.7B, is the high degree of scatter, with a correlation coefficient for the linear fit of only 0.6 and a slope of 0.46, very different from the "theoretical slope" of unity. Perhaps of greatest significance are the many positive deviations from the general pattern. These appear to show that antibody catalysis can fortuitously achieve more than is predicted from catalysis through transition state stabilization alone.

Tantillo and Houk [75] have recently developed a computer simulation model to search for optimal haptens to elicit efficient catalytic antibody esterases. Their study evaluated the fidelity of the structures and charge distributions of $TS^{\ddagger}$s and already utilized TSAs by comparing the molecular electrostatic potentials (MESPs) on the van der Waals surfaces of the $TS^{\ddagger}$s and TSAs. It was concluded that phosphorus (V)-based haptens, such as those shown in Fig. 16.4, do not reproduce a perfect mimic of the $TS^{\ddagger}$s and that more proficient catalytic antibodies may be obtained by hapten redesign.

There is accumulating evidence from simulation studies that complementary electrostatic effects are the primary source of enzymatic catalytic power [76–78]. Although the binding step usually involves hydrophobic effects, these effects do not help in the chemical step, which involves mainly changes in the charge distribution in the reaction locus. To illustrate this point, Warshel and coworkers [79] utilized ab initio calculations of the electrostatic potentials of the $TS^{\ddagger}$ **11** and the oxabicyclic TSA **12** of the Claisen rearrangement of chorismate **13** to prephenate **14** and then examined the complementary molecular interaction potentials of the Bacillus subtilis chorismate mutases ($k_{cat}/k_{uncat} > 10^6$) [80] and an antibody chorismate mutase, 1F7 ($k_{cat}/k_{uncat} > 200$) [81–83] (Fig. 16.8). This analysis revealed that the efficiency of the two biocatalysts for this pericyclic reaction correlated well with their degree of complementarity to the electrostatic charges of **11**.

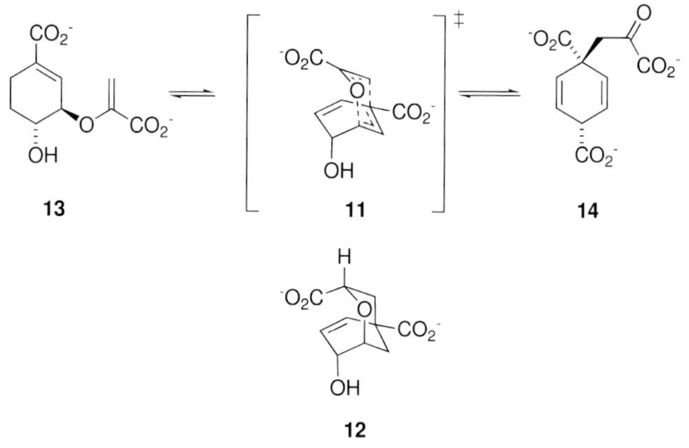

Fig. 16.8 Claisen rearrangement of chorismate **13** to prephenate **14**. The transition state **11** and transition state analog **12** are also shown.

Our imperfect ability to devise a stable mimic for the mercurial transition state structure will always be a limitation to the generation of the most efficient catalytic antibody. The fractional bond orders, expanded valences, extended bond lengths, distorted bond angles, and high degree of charge separation in $TS^{\ddagger}$s cannot be reproduced in stable molecules. To put this in perspective, even the best "transition state analog" enzyme inhibitors developed over the past three decades fall well short of the theoretical enhancement in binding affinity, relative to substrate, that this designation should bestow [84].

While hapten fidelity to the TS‡ can be a serious limitation, sometimes an unexpected benefit may be observed in which the antibody combining site inaccurately complements the hapten. For example, antibody 43C9 elicited to the TSA phosphonamidate **9** is a potent anilidase (Fig. 16.6). However, its mechanism has been shown to proceed via a ping-pong mechanism, with initial acylation of His^{L91} being followed by deacylation by attack of water [54]. This nucleophilic mechanism is not programmed by any obvious component of the TSA **9**. However, it is reasonable to speculate that the nucleophilic mechanism of 43C9 may have been programmed fortuitously via a unique reactive immunization approach [85]. Given the known propensity for phosphonamidate esters to undergo acid-catalyzed hydrolysis, it is possible that during the evolution of the immune response to **9**, the conjugate acid form of His^{L91} may have been selected via a general-acid-catalyzed hydrolysis reaction of **9** – a question easily answered by a combination of hapten analog binding studies and kinetic investigations with **9**.

An additional reason for the limitation in the catalytic efficiency of antibodies elicited by TSAs is simply that the diversity of antibody sequences sampled may be too few. The genetic diversity of the antibody response to a number of model haptens has been studied by sequence analysis, and, in general, the response to simple antigens is *very* restricted [86]. Catalytic antibodies are identified from those that bind tightly to the TSA hapten, and therefore their diversity of structure is expected to be even more restricted than is the case for simple hapten-binding antibodies. Knossow has analyzed the structures of three esterase antibodies, each elicited in response to the same phosphonate TSA hapten [47]. Despite significant structural differences in their combining sites, the catalytic activity of all three immunoglobulins is accounted for by TS‡ stabilization comprising similar oxyanion holes, each incorporating a tyrosine residue. These structural data strongly supports the notion that evolution of immunoglobulins for binding to a single TSA hapten, followed by selection from a large hybridoma repertoire by screening for catalysis, leads to antibodies with a high degree of structural convergence.

The question what can be done to improve the catalytic efficiency of antibodies necessarily arises, given that TSA designs will always be flawed and that establishing differential affinity for the TS‡ over the ground state is not a first-order problem? A number of solutions have been tested, and each involves modification of the immunization/hapten strategy to allow the incorporation of chemical mechanisms additional to TS‡ stabilization into antibody combining sites. It is well known that enzymes utilize a wide range of devices to achieve catalysis as well as dynamic interactions to guide the substrate toward the transition state, which is then selectively stabilized. "Bait-and-switch" [87] "reactive immunization" [85, 88–90], and "heterologous immunization" [91, 92] are a few examples of such approaches that have been utilized with considerable success. Recently, the problem has been approached in reverse, by asking the question how can we select for an antibody sequence that already possesses efficient catalytic activity for a given reaction? The problem then moves away from how to program a catalytic mechanism to how to select for it. Janda and coworkers [52, 93, 94] and Blackburn and coworkers [95] have developed elegant selection criteria that involve affinity trapping of catalytic antibodies from within antibody libraries expressed on the surface of filamentous phage.

In conclusion, transition state mimicry has received broad-based utility throughout the catalytic antibody field since its invention in the mid-1980s. However, the catalytic activity of antibodies elicited in this manner has rarely approached that of enzymes. From a pragmatic and fundamentally biased perspective, one might ask why the elicitation of enzyme-like catalytic activity is a necessary prerequisite to establish antibodies as "valid" biocatalysts? Clearly, the exceptional catalytic level of enzymes is not an *a priori* requirement for functional activity either *in vitro* or *in vivo*. From a chemist's point of view, the ability to program an antibody to catalyze reactions that enzymes cannot catalyze, such as the Diels-Alder cycloaddition [96, 97], the 1,3-dipolar cycloaddition [98], or the Bergman cycloaromatization reaction [99], with enhancement rations of ca. 10^3 and with exquisite control of regio- and stereo-selectivity [28], is reason enough to continue exploring the scope of these remarkable biocatalysts.

References

1 HALDANE, J. B. S., *Enzymes*, Longmans, Green, London **1930**
2 PAULING, L., *Nature* 161 (**1948**), p. 707
3 PAULING, L., *Chem. Eng. News* 24 (**1946**), p. 1375
4 EYRING, H., *Chem. Rev.* 27 (**1935**), p. 65
5 ALBERY, J., KNOWLES, J. R., *Biochemistry* 15 (**1976**), p. 5631
6 ALBERY, J., KNOWLES, J. R., *Angew. Chem., Int. Ed. Engl.* 16 (**1977**), p. 285
7 KURZ, J. L., *J. Am. Chem. Soc.* 85 (**1963**), p. 987
8 WOLFENDEN, R., *Nature* 223 (**1969**), p. 704
9 WOLFENDEN, R., *Acc. Chem. Res.* 5 (**1972**), p. 10
10 LIENHARD, G. E., *Science* 180 (**1973**), p. 149
11 ZEWAIL, A. H., BERNSTEIN, R. B., *Chem. Eng. News* 66 (**1988**), p. 24
12 POLANYI, J. C., ZEWAIL, A. H., *Acc. Chem. Res.* 28 (**1995**), p. 119
13 HOUK, K. N., GONZALEZ, J., LI, Y., *Acc. Chem. Res.* 28 (**1995**), p. 81
14 HAMMOND, G. S., *J. Am. Chem. Soc.* 77 (**1955**), p. 334
15 BARTLETT, P. A., LAMDEN, L. A., *Bioorg. Chem.* 14 (**1986**), p. 356
16 ALBERG, D. G., LAUHON, C. T., NYFELER, R., FASSLER, A., BARTLETT, P. A., *J. Am. Chem. Soc.* 114 (**1992**), p. 3535
17 JENCKS, W. P., In *Catalysis in Chemistry and Enzymology*, McGraw Hill, New York **1969**, p. 3
18 RASO, V., STOLLAR, B. D., *Biochemistry* 14 (**1975**), p. 584
19 RASO, V., STOLLAR, B. D., *Biochemistry* 14 (**1975**), p. 591
20 BEREK, C., MILSTEIN, C., *Immunol. Rev.* 105 (**1988**), p. 5
21 BURTON, D. R., *Acc. Chem. Res.* 26 (**1993**), p. 405
22 KÖHLER, G., MILSTEIN, C., *Nature* 256 (**1975**), p. 495
23 KÖHLER, G., HOWE, S. C., MILSTEIN, C., *Eur. J. Immunol.* 6 (**1976**), p. 292
24 TRAMONTANO, A., JANDA, K. D., LERNER, R. A., *Proc. Natl. Acad. Sci. U.S.A.* 83 (**1986**), p. 6736
25 TRAMONTANO, A., JANDA, K. D., LERNER, R. A., *Science* 234 (**1986**), p. 1566
26 POLLACK, S. J., JACOBS, J. W., SCHULTZ, P. G., *Science* 234 (**1986**), p. 1570
27 WENTWORTH JR., P., JANDA, K. D., *Curr. Opin. Chem. Biol.* 2 (**1998**), p. 138
28 WENTWORTH JR., P., JANDA, K. D., In *Comprehensive Asymmetric Catalysis*, JACOBSEN, E. N., PFALTZ, A., YAMAMOTO, H., Eds., Springer-Verlag, New York **1999**, Vol. I–III, p. 1403
29 WENTWORTH JR., P., *Science* 296 (**2002**), p. 2247
30 WENTWORTH JR., P., JANDA, K. D., *Cell Biochem. Biophys.* 35 (**2001**), p. 63
31 DATTA, A., PARTRIDGE, L. J., BLACKBURN, G. M., *European network on antibody catalysis 1993–95*, CSC-EC-EAEC, **1996**
32 BLACKBURN, G. M., DATTA, A., DENHAM, H., WENTWORTH JR., P., *Adv. Phys. Org. Chem.* 31 (**1998**), p. 249

33 JENCKS, W. P., In *Catalysis in Chemistry and Enzymology*, McGraw-Hill, New York **1969**, p. 288
34 SHEARER, E. C., TURNER, G. M., SHIELDS, G. C., *Int. J. Quantum Chem., Quantum Biology Symposium* 22 (**1995**), p. 83
35 BLACKBURN, G. M., DATTA, A., DENHAM, H., WENTWORTH, P., In *Advances in Physical Organic Chemistry, Vol 31*, **1998**, p. 249
36 JACOBSEN, N. E., BARTLETT, P. A., *J. Am. Chem. Soc.* 103 (**1981**), p. 654
37 BARTLETT, P. A., MARLOWE, C. K., *Biochemistry* 22 (**1983**), p. 4618
38 SHEN, J.-Q., Ph.D., Sheffield University **1995**
39 BENEDETTI, F., BERTI, F., COLOMBATTI, A., EBERT, C., LINDA, P., TONIZZO, F., *Chem. Commun.* 1417 (**1996**)
40 SHOKAT, K. M., SCHULTZ, P. G., *Annu. Rev. Immunol.* 8 (**1990**), p. 335
41 KITAZUME, T., TSUKAMOTO, T., YOSHIMURA, K., *J. Chem. Soc. Chem. Commun.* (**1994**), p. 1355
42 MACBEATH, G., HILVERT, D., 3 *Chem. Biol.* (**1996**), p. 433
43 WADE, H., SCANLAN, T. S., *Annu. Rev. Biophys. Biomol. Struct.* 22 (**1997**), p. 461
44 LESLEY, S. A., PATTEN, P. A., SCHULTZ, P. G., *Proc. Natl. Acad. Sci. U.S.A.* 90 (**1993**), p. 1160
45 WEDEMAYER, G. J., WANG, L. H., PATTEN, P. A., SCHULTZ, P. G., STEVENS, R. C., *J. Mol. Biol.* 268 (**1997**), p. 390
46 WEDEMAYER, G. J., PATTEN, P. A., WANG, L. H., SCHULTZ, P. G., STEVENS, R. C., *Science* 276 (**1997**), p. 1665
47 CHARBONNIER, J.-P., GOLINELLI-PIMPANEAU, B., GIGANT, B., TAWFIK, D. S., CHAP, R., SCHINDLER, D. G., KIM, S.-H., GREEN, B. S., ESHHAR, Z., KNOSSOW, M. *Science* 275 (**1997**), p. 1140
48 GUO, J., HUANG, W., SCANLAN, T. S., *J. Am. Chem. Soc.* 116 (**1994**), p. 6062
49 ZHOU, G. W., GUO, J., HUANG, W., FLETTERICK, R. J., SCANLAN, T. S., *Science* 265 (**1994**), p. 1059
50 BACA, M., SCANLAN, T. S., STEPHENSON, R. C., WELLS, J. A., *Proc. Natl. Acad. Sci. U.S.A.* 94 (**1997**), p. 10063

51 BUCHBINDER, J. L., STEPHENSON, R. C., SCANLAN, T. S., FLETTERICK, R. J., *J. Mol. Biol.* 282 (**1998**), p. 1033
52 GAO, C., LAVEY, B. J., LO, C.-H. L., DATTA, A., WENTWORTH JR., P., JANDA, K. D., *J. Am. Chem. Soc.* 120 (**1998**), p. 2211
53 JANDA, K. D., SCHLOEDER, D., BENKOVIC, S. J., LERNER, R. A., *Science* 241 (**1988**), p. 1188
54 STEWART, J. D., KREBS, J. F., SIUZDAK, G., BERDIS, A. J., SMITHRUD, D. B., BENKOVIC, S. J., *Proc. Natl. Acad. Sci. U.S.A.* 91 (**1994**), p. 7404
55 BENKOVIC, S. J., ADAMS, J. A., BORDERS JR., C. L., JANDA, K. D., LERNER, R. A., *Science* 250 (**1990**), p. 1135
56 GIBBS, R. A., BENKOVIC, P. A., JANDA, K. D., LERNER, R. A., BENKOVIC, S J., *J. Am. Chem. Soc.* 114 (**1992**), p. 3528
57 KREBS, J. F., SIUZDAK, G., DYSON, H. J., STEWART, J. D., BENKOVIC, S. J., *Biochemistry* 34 (**1995**), p. 720
58 ROBERTS, V. A., STEWART, J., BENKOVIC, S. J., GETZOFF, E. D., *Protein Engineering* 6 (**1993**), p. 85
59 ROBERTS, V. A., STEWART, J., BENKOVIC, S. J., GETZOFF, E. D., *J. Mol. Biol.* 235 (**1994**), p. 1098
60 STEWART, J. D., ROBERTS, V. A., THOMAS, N. R., GETZOFF, E. D., BENKOVIC, S. J., *Biochemistry* 33 (**1994**), p. 1994
61 THAYER, M. M., OLENDER, E. H., ARVAI, A. S., LOIKE, C. K., CANESTRELLI, I. L., STEWART, J. D., BENKOVIC, S. J., GETZOFF, E. D., ROBERTS, V. A., *J. Mol. Biol.* 291 (**1999**), p. 329
62 JACOBS, J. W., *Bio./Technology* 9 (**1991**), p. 258
63 WESTERICK, J. O., WOLFENDEN, R., *J. Biol. Chem.* 247 (**1972**), p. 8195
64 THOMPSON, R. C., *Biochemistry* 12 (**1973**), p. 47
65 BARTLETT, P. A., GIANGIORDANO, M. A., *J. Org. Chem.* 61 (**1996**), p. 3433
66 JENCKS, W. P., *Adv. Enzymol. Relat. Areas Mol. Biol.* 43 (**1975**), p. 219
67 BENKOVIC, S. J., NAPPER, A. D., LERNER, R. A., *Proc. Natl. Acad. Sci. U.S.A.* 85 (**1988**), p. 5355
68 TRAMONTANO, A., AMMANN, A. A., LERNER, R. A., *J. Am. Chem. Soc.* 110 (**1988**), p. 2282

69 Stewart, J. D., Benkovic, S. J. *Nature* 375 (**1995**), p. 388

70 Janda, K. D., Shevlin, C. G., Lerner, R. A., *Science* 259 (**1993**), p. 490

71 Cravatt, B. F., Ashley, J. A., Janda, K. D., Boger, D. L., Lerner, R. A., *J. Am. Chem. Soc.* 116 (**1994**), p. 6013

72 Wentworth Jr., P., Datta, A., Blakey, D., Boyle, T., Partridge, L. J., Blackburn, G. M., *Proc. Natl. Acad. Sci. U.S.A.* 93 (**1996**), p. 799

73 Wentworth Jr., P., Datta, A., Smith, S., Marshall, A., Partridge, L. J., Blackburn, G. M., *J. Am. Chem. Soc.* 119 (**1997**), p. 2315

74 Phillips, M. A., Kaplan, A. P., Rutter, W. J., Bartlett, P. A., *Biochemistry* 31 (**1992**), p. 959

75 Tantillo, D., Houk, K., *J. Org. Chem.* 64 (**1999**), p. 3066

76 Warshel, A., *Computer Modeling of Chemical Reactions in Enzymes and Solutions*, John Wiley & Sons, New York **1991**

77 Warshel, A., *J. Biol. Chem.* 273 (**1998**), p. 27035

78 Villa, J., Warshel, A., *J. Phys. Chem. B* 105 (**2001**), p. 7887

79 Barbany, M., Gutiérrez-de-Terán, Sanz, F., Villá-Freixa, J., Warshel, A., *ChemBiochem* 4 (**2003**), p. 277

80 Chook, Y. M., Ke, H. M., Lipscomb, W. N., *Proc. Natl. Acad. Sci. U.S.A.* 90 (**1993**), p. 8600

81 Hilvert, D., Carpenter, S. H., Nared, K. D., Auditor, M.-T. M., *Proc. Natl. Acad. Sci. U.S.A.* 85 (**1988**), p. 4953

82 Bowdish, K., Tang, Y., Hicks, J. B., Hilvert, D., *J. Biol. Chem.* 18 (**1991**), p. 11901

83 Haynes, M. R., Stura, E. A., Hilvert, D., Wilson, I. A., *Science* 263 (**1994**), p. 646

84 Mader, M. M., Bartlett, P. A., *Chem. Rev.* 97 (**1997**), p. 1281

85 Wirsching, P., Ashley, J. A., Lo, C.-H. L., Janda, K. D., Lerner, R. A., *Science* 270 (**1995**), p. 1775

86 Milstein, C., Neuberger, M. S., *Adv. Protein Chem.* 49 (**1996**), p. 451

87 Janda, K. D., Weinhouse, M. I., Schloeder, D. M., Lerner, R. A., Benkovic, S. J., *J. Am. Chem. Soc.* 112 (**1990**), p. 1274

88 Lo, C.-H. L., Wentworth Jr., P., Jung, K. W., Yoon, J., Ashley, J. A., Janda, K. D., *J. Am. Chem. Soc.* 119 (**1997**), p. 10251

89 Wagner, J., Lerner, R. A., Barbas III, C. F., *Science* 270 (**1995**), p. 1797

90 Datta, A., Wentworth Jr., P., Shaw, J. P., Janda, K. D., *J. Am. Chem. Soc.* 121 (**1999**), p. 10461

91 Suga, H., Ersoy, O., Williams, S. F., Tsumuraya, T., Margolies, M. N., Sinskey, A. J., Masamune, S., *J. Am. Chem. Soc.* 116 (**1994**), p. 6025

92 Tsumuraya, T., Suga, H., Meguro, S., Tsunakawa, A., Masamune, S., *J. Am. Chem. Soc.* 117 (**1995**), p. 11390

93 Janda, K. D., Lo, C.-H. L., Li, T., Barbas III, C. F., Wirsching, P., Lerner, R. A., *Proc. Natl. Acad. Sci. U.S.A.* 91 (**1994**), p. 2532

94 Lo, L.-C., Lo, C.-H. L., Janda, K. D. *Bioorg. Med. Chem. Lett.* 6 (**1996**), p. 2117

95 Cesaro-Tadic, S., Lagos, D., Honegger, A., Rickard, J. H., Partridge, L J., Blackburn, G. M., Plückthun, A., *Nat. Biotech.* 21 (**2003**), p. 679

96 Gouverneur, V. E., Houk, K. N., De Pascual-Teresa, B., Beno, B., Janda, K. D., Lerner, R. A., *Science* 262 (**1993**), p. 204

97 Hilvert, D., Hill, K. W., Nared, K. D., Auditor, M.-T. M., *J. Am. Chem. Soc.* 111 (**1989**), p. 9261

98 Toker, J. D., Wentworth Jr., P., Hu, Y., Houk, K. N., Janda, K. D., *J. Am. Chem. Soc.* 122 (**2000**), p. 3244

99 Jones, L. H., Harwig, C. W., Wentworth Jr., P., Simeonov, A., Wentworth, A. D., Py, S., Ashley, J. A., Lerner, R. A., Janda, K. D. *J. Am. Chem. Soc.* 123 (**2001**), p. 3607

17
Polyclonal Catalytic Antibodies

Marina Resmini, Elizabeth L. Ostler, Keith Brocklehurst, Gerard Gallacher

17.1
Introduction

The production of receptors that can achieve recognition at the molecular level is a very important goal of organic and bio-organic chemistry. The study of catalytic antibodies is certainly at the forefront of the search for a wide range of novel enzyme-like mimics and aims to bring together the basic concepts of enzyme catalysis and the potential of the immune system to generate antibodies against virtually any molecule. The foundation for the development of this field was laid by Jencks in 1969 [1] with his proposal that the use of a stable analog of a reaction transition state as a hapten might select IgG molecules with catalytic properties. Initially, several groups worked on this new idea, but they failed to isolate any antibody with catalytic activity. Raso and Stollar [2] reported that polyclonal antibodies generated by immunization with a phosphopyridoxyltyrosine derivative and isolated by using affinity chromatography failed to show any catalytic activity in the formation of a Schiff base. In 1980, a short communication by Kohen et al. [3] reported on the esterase-like activity of monoclonal IgG, elicited against the 2,4-dinitro-phenyl group, toward substrates containing the same haptenic moiety.

Following the publication in 1975 of Kohler and Milstein's [4] report on their revolutionary technology for the production of monoclonal antibodies, further studies were carried out. In 1986, two groups led by Lerner [5] (Scripps Research Institute, La Jolla) and Schultz [6] (Berkeley) independently reported the successful isolation for the first time of monoclonal catalytic antibodies. At that time the use of the monoclonal technology was viewed by many as essential to success, although subsequent research proved that in fact hapten design and the choice of reaction had been the determining factors. Since then the field has developed rapidly in a number of directions. Most investigations have been concerned with monoclonal catalytic antibodies that can certainly provide specific advantages when successfully isolated. Nevertheless, a number of researchers have spent the last 12 years successfully working with polyclonal catalytic antibodies (PCAs), and their results have complemented those obtained with monoclonals. The aim of this chapter is to review the work reported

Catalytic Antibodies. Edited by Ehud Keinan
Copyright © 2005 WILEY-VCH Verlag GmbH & Co. KGaA, Weinheim
ISBN: 3-527-30688-9

since the early discouraging results and to highlight the applications and potential advantages that PCAs have over their monoclonal counterparts.

17.2
The Importance of Polyclonal Catalytic Antibodies

The polyclonal approach offers a number of advantages of importance for the study of antibody catalysis. One of the unique features is that polyclonal antibodies represent the entirety of the immune response, since none of the IgG species produced are lost during the isolation step. This is in contrast with antibodies produced by hybridoma technology, where the processes of cell fusion, culture, and clone selection lead inevitably to loss of the initial IgG repertoire. For this reason, polyclonal antibodies can provide a more efficient way to screen new haptens designed for their ability to generate catalytic antibodies and also to attempt to identify general trends relating hapten structure to antibody catalytic activity. This is particularly important since all experts in the field agree that one of the keys to high catalytic efficiency lies in hapten design.

Polyclonal antibody preparations are obtained and analyzed without sacrificing the animal. Samples of serum are taken at different stages of the immunization programme, and this allows monitoring of the maturation of the catalytic activity during the whole period. A further advantage is that the immunization procedures and schedules can be systematically studied and optimized.

Polyclonal antibodies are cheap and simple to produce compared to monoclonals. They do not require highly specialized equipment or dedicated technicians. The first serum sample is available for evaluation within 8 to 12 weeks from the first immunization, thus shortening considerably the time required for gaining initial information on the quality of the immunogen.

Although with mice and rabbits the amounts of antibodies obtained from each bleed can be a limiting factor, in the case of sheep polyclonal antibodies, each bleed (up to 350 mL) provides large quantities of IgG, in the multi-gram range. This allows a large number of experiments to be performed, as illustrated by the extensive work carried out by the group led by Gallacher and Brocklehurst. Over 10 years this group has published more than 10 papers reporting the results of studies into sheep polyclonal catalytic antibodies generated by a single immunogen.

Another important advantage of working with polyclonal antibodies is the possibility of investigating how the different animal immune systems respond to immunization. The same immunogen can be used to generate a response in several different animals of the same or different species. Characterization of the catalytic activity of the different preparations may provide information about the role played by genetic diversity in the optimization of antibody catalysis and insight into the best possible combination of hapten design and immune system genetics. In the case of monoclonal antibodies, this particular aspect is restricted by the hybridoma technology that is applied only to mice and rats, therefore limiting the different animals that can be utilized.

All of the advantages mentioned so far highlight the type of information that the study and development of polyclonal catalytic antibodies can offer. One ultimate goal is represented by the potential therapeutic applications that this technique could bring. A number of groups have worked on possible strategies involving the generation of catalytic antibody activity in patients through immunization, and the preliminary results are very encouraging. These are discussed later in the chapter.

17.3
Demonstration of Polyclonal Antibody Catalysis

The group led by Brocklehurst and Gallacher was the first to describe the generation and characterization of polyclonal catalytic antibodies [7, 8]. A preparation of sheep IgG was obtained by immunization with phosphate (1) and shown to catalyze the hydrolysis of the mixed carbonate (2). The reaction was chosen because the products, two alcohols and carbon dioxide, are less likely to give product inhibition.

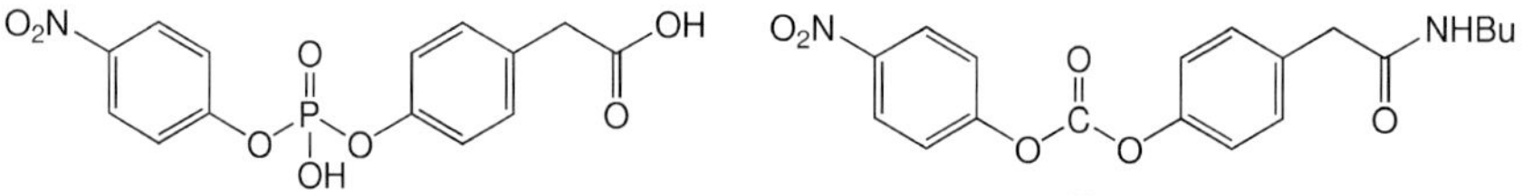

Fig. 17.1

The design of the substrate aimed to maximize hydrogen bonding and polar and hydrophobic interactions on both sides of the planar carbonyl and to provide a chromogenic leaving group that would facilitate characterization of the hydrolysis reaction. The polyclonal IgGs were isolated from the antiserum by salt precipitation followed by Protein G-Sepharose chromatography. The catalyzed reaction exhibited single-site saturation behavior (Michaelis-Menten kinetics) with K_m = 3.34 µM and k_{cat} = 0.029 s^{-1}. The latter was calculated assuming all the IgG present to be active and two active sites per antibody molecule. In fact it is more likely that the proportion of catalytically active IgG contained in the polyclonal mixture is less than 10% [8], therefore giving an activity that compares favorably with those obtained with monoclonal catalytic antibodies raised to similar immunogens [5, 6].

Shortly after this, Iverson et al. [9] in 1993 reported the production and characterization of rabbit polyclonal antibodies that catalyze the hydrolysis of the triphenylmethyl ether substrate (3). The IgG preparations were obtained following immunization with the phosphonium hapten (4) that mimics the positive charge of the protonated species generated during the acid-catalyzed hydrolysis of (3) and is expected to generate a complementary negative charge in the antibody binding pocket.

This reaction was chosen because the authors had previously generated monoclonal catalytic antibodies using the same hapten [10], which provided a basis for comparison. Since there is no natural enzyme activity known for this reaction, the problem of enzyme contamination, often mentioned by critics of the polyclonal approach, was completely avoided. The polyclonal antibodies isolated displayed satura-

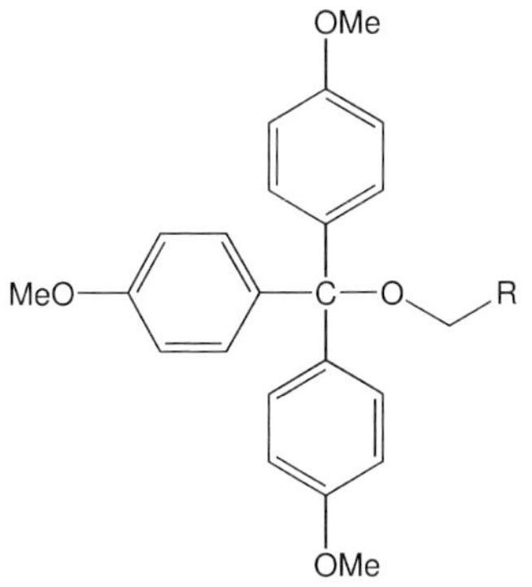

(3) R = CH₂CH₂(OCH₂CH₂)₅OH 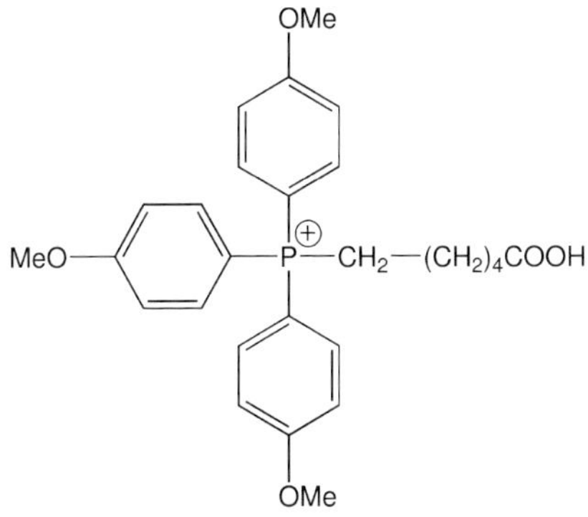   (4)

Fig. 17.2

tion behavior consistent with classic Michaelis-Menten kinetics. The kinetic parameters obtained compared well with those obtained previously by using monoclonal antibodies. One interesting finding was that catalytic activity appeared later in the immune response than the antibody binding. The authors suggested that the affinity maturation process occurring at the later stage is the significant factor leading to catalytically active antibodies.

In the following years a number of groups carried out further work on hydrolytic reactions. In 1995, a report [11] appeared in the literature on the generation of rabbit polyclonal antibodies catalyzing the hydrolysis of naproxen ethyl ester (5). These were obtained by immunization using the corresponding racemic ethyl phosphonate (6) as hapten.

Fig. 17.3

The purpose of this work was to obtain a catalyst that would allow the preparation of optically pure naproxen by kinetic resolution. Initial results showed a rate acceleration of 1240-fold. Subsequent work [12] on the enantioselectivity of these polyclonal preparations showed that they were highly selective for the R-isomer, catalyzing its hydrolysis, but not that of the corresponding S-isomer. Interestingly, the same experiment performed using the commercial and semi-purified lipase of *Candida cylindracea* found that only the hydrolysis of the S-isomer is catalyzed.

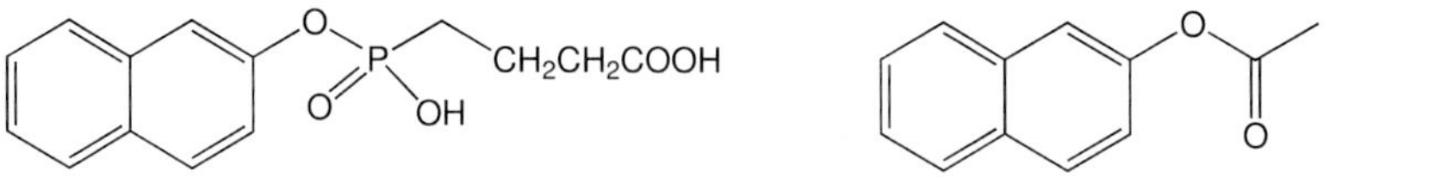

(7)

(8)

Fig. 17.4

The same group reported [13] the generation of rabbit polyclonal antibodies, using the phosphonate hapten (7), which catalyzes the hydrolysis of β-naphthyl acetate (8) to β-naphthol with a rate enhancement of $k_{cat}/k_{uncat} = 4.6 \times 10^4$.

In the same year, 1997, another report [14] appeared on the generation of rabbit polyclonal antibodies catalyzing ester hydrolysis. The designs of the substrates and phosphonate hapten are very similar to the ones previously used for monoclonal antibodies by other groups [15], and the value of k_{cat} (0.22 min^{-1}; 100% IgG) is comparable when allowance is made for the fact that only a small percentage ($\leq 10\%$) of the mixture is in fact catalytically active.

Following previous work done with monoclonal antibodies [16], rabbit polyclonal antibodies have been generated also for pericyclic reactions such as the hetero-Diels-Alder and the oxy-Cope rearrangement, two reactions for which there are no known enzymes available.

In the first case [17], immunization with the product analog (9) generated antibodies that selectively catalyze the formation of the *endo* adduct resulting from the reaction between diene (10) and dienophile (11).

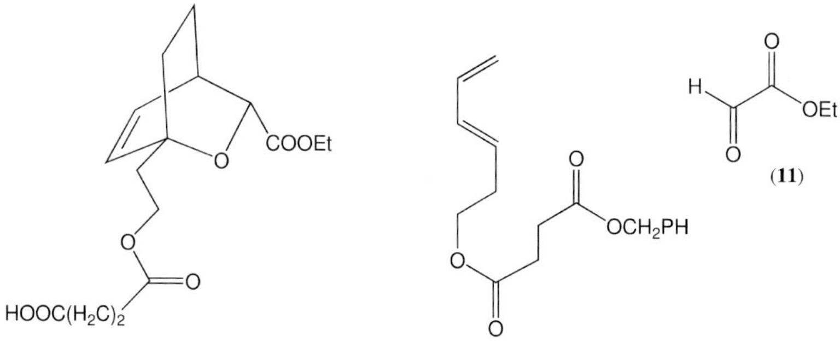

(9)

(10)

(11)

Fig. 17.5

The reaction followed Michaelis Menten kinetics with $K_m = 96$ µM and $k_{cat} = 1.7$ min^{-1}, and addition of a hapten analog resulted in essentially complete inhibition of the catalyzed reaction. The authors demonstrated that the reaction was antibody mediated, because in the presence of normal rabbit IgGs or in the absence of catalyst no cycloaddition reaction could be detected. The aim of this work was to develop a stereoselective catalyst that could be used in the synthesis of the biologically important ulosonic acid and its analogs.

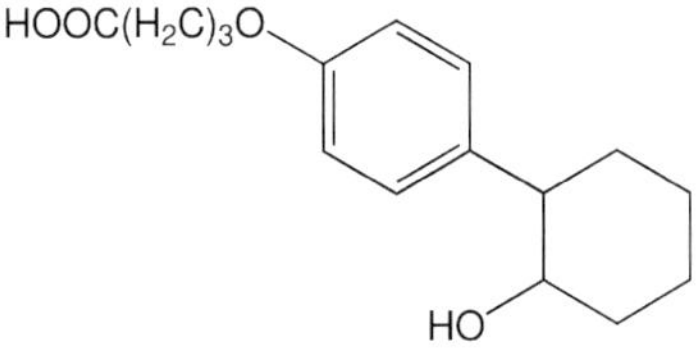

Fig. 17.6

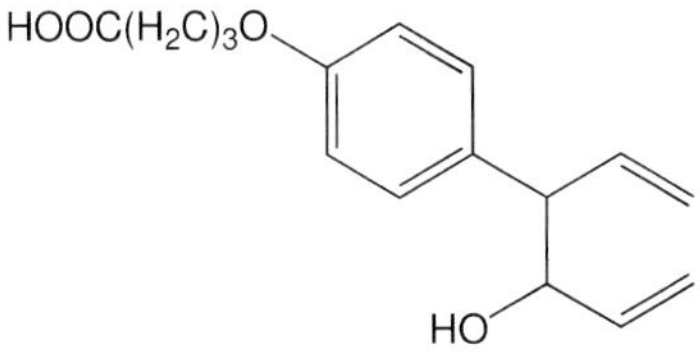

More recently [18], a polyclonal antibody preparation was generated by using the cyclohexanol transition-state analog (12) that catalyzes the oxy-Cope rearrangement of substrate (13) with $K_m = 1.025$ mM and $k_{cat} = 10$ min^{-1}.

17.4
Hapten Design and Catalytic Antibody Activity

Since the initial reports on the first catalytic antibody, more than 100 different reactions have been shown to be catalyzed by antibodies. Although a great variety of substrates and haptens have been used, the catalytic efficiency of the isolated IgGs, with a few exceptions [19], has yet to equal that of enzymes. Workers in the field agree that the role played by the structure of the hapten is of fundamental importance in determining the activity of the antibodies. As mentioned earlier, polyclonal antibodies can provide an important tool for assessing large numbers of haptens. Despite this great potential, the number of reports using polyclonal antibodies for this purpose has been remarkably small and confined mainly to the work carried out by Iverson and coworkers.

The first report on this topic was by Gallacher and Brocklehurst in 1993 [20]. While investigating the polyclonal preparations generated by phosphate immunogen (1), this group used the sulfone hapten (14) to immunize three sheep. None of the antibody preparations obtained showed any catalytic activity towards either the corresponding carbonate substrate (15) or the original carbonate (2).

This result confirmed that the catalytic activity observed previously [8] was not due to enzyme contamination, and demonstrates that the sulfone group is substantially poorer as an analog of the postulated transition state for the hydrolysis of carbonates.

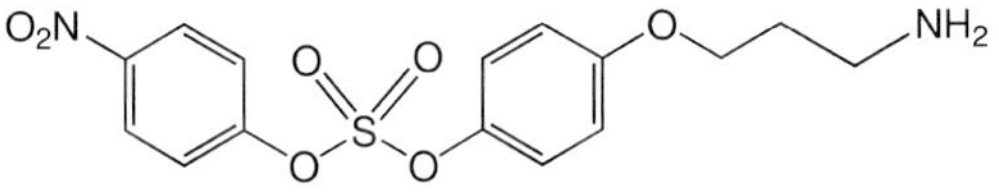

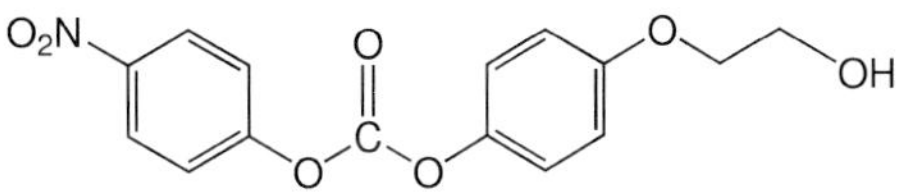

Fig. 17.7

In 1994 Wilmore and Iverson [21] carried out a comparative study by generating rabbit polyclonal antibodies against the phosphate (**16**) and the phosphorothioate (**17**) and studying their catalytic activity and specificity towards esters, thioesters, carbonates, and amides.

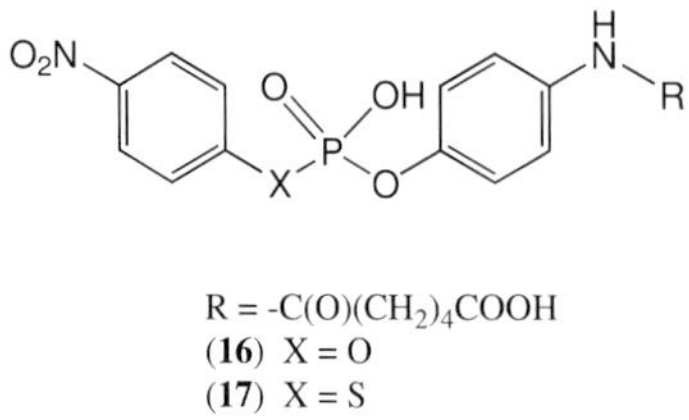

R = -C(O)(CH$_2$)$_4$COOH
(**16**) X = O
(**17**) X = S

Fig. 17.8

The hypothesis was that the phosphorus-sulfur bond, being 0.5 Å longer than the analogous phosphorus-oxygen bond, would provide a better mimic for a more expanded transition state, leading to enhanced catalytic activity [22]. Characterization of the isolated preparations showed that (i) the phosphate hapten produced antibodies catalyzing the ester and carbonate substrate more efficiently, (ii) both haptens produced antibodies with similar activities towards a thioester substrate, and (iii) none of the isolated preparations showed any catalytic activity towards the amide substrate. The conclusion was that substitution of the oxygen by a sulfur atom in the phosphate did not lead to any significant advantage.

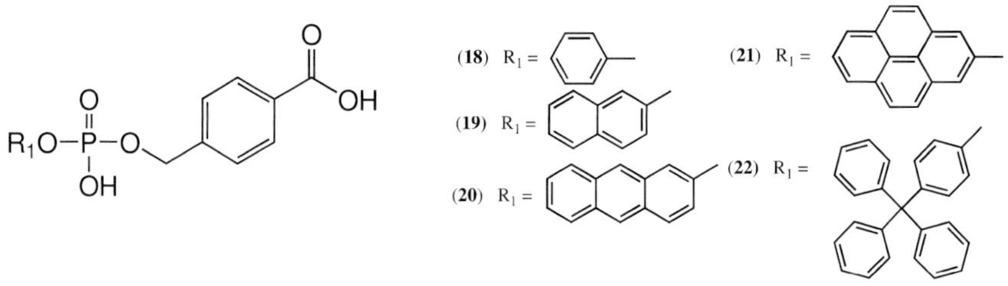

Fig. 17.9

Wallace and Iverson [23] subsequently extended this work to investigate the importance of hapten size and hydrophobicity in eliciting catalytic activity. A series of haptens (**18–22**) containing aromatic groups increasing in size attached to the phosphate moiety were used to elicit mice polyclonal antibodies. Kinetic studies on the hydrolysis of the corresponding carbonate substrates revealed that only the two smaller and less hydrophobic haptens (**18**) and (**19**) produced antibodies with some catalytic activity [k_{cat}/k_{uncat} = 4300 for (**18**) and k_{cat}/k_{uncat} = 1000 for (**19**)]. Affinity studies revealed that all of the haptens elicited good immune responses and that hapten affinity increased along with increase in the size of the planar aromatic group in haptens (**18**)–(**21**). Even the non-catalytic polyclonal preparations elicited by haptens (**20**) and (**21**) showed good binding for their substrates. One possible explanation is that the larger size of the aromatic portions of haptens (**20**), (**21**), and (**22**) led the immune response to divert the binding energy away from the phosphate group, therefore precluding transition state stabilization and thus catalysis. This work has demonstrated that catalytic activity can depend on structural features distinct from those that mimic the rate-limiting transition state.

The most recent report [24] on this topic describes an investigation of the effects of small modifications in haptens on the resultant catalytic activity. A total of 12 phosphate and phosphonate haptens were used to elicit polyclonal antibodies. Analysis of the isolated antibodies showed that six haptens containing benzyl phosphonate moieties failed to elicit any catalytic activity, despite a high affinity immune response. Investigation of a series of benzyl phosphate haptens using different animal models showed that the elicited catalytic activity was in each case significantly lower than the ones previously obtained using a phenyl phosphate hapten. A possible explanation highlighted by the authors is concerned with the nature of the leaving group in the corresponding substrates. When a phenyl or 4-nitro phenyl group is present, the hapten appears to be successful. On the contrary, if 4-alkyphenyl or benzyl leaving groups are present, the hapten is unsuccessful. Furthermore, the phenyl or 4-nitrophenyl group needs to be located on the side of the hapten opposite the linker. It is possible to hypothesize that flexibility in the region of the hapten opposite the linker decreases the effectiveness in eliciting high catalytic activity by disrupting catalysis of a bound substrate, although the "flexibility hypothesis" is complicated by the use of substrates with different leaving group abilities. Flexibility may lead a hapten to rearrange into a low-energy conformation that may not be compatible with catalysis. These arguments are in accord with the results of the studies conducted by Linder et al. [25] using hydrolytic monoclonal antibodies.

17.5
Kinetic Activity, Homogeneity, and Variability

All catalytic polyclonal antibodies isolated so far have been shown to obey Michaelis-Menten kinetics, displaying single-site saturation behaviour. As first reported by Gallacher et al. [8] and subsequently confirmed by all others, "whatever structural heterogeneity exists in the IgG, the catalytic characteristics of the active antibodies are all sufficiently similar that differences between them are not readily detected as deviations from a single-site saturation model".

Given that the initial work by Gallacher et al. using the carbonate substrate (2) was confined to a small range of substrate concentration (0.2–17 µM), it was decided to extend this by using (23), which has the same reaction center as substrate (2) but a much higher solubility [26].

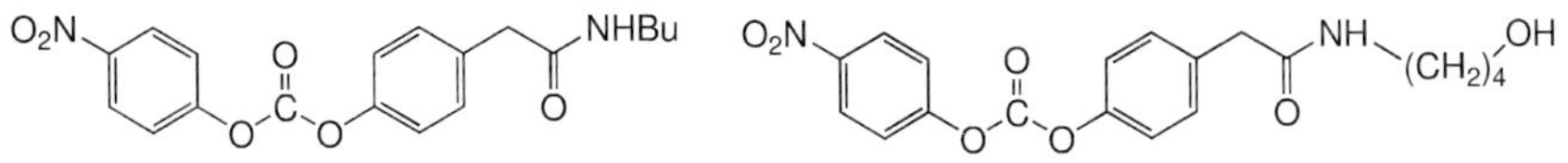

Fig. 17.10

This allowed kinetic studies to be carried out using substrate concentrations up to 85 µM without increasing the percentage of organic solvent. Plots of v vs [S] and the linearity of the [S]/v vs [S] plots showed again good adherence to the Michaelis-Menten equation and no evidence of functional heterogeneity.

A topic of particular interest is the study of antibody variability. This includes both monitoring the immune response in the same animal over a long period of time and investigating the response of the same hapten in different animals. The former is of particular interest since it can be studied only using the polyclonal approach, as highlighted earlier.

Gallacher and Brocklehurst were the first to publish results on the variability of polyclonal catalytic antibodies [20]. Their initial immunization program on three sheep using the same immunogen continued for over two years. Overall, the three sheep led to thirteen bleeds, each displaying catalytic activity and containing antibodies that showed similar catalytic parameters (only 4-fold variation in K_m and 31-fold variation in k_{cat}). This demonstrated that not only it is possible to routinely elicit a catalytic antibody response in a host animal but also that within the same animal the catalytic fraction of IgG present is relatively constant. It follows that the immune systems of the animals responds with relative uniformity to the same hapten.

These results were further confirmed and extended by the work of Iverson on polyclonal antibody catalytic variability [27]. Hapten (**4**) discussed earlier was used to immunize five rabbits, and analysis of the different bleeds showed that (i) the catalytic activity was remarkably similar for all the rabbits, (ii) the level of activity was similar to that obtained from the best murine monoclonal antibody elicited by the same hapten (when k_{cat} was calculated using [IgG] obtained from inhibition studies), and (iii) the binding and catalytic properties are relatively homogeneous within the polyclonal samples from each animal, as demonstrated by analysis of the eluted fractions of antibody from a substrate column. Previous work [28] carried out by the same group using binding antibodies had already provided evidence of the remarkably homogeneous nature of the immune response.

17.6
Assessment of Contamination

One important issue related to all catalytic antibodies is enzyme contamination. This is particularly relevant in the case of hydrolytic reactions, since hydrolytic enzymes are known to be present in the serum, and therefore the purification protocols must be rigorous and special controls must be in place to ensure that catalysis is antibody-mediated. Hapten inhibition is certainly the most common control used to verify that catalysis is occurring in the antibody binding site, but this on its own might not be sufficient. With polyclonal antibodies these issues have been raised many times.

A particularly convincing approach to the assessment of contamination was published by Gallacher and Brocklehurst [8]. This makes use of the strict specificity exhibited by antibodies that is not shared by well-known hydrolytic enzymes. Thus the *anti*-phosphate polyclonal catalytic antibody preparation 270-22, which showed the highest catalytic activity against the substrate (2), did not catalyze the hydrolysis of the corresponding 2-nitrophenyl isomer (24). In marked contrast, two different carboxylesterase enzymes, pig liver carboxylesterase and rabbit liver carboxylesterase, failed to discriminate substantially between the two substrates, with similar values of K_m and k_{cat}. Perhaps more importantly, the same failure to discriminate was obtained when the whole serum from a non-immunized sheep or the whole serum from a sheep immunized with an unrelated hapten were assayed with both substrates. These results provide compelling evidence that the observed catalytic activity with substrate was undoubtedly due to the elicited IgGs and not to any contaminant enzymes.

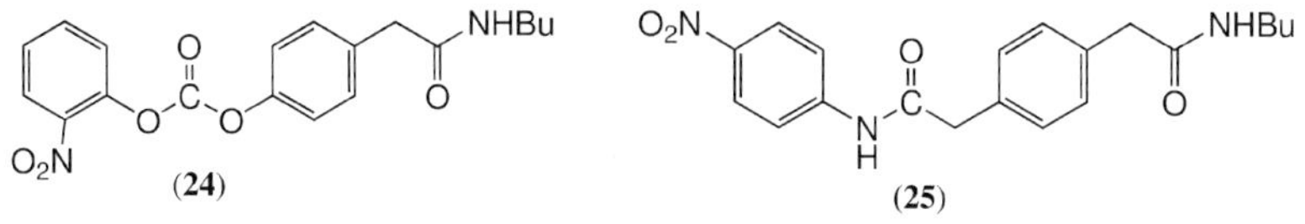

Fig. 17.11

An early investigation [29] of the possibility that these antibodies might be able to catalyze the hydrolysis of the related anilide (25) was hindered by low rates of reaction. The latest published result [30], however, shows more clearly that the same *anti*-phosphate antibodies catalyze the hydrolysis of the β-lactam substrate (26). The choice of this substrate was prompted by results obtained from unpublished preliminary specificity studies, in particular the finding that the second aryl ring is not always required for these antibodies to mediate catalysis [31].

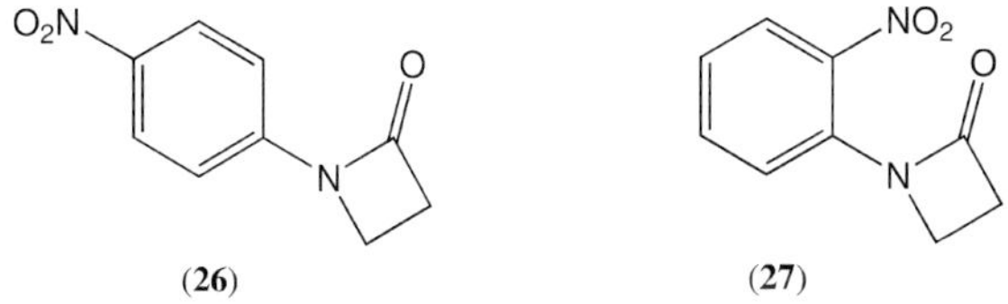

Fig. 17.12

IgG purified from bleed 270-26 catalyzes the hydrolysis of (26) with $k_{cat} = 1.3 \times 10^{-5}$ s^{-1} and $K_m = 30$ μM at pH 9 and 37 °C. Whilst the activity is moderate, it nevertheless demonstrates the suitability of N-aryl-β-lactams as catalytic antibody substrates. As expected, the 2-nitrophenyl-β-lactam (27) is not a substrate for the antibodies. More importantly, these compounds are sufficiently stable for medical applications and open up exciting possibilities in the area of prodrug activation. It should be pointed out that these experiments were carried out using the same sheep polyclonal antisera originally generated in 1987 and stored as serum at −20°C since that time.

17.7
Mechanistic Studies

Polyclonal antibody preparations are a mixture of IgGs, and as such, any detailed mechanistic studies are deemed to be difficult and certainly more complicated than in the case of monoclonal antibodies. Nevertheless, following the extensive investigation of sheep polyclonal antibodies and the resulting conclusion that functional homogeneity was consistent and reproducible, Resmini et al. [26] focused attention on mechanistic studies using PCA 270-26 and the more soluble carbonate (23).

A study of the pH dependence of k_{cat} and k_{cat}/K_m of these antibodies identified a pK_a value of 9 for a kinetically influential ionization, suggesting the involvement in the catalytic act of a tyrosine or a lysine residue. Group-selective chemical modification studies established that the side chains of tyrosine and arginine residues are essential for catalytic activity and provided no evidence for the involvement of side chains of lysine, histidine, or cysteine residues. Moreover the results obtained from binding studies with ELISA experiments were used to define the transition state of the catalyzed reaction as containing the undissociated tyrosine side chain and hydroxide ion. As is well known, it is not possible to distinguish alternative ionic forms of transition states by pH-dependent kinetics data alone.

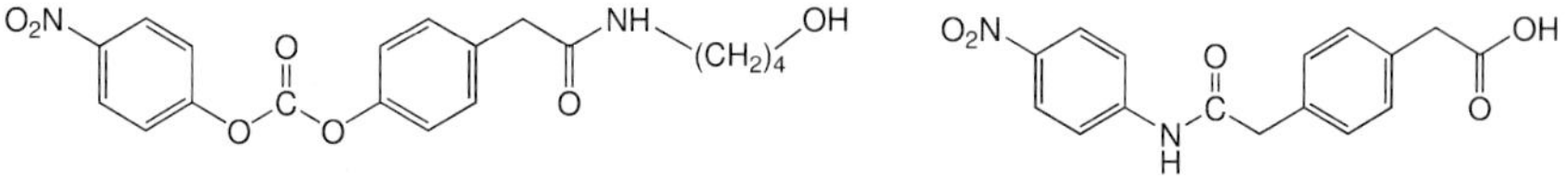

Fig. 17.13

Binding to the analogous substrate (28) was shown to be weaker at pH 9 than at pH 6, and when the antibody tyrosine side-chains were nitrated using tetranitromethane, thereby reducing the pK_a value of the tyrosine from 9 to 7, the binding activity was totally lost at pH 9 but retained at pH 6. This suggests that the tyrosine is required for catalysis in its undissociated form. The combination of the kinetic and binding data suggests that catalysis involves assistance to the reaction of the substrate with hydroxide ions by hydrogen bond donation at the reaction center by tyrosine side chains, arginine side chains also being involved. Interestingly, these same residues have been found to be important in a number of hydrolytic monoclonal antibodies and to have clearly similar roles.

17.8
Investigations of Active-Site Availability

One of the major factors preventing a detailed kinetic characterization of polyclonal catalytic antibodies is that the percentage of catalytic IgG in the protein mixture has always been uncertain. The IgG also contains binding-only IgG and non-binding/non-catalytic IgG. A number of research groups have used inhibition experiments to obtain an estimate of the percentage of hapten-binding antibodies, and their values have always ranged between 8 and 12%. Of course, while the functional homogeneity of the polyclonal catalytic preparations has been widely demonstrated, the same cannot be said for the hapten-binding IgGs. The heterogeneity at that level can certainly be expected to be much higher, and therefore the inhibition experiments will only provide an indication of the content of binding antibodies, because much will depend on the affinity of the different binding IgG populations for the hapten. It is important to remember that only a small percentage of binding antibodies will be catalytically active.

Attempts to isolate a catalytic fraction by affinity chromatography have to date been disappointing. Stephens et al. [27] reported that after loading the rabbit polyclonal IgG samples onto affinity columns all attempted elutions (including pH changes, different ionic strengths, and organic solvents) failed to elute active antibodies. The same result was obtained when they used as a positive control their catalytic monoclonal antibody. It appears that the antibody-antigen interactions cannot be easily dissociated under the conditions that allow retention of the antibody's activity.

Resmini et al. [32] decided to approach the problem from a different point of view and use a kinetic approach to determine the concentration of active sites. A combination of steady-state (excess substrate) and single-turnover (excess catalyst) kinetics was used to estimate the proportion of hapten-binding antibodies that are potentially catalytic. This method allows the determination of the catalytic fraction within a mixture containing: non-binding IgG, binding but non-catalytic IgGs, and catalytic IgGs, which may bind substrate in a productive (catalytic) or non-productive way. This approach was designed as an extension of the treatment of the simple two-step model of enzyme catalysis that would be applicable to the circumstances found both with catalytic antibodies and with other macromolecular catalysts. The kinetic method was validated by using α-chymotrypsin. In the case of catalytic antibody preparations, the non-binding fraction was required for this calculation and was estimated by measuring the breakthrough fraction from hapten analog affinity chromatography. This estimate was used to show that the catalytic fraction of the antibodies purified from bleed 271-22 was at most 8% and probably less than 1%. When these values are applied to the experimental data to calculate k_{cat} for various PCAs, it becomes evident that the rate accelerations compare very favorably with the ones obtained using monoclonal catalytic antibodies.

17.9
Therapeutic Applications

One of the main targets of catalytic antibody research is the development of therapeutic agents. Among the obvious advantages are: (i) antibodies showing catalytic activity and having turnover capabilities could be used in much lower doses than binding antibodies and drugs; (ii) antibodies are biocompatible and have long half-lives in serum; (iii) the progress in antibody engineering has led to a reduction of the immunogenicity of xeno-antibodies [33]. Two therapeutic approaches are envisaged:

the first one, termed passive immunization, is when catalytic antibodies are administered to patients to exert a designed therapeutic effect; the second one, termed active immunization, is when a hapten is used to directly immunize patients with the aim of generating their own catalytic antibodies. Therapeutic antibodies are reviewed elsewhere in this book and therefore only the active immunization involving polyclonal antibodies will be covered in this chapter.

Basmadjian et al. [34] were the first to report an attempt to use active immunization of mice to generate polyclonal antibodies capable of degrading cocaine *in vivo* by catalyzing its hydrolysis to the inactive metabolite benzoylecgonine. They used phosphonate transition state analogs similar to those used successfully by Landry to generate monoclonal catalytic antibodies [35]. The haptens were coupled to the immunogenic protein diphtheria toxoid and used to immunize over 160 mice. Unfortunately, all of the published data were obtained using the whole sera, which is known to contain many esterases, therefore making the interpretation of the results more complex. No follow-up work by this group on the polyclonal approach appears to have been published to date.

A significant step forward in the development of therapeutic catalytic antibodies is represented by the work of Renard and Mioskowski on the generation of antibodies that detoxify organophosphorus nerve agents by catalyzing their hydrolysis [36]. Among the chemical warfare nerve agents, the poisonous VX (**29**) was chosen as a target because of its high toxicity. The alpha-hydroxyphosphinate analog (**30**) was chosen as hapten, its design aiming to bring a water molecule close to the reactive phosphate group in the antibody binding site. Moreover, the presence of the free terminal amino group enabled conjugation to the carrier protein. Initially, immunogen (**30**) was used to elicit polyclonal antibodies in mice in order to minimize the amount of (**29**) required for screening. Preliminary results were interesting and showed that the generated IgGs were able to neutralize the highly toxic agent VX in a very specific way and were not active toward other structurally related phosphorus chemical poisons. The small volume of antisera obtained from each bleed prevented further characterization of the activity, and therefore the authors decided to use the monoclonal approach.

In order to avoid using dangerously high quantities of the toxic compound (**29**), hapten (**31**) containing an aromatic moiety instead of the methyl group was used.

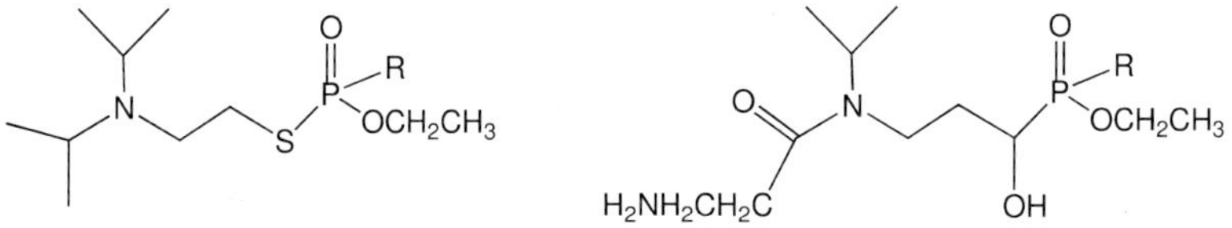

Substrates
(**29**) : R = Me
(**32**) : R = Phe

Haptens
(**30**) : R = Me
(**31**) : R = Phe

Fig. 17.14

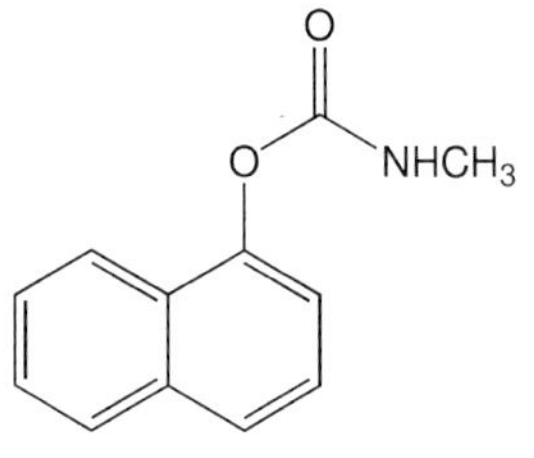

(33) 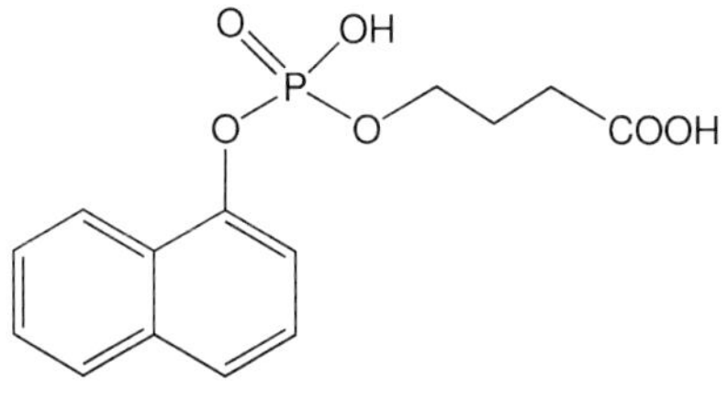   (34)   Fig. 17.15

The activity against (29) was very poor, as expected, since the affinity was substantially decreased. Nevertheless, one antibody showed significant activity when tested against compound (32), a less toxic VX analog. This work successfully proved the viability of the *in vivo* detoxification process via both passive and active immunization, although in order to envisage any animal or clinical applications, antibodies with higher affinities and catalytic activities will have to be elicited.

A further advance in demonstrating the potential of active immunization was recently provided by work carried out by Wilkinson's group at the University of Texas [37]. This group reported the generation of mice polyclonal antibodies that catalyze the hydrolysis of carbaryl (33), a widely used broad-spectrum carbamate insecticide that has high toxicity for animals and humans. The phosphate hapten (34) was conjugated to BSA and used as the immunogen. Polyclonal catalytic antibodies were isolated and purified from the antisera, and the catalyzed reaction was shown to obey Michaelis-Menten kinetics *in vitro* with $K_m = 8$ µM and $k_{cat} = 0.5$ s^{-1} (assuming only 1% of IgG to be catalytically active) and was inhibited by the free hapten. The whole sera were also tested under physiological conditions and shown to be active when compared to non-immunized sera. Furthermore the capability of these antibodies to decrease the concentration of carbaryl *in vivo* was also tested by immunizing mice with hapten (34) and determining the carbaryl concentration in the blood 1 h after administering fixed amounts of carbaryl.

To evaluate the efficacy of the polyclonal antibodies induced *in vivo*, the ED$_{50}$ value was determined, that is the dose of carbaryl that produces lowest-grade tremors in 50% of the animals. Results showed that the ED$_{50}$ for the immunized animals was 43% higher than for the non-immunized ones, indicating that, for the former group, part of the carbaryl is being hydrolyzed by the PCAs. These are the first characterized animal data showing that *in vivo*-induced polyclonal antibodies can prevent the toxic effects of insecticides. These results further confirm the great potential of active immunization as a means to protect against pesticides, toxic agents, and narcotic drugs.

17.10
Antiidiotypic Antibodies

The antiidiotypic approach represents an alternative method for the generation of antibody catalysts. It is based on the idea proposed initially by Niels Jerne [38] and later supported by experimental results [39] showing that *anti*-idiotypic antibodies can recognize antigenic determinants in an antibody's combining site that was in contact with the original hapten. It follows that the binding site of the *anti*-idiotypic antibody could carry an "internal image" of the antigen. This concept has been applied to obtain *anti*-idiotypic antibodies mimicking the activity of different enzymes, such as acetylcholinesterase [40], carboxypeptidase [41], β-lactamase, and protease [42]. Both monoclonal and polyclonal antibodies have been used in this method, in some cases within the same project, demonstrating once more the complementarity of the two approaches. An example of this is represented by the work carried out by Friboulet's group in Compiegne, working on the generation of antibodies with cholinesterase activity. Their initial work focused on using an *anti*-cholinesterase monoclonal antibody as antigen to generate rabbit polyclonal catalytic antibodies [43]. The advantage here is that the entire immune response is sampled, and therefore different modes of action can be distinguished and also novel activity can be identified. The antisera were characterized and shown to have different catalytic constants from those of a acetylcholinesterase but similar specificity toward substrates and inhibitors. Following these positive results, the same immunogen was successfully used for the generation of monoclonal antibodies [40]. The isolation of a highly active antibody has allowed extensive characterization and studies of structure-activity relationships, with particular regard to comparisons with the analogous enzymes and the monoclonal antibody used as the antigen. Similarly, the polyclonal antiidiotypic approach has yielded catalytic antibodies with DNAse [44] and β-lactamase activity [45].

The possibility of generating catalytic antibodies using the *anti*-idiotypic approach opens up new routes to both the study of structure-activity relationships of enzymes and the production of new catalytic activities. Such results promise contributions to the understanding of catalysis in general and of enzyme mechanisms in particular.

17.11
Naturally-Occurring Catalytic Antibodies

Generation of catalytic activity in an antibody was thought to require immunization with particular man-made immunogens. The discovery by Paul [46], however, that certain autoantibodies to vasoactive intestinal peptide (VIP), found in the sera of patients with autoimmune diseases, can accelerate the hydrolysis of a peptide bond provided the first example of natural catalytic antibodies. Subsequently L-chains derived from these autoantibodies were found to accelerate the hydrolysis of biologically active peptides as well as synthetic amidase substrates [47]. Previously, IgGs with DNase activity had been found in systemic lupus erythematosus (SLE) patients by Gabibov and coworkers [48], and RNA-hydrolysing IgGs in SLE patients were discovered after that [49]. It appears that the sera of patients with several autoimmune and viral pathologies as well as lymphoproliferative diseases contain a number of these catalytic autoantibodies. The literature on this topic is extensive and would require a dedicated chapter to be properly described. It is mentioned within this chapter because all these autoantibodies are polyclonal in nature and are generated by the immune system within itself. The origin of these natural catalytic antibodies is still not clear, and a number of different hypotheses are currently being investigated. It is interesting to mention that one of the possible explanations involves the mimicry of enzymatic sites by *anti*-idiotypic antibodies resulting from immune regulation dysfunction. Evidence for this hypothesis is that sequence homology has been observed between a catalytic antiidiotypic antibody and other autoantibodies [50]. Another interesting result is the finding that when autoimmune mice have been used for eliciting catalytic antibodies via the standard *anti*-hapten approach, a larger number of catalytic antibodies have been obtained [51]. This is further evidence of a relationship between and catalytic antibodies.

17.12
Conclusions

Polyclonal catalytic antibodies have been used to investigate a variety of reactions and to study the steric and electronic features required in the hapten design to elicit successful catalysis. It has been suggested that structural rigidity and hydrophobicity play a major role and that the less flexible the hapten is, the greater are the chances of generating high catalytic activity. All the published data support the view that when a well-designed immunogen is used, the catalytic antibody response is reproducible in all animals and is functionally homogenous. In all cases, single-site saturation behavior has been observed, and the kinetic parameters compare well with the ones obtained with monoclonal antibodies when the catalytically active [IgG] (1–8% of the total) is taken into account. Polyclonal antibodies are therefore an excellent tool for evaluating different immunogens designed to generate catalytic activity, especially if we consider the advantage that the entire immune system of individual animals is sampled.

The polyclonal approach allows monitoring of antibody activity, in terms of both hapten binding and catalysis, during the immunization process. More than one group has observed that development of hapten-binding ability precedes development of catalytic activity, and thus the former is not always a good indication of the latter. It would be interesting to extend this type of study using a greater variety of haptens and to complement it by using different animal systems including non-mammals such as chickens.

The study of polyclonal catalytic antibodies has already contributed significantly to the understanding of how the immune system works and of the requirements for successful hapten design. Recently, this has led to interesting developments in therapeutic applications, both by active and passive immunization. More work is required, however, to develop an effective system, and polyclonal catalytic antibodies provide a cheap and simple, yet powerful, tool for obtaining the information we need to achieve this effectively and reliably.

Acknowledgements

We thank the EC (ERBFM-RXCT 980 193) and the London Central Research Fund (equipment grant to MR) for financial support.

References

1 JENCKS, W. P., *Catalysis in Chemistry and Enzymology*, McGraw-Hill, New York **1969**, p. 1282–1320
2 RASO, V., STOLLAR, B. D., The Antibody-Enzyme Analogy. Characterization of Antibodies to Phosphopyridoxyltyrosine Derivatives, *Biochemistry* 14 (**1975**), p. 584–591
3 KOHEN, F., KIM, J. B., LINDNER, H. R., HESHHAR, Z., GREEN, B. S., Monoclonal immunoglobulin G augments hydrolysis of an ester of the homologous hapten, *FEBS Lett.* 111 (**1980**), p. 427–431
4 KOHLER, G., MILSTEIN, Continuous cultures of fused cells secreting antibody predefined specificity, *Nature (London)* 256 (**1975**), p. 495–497
5 TRAMONTANO, A., JANDA, K. D., LERNER, R. A., Catalytic antibodies, *Science* 234 (**1986**), p. 1566–1570
6 POLLACK, S. J., JACOBS, J. W., SCHULTZ, P. G., Selective chemical catalysis by an antibody, *Science* 234 (**1986**), p. 1570–1573
7 GALLACHER, G., JACKSON, C. S., TOPHAM, C. M., SEARCEY, M., TURNER, B. C., BADMAN, G. T., BROCKLEHURST, K., Polyclonal-antibody-catalyzed hydrolysis of an aryl nitrophenyl carbonate, *Biochem. Soc. Trans.* 18 (**1990**), p. 600–601
8 GALLACHER, G., JACKSON, C. S., SEARCEY, M., BADMAN, G. T., GOEL, R., TOPHAM, C. M., MELLOR, G. W., BROCKLEHURST, K., A polyclonal antibody preparation with Michaelian catalytic properties, *Biochem. J.* 279 (**1991**), p. 871–881
9 STEPHENS, D. B., IVERSON, B. L., Catalytic polyclonal antibodies, *Biochem. Biophys. Res. Commun.* 192 (**1993**), p. 1439–1444
10 IVERSON, B. L., CAMERON, K. E., JAHANGIRI, G. K., PASTERNAK D. S., Selective cleavage of trityl protecting groups catalyzed by an antibody, *J. Am. Chem. Soc.* 112 (**1990**), p. 5320–5323
11 ZHAO, H., JI, Y., YANG, B., YE, M., HU, Y., WU, Y. Polyclonal antibody-catalyzed hydrolysis of naproxen ethyl ester, *Chin. Sci. Bull.* 40(21) (**1995**), p. 1794–1796
12 HU, Y., YANG, B., ZHAO, H., WU, Y., JI, Y., YE, M., Enantioselective hydrolysis of naproxen ethyl ester catalyzed by polyclonal antibodies, *Chin. Sci. Bull.* 42(9) (**1997**), p. 741–744
13 YANG, B. H., KONG, H. K., JIANG, J. Q., JI, Y. Y., YEH, M., Polyclonal antibody-catalyzed hydrolysis of beta-naphthol acetate, *Chin. Chem. Lett.* 7(2) (**1996**), p. 123–126
14 LIU, D., YU, Y., WONG, W.-K. R., ZHAO, X., ZHANG, J., An enzyme-like polyclonal antibody capable of catalyzing ester hydrolysis, *Enzymol. Microb. Technol.* 20 (**1997**), p. 24–31
15 TAWFIK, D. S., ZEMEL, R. R., ARAD-YELLIN, R., GREEN, B. S., ESHHAR Z., Simple method for selecting catalytic monoclonal antibodies that exhibit turnover and specificity, *Biochemistry* 29 (**1990**), p. 9916–9921; JACOBS,

J. W., Schultz, P. G., Sugasawara, R., Powell, M., Catalytic Antibodies, *J. Am. Chem. Soc.* 109(7) (**1987**), p. 2174–2176

16 Meekel, A. A. P., Resmini, M., Pandit U. K., First example of an antibody-catalyzed hetero-Diels-Alder reaction, *Chem. Commun.* 5 (**1995**), p. 571–572; Meekel, A. A. P., Resmini M., Pandit U. K., Regioselectivity and enantioselectivity in an antibody catalyzed hetero Diels-Alder reaction. *Bioorg. Med. Chem.* 4 (**1996**), p. 1051–1057; Resmini, M., Meekel, A. A. P., Pandit U. K., Catalytic antibodies: Regio- and enantioselectivity in a hetero Diels-Alder reaction. *Pure Appl. Chem.* 68 (**1996**), p. 2025–2028

17 Hu, Y.-J., Ji, Y.-Y., Wu, Y.-L., Yang, B.-H., Yeh, M., Polyclonal catalytic antibody for hetero-cycloaddition of hepta-1,3-diene with ethyl glyoxylate – an approach to the synthesis of 2-nonulosonic acid analogs, *Bioorg. Med. Chem. Lett.* 7(13) (**1997**), p. 1601–1606

18 Yang, B.-H., Zhao, J.-J., Chen G.-N., Ji, Y.-Y., Yeh, M., Polyclonal antibody with catalytic activity for Oxy-Cope rearrangement, *Chin. J. Org. Chem.* 20(5) (**2000**), p. 726–730

19 Barbas, C. F., Heine, A., Zhong, G. F., Hoffmann, T., Gramatikova, S., Bjornestedt, R., List, B., Anderson, J., Stura, E. A., Wilson, I. A., Lerner, R. A., Immune versus natural selection: antibody aldolases with enzymic rates but broader scope, *Science* 278 (**1997**), p. 2085–2092

20 Gallacher, G, Jackson, C. S., Searcey, M., Goel R., Mellor, G. W., Smith, C. Z, Brocklehurst, K., Catalytic antibody-activity elicited by active immunization – evidence for natural variation involving preferential stabilization of the transition-state, *Eur. J. Biochem.* 214 (**1993**), p. 197–207

21 Wilmore, B. H., Iverson, B. L., Phosphate versus phosphorothioate haptens for the production of catalytic polyclonal antibodies, *J. Am. Chem. Soc.* 116 (**1994**), p. 2181–2182

22 Blackburn, G. M., Kingsbury, G., Jayaweera, S., Burton, D. R., in *Catalytic Antibodies*, Ciba Foundation Symposium 159, John Wiley and Sons, Chichester **1991**, p. 211–226

23 Wallace, M. B., Iverson, B. L., The influence of hapten size and hydrophobicity on the catalytic activity of elicited polyclonal antibodies, *J. Am. Chem. Soc.* 118 (**1996**), p. 251–252

24 Odenbaugh, A. L., Helms, E. D., Iverson, B. L., An investigation of antibody acyl hydrolysis catalysis using a large set of related haptens, *Bioorg. Med. Chem.* 8 (**2000**), p. 413–426

25 Lindner, A. B., Eshar, Z., Tawfik, D. S., Conformational changes affect binding and catalysis by ester-hydrolysing antibodies, *J. Mol. Biol.* 285 (**1999**), p. 421–430

26 Resmini, M., Vigna, R., Simms, C., Barber, N. J., Hagi-Pavli, E. P., Watts, A. B, Verma, C., Gallacher, G., Brocklehurst, K., Characterization of the hydrolytic activity of a polyclonal catalytic antibody preparation by pH-dependence and chemical modification studies: evidence for the involvement of Tyr and Arg side chains as hydrogen-bond donors, *Biochem. J.* 326 (**1997**), p. 279–287.

27 Stephens, D. B., Thomas, R. E., Stanton, J. F., Iverson B. L., Polyclonal antibody catalytic variability, *Biochem. J.* 332 (**1998**), p. 127–134

28 Shreder, K., Harriman, A., Iverson, B. L., Polyclonal antibodies elicited via immunization with a ru(bpy)(3)(2+)-methyl viologen conjugate – is a polyclonal antibody immune-response always heterogeneous, *J. Am. Chem. Soc.* 117 (**1995**), p. 2673–2674

29 Gallacher, G., Searcey, M., Jackson, C. S., Brocklehurst, K., Polyclonal antibody-catalyzed amide hydrolysis. *Biochem. J.* 284 (**1992**), p. 675–680

30 Ostler, E. L., Resmini, M., Boucher, G., Romanov, N., Brocklehurst, K., Gallacher, G., Polyclonal antibody-catalyzed hydrolysis of a β-lactam, *Chem. Commun.* 3 (**2002**), p. 226–227

31 Gallacher, G., et al., Unpublished results

32 Resmini, M., Gul, S., Carter, S., Sonkaria, S., Topham, C. M., Gallacher, G., Brocklehurst, K., A general kinetic approach to investigation of active-site availability in macromolecular catalysts, *Biochem. J.* 346 (2000), p. 117–125; Topham, C. M., Gul, S., Resmini, M., Sonkaria, S., Gallacher, G., Brocklehurst, K., The kinetic basis of a general method for the investigation of active site content of enzymes and catalytic antibodies: first order behaviour under single turnover and cycling conditions, *J. Theoret. Biol.* 204 (2000), p. 239–256; Brocklehurst, K., Resmini, M., Topham, C. M., Kinetic and titration methods for determination of active site contents of enzyme and catalytic antibody preparations, *Methods* 24 (2001), p. 153–167

33 Tellier, C., Exploiting antibodies as catalysts: potential therapeutic applications, *Transfus. Clin. Biol.* 9 (2002), p. 1–8

34 Basmadjian, G. P., Singh, S., Sastrodjojo, B., Smith, B. T., Avor, K. S., Chang, F., Mills, S. L., Seale, T. W., Generation of polyclonal catalytic antibodies against cocaine using transition-state analogs of cocaine conjugated to diphtheria toxoid, *Chem. Pharm. Bull.* 43(11) (1995), p. 1902–1911

35 Landry, D. W., Zhao, K., Yang, G. X. Q., Glickman, M., Gerogiadis, T. M., Antibody-catalyzed degradation of cocaine, *Science* 259 (1993), p. 1899–1901

36 Renard, P.-Y., Vayron, P., Taran, F., Mioskowski, C., Design and synthesis of haptens for antibody catalyzed hydrolysis of organophosphorus nerve agents *Tetrahedron Lett.* 40 (1999), p. 281–284; Vayron, P., Renard, P.-Y., Taran, F., Creminon, C., Frobert, Y., Grassi, J., Mioskowski, C., Toward antibody-catalyzed hydrolysis of organophosphorus poisons, *Proc. Natl. Acad. Sci. USA* 97(13) (2000), p. 7058–7063

37 Wang, J., Han, Y., Wilkinson, M. F., An active immunization approach to generate protective catalytic antibodies, *Biochem. J.* 360 (2001), p. 151–157; Wang, J., Han, Y., Liang, S., and Wilkinson, M. F., Catalytic antibody therapy against the insecticide carbaryl, *Biochem. Biophys. Res. Commun.* 291 (2002), p. 605–610

38 Jerne, N. K. Towards a network theory of the immune system, *Ann. Immunol.* 125C (1974), p. 373–389

39 Pan, Y., Yuhasz, S. C., Amzel, L. M., Antiidiotypic antibodies – biological function and structural studies, *FASEB J.* 9 (1995), p. 43–49

40 Izadyar, L. Friboulet, A., Remy, M. H., Roseto, A., Thomas, D., Monoclonal antiidiotypic antibodies as functional internal images of enzyme active-sites – production of a catalytic antibody with a cholinesterase activity, *Proc. Natl. Acad. Sci. USA* 90 (1993), p. 8876–8880

41 Hu, R., Xie, G. Y., Zhang, X., Guo, Z. Q., Jin, S., Production and characterization of monoclonal *anti*-idiotypic antibody exhibiting a catalytic activity similar to carboxypeptidase A, *J. Biotechnol.* (1998), p. 109–115

42 Lefevre, S., Debat, H., Thomas, D., Friboulet, A., Avalle, B., A suicide-substrate mechanism for hydrolysis of β-lactams by an *anti*-idiotypic catalytic antibody, *FEBS Lett.* (2001), p. 25–28; Avalle, B., Thomas, D., Friboulet, A., Functional mimicry: elicitation of a monoclonal *anti*-idiotypic antibody hydrolyzing β-lactams, *FASEB J.* 12 (1998), p. 1055–1060

43 Joron, L., Izadyar, L., Friboulet, A. Remy, M.-H., Pancino, G., Roseto, A., Thomas, D., Antiidiotypic antibodies exhibiting an acetylcholinesterase abzyme activity, *Ann. N.Y. Acad. Sci.* 672, (1992), p. 216–223

44 Crespeau, H., Laouar, A., Rochu, D., Polyclonal DNAse abzyme formed by the antiidiotypic internal image method, *C. R. Acad. Sci. Ser. Iii-Sciences de la Vie-Life Sciences* 317, (1994), p. 819–823

45 Avalle, B., Debat, H., Friboulet, A., Thomas, D., Catalytic mechanism of an abzyme displaying a β-lactamase-like activity, *Appl. Biochem. Biotechnol.* 83 (2000), p. 163–169

46 Paul, S., Mei, S., Mody, B., Eklund, S. H., Beach C. M., Massey, R. J., Hamel, F., Cleavage of vasoactive-intestinal-peptide at multiple sites by autoantibodies, *J. Biol. Chem.* 266 (**1991**), p. 16128–16134

47 Paul, S., Li, L., Kalaga, R., Wilkinsstevens, P., Stevens, F. J., Solomon, A. Natural catalytic antibodies – peptide-hydrolyzing activities of Bence-Jones proteins and v-l fragment. *J. Biol. Chem.* 270 (**1995**), p. 57–61; Sun, M., Gao, Q. S., Kirnarskiy, L., Rees, A., Paul, S., Cleavage specificity of a proteolytic antibody light chain and effects of the heavy chain variable domain. *J. Mol. Biol.* 271 (**1997**), p. 374–385

48 Shuster, A. M., Gololobov, G. V., Kvashuk, O. A., Bogomolova, A. E., Smirnov, I. V., Gabibov, A. G., DNA hydrolyzing autoantibodies. *Science* 256 (**1992**), p. 665–667

49 Buneva, V. N., Andrievskaya, O. A., Romannikova, I. V., Gololobov, G. V., Yadav, R. P., Yamkovoi, V. I., Nevinskii, G. A., Interaction of catalytically active antibodies with oligoribonucleotides, *Mol. Biol. (Moscow)* 28 (**1994**), p. 738–743

50 Debat, H., Avalle, B., Chose, O., Sarde, C. O., Friboulet, A., Thomas, D., Overpassing an aberrant V(kappa) gene to sequence an *anti*-iciotypic abzyme with (beta)-lactamase-like activity that could have a linkage with autoimmune diseases *FASEB J.* 15 (**2001**), p. 815–822

51 Takahashi, N., Kakinuma, H., Hamada, K., Improved generation of catalytic antibodies by MRL/MPJ-Ipr/Ipr autoimmune mice, *J. Immunol. Methods* 235 (**2000**), p. 113–120

18
Production of Monoclonal Catalytic Antibodies: Principles and Practice
Diane Kubitz, Ehud Keinan

18.1
Introduction

Antibody catalysis is an experimental science. Although success in this field is dependent on conceptual insight and creative design, success reflects the quality of the experimental procedures. There are many conceptual steps on the way toward the realization of a new antibody catalyst, including mechanistic understanding of the specific reaction to be catalyzed, scholarly prediction of the transition state of highest energy, design of a chemically stable transition state analog, and planning of synthetic schemes for hapten, substrates, products, etc. Yet, the critical components of the entire effort are the experimental procedures, such as organic synthesis, immunization, screening, production of monoclonal antibodies, kinetic experiments, crystallographic studies, etc. Accordingly, we found it appropriate to devote one chapter in this book to the experimental work that is needed for the general production and purification of catalytic monoclonal antibodies via the hybridoma technology. No attempt is made here to cover any aspect of organic synthesis and kinetic studies which are specific to the chosen reaction to be catalyzed. We have focused on the experimental procedures that follow conjugation of the appropriate hapten to the carrier protein.

Although hybridoma technology is a mature and well-established method, it involves a variety of independent variables and steps. Each step can be carried out in many different ways, and the diversity of the published approaches reflects specific biological problems and experimental tradition that characterize any given laboratory. The literature offers a broad variety of methods, which differ from one another in speed, convenience, reproducibility, and cost. There are no right or wrong approaches, because every laboratory chooses the published strategies and adapts them to their specific needs. In this chapter we describe the methods of producing catalytic antibodies that have been used continuously over the past two decades in our laboratories at The Scripps Research Institute, without claiming that these methods are superior to others. For example, while the experimental literature offers many different immunization procedures with various antigen doses, time frames, and adjuvants, this account covers the procedures that have been working for us success-

Catalytic Antibodies. Edited by Ehud Keinan
Copyright © 2005 WILEY-VCH Verlag GmbH & Co. KGaA, Weinheim
ISBN: 3-527-30688-9

fully over the years. These procedures are based on a number of leading references dealing with the production of monoclonal antibodies [1–3] and their purification [4, 5], as well as on the accumulating experience. Expectedly, the significant advance of chemical immunology within the past 15 years is reflected by the major improvements of protocols since an earlier description of these procedures was published in 1989 [6].

The hybridoma library obtained from immunization against a given hapten comprises a diverse population of binding proteins. An efficient method of screening this library is one of the keys to success in the field of antibody catalysis. The screening issue is not less important than either the understanding of the mechanistic details of the given reaction to be catalyzed or the rational design of hapten. The critical importance of good screening methods is highlighted by J.-L. Reymond in Chapter 8 of this book. Unlike strategies of rational design, including *de novo* synthesis of biocatalysts, as well as the preparation of organic and organometallic catalysts, the catalytic antibodies approach is characterized by a significant element of unpredictability. There are many facets of biocatalysis that are not yet fully understood, particularly those related to protein dynamics along the reaction coordinate. Even when some of this is understood, we do not yet know how to design the antibody active site accordingly. Therefore, the use of a transition state analog, even an optimal one, which is often impossible to make, represents only a "snapshot" of a continuous, dynamic process and is only a general guideline for the immune system. The resultant broadly diverse population of antibodies reflects the variety of ways in which the immune system can respond to this transition state analog. The beauty of this approach is the element of unpredictability, which is reminiscent of browsing scientific journals – a method by which one can find valuable items that were not looked for. In contrast, other methods, such as the imprinted polymers methodology, at their best provide the experimenter with what he or she has been searching for. Transition state binding is a parameter that can be well designed by immunization with an appropriate hapten. However, since there are several other parameters that lead to biocatalysis, the best hapten binder in a given antibody library may not necessarily be the best catalyst. Thus, although screening for binding is convenient and well established by traditional immunological methods, one should remember that in many ways it seems like searching for a lost coin under the street lamp. A much more efficient approach would be the screening for actual catalysis. Several methods have been used for such screening, including catElisa and the use of chromogenic and fluorogenic assays, all of which are covered by Chapter 8 in this book. The screening methods described below represent those more generally used for hapten binding. These are the basics upon which improvements and screening for catalysis may be designed.

18.2
Immunization

Not all hapten conjugates will generate a strong antibody response even if the exact immunization protocol is followed. Variables that can influence the *anti*-hapten response include the length of the spacer group or linker, the hapten density on the carrier protein, the choice of carrier protein, the stability of the conjugate, the strain of mice, and the adjuvants used. We generally use 8–12 week old mice for immunization. These 129GIX+ mice have routinely given us the best response to small organic hapten and peptide conjugates. However, we have also successfully used Balb/c, Swiss Webster, A/J, and C57 mice. We use RIBI's Adjuvant System (RAS) MPL+TDM (# R700, Corixa), which gives as good a response as Complete Freunds Adjuvant (CFA) and sometimes better, and has none of the adverse side affects associated with CFA. We have also been very successful using ALUM (#77161, Pierce) for boosts.

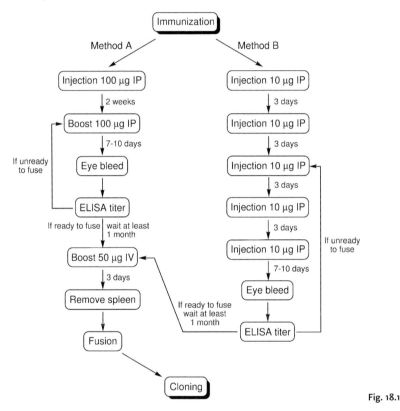

Fig. 18.1

In general, we immunize between 4 and 8 mice per hapten using two different immunization strategies in an attempt to get a stronger response to weaker immunogens. The first method is the one we have used for many years (Fig. 18.1). Each mouse receives a solution of 100 µg hapten-KLH conjugate in 100 µL phosphate-buffered saline (PBS: 10 mM sodium phosphate, 160 mM sodium chloride, pH 7.4) mixed with an equal volume of RAS reconstituted in PBS. The mice are injected intraperitoneally (IP) using a 23 gauge needle. After two weeks, they are given a second injection exactly similar to the first (100 µg per mouse in RAS, IP). Approximately 7–10 days later, the mice are bled and the serum titer determined by ELISA. If the titer (dilution giving 50% of the maximum absorbance) is less than 6400, the mice are given additional boosts of 50–100 µg after 2–3 weeks until the titer is 12 800 to 25 600 or more. Once a sufficient titer is reached, we wait at least 1 month for the circulating antibody levels to decrease before giving the mouse a final injection of 50 µg KLH conjugate in 100 µL PBS intravenously (IV) in the lateral tail vein. The spleen is removed 3 days later for fusion.

The second method, which we have been using additionally throughout the last year, uses lower antigen amounts, with more frequent injections. 10 µg hapten conjugate in 50 µL PBS per mouse mixed with an equal volume of RAS is injected IP, and this is repeated every 3 days (4 more times) over a 2 week period. About 7–10 days later, the mice are bled, and the serum titer is determined by ELISA. If it is found that additional boosts are required, the mice are boosted with three 10 µg boosts, 3 days apart, in RAS. As with the higher-dose immunization, once a sufficient titer is reached, we wait at least 1 month for the circulating antibody levels to decrease before giving each mouse a final injection of 50 µg KLH conjugate in 100 µL PBS intravenously (IV) in the lateral tail vein. The spleen is removed 3 days later for fusion.

18.3
Hybridoma Production and Screening

At least one week prior to the fusion, we thaw a fresh vial of myeloma cells. It is critical that the myeloma cells be healthy and in log phase on the day of the fusion in order for it to be successful. Cells that have been growing too long and have come out of log phase or cells that have not been growing long enough to achieve log growth will not fuse optimally. We try to keep the cell density between 3 and 6×10^5/mL. It is important to choose a myeloma line that has been selected to be a non-producer of IgG. We use the X63-Ag8.653 line because it is a non-producer and has high fusion efficiency.

The spleen cells from the hyper-immunized mouse and myeloma cells are washed 3 times with 30 mL RPMI 1640 media (supplemented with 2 mM L-glutamine, 1 mM sodium pyruvate, 10 mM HEPES, and 50 µg/mL gentamycin) and mixed together in a 5:1 ratio (spleen:myeloma) in a 50 mL conical tube. After the final spin, it is important to aspirate the media completely, then to spread the pellet out by gently tapping the tube. It is also important that the cells be spread along the bottom edge of the conical so that all the cells have equal access to the PEG. Addition of 1 mL 50% PEG 1500 (BMB #783-641) that is pre-warmed to 37 °C is done drop-wise over 1 min using a 1 mL syringe and an 18 g needle, while gently rotating and tapping the tube to re-suspend the cells. The PEG is then slowly diluted out with 1 mL RPMI-1640 media over 1 min, and then 8 mL over 2 min. The cell membranes are very fragile at this point. The cells are then placed in a 37 °C water bath for 10 min and centrifuged. The supernatant is decanted and the cells are gently re-suspended in 5 mL complete media (supplemented RPMI-1640 with 10% FCS). The cells are again placed in a 37 °C water bath for 10 min and are then added to 225 mL HAT media (RPMI-1640 supplemented with 2 mM L-glutamine, 1 mM sodium pyruvate, 10 mM HEPES, 50 μg/mL gentamycin, 10% FCS, 0.1 mM hypoxanthine, 0.4 μM aminopterin, and 16 μM thymidine). They are plated into fifteen 96-well plates (Corning 3596, 150 μL/well). The fusion is fed 50 μL/well with HT media (RPMI-1640 supplemented with 2 mM L-glutamine, 1 mM sodium pyruvate, 10 mM HEPES, 50 μg/mL gentamycin, 10% FCS, 0.1 mM hypoxanthine, and 16 μM thymidine) on days 4, 8, and 12. By this time, macroscopic colonies should be seen. We generally get growth in 40–60 % of the wells, so we assay for antigen binding directly from the 96 well plates.

Having a sensitive enzymatic assay that could detect catalytic activity at the antibody concentrations in cell supernatant (1–20 μg/mL) would greatly facilitate the screening process and eliminate the labor-intensive procedures of subcloning, producing, and purifying large numbers of antigen binders that are non-catalytic. If it is possible to determine background rates and sensitivity levels of the assay ahead of time so that screening is done directly from the 96-well plate, this greatly increases the chances of finding catalysts. We initially screen 600–900 hybridomas for binding, but only a small subset of these is carried through subcloning and large-scale antibody production and purification to be tested for catalytic activity. As the hybridomas are expanded and moved from the 96-well plate to the 48-well, 24-well, etc., we continually monitor their binding by ELISA to the hapten coupled to BSA. We characterize the hybridomas by titer and isotype, keeping only IgG subclasses that have stable titers of 16 or above. We also rank them by their affinity for related ligands and substrates using a competitive inhibition assay.

Our ELISA procedure can be modified or adapted if necessary, depending upon the nature of the hapten being used. Since we regularly immunize with a KLH-conjugate, it is important to have another carrier protein for screening purposes. BSA is easy to conjugate and does not aggregate and precipitate easily, so that the plate is evenly coated. The hapten-BSA conjugate is diluted to 50 μg/mL in PBS and plated at 25 μL/well into a 96-well 1/2-area EIA plate (Corning 3696) and allowed to dry overnight at 37 °C. The antigen is fixed to the plate with 50 μL/well methanol for 5 min. The methanol is then shaken out and the plates allowed to air dry for 5–10 min. Non-specific binding sites are blocked with 50 μL/well of "BLOTTO" (5% w/v non-fat powdered milk in PBS) for 30 min at room temperature. The excess BLOTTO is shaken out, and the primary antibody is immediately added. The plates should not be allowed to dry out from this point on, because this may lead to non-specific binding. Addition of 25 μL/well of the primary antibody diluted in BLOTTO, (1:1 for tissue culture supernatant, 1:100 for mouse sera, and 1:1000 for ascites) is followed by incubation in a moist chamber for 1–2 h at 37 °C. The plates are washed 10 times with de-ionized water using a showerhead. It is recommended to alternate the direction of wash in order not to miss any well. The excess water is shaken out and 25 μL/well of the secondary antibody are added, followed by goat *anti*-mouse-horseradish peroxidase conjugate (Southern Biotech) diluted 1:2000 in BLOTTO, and the plates are then incubated for 1 h at 37 °C in a moist chamber. When isotyping the hybridomas, we use the HRP clonotyping kit (Southern Biotech) with conjugates specific for each of the murine heavy and light chains. The plates are washed 10 times with de-ionized water and 50 µl/well of the developer (30 mL 0.1 M citrate buffer, pH 4.0, with 9 µL 30% H_2O_2, 200 µL 45 mg/mL ABTS) is added. The plates are then read in an ELISA reader (Molecular Devices V-max kinetic plate reader) at 414 nM after 30 min.

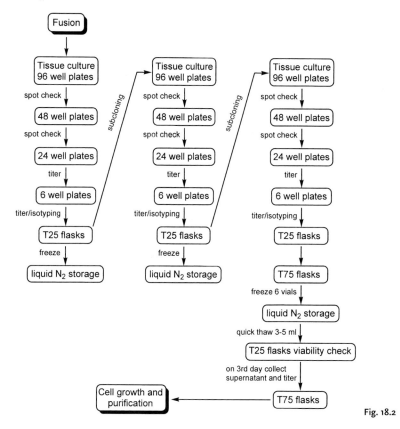

Fig. 18.2

The hybridomas are subcloned through at least 2 generations to guarantee monoclonality (Fig. 18.2). We subclone by limiting dilution sometimes 3 or 4 times if the hybridoma does not appear to be stable. Newly established hybridomas that have not been growing long in culture do not grow well when plated out at limiting dilutions, but it is important to start subcloning the hybridomas as soon as possible to avoid the cells being taken over by non-producers. We take both of these issues into consideration by titering the individual wells repeatedly while they are being expanded, so that we can tell when they are beginning to lose activity. In this way we can wait to subclone until the cells are dividing well in culture. We generally have a large number of hybridomas to subclone, so we clone by limiting dilution without counting the cells. We make a series of dilutions that result in plating out approximately 0.5 and 5 cells per well in 200 µL in a 96 well plate (48 wells at each density). It takes 10–14 days for macroscopic colonies to show up. We pick 8–16 different clones per hybridoma to move up into a 48 well (depending upon the initial titer of the hybridoma), trying to take wells that appear to have only one clone. At this point many of the clones will not be positive by ELISA. We pick the 6 best to move up to a 24-well plate, and at this point we titer all the subclones. We recheck the titer and isotype at the 6-well stage and pick the best one to freeze down and subclone again. The whole process is repeated a second and third time until all clones from a 96-well plate show positive binding, with identical isotypes and equal titers. At this point we feel relatively secure that we have a monoclonal stable line, and we do a large-scale multiple freeze-down of the hybridoma with 6 identical aliquots of cells.

In order to have a frozen stock of cells with greater than 95% viability it is important to have the cells dividing well and in log phase. We freeze the cells when we have 50 mL of 4 to 6×10^5 cells/mL. The cells are pelleted, re-suspended in 3 mL of freezing media (90% fetal calf serum and 10% DMSO), and aliquoted 0.5 mL per cryule vial. The vials are placed immediately into a cryo-container (Nalge 5100-0001) at -80 °C, which allows for a controlled, slow rate of freezing (1 °C per min). After 4–6 hours the cells are then placed in a liquid nitrogen storage tank. After 2–3 days we thaw one of the vials to make sure it is viable and not contaminated. The vials are thawed quickly in a 37 °C water bath and the cells placed in 10 mL of complete RPMI-10% FCS. The cells are then pelleted to remove any traces of the DMSO and then re-suspended in 5 mL media. We culture the cells for at least 48 h to make sure they are viable and start dividing, and then we test the supernatant for antibody titer. Once we have the cloned hybridoma cells backed up in liquid nitrogen, we can start large-scale antibody production.

18.4
Large-Scale Antibody Production

There are a number of different techniques that can be used to produce the monoclonal antibodies, depending upon the amount of antibody required, the facilities available, the time frame, and the available budget. In the past we have grown our hybridomas exclusively in mice for ascites production. We have made an effort in recent years to find alternatives to ascites production in order to reduce painful procedures using animals. At this time we only use ascites when we have tried a few different methods and have been unsuccessful in producing our antibody *in vitro*.

If we need 30 mg or less of purified antibody we produce it in T-flasks (Fig. 18.3). Hybridoma cells will typically produce 2–50 µg/mL of antibody, depending upon the isotype of the clone and the cell density. Antibody levels can be maximized by growing up to 500–1000 mL of cells and allowing them to grow until they start dying (sometimes 10–14 days). The supernatant is then collected, centrifuged, filtered through a 0.2 µm filter, concentrated by ultra-filtration, and purified. We have been able to obtain as much as 60 mg this way, but 30–40 mg is the average. On traditionally low producers like IgG3 antibodies, the yield does not exceed 20 mg.

For larger amounts of antibody we have been very successful in growing the hybridomas in Integra Bioscience's CELLine flasks. Once established, we routinely get 15–20 mg/harvest from each CL-1000 flask, and the flasks can be grown for 2–3

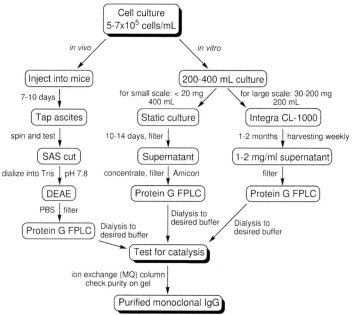

Fig. 18.3

months and harvested once or twice per week, resulting in 150–300 mg of antibody. When we need 1–2 g antibody, we simply inoculate 5–6 CL-1000s. The initial cost of the *in vitro* antibody production in the CELLine flasks higher than in antibody production in ascites, but, barring any contamination, the flasks can be grown for months with continual harvests, bringing the overall cost per mg of antibody down. Purification of the antibody is easier as well.

Before starting any large-scale antibody production it is critical that the hybridomas are well cloned with a stable titer and are in the log phase. When inoculating a CL-1000 flask, we have had the best results using 200 mL of 5 to 7×10^5 cells/mL, ($\sim 1 \times 10^8$ cells). The cells are centrifuged at 1200 rpm and re-suspended in 15 mL of complete RPMI supplemented with 20% FCS, 50 I.U./mL penicillin, and 50 µg/mL streptomycin. One liter of complete RPMI, with "pen-strep" (but without FCS)" is placed in the basal chamber of the flask. Then the 15 mL of cells are pipetted into the cell compartment, removing any trapped air bubbles. After one week we check the viability of the cells with trypan blue. If the viability is less than 50%, we generally grow up more cells to re-inoculate. If viability is 50% or greater, we will do a partial harvest of the cells (the exact amount will vary with the total number of cells). We try to put $\sim 5 \times 10^7$ viable cells in 15 mL of 20% FCS back into the cell compartment. The rest of the harvested cells are spun, and the supernatant is collected and filtered through a 0.2 µM filter and purified. We change the basal media (one liter of complete RPMI with pen-strep without FCS) every week and harvest the cell compartment. After approximately 4 weeks we can usually harvest the cell compartment twice per week, resulting in close to 30–40 mg of antibody per week for good producers.

Occasionally we find a cell line that will not adapt to growing at high densities in the CL-1000. If we need to produce over 50 mg we will use ascites production. Since we generally immunize 129GIX+ mice and the myeloma cell line we use (X63 Ag8.653) was isolated from Balb/c mice, we need make our ascites in a hybrid strain, Balb/c X 129GIX+. The mice are primed with 0.2 mL of pristane (2,6,10,14-tetramethyl pentadecane, Aldrich) intraperitoneally at least one week prior to injecting the cells. An amount of 50 mL of 5×10^6 cells/mL are spun, washed once in PBS, and re-suspended into 2 mL of PBS. We then inject 500 mL of cells per mouse intraperitoneally into 4 mice. After 7–14 days the ascites should begin to form. The mice can be "tapped" sometimes as many as 3 times in a 7 day period, resulting in 10–20 mL of ascites per mouse. The ascites fluid is centrifuged at 3000 RPM for 10 min to remove the blood cells and stored at −20 °C until it is purified. We get between 0.5 and 15 mg of antibody per mL of ascites depending on the hybridoma. We have found that most IgG1s produce 10–15 mg/mL, IgG2as produce 5–7 mg/mL, IgG2bs produce 2–4 mg/mL, and IgG3s produce 0.5–2 mg/mL.

18.5
Antibody Purification

For the purification of antibodies made in T-flasks or CL-1000s, we affinity purify the supernatant over Protein G (Fig. 18.3). We use Gammabind™ plus Sepharose™ (Amersham Pharmacia Biotech AB, 17-0886-04) and Hi Trap Protein G HP, 5 mL pre-packed columns (Amersham Pharmacia Biotech AB, 17-0405-01), depending upon the amount of antibody being purified and the machine (FPLC or AKTA Prime) being used. We bind the antibody in PBS, pH 7.4, and wash with 3 column volumes to make sure all unbound proteins are removed. We then elute with 0.1 M acetic acid (pH 3.0) and neutralize the eluted antibody with 1M TRIS, pH 9.0. The antibody is then dialyzed into PBS pH 7.4, concentrated by ultra-filtration, sterile filtered, and stored at 4 °C. For long-term storage, we aliquot the antibody and freeze it at −20 °C.

For purification of ascites, we start by precipitating the antibody using saturated ammonium sulfate (SAS). We slowly add the ice-cold SAS dropwise to the ascites while stirring on ice and let it sit on ice for at least 15–20 min, then centrifuge at 9000–10000 rpm for 15 min. Doing a 50% cut gives a slightly higher antibody yield, but it can precipitate additional serum proteins as well. A 45% cut is generally cleaner, but more antibodies are left behind. The pellet is re-suspended in PBS and extensively dialyzed to remove the excess salt. We switch the dialysis buffer from PBS to 50 mM TRIS, pH 7.8 (or 8.0 for IgG2as), and for our second purification step we use an ion-exchange column. We use DEAE- Sephacel™ (Amersham Pharmacia Biotech AB, 17-0500-01), which has a strong positive charge at pH 7.8 and 8.0. Since immunoglobulins are the most basic of all serum proteins (with isoelectric points between 6 and 8), at pH 7.8 the IgG will be eluted first. We elute using a step gradient of increasing NaCl concentration, (50 mM, 75 mM, 100 mM, 150 mM, 250 mM, and 500 mM). We test all fractions by ELISA and concentrate the antibody-containing ones by ultra-filtration. The antibody is then dialyzed into PBS to be affinity purified on Protein G (as above).

18.6
Testing for Catalytic Activity

Once we have made a panel of purified antibodies, we dialyze them into the appropriate buffer for each catalytic reaction. Most antibodies will tolerate a pH of between 5 and 9 without too much trouble. If it is necessary to use a buffer without any salt, it helps to keep the antibody in solution if the molarity of the buffer is increased to 100 mM or more. For pH extremes or low-ionic-strength buffers, we dialyze a very small amount of antibody first using a slide-a-lyzer (Pierce, 66415). This lets us know if our antibodies will tolerate the chosen buffer system. If we see any precipitation of the antibody, we have the opportunity to modify the buffer without losing our entire antibody.

Once we have identified potential catalysts, we purify the antibody through additional columns to rule out any enzyme contamination. We will use Source S and Source Q anionic and cationic exchange columns (Amersham Pharmacia Biotech AB) with a linear salt elution, and gel-filtration columns using HiPrep 26/60 Sephacryl S-200 High Resolution (Amersham Pharmacia Biotech AB, 17-1195-01). The purified antibody is analyzed by SDS-PAGE to check for contaminating proteins. If the antibody retains its catalytic activity throughout all of the purification tests and can be inhibited, we grow it up a second time, going through all of the purification steps, to make sure it is reproducible.

18.7
Preparation of Fab, F(ab')$_2$, and Fab' Fragments

Antibody fragments, mainly Fab (50 000 dalton), are prepared primarily for structure determination by X-ray crystallography. In some cases they may be prepared for catalysis when either a monovalent catalyst is required or in order to verify that catalysis occurred exclusively in the binding site.

Antibodies may be conveniently considered as having three protein domains, two identical antigen binders, Fab, and an effector domain, Fc. Antibodies can be fragmented by partial digestion with either papain or pepsin (Fig. 18.4) [4]. Papain treatment produces two Fab fragments and one Fc fragment. In contrast, pepsin treatment can be used to release two antigen binding domains still bound together, F(ab')$_2$. The different mouse subclasses vary in their susceptibility to enzyme digestion to form antibody fragments. Mouse IgG2a and IgG2b are easily digested with papain to form two Fab fragments and one Fc fragment (Fig. 18.4). As mouse IgG1s are more resistant to papain, we generally use pepsin to form the F(ab')$_2$, which can then be reduced by cysteine to form the Fab' fragment.

18.7 Preparation of Fab, F(ab')₂, and Fab' Fragments

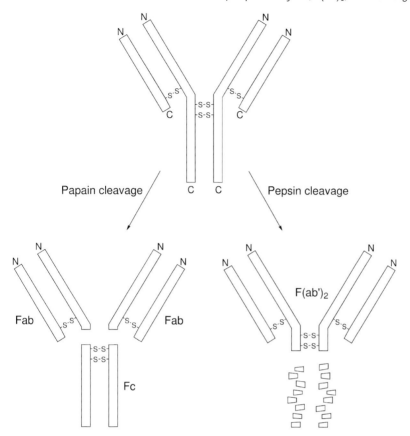

Fig. 18.4

Before starting any large-scale digestion, it is important to run small test digests, varying the cysteine concentration, enzyme concentration, and digestion time, to determine the optimal digestion conditions using all fresh reagents. The sample is dialyzed into sodium acetate (0.1 M, pH 5.5). When using mercuripapain (Sigma P-9886), the enzyme needs to be pre-activated for 30 min at 37 °C in Buffer A (100 mM sodium acetate, pH 5.5, 1 mM EDTA, 50 mM cysteine). For IgG2b we have found that 40 µg of papain per mg of antibody is a good starting enzyme concentration. For IgG2a we have been able to get complete digestion with as low as 20 µg of papain per mg of antibody.

When there is a good supply of antibody it saves time to set up 3 or 4 different tubes with different amounts of enzyme (20, 30, 40, and 50 µg) using 5 mg of antibody per test digest in 1 mL total volume. EDTA (1/100 volume of 100 mM) and 50 µl of pre-activated papain solution are then added. The tubes are sealed, mixed well, and placed in a 37 °C water bath. Aliquots (20 µl each) are taken at 0.5, 1, 2, 4, 6, and 24 h to determine the optimal digestion time. Iodoacetamide (2 µl, 0.75 M, 1/10 volume) is then added to inactivate the papain, and the mixture is incubated for 30 min at room temperature. Samples are stored at 4 °C until the digestion mixture is analyzed using SDS-PAGE. If the digestion of the IgG is still incomplete, the test digest is repeated with increased concentration of cysteine. When the optimal conditions are achieved, the digest reaction is scaled up.

To purify the Fc and Fab, the sample is first dialyzed into PBS and then purified on a Protein A column. The Fc (and any remaining IgG) are retained on the column while the Fab and enzyme are eluted. The Fc can then be eluted with 0.1 M acetic acid, pH 3.0. The crude Fab can then be loaded onto a protein G column and eluted with 0.1 M acetic acid, pH 3.0. All fractions are analyzed by SDS-PAGE. If the Fab is still not sufficiently pure, it is further purified on an IEF gel to determine its pI. Dialysis into the appropriate pH buffer is carried out before purification on either an anionic or cationic exchange column. In most cases we use 20 mM PB, pH 7.0, to bind the protein on a cationic exchange column (Source S) and elute it with a linear gradient from 0 to 500 mM NaCl in 20 mM PB, pH 7.0. Finally, the sample is dialyzed into PBS, pH 7.4.

For IgG1 antibodies, the sample is dialyzed into 0.1 M sodium acetate, pH 7.0, and then into 0.1 M sodium acetate, pH 3.8, immediately before digesting. Fresh pepsin (Sigma, P-6887) stock solution is prepared at 10 mg/mL in 0.1 M sodium acetate, pH 3.8. The test digestion is carried out at 37 °C with 5 mg/mL antibody, various pepsin concentrations (1–4% by weight), and various incubation times (0.5, 1, 2, 4, 6, and 24 h). Samples of 20 µL each are taken at each time point, the reaction is stopped by adding 1/10 volume of 2 M TRIS pH 8.0, and the degree of digestion is analyzed by SDS-PAGE. Too long a digestion time may result in decreased yields. If the digestion of IgG is incomplete, lowering of the pH may facilitate it. Once the digestion conditions are optimized, the digestion can be scaled up. The digestion mixture is dialyzed into PBS, pH 7.4, and loaded on a Protein A column to bind any undigested IgG. Immunopure A Binding Buffer (Pierce, 21001) will increase IgG1 binding to Protein A. This works with Immunopure Mouse IgG1 Mild Elution Buffer (Pierce, 21022). Once the IgG is removed, the mixture is loaded onto a Protein G column in PBS and eluted with 0.1 M acetic acid to remove the enzyme. At this point the mixture should contain only $F(ab')_2$ with a single band of 100 000 dalton on SDS-PAGE. If further purification is required, the pI of the sample should be determined, and it is then purified on an ion-exchange column. Most of our $F(ab')_2$s have had a pI of 7.5–8.0, and we have run a Source S cationic exchange column in 20 mM PB, pH 6.5. To generate the Fab', the sample needs to be reduced with cysteine. For that purpose, 100 mM cysteine solution is prepared in dH_2O. A set of experiments is carried out with the antibody sample (3 mg/mL) using multiple tubes with different cysteine concentrations (2.5–10 mM) and various times (1–4 h). The reaction is stopped by the addition of iodoacetamide (20 µl, 0.75 M) per 1 mL sample and the mixtures are analyzed by SDS-PAGE. After achieving satisfactory conditions, the reaction is scaled up and the resultant Fab' is then dialyzed into PBS, pH 7.4.

18.8
Conclusion

We are continually exploring new ways of making and identifying catalytic antibodies. These include, for example, improving our immunization procedures to elicit tighter binders with higher affinities for the substrates, testing out new media formulations and growth conditions to improve subcloning efficiencies, and developing more sensitive catalytic assays that can be used to detect catalytic activity from cell supernatant instead of purified antibody. All of these would shorten the time between hapten design and synthesis and identification of antibody catalysts.

References

1 GODING, J. E., *Monoclonal Antibodies: Principles and Practice*, 3rd Edition, Academic Press **1996**, Chapter 8, p. 141–191
2 HARLOW, E., LANE, D., *Antibodies: A Laboratory Manual*, Cold Spring Harbor Laboratory **1988**
3 PETERS, J. H., BAUMGARTEN, H. (eds), *Monoclonal Antibodies*, Springer-Verlag **1992**
4 GODING, J. E., *Monoclonal Antibodies: Principles and Practice*, 3rd Edition, Academic Press **1996**, Chapter 9, p. 192–233
5 PROTEIN PURIFICATION HANDBOOK, Amersham Pharmacia Biotech **1999**
6 TRAMONTANO, A., SCHLOEDER, D., *Production of Antibodies that Mimic Enzyme Catalytic Activity*, in: Methods in Enzymology, Langone, J. J., (ed), Vol. 178 (**1989**), p. 531–550

19
Natural Catalytic Antibodies – Abzymes
Georgy A. Nevinsky, Valentina N. Buneva

19.1
Abbreviations

Abs, antibodies; Abzs, abzymes or catalytically active antibodies; SLE, systemic lupus erythematosus; HT, Hashimoto's thyroiditis; MS, Multiple sclerosis; ODN, oligodeoxyribonucleotide; ON, oligonucleotide; SDS-PAGE, SDS-polyacrylamide gel electrophoresis; VIP, intestinal vasoactive peptide; hMBP, human myelin basic protein; AI, autoimmune; AD, autoimmune disease; AIDS, autoimmune deficiency syndrome; PK, protein kinase

19.2
Introduction

The possibility of the induction of catalytic Abs was initially suggested by Pauling in 1948; he noted similarities between the mechanisms of antibody-antigen recognition and the interaction of a transition state with an enzyme [1]. The first polyclonal Abz was reported in 1966 by Slobin [2]. This IgG was obtained by immunization of rabbits with a conjugate of *p*-nitrocarbobenzoxy with bovine serum albumin and hydrolyzed *p*-nitrophenyl acetate.

In 1969, Jencks proposed the hypothesis that Abs obtained during immunization with chemically stable transition state analogs could catalyze the corresponding chemical reactions [3]. In 1985, a general method for generating catalytic monoclonal Abzs against transition state analogs and a way to use these Abs to accelerate chemical reactions was first described [4]. One year later, two groups were able to produce the first monoclonal Abs with catalytic properties. These were generated against hapten analogs of the transition states for *p*-nitrophenylphosphorylcholine [5] or for monoaryl phosphonate esters [6, 7].

As mentioned above, later different natural catalytic Abs were detected in sera of patients with several autoimmune (AI) and viral pathologies as well as in healthy human milk. In this chapter, for simplicity and in order to distinguish natural catalytic

Catalytic Antibodies. Edited by Ehud Keinan
Copyright © 2005 WILEY-VCH Verlag GmbH & Co. KGaA, Weinheim
ISBN: 3-527-30688-9

Abs from those against transition state analogs, the latter will be termed "artificial abzymes".

Artificial Abzs catalyzing the hydrolysis of amides and esters, as well as reactions of cyclization [8–11] decarboxylation [12–14] lactonization [15] peroxidation [16], photochemical thymine dimer cleavage, bimolecular amide bond formation, and other reactions not known to be catalyzed by known enzymes have been described. A number of papers have described Abzs performing other specific functions directed against proteins, including formation of cyclic peptides [17], catalysis of peptidyl-prolyl *cis-trans* isomerization in protein folding [18], and development of a novel enzymatic activity cleaving the bacterial protein HPr [19]. Some Abzs have been described that require cofactors for activity similar to standard enzymes [20]. The field of artificial Abzs has been amply reviewed recently (see [21–24] and references therein for more detailed description of the relevant reactions).

The evolution of the technology of artificial Abzs during about the last two decades has led not only to the rapid development of direct approaches for the generation of Abs with specified properties, but also to the creation of strategies to revise the targeting specificity of individual Abzs. Such modifications of antigen binding specificity can be achieved genetically *in vitro* by the application of site-directed mutagenesis or genetic selection or screening (using approaches such as phage display, as detailed in other chapters). Alternatively, modification can be induced directly on purified antibody via selective chemical modification by the direct introduction of catalytic groups into the Ab combining site. Some studies describing these approaches are reported in [25–29] and references therein. The employment of the approaches has demonstrated that the substrate specificity (and/or the specific activity) of some artificial Abzs is comparable to or even higher than that of enzymes with the same catalytic activity [12, 30, 31]. The mechanistic basis for the activity of such Abzs is becoming well understood (see [32] and the discussion below).

19.3
Natural Catalytic Antibodies

In contrast to the artificially designed transition state-directed Abs described in the preceding section, an alternative means of developing catalytic Abs is through the study and purification of auto-Abs produced naturally in human sera. As noted by Suzuki, naturally occurring Abs can (a) be quickly and easily purified at low cost, (b) possess novel catalytic activities of interest, (c) be informative for the development of therapeutic strategies involving Abs administered in serum, and (d) be useful for the design of haptens for the production of artificial Abzs [24]. Other issues relevant to the comparison of artificial with natural and *in vitro* with *in vivo* Abz generation approaches are discussed in [33]. The present chapter collates rigid criteria indicating that efficient naturally occurring catalytic Abs do exist and describes methods of natural Abzs purification.

The first example of a natural Abz was an IgG which is found in bronchial asthma patients and which hydrolyzes intestinal vasoactive peptide [34], the second was an

IgG with DNase activity in SLE [35, 36], and the third was an IgG with RNase activity in SLE [37]. However, to be precise, the first natural Abz was reported in 1969, when Kulberg [38] found proteolytic activity in highly purified rabbit Abs, and the second was catalytic IgGs from the human sera [39] and complexes of mice IgG with antigen demonstrating superoxide dismutase activity [40]. But although these Abs were isolated by affinity chromatography and some of the criteria assigning catalytic activity to Abs (see Section 19.7) were studied, the authors [39–41] could not exclude the possibility that the catalytic activity was due to contaminating enzymes. A similar situation occurred with the human milk sIgA possessing protein kinase activity [42]; the existence of protein kinase Abzs using different special rigid criteria (see Section 19.7) was confirmed later [43].

The occurrence of auto-Abs is a feature of AI diseases. Abs to DNA, to nucleoprotein complexes, and to enzymes that participate in nucleic acid metabolism have been described: these also occur in certain infections [44–50]. There can be spontaneous induction of *anti*-idiotypic Abs elicited by a primary antigen, including some with catalytic activity, or a transition from polyreactive catalytic activity to autoantigen-directed activity.

Catalytic IgGs and/or IgMs hydrolyzing RNA [37, 51–55], DNA [51, 52, 54–56], peptides and proteins [57, 58], or polysaccharides [59] have been described in the sera of patients with AI pathologies including SLE [37, 54, 55], Hashimoto's thyroiditis and polyarthritis [53, 56, 57], rheumatoid arthritis [58, 59], multiple sclerosis [52], lymphoproliferative diseases [52, 60, 61], polynephritis and malignant tumors [57, 59], and two viral diseases, viral hepatitis [51] and acquired immunodeficiency syndrome [62]. We could not detect DNase or RNase activities in Abs from the sera of 50 normal men and women or from patients with influenza, pneumonia, tuberculosis, tonsillitis, duodenal ulcer, and some types of cancer [51, 52]. Some healthy patients demonstrated Abzs with low proteolytic [34] and polysaccharide-hydrolyzing activities [59]. We also detected low DNase and/or RNase activity of Abs from a small percentage of healthy donors, but these activities were usually very low – on the borderline of the sensitivity of the methods used.

Amidase and peptidase activities were found in IgGs from the sera of patients with rheumatoid arthritis [51, 52, 63], factor VIII-cleaving allo-Abs in the sera of patients with severe hemophilia [64], and DNA-hydrolyzing, amidolytic, and peptidolytic activities in Bence-Jones proteins from patients with multiple myeloma (reviewed in [65]). The recent identification in multiple myeloma patients of an Ab light chain which cleaves the HIV protein gp120 demonstrates that natural Abzs are not restricted to autoantigenic substrates [66].

Natural Abzs also occur in the milk of normal human mothers, including sIgA and/or IgG possessing DNase and RNase [67–70] or nucleotidase activities [71], and protein kinase [42, 43, 72, 73] and lipid kinase activities [74, 75], which are the first examples of natural Abzs with synthetic activity. These Abzs show specific activities significantly higher than those found in most ADs. This is in agreement with the results of Neustroev's group [59, 76], that the specific activities of milk amylolytic Abzs were 5–50 times higher than those of Abzs in AI patients.

The idea that Abzs are involved in natural immunity arose after the discovery, by Paul et al. [34], of Abzs which specifically hydrolyze VIP, which are currently the most studied natural Abzs [26, 77–79] and whose mechanism and functional roles have been reviewed [47, 48, 80]. Some interesting aspects of nucleophilic proteolytic Abs were discussed by Gololobov et al. [81, 82]. Other relevant issues, including comparisons of artificial with natural and *in vitro* with *in vivo* Abz generation and analytical approaches, have been discussed [24, 33, 49, 50, 83, 84].

19.4
Peculiarities of the Immune Status of Patients with Various Autoimmune Diseases

A special feature of autoimmune diseases (ADs) is high concentrations of auto-Abs (Abs to endogenous antigens) [85, 86]. The development of ADs is characterized by the spontaneous generation of primary Abs to proteins, nucleic acids (NA), and their complexes, followed by the generation of secondary Abs to the primary ones, etc. [44, 87, 88].

Concentrations of DNA and *anti*-DNA Abs are especially high in patients with systemic lupus erythematosus (36% of SLE patients) [61, 89, 90]. Increased concentrations of DNA and *anti*-DNA Abs were also found in the blood of patients with other different ADs: polymyositis (49%), MS (17%), primary Sjogren's syndrome (18%), AI thyroid diseases (23%), myasthenia gravis (6%), and rheumatoid arthritis (7%). AI hepatitis, [61, 89, 90], and also with lympho-proliferative [61] and some viral diseases (e.g., viral hepatitis and AIDS) [62]. Although some viral diseases like AIDS and viral hepatitis are not related to ADs, they significantly influence the immune status of patients. For example, the development of both viral hepatitis [91] and AIDS [62, 92] is accompanied by AI humoral and cellular reactions; as in the case of AD, tissue-specific and organ-non-specific Abs were found in the blood of patients with hepatitis.

In the case of bronchial asthma, vasoactive intestinal peptide (VIP), widely distributed in the central and peripheral nervous systems, acts as auto-antigen [34].

The mechanism of AI damage of the thyroid gland has not been adequately studied, and its pathogenesis remains unclear; different Abs are believed to play a significant role in pathogenesis of Hashimoto's thyroiditis (HT). There are several types of Abs in thyroid gland diseases: to thyroglobulin (Tg), to the microsomal fraction of tyrocytes (MFT), superficial Ag, the second colloidal Ag, thyrostimulating Abs, and some others. There was the suggestion of a damaging action of Abs on thyrocytes [93], and a pathogenic role of Abs to Tg and MFT has been proved [94].

Auto-Abs to Tg, a 660 kDa precursor of thyroid hormones, are found in nearly all patients with HT and a smaller proportion of healthy individuals [95]. A high concentration of *anti*-Tg Abs is one of the main indicators of a systemic AI process in patients with HT. Recently it was shown that *anti*-Tg specific auto-Abs from sera of patients with HT are capable of hydrolyzing thyroglobulin [96].

Multiple sclerosis (MS) is a chronic demyelinating disease of the central nervous system. Its etiology remains unclear, and the most widely accepted theory of MS pathogenesis assigns the main role in the destruction of myelin-proteolipid shell of axons to inflammation related to AI reactions [97]. Although the T-cell immune system plays a leading role in MS pathogenesis, normal functioning of the B-cell system is the necessary condition for development of the disease. Recent data suggest an important dual role of autoreactive Abs in the pathogenesis of MS: they may be harmful in lesion formation but also potentially beneficial in repair [98]. Elevated Ab titers against a variety of antigens have been described in the cerebrospinal fluid and sera of MS patients [99]. Current evidence from animal models and clinical studies suggests that a crucial role in MS immunopathogenesis belongs to auto-Abs against myelin autoantigens which are involved in Ab-mediated demyelination [100], and to auto-Abs against oligodendrocyte progenitor cell surface protein, which could block remyelination by eliminating or impeding these cells [101].

Interestingly, we have recently shown that highly purified MS IgGs (but not Igs from the sera of healthy individuals) catalyze the hydrolysis of human myelin basic protein (hMBP). The enzymatic properties of IgG proteolytic activity distinguish it from other known mammalian proteases. In contrast to known human Ca^{2+}-dependent protease specific to hMBP from axon myelin-protein-lipid envelopes, one subfraction of MS IgGs proteolytic activity is Mg^{2+}-dependent (Polosukhina, Favorova, Nevinsky et al., personal communication), while another Ab subfraction is serine protease-like (Garmashova, Polosukhina, Nevinsky et al., personal communication). Both IgG subfractions demonstrate high affinity for MBP and hydrolyze specifically only hMBP but not any other tested protein of human blood or any of several dozen tested proteins of mammalian, bacterial, or viral origin.

It is appropriate to mention here that the main targets of both above-mentioned auto-Abs are glycoproteins: these are myelin oligodendrocyte glycoprotein (MOG), which is expressed preferentially on the outermost surface of the myelin sheath [102] and a progenitor cell-specific surface glycoprotein AN2 [101], respectively. As far as we know, the possibility that carbohydrate groups of these glycoproteins participate in epitope formation has not yet been investigated.

In patients with MS, increased levels of Ig (usually IgG) and free light chains (see [57, 97]) were observed in the brain tissue and cerebrospinal fluid. As in the case of other Ads, poly-specific DNA-binding Abs interacting with phospholipid have also been detected in MS patients [57, 91, 103]. New keys in the understanding of MS pathogenesis have arisen after cloning the IgG repertoire directly from active plaque and periplaque regions in MS brain and from B-cells recovered from the cerebrospinal fluid of a patient with MS with subacute disease [104]. It was found that high-affinity *anti*-DNA Abs were a major component of the intrathecal IgG response in the patients with MS. Furthermore, DNA-specific monoclonal Abs were shown to rescue from two individuals with MS, and a DNA-specific Ab was shown to rescue from an individual suffering from SLE bound efficiently to the surface of neuronal cells and oligodendrocytes. For two of these Abs, cell-surface recognition was DNA dependent. The findings indicate that *anti*-DNA Abs may promote important neuropathological mechanisms in chronic inflammatory disorders such as MS and SLE [104].

In addition, we have shown that sera of > 90% of MS patients contain catalytically active IgGs hydrolyzing DNA and RNA [52]. Moreover, IgGs and IgMs from the sera of patients with MS were found to possess amylolytic activity, hydrolyzing maltooligosaccharides, glycogen, and several artificial substrates [105]. Thus, these catalytic auto-Abs may be an important part of AI response in MS [105].

Among all these pathologies, only SLE is usually considered to be related to patients' autoimmunization with DNA, since the sera of such patients usually contain DNA at a high concentration. Therefore, at first glance it is reasonable to suggest that *anti*-DNA Abs do not play a significant leading role in the pathogenesis of most AI or viral diseases except for SLE. The chances are very high that some specific Abzs which are considered to be related to certain diseases (for example, Igs-degrading VIP and thyroglobulin in the case of bronchial asthma [34]. and HT [96], respectively) may be more important for the development of AI processes. However, although *anti*-DNA Abs may be considered as secondary or accompanied Abs, from our point of view, non-specific Abs against nucleic acids and DNA-, RNA-, and nucleotide-hydrolyzing Abs, which may occur in any AD, also play a remarkable role during AI processes.

A perpetually growing number of observations suggest that ADs originate from defects in hematopoietic stem cells (HSCs) [106]. Increased numbers of actively proliferating HSCs have been found in NZB mice characterized by spontaneous development of SLE (SLE) [107]. MRL*lpr/lpr* mice have radioresistant abnormal stem cells: an age-dependent increase in splenic colony-forming units (CFUs) in *lpr/lpr* has been described [108]. Abnormal bone marrow stem cell differentiation (numbers of bone marrow-committed precursors) in NZB mice was shown earlier: augmentation of erythroid differentiation is accompanied by a decline in myeloid precursors [109].

We have recently analyzed lymphocyte proliferation and apoptosis at different stages of the AI condition in MRLMpJ*lpr* mice [110]. Hematopoietic progenitors colony formation in the course of the disease was characterized. A detectable difference at the level of lymphocyte proliferation, cell apoptosis, and the relative amount of BFU-E and CFU-GEMM cell colonies was revealed between healthy young mice and animals spontaneously developing pronounced symptoms of the AI disorder. In addition, we have observed a correlation of production in MRLMpJ*lpr* mice of Abs-hydrolyzing DNA with disturbance of the bone marrow stem cell differentiation. Thus, the study of hematopoietic disturbances in experimental animals may be very useful for understanding the mechanism of AD development and the production of Abzs.

19.5
The Origin of Natural Autoimmune Abzymes

The origin of natural Abzs is complex. On the one hand they may be directed against analogs of transition states of catalytic reactions or even against substrates of enzymes acting as haptens. For example, Abzs-hydrolyzing VIP and thyroglobulin in the sera of patients with asthma and HT, respectively, are Abs against these proteins [48, 57]. On the other hand, in ADs, *anti*-idiotypic Abs can be induced by a primary antigen and may show some of its characteristics, including catalytic activity [44, 45, 48, 111]. It is commonly believed that Abzs are associated with autoimmunization and that auto-

19.5 The Origin of Natural Autoimmune Abzymes

Abs are most probably of *anti*-idiotypic nature. Building on earlier observations on the existence of idiotypic determinants related to the antigen, Jerne proposed that the immune system is self-regulated by a network of idiotype/*anti*-idiotype interactions [112]. The simplified model for the realization of this network may be presented as shown in Scheme 19.1.

Ag1   Ag2   Ag3   Ag4

         Ab1   Ab2   Ab3   Scheme 19.1

Antibody 1 can effectively bind antigen 1, Ag1. At the same time, the former can also be an antigen-inducing generation of secondary Ab2. The latter may also be an antigen causing generation of Ab3, and this can be continued (Ab4, Ab5, etc.). The scheme shows the great similarity between the *anti*-gene determinant of Ag1 and the antigen-binding site of Ab2 (the same is true for the couple Ab1-Ab3). Antibodies 1 and 2 are denominated as idiotype and *anti*-idiotype, respectively, etc. There is convincing evidence that such idiotype-*anti*-idiotype networks are actually present in the body. The presence of blood serum Ab4 (notation used in Scheme 19.1 is used) has been recognized in experimental animals [112].

If the active site of an enzyme plays the role of antigen triggering this *anti*-idiotypic chain, it is logical to suggest that the secondary *anti*-idiotypic Ab2 will possess a protein structure, part of which represents an "internal image or mould" of the active site of this enzyme, and consequently these Abs may possess some properties of this enzyme. It is suggested that DNA-hydrolyzing Abs generated in patients with SLE have *anti*-idiotypic nature [36, 90]. The authors suggest that these DNA-hydrolyzing Abs are *anti*-idiotypic Abs to topoisomerase I, because in the blood serum of patients with SLE an increased level of Abs against this enzyme was noted. Data of the other study indicate that DNA-hydrolyzing Abs are *anti*-idiotypes to DNAse I [113].

The remarkable property of idiotypic mimicry has been exploited to develop *anti*-idiotypic Abzs (see [111] and references therein); monoclonal Abs were selected by immunizing mice with a monoclonal Ab against the active center of acetylcholinesterase [114–117], and a monoclonal *anti*-idiotypic Ab against carboxypeptidase showed catalytic activity similar to that of the original antigen [118, 119]. Overpassing an aberrant Vκ gene to sequence an *anti*-idiotypic Abz with β-lactamase-like activity was demonstrated [120].

The concept of antigen internal image was applied to the production of catalytic rabbit Abzs with DNase activity [121]. Ab to DNase (Ab1) was acting as a competitive inhibitor of the catalysis, and thus was assumed to contain *anti*-active site. This Ab1 was used to elicit a polyclonal *anti*-idiotypic Ab2. This later exhibited DNA recognition specificity, suggesting the existence of structural internal images mimicking the conformation of the active site. Moreover, Ab2 was able to hydrolyze DNA, indicating the existence of internal images mimicking the enzymic activity of DNase.

As noted above, IgGs and IgMs from the sera of patients with AI or viral diseases, and IgGs and sIgAs from human milk, possess DNase and RNase activities, raising the question of whether the same Ab molecule can show both activities, although RNases usually cannot hydrolyze DNA and *vice versa*. Efficient separation of Abs with only DNase or only RNase activity was not achieved using different chromatographic adsorbents in our studies. Using the same electrophoretically homogeneous preparations of catalytic IgGs, IgMs, and sIgAs purified by several chromatographies including affinity chromatography on DNA- or RNA-cellulose (Section 19.6), except that a separation of small subfractions of MS IgG possessing only one activity was sometimes possible by high-efficiency ion-exchange chromatography [52]. Since monoclonal lupus IgGs which recognize specific DNA sequences show both DNase and RNase activities [122], it is plausible that in subfractions of polyclonal IgG, IgM, and sIgA both activities can also reside in the same protein, an idea supported by our finding that ribo-oligonucleotides inhibit the hydrolysis of DNA and *vice versa* [50]. The RNase activity of IgG and IgM from patients is often 5–400 times higher than that of DNase, but each is different for individual AD patients or milk donors; DNase and RNase Abzs from patients with different diseases or from human milk are correlated [37, 50–52, 54–56, 68–70], and recently we measured a correlation coefficient of 0.75 for IgGs from 90 MS patients (Ershova, Garmashova et al., personal communication). The nature of AI patients' natural polyclonal Abs with DNase and RNase activities and whether Abs directed against the active centers of DNases and RNases are capable of hydrolyzing both RNA and DNA are still open questions.

There are good reasons to believe that an Ab directed against DNA could in principle hydrolyze both DNA and RNA, since we showed earlier [122] that several mouse SLE monoclonal IgGs to DNA hydrolyze RNAs 30–100 times faster than DNA. DNase Abs could also be *anti*-idiotypes to DNAse I [113]. However, our recent data support this idea only partially, because < 10% of the total DNase activity of purified IgG from AI patients binds to immobilized DNase I or DNase II (Matushin et al., personal communication). In addition, our data show no good correlation between the relative levels of *anti*-DNA Abs and DNase Abzs from patients with SLE and MS; the major fraction of Abs with high affinity for DNA is usually not catalytically active, while in patients with a low concentration of *anti*-DNA Abs the total Abz DNase can be very high (Ershova, Garmashova et al., personal communication).

The existence of additional pathways of Abz generation in AI patients cannot be excluded. For example, Bronshtein et al. have suggested that some DNase Abzs of SLE patients are *anti*-idiotypic Abs to topoisomerase I [123–125]. From these data and our results on the extreme polyclonality of DNase and RNase Abzs from AI patients (see below), it seems reasonable to conclude that the polyclonal Abz pool contains different types: (a) Abs to DNA or RNA, (b) *anti*-idiotypic Igs to DNases and RNases, and probably (c) to other enzymes such as topoisomerases, or to their complexes with nucleic acids, or with other ligands including allosteric enzyme regulators. Some antigens may change conformation when they associate with other proteins, and their structure in such complexes could mimic that of a transition state of the antigen's reaction. One can imagine that more detailed study of Abzs could provide new insights into fundamental aspects of biological catalysis.

Data discussed in this section suggest the possibility of Abz production via the pathway described for artificial Abs induced by transition state analogs. In this case, a certain conformer of some endogenous components might play the role of transition state analog. ADs are accompanied by intensive cell apoptosis. This may result in abnormally high concentrations in the blood of both native and partially destroyed proteins, NAs, other cell components, etc. The literature data and our results [50, 83, 110, 126] suggest that both pathways of Abz generation (production of *anti*-idiotypic Abs and Ig to substrate itself) can be realized in ADs.

19.6
Peculiarities of the Immune Status of Pregnant and Lactating Women and the Origin of Natural Abzymes from Human Milk

Human milk contains various types of Abs (IgG, IgM, IgA and sIgA), of which sIgA is the major component (> 85–90%) [127, 128]. Today the source of IgG in milk is still debated; it may be partially synthesized locally by specific cells of the mammary gland and partially transferred from the mother's blood circulation system [127]. On the other hand, sIgA is produced by B-lymphocytes of the local immune system of the mammary gland [128] and is present in Payer's patch lymphoid cells, which migrate to mucosal sites and generate local secretory IgA (sIgA) Ab responses. The origins of human milk catalytic IgG and sIgA may thus be different, and at this time it is of interest to compare different IgGs and sIgAs, because answering this question would help to explain how milk catalytic Abs arise and what functions they may play in healthy organisms (see below).

As mentioned above, a number of Abzs have been detected recently in the sera of patients with ADs. These sera contain NAs and certain proteins at higher concentrations than those in the sera of normal humans. These components of sera could cause autoimmunization, and therefore it is commonly believed that the presence of catalytic Abs is associated with autoimmunization (see Section 19.4). In spite of the absence of obvious immunizing factors that are found in normal humans, we have recently presented evidence that the milk of healthy human mothers contains sIgA and IgG possessing different catalytic activities. In addition to IgGs and sIgAs with DNase, RNase [67–70], ATPase [71], and amylolytic [76] activities, human milk also contains very unusual IgGs and sIgAs with synthetic protein [43, 72, 73], lipid [74, 75], and polysaccharide kinase activities (see below). In this connection it should be emphasized that we did not find detectable nuclease activity in Abs from the sera of 50 normal donors (men and women) or in patients with pathologies proceeding without disorder of immuno status [51, 52] [12–15]. In addition, we have not found Abz with unusual synthetic activities in the blood of AD patients. Specific activities of DNase and RNase IgGs from human milk were significantly higher than that from the blood of patients with different AI [49, 83, 126]. Moreover, the specific activities of milk sIgAs in the hydrolysis of DNA and RNA are usually higher than those of IgGs from the same samples of human milk.

As noted above, the specific human milk IgGs and sIgA with amylolytic activity were detected [76]. Interestingly, the specific activities of IgG and sIgA from human milk is 2- to 3-fold higher than those of IgG from pregnant women's blood and approx. 10-fold higher than those from oncological patients [67–70]. Only in the case of such ADs as MS and SLE does the specific α-amylolytic activity of some donors reach the level of IgG and sIgA auto-Ab fractions from human milk [105]. Thus, human milk can be considered as a "potential reservoir" of Abzs with a number of different catalytic activities which have not so far been revealed in AI patients' blood. In addition, we suppose that human milk contains not only the Abzs listed above but also other catalytic Abs hydrolyzing peptides, proteins, etc. as well as Igs with different synthetic activities which have not so far been studied. Interestingly, the substrate specificities of milk Abzs, including DNase and RNase Abs, are different from those from patients with AI pathologies (see below). At the same time, the origin of the natural Abzs of human milk is unknown but is very intriguing.

From an analysis of published data, a possible process by which pregnant women are directly immunized may be a result of a specific response of their immune system to certain components of viruses, bacteria, or food, which most probably can effectively stimulate production of different Abs. Thus, the immunization of animals by the injection of antigens (mainly various proteins) into the blood or by oral administration no more than 1–3 months before the birth of neonates leads to the production of *anti*-protein Abs, which may then be detected in the milk at high concentration [105, 122, 129, 130]. This means that, in contrast to normal humans, pregnant women may be effectively immunized by components of various viruses and bacteria when they are in contact with such compounds. Moreover, the production of milk Abs 1–3 months after immunization speaks in favor of the existence of a specific "immunomemory" in pregnant females. In this connection it is interesting to mention that recently we have observed a significant increase of RNA- and DNA-hydrolyzing Abs in the milk and sera of normal mothers who suffered from viral or allergic diseases during their pregnancy [69, 70]. Since we did not detect catalytic Abs in the sera of patients with influenza, pneumonia, tuberculosis, tonsillitis, duodenal ulcer, or some types of cancer [131], it is quite possible that the immune system of normal humans (men and women) cannot be stimulated by proteins and NAs in the same way as that of pregnant women. Moreover, most probably there may be autoimmunization of mothers during pregnancy similar to that in AD patients; an increased level of NAs in the sera of normal women during the 3 first months of pregnancy, like that in cases of ADs, was revealed recently [132]. In addition, the apoptosis of several types of cells during the last three months of pregnancy was recently demonstrated [133], as was a very low concentration of embryo cells in the blood of pregnant women [134]. Taken together, we believe that during pregnancy women are subjected to immunization with various components of viruses and bacteria, including those of an *anti*-idiotypic nature as in ADs.

There are many published data obviously demonstrating that during pregnancy, immediately away after delivery, and at the beginning of lactation women are very often characterized by immune processes similar to those for AD patients [135, 136]. Interestingly, the pregnancy can "activate" an appearance of some manifestations of different AI pathologies in previously healthy women. A sharp exacerbation of AI reactions (so called "AI shock") can occur immediately after childbirth. Independently of an existence or absence of AI processes in women during pregnancy, sometimes there may be several different post-natal AI pathologies: SLE, HT, phospholipids syndrome, polymiosit, AI myocardium etc. [135, 136]. Post-natal AD can occur immediately after delivery or later. One AD (often post-natal) is HT, which according to the literature data can be found among 1.9–16.7% of human mothers [137, 138]. Different indicators of these ADs may often be revealed during the first 3–6 months of the post-natal period, although there may be such events even at a later time but usually no later than 1 year.

Thus, one cannot exclude the possibility that molecular mechanisms of activation of immune system leading to the production of auto-Abs in AD patients and in human mothers may be to some extent similar or overlapping. The question is what are the possible similarities and possible differences between AI processes in AD patients and those in human mothers. In this connection it should be noted that patients with various typical ADs are practically incurable, and such diseases have often a chronic character with periodic exacerbations and remissions of the pathologies. At the same time, different appearances of various ADs which can be detected during the pregnancy of women are often restricted to the time of a pregnancy and a short post-natal period. For example, during the first 3 months of the pregnancy an increased level of Abs to thyroid hormone was demonstrated only in 102 [18,7%] of 545 women [139]. Only 33% of women with increased titers of these Abs began to ache with post-natal HT, while only 3% of pregnant women demonstrating no increased level of auto-Abs became ill [139]. In all the other women there were no apparent post-natal ADs. Thus, a disappearance of processes of AI reactions in the post-natal period can be considered as a norm for human mothers, even if in some single cases "a temporary activation of AI processes" in pregnant women can sometimes gradually or suddenly (though AI shock) transform to typical chronic AI process.

As discussed above, the milk of healthy human mothers contains many Abzs with different activities. However, there was no data about the presence of Abzs in the blood of women during their pregnancy and in early post-natal period. Taking this into account, we have recently studied the dynamics of RNase and DNAse activities of Abs from the blood of pregnant women [140]. It was shown that the relative catalytic activity of Abs in the hydrolysis of DNA and RNA from the blood of different pregnant women could vary over a large range. The dynamics of changes of Abs DNase activity in the course of pregnancy has an individual character for each woman (Fig. 19.1). At the same time, a rising tendency to activity in the first and/or third terms of pregnancy takes place. Interestingly, a relative activity of blood's Abs significantly increases after delivery with the beginning of lactation. However, Abzs from milk have remarkably higher activities than those from the blood of the same women (Fig. 19.2). The data obtained demonstrated unambiguously that, in principle, in contrast to females before pregnancy, the blood of pregnant women can contain catalytic Abs, but the relative catalytic activity of such Abzs is significantly increased in the early post-natal period, e.g., immediately after the beginning of the lactation.

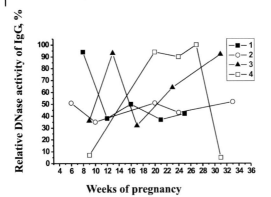

Fig. 19.1 The dynamics of a change in relative DNase activity of IgGs from the sera of four women (1–4) during the course of pregnancy. The maximal activity of one of the samples used was taken as 100%. The details of the analysis are given in [140]. As mentioned above, we have shown that the blood of patients with HT contains DNase and RNase Abzs [53, 56, 141, 142]. In addition, HT is known as an AI pathology which can be often induced by the pregnancy. Therefore, we have compared the levels of DNase activity of IgG and IgM from the blood of pregnant healthy females without pathologies and women with an acute form of HT. It was shown that DNAse activity of Abzs in women with AD induced by pregnancy is overall about 4–5 times higher than this activity in the case of healthy pregnant women [140].

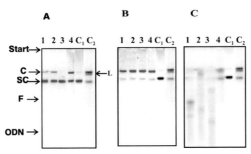

Fig. 19.2 DNase activities of catalytic IgGs in the cleavage of supercoiled (sc) pBluescript DNA (15 µg) by IgG (0.5 µM) at 37 °C for 2 h from the blood sera (A and B) and milk (C) of different women before (A) and after the beginning of lactation (B and C); in all panels (A–C) lanes 1–4 correspond to Abs from the sera or milk of the same four women, the lanes C_1 and C_2 correspond to DNA hydrolysis under the same conditions with control IgG from healthy men (negative control) and with IgG from an AI patient (positive control), respectively. Different forms of plasmid DNA are marked as: C, circular; L, linear; F and ODN, fragments of DNA of different length including ODNs. The details of the analysis are given in [140].

Interestingly, after delivery the sera Abz hydrolyzing NA has high activity even in the case of women no demonstrating such activity during pregnancy (Fig. 19.2, compare lanes 3 in panels A, B and C). As mentioned above, these activities of Abs from human milk were significantly higher than those of Abs from the blood of the same women. The level of enzymic activities of Abzs from milk and blood did not change significantly during the first month after delivery, and then it slowly decreased during months 2–4 and sometimes 2–8 of the post-natal period.

Taken together, in contrast to IgG from normal women before pregnancy and IgG from men, IgG from the serum of healthy pregnant women has DNAse and RNase activities which increase remarkably after the beginning of lactation [140], and this increase is much more significant in mothers who suffered from viral direases, for example, influenza or allergic diseases [140], or in females with induced AD during pregnancy [69]. Our data and the data in the literature show that the immune system of pregnant women may be stimulated by proteins and NAs in a different manner to that of normal humans.

Nevertheless, all the above data do not tell us about the nature of milk Abzs and the possible antigens stimulating the production of milk Abzs with various catalytic activities. As is known from the literature and follows from our results, *anti*-nucleic acid Abs and Abzs with nuclease activities from AD patients are first of all anti-DNA Abs (for review see [126]). Interestingly, on immunization of rabbits with DNA and RNA and with their complexes with different proteins, DNA stimulates a better an AI answer than RNA (Krasnorudskii, personal communication). In addition, immunization of the animals with DNA-protein complexes leads to the production of Abzs demonstrating higher activity in the hydrolysis of NAs than that of catalytic Abs obtained on immunization with RNA-protein complexes. At the same time, as was recently shown by Semenov et al. (personal communication), the preparations of total NAs isolated from human milk contain ~ 80–90% of RNA and only ~ 10–20% of DNA. In addition, total milk RNA contains some ribosomal RNA and tRNA of specific sequences in increased concentration. Therefore, one cannot exclude the possibility that the human milk *anti*-NA abzymes may be Abs not only against DNA but also against RNA.

The milk IgGs and sIgAs are polyclonal in origin and consist of many subfractions with different affinities for immobilized DNA [49, 67, 68, 83, 126]. As mentioned above, milk IgG may be partially synthesized locally by specific cells of the mammary gland. Therefore, one cannot exclude the possibility that some of specific milk RNAs can serve as antigens stimulating the production of specific mammary gland IgG subfractions with catalytic activities. Another part of milk IgG is usually transferred from the mother's blood circulation system [144], and sIgA is produced from IgA migrating to the mucosal site from Payer's patch lymphoid cells (duodenum). In Payer's patch, the cell precursors of B-lymphocytes commit to antigens of intestinal flora and food antigens. During maturing, these cells leave the intestine and settle in mucosal and other glands [132]. Thus, the repertoire of milk Abzs should be determined to a great extent by antigens which are presented in the intestine. As known, *Escherichia coli* is the major component of the human intestine, and human milk contains many different Abs against proteins and other polymeric molecules of *E. coli* [128]. Taking these data into account one cannot exclude the possibility that specifically activated immune systems of lactating women can produce not only classical Abs to *E. coli* components but also catalytically active Abs. Therefore, we have tried to elucidate whether some of milk Abzs are Igs against proteins of *E. coli*.

First we have shown that proteins of *E. coli* can be adsorbed on the agarose-bearing immobilized milk sIgA fraction. As mentioned above, milk IgGs and sIgAs fractions catalyze the hydrolysis of ATP and the phosphorylation of milk proteins; both reactions lead to the formation of free *ortho*-phosphate (P_i). We have developed an *in situ* method for the detection of the Ab-dependent formation of P_i [71]. The samples of *E. coli* proteins (crude extract) were subjected to SDS-PAGE, and the proteins were then transferred from the gel to a nitrocellulose membrane. This membrane was treated with a mixture of electrophoretically homogeneous human milk sIgA having an affinity for ATP-Sepharose (see Section 19.6). Nitrocellulose replicas of *E. coli* proteins prepared in the same way without further treatment with Abs or treated with human Abs having no affinity for ATP were used as controls. To reveal ATPase activity of human Abs bound with *E. coli* proteins, the membranes were treated with γ-[^{32}P]ATP in the presence of $Pb(NO_3)_2$. Because of the catalytic activity of Abs, the formation of free *ortho*-phosphate leads to the formation of an insoluble precipitate of ^{32}P-labeled $Pb_3(PO_4)_2$ (Fig. 19.3). Comparison of lanes 2 and 5 or 6 (Fig. 19.3) as well as 3 and 5 or 6 demonstrates that only a few *E. coli* proteins forming complexes with milk Abs produce free *ortho*-phosphate, and there is no precipitation of ^{32}P-labeled $Pb_3(PO_4)_2$ in some of the protein bands in the control lane 4 in contrast to lanes 5 and 6. Interestingly, IgG and sIgA abzymes from milk of different donors can differ very much in their specific enzymic activities, affinity to immobilized ATP, and K_m values for ATP [71]. As seen from Fig. 19.3, milk sIgAs of the first donor (lane 5) interact mainly with *E. coli* proteins of ~ 20–23, ~ 46 and ~ 70 kDa, while Abzs of the second donor react with ~ 23 kDa protein (lane 6). ATPase sIgA of some other donors also demonstrated major interaction with these 23, 46 and 76 kDa proteins and some other minor proteins from ~ 14 to ~ 20 kDa (data not shown). Thus, the repertoire of *E. coli* proteins, which are capable of interacting with milk ATPase Abs, can be changed from donor to donor. In this connection it should be mentioned that milk IgGs and sIgA abzymes possess an affinity for some milk proteins, for example, casein, and the latter stimulates remarkably the IgG- and sIgA-dependent hydrolysis of ATP [43]. Taking into account published data that some Abs can be polyreactive in

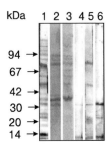

Fig. 19.3 Analysis of human milk Abs interaction with total proteins of *E. coli* after SDS-PAGE of proteins of bacterial lysates and following transfer of the proteins onto a nitrocellulose membrane. Lane 1 corresponds to colloid silver staining of *E. coli* proteins; lane 2, immunoblot of total *E. coli* proteins (alkaline phosphatase conjugated with milk sIgA); lane 3, colloid silver staining of *E. coli* proteins purified on the adsorbent bearing immobilized human milk sIgA; lane 4, analysis of an intrinsic ATPase activity of total *E. coli* proteins by incubation of the nitrocellulose with γ-[^{32}P]ATP in the presence of $Pb(NO_3)_2$; lanes 5 and 6, detection of *E. coli* proteins reacting with ATPase sIgAs from two different donors of milk, after sequential treatments of the proteins with catalytic milk sIgA and γ-[^{32}P]ATP in the presence of $Pb(NO_3)_2$. The details of the analysis of different Ab's activities are given in [143].

their nature, one cannot exclude the possibility that some fractions of milk Abz can also be polyreactive Abs.

Using affinity chromatography of *E. coli* proteins on adsorbents bearing immobilized total milk sIgAs and ELISA analysis, we have shown that milk ATPase Abs interacting with *E. coli* proteins are only a very small part of total human milk Abs binding bacterial proteins. These data support the idea that intestinal bacteria proteins are potential antigens for the production of different Abzs in organisms of human mothers. But the data obtained do not permit us to conclude that components of *E. coli* are a single incentive for the production of catalytic Abs. As mentioned above, in contrast to ordinary donors (men and women), pregnant females may be directly immunized by different components of viruses, bacteria, or food, even on oral administration. Thus, we cannot exclude the possibility that some of these components effectively stimulate the production of different Abzs.

Taken together, one can suppose that the production of human milk and blood Abz by organisms of normal woman can occur to some extent in a similar way to that in the case of patients with ADs, but this process is most probably more complicated and includes additional mechanisms like "immuno memory", protection from development of typical AI-processes, immuno response to many components of an aggressive environment, etc. There is one additional and very important difference between the production of Abz in organisms during typical ADs and that during pregnancy. As mentioned above, we have observed a correlation of production in MRLMpJ*lpr* mice of IgGs and IgMs hydrolyzing DNA with disturbance of the bone marrow stem cell differentiation. The disturbance of marrow stem cell differentiation in animals spontaneously developing pronounced symptoms of AI disorder usually leads not only to an increase in auto-Abs and Abzs but also to the appearance of other different characteristics of AI processes like proteinuria etc. Interestingly, the appearance of auto-Abs and Abzs in pregnant and lactating mice occurs without the appearance of typical AI pathologies, and, in contrast to mice with spontaneous or induced AD, no very drastic change of bone marrow stem cell differentiation for such animals is revealed (Orlovskaya, Nevinsky; unpublished results).

Finally, it is obvious that peculiarities of the immune system of pregnant and lactating women are extremely unusual, and therefore they are of great interest for many scientists. Progress in immunology in pregnancy most probably can help to develop methods of transplantation, and this requires a study of immunological tolerance of mother and child. It is obvious that during pregnancy a very strong reorganization of the immune system of women takes place, which not only does not prevent growth of the baby but also protects the organisms of women and babies from infection and launches the process of embryo implantation.

19.7
Purification of Natural Abzymes

Abz-mediated catalysis is often characterized by low reaction rates, and it is therefore important to prove that catalytic activities are not due to contaminating enzymes.

However, there are few studies concerning human monoclonal Abs. In contrast, natural Abzs from the sera of patients are usually polyclonal in origin and are products of different immuno-competent cells [50]. Their purification is one of the most complicated aspects of their study, and here we discuss in detail some specific methods of Abz purification and characterization illustrated by our own work.

Affinity chromatography on immobilized hapten usually produces homogeneous monoclonal Abzs, but polyclonal Abs from patients are very heterogeneous [52, 55, 69, 145], and their affinity for an immobilized substrate analogs can differ by several orders of magnitude [50]. The first step of purification should separate Abs efficiently from other proteins (NAs etc.), so that a fraction with high affinity for potential substrates is obtained. Practically all published methods for purifying natural Abzs include purification up to the step of removing Abs with affinity to the substrate but no catalytic activity, but as yet there are no methods which include this final separation step. Specific activities are difficult to quantify because nuclease Abzs are often only about 0.001–10% of the total Abs with affinity for immobilized DNA [50, 52, 55, 69, 145], and acidic treatment during their purification may lead to loss of activity. The same situation was observed for other human sera and milk Abzs with different activities [50].

In the first step of Abz purification, affinity chromatography on adsorbents bearing *anti*-IgG, *anti*-IgM, or *anti*-IgA Abs is often used in conditions to remove non-specifically bound proteins (buffers containing 1% Triton X-100 and 0.3-1.0 M NaCl) (see [35, 50] and references therein; also [52, 55, 70, 72]). Some Ab fractions may be eluted by buffer at pH 4.6, but they usually (except in rare cases) do not contain Abzs [50, 52, 70], and the main fraction containing all types of Abzs is usually eluted in glycine-HCl, pH 2.6, leading to a preparation homogeneous by SDS-PAGE revealed by silver-staining and by Western blot.

Adsorbents bearing immobilized Abs are often not convenient for Abz purification, and we have found that protein A-Sepharose is an optimal resin for the first step because, for unknown reasons, it has a higher affinity for Abzs than for Abs without catalytic activity (see [50] and references therein; also [52, 55, 70, 72]). Only 4–10% of the total Abs in human milk were eluted from protein A-Sepharose in buffer at pH 2.6, but these contained 80–95% of the initial Abz DNase activity and 80–95% of the Abz protein kinase [72], lipid kinase [74, 146], RNase, amylolytic [69, 70], and ATPase activities [71] (for example, Fig. 19.4). Similar results were obtained for autoimmune sera [52, 54, 55]. Our protocol to obtain human sera IgG or milk IgG and sIgA Abzs with different activities includes practically all chromatographic steps usually used for purification of Abzs including sequential chromatography on protein A-Sepharose and DEAE-cellulose, gel filtration on Toyopearl HW-55 or other adsorbents allowing high resolution in conditions of "acidic shock", and then chromatography on DNA-cellulose (or other affinity adsorbents bearing immobilized ligands: ATP, protein, etc.) to obtain a fraction with high affinity for nucleic acids [50, 70].

Fig. 19.5 demonstrates a typical example of a separation of IgG and IgM fractions of purified SLE Abs using gel filtration. The following chromatography of each of purified IgG and sIgA fractions on DEAE-cellulose provides additional purification of IgGs and sIgAs from the possible minimal admixtures of other types of Igs. In some

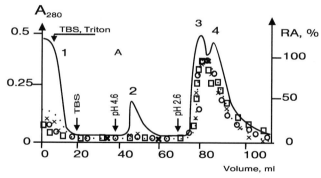

Fig. 19.4 Purification of Abs from human milk by chromatography on protein A-Sepharose. (–): absorption at 280 nm; RA: relative activity of Abs in the hydrolysis of supercoiled DNA (·), ATP (□), polysaccharide (×), and phosphorylation of casein (○). These activities were assayed using 3 µL of each fraction in 30–50 µL of reaction mixture and expressed relative to the activity of the fraction having maximal activity (100%). The details of the analysis are given in [71, 73, 76, 105, 143].

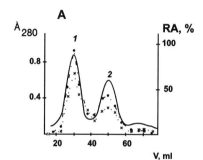

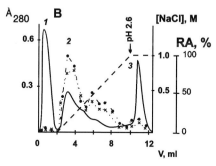

Fig. 19.5 Purification of RNase IgM from serum of SLE patient by sequential chromatography on Toyopearl HW 60 (**A**, after protein A-Sepharose), and DNA-cellulose (**B**, after gel filtration of Toyopearl HW 60 and chromatography on DEAE-cellulose). (–): absorption at 280 nm, (·) and (×): relative activity (RA) of Abs in $[5'-^{32}P](pA)_{13}$ and $[5'-^{32}P]d(pA)_{13}$ hydrolysis, respectively, assayed using 1–5 µL of each fraction in 100 µL of reaction mixture relative to the activity of the fraction having maximal activity (100%). The details of the analysis are given in [55].

cases it was more convenient first to carry out chromatography on DEAE-cellulose and then to perform gel filtration, as shown in Fig. 19.6.

Although these chromatographies, including affinity chromatography, enrich Abs with substrate-binding activity, they cannot separate Abzs from catalytically inactive Abs.

The sera of patients in the initial stages of ADs usually contain auto-IgM, and later auto-IgG may be detected. Essentially homogeneous IgG can be obtained by the stan-

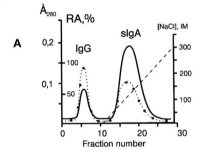
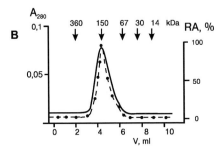
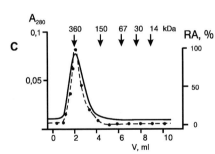

Fig. 19.6 Separation of human milk IgG and sIgA by sequential chromatography on DEAE-cellulose (**A**) and gel filtration on Toyopearl HW 60 of purified IgG (**B**) and sIgA (**C**). (–): absorption at 280 nm; RA, relative activity of Abs in the hydrolysis of ATP (·). This activity was assayed using 1 μL of each fraction in 20 μL of reaction mixture relative to the activity of the fraction having maximal activity (100%). The details of the analysis are given in [69–73, 143].

dard method on protein A-Sepharose when the concentration of IgG is significantly higher than that of IgM [54, 55]. To separate IgM, the combined IgG + IgM fraction was first enriched, and then IgM and immunocomplexes were precipitated essentially completely with 6% polyethylene glycol [55]; the precipitate contained only a small amount of IgG, and, after chromatography on protein A-Sepharose under conditions to remove non-specifically bound proteins, the fraction eluted by pH 2.6 buffer contained IgM and a small amount of IgG, and no contaminating proteins detectable by silver staining. IgM was then separated from the remaining IgG by gel filtration at pH 2.6, and each fraction was further purified by chromatography on DEAE-cellulose and DNA-cellulose [55]. Chromatography of Abs with DNase [50, 70] and other activities [43, 71–73] on affinity adsorbents usually leads to many subfractions, because natural Abzs are polyclonal in origin.

Interestingly, an elution of all subfractions of polyclonal Abzs from adsorbent bearing immobilized specific ligand (NA, ATP, protein, etc.) requires very drastic

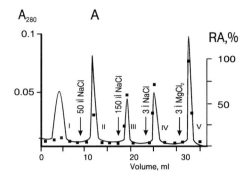
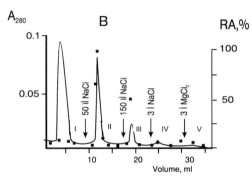

Fig. 19.7 Purification and analysis of milk protein kinase sIgA (**A**) and IgG (**B**) to ATP by chromatography of purified preparations of these Abs on ATP-Sepharose. (–): absorption at 280 nm, (●): relative activity (RA) in casein phosphorylation. This activity was assayed using 2 µL of each fraction in 20 µL of reaction mixture relative to the activity of the fraction having maximal activity (100%). The ratio of protein in peaks I–V eluted by different concentrations of the salt was dependent very much on the donor's milk. The PK-activity was analyzed by ^{32}P-label accumulation in the acid-insoluble fractions using 0.5 mg/mL of casein, 10 nM γ-[^{32}P]ATP, and optimal conditions [72, 73].

conditions: 3 M MgCl$_2$ or acidic buffer (pH 2.3–2.6). For example, affinity chromatography of milk sIgA or IgG Abzs on ATP-Sepharose demonstrated that Abs possessed a heterogeneous affinity for ATP (Fig. 19.7) [69]. The ratio of protein in peaks I-V eluted by different concentrations of the salt was dependent on the donor's milk. Nevertheless, in all 50 samples the adsorption of 70–90% of sIgA (previously purified as described above) took place. All fractions of sIgA (peaks II-V, but not I) had kinase activity (Fig. 19.7A). A fraction of sIgA was eluted at a high concentration of NaCl (3 M) or MgCl$_2$ (3 M) under conditions required for disruption of very stable immune complexes. It should be emphasized that all milk kinases of a non-immunoglobulin nature were eluted from ATP-Sepharose at a concentration of NaCl lower than 0.1–0.2 M.

Fig. 19.8 demonstrates that 70–80% of sIgA purified on ATP-Sepharose possesses a different (high) affinity to casein-Sepharose, and all subfractions are catalytically active [69]. It should be noted that the relative amounts of Abs (and their catalytic activities) eluted from adsorbents of various affinities by various concentrations of NaCl strongly depend on the donor's milk.

A similar situation was observed in the case of affinity chromatography of purified IgG and IgM Abzs from the sera of patients with different ADs on DNA-Sepharose (for example, Fig. 19.5B). All human nucleases of a non-immunoglobulin nature were eluted from DNA-cellulose at a concentration of NaCl lower than 0.1–0.5 M, but in the case of Abzs several fractions were eluted only at a high concentration of NaCl (3 M) or MgCl$_2$ (3 M) under conditions required for the disruption of very stable

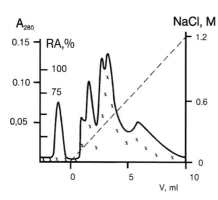

Fig. 19.8 Purification and analysis of milk protein kinase sIgA affinity to casein by chromatography of purified preparations of these Abs on Casein-Sepharose. (–): absorption at 280 nm, (×): relative activity (RA) in casein phosphorylation. This activity was assayed using 2 µL of each fraction in 20 µL of reaction mixture relative to the activity of the fraction having maximal activity (100%). The ratio of protein in different peaks eluted by different concentrations of the salt was dependent very much on the donor's milk. The PK-activity was analyzed by ^{32}P-label accumulation in the acid-insoluble fractions using 0.5 mg/mL of casein, 10 nM γ-[^{32}P]ATP, and optimal conditions [72, 73].

immune complexes. We have observed variations in the affinity for immobilized nucleic acids in the case of IgG and IgM Abzs from the sera of AI patients [52, 54, 55].

Chromatography of Abzs on protein A-Sepharose and subsequent gel filtration in acidic buffer (pH 2.6) are extremely effective and possibly irreplaceable, since they practically guarantee homogeneous preparations with no contaminating proteins or enzymes. However, these procedures are potentially harmful to catalytic activity because of the so-called "acidic shock" which many enzymes do not survive; for example, the protein kinase of a non-immunoglobulin nature in human milk loses its activity completely [50]. The stability of Abzs to "acidic shock" or "alkaline shock" is relatively low; DNase Abzs from the sera of patients with MS eluted at pH 2.6 and immediately neutralized with Tris-HCl, pH 9.0 were inactive but recovered their activity slowly after dialysis against Tris-HCl buffer, pH 7.5, and storage at 4 °C for 1–5 weeks [52]. Different human RNases and DNases usually recover their activity rapidly. At the same time, different Abzs, for example human IgG, IgM, and sIgA RNase Abzs from the sera of AI patients and from human milk, also RNase Abzs of patients with SLE, HT, polyarthritis, and viral hepatitis, recover activity relatively slowly and at different rates [147]. Rapid "acidic shock" of polyclonal IgGs from AI patients, which possess two different RNase activities detected by cleavage of [^{32}P]tRNA^{Phe}, one active in low salt conditions with specificity similar to RNase A and another stimulated specifically by Mg^{2+} ions (which is unique and different from those of known pro- and eucaryotic RNases [53, 147]), usually results in a low activity similar to that of RNase A in low-salt conditions; this slowly increases during storage of the dialyzed solution, and during this increase new sites in tRNA are cleaved in addition to those at CA and UA, and a unique activity specifically stimulated by Mg^{2+} ions appears. The rate of recovery of each activity varies from patient to patient and from one disease to another, some requiring 3–6 weeks of storage of neutral solutions at 4 °C. These results, together with those in Section 19.10, show that natural Abzs from the sera of patients consist of different sets of Abz subfractions with quite distinct enzymic properties; different subfractions are denatured to a variable degree in drastic conditions and renature at different rates, although complete renaturation most probably can never be achieved. Thus, the use of drastic conditions for Abz purification appears at the present time to be a very useful method for obtaining homogeneous preparations, but does not provide highly active Abzs. Although such preparations are usually homogeneous by SDS-PAGE revealed by silver staining, this does not prove that the catalytic activities are an intrinsic property of Abs, and as a rule further experiments to rule out artefacts due to co-purifying enzymes should be performed by methods described in the next Section.

19.8
Criteria to Establish that Catalytic Activity is Intrinsic to Antibodies

The application of rigid criteria allowed the authors of the first article concerning natural Abzs [34] to conclude that VIP-hydrolyzing activity is an intrinsic property of Abs from the sera of patients with asthma. The most important of these criteria are: (1) electrophoretic homogeneity of the IgG by silver staining, (2) catalytic activity of Fab fragments of the Ab, (3) complete adsorption of activity on *anti*-IgG-Sepharose and its elution with buffer of low pH, (4) precipitation of activity by *anti*-IgG Abs, (5) activity only in IgG of patients with asthma and not in healthy donors, (6) activity not lost upon gel filtration of IgG in conditions of "acidic shock" and the eluted peak of activity coinciding exactly with 150 kDa IgG, (7) the K_m value for hydrolysis of VIP [38 nM] showing a high affinity for the substrate comparable to the K_d of antigen-Ab complexes, and (8) the substrate specificity of VIP hydrolysis different from that of all known proteases.

Similar arguments were presented in the first reports of natural DNase and RNase IgGs (see [35, 37, 50] and references therein).

Our studies of catalytic Abs have led us to add further criteria to the above list:

1. Preservation of Ab's catalytic activity is demonstrated after gel filtration not only under conditions of extremely low pH but also in other severe conditions that effectively dissociate non-covalent complexes: examples of such conditions include the use of buffers with strongly alkaline pH (pH 10–11) or neutral buffers containing 5 M thiocyanate or 6 M guanidine chloride [43, 73].

2. We examined the efficiency with which Abzs are separated during purification from known serum enzymes with the same activity using mixtures of homogeneous Abzs and blood nucleases [53, 55, 148] and found that chromatography on protein A-Sepharose and gel filtration in drastic conditions led to efficient separation of Abs from added RNases.

3. All known human serum RNases ($M_r \approx 32$ kDa) can be separated from RNase Abzs on Centricon-100 devices, which retain 100 kDa molecules including IgGs (150 kDa). Only the Abz identified by its Mg^{2+}-stimulated activity was retained after filtration of a mixture of Abzs and RNase A in the absence or presence of 0.1% Triton X-100 and a high concentration of salts to dissociate non-covalent protein complexes, showing that this method can efficiently separate Abzs from enzyme contaminants [53].

4. Finally, thermal denaturation experiments were used to show that catalytic IgG and IgM can be distinguished by their lower thermostability from five RNases and two DNases of human blood [53–5, 148, 149]; they lose activity after 30–40 min incubation at only 45–50 °C, in parallel with loss of antigen-binding activity [150].

5. We have developed other approaches to provide direct evidence that an Abz possesses intrinsic enzymic activity. The first is a very sensitive and specific method based on affinity modification of an Abz by a chemically reactive [^{32}P]substrate analog whose covalent binding to the active site is detected after SDS-PAGE in non-reducing conditions and/or to the L- or H-subunits in reducing conditions [43, 50, 67–73]. No contamination of catalytic Abs with enzymes was detected by this very sensitive method. For example, the affinity modification of milk sIgA with protein kinase activity by the 2',3'-dialdehyde derivative of ATP led to inactivation of Abs enzymic activity: α-[^{32}P]oxATP modified only the L-chain of sIgA with stoichiometry 2.0–3.0 mole of reagent per mole of sIgA (Fig. 19.9). ATP protected sIgA from both enzymatic activity inactivation and covalent binding of the reagent. Labelled ^{32}P-sIgA had a positive response to Abs against sIgA. The stoichiometry of the affinity modification is a strong argument for the reagent covalent binding with the L-subunit of sIgA, but not with any protein kinases co-purifying with Abs.

Affinity modification of IgG from SLE patients and human milk by chemically active derivatives of ODNs led to preferential modification of the L-chain (Fig. 19.9) [67, 68, 70], but both L- and H-chains became labelled after incubation of sIgA with an affinity probe for DNA-binding sites (Fig. 19.9) [69].

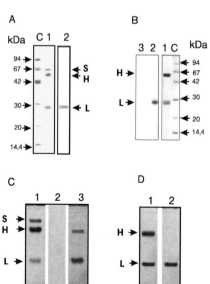

Fig. 19.9 Electrophoretic analysis (radioautograph) of affinity modification of milk PK-sIgA (A, lane 2), milk ATPase IgG (B) by 1 μM 2',3'-dialdehyde derivative of α-[^{32}P]ATP nucleotide oxidized by NaIO$_4$ in the absence (lane 2, panels A and B) and in the presence (lane 3, panel B) of 100 μM ATP as well as milk DNase sIgA by 0.1 μM alkylating 4-(N-2-chloroethyl-N-methylamino)-5'-phosphoamide derivative of 5'-[^{32}P]d(pT)$_{10}$ (C) in the absence (lane 3) and in the presence of 10 μM d(pT)$_{10}$ (lane 2), and milk RNase IgG by 0.1 μM 5'-[^{32}P](pU)$_{10}$ oxidized by NaIO$_4$ (D, lane 2). The reduction of the Schiff's base formation from the condensation of oxidized ATP and (pU)$_{10}$ analogs with Abs was carried out with NaBH$_4$. The protein analyzed by SDS-PAGE in reducing conditions in a Laemmli system. Gels stained with silver show the position of different subunits of Abs (lanes 1). Arrows indicate the positions of molecular mass markers (lanes C). The details of the above different types of analysis are given in [67, 69, 70, 72, 73].

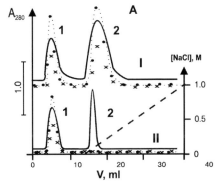

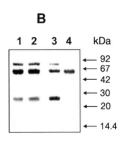

Fig. 19.10 A Affinity chromatography of purified milk nuclease sIgA on DNA-cellulose before (curve I) and after dissociation into subunits (curve II) using 0.3 M DTT and 5 M urea. (–): absorption at 280 nm, (•) and (×)- relative activity of Abs in $5'\text{-}[^{32}P](pU)_{10}$ and $5'\text{-}[^{32}P]d(pT)_{10}$ hydrolysis, respectively, relative to the activity of the fraction having maximal activity in hydrolysis of $(pU)_{10}$ (100%). B Silver-stained SDS-PAGE of peak fractions from panel A in a reducing 12% gel: lanes 1 and 2 correspond to peaks 1 and 2 of curve I and lanes 3 and 4 to peaks 1 and 2 of curve II. Arrows indicate the positions of molecular mass markers. The details of the experiments are given in [67–70].

6. A second direct approach is to analyze the activity of isolated L- and H-subunits of Ig after separation by affinity chromatography in mild dissociating conditions (1 mM mercaptoethanol and 4 M urea) [67–70]. As an example, DNA-cellulose adsorbed only the light chain of milk IgG, which showed DNase and RNase activity, whereas the heavy chain of sIgA was adsorbed and the catalytic light chain had a low affinity for DNA (Fig. 19.10) [69].

ATP-Sepharose adsorbed only the L-chain of milk sIgA with protein kinase activity following separation of the subunits of sIgA after the Abs complex dissociation using DTT and urea. The protein adsorbed by affinity adsorbent had electrophoretic mobility of the L-chain and reacted positively with Abs against the sIgA L-chain [42, 43, 72, 146].

7. A further strong criterion is the detection of an Ab's catalytic activity and that of its Fab fragments *in situ* after SDS-PAGE in non-reducing or reducing conditions in a gel containing its substrate (RNA, DNA, or ATP) [54, 55, 67–71, 149]. We applied this method to establish the DNase activity of IgG from human milk [67, 68]; after electrophoresis, the gel was incubated to allow protein renaturation and stained with ethidium bromide to reveal DNase activity as a sharp dark band at the position of the Ab on a background of fluorescent DNA; the positions of IgG and its subunits were then revealed by Coomassie blue staining (for example, Fig. 19.11). DNase activity was detected only in the bands corresponding to the IgG and its Fab fragments in non-reducing conditions, and only in the band containing light chains in dissociating condi-

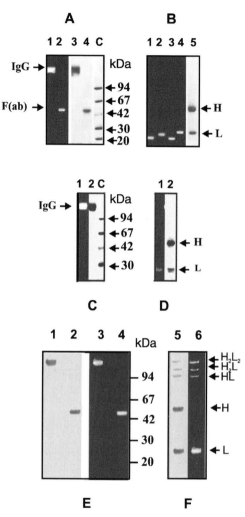

Fig. 19.11 *In situ* gel assay of RNase (**A** and **B**) and DNase (**C** and **D**) activity of catalytic SLE IgG (**A**, lanes 1 and 3; **C** lanes 1 and 2) and its Fab fragments (**A**, lanes 2 and 4) in non-reducing conditions or of separated chains of IgG (**B**, lanes 4 and 5; **D**, lanes 1 and 2) and pancreatic ribonuclease A (**B**, lane 1), human sera RNase 4 (**B**, lane 2), human sera RNase 3 (**B**, lane 3) by SDS-PAGE in reducing conditions in a gel containing RNA (**A** and **B**) or DNA (**C** and **D**). RNase and DNase activities were revealed by ethidium bromide staining as a sharp dark band on the fluorescent background; negatives shown in **A**, lanes 1, 2; **B**, 1–4; **C**, lane 1; **D**, lane 1. Gels stained with Coomassie R250 show IgG (**A**, lane 3; **C**, lane 2), its Fab fragments (**A**, lane 4), or separated L and H chains (**B**, lane 5; **D**, lane 2). Lane C (**A**) corresponds to control proteins with known molecular masses. The details of the analysis are given in [55, 149].

In situ gel assay of DNase activity of milk IgG (lanes 1, 3, 5–6) and its Fab fragments (lanes 2, 4) by SDS-PAGE in a gel containing DNA. Before electrophoresis the samples were incubated in non-reducing (**E**), or reducing conditions (**F**). DNase activity was revealed after ethidium bromide staining (lanes 3, 4, and 6) (the negatives of the films are shown). Gels stained with Coomassie R250 show positions of IgG (lane 1), its Fab fragments (lane 2) and separated H- and L-chains (lane 5). Arrows indicate the positions of molecular mass markers. The details of the experiments are given in [70].

tions (2-mercaptoethanol). Since contaminating enzymes could not migrate simultaneously with IgG, its Fab fragments, and its separated L-chains, these results prove unambiguously that catalytic activity is a property of the IgG. By the same approach we showed that the L-chain is responsible for DNase and/or RNase activities of IgG and IgM of SLE [54, 55, 149] and that DNase activity is an intrinsic property of IgG from MS [52, 145], and HT patients [84, 142].

We used an in-gel ATPase assay adapted from Gomori's method of P_i precipitation on thin sections of mammalian tissues, which was used earlier for a histochemical study of ATPase activity [151] for *in situ* detection of Abz-dependent formation of P_i in

SDS-PAGE gels [71, 143]. We established suitable conditions for effective precipitation of $Pb_2(PO_4)_3$ in regions of gels containing orthophosphate and for efficient removal of Pb salts non-specifically adsorbed to proteins. After ATPase Ab-dependent formation of $Pb_2(PO_4)_3$, gels were stained with Na_2S, causing the formation of black-brown $Pb_3(S)_2$, or were examined by autoradiography to reveal $[^{32}P]Pb_2(PO_4)_3$.

SDS-PAGE of native milk IgG or purified Fab fragments in nonreducing conditions revealed a single protein band catalyzing formation of orthophosphate from ATP (Fig. 19.12). After complete or partial reduction of IgG using a high (1%) or low [50 mM] concentration of 2-mercaptoethanol, the in-gel assay showed the absence of ATP-hydrolyzing activity in the separated L- or H-chains (Fig. 19.12). However, mild treatment of IgG with 50 mM 2-mercaptoethanol followed by SDS-PAGE in non-reducing or mildly reducing conditions revealed ATP-hydrolyzing activity in the bands corresponding to the original IgG (H_2L_2) and to its partially reduced forms, H_2L and HL (Fig. 19.12). In the same conditions there was no $Pb_2(PO_4)_3$ formation in lanes containing bovine serum albumin or milk casein and other control proteins, which have no ATP-hydrolyzing activity (Fig. 19.12). However, formation of $Pb_3(S)_2$ or $[^{32}P]Pb_2(PO_4)_3$ took place at positions corresponding to alkaline phosphatase and to several nucleotide-hydrolyzing enzymes of E. coli cells extract (Fig. 19.12). The above data strongly indicate that ATP-hydrolyzing activity is an intrinsic property of different oligomeric forms of L- and H-subunits of IgG.

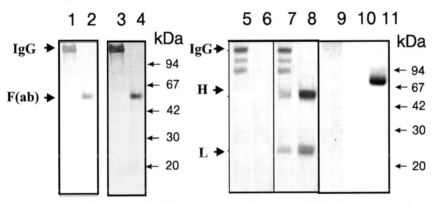

Fig. 19.12 *In situ* gel assay of ATP-hydrolyzing activity of human milk IgG, its Fab fragments, oligomeric forms of H and L chains, and control proteins (7–10 μg per sample) after separation by SDS-PAGE in a mild reducing condition (50 mM 2–mercaptoethanol). Before electrophoresis the samples were incubated in non-reducing (lanes 1 and 2), mild (lanes 5, 9–12) or drastically reducing (lane 6) conditions (1% 2–mercaptoethanol). ATP-hydrolyzing activity was revealed by staining of orthophosphate formed by sequential treatment of the gel with $Pb(NO_3)_2$ and Na_2S. The proteins assayed were catalytic IgG (lanes 1, 5, 6) and its Fab fragments (lane 2), bovine serum albumin (lane 9), milk casein (lane 10), and alkaline phosphatase (lane 11). Gels stained with Coomassie blue show IgG (lane 3) and its Fab fragments (lane 4) after incubation in a non-reducing condition; IgG after incubation in mildly reducing (lane 7) or strongly reducing conditions (lane 8). Arrows indicate the positions of molecular mass markers and control proteins. The details of the experiments are given in [143].

The subunit of an Abz which is responsible for catalysis can therefore be identified by these methods. DNase activity was found in the L-chain of milk IgG [67, 68], whereas in milk sIgA the heavy chain showed high affinity for DNA-cellulose (see above) but no DNase activity [69], suggesting that both light and heavy subunits contribute to the catalytic center. An affinity derivative of $d(pN)_{10}$ (see above) preferentially modified the L-chain of IgG, but both L- and H-chains of sIgA Abzs bound an affinity probe for DNA-binding sites.

Interestingly, the separated L-subunits of ATP-hydrolyzing milk IgG are not capable of hydrolyzing ATP [71, 143]. This activity can be detected only for the original milk ATPase IgG (H_2L_2) and its partially reduced forms, H_2L and HL.

The crystal structure of an Abz with esterase-like activity shows that the ligand *p*-nitrophenyl ester interacts with amino acid residues of both light and heavy chains and that both are required for catalysis [152], and most probably a similar situation occurs in the case of DNase sIgA and ATPase IgG from milk.

In conjunction with Paul's original list (see above), these tests provide strong evidence supporting the catalysis of various reactions by natural Abzs. Some of these tests, and particularly the *in situ* observation of catalytic activity induced by an Ab, its Fab fragment, and one of its individual separated chains, may be considered as the most rigid argument unambiguously assigning observed catalytic activity to Abs versus contaminating proteins, while also providing the basis for subsequent structural and functional analysis of abzyme activity.

19.9
Catalytic Antibodies Catalyzing Transformations of Water and Oxygen Radicals

As mentioned above (Section 19.2), one of the first natural Abs, IgG from the human sera [153] and complexes of mice IgG with antigen [39], reported by Kulberg in the end of 80s, demonstrated superoxide dismutase activity. It was shown that human IgG in the complex with products of catabolic destruction of cell receptors (R-proteins) possesses superoxide dismutase activity and that the specific activity of this complex ($k_{cat} = 2.4 \times 10^9$ M^{-1} c^{-1}) is comparable with that for superoxide dismutase from bovine erythrocytes ($k_{cat} = 0.9 \times 10^9$ M^{-1} c^{-1}). This activity was significantly lower for isolated IgG (37-fold) and R-protein (6.7-fold) [153].

In addition, these authors have shown that antireceptors as components of cellular receptors possess the ability to catalyze superoxide radical production [39]. The ligand-binding components of this receptor, on the contrary, catalyzed superoxide dismutation. Addition of ligand to such a complex made it a superoxide generator. The complex with an excessive ligand possessed reciprocal catalytic properties expressed in dismutation of this radical. *Anti*-idiotypic antibodies and Ab proper had no capacity for catalyzing the superoxide-dependent reaction. In contrast to the former, the Abs were capable of superoxide radical binding. However, in the presence of *anti*-idiotypic Abs they lost these properties. An excessive specific antigen induced superoxide dismutase activity in its complex with the Abs. The antigen-Ab complex with excessive *anti*-idiotypic Abs generated superoxide radical.

Recently, these first findings concerning Abs catalyzing conversion of superoxide radical were continued by [154] using modern methods and approaches. It has been shown that different mono- and polyclonal Abs from various sources as well as their Fab fragments have the intrinsic ability to intercept $^1O_2$ and efficiently reduce it to $O_2^{\bullet-}$, thus offering a mechanism by which oxygen can be reduced and recycled during phagocyte action, thereby potentiating the microbial action of the immune system. These intact Igs and their fragments were shown to catalyze the formation of hydrogen peroxide.

Later, the same authors unambiguously showed that hydrogen peroxide formation is a catalytic process, and they identify the electron source for a quasi-unlimited generation of H_2O_2 [155]. Abs produce up to 500 mole equivalents of H_2O_2 from singlet $^1O_2^*$, without a reduction in rate, and metals or Cl^- were excluded as the electron source. On the basis of isotope incorporation experiments and kinetic data, they propose that Abs use H_2O as an electron source, facilitating its addition to $^1O_2^*$ to form H_2O_3 as the first intermediate in a reaction cascade that eventually leads to H_2O_2. X-ray crystallographic studies with xenon point to putative conserved oxygen binding sites within the Ab fold where this chemistry could be initiated. These findings suggest a protective function of Igs against $^1O_2^*$ and raise the question of whether the need to detoxify $^1O_2^*$ has played a decisive role in the evolution of the Ig fold.

Recently, it was shown that different R or RO radicals, including those generated from the saccharide-containing Fe-XO/HX system, are effectively scavenged by Abzs containing selenium, and these can be considered as promising antioxidants [156]. Two types of selenium-containing monoclonal Abzs demonstrating glutathione peroxidase activity were developed. Ab-dependent scavenging of OH was examined in an iron-containing xanthine oxidase/hypoxanthine system by using a spin trapping method. It was found that the abilities of these two Se-Abzs to scavenge OH surpassed the level of native glutathione peroxidase. M1c8 Abz showed the greatest OH radical scavenger effect [157].().

The interaction of human sera IgG (subclasses G1, G2, G4) with different metal ions (Mg^{2+}, Ca^{2+}, Cu^{2+}, Zn^{2+}, Ni^{2+}, Co^{2+}, Fe^{2+}, and Cr^{3+}) using metal-chelating adsorbent was investigated. The IgG-Cu^{2+} complex-dependent hydrolysis of H_2O_2 was demonstrated [158].

Thus, it seems likely that some Abzs can convert oxygen, via reducing it to $O_2^{\bullet-}$, into hydrogen peroxide, while other Abs hydrolyze H_2O_2. Taken together, we can exclude the possibility that a specific repertoire of different polyclonal Abs can serve as an additional natural system of detoxication of different active forms of oxygen.

19.10
Antibodies with Proteolytic Activities

As mentioned above, most probably the first natural IgG with proteolytic activity was revealed in rabbit sera by Kulberg et al. [41]. At least we could not find earlier publications concerning natural Abzs. Since at that time the study of Abzs was very rare event, the author could not assert uniquely that the observed activity is an intrinsic property of IgGs in spite of checking some of the strict criteria listed above (Section 19.2).

19.10.1
Antibodies Hydrolyzing Vasoactive Intestinal Peptide

The VIP-hydrolyzing Abz from asthma patient plasma was the first natural polyclonal abzyme, the existence of which was shown by various modern methods [34]. VIP-binding auto-Abs were observed in the plasma of 18% of asthma patients and 16% of healthy individuals [159]. The auto-Abs from asthma patients exhibited a larger VIP-binding affinity than those from healthy donors ($K_a = 7.8 \times 10^9$ M^{-1} and 0.13×10^9 M^{-1}, respectively).

Ten fragments of VIP were tested for reactivity with a proteolytic IgG that cleaves full-length VIP(1–28) at the Gln16-Met17 peptide bond. A large COOH-terminal subsequence, VIP(15–28), was bound by the Ab with high affinity ($K_I = 1.25$ nM), suggesting that it is the Ab-binding epitope. VIP(22–28), a short subsequence distant from the scissile bond, inhibited the binding ($K_I = 242$ µM) and hydrolysis ($K_I = 260$ µM) of full-length VIP by this Abz in a competitive fashion. The auto-Ab did not show detectable binding of short VIP subsequences that encompass the scissile bond, VIP(15–21), VIP(11–17), and VIP(13–20). These data show that residues 22–28, located four amino acids distant from the scissile bond, contribute in recognition of VIP by the catalytic Ab [160, 161].

In functional studies, chromatographically separated light chains of these VIP Abzs were found to be active in the hydrolysis of VIP [162, 163]. The light chains hydrolyzed VIP with specific activity 32-fold greater than that of Fab; the hydrolytic activity was saturable ($V_{max} = 0.19$ pmol/min/µg light chains) [164].

The monoclonal Ab raised by immunization with a VIP conjugate with protein selectively hydrolyzes one peptide bond VIP (residues 14–22) [165].

The light chain of the VIP abzyme was expressed in bacteria, purified, and found to possess independent catalytic activity [26, 79]. Subsequently, single-chain Fv constructs containing the VL domain of the *anti*-VIP light chain linked via a 14-residue peptide to its natural VH domain partner possessed an increased affinity for the substrate ground state. From these and other data, a model of catalysis by the *anti*-VIP Abs was proposed. According to the model, the essential catalytic residues are located in the VL domain, and additional residues from the VH domain are involved in high-affinity binding of the substrate [163].

Molecular modeling suggested the presence of a serine protease-like site in the light chain of VIP-hydrolyzing Abz. This assumption was supported by inhibition of the hydrolytic activity of recombinant L-chain by serine protease inhibitors but not by inhibitors of other classes of proteases. The serine protease mechanism was further supported by the observation that catalytic activity was lost following site-directed mutagenesis at a framework region residue, Asp1 [81], and at two complementarity-determining region residues, Ser27a and His93 [26], residues forming a catalytic triad modeled to be similar to that found in serine proteases. The effect of these mutations was specific to catalysis rather than binding, as the affinity of the light chain for the substrate ground state was almost unaffected by mutations at Ser27a or His93. In contrast, a Ser26 single mutant and His27d/Asp28 double mutant displayed increased K_m (by about tenfold) and increased turnover (by about tenfold). Thus, two types of light-chain amino acid residues participating in catalysis were suggested: those essential for catalysis and those participating in VIP binding, and indirectly limiting Abz/substrate turnover. It may be noted that all three critical catalytic residues (Ser27, His93, Asp1) were present in the germline counterpart of the mature V(L), while the mature and germline sequences differed by only four amino acids remote from the catalytic site [81]. Differences between the kinetic constants of the mature and germline light chains were marginal. These data show that catalytic activity of VIP-hydrolyzing Abz is encoded by a germline VL gene, but can potentially be improved by somatic sequence diversification and pairing of the L-chain with the appropriate heavy chain (for review see [47, 48, 82]. Some of the data on the study of VIP-hydrolyzing Abs are discussed in several reviews [165, 166].

19.10.2
Abzymes Hydrolyzing Thyroglobulin, Prothrombin, and Factor VIII

Following the initial discovery of *anti*-VIP Abzs, a number of natural catalytic Abs with diverse activities were subsequently detected in the sera of patients with different AI pathologies. Proteolytic IgG directed at thyroglobulin (Tg), a precursor of thyroid hormone, were found in the sera of patients with HT and SLE [57, 96, 167]. The apparent K_m value for Tg was in the nanomolar range, a property typical of an Ab combining site [57]. The Tg Abz also hydrolyzed tripeptide substrates with lower affinity. No cleavage of Tg by IgG from subjects with HIV-1 infection, or from mice hyperimmunized with an albumin-hapten conjugate was evident, suggesting that generation of Tg-cleaving Abs does not accompany V region affinity maturation in response to unrelated Ags [96].

Prothrombin is the precursor of thrombin, a central enzyme in coagulation. Auto-Abs to prothrombin are associated with thromboembolism, but the mechanisms by which the Abs modulate the coagulation processes are not understood. A panel of 34 monoclonal Ab light chains isolated from patients with multiple myeloma was screened for prothrombinase activity by an electrophoresis method [168, 169]. Two light chains with the activity were identified, and one of the light chains was characterized further. Four cleavage sites in prothrombin were identified by N-terminal sequencing of the fragments: Arg155-Ser156, Arg271-Thr272, Arg284-Thr285, and Arg393-Ser394. The light chain did not cleave albumin, Tg, and annexin V under conditions that readily permitted detectable prothrombin cleavage.

Others have reported factor VIII-cleaving alloantibodies in the sera of patients with severe hemophilia [64]. The identified proteolytic Abs cleaved their protein substrates at several fixed positions, indicating specificity of action. The K_m of Tg- and prothrombin-directed Abs for Tg and a peptide 268–271 of prothrombin were 39 nM and 103 µM, respectively, demonstrating high affinity substrate recognition. In a study of the development of catalytic specificity by Abzs, the polyreactive (non-specific) peptidase activity of serum IgGs from healthy individuals and patients with AI disease was compared, based on the extent to which these sera cleaved a synthetic protease substrate, Pro-Phe-Arg-methylcoumarinamide [96].

A transition from a polyreactive proteolytic activity to autoantigen-directed activity in AI disease was suggested [96]. Finally, the identification of an HIV gp120-cleaving Ab light chain from multiple myeloma patients has recently demonstrated that natural catalytic immunity is not restricted to autoantigenic substrates [66].

In the context of this section, some data concerning monoclonal Abzs may be interesting. Thus, the reactivity of phosphonate ester probes with several available proteolytic Ab fragments was characterized [170]. Somewhat unexpectedly, a phosphonate monoester also formed stable adducts with the Abs. Improved catalytic activity of phage Abs selected by monoester binding was evident. Turnover values (k_{cat}) for a selected Fv construct and a light chain against their preferred model peptide substrates were 0.5 and 0.2 min^{-1}, respectively, and the corresponding K_m values were 8–10 µM. The covalent reactivity of Abs with phosphonate esters suggests their ability to recapitulate the catalytic mechanism utilized by classical serine proteases.

19.11
Human and Animal RNase and DNase Abzymes

As mentioned above (Section 19.2), DNase [35] and RNase [37]. Abzs were first found in the sera of patients with SLE and later in patients with different ADs and the milk of healthy women [50, 83, 126]. Recently Abzs hydrolyzing DNA were detected in cow's milk [171]. In addition, we have shown that DNase and RNase Abzs are present in the blood sera of MRL*lpr/lpr* mice showing pronounced evidence of AI disorder [172] and rabbits immunized with a complex of DNA (or RNA) with proteins or human DNase I and II (Krasnorudskii, personal communication).

The DNase and RNase activities of Abzs have been studied by cleavage of supercoiled plasmid DNA [35, 173], cleavage of single- and double-stranded [^{32}P]ON assayed by PAGE in gels containing urea [37, 174], estimation of acid-insoluble [^{32}P]DNA, flow linear dichroism, and increase in optical density at 260 nm of solutions containing homopolymeric poly(N) or cCMP [37, 51, 52, 54, 55, 122, 145]. Fab fragments of DNase Abzs from SLE patients showed a high affinity (K_m = 43 nM) and a considerable catalytic efficiency (k_{cat} = 14 min^{-1}) evaluated using plasmid DNA and linear dichroism methods, and behaved as monoclonal Igs with a single value of K_m [173].

Polyclonal DNase Abs from patients with AI diseases and from human milk usually demonstrate relatively low specific activities, 0.001–5% of those of known human DNases and RNases, those from human milk usually being more active [50]. However, the relative activities of Abs from the sera varied markedly from patient to patient. Fig. 19.13 illustrates a cleavage of plasmid DNA by Abs from various patients with different ADs after 2 h of incubation. One can see that in this period some Abs cause only single breaks in one strand of supercoiled DNA (lanes 1–3), whereas others cause multiple breaks and as a result the formation of linear DNA (lanes 4–6). The most active Abs hydrolyze DNA into short and medium length ODNs (lanes 7–10).

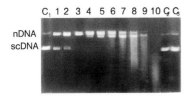

Fig. 19.13 DNase activities of catalytic IgGs from different AD patients in cleavage of supercoiled (sc) and nicked (n) pBR322 DNA: lanes 1–10 correspond to Abs from the sera of 10 different patients; lane C_1 – DNA incubated alone; lanes C_2 and C_3 – DNA incubated with Ab from the sera of two healthy donors. The estimation of the possible relative DNase activities of Abs from human milk and from the sera of patients with different ADs were carried out as described in [51, 52, 56, 70, 73, 126].

It should be mentioned that Fig. 19.13 illustrates a range of possible changes of the relative DNase activities for patients with different ADs and viral pathologies. When passing from one pathology to another, only the relative percentages of patients with low, middle and high DNase activities are usually changed.

Interestingly, the specific DNase and RNase activities of IgM from the sera of SLE patients were usually about 5–10 times higher than those of IgG [55, 149], and we recently observed a similar situation for Abzs from the sera of patients with other AI diseases (data in preparation). The switch from the synthesis of IgM to IgG antibodies may be accompanied by the selection of Abzs possessing weaker catalytic properties [149].

IgG and sIgA Abzs from human milk behave mainly as monoclonal Igs in the hydrolysis of plasmid DNA and ODNs [69, 70] in spite of the possibility of their separation as several subfractions under affinity chromatography on DNA-cellulose (see Section 19.6). At the same time, we recently showed diverse affinity and substrate specificities toward ODNs of IgGs from patients with SLE and MS [145, 149]. The affinity of MS IgGs for plasmid DNA was extremely high and did not allow estimation of K_m using different concentrations of DNA; we therefore determined K_I using plasmid DNA (K_I = 0.34 nM) as an inhibitor of Ab-dependent hydrolysis of $d(pT)_{10}$ and found that it is about 120 times lower than the K_m for DNA (43 nM) for IgG from SLE patients [62].

The ODN cleavage patterns were distinct for different patients [145] (Fig. 19.14); some Abzs were sequence-dependent, whereas others produced both 5′-phosphate-terminated products like those of DNase I and 3′-phosphate terminated products typical of DNase II, or hydrolyzed hetero-ODN in a sequence-independent manner. Hydrolysis of ODNs was also strongly dependent on reaction conditions, and, on addition of EGTA, $MgCl_2$, or NaCl. MS IgGs, demonstrated different combinations of endo- and exonuclease activities, but the properties of the DNase Abzs distinguished them from other known DNases [145].

Interestingly, polyclonal Abzs from the sera of patients with different typical ADs usually hydrolyze both single- and double-stranded DNA with comparable efficiency, or hydrolysis of double-stranded plasmid DNA may be remarkably faster [50, 83, 126]. As we have recently shown, Abs of patients with atherosclerosis, which could not be assigned to pathologies with typical disorder of immune status, also possess DNAse activity, but these Abzs effectively hydrolyze only single-stranded DNA. In addition, IgGs from the sera of patients with one of AI diabetes demonstrate mainly

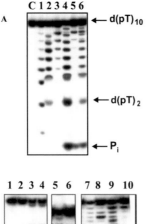
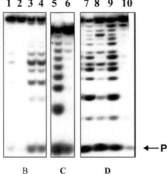

Fig. 19.14 Demonstration of catalytic diversity of MS IgGs in the hydrolysis of different ODNs in comparison with substrate hydrolysis by human sera DNases. Cleavage patterns of $[^{32}P]d(pT)_{10}$ (5 nM) by IgGs from sera of MS patients and by different DNases (A). The ODN was digested with DNase II (lane 1; 50 μmg/mL, 50 mM sodium acetate (pH 5.0), 0.8 mM $MgCl_2$), DNase I (lane 2; 50 μmg/mL, 20 mM Tris-Hepes (pH 7.4), 4.8 mM $MgCl_2$, and 0.5 mM $CaCl_2$), or with Abs from different patients (lanes 3–6, 30–90 μmg/mL, 20 mM Tris-Hepes (pH 7.4), 5 mM $MgCl_2$). Lane 3 corresponds to Abz4, lane 4 to Abz1, lane 5 to Abz5, lane 6 to Abz2; lane C to ODN alone.
Influence of some reaction components on the hydrolysis pattern of various 5 nM $[5'-^{32}P]$ODNs (B, dAGCAGTGGCGCCCGA; C, $d(pT)_{10}$; D, dCCAGTCAC-GACGTT) by Abs from sera of different MS patients (lanes 1–4 – Abz3; lanes 5–8 – Abz1; lanes 9–10 - Abz2) in 20 mM Tris-Hepes, pH 7.4, containing (with the exception of lane 10) 5 mM $MgCl_2$ and: lane 1 – 7 nM plasmid DNA; lane 2 – 0.2 μmM none-labeled ODN of the same sequence; lane 3, 5, 7, 9 – no additions; lane 4 – 100 mM NaCl; lane 6 – 2 mM potassium phosphate; lane 8 – 0.3 mM EGTA; lane 10 – 1 mM EDTA. The details of the analysis are given in [145].

very low activities in the hydrolysis of ss ODNs, and a detectable level of plasmid DNA hydrolysis by these Igs is a very rare event.

We have shown that the specificity of IgG Abzs from HT and polyarthritis patients [56], SLE [37], MS [52], and different types of hepatitis [51] for classic homopolynucleotide poly(N) substrates, cCMP, and $tRNA^{Phe}$ with a compact and stable structure [53, 56, 147, 148] differed but was correlated with the type of disease and was distinguishable from those of known pro- and eucaryotic RNases. The activity was strongly dependent on the patient, but in general increased in the order diabetes < hepatitis < polyarthritis < AID ≤ HT < SLE ≤ MS.

Table 19.1 summarizes the relative rates of various hydrolytic reactions catalyzed by human RNases and Abzs isolated from blood of patients with various ADs. Interestingly, all IgG- and IgM-abzymes effectively hydrolyzed poly(A), whereas RNase A and related human blood RNases were almost inactive with this substrate [37, 52–54, 56, 122, 147, 148, 174, 175]. Using the substrate specificity of the hydrolytic reaction catalyzed by Abzs, it is possible to discriminate each AD studied [53, 56]. The ratio of rates of poly(A) and poly(C) hydrolytic reactions catalyzed by RNases and Abzs is especially demonstrable (Table 19.1).

The specific activity was about 1–20% of that of RNase A and of six known human sera RNases, while poly(A) was hydrolyzed 2–10 times faster than by RNase A, one of the most active RNases known [37, 51, 52, 56]. The specific activity of homogeneous Abs of several SLE and MS patients was about 40–400% of that of RNase A [52]. The

Tab. 19.1: Relative specific activities of IgG preparations isolated from blood of patients with various ADs and human plasma RNAse A and nucleases evaluated in hydrolysis of polyribonucleotides and cCMP.

Catalyst	Relative specific activity, %				
	cCMP	poly(U)	poly(A)	poly(C)	Total RNA
RNAse A	100	2 (200)**	0.01	100 (10000)	15 (1500)
Human blood RNases	–	3–16	0	100	5–10
IgG-Hepatitis B	4–5	0.02 (0,3)	0.06	0	
IgG-Polyarthritis	4	0.2 (10)	0.02	14 (700)	3 (150)
IgG-H. thyroiditis	10	0.2 (13)	0.016	10 (625)	2 (125)
IgG-SLE	35	0.5–0.8 (4)	0.2	4 (20)	5.0 (25)
IgG-Multiple sclerosis	4–8	0.6–4.0 (8)	0.1–0.2	5–7 (43)	10–15 (83)

* Specific activities were evaluated per mole of protein. Specific activities of RNase A catalyzing reactions of poly(C) and cCMP hydrolysis were taken as 100%, and specific activities of each catalyst (with each substrate) were normalized to RNase activity. Numbers in brackets indicate reaction rate ratio of hydrolysis of each polymer and poly(A).

pH dependencies and salt effects also varied in different patients, but activity was specifically stimulated by Mg^{2+} ions, which essentially completely inhibit all known human RNases. Thus, RNA-hydrolyzing Abzs may be the most active among natural Abzs known at present.

According to our data, the diversity of RNase activity of polyclonal Abs is much greater than that of DNase activity. We investigated hydrolysis by IgGs from the sera of 50 patients with different diseases of the *in vitro* transcript of human mitochondrial $tRNA^{Lys}$, which has a less stable structure than $tRNA^{Phe}$ [53, 56, 147, 148]. In contrast to the classical substrates and $tRNA^{Phe}$, there was no correlation between a patient's IgG cleavage specificity for $tRNA^{Lys}$ and a specific disease, and each patient demonstrated an individual repertoire of polyclonal RNA-hydrolyzing IgGs independently of the disease. Fig. 19.15 demonstrates some patterns of $tRNA^{Lys}$ cleavage by Abs from patients with several different pathologies, although if we take into account all ADs analyzed, the variation of substrate specificities of RNase Abs is significantly wider [53, 56, 147, 148, 175].

Polyclonal Abzs usually contain a major subfraction characteristic of each pathology and share common properties with classical substrates in each disease, which distinguishes them from Abs of other diseases as seen by their specificity, kinetics and optimal conditions (see Table 19.1). However, RNase Abzs from patients with the same AI disease can demonstrate pronounced catalytic diversity. For example, the pH dependencies of the initial rates of $ribo(pA)_{13}$ hydrolysis by 10 IgMs (Fig. 19.16) [55] had individual features that contrast with all human RNases, which show a single pH optimum; one had a pronounced optimum at pH 8.5, two had two pH optima (for example, at pH 7.8 and 8.5), three had several pH optima, while five had no clear pH optimum between 6 and 9. Only one IgM having a pronounced optimum at pH 8.5 demonstrated a single K_m for $(pA)_{13}$ substrate, and two with two pH optima demonstrated two K_m and V_{max} values. Five IgMs with comparable activity at pHs

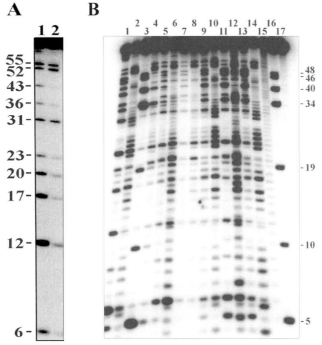

Fig. 19.15 Cleavage of the 5'-[^{32}P] in vitro transcript of human mitochondrial tRNA^{Lys} with a point mutation A9C by RNase A (**A**) and Abs isolated from the sera of different patients (**B**). Autoradiography of 15% denaturing polyacrylamide gels. **A** Lane 1 – hydrolysis in the low salt conditions; lane 2 – hydrolysis in the presence of 5 mM MgCl$_2$. **B** Lanes 1–2 and 3–4 correspond to cleavage patterns by two unrelated patients with Hashimoto's thyroiditis; lanes 5–6 and 7–8 to two patients with hepatitis B; lanes 9–10, 11–12 and 13–14, to three patients with SLE. Lanes 1, 3, 5, 7, 9, 11, 13 – hydrolysis in the low salt conditions; lanes 2, 4, 6, 8, 10, 12, 14 – hydrolysis in the presence of 5 mM MgCl$_2$. Lane 15, partial hydrolysis by imidazole; lane 16, partial hydrolysis by RNase T1; lane 17, control (incubation of RNA without Abs). The details of the analysis are given in [53].

from 6.0 to 9.0 demonstrated fan-like Cornish-Bowden dependencies and smooth changes of apparent K_m and V_{max} with increasing substrate concentration and no evident intersection points (Fig. 19.17). The remaining 2 IgMs showed intermediate characteristics and many intersection points with a change in apparent K_m in the range 40–500 nM. RNase A and several RNases from SLE patients had classical Cornish-Bowden dependencies and a single K_m with a substrate affinity about 2–3 orders of magnitude lower than IgMs (Table 19.2) [55].

Nearly the same situation was revealed for SLE IgGs (Table 19.2) [149]. Substrate specificity of SLE IgG RNases is unique within certain limits for practically every SLE patient. For example, the pH dependence of the initial rate of [5'-^{32}P](pA)$_{13}$ hydrolysis by IgGs showed individual features in all 6 IgGs studied. Fig. 19.16 demonstrates three pH dependencies for sera IgGs of three different patients. In contrast to all human RNases, which have one pronounced pH optimum [6.8–7.5] for hydrolysis of

Tab. 19.2: Kinetic parameters for hydrolysis of different ribo- and deoxyoligonucleotides by catalytic SLE-Ab, human serum RNases and pancreatic ribonuclease A.

Substrate	Catalyst	Apparent K_m^*, M	k_{cat}, min^{-1}	k_{cat}/K_m, M^{-1}min^{-1}
		IgG antibodies		
d(pA)$_{13}$	IgG-1**	7×10^{-8}	2.0×10^{-2}	2.9×10^5
(pA)$_{13}$		4×10^{-8}	1.4	3.5×10^7
d(pA)$_{13}$	IgG-2	$4.7 \times 10^{-8} - 3.0 \times 10^{-7***}$	$2.0 \times 10^{-3} - 7.1 \times 10^{-2}$	$0.43 - 2.4 \times 10^5$
(pA)$_{13}$		$5.1 \times 10^{-8} - 4.4 \times 10^{-7}$	$0.12 - 0.84$	$1.9 - 2.3 \times 10^6$
p(U)$_{10}$		$9.0 \times 10^{-8} - 4.1 \times 10^{-7}$	$3.2 \times 10^{-3} - 1.3 \times 10^{-2}$	$3.1 - 3.6 \times 10^4$
(pA)$_{13}$	IgG-3 – IgG-4	$1 \times 10^{-8} - 2 \times 10^{-6***}$	$1.0 \times 10^{-2} - 2.5$	$1 - 1.25 \times 10^6$
		IgM antibodies		
d(pA)$_{13}$	IgM-1	5.4×10^{-8}	2.7×10^{-2}	5.0×10^5
(pA)$_{13}$		4.3×10^{-8}	1.5	3.4×10^7
p(U)$_{10}$		1.3×10^{-7}	5.8	4.5×10^7
p(A)$_{13}$	IgM-2	8.7×10^{-8} (K_{m1})	1.8	2.1×10^7
		10.2×10^{-8} (K_{m2})	3.2	3.1×10^7
p(A)$_{13}$	IgM-5	$0.8 \times 10^{-8} - 8 \times 10^{-6***}$	$2.0 \times 10^{-2} - 8.0$	$1.0 - 2.7 \times 10^6$
		Human RNases		
p(A)$_{13}$	RNase A	3.4×10^{-6}	2.2×10^{-2}	6.4×10^3
p(A)$_{13}$	RNase 3	4.9×10^{-6}	1.7×10^{-2}	3.5×10^3
p(U)$_{10}$		2.1×10^{-6}	37	1.8×10^7
p(A)$_{13}$	RNase 4	7.2×10^{-6}	5.2×10^{-2}	7.2×10^3
p(U)$_{10}$		5.6×10^{-6}	26	4.6×10^6

* The errors of the values were within ±10–30%. The same number of IgGs and IgMs correspond to the same number of SLE patient.
*** Several K_m and V_{max} values are lying in the indicated ranges.

poly(A) [176–178] (Fig. 19.16), catalytic IgGs usually showed high activity over a wide range of pH values between 6.0 and 9.5. Interestingly, the mode of [5′–^{32}P](pA)$_{13}$ hydrolysis slightly changes over a wide range of pH values (6.0–9.0) in the case of the above listed Abs. Interestingly, in contrast to the hydrolysis of plasmid DNA, which is characterized by one K_m (and V_{max}) value [173], we have observed for some SLE IgGs and IgMs samples two or more K_m (and V_{max}) values for ODN substrate.

Very different catalytic properties of IgG subfractions entering into the composition of the analyzed total polyclonal SLE IgGs and their variations between patients were also demonstrated by kinetic and thermodynamic studies [149]. In addition, IgG purified on DNA-cellulose was fractionated on Sepharose bearing immobilized monoclonal *anti*-λ or *anti*-κ-L-chains of human Ig, 60–70% of IgG was adsorbed by Abs against κ- and 30–40% by Abs against λ-light chains (Fig. 19.18) [55, 149]. The fraction containing the κ-light chain was about 10–50 times more active in thehydrolysis of both RNA and DNA. Analysis of subfractions of Fab fragments by the same method led to the same conclusion. A similar situation was observed for SLE IgM and Abs from patients with other AD pathologies and for Abzs from human milk [69, 70].

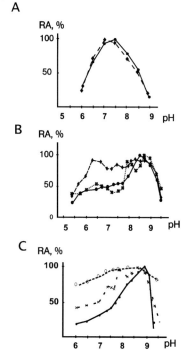

Fig. 19.16 Influence of pH on the relative activity (RA) of human sera RNase 3 (•) and RNase4 (♦) (**A**), IgM (**B**) and IgG (**C**) preparations from sera of several different SLE patients in the hydrolysis of $[5'-^{32}P](pA)_{13}$: three IgMs, (•) – IgM-1, (×) – IgM-2 and (♦) – IgM-3 (**B**), and three IgGs: (•) – IgG-1, (×) – IgG-2, and (○) – IgG-4 (**C**), were used. The details of the analysis are given in [55, 149]. IgM and IgG samples having the same number correspond the same SLE patient.

In contrast to normal enzymes, for example, known human sera DNases of I and II types, different subfractions of catalytic Abs can differ in their isoelectric points. Therefore, we recently analyzed DNase activity of electrophoretically homogeneous preparations of IgGs from the sera of SLE patients after their separation using the method of electrofocusing (Kuznetsova, personal communication). It was shown that DNA-hydrolyzing activity was associated with many IgGs having isoelectric points from 3 to 8. The same situation was observed for DNase IgGs from the sera of MRL*lpr/lpr* mice which spontaneously developed pronounced symptoms of AI disorder. At the same time, DNA-hydrolyzing activities of DNases from human and mouse blood were neatly localized in agarose gel after electrofocusing. Thus, the data of this approach can also be used for additional characterization of the diversity of polyclonal Abzs.

Taken together, these data demonstrate that subfractions of polyclonal Abzs from the sera of AD patients and from milk of healthy human mothers display a multitude of different enzymic specificities for DNA, RNA or tRNA substrates, pH, and salt concentration [50, 52–55, 67–70, 83, 126, 145, 147, 149]. They differ in the polyclonal complexity of patients, who may have a relatively small or extremely large pool of DNase and RNase Abzs with differences in relative amounts of light chains of κ- and λ-types, optimal pH, net charge, activation by Mg^{2+} ions, substrate specificities, and apparent K_m and V_{max}. These data lead to new questions concerning DNase and especially RNase Abzs: why do humans have such activities, why are there so many types, and are they advantageous or not? Our findings that AI patients and human

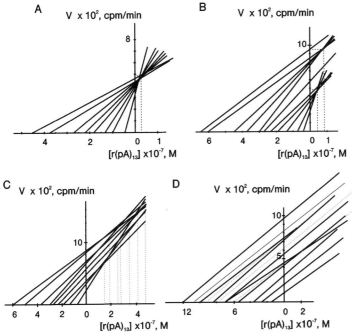

Fig. 19.17 Initial rates of IgM-dependent hydrolysis of $[5'-^{32}P](pA)_{13}$ as a function of its concentration for Ab preparations from sera of four different SLE patients. The K_m and V_{max} values were determined according to the representation of Cornish-Bowden (**A–D**). The common intersection point gives the values of K_m and V_{max} in the case of IgM-1 (**A**); two values of both K_m and V_{max} values for IgM-2 (**B**); a set of intersection points for $(pA)_{13}$ shows a range of K_m and V_{max} values of IgM-3 (**C**); the absence of obvious intersection points in the IgM-4 preparation (**D**) show stepwise change of the K_m and V_{max} values as a result of the functioning of many subfractions of IgM having different kinetic and thermodynamic characteristics of their interaction with the substrate. The details of the experiments are given in [55].

mothers may be considered as large "reservoirs" of natural DNase and RNase Abzs of new types may have applications in the scientific, medical, and biotechnological fields, some of which are discussed in Section 19.13.

It should be mentioned that one of the possible ways of more detailed investigation of natural polyclonal Abzs is to compare them with monoclonal Abs with the same activities. However, there are not yet any examples in the literature of human monoclonal DNase and RNase Abzs.

The mouse monoclonal *anti*-DNA IgG BV 04-01 and its Fab fragment catalyzed hydrolysis of both single- and double-stranded DNA in the presence of Mg^{2+} ions [179] with efficient hydrolysis of the C-rich region of the ODN A7C7ATATAGCGCGT7 as well as preference for cleavage within CG-rich regions of double-stranded DNA. Data on the specificity of ss DNA hydrolysis were used to model the catalytically active site utilizing the previously solved X-ray structure of $(dT)_3$-liganded Fab BV 04-01 [180]. The resulting model suggested that BV 04-01 activates the target phosphodiester bond by induction of conformational strain. In addition, the Abz-DNA complex contained

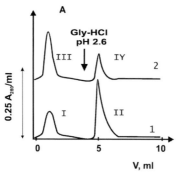

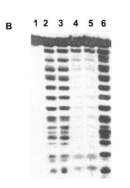

Fig. 19.18 A Separation of IgG from serum of an SLE patient by chromatography on Sepharose bearing immobilized monoclonal IgGs against κ- (curve 1) or λ (curve 2) L-chain of human Ig. (–): absorption at 280 nm. B Cleavage of $[5'-^{32}P](pA)_{13}$ by the initial SLE-IgG (lane 6) and its subfractions separated on adsorbents with immobilized monoclonal IgGs against κ- (lanes 2 and 4) and against λ (lanes 3 and 5) L-chains. Lanes 2 and 3 correspond to IgGs interacting with anti-κ-Abs (peaks II and III), lanes 4 and 5 interacting with anti-λ-Abs (peaks I and IY). IgG and its subfractions (10 μg/mL) were incubated with substrate at 37 °C for 1 h and the products were analyzed using urea-electrophoresis in 20% polyacrylamide gel (autoradiograph). The details of the analysis are given in [149].

a potential Mg^{2+} ion coordination site composed of the L32Tyr and L27dHis amino acid side chains and a DNA 3'-phosphodiester group, suggesting that induction of strain and metal coordination could be constituents of the mechanism by which this Ab catalyzed DNA hydrolysis. Sequence data for the BV 04-01 VH and VL genes suggested that the proposed catalytic Ab active site was germline encoded, suggesting that catalytic activity represents an important function of some Ab molecules.

19.12
Human Milk Abzymes with Various Activities

19.12.1
Abzymes with Protein Kinase Activity

As mentioned above, the milk of healthy parturient women contains small subfractions of polyclonal IgGs and sIgAs catalyzing reactions of hydrolysis of DNA and RNA [67–70]; it also exhibits phosphatase activity by catalyzing the cleavage of 5'-terminal phosphate of DNA and RNA [67–70]. The catalytic properties of these Abzs are different from those of Abzs from the sera of AD patients, but all these Abs catalyze chemical reactions of the same types (Section 19.10). At the same time, human milk contains extremely unusual Abzs, which have not yet been found in the sera of AD patients.

The first example of such Abzs was milk sIgA with protein kinase activity [43, 72, 73, 146, 181–183]. This Abz is interesting from many points of views. Thus, it is a first example of the catalytic activity of IgA Igs, of Abzs catalyzing a bisubstrate reaction, and of natural Abs with synthetic activity. Secondly, enzymatic phosphorylation of proteins plays a fundamental role in the regulation of key physiological processes. Therefore a biological function of such Abzs may be very important.

As is known, most artificial Abzs possess hydrolytic activity, and there are only a few examples of such Abzs with synthetic activities [24, 49, 184–187]. Therefore, it was difficult to imagine that natural catalytic Abzs with synthetic activity could in principle exist. But the ability of sIgA to phosphorylate serine residues of various milk proteins (approx. 15) in the presence of γ-[^{32}P]-dATP was shown by us to be an intrinsic property of this Ab [43, 72, 73, 146]. In order to prove the existence of this catalytic activity we have purified milk sIgA using many different sequential chromatographies, including ones with very drastic conditions (Section 19.6), and have checked all the rigid criteria allowed by Paul et al. [34] as well as all additional criteria elaborated by us (Section 19.6) [43, 72, 73, 146].

As mentioned above (Section 19.6), chromatography on ATP-Sepharose showed that the bound sIgA was composed of fractions having different affinities for ATP; the amount of protein in the peaks eluted by different concentrations of NaCl and of 3.0 M $MgCl_2$ depends on the salt concentration used [43, 72, 73, 146]. Moreover, the use of a gradient of concentration of NaCl and/or $MgCl_2$ usually leads to a distribution of sIgA along the whole volume of the eluates. Consequently, the preparations of polyclonal sIgA consist of a very large number of different subfractions of Abzs having different affinities for ATP. The same situation was observed for the affinity of sIgA-Abz for casein [43, 72, 73, 146]. All the data obtained have unambiguously shown that Abzs with protein kinase activity are generated in the tissue of healthy mothers.

Interestingly, the substrate specificity of this Abz is quite unique and different from all known viral-, pro- or eucaryotic protein kinases [73]. For example, in contrast to these protein kinases, the sIgA fraction transfers [^{32}P]phosphate only onto serine residues of proteins, and utilizes as substrate not only ATP but also other NTPs and dNTPs with comparable efficiency: ATP (100%), dATP (70–80%), GTP and dGTP (200–300%), and UTP and dTTP (30–60%) [73]. We have found only one protein kinase of non-immunoglobulin (PK) in human milk – a 71 kDa kinase which specifically phosphorylates only casein and uses as substrate only ATP ($K_m = 10^{-4}$ M). In addition, this non-immunoglobulin PK completely loses its activity under drastic conditions of Abzs purification, and it can be purified using only very mild conditions (Section 19.6). In contrast to this kinase, polyclonal fractions of sIgA-PK (having different affinities for ATP-Sepharose, see Section 19.6) phosphorylate about 15 proteins of human milk (Fig. 19.19) and utilize as substrates all NTPs with K_Ms from 10^{-7} to 10^{-5}M [13, 14]. It should be mentioned that in contrast to 71 kDa kinase and some different commercial protein kinases, milk Abzs could not effectively phosphorylate thermo-inactivated human casein, and the drastic conditions dephosphorylated hu-

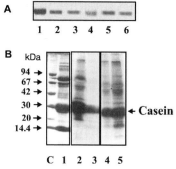

Fig. 19.19 A SDS-PAGE analysis of two different human milk sIgAs (lanes 1–4) and one IgG (lanes 5–6) for relative enzymic activity in casein phosphorylation (radioautograph). The PK activity was analyzed using 0.03 mg/mL of Abs, 0.5 mg/mL of human milk casein, 1 µM γ[^{32}P]ATP or 10 µM [^{32}P]ortho-phosphate of the same specific radioactivity and optimal conditions: 50 mM Tris-HCl buffer (pH 6.8), 1 mM MgCl$_2$, 0.1 mM EDTA, 50 mM NaCl: lanes 1, 3, and 5 correspond to ATP; lanes 2, 4, and 6 to ortho-phosphate. **B** SDS-PAGE analysis of total milk protein phosphorylation (radioautograph) after addition of 1 µM γt[^{32}P]ATP to fresh defatted milk (lane 2), after removal of PK Abz from the defatted milk using protein A-Sepharose (lane 3), and further addition to the latter milk protein preparation of purified preparation of IgG (lane 4), or sIgA (lane 5). Lane 1 corresponds to gel stained with Comassie R250. Arrows indicate the positions of molecular mass markers (lane C). The details of the experiments are given in [72–74].

man or bovine milk casein. The best substrate for milk Abzs is human milk casein purified using very mild conditions, although it can utilize native bovine milk casein.

Interestingly, all known kinases usually use as a donor of phosphate group different nucleotides, like ATP, whose phosphoanhydride bonds are rich in energy [181–183]. In the literature, there are no examples of kinases that are capable of directly using inorganic ortho-phosphate as a donor of phosphate group. The interaction of proteins with ortho-phosphate leading to the formation of a covalent bond is an extremely rare event. For example, it was shown that incubation of pyrophosphatase with ortho-phosphate led to protein phosphorylation [188–190]. However, we have not found in the literature any examples of enzymes catalyzing the transfer of ortho-phosphate from solution to any substrate (protein, lipid, etc.). At the same time, milk catalytic IgG and sIgA possess this unique capability. Incubation of electrophoretically homogeneous human milk casein with very carefully purified IgG or sIgA in the presence of [^{32}P]ortho-phosphate leads to effective phosphorylation of casein (Fig. 19.19). The affinity of ortho-phosphate for Abzs (in terms of apparent K_m) was ~ 10 times lower than ATP, but the relative value of V_{max} of casein phosphorylation, depending on the Abs preparation used, varies from 20% of that for ATP to 50%.

According to our preliminary data, milk catalytic sIgA and IgG can react with one of the substrates (NTPs, dNTPs, orthophosphate), forming a covalently modified Abz, and can then transfer the activated phosphoryl group to the casein. In another words, these Abs can be considered as an example of Abzs demonstrating a ping-pong reaction mechanism.

As follows from the thermodynamics of an aqueous medium, phosphorylation of proteins directly with *ortho*-phosphate is not possible, since it demands quite a large consumption of energy. In order to explain the possibility of pyrophosphatase phosphorylation directly with *ortho*-phosphate, the authors analyzed X-ray analysis data of this enzyme. It was revealed that the phosphorylated residue of pyrophosphatase is located in a very hydrophobic "pocket" of the enzyme, which is not accessible by molecules of water. Finally, the possibility of this reaction was explained by a very high hydrophobicity of this pocket, which makes it to some extent comparable to the gas phase [191, 192]. As is known, in some cases estimation of active center hydrophobicity shows that this exceeds that of many organic solvents. In this case, enzyme-dependent chemical reaction can proceed according to the thermodynamics of a gas phase, where such direct transfer of phosphate with formation of a covalent bond is not forbidden. Taken together, one cannot exclude a similar mechanism for milk Abs phosphorylation directly by *ortho*-phosphate with subsequent transfer of the activated phosphoryl group to casein.

Interestingly, proteolytic hydrolysis of casein phosphorylated by 71 kDa milk kinase resulted in the formation of many very short labeled polypeptides, whereas digestion of casein phosphorylated by sIgA fractions led to the preferential formation of two high-molecular-mass labeled polypeptides [73]. These results are evidence in favor of the phosphorylation of casein at different sites by sIgA-PK fractions and by 71 kDa protein kinase.

The optimal conditions and pH optimum (6.8) for the PK activity of sIgA were found [43, 72, 73]. In contrast to known kinases [181–183], sIgA catalyzes the reactions at about 10–20 times lower concentration of $MgCl_2$ and has a lower pH optimum (pH optimum for PKs is about 8.0). Typically of PK's phenomena of autophosphorylation of the proteins, phosphorylation of histones, and activation of PKs by iron ions [181–183] was not observed in case of sIgA [73].

The specific activity of sIgAs was often higher than that for IgG Abzs and, depending on the donor's milk, was estimated to be 2–10% (4–25 units/mg; non-separated on ATP-Sepharose sIgA, average of 17 donors' milk) [73] in comparison with that for known PKs [181–183]. The K_m values for ATP in the case of sIgA fractions (polyclonal in origin) having different affinities for ATP-Sepharose were estimated to be in the range: 15.0 µM. (peak 2) to 0.1 µM (peak 5, Fig. 19.7, Section 19.6).

Taken together, small fractions of IgG and sIgA from human milk are capable of phosphorylating many proteins of human milk.

19.12.2
Abzymes with Lipid Kinase Activity

A study of homogeneous milk IgGs and sIgAs with protein kinase activity revealed ^{32}P-labeling of low- (and middle-, see below) molecular-weight compounds which gave a visible radioactive background of the gel after SDS-PAGE [74, 146]. Interestingly, these ^{32}P-labeled compounds were not removed from Abs by dialysis, precipitation with 5–10% trichloroacetic acid or ammonium sulfate, gel filtration of Abs in salt-containing solutions, and washing of protein A-Sepharose-bound Abs with 1%

Triton X-100. These data are evidence in favor of a very high affinity of sIgA for these endogenous milk compounds of low molecular weight. Separation of sIgA from IgG using different chromatographic procedures including the use of a DEAE-cellulose with subsequent gel filtration in 0.05 M NaOH or acidic conditions (Sections 19.6 and 19.7) indicated that the low-molecular-weight compounds that are tightly bound with these Abs can be partially separated from the Abs (Fig. 19.20). Only gel filtration of Abs in a buffer containing 5% dioxane (non-denaturing resolution) led to effective removal of these compounds, and they were completely separated in denaturing conditions by extraction of the Ab pellets (or solutions) with a chloroform-methanol mixture (2:1) (Fig. 19.20) [74, 146].

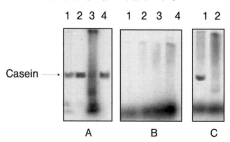

Fig. 19.20 12% SDS-PAGE analysis of phosphorylation of casein and lipids by different sIgA preparations. **A** Phosphorylation of casein by sIgA from two different donors: lanes 1, 2 and 3, 4, respectively. ^{32}P-labeled casein was precipitated with trichloroacetic acid and then was electrophoresed before (lanes 1 and 3) and after treatment of the pellet with methanol-chloroform mixture (lanes 2 and 4). **B** The decrease in the level of ^{32}P background conditioned by phosphorylation of tightly bound lipids after different steps of sIgA purification: lane 4 – protein A-Sepharose; lane 3 – DEAE-cellulose; lane 2 – gel filtration of Abs using acidic buffer; lane 4 – ATP- Sepharose. **C** Influence of casein (1 mg/mL) on a phosphorylation of endogenous lipids; incubation of sIgA in the absence (lane 2) and in the presence (lane 1) of casein. The details of the analysis are given in [74].

In some cases, the relative amount of ^{32}P-labeled low-molecular-weight compounds was so great that it was impossible to reveal ^{32}P-labeled casein because of very high background (Fig. 19.20). Therefore, after phosphorylation of casein for analysis of its phosphorylation before SDS-PAGE, these compounds were removed from protein solutions by extraction with the chloroform-methanol mixture [Fig. 19.20] [74, 146]. TLC analysis of ^{32}P-labeled products in a system usually used for the study of small compounds of a different nature has shown that the labeled compounds are phospholipids [74, 146].

It is known that different lipids are very important for leaving cells (for review see [193, 194]). Different lipids are now recognized to play a dual role in cellular signaling, acting as intracellular as well as extracellular signaling molecules. They are regulators of heart rate, oxidative burst, neurite retraction or platelet activation.

The main lipids of human milk are di- and triacylglycerols (98%) [194–196]. Phospholipids are mostly also glycerolipids. The relative amounts of other phospholipids in human milk was estimated as (weight percent of total lipids): phosphatidyl choline (30.0) phosphatidyl ethanolamine (28.0), phosphatidylserine (5.3), phosphatidyl inositol (4.0), sphingomyelin (31.0) [194–196]. Human milk also contains small amounts of some other lipids. The total amount of different phospholipids was estimated by different authors to be about 0.27–0.4 g/L [194–196].

Milk contains sphingomyelins, acidic glycosphingolipids or gangliosides, and neutral glycosylceramides [194–196]. The amount of gangliosides GM1, GM3, and GD3 is about 11 mg/L. Gangliosides contain one (GM1, GM3) or two (GD3) sialic acids (N-acetylneuraminic acid) as part of the carbohydrate group and are acidic. They differ also in the number of sugar moieties.

The ganglioside GM1 binds the enterotoxins of *Vibrio cholerae, E. coli,* and *Campylobacter jejuni* [193–195, 197]. Gangliosides were postulated to promote fusion of microdroplets of lipid into globules in the mammary gland secreting cell (reviewed in [193–195, 197]). It was found that series of glycolipids bound shigatoxin and the shiga-like toxin receptors. These compounds may be relevant to the protection of the breast-fed infant from several pathogens including HIV.

Taking the above data into account, we aimed to find out what kind of lipids are tightly bound with milk IgG and sIgA [74, 75]. We have present evidence that a small fraction of sIgA from the milk of healthy mothers is tightly bound to unusual minor lipids which may be phosphorylated after the addition of γ-[^{32}P]ATP to preparations of electrophoretically homogeneous Abs or directly to native milk (Fig. 19.21). Like known gangliosides GM1, GM3, and GD3, two minor lipids were shown to contain residues of neuraminic acid (Fig. 19.21). However, in contrast to the gangliosides containing one or two residues of fatty acid, minor lipids show four or five of such residues, and these lipids cannot be oxidized with periodate (Fig. 19.21). Thus, the minor lipids cannot be assigning to known lipids containing neuraminic acid.

Application of a set of rigid criteria worked out previously and some additional criteria (Section 19.7) allowed us to conclude that the observed minor lipid kinase activity is an intrinsic property of sIgA and IgG from human milk and is not due to copurifying enzymes (personal communication, Gorbunov, Karataeva). Interestingly, human milk minor lipids are bound to IgG and sIgA so tightly that a small part of them remains bound to Abs even after SDS-PAGE of a mixture of IgG and sIgA purified by chromatography on protein A-Sepharose (Fig. 19.21E). To allow Abz renaturation, the gel was washed and incubated for the removal of SDS using special conditions. To reveal the products of the lipid phosphorylation, the gel was incubated with γ-[^{32}P]ATP. The positions of Abs on the gel were revealed by Coomassie blue staining and coincided with positions of [^{32}P]lipids (Fig. 19.21E) [74, 75].

Interestingly, like milk Abs with protein kinase activity (see above), IgGs and sIgAs phosphorylating minor lipids can also use as a substrate not only ATP, but also other different NTPs and *ortho*-phosphate [74, 75]. At the same time, substrate specificity of lipid-Abz is different from that of PK-abzymes. For example, [γ–^{32}P]GTP is the best donor of phosphate group in the phosphorylation of proteins, but detectable phosphorylation of lipids with this nucleotide was not revealed. On the contrary, lipid-sIgA effectively used [γ–^{32}P]UTP (~ 50% as compared with ATP, 100%), which is a poor substrate for PK-sIgA [73]. The levels of lipids phosphorylation in the case of ATP (100%) and *ortho*-phosphate (60–80%) were comparable.

We have used Ab-dependent phosphorylation of tightly bound lipids for the estimation of apparent K_m values for *ortho*-phosphate as a very unusual substrate of

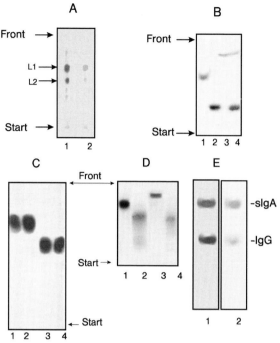

Fig. 19.21 Separation of ^{32}P phosphorylated lipids tightly bound with sIgA fractions, which are soluble in chloroform-methanol mixture, by TLC on a Kieselgel plates in different systems (autoradiographs). **A** Lipids from purified sIgA (lane 1) and from native human milk (lane 2); system A (chloroform-methanol-7 M NH$_4$OH; 60:35:5). **B** TLC in system B (chloroform-methanol-H$_2$O; 65:25:4) of methanolyzed L1 and L2 [^{32}P]lipids before (lanes 1 and 3) and after their hydrolysis with neuraminidase (lanes 2 and 4), respectively. **C** TLC analysis in system B (Kieselgel plates) of products of oxidation with NaIO$_4$ of L1 and L2 [^{32}P]lipids (lanes 1–2 and 3–4, respectively); **D** TLC analysis in system C (isopropanol-H$_2$O-7 M NH$_4$OH; 6:2:0.8) of products of L1 and L2 lipids methanolysis (lanes 1–2 and 3–4, respectively). Lanes 1 and 3 correspond to the compounds before, and lanes 2 and 4 after, incubation with sodium periodate. **E** *In situ* gel assay of milk Ab lipid kinase activity (lane 2) after SDS-PAGE of the mixture of milk sIgA and IgG (purified on protein A-Sepharose, 10 μg of each Abs) in non-reducing conditions. The gel was incubated with 100 nM γ[^{32}P]ATP, and phosphorylation of tightly bound lipids was revealed by radio-autography (lane 2). Gel stained with Coomassie R250 shows positions of IgG and sIgA antibodies (lane 1). The details of the analysis are given in [74, 75].

lipid-Abzs. But, as in the case of DNase and RNase Abzs (Section 19.10), a situation where the initial rate data obtained at increasing substrate concentrations was consistent with Michaelis-Menten kinetics was observed only for single Ab preparations. The major part of IgGs and sIgAs demonstrated two or more of both apparent K_m and V_{max} values. But this was not surprising, since lipid kinase Abzs, like Abs with protein kinase activity, can be separated into many subfractions demonstrating different affinities for ATP-Sepharose (see Section 19.6). Small subfractions of Abz demonstrating maximal affinity for Pi were characterized by the apparent $K_m(I)$ values for [^{32}P]*ortho*-phosphate, varying over a range from 1.6 to 5.6 μM, depending on donor's milk. The increase of [^{32}P]*ortho*-phosphate concentration reveals the second $K_m(2)$ values (3.5–42 µM), and sometimes the third $K_m(3)$ values (7–130 µM) [personal communication, Gorbunov, Karataeva et al.]

Using complete phosphorylation of lipids at saturated concentration of [^{32}P]*ortho*-phosphate or γ-[^{32}P]ATP, we have calculated that, depending on donor's milk, different purified IgG and sIgA preparations contain only ~ 2–7% of Abz tightly bound to minor lipids. As mentioned above (Section 19.6), protein A-Sepharose adsorbs only 4-10% of the pool of milk Abs which contain ~ 80–95% of the initial milk sIgA with different kinase activities, while 90–96% of total milk sIgAs do not interact with this adsorbent (these values varied between donors). This means that only ~ 0.05–1% of total milk sIgAs are bound to minor lipids which can be phosphorylated in the presence of ATP. We observed nearly the same low percentage of catalytic Abs in the case of other Abzs from blood sera of AD patients and human milk (see above).

19.12.3
Abzymes with Oligosaccharide Kinase Activity

Interestingly, the mixture of IgG and sIgA purified by chromatography on protein A-Sepharose contain not only Abs with protein and lipid kinase activity, but also Abzs which are capable of phosphorylating tightly bound oligosaccharides (unpublished results of coworkers of our laboratory, Karataeva and Gorbunov). These ^{32}P-labeled oligosaccharides like [^{32}P]lipids cannot be removed from Abs by dialysis or precipitation of Abs by ammonium sulphate, gel filtration of Abs in salt containing solutions, or washing of protein A-Sepharose-bound Abs with 1% Triton X-100. Similarly to tightly bound lipids, these [^{32}P]polysaccharides can be lost stepwise by Abs at different stages of Abz purification. However, even after all the stages of their purification described above (see Section 19.6), milk Abs usually contain small fractions of IgG and sIgA complexes with oligo and polysaccharides. Interestingly, a ratio of [^{32}P]oligo- and polysaccharides in each sample of Abs depends very much on the donor's milk. Phosphorylation of polysaccharides can be analyzed using SDS-PAGE similar to that used for the study of proteins labeling, or even in parallel with analysis of protein phosphorylation, since usually these polysaccharides demonstrate faster electrophoretic mobility than proteins. In addition, for synchronous analysis of phosphorylation of protein, lipids and polysaccharides, lipids can first be removed from reaction mixture by extraction with a mixture of chloroform and methanol, and then proteins can be precipitated before SDS-PAGE with 5–10% trichloroacetic acid or denatured by boiling. The latter solution of trichloroacetic acid usually contains ^{32}P-labeled oligo- and polysaccharides. The phosphorylation of oligosaccharides after Abs incubation with γ-[^{32}P]ATP or [^{32}P]*ortho*-phosphate can easily be analyzed by TLC using special conditions for this chromatography (Fig. 19.22).

As we have recently shown, electrophoretically homogeneous milk IgGs and sIgAs contain several [^{32}P]polysaccharides of different lengths (according to SDS-PAGE data and following autoradiography) and several [^{32}P]oligosaccharides, which were revealed by autoradiography of plates after TLC of these compounds (for example, Fig. 19.22). We have begun (together with the group of K. Neustroev) to analyze the

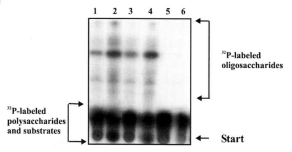

Fig. 19.22 Separation of ^{32}P phosphorylated oligosaccharides tightly bound with sIgA fractions and soluble in aqueous medium (after extraction of [^{32}P]lipids from the reaction samples with methanol chloroform mixture) by TLC on a Kieselgel plates (autoradiographs); a system dioxane/7M NH$_4$OH/H$_2$O, 6:1:4, was used. [^{32}P]Oligosaccharides of sIgAs from two donors were analyzed; lanes 1 and 3 – first donor, lanes 2 and 4 – second donor. γ[^{32}P]ATP (lanes 1 and 2) and [^{32}P]ortho-phosphate (lanes 3 and 4) were used as substrates.
The reaction mixtures for phosphorylation of sIgA (0.2 mg/mL) containing 20 mM Tris-HCl (pH 7.5), 30 mM NaCl, 2 mM MgCl$_2$, and 1 μM γ[^{32}P]ATP or [^{32}P]Pi were incubated at 37 °C for 7 h. The reaction mixtures were boiled for 5 min, denatured Abs and [^{32}P]lipids were removed by sequential centrifugation and extraction of the mixture with methanol chloroform mixture, respectively, and 2 μL of reaction mixture were used for TLC analysis. Lanes 5 and 6 correspond to incubation of sIgA for 5 c in standard conditions with γ[^{32}P]ATP or [^{32}P]P$_i$, respectively. The excess of radioactive substrate and phosphorylated polysaccharides demonstrated very low mobility in the TLC conditions used (see Figure). We are grateful to coworkers of our laboratory Karataeva N. and Gogbunov D. for the data cited above.

structure of these oligo- and polysaccharides using nine different enzymes hydrolyzing polysaccharides of different structures. According to our preliminary data, none of the individual [^{32}P]oligo- or polysaccharides, purified by TLC or SDS-PAGE, can be completely hydrolyzed by only one of the different enzymes used. In addition, even the use of two or three of these enzymes did not lead to the formation of very short fragments of oligo- or polysaccharides. This means that these compounds are not simple or regular and have a very unusual and complicated branchy structure which is not characteristic of the major oligo- or polysaccharides of human milk. A more detailed analysis of the structure of the [^{32}P]oligo- or polysaccharides is in a progress. However, it seems probable that oligo- and polysaccharides which can be phosphorylated by milk Abzs like lipids are minor and have unusual structure.

So far, we have checked some of general rigid criteria (Section 19.7), allowing us to assign uniquely polysaccharide phosphorylation activity to milk catalytic IgG and sIgA. All of the applied criteria support the idea that catalytic activity is an intrinsic property of Abs. In addition, according to our preliminary data, these catalytic Abs, like Abzs with protein and lipid kinase activities, are very heterogeneous and demonstrate several K_m values for ATP and *ortho*-phosphate as substrates/donors of phosphate group in the phosphorylation reaction.

19.12.4
Nucleotide-Hydrolyzing Abzymes of Human Milk

Other very interesting Abzs in human milk are IgG and sIgA that catalyze the hydrolysis of ATP and other nucleotides [71, 143]. These are very different from known phosphatases and ATPases [198, 199]. In contrast to known ATPases [200], IgG utilizes as a substrate not only ATP but also other NTPs and dNTPs with comparable efficiency [71, 143]. In addition, catalytic IgG was shown to hydrolyze not only NTPs, but also ADPs and NMPs. It is known that some phosphatases are also capable of hydrolyzing not only NMPs but also NDPs and NTPs at a significantly lower rate [152]. Nevertheless, in contrast to known phosphatases, catalytic IgG hydrolyzes NTPs faster than NMPs and is activated by Mg^{2+} ions [71, 143]. The affinity of IgG for ATP is about 1–2 orders of magnitude higher than that of other ATPases and phosphatases [130, 152, 200]. Moreover, catalytic IgG did not show hydrolysis of p-nitrophenylphosphate or α-naphthylphosphate [71, 143], which are typical substrates of known phosphatases [130, 152, 200]. Thus, the substrate specificity of catalytic IgG is completely different from those of known ATPases and phosphatases. All observations demonstrate that nucleotide-hydrolyzing catalytic Abs like IgG and sIgA-kinases (see above) are very heterogeneous in their affinity for immobilized ATP and in other enzymic parameters. A very wide polyclonality of this Abs can be seen from different data [71, 143]. For example, 80–90% of IgG-ATPase activity was adsorbed by Sepharose bearing immobilized monoclonal Abs against κ-light chains, and 10–20% by Sepharose bearing Abs against λ-light chains of IgG [71, 143].

In spite of the polyclonal nature of catalytic IgG, the initial rate data obtained at increasing substrate concentrations (NTPs and dNTPs) were consistent with Michaelis-Menten kinetics [71, 143]. The apparent K_m (44–79 µM), V_{max} (0.57–1.1 µM/min), and k_{cat} (0.52–1.0 min^{-1}) values for hydrolysis of different NTPs and dNTPs were comparable in all cases. The apparent k_{cat} values for nucleoside-5′-triphosphate hydrolysis were comparable with (or even higher than) those for known catalytic Abs including anti-thyroglobulin Abs [57], DNA-hydrolyzing Abs [67], and some artificial abzymes against transition state complexes [83].

In contrast to NTP and dNTP, hydrolysis of AMP by the same preparation of catalytic IgG was not consistent with Michaelis-Menten kinetics. Analysis of initial velocities as a function of AMP concentration allowed us to estimate the range of apparent K_m and V_{max} values to be 70–300 µM and 0.15–0.5 µM/min [71, 143]. These results are in agreement with the polyclonality of the IgG, which was demonstrated by its fractionation on ATP-Sepharose, protein A-Sepharose, and Sepharose bearing immobilized anti-κ and anti-λ L chain antibodies.

The specific activity of Abs is ~2 orders of magnitude lower than those for ATPases and phosphatases. But several considerations allow us to suggest that the specific activities of ATPase IgG and sIgA fractions from certain donors are significantly higher than those measured in experiments. Firstly, the specific activities were calculated relative to the total concentration of protein, although it is obvious that Ab with enzyme activity is associated with a minor component of the total pool of analyzed Ab fractions (Section 19.6). As mentioned above, the amount of DNA, RNA, and other

Abzs in purified preparations does not usually exceed 1–5% of the total Abs [50, 126]. Secondly, as was shown above (Section 19.6), several acidic treatments of Abs during their purification led to partial inactivation of their catalytic activity [57]. Thus, the real V_{max} values for nucleotide-hydrolyzing Abs may be at least 10–100 times higher.

19.12.5
Human Blood Sera and Milk Abzymes Hydrolyzing Polysaccharides

Catalytic IgGs with glycosidase activity were generated for the first time using an approach of Ab inducing to a half-chair transition state analog, based on the idea that glycosidases stabilize the intermediate half-chair conformers along the hydrolysis pathway [201]. The data reported later by Neustroev's group [59, 76] provided strong evidence that amylolytic activity is an intrinsic property of natural IgG and sIgA derived from milk of healthy mothers and from the sera of patients with some oncological diseases: it is not due to copurifying enzymes and shows biochemical properties quite different from known enzymes. Interestingly, a very low but detectable amylolytic activity was revealed for IgGs from the sera of some healthy donors [59, 76]. A similar situation was observed previously for VIP- [34] and thyroglobulin-hydrolyzing IgGs [57, 96, 167] from normal humans.

In contrast to human amylases, the amylolytic activity of milk IgG and sIgA Abzs from all donors did not depend on Cl^- ions. The pH optima of 4-nitrophenyl 4,6-O-ethylidene-α-D-maltoheptoside (EPS) and maltopentoside hydrolyses were in the range 6.2–7.5 for the IgG and sIgA fractions [59, 76]. The K_M (~ 0.2–0.3 mM) and V_{max} (3–3.2 kat/kg) values for EPS for all IgG and sIgA (and IgG's Fab fragments) were comparable. Interestingly, the specific activities of IgG and sIgA from human milk were 2–3 times higher than those of IgG from blood of pregnant women, and ~ 10 times higher than those from oncological patients [59, 76].

It should be pointed out that diversity of milk Abzs depends very much on the substrate used; a small number of Abzs possess α-glycosidase activity in the hydrolysis of p-nitrophenyl-α-D-glucopyranoside and maltose [76]. Interestingly, specific activity of the same IgG and sIgA fractions toward PNP-α-glucopyranoside varied significantly widely in comparison with that for amylolytic activity. The K_m and V_{max} values characterizing the hydrolysis of p-nitrophenyl α-D-maltooligosides catalyzed by several different milk IgGs and sIgAs varied in the range 0.2–0.5 mM and 13–19 nM min^{-1} µg^{-1}) [76].

The specific amylolytic activity of Abs is about two orders of magnitude lower than that for known α-amylases including α-amylases from human blood [202]. However, the specific activities of amylolytic IgG and sIgA are comparable with that for catalytic *anti*-thyroglobulin Abs [57], SLE DNase IgG [173], and some artificial Abzs against transition state complexes [24]. In addition, taking into accounts several different reasons for decreased activity of natural Abzs (Section 19.6), these observations suggested that the specific activities of the catalytic fraction of Abs in fact are significantly (at least 10–50 times) higher than the found values.

Analysis of 15 IgG and 12 sIgA preparations from human milk of different donors shows quite major differences in the mode of action of various Abs preparations in the hydrolysis of maltohexaose and EPS. Interestingly, most Abs did not hydrolyze the short oligosaccharides maltose and maltotriose (except one sIgA preparation), but demonstrated amylolytic activity toward longer maltooligosaccharides, but final products of the hydrolysis of oligomers containing 4–6 links were different for various IgGs and sIgAs. Different Abs giving similar sets of final products from the hydrolysis of maltohexaose gave different ratios of maltose and maltotriose at the final reaction stage (Table 19.3). This undoubtedly shows differences in the mode of action of these Abzs.

Tab. 19.3: Products of maltooligosaccharide hydrolysis catalyzed by several different preparations of IgG and sIgA*

Substrate	IgG^3		$sIgA^6$		$sIgA^9$		IgG^6	
	Initial product(s)	Final products	Initial product(s)	Final products	Initial product(s)	Final products	Initial product(s)	Final products
Maltose	–	–	–	–	G1**	G1	–	–
Maltotriose	–	–	–	–	G1, G2	G1	–	G2
Maltotetraose	G2	G2	G2	G2	G2	G1	G2	G2
Maltopentaose	G2, G3	G2, G3	G2, G3	G2, G3	G2, G3	G1	G1, G2, G3	G1, G2, G3
Maltohexaose	G2, G3, G4	G2, G3	G2, G3, G4	G2, G3	G1, G2, G3, G4, G5	G1	G1, G2, G3, G4	G1, G2, G3

* Incubations were carried out in 30 mM/l Tris HCl, pH 7.5, 1 mM/l NaN$_3$, at 37°C for 24 and 48 hrs. Concentration of maltooligosaccharides was 5 mg/ml. The initial and final products were detected after 24 h and 48 h of incubation.
** G1, glucose; G2, maltose; G3, maltotriose; G4, maltotetraose, G5, maltopentaose

Recently, IgGs and IgMs from the sera of patients with MS were found to possess amylolytic activity, hydrolyzing α-(1 → 4)-glucosyl linkages of maltooligosaccharides, glycogen, and several artificial substrates [105]. Individual IgM fractions isolated from 54 MS patients had approximately three orders of magnitude higher specific amylolytic activity than that for healthy donors, whereas IgG from only a few patients had high amylolytic activity. Strict criteria were used to prove that the amylolytic activity of IgMs and IgGs is their intrinsic property and is not due to any enzyme contamination. Fab fragments produced from IgM and IgG fractions of the MS patients displayed the same amylolytic activity. IgMs from various patients demonstrated different modes of action in hydrolyzing maltooligosaccharides: exo-amylolytic pattern and exo-amylolytic activity combined with exo-glycosidase one.

The K_m values in the hydrolysis of PNP-maltopentaose and EPS were in the range of 0.1–0.3 mM for IgMs. The optimal pH for all samples tested was in the range of 6.5–7.5. The initial rates of hydrolysis of maltohexaose by IgM fractions of 16 MS patients were 4–5 times higher than that for glycogen. This is an essential difference between IgM kinetic parameters and a human salivary α-amylase, where an increase in the initial rate of hydrolysis with increasing length of substrate is characteristic [69].

Samples of IgMs isolated from different MS patients using a standard procedure demonstrated a remarkable variety of modes of action in the hydrolysis of oligosaccharides. Note that a difference in mode of action was also found in IgG and sIgA fractions from human milk, but the presence of exo-hydrolytic activity in these Abs is rare. Such a variety of modes of action was not found for Abzs with proteolytic activity from patients with HT, rheumatoid arthritis, and asthma [57, 58, 159] and for DNase IgGs from SLE patients [173]. However, as mentioned above, an extreme diversity of RNA-hydrolyzing IgGs and/or IgMs from the sera of patients with SLE and especially DNase IgGs from MS patients was observed (Section 19.10). Thus, individual MS patients may have a relatively small or an extremely large pool of not only polyclonal nuclease Abs but also Abs hydrolyzing polysaccharides.

Since according to published data [105] nearly 100% of the MS patients were characterized by high amylolytic activity of IgMs, analysis of this activity could be used as an additional criterion of this pathology.

19.12.6
Peculiarities of Milk Abzymes

It is obvious that human milk contains different NA-, nucleotide-, and polysaccharide-hydrolyzing Abzs and some very unusual catalytic Abs with synthetic activity. Taking into account the data regarding potential mechanisms by which pregnant women could be directly immunized as a result of a specific response of their immune system to certain components of viruses, bacteria, or food, which most probably can effectively stimulate the production of different Abz (see above), this is not surprising. Nobody has yet studied the possibility of the presence of Abs with proteolytic activity in the milk and blood of lactating woman, but we believe that such Abzs are present in human milk. In addition, it is obvious that a repertoire of Abzs with various activities in the sera and milk of human mothers must depend very much on individual peculiarities of their immune systems, on different diseases during the pregnancy (especially during last 3 month of pregnancy) including different viral and bacterial pathologies, and on peculiar properties of the environment. In this connection it should be mentioned that according to our preliminary data severe technological pollution very much influences the properties of Abzs of human mothers. We have observed particularly large changes in Abzs in the case of mothers who were under the constant and prolonged influence of incorporated radioactive isotopes (for example, ^{137}Cs and ^{90}Sr). Interestingly, the milk of such mothers contained unusual IgG and sIgA Abzs without covalent bonds between different subunits of these oligomeric molecules; Abzs were presented by free catalytically active light or even heavy chains of Abs, and such incomplete Abs were more catalytically active than complete Abzs of healthy mothers (our unpublished data). One cannot exclude the possibility that production of Abzs in organisms of healthy mothers could have a

19.13
Biological Roles of Abzymes

Abzs have been studied primarily in the context of AI diseases, where their biological role remains unknown: do they have a function or do they represent a dysfunction? One cannot exclude the possibility that some may play positive and others negative roles. In bronchial asthma it has been suggested that respiratory tract dysfunction may stem from the protease activity of auto-Abzs, resulting in a deficit of VIP, which plays a major role in the pathophysiology [46, 159]. In HT, Abzs hydrolyzing thyroglobulin have been considered a positive factor, since they could minimize AI responses to thyroglobulin and prevent the formation of immune precipitates [57], but on the other hand *anti*-VIP Abs are cytotoxic and mice immunized with *anti*-hormone IgGs from human sera develop asthma [48]. Abzs hydrolyzing DNA or RNA could at first glance be considered as non-specific side-products of the AI process since they occur in the sera of patients with many AI and viral diseases (see above), but, in SLE, DNase Abzs are cytotoxic like the tumor necrosis factor [61, 203, 204], which induces cell death via apoptotic mechanism. The DNase and cytotoxic properties of *anti*-DNA auto-Abs from SLE patients and from autoimmune MRL-lpr/lpr, SJL/J, and (NZBx NZW)F(1) mice are correlated, and the catalytic and cytotoxic properties are related to the stage of SLE [205, 206]; treatment of cells with *anti*-DNA Abs induced internucleosomal DNA fragmentation, a characteristic of apoptotic cell death. It was shown that incubation of milk DNase sIgA with nuclei of porcine embryo kidney cells permeabilized by Triton X-100 causes the formation of electrophoretically mobile forms of nuclear DNA and inhibition of phosphorylation of nuclear proteins [207].

Complement-independent cytotoxicity of Abzs is suggested by further evidence. DNase Abzs from lymphoproliferative patients are cytotoxic [61], and DNA-hydrolyzing Bence-Jones proteins from multiple myeloma patients enter the nucleus and cause DNA fragmentation; catalytically active protein preparations were significantly cytotoxic and their activity was related to the progressive deterioration of clinical status [65]. In MS, the theory of pathogenesis focused until recently on the degradation of myelin by AI inflammatory processes [97], but recently high-affinity *anti*-DNA Abs have been identified as a major component of the intrathecal IgG in patients' brains and cerebrospinal fluid cells [104]. DNase activity was found in Abs from ~ 90% of MS patients [52], and using a new highly sensitive method we found recently that the blood of ~ 80% of MS patients contained *anti*-DNA Abs in a higher concentration than that in normal humans (Garmashova, Ershova et al., personal communication). Our previous data showed that DNase/RNase Abzs from some MS, HT, polyarthritis, and viral hepatitis patients were cytotoxic, and this suggests that Abzs from patients with other AI diseases are also cytotoxic and play an important role in pathogenesis. It is therefore reasonable to propose that even apparently "non-specific" Abzs from patients with AI diseases are cytotoxic and therefore important for the development of each type of disease.

Some Abzs of healthy donors may also play an important role for humans. For example, Abs generating H_2O_2 from singlet oxygen may participate in Ab-mediated cell killing by event-related production of hydrogen peroxide [155]. Alternatively, Abs may function to defend an organism against oxygen.

The work of Ehrlich clearly demonstrated that passive immunity in mice could be acquired both via the placenta before birth and via the milk after birth [208]. The immune status of the newborn varies from species to species, but all mammalian neonates are essentially agammaglobulinemic at their mucosal surfaces at birth. Milk contains a wide array of Abs to bacterial, viral, and protozoal antigens [198, 199, 208, 209] with antimicrobial activities, which reach mucosal surfaces of intestinal and respiratory tracts and as a result protect infants from infection and disease. The Abs are active at the mucosal surface in dealing with the replication and colonization of pathogenic microorganisms, as well as in limiting the access of environmental antigens [198, 199, 208, 209]. Passive immunity of the child may also be acquired via mother's milk IgG after the transfer of Abs across the epithelium of intestinal surfaces to the newborn's circulatory system [210].

As mentioned above, during the last 2–4 months of pregnancy the "immunomemory" of women "collects information" about all inside and outside compounds which can be harmful for infants and produce Abs to these compounds after the beginning of lactation. Therefore, it appears likely that catalytic Abs of the mother's milk may have a positive function during breast-feeding of the newborn. For example, it is known that increased amounts of DNases and RNases in human blood or therapy of patients with nucleases leads to protection from different viral and bacterial diseases (see [50] and references therein). Recently, an inverse correlation between mammary tumor incidence and the amount of RNase activity in human milk was revealed [211]. One might well suppose that the milk DNA- and RNA-hydrolyzing Abs are capable not only of neutralizing viral and bacterial nucleic acids by complex formation with them, but additionally of hydrolyzing them; the DNA-hydrolyzing activity of Abs raises the possibility that these Abzs may provide protective functions for the newborn through the hydrolysis of viral and bacterial nucleic acids. In addition, according to our preliminary data, milk DNase/RNase Abzs are cytotoxic and effectively kill tumor cells of different lines (L929, mouse fibroblasts, and HL-60, human promyelocytes). It is possible that ATP-hydrolyzing IgG and sIgA immunoglobulins with protein, lipid, and polysaccharide kinase activities in human milk may also be important in the protective role of milk. Specific stimulation of the production of various Abs, including catalytic Abs, by the immune system of mothers as a result of autoimmunization and/or viral and bacterial infection may be a way of strengthening the protective function of breast milk and may play a very important role for passive immunity of neonates and contribute to mucosal immunity by policing the function of some cells [69].

19.14
Natural Abzymes as Tools for Investigating RNA Structure

Chemical and enzymatic probes to cleave RNA at specific sequences or structural motifs are crucial tools for the investigation of RNA structure. Although a number of such probes are presently available, there is a need to expand this arsenal of tools in order to detect structural motifs in RNA molecules in solution (see [56] and references therein). Nucleases are specific either for sequences (e.g., RNase TI is specific for guanosines and RNase A for Py-A sequences) or for structural features (e.g., nuclease S1 cleaves exclusively single-stranded domains of RNA). As noted above, Abzs of AI patients show novel RNase activities, including some stimulated by Mg^{2+}, which are not sequence-specific but sensitive to subtle and/or drastic folding changes as evidenced by work with structurally well-studied tRNA substrates [53, 56, 147, 148]. An interesting example of the utility of such RNases is provided by a study of cleavage by SLE Abzs of two $tRNA^{Lys}$ molecules [53, 147], one from human mitochondria and the other a mutant found in the serious neuromuscular disorder myoclonic epilepsy with ragged-red fibers (MERRF) in which A at position 50 is replaced by G, a mutation believed to change the tRNA structure and cause the functional disorder. RNase A and other RNases used for probing structure showed no difference in their cleavage patterns [53, 147], but RNase Abzs in the presence of Mg^{2+} produced new cleavage sites with different patterns: the mutant tRNA displayed a significantly different sensitivity in the mutated region and cleavage also occurred at new positions indicating local structural or conformational changes.

Most Mg^{2+}-dependent Abzs from AI and viral patients display no sequence specificity but are rather sensitive to structural features of tRNAs specific for Phe, Lys, Asp, and Gln [53, 56, 147, 148]. Abzs from some patients demonstrate a major RNase A-type specificity with minor differences (preference for CpA and UpA sequences), while others contain a major subfraction possessing an RNase TI-type specificity. The Mg^{2+}-stimulated IgG-associated RNase demonstrates in most cases a cobra venom RNase VI-like action on $tRNA^{Phe}$ with a unique Mg^{2+}-activated specificity for double-stranded regions [53, 147]. SLE and other ADs Abzs show specificities quite different from those of RNase VI in spite of some similarities, and the observed specificity differs remarkably from patient to patient. Thus, Abzs can discriminate between sequences and subtle or large structural changes, including stability and folding; they may become tools for investigating RNA structures in solution, but, since their specificities are multiple, further applications will require the development of monoclonal Abzs.

19.15
Abzymes as Diagnostic Tools and Tools for Biological Manipulations

Studies of Abzs in AI patients and human mothers show the very wide spectrum of natural Abzs which can be formed. Their application in medicine, science, and biotechnology requires the production of monoclonal preparations by hybridoma technology and subsequent screening for interaction with hapten and for Ab-dependent catalysis. Practically all types of Abzs described here could in principle be used as diagnostic tools for AI diseases; for example, according to our data, the serum of ~ 90% of MS patients contains DNase Abzs. Further, comparison of the substrate specificity of Abzs, for example, those with RNase activity, should allow us to distinguish between different types of AI (Section 19.10).

We have analyzed the relative activities of DNase and RNase Abs from the sera of patients with different diseases: SLE, MS, HT, polyarthritis, and viral hepatitis. It was shown that in all cases the relative level of these activities increases during an exacerbation and decreases at a remission of disease [50, 126]. Interestingly, we have found good correlation between activities of Abz in the hydrolysis of single- and double-stranded DNA, as well as DNase and RNase activities of Abzs of different AD patients. At the same time there was no clearly expressed correlation between the total titers of *anti*-DNA Abs and their ability to hydrolyze NAs. Nevertheless, all these results can be considered as preliminary, since they do not provide a common picture of a possible connection of Abz activities with other immunological and biochemical characteristics of different ADs.

Up to the present time, there have been no published statistical data concerning possible correlations of Abz activities with the onset, progress, or remission of human diseases or with biochemical and immunological indices which characterize them, and we have therefore recently carried out the first detailed investigation of this type on 120 HT patients [141, 142].

Abs are believed to play a significant role in the pathogenesis of HT (see above) [93]. IgG from 65% of patients with HT possessed DNase and RNase activities, and to correlate these with biochemical and immunological indices we compared them with thyroid hormone status. We found practically linear relationships between the proportion of patients having Abzs and an increase of thyrotropic hormone, and between the percentage of such patients with a decreased concentration of thyroid hormones (Fig. 19.23). All patients with an increase in thyrotropic hormone (13 ME/L) and/or a very low concentration of thyroxin (50 nmole/mL), which characterize typical hypothyrosis, contained serum Abzs and showed a correlation of Abz activity with exacerbation of thyroid gland damage, while patients at the initial stage or during remission of HT had no Abz activity or a reduced amount. The specific activity of DNase IgGs increased with the relative amount of *anti*-thyroglobulin Abs and therefore provides a good indicator of the progress of AI pathology.

The very widely used therapy with thyroxin led to a temporary change of blood hormone concentration but not of Abz levels or AI status, but the immunosuppressive drug plaquenil significantly reduced the DNase activity of Abs, and this correlated with enhancement of thyroid hormone concentrations (Fig. 19.24), elevation of functional activity of the thyroid, and improvement of the patient's clinical state.

Recently we have performed a similar study of the relative activity of DNase and RNase Abz of MS patients and have shown that there are correlations of these two activities with some of the ~ 13 different clinical criteria of MS proposed by Poser [212, 213], various types of this pathology, and some biochemical and immunological indexes of patients analyzed (to be published elsewhere).

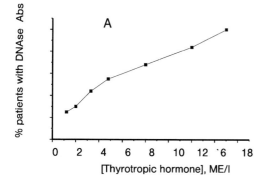

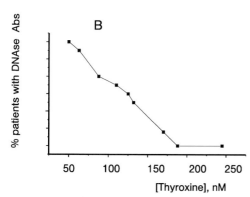

Fig. 19.23 Dependence of the relative number (%) of HT patients containing DNA-hydrolyzing activity upon the patients' serum concentration of thyrotropic hormone (**A**), or of thyroxin (**B**). Most HT patients (83 of 120) can be placed in groups demonstrating comparable concentration ([C] ±20%, see Figure) of thyrotropic or thyroxin hormone (each group contains 6–16 patients); % value is expressed as the % of patients with Abzs out of all patients in each group demonstrating a comparable concentration of one of the hormones. The details of the analysis are given in [142].

We consequently believe that Abzs offer good indicators of alteration of AI processes in different AI diseases. Study of Abzs shows the extremely wide potentialities of the immune system in producing Abs possessing very different enzymic activities, which very often are not comparable with those of known enzymes, and natural Abs with specified and novel functions may have wide potential for biotechnology and medicine.

In order to develop Abzs as tools for biological manipulations, it is necessary to understand the structural basis of their activity in detail. Although the study of polyclonal Abzs provides important information about the potential repertoire of catalytic activities of Abs possible repertoire of Abzs in the blood of patients with different ADs for different prospects for immunotherapeutic Abzs, structural studies require pure uniform material concentrated to high levels. Monoclonal and recombinant Abzs corresponding to specific polyclonal AI Abs potentially represent a new generation of therapeutics with enhanced antigen inactivation capability.

The AI repertoire is well known from previous studies to be capable of producing catalytic Abs directed to self-antigens. The ability of 26 monoclonal light chains from multiple myeloma patients to cleave gp120, a foreign protein, was analyzed [66]. One L chain specifically hydrolyzed gp120 protein (K_d = 130 nM). The toxic effect of gp120 in neuronal cultures was reduced by about 100-fold by pretreatment

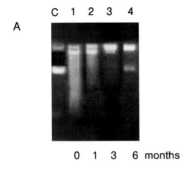

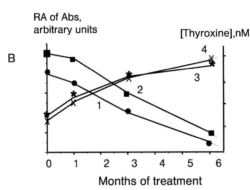

Fig. 19.24 Time dependencies of the decrease of relative DNase activity of IgG (lanes 1 and 2) and of increase of thyroxin concentration (lanes 3 and 4) of two patients treated with plaquenil. The relative activity of Abs is expressed in arbitrary units (arbitrary units are equal to the number of the lanes, which are given on Fig 19.13); curves 1 and 3 correspond to Abs of patients demonstrating an initial activity of about 8 arbitrary units, curves 2 and 4 to other patient with an initial activity of about 10 arbitrary units. The details of the analysis of 30 HT patients treated with plaquenil are given in [142].

of the protein with the L chain. These observations open up the possibility of utilizing gp120-cleaving Abs in the treatment of AIDS.

Thus, SLE patients were found to develop Abs to a conserved component of the CD4 binding site of gp120, potentially offering a means to obtain human Abs expressing broad reactivity with various HIV strains [214–216]. Covalently reactive antigen analogs capable of selective recognition of nucleophilic Abs were synthesized and applied to isolate Fv and L chain catalysts from lupus phage repertoires. CRA binding by the recombinant Ab fragments was statistically correlated with catalytic cleavage of model peptide substrates. A peptidyl CRA composed of residues 421–431 with a phosphonate diester moiety at its C terminus was validated as a reagent that combines non-covalent and covalent binding interactions in recognition of a gp120ase L chain. A general challenge in the field is the apparent instability of the catalytic conformation of the Abs. In reference to therapy of HIV infection, assurance is required that the Abs recognize the native conformation of gp120 expressed as a trimer on the virus surface.

Some additional different medical potentials of catalytic antibodies are discussed in several articles and reviews [217, 218].

19.16
Conclusion

A growing body of data suggests that catalytic Abs are important mediators of immunological defence, regulation, and AI dysfunction. The degree to which Abzs contribute to these biological phenomena will require continuing studies of Ab-mediated catalysis following experimental immunization and AI disease, as well as mechanistic investigation of catalysis by Abs and their subunits. From a technological point of view, the further study of both artificial and natural Abzs with the goal of understanding structure-function relationships in order to enable tailor-made catalysts of potential therapeutic application to be produced is of high importance. Because the catalytic activity of certain Abs is an innate function [81], catalysts with specificity for virtually any target polypeptide could potentially be developed and then improved by *in vitro* or *in vivo* affinity selection. The increasing number of available X-ray structures of catalytic Abs shows the multiplicity of solutions to the question of how an Ab can catalyze an enzyme-like reaction. These strategies include amino acid arrangements analogous to those in enzymes [219], but also arrangements completely different from those selected in enzymes by natural evolution [220]. *Anti*-idiotype approaches may enable Abzs to mimic many useful enzymes [115], enhancing their possible functions. The phenomenon of Abz catalysis can potentially be applied to isolate efficient catalysts suitable for passive immunotherapy for the treatment of major diseases. For example, cocaine-hydrolyzing abzymes have been developed, and these may provide a novel approach to the problems of drug addiction [221]. Abzs that cleave the gp120 protein of HIV may be of use in the treatment of AIDS [66]. Through the rational design of Abzs with specified and novel catalytic functions allowing the selective cleavage of surface proteins, it is to be hoped that an unprecedented level of control over *in vivo* protein-protein associations may be achieved.

Acknowledgements

Our study of natural catalytic Abs was supported by grants from the Russian Foundation of Basic Research (1990–2002), the Russian Ministry of General and Professional Education (1992–2002), and the Siberian Division of the Russian Academy of Sciences (1990–2002).

References

1 Pauling, L., *Chem. Eng. News* 24 (**1946**), p. 1375–1377
2 Slobin, L. I., *Biochemistry* 5 (**1966**), p. 2836–2841
3 Jencks, W., *Catalysis in chemistry and enzymology*, McGraw-Hill, New York **1969**
4 Schochetman, G., Massey, R., International Patent no. W085/02414 (**1985**)
5 Pollack, S. J., Jacobs, J. W., Schultz, P. G., *Science* 234 (**1986**), p. 1570–3
6 Tramontano, A., Janda, K. D., Lerner, R. A., *Science* 234 (**1986**), p. 1566–70
7 Tramontano, A., Janda, K. D., Lerner, R. A., *Proc. Natl. Acad. Sci. USA* 83 (**1986**), p. 6736–40
8 Janda, K. D., Shevlin, C. G., Lerner, R. A., *Science* 259 (**1993**), p. 490–3
9 Li, T., Janda, K. D., Ashley, J. A., Lerner, R. A., *Science* 264 (**1994**), p. 1289–93
10 Li, T., Janda, K. D., Lerner, R. A., *Nature* 379 (**1996**), p. 326–7
11 Wentworth, P., Jr., Liu, Y., Wentworth, A. D., Fan, P., Foley, M. J., Janda, K. D., *Proc. Natl. Acad. Sci. USA* 95 (**1998**), p. 5971–5
12 Barbas, C. F., 3rd, Heine, A., Zhong, G., Hoffmann, T., Gramatikova, S., Bjornestedt, R., List, B., Anderson, J., Stura, E. A., Wilson, I. A., Lerner, R. A., *Science* 278 (**1997**), p. 2085–92
13 Hotta, K., Lange, H., Tantillo, D. J., Houk, K. N., Hilvert, D., Wilson, I. A., *J. Mol. Biol.* 302 (**2000**), p. 1213–25
14 Smiley, J. A., Benkovic, S. J., *Proc. Natl. Acad. Sci. USA* 91 (**1994**), p. 8319–23
15 Napper, A. D., Benkovic, S. J., Tramontano, A., Lerner, R. A., *Science* 237 (**1987**), p. 1041–3
16 Ding, L., Liu, Z., Zhu, Z., Luo, G., Zhao, D., Ni, J., *Biochem. J.* 332 (Pt 1) (**1998**), p. 251–5
17 Smithrud, D. B., Benkovic, P. A., Benkovic, S. J., Roberts, V., Liu, J., Neagu, I., Iwama, S., Phillips, B. W., Smith, A. B., 3rd, Hirschmann, R., *Proc. Natl. Acad. Sci. USA* 97 (**2000**), p. 1953–8
18 Ma, L., Hsieh-Wilson, L. C., Schultz, P. G., *Proc. Natl. Acad. Sci. USA* 95 (**1998**), p. 7251–6
19 Liu, E., Prasad, L., Delbaere, L. T., Waygood, E. B., Lee, J. S., *Mol. Immunol.* 35 (**1998**), p. 1069–77
20 Iverson, B. L. Lerner, R. A., *Science* 243 (**1989**), p. 1184–8
21 Lerner, R. A., Tramontano, A., *Trends Biochem. Sci.* 12 (**1987**), p. 427–38
22 , *Trends Cell Biol.* 9 (**1999**), p. 24–8
23 Stewart, J. D., Benkovic, S. J., *Int. Rev. Immunol.* 10 (**1993**), p. 229–40
24 Suzuki, H., *J. Biochem. (Tokyo)* 115 (**1994**), p. 623–8
25 Ersoy, O., Fleck, R., Blanco, M. J., Masamune, S., *Bioorg. Med. Chem.* 7 (**1999**), p. 279–86
26 Gao, Q. S., Paul, S., *Methods Mol. Biol* 51 (**1995**), p. 319–27

27 Miller, G. P., Posner, B. A., Benkovic, S. J., *Bioorg. Med. Chem.* 5 (**1997**), p. 581–90
28 Roberts, V. A., Stewart, J., Benkovic, S. J., Getzoff, E. D., *J. Mol. Biol.* 235 (**1994**), p. 1098–116
29 Stewart, J. D., Krebs, J. F., Siuzdak, G., Berdis, A. J., Smithrud, D. B., Benkovic, S. J., *Proc. Natl. Acad. Sci. USA* 91 (**1994**), p. 7404–9
30 Gouverneur, V. E., Houk, K. N., de Pascual-Teresa, B., Beno, B., Janda, K. D., Lerner, R. A., *Science* 262 (**1993**), p. 204–8
31 Janda, K. D., Lo, L. C., Lo, C. H., Sim, M. M., Wang, R., Wong, C. H., Lerner, R. A., *Science* 275 (**1997**), p. 945–8
32 Thayer, M. M., Olender, E. H., Arvai, A. S., Koike, C. K., Canestrelli, I. L., Stewart, J. D., Benkovic, S. J., Getzoff, E. D., Roberts, V. A., *J. Mol. Biol.* 291 (**1999**), p. 329–45
33 Fastrez, J., *Mol. Biotechnol.* 7 (**1997**), p. 37–55
34 Paul, S., Volle, D. J., Beach, C. M., Johnson, D. R., Powell, M. J., Massey, R. J., *Science* 244 (**1989**), p. 1158–62
35 Shuster, A. M., Gololobov, G. V., Kvashuk, O. A., Bogomolova, A. E., Smirnov, I. V., Gabibov, A. G. *Science* 256 (**1992**), p. 665–7
36 Shuster, A. M., Gololobov, G. V., Kvashuk, O. A., Gabibov, A. G., *Dokl. Akad. Nauk. SSSR* 318 (**1991**), p. 1262–4
37 Buneva, V. N., Andrievskaia, O. A., Romannikova, I. V., Gololobov, G. V., Iadav, R. P., Iamkovoi, V. I., Nevinskii, G. A., *Mol. Biol. (Mosk.)* 28 (**1994**), p. 738–43
38 Kulberg, A., Docheva Iu, V., Tarkhanova, I. A., Spivak, V. A., *Biokhimiia* 34 (**1969**), p. 1178–83
39 Kulberg, A. Y. Petyaev, I. M., *Immunology (Moscow)* 6 (**1988**), p. 10–13
40 Kulberg, A. Y., Petyaev, I. M., Zamotaeva, N. G., *Immunology (Moscow)* 3 (**1988**), p. 37–40
41 Kulberg, A., Docheva Iu, V., Tarkhanova, I. A., Spivak, V. A., *Biokhimiia* 34 (**1969**), p. 1178–83
42 Kit, Y. Y., Kim, A. A., Sidorov, V. N., *BioMed. Sci.* 2 (**1991**), p. 201–4
43 Kit, Y. Y., Semenov, D. V., Nevinskii, G. A., *Mol. Biol. (Moscow)* 29 (**1995**), p. 893–906
44 Earnshaw, W. C., Rothfield, N., *Chromosoma* 91 (**1985**), p. 313–21
45 Reimer, G., Raska, I., Tan, E. M., Scheer, U., *Virchows Arch. B Cell Pathol. Incl. Mol. Pathol.* 54 (**1987**), p. 131–43
46 Paul, S. *Appl. Biochem. Biotechnol.* 47 (**1994**), p. 241–53; discussion p. 253–5
47 Paul, S., *Ann. N Y Acad. Sci.* 865 (**1998**), p. 238–46
48 Paul, S., *Appl. Biochem. Biotechnol.* 75 (**1998**), p. 13–24
49 Nevinsky, G. A., Semenov, D. V., Buneva, V. N., *Biochemistry (Mosc)* 65 (**2000**), p. 1233–44
50 Nevinsky, G. A., Kanyshkova, T. G., Buneva, V. N., *Biochemistry (Mosc)* 65 (**2000**), p. 1245–55
51 Baranovsky, A. G., Matushin, V. G., Vlassov, A. V., Zabara, V. G., Naumov, V. A., Giege, R., Buneva, V. N., Nevinsky, G. A., *Biochemistry (Mosc.)* 62 (**1997**), p. 1358–66
52 Baranovskii, A. G., Kanyshkova, T. G., Mogelnitskii, A. S., Naumov, V. A., Buneva, V. N., Gusev, E. I., Boiko, A. N., Zargarova, T. A., Favorova, O. O., Nevinsky, G. A., *Biochemistry (Mosc.)* 63 (**1998**), p. 1239–48
53 Vlassov, A., Florentz, C., Helm, M., Naumov, V., Buneva, V., Nevinsky, G., Giege, R., *Nucleic Acids Res.* 26 (**1998**), p. 5243–50
54 Andrievskaia, O. A., Buneva, V. N., Zabara, V. G., Naumov, V. A., Iamkovoi, V. I., Nevinskii, G. A., *Mol. Biol. (Mosk)* 32 (**1998**), p. 908–15
55 Andrievskaia, O. A., Buneva, V. N., Zabara, V. G., Naumov, V. A., Iamkovoi, V. I., Nevinskii, G. A., *Med. Sci. Monit.* 6, (**2000**), p. 460–70
56 Vlasov, A. V., Baranovskii, A. G., Kanyshkova, T. G., Prints, A. V., Zabara, V. G., Naumov, V. A., Breusov, A. A., Giege, R., Buneva, V. N., Nevinskii, G. A., *Mol. Biol. (Mosk.)* 32 (**1998**), p. 559–69

57 Li, L., Paul, S., Tyutyulkova, S., Kazatchkine, M. D., Kaveri, S., *J. Immunol.* 154 (**1995**), p. 3328–32

58 Kalaga, R., Li, L., O'Dell, J. R., Paul, S., *J. Immunol.* 155 (**1995**), p. 2695–702

59 Savel'ev, A. N., *Protein Peptide Lett.* 6 (**1999**), p. 179–184

60 Kozyr, A. V., Kolesnikov, A. V., Iakhnina, E. I., Astsaturov, I. A., Varlamova, E., Kirillov, E. V., Gabibov, A. G., *Biulet. Eksped. Biol. Med.* 121 (**1996**), p. 204–6

61 Kozyr, A. V., Kolesnikov, A. V., Aleksandrova, E. S., Sashchenko, L. P., Gnuchev, N. V., Favorov, P. V., Kotelnikov, M. A., Iakhnina, E. I., Astsaturov, I. A., Prokaeva, T. B., Alekberova, Z. S., Suchkov, S. V., Gabibov, A. G., *Appl. Biochem. Biotechnol.* 75 (**1998**), p. 45–61

62 Gololobov, G. V., Mikhalap, S. V., Starov, A. V., Kolesnikov, A. F., Gabibov, A. G., *Appl. Biochem. Biotechnol.* 47 (**1994**), p. 305–14; discussion p. 314–5

63 Matsuura, K., Ikoma, S., Sugiyama, M., Funauchi, M., Sinohara, H., *Appl. Biochem. Biotechnol.* 83 (**2000**), p. 107–13; discussion p. 113–4, p. 145–53.

64 Lacroix-Desmazes, S., Moreau, A., Sooryanarayana, Bonnemain, C., Stieltjes, N., Pashov, A., Sultan, Y., Hoebeke, J., Kazatchkine, M. D., Kaveri, S. V. *Nat. Med.* 5 (**1999**), p. 1044–7

65 Sinohara, H. Matsuura, K., *Appl. Biochem. Biotechnol.* 83 (**2000**), p. 85–94

66 Paul, S., Kalaga, R., Gololobov, G. V., D., B., *Appl. Biochem. Biotechnol.* 83 (**2000**), p. 71–84

67 Kanyshkova, T. G., Semenov, D. V., Khlimankov, D., Buneva, V. N., Nevinsky, G. A., *FEBS Lett.* 416 (**1997**), p. 23–6

68 Kanyshkova, T. G., Semenov, D. V., Vlasov, A. V., Khlimankov, D., Baranovskii, A. G., Shipitsyn, M. V., Iamkovoi, V. I., Buneva, B. N., Nevinskii, G. A., *Mol. Biol. (Mosk.)* 31 (**1997**), p. 1082–91

69 Nevinsky, G. A., Kanyshkova, T. G., Semenov, D. V., Vlassov, A. V., Gal'vita, A. V., Buneva, V. N., *Appl. Biochem. Biotechnol.* 83 (**2000**), p. 115–29; discussion p. 129–30, p. 145–53

70 Buneva, V. N., Kanyshkova, T. G., Vlassov, A. V., Semenov, D. V., Khlimankov, D., Breusova, L. R., Nevinsky, G. A. *Appl. Biochem. Biotechnol.* 75 (**1998**), p. 63–76

71 Semenov, D. V., Kanyshkova, T. G., Kit, Y. Y., Khlimankov, D. Y., Akimzhanov, A. M., Gorbunov, D. A., Buneva, V. N., Nevinsky, G. A., *Biochemistry (Mosc.)* 63 (**1998**), p. 935–43

72 Kit, Y., Semenov, D. V., Nevinsky, G. A., *Biochem. Mol. Biol. Int.* 39 (**1996**), p. 521–7

73 Nevinsky, G. A., Kit, Y., Semenov, D. V., Khlimankov, D., Buneva, V. N., *Appl. Biochem. Biotechnol.* 75 (**1998**), p. 77–91

74 Gorbunov, D. A., Semenov, D. V., Shipitsyn, M. V., Kit, Y. Y., Kanyshkova, T. G., Buneva, V. N., Nevinsky, G. A. *J. Russ. Immunol.* 5 (**2000**), p. 267–78

75 Gorbunov, D. A., Semenov, D. V., Shipitsin, M. V., Nevinsky, G. A., *Doklad. Biochem. Biophys.* 377 (**2001**), p. 62–4

76 Savel'ev, A. N., Kanyshkova, T. G., Kulminskaya, A. A., Buneva, V. N., Eneyskaya, E. V., Filatov, M. V., Nevinsky, G. A., Neustroev, K. N., *Clin. Chim. Acta* 314 (**2001**), p. 141–52

77 Gao, Q. S., Sun, M., Tyutyulkova, S., Webster, D., Rees, A., Tramontano, A., Massey, R. J., Paul, S., *Ann. N Y Acad. Sci.* 764 (**1995**), p. 567–9

78 Li, L., Kalaga, R., Paul, S., *Clin. Exp. Immunol.* 120 (**2000**), p. 261–6

79 Tyutyulkova, S., Gao, Q. S., Thompson, A., Rennard, S., Paul, S., *Biochim. Biophys. Acta* 1316 (**1996**), p. 217–23

80 Paul, S., *Mol. Biotechnol.* 5 (**1996**), p. 197–207

81 Gololobov, G., Sun, M., Paul, S., *Mol. Immunol.* 36 (**1999**), p. 1215–22

82 Gololobov, G., Tramontano, A., Paul, S. *Appl. Biochem. Biotechnol.* 83

(2000), p. 221–31; discussion p. 231–2, p. 297–313

83 NEVINSKII, G. A., FAVOROVA, O. O., BUNEVA, B. N., in *Protein-protein interactions. A molecular cloning manual* (Golemis, Ed.) Cold Spring Harbor Lab. Press, Cold Spring Harbor, New York 2002, p. 523–534

84 NEVINSKII, G. A., KANYSHKOVA, T. G., SEMENOV, D. V., BUNEVA, V. N., *Vestn. Ross. Akad. Med. Nauk* (2001), p. 38–45

85 ZOUALI, M., *Arch. Immunol. Ther. Exp. (Warsz.)* 49, (2001), p. 361–5

86 PISETSKY, D. S., *Isr. Med. Assoc J.* 3 (2001), p. 850–3

87 RAPTIS, L. MENARD, H. A. *J. Clin. Invest.* 66 (1980), p. 1391–9

88 REIMER, C., RASKA, I., TAN, E. M., SHEER, U., *Virchows Arch.* 54 (1987), p. 131–36

89 SHOENFELD, Y., TEPLIZKI, H. A., MENDLOVIC, S., BLANK, M., MOZES, E., ISENBERG, D. A., *Clin. Immunol. Immunopathol.* 51 (1989), p. 313–25

90 GABIBOV, A. G., GOLOLOBOV, G. V., MAKAREVICH, O. I., SCHOUROV, D. V., CHERNOVA, E. A., YADAV, R. P. *Appl. Biochem. Biotechnol.* 47 (1994), p. 293–302; discussion p. 303

91 SUGIAMA, Y., YAMAMOTA, T. *Tohoku J. Exp. Med.* 178 (1996), p. 203–215

92 BIGAZZI, P. E., in *Mechanisms of Immunology* (Cohen, S., Wozd, P. E, McKlaskie, R. T., Ed.) Meditsina, Moscow 1983, p. 181–206

93 JAGER, L., Jena (1988), p. 270–279

94 BOGNER, U., HEGEDUS, L., HANSEN, J. M., FINKE, R., SCHLEUSENER, H., *Eur. J. Endocrinol.* 132 (1995), p. 69–74

95 SENDA, Y., NISHIBU, M., KAWAI, K., MIZUKAMI, Y., HASHIMOTO, T., *Rinsho Byori* 43 (1995), p. 1243–50

96 PAUL, S., LI, L., KALAGA, R., O'DELL, J., DANNENBRING, R. E., JR, SWINDELLS, S., HINRICHS, S., CATUREGLI, P., ROSE, N. R., *J. Immunol.* 159 (1997), p. 1530–6

97 BOIKO, A. N. FAVOROVA, O. O., *Mol. Biol. (Mosk.)* 29, (1995), p. 727–49

98 ARCHELOS, J. J., STORCH, M. K., HARTUNG, H. P., *Ann. Neurol.* 47 (2000), p. 694–706

99 TROTTER, J. L., CROSS, A. N., (Cook, S. D., Ed.) Marcel-Dekker, New York 1996, p. 187–199

100 HEMMER, B., ARCHELOS J. J., HARTUNG H. P. *Nat. Rev. Neurosci.* 3 (2002), p. 291–301

101 NIEHAUS, A., SHI, J., GRZENKOWSKI, M., DIERS-FENGER, M., ARCHELOS, J., HARTUNG, H. P., TOYKA, K., BRUCK, W., TROTTER, J. *Ann. Neurol.* 48 (2000), p. 362–71

102 IGLESIAS, A., BAUER, J., LITZENBURGER, T., SCHUBART, A., LININGTON, C., *Glia* 36 (2001), p. 220–34

103 SHOENFELD, Y., BEN-YEHUDA, O., MESSINGER, Y., BENTWITCH, Z., RAUCH, J., ISENBERG, D. I., GADOTH, N., *Immunol. Lett.* 17 (1988), p. 285–91

104 WILLIAMSON, R. A., BURGOON, M. P., OWENS, G. P., GHAUSI, O., LECLERC, E., FIRME, L., CARLSON, S., CORBOY, J., PARREN, P. W., SANNA, P. P., GILDEN, D. H., BURTON, D. R., *Proc. Natl. Acad. Sci. USA* 98 (2001), p. 1793–8

105 SAVEL'EV, A. N., IVANEN, D. R., KULMINSKAYA, A. A., ERSHOVA, N. A., KANYSHKOVA, T. G., BUNEVA, V. N., MOGELNITSKII, A. S., M., D. B., FAVOROVA, O. O., NEVINSKY, G. A., NEUSTROEV, K. N., *Immunol. Lett.* (2002)

106 IKEHARA, S., KAWAMURA, M., TAKAO, F., INABA, M., YASUMIZU, R., THAN, S., HISHA, H., SUGIURA, K., KOIDE, Y., YOSHIDA, T. O., ET AL., *Proc. Natl. Acad. Sci. USA* 87 (1990), p. 8341–4

107 MORTON, J. I., SIEGEL, B. V., *Immunology* 34 (1978), p. 863–8

108 ISHIDA, T., INABA, M., HISHA, H., SUGIURA, K., ADACHI, Y., NAGATA, N., OGAWA, R., GOOD, R. A., IKEHARA, S., *J. Immunol.* 152 (1994), p. 3119–27

109 ORLOVSKAYA, I. A., KOZLOV, V. A., *Russ. J. Immunol.* 6(2) (2001), p. 168–175

110 ORLOVSKAYA, I. A., DUBROVSKAYA, V. V., TOPORKOVA, L. B., CHERNYKH, E. R., TIKHONOVA, M. A., SAKHNO, L. V., BUNEVA, V. N., KOZLOV, V. A., NEVINSKY, G. A., in *Third International Conference on Bioinformatics of Genome Regulation and Structure (BGRS" 2002)* Novosibirsk 2002, p. 49–51

111 Avalle, B., Zanin, V., Thomas, D., Friboulet, A., *Appl. Biochem. Biotechnol.* 75 (1998), p. 3–12

112 Jerne, N. K., *Ann. Immunol. (Paris)* 125C (1974), p. 373–89

113 Puccetti, A., Madaio, M. P., Bellese, G., Migliorini, P., *J. Exp. Med.* 181 (1995) p. 1797–804

114 Izadyar, L., Friboulet, A., Remy, M. H., Roseto, A., Thomas, D., *Proc. Natl. Acad. Sci. USA* 90 (1993), p. 8876–80

115 Kolesnikov, A. V., Kozyr, A. V., Alexandrova, E. S., Koralewski, F., Demin, A. V., Titov, M. I., Avalle, B., Tramontano, A., Paul, S., Thomas, D., Gabibov, A. G., Friboulet, A. *Proc. Natl. Acad. Sci. USA* 97 (2000), p. 13526–31

116 Alexandrova, E. S., Koralewski, F., Titov, M. I., Demin, A. V., Ignatova, A. N., Kozyr, A. V., Kolesnikov, A. V., Tramontano, A., Paul, S., Thomas, D., Gabibov, A. G., Friboulet, A., *Doklad. Biochem. Biophys* 377 (2001), p. 75–8

117 Aleksandrova, E. S., Koralevski, F., Titov, M. I., Demin, A. V., Kozyr, A. V., Kolesnikov, A. V., Tramontano, A., Paul, S., Thomas, D., Gabibov, A. G., Gnuchev, N. V., Friboulet, A., *Bioorg. Khim.* 28 (2002), p. 118–25

118 Hu, R., Xie, G. Y., Zhang, X., Guo, Z. Q., Jin, S., *J. Biotechnol.* 61 (1998), p. 109–15

119 Friboulet, A., Izadyar, L., Avalle, B., Roseto, A., Thomas, D., *Appl. Biochem. Biotechnol.* 47 (1994), p. 229–37; discussion p. 237–9

120 Debat, H., Avalle, B., Chose, O., Sarde, C.-O., Friboulet, A., Thomas, D., *FASEB J.* 15 (2001), p. 815–822

121 Crespeau, H., Laouar, A., Rochu, D., *C. R. Acad. Sci. III* 317 (1994), p. 819–23

122 Andrievskaia, O. A., Kanyshkova, T. G., Iamkovoi, V. I., Buneva, V. N., Nevinskii, G. A., *Doklad. Akad. Nauk.* 355 (1997), p. 401–3

123 Bronshtein, I. B., Shuster, A. M., Gololobov, G. V., Gromova, II, Kvashuk, O. A., Belostotskaya, K. M., Alekberova, Z. S., Prokaeva, T. B., Gabibov, A. G., *FEBS Lett.* 314 (1992), p. 259–63

124 Bronshtein, I. B., Shuster, A. M., Gromova, II, Krashuk, O. A., Geva, O. N., Alekberova, Z. S., Gabibov, A. G., *Doklad. Akad. Nauk. SSSR* 318 (1991), p. 1496–500

125 Bronshtein, I. B., Shuster, A. M., Gromova, II, Kvashchuk, O. A., Geva, O. N., Gololobov, G. V., Prokaeva, T. B., Alekberova, Z. S., Gabibov, A. G., *Biuet. Eksped. Biol. Med.* 110 (1990), p. 598–600

126 Nevinsky, G. A., Buneva, B. N., *J. Immunol. Methods* 269 (2002), p. 235–

127 Mestecky, J., Russell, M. W., Jackson, S., Brown, T. A., *Clin. Immunol. Immunopathol.* 40 (1986), p. 105–14

128 Hanson, L. A., Hahn-Zoric, M., Berndes, M., Ashraf, R., Herias, V., Jalil, F., Bhutta, T. I., Laeeq, A., Mattsby-Baltzer, I., *Acta Paediatr. Jpn.* 36 (1994), p. 557–61

129 Hardonk, M. J., Koudstaal, J., *Prog. Histochem. Cytochem.* 8 (1976), p. 1–68

130 Komoda, T., Koyama, I., Hasegawa, M., Sakagishi, Y., *Rinsho Byori* 31 (1983), p. 682–91

131 Fey, H., Butler, R., Marti, F., *Vox Sang.* 25 (1973), p. 245–53

132 Mestecky, J., McGhee, J. R., *Adv. Immunol.* 40 (1987), p. 153–245

133 Kazakov, V. I., Bozhkov, V. M., Linde, V. A., Repina, M. A., Mikhailov, V. M., *Tsitologiia* 37 (1995), p. 232–6

134 Mikhailov, V. M., Linde, V. A., Rozanov Iu, M., Kottsova, N. A., Susloparov, L. A., Konycheva, E. A., *Tsitologiia* 34 (1992), p. 67–73

135 Dayan, C. M., Daniels, G. H., *North. Engl. J. Med.* 335 (1996), p. 99–107

136 Amino, N., Tada, H., Hidaka, Y., *Thyroid* 9 (1999), p. 705–13

137 Freeman, R., Rosen, H., Thysen, B., *Arch. Intern. Med.* 146 (1986), p. 1361–4

138 Tanaka, A., Lindor, K., Ansari, A., Gershwin, M. E., *Liver Transpl.* 6 (2000), p. 138–43

139 Amino, N., Mori, H., Iwatani, Y., et al., 306 (1986), p. 849–852

140 Buneva, B. N., Kudryavtseva, A. N., Gal' vita, A. V., Dubrovskaya, V. V., V., K. O., Kalinina, I. A., Galenok, V. A., Nevinskii, G. A., *Biochemistry (Moscow)* (2002), in press

141 Breusov, A. A., Gal' vita, A. V., Benzo, E. S., Baranovskii, A. G., Prints, A. V., Naumov, V. A., Buneva, V. N., Nevinsky, G. A., *Russ. J. Immunol.* 6 (2001), p. 17–28

142 Nevinsky, G. A., Breusov, A. A., Baranovskii, A. G., Prints, A. V., Kanyshkova, T. G., Galvita, A. V., Naumov, V. A., Buneva, V. N., *Med. Sci. Monit.* 7 (2001), p. 201–11

143 Semenov, D. V., Kanyshkova, T. G., Karataeva, N. A., Gorbunov, D. A., Kuznetsova, I. A., Buneva, V. N., Nevinsky, G. A., *Med. Sci. Monit.* (2002)

144 Mohan, C., Adams, S., Stanik, V., Datta, S. K. *J. Exp. Med.* 177 (1993), p. 1367–81

145 Baranovskii, A. G., Ershova, N. A., Buneva, V. N., Kanyshkova, T. G., Mogelnitskii, A. S., Doronin, B. M., Boiko, A. N., Gusev, E. I., Favorova, O. O., Nevinsky, G. A., *Immunol. Lett.* 76 (2001), p. 163–7

146 Kit, Y. Y., Shipitsin, M. V., Semenov, D. V., Richter, V. A., Nevinsky, G. A., *Biochemistry (Mosc.)* 63 (1998), p. 719–24

147 Vlassov, A. V., M., H., Florentz, C., Naumov, V., Breusov, A. A., Buneva, V. N., Giege, R., Nevinsky, G. A., *Russ. J. Immunol.* 4 (1999), p. 25–32

148 Vlasov, A. V., Helm, M., Naumov, V. A., Breusov, A. A., Buneva, V. N., Florentz, C., Giege, R., Nevinskii, G. A., *Mol. Biol. (Mosk.)* 33 (1999), p. 866–72

149 Andrievskaya, O. A., Buneva, V. N., Baranovskii, A. G., Gal'vita, A. V., Benzo, E. S., Naumov, V. A., Nevinsky, G. A., *Immunol. Lett.* 81 (2002), p. 191–8

150 Rosenqvist, E., Jossang, T., Feder, J., *Mol. Immunol.* 24 (1987), p. 495–501

151 Burstone, M. S., Academic press, New York and London (1962)

152 Golinelli-Pimpaneau, B., Gigant, B., Bizebard, T., Navaza, J., Saludjian, P., Zemel, R., Tawfik, D. S., Eshhar, Z., Green, B. S., Knossow, M., *Structure* 2 (1994), p. 175–83

153 Kylberg, A. Y., Petyaev, I. M., Zamotaeva, N. G., *Immunology (Moscow)* 3 (1988), p. 37–40

154 Wentworth, A. D., Jones, L. H., Wentworth, P., Jr., Janda, K. D., Lerner, R. A., *Proc. Natl. Acad. Sci. USA* 97 (2000), p. 10930–5

155 Wentworth, P., Jr., Jones, L. H., Wentworth, A. D., Zhu, X., Larsen, N. A., Wilson, I. A., Xu, X., Goddard, W. A., 3rd, Janda, K. D., Eschenmoser, A., Lerner, R. A., *Science* 293 (2001), p. 1806–11

156 Luo, G. M., Qi, D. H., Zheng, Y. G., Mu, Y., Yan, G. L., Yang, T. S., Shen, J. C., *FEBS Lett.* 492 (2001), p. 29–32

157 Dong, F.-x., Zheng, Y.-g., Qi, D.-H., Luo, G.-m., Yang, T.-s., *Bopuxue Zazhi (China)* 17 (2000), p. 401–406

158 Generalov, I. I., Novikov, D. K., Zhiltsov, I. V., *Vestsi Natsyyanal'nai Akademii Navuk Belarusi, Seryya Biyalagichnykh Navuk* 1 (1999), p. 90–96

159 Paul, S., Said, S. I., Thompson, A. B., Volle, D. J., Agrawal, D. K., Foda, H., de la Rocha, S., *J. Neuroimmunol.* 23 (1989), p. 133–42

160 Paul, S., Volle, D. J., Powell, M. J., Massey, R. J., *J. Biol. Chem.* 265 (1990), 11910–3

161 Paul, S., Johnson, D. R., Massey, R., *Ciba Found. Symp.* 159 (1991), p. 156–67; discussion p. 167–73.

162 Sun, M., Li, L., Gao, Q. S., Paul, S., *J. Biol. Chem.* 269 (1994), p. 734–8

163 Sun, M., Gao, Q. S., Kirnarskiy, L., Rees, A., Paul, S., *J. Mol. Biol.* 271 (1997), p. 374–85

164 Mei, S., Mody, B., Eklund, S. H., Paul, S., *J. Biol. Chem.* 266 (1991), p. 15571–4

165 Paul, S., Sun, M., Mody, R., Tewary, H. K., Stemmer, P., Massey, R. J., Gianferrara, T., Mehrotra, S., Dreyer, T., Meldal, M., et al., *J. Biol. Chem.* 267 (1992), p. 13142–5

166 Sun, M., Gao, Q. S., Li, L., Paul, S., *J. Immunol.* 153 (1994), p. 5121–6

167 Li, L., Kaveri, S., Tyutyulkova, S., Kazatchkine, M. D., Paul, S., *Ann. N Y Acad. Sci.* 764 (1995), p. 570–2.

168 Thiagarajan, P., Paul, S., *Chem. Immunol.* 77 (2000), p. 115–29
169 Thiagarajan, P., Dannenbring, R., Matsuura, K., Tramontano, A., Gololobov, G., Paul, S., *Biochemistry* 39 (2000), p. 6459–65
170 Paul, S., Tramontano, A., Gololobov, G., Zhou, Y. X., Taguchi, H., Karle, S., Nishiyama, Y., Planque, S., George, S., *J. Biol. Chem.* 276 (2001), p. 28314–20
171 Stepaniak, L., *Prep. Biochem. Biotechnol.* 32 (2002), p. 17–28
172 Toporkova, L. V., Dubrovskaya, V. V., Sakhno, L. V., Tikhonova, M. A., Chernykh, E. R., Buneva, V. N., Nevinsky, G. A., Kozlov, V. A., Orlovskaya, I. A., *Russ. J. Immunol.* (2002)
173 Gololobov, G. V., Chernova, E. A., Schourov, D. V., Smirnov, I. V., Kudelina, I. A., Gabibov, A. G., *Proc. Natl. Acad. Sci. USA* 92 (1995), p. 254–7
174 Shchurov, D. V., Makarevich, O. I., Lopaeva, O. A., Buneva, V. N., Nevinskii, G. A., Gabibov, A. G., *Doklad. Akad. Nauk.* 337 (1994), p. 407–10
175 Vlassov, A. V., Andrievskaya, O. A., Kanyshkova, T. G., Baranovsky, A. G., Naumov, V. A., Breusov, A. A., Giege, R., Buneva, V. N., Nevinsky, G. A., *Biochemistry (Mosc)* 62 (1997), p. 474–9
176 Sorrentino, S., Libonati, M., *FEBS Lett.* 404 (1997), p. 1–5
177 Sorrentino, S., Libonati, M., *Arch. Biochem. Biophys.* 312 (1994), p. 340–8
178 Akagi, K., Murai, K., Hirao, N., Yamanaka, M., *Biochim. Biophys. Acta.* 442 (1976), p. 368–78
179 Gololobov, G. V., Rumbley, C. A., Rumbley, J. N., Schourov, D. V., Makarevich, O. I., Gabibov, A. G., Voss, E. W., Jr, Rodkey, L. S., *Mol. Immunol.* 34 (1997), p. 1083–93
180 Herron, J. N., He, X. M., Ballard, D. W., Blier, P. R., Pace, P. E., Bothwell, A. L., Voss, E. W., Jr, Edmundson, A. B., *Proteins* 11 (1991), p. 159–75
181 Hardie, G., Hanks, S., *The protein kinase. Protein-serine kinases*, Vol. 1, Academic Press, Harcourt Brace and Company, London, San Diego, New York, Boston, Sydney, Tokyo, Toronto **1995**
182 Hardie, G., Hanks, S., *The protein kinase. Protein-tyrosine kinases*, Vol. 2, Academic Press, Harcourt Brace and Company, London, San Diego, New York, Boston, Sydney, Tokyo, Toronto **1995**
183 Severin, E. S., Kochetkova, M. N. in: *Role of the phosphorylation in the cell activity regulation*, Science Press, Moscow **1985**
184 Bolon, D. N., Voigt, C. A., Mayo, S. L., *Curr. Opin. Chem. Biol.* 6 (2002), p. 125–9
185 Blackburn, G. M., Kang, A. S., Kingsbury, G. A., Burton, D. R., *Biochem. J.* 262 (1989), p. 381–90
186 Benkovic, S. J., *Annu. Rev. Biochem.* 61 (1992), p. 29–54
187 Shokat, K. M., Schultz, P. G., *Annu. Rev. Immunol.* 8 (1990), p. 335–63
188 Kasho, V. N., Baykov, A. A. *Biochem. Biophys. Res. Commun.* 161 (1989), p. 475–80
189 Shestakov, A. A., Baykov, A. A., Avaeva, S. M., *FEBS Lett.* 262 (1990), p. 194–6
190 Smirnova, I. N., Shestakov, A. S., Dubnova, E. B., Baykov, A. A., *Eur. J. Biochem.* 182 (1989), p. 451–6
191 Dewai, M. J. S. *Enzymes* 36 (1986), p. 8–
192 Warshel, A., Aqvist, J., Creighton, S. *Proc. Natl. Acad. Sci. USA* 86 (1989), p. 5820–4
193 Kates, M., *Techniques of lipidology*, Elsevier, New York **1972**
194 zu Meyer Heringdorf, D., van Koppen, C. J., Jakobs, K. H., *FEBS Lett.* 410 (1997), p. 34–8
195 H., W., Elsevier, Amsterdam **1985**
196 Wiegandt, H., Elsevier, Amsterdam **1985**
197 M., K., *Techniques of lipidology*, Elsevier, New York **1972**
198 Fiat, A. M., Jolles, P., *Mol. Cell Biochem.* 87 (1989), p. 5–30
199 Gillin, F. D., Reiner, D. S., Wang, C. S. *Science* 221 (1983), p. 1290–2
200 Balint, J. P., Jr, Ikeda, Y., Nagai, T., Terman, D. S., *Immunol. Commun.* 10 (1981), p. 533–40

201 Suga, H., Tanimoto, N., Sinskey, A. J., Masamune, S., *J. Am. Chem. Soc.* 116 (**1994**), p. 11197–98
202 Marshall, J. J., Miwa, I., *Biochim. Biophys. Acta* 661 (**1981**), p. 142–47
203 Gabibov, A. G., Kozyr, A. V., Kolesnikov, A. V., *Chem. Immunol.* 77 (**2000**), p. 130–56
204 Kozyr, A. V., Sashchenko, L. P., Kolesnikov, A. V., Zelenova, N. A., Khaidukov, S. V., Ignatova, A. N., Bobik, T. V., Gabibov, A. G., Alekberova, Z. S., Suchkov, S. V., Gnuchev, N. V., *Immunol. Lett.* 80 (**2002**), p. 41–7
205 Sushkov, S. V., Gabibov, A. G., Alekberova, Z. S., Gnuchev, N. V., *Ter. Arkh. (Moscow)* 73 (**2001**), p. 58–65
206 Suchkov, S. V., Gabibov, A. G., Gnuchev, N. V., *Ontogenez* 32 (**2001**), p. 348–52
207 Kit, Y. Y., Mitrofanova, E. E., Shestova, O. E., Kuligina, E. V., Richter, V. A., *Ukr. Biokhim.* 72 (**2000**), p. 73–76
208 Tomasi, T. B. in *Immunology of Breast milk* (Ogra, P. L. and Dayton, D. H., Eds.), Raven Press, New York **1979**, p. 1–5
209 Redhead, K., Hill, T., Mulloy, B., *FEMS Microbiol. Lett.* 58 (**1990**), p. 269–73
210 Brambell, F. W., *The Transmission of Passive Immunity from Mother to Young*, North-Holland, Amsterdam, London **1970**

211 Ramaswamy, H., Swamy, C. V., Das, M. R., *J. Biol. Chem.* 268 (**1993**), p. 4181–7
212 Poser, C. M., Paty, D. W., Scheinberg, L., McDonald, W. I., Davis, F. A., Ebers, G. C., Johnson, K. P., Sibley, W. A., Silberberg, D. H., Tourtellotte, W. W., *Ann. Neurol.* 13, (**1983**), p. 227–31
213 Kurtzke, J. F. *Neurology* 33 (**1983**), p. 1444–52
214 Zhou, Y., Karle, S., Taguchi, H., Planque, S., Nishiyama, Y., Paul, S., *J. Immunol. Methods* 269 (**2002**), p. 257
215 Paul, S., Baukloh, V., Mettler, L., *J. Immunol. Methods* 56 (**1983**), p. 193–9
216 Paul, S. M., Liberti, P. A., *J. Immunol. Methods* 21 (**1978**), p. 341–53
217 Blackburn, G. M., Datta, A., Patridge, L. J., *Pure Appl. Chem.* 68 (**1996**), p. 2009–2016
218 Wentworth, P., Datta, A., Blakey, D., Boyle, T., Partridge, L. J., Blackburn, G. M., *Proc. Natl. Acad. Sci. USA* 93 (**1996**), p. 799–803
219 Zhou, G. W., Guo, J., Huang, W., Fletterick, R. J., Scanlan, T. S., *Science* 265 (**1994**), p. 1059–64
220 Charbonnier, J. B., Carpenter, E., Gigant, B., Golinelli-Pimpaneau, B., Eshhar, Z., Green, B. S., Knossow, M. *Proc. Natl. Acad. Sci. USA* 92 (**1995**), p. 11721–5
221 De Prada, P., Winger, G., Landry, D. W. *Ann. NY Acad. Sci.* 909 (**2000**), p. 159–169

Index

a

A-sepharose 520
AbM 74
absorbance spectrum 359
absorption maximum 359
acetal 134
acetylcholinesterase (AChE) 229, 427, 428, 443, 511
AChE-tagged 229
acid-limb pH profile 433
acidic shock 524
activation energy 14
active immunization 482
active site 481
active-site general base 432
acyl enzyme 158
acyl migration 376
acyl transfer 439
acyl transfer reactions 371
acyl-antibody 51
acyl-antibody intermediate 209, 210
acyl-enzyme intermediate 431–434
addition 405
ADEPT 290, 291, 295
adjuvants 493
affinity ceiling 429
affinity chromatography 481, 507, 512, 519, 520, 523, 527
affinity maturation 2, 7, 10, 12, 15, 56, 57, 160, 161, 165, 177, 445, 473
affinity selection 250

aggregation pheromone 134
aldol 139
aldol acceptor 175
aldol condensation 30, 46, 432
aldol donor 175
aldol reaction 136, 306, 308, 311, 314, 392
aldolase 54, 175–177
aldolase antibody 136, 137, 142, 284, 392
aldolization 392
alkenes 386
4-alkyphenyl 477
allergic diseases 514
allosteric binding 17
allylic isomerization 161
alpha lytic protease 443
α-agglutinin 275
α-amylase 554
alpha-cubebene 136
α-D-maltooligosides 552
α-multistriatin 134, 136, 149, 409
AMBER 102
amidase 52, 381, 507
amidase activity 230
amide bond 381
amine oxide 403
aminopterin 495
amplification 243, 250, 264, 279
amylolytic activity 510, 514, 553
anchimeric assistance 173
   antibody catalysis 199
   cationic cyclization 199

Catalytic Antibodies. Edited by Ehud Keinan
Copyright © 2005 WILEY-VCH Verlag GmbH & Co. KGaA, Weinheim
ISBN: 3-527-30688-9

animal models   477
anionic binding motifs   155
anionic binding site   157, 159, 160
anthropological research   118
anti-Baldwin epoxide cyclization
      224
anti-dansyl antibody   402
anti-hapten approach   485
anti-idiotypic antibody   484, 507
anti-phosphonate antibody   50, 432
anti-thyroglobulin   551, 552
antibody
      1D4   161, 169, 206, 440, 444
      1E9   38–40, 42, 43, 91, 168,
           169
      1F7   34, 35, 37, 38, 106, 163,
           386, 464
      2D5   439, 440
      2H6   447
      3G2   377
      4B2   161, 162, 440
      4C6   172, 344
      4D5   388
      5C8   189
      6D6   50
      6D9   82, 156, 422–425, 427,
           429, 436, 437, 442
      6F11   432
      7B9   378
      7C8   156, 422–425, 433, 436
      7D4   98, 388
      7F11   440
      7G12   3, 4, 7, 9, 13, 44, 45, 62,
           377
      8C7   359
      9D9   100
      10F11   91, 100, 165, 444
      11F1-2E11   386
      12B4   360
      12F12   447
      13G5   91, 98, 167, 388
      14-10   437, 438
      14B9   189
      14D9   134, 230, 231
      16G3   81
      17E8   49, 86, 156–158, 374,
           422, 423, 425, 433, 442,
           445, 460
      17G8   197
      19A4   172, 173
      19C9   231
      20B11   404
      20F10   360, 362
      20G9   433
      21D8   107, 159
      21H3   51, 52, 379, 447
      21H9   360
      22C8   96, 98, 388
      22H25   231
      24H6   309
      25A10   202
      25D11   340
      26D9   399
      27A6   439, 440
      28B4   3, 12, 18, 26, 160, 161,
           445
      29G11   49, 86, 156, 157, 422,
           423, 460
      29G12   105
      30C6   440
      30C8   439
      33F12   54, 175, 177, 308, 310
      38C2   54, 142, 143, 145, 147,
           177, 284, 288, 293, 301,
           308, 310, 321, 395
      38C6   234
      38G2   441
      39-A11   40–42, 167, 445
      39A11   3, 25, 91, 93, 169
      40F12   177, 314
      42F1   314
      43C9   51, 83, 156, 421, 423,
           432–435, 442, 444, 461
      48G7   3, 12, 21, 47–49, 85, 156,
           157, 423–425, 429, 430,
           444, 460
      84A3   441
      84G3   177, 309, 316, 397
      87D7   195
      93F3   177, 309, 316, 396

AZ-28   36–38, 101, 164, 445
AZ28   3, 12, 22
BV 04-01   541
CNJ206   49, 84, 156, 157, 423–425, 444
conformational effects   350
D2.3   156, 157, 422–426, 433, 436, 442, 444, 446, 460
D2.4   422–424, 426, 433, 436, 446
D2.5   422, 423, 446
DB3   42, 169
induced fit binding   161
MOPC167   226
NaBH$_3$CN   403
PCP21H3   210, 432
SPO50C1   209
UD4C3.5   358
antibody affinity   429
antibody binding site   357
antibody fragments   *see* Fab, Fc
antibody library   306, 325
antibody oxyanion holes   425
antibody production   495, 497
antibody purification   499
antibody variability   478
antibody-catalyzed cationic polyene cyclization   197
anticancer drug   285
antioxidants   531
antiperiplanar elimination   407
apoptosis   510, 513, 555
approximation   162, 165, 169
Aristotle   120
artificial selection   246
aryl carbamate hydrolysis   207
arylnitroso   389
arylsulfonate   400
ascites fluid   135, 497, 499
aspartic hydrolases
   mechanism   420
aspartyl hydrolases   418
association rate   429
asthma   510, 554
asymmetric building block   134

asymmetric synthesis   149
ATP   543
ATP-sepharose   518, 527, 543
ATPase   519
autoantibody   484, 506, 507
AutoDock   85, 109
autoimmune disease   507, 514
autoimmune shock   515
autoimmunity   61, 346, 505
   autoimmune mice   485
autoimmunization   513, 514
autoradiography   549
aza Diels-Alder reaction   390
4-aza-steroid *N*-oxide   402

b
B-cell   509
B-lymphocytes   513
B$_{Ac}$2 pathway   207
backbone NH group   427
backcrossing   268
bactericidal activity   340
bait-and-switch   185, 190, 191, 200, 371, 421, 439, 440, 465
baiting strategy   161, 162
Baldwin's rules   171, 185
bark beetle   142
benzoic acid   342
benzonitrile *N*-oxide   391
benzoylecgonine   482
benzyl phosphonate   477
Bergman cycloaromatization reaction   466
β-amino alcohol   439
β-diketosulfone   395
β-elimination   234
β-galactosidase   235
β-hydrogen abstraction   362
β-hydroxyketones   395
β-lactam substrate   479
β-lactamase   17, 256, 268
β-naphthyl acetate   474
bicyclic hapten   388
bimolecular binding   446
bimolecular process   380, 388

bimolecular rearrangements 165
binding energy 14, 60, 208, 217
binding mode 365
biocatalysis 366
bispecific antibody 291
bisubstrate reaction 543
brain tissue 509
brevicomins 142, 143, 149, 409
bronchial asthma 506, 555
burst kinetics 432

c

calmodulin 261
calmodulin-binding peptide 261
calmodulin-tagged phage 261
camptothecin 289, 323
Campylobacter jejuni 547
cancer 285, 290
canonical binding array 424
carbamate insecticide 483
carbaryl 483
carbocation 191
carbohydrates 376, 383
carbonate 472, 476
carboxybenzisoxazole decarboxylation 107
carboxylesterases 437
carboxypeptidase 511
carboxypeptidase A 428
caribou 119
caribou hunting 125
casein 518, 543
cat-ELISA 217, 222
catalase 347
catalysis 454
catalytic activity 470
catalytic ceiling 429
catalytic efficiency 361
catalytic triad 429
cation stabilization 161
cationic cyclization 147, 191, 400
cationic cyclopropanation 196
cationic exchange column 500
cationic peptide 260
CD4 560

cDNA 257
CDR 218, 269
cell-culture cytotoxicity 138
cellular immunity 339
cephalosporinase 269
cerebrospinal fluid 509
chair-like transition state 387
chemical adaptor 296
chemical evolution 245
chemical instruction theory 14
chemical rescue 421, 431, 435, 437
chemoselectivity 133, 370, 403
chloramphenicol ester 437
cholinesterase 484
chorismate 386
chorismate mutase 22, 46, 106, 163, 226, 386, 464
chorismate mutase antibody 38
chromogenic assay 217
circle of life 119
cis-cyclobutanol 360
Claisen rearrangement 163, 386, 464
cocaine 235, 300, 482
cocaine esterase (cocE) 427, 437
cocaine-hydrolyzing abzymes 561
cofactor-dependent reactions 60
combinatorial library 249
combinatorial mutagenesis 43
combinatorial phage display library 249
combinatorial protein library 279
compartmentalization 271
complementarity-determining regions *see* CDR
computational analysis 186, 204
confocal fluorescence microscopy 275
conformational change 7, 388, 421, 437, 443–445, 447
conformational constraints 387
conformational diversity 14, 445
conformational flexibility 10, 24, 27
conformational isomerism 448
conformational plasticity 14

conformational requirement 365
conserved region 344
continuous-flow reactor 149
convergence 424
Coomassie blue 527, 547
cooperativity 18, 21, 27
covalent catalysis 175
covalently reactive antigen 560
CRA 560
Cree Indians 119
cross-reactivity 41
crystal structure 84, 90, 107, 424, 530
cultural changes 120
cutinase 428
cyclization 365, 398, 506
cyclization transition state 365
cycloaddition 25, 38, 354, 388
[4+2] cycloaddition 38
cyclopropanol 362

d

D-amino acids 373
D-antibodies 419, 425, 428, 437, 446
de-acylation 434
DEAE-cellulose 520
debenzoylation 235
decalin 401
decarboxylase antibody 62
decarboxylation 46, 137, 159, 319, 506
degradation 141
dehydratation 226
dehydration 147
dehydrofluorination 226
dehydrogenase 227
δ-lactone 399
2-deoxyribose-5-phosphate aldolase 176, 177
dephosphonylation 432
DERA 177
Dess-Martin reagent 141
destabilization 43

desymmetrization 371, 379
  of *meso* compounds 408
diagnostic tools 558
Diels-Alder cycloaddition 25, 39, 40, 165, 167, 203, 204, 386, 388, 466
Diels-Alderase 3, 43, 50, 90, 93, 96, 98, 100, 224, 445
digestion 502
dihydrofolate reductase 17
dihydrogen trioxide 336
diketone 306, 308
1,3-dipolar cycloaddition 391, 466
diradical 211
directed evolution 38, 132, 244, 251, 262, 269, 271, 272, 274, 276, 279
disfavored processes 185
disfavored reactions 184, 187
disfavored ring closure 171
disfavored transformations 184
dissociation rate 429
disulfide-stabilized Fv   see dsFv
diversity 243, 245, 246, 263, 270
DNA catalysts 63
DNA photolyase 350, 358
DNA photoreactivating enzymes 352
DNA polymerase 272
DNA shuffling 61, 262, 267–271, 276
DNA-cellulose 520
DNase 507, 514, 515, 534, 535
DNase II 535
docking 100
docking studies 42, 189, 206
domain swapping 61
doxorubicin 289, 323
drug-masking linker 286
dsFv 249
dsFv phage 250
dsFv phage library 250
Dutch elm disease 134
dynamic kinetic resolution 408

e

E. coli   517, 529, 547
E1cB (elimination-addition) process   207
eclipsed conformation   206
economic changes   120
EDTA   261
effective molarity   39, 205, 210
electrofocusing   540
electronically excited molecule   213
electrophilic addition   172
electrophilic attack   173
electrophoretic analysis   526
electrostatic complementarity   155, 156
electrostatic interactions   17
electrostatic stabilization   459
elimination   405, 406, 419
ELISA   228, 238, 239, 255, 493–495, 497, 499
ELISA reader   496
enamine   137, 306, 307, 309
enamine intermediate   314
enantiofacial selectivity   158
enantiomer   187
enantiomeric excess   186, 187, 204, 359, 381
enantiomeric purity   139, 141, 142
enantioselective antibody   360
enantioselective protonolysis   136
enantioselective synthesis   146
enantioselectivity   133, 136, 143, 149, 188, 313, 447
enantiotopic group   378
endo adduct   474
endo isomer   388
endo transition state   41
endogenous nucleophile   421, 430, 432, 433, 448
endoperoxides   336
enhanced permeability and retention   295
enol ether   134, 141
entropy trap   164
environment   119

enzyme contamination   475
enzyme digestion   500
enzyme-calmodulin construct   261
enzyme-coupled assay   222
enzyme-like mimics   470
epothilone   31, 136, 149, 409
epothilone A   137, 139, 141, 142
epothilone B   137
epothilone C   139, 141
epoxide   134, 401
epPCR   267, 269–274
epPCR mutagenesis   276
EPR   295
EPR effect   295
error-prone PCR   *see* epPCR
ester hydrolysis   148, 155, 418
    mechanism   419
esterase   15, 50, 157
esterification   141
esterolysis   235
esterolytic antibody   49, 50, 61, 428
etoposide   292, 298, 299, 323
European elm bark beetles   136
evolution   1, 7, 356
evolutionary force   272
evolutionary pressure   251, 272, 275, 279
excision repair mechanism   355
exo Diels-Alder product   388
exogenous amine nucleophile   437
exogenous hydroxide   419
exogenous hydroxide anion   435
exogenous nucleophile   421, 438
exogenous nucleophilic attack   431

f

Fab   461, 500
    structure   189
Fab'   500
Fab-phage library   257, 264
Fab-presenting phages   252, 256
F(ab')$_2$ 500, 502
FACS   273, 274
factor D   443
factor VIII   533

factor VIII cleavage 507
FADH 355
family shuffling 269
Fc fragment 14
FDP aldolase 56
fermentation 59
ferrocene 204, 389
ferrochelatase 3, 43, 45
filamentous phage 244, 255
Fischer, E. 7
flavin 353
flexibility hypothesis 477
flow cytometry 273, 275
fluorescence 4, 236, 321
fluorescence quenching 446
fluorescence-activated cell sorter
       see FACS
fluoroacetone 394
fluorogenic assay 217
formamide-hydrogen peroxide 404
fragmentation transition states 365
free energies 365
free energy perturbation (FEP) 74, 109
free-energy perturbation methods 87
FRET 273
Freunds adjuvant 493
(+)-frontalin 146, 149
functional homogeneity 480
Fv 461

g
gangliosides 547
GC 225
gel filtration 520
gel-filtration column 500
general acid 189
general acid/base mechanism 421, 439
general acid/general base 190, 420
general base 431, 434
general repair system 355
general-acid catalysis 420
general-base catalysis 431, 438

general-base catalyst 440
general-base residue 439
genetic engineering 127
genetic selection 328
gentamycin 494, 495
germline 7, 9, 12, 27, 160, 165, 168, 169, 175
germline antibody 2
germline precursor 37, 47
glutathione peroxidase 531
glycerolipids 546
glycoproteins 509
glycosidase 383
glycosidic bond 383
glycosphingolipids 547
Gomori's method 528
gp120 559
gradualism 22
GRASP 37
growth inhibition 286
Grubb's catalyst 141

h
H_2O_2 336
Haldane, J. B. S. 7
hapten 78, 81, 86, 88, 94, 471
    anti-hapten 220
    aryl phosphates 60
    β-diketone 137
    cationic 53
    diketone 307
    homochiral 373
    hydrophobicity 476
    phosphonamidate 52
    phosphonate 47
    phosphonium 472
    piperidinium 392, 404
    racemic 371
    reactive 373
    size 476
    sulfone 475
hapten conjugate 492, 494
hapten design 358
hapten flexibility 205
hapten inhibition 478

Hashimoto's thyroiditis 508
Haurowitz, F. 14
heart rate 546
helper phage 253
hematopoietic stem cells 510
hematoporphyrin 336
hemophilia 533
Henry type *retro*-aldol reaction 397
heparin resistance 272
hepatitis 536
hetero-Diels-Alder reaction 390, 474
heterologous immunization 371, 437, 465
hexokinase 14, 24
high-efficiency ion-exchange chromatography 512
high-throughput screening 239
historical evolution 118
historical perspective 118, 119
history of science 118, 122
HIV 507
hMBP 509
homology 83
homology modeling 80
hormone 555
horseradish peroxidase 495
HP-TLC 231
HPLC 222
HPMA 295
HT 510, 558
human behavior 118
human growth hormone 14
human growth hormone receptor 14
human milk 505, 507, 513, 534, 535, 543
human myelin basic protein (hMBP) 509
human serum 506, 507
humanism 120
humoral immunity 339
hybridoma 221
hybridoma cells 135, 497
hybridoma production 494

hybridoma screening 494
hybridoma technology 126, 250, 491
hydrogen bond 459
hydrogen bonding 17
hydrogen peroxide 178, 336, 404, 531
hydrogen-bond donor 424
hydrogenation 141
hydrolysis 46, 286, 536
hydrolytic antibody 62, 429, 442, 448
hydrolytic enzymes 46, 418, 439
hydrolytic mechanism 49, 77
hydrolytic reactions 536
hydrophobicity 485
hydroxide ion 430
hydroxyacetone 394
hydroxybrevicomins 142, 143
hydroxyl radical 342
hydroxylamine 436, 437
hyperconjugation 23, 164, 172
hypothyrosis 558
hypoxanthine 495, 531
hysteretic 445

i

idiotype 511
idiotypic mimicry 511
imine 226
iminium ion 137
immobilization 149
immobilized phage 267
immune response 471, 478
immune system 346
immunization 492, 503, 514
immuno-memory 514, 519
immunoglobulin fold 343
immunoglobulin G 14
immunology 336
immunotherapeutics 559
in vitro selection 325
in vitro evolution 126, 250, 275
in vitro immunization 384
in vitro recombination 269

in vitro selection   254
in vivo detoxification   483
induced fit   8, 10, 27, 165, 195, 209, 444, 446, 447
    mechanism   443
    model   443
inhibition experiments   480
innate immunity   339
insulin   299
intestinal vasoactive peptide   506
intramolecular cyclization   185
intramolecular hydrogen transfer   359
intramolecular transacylation   380
intrinsic catalytic activity   525
intrinsic enzymic activity   526
irradiation   358
irrational design   243
isoelectric point   540
isomerization   446
isotyping   495

j

jeffamine   9
Jencks, W. P.   30, 122, 505
Jerne, N. K.   511

k

KDPG aldolase   56
Kemp elimination   387
Kemp, D. S.   232
ketal   134
ketal formation   210
ketalization   143
kinase activity   523
kinetic control   185, 186
kinetic resolution   145, 148, 309, 316, 371, 373, 377, 393, 399, 473
KLH conjugate   494, 495
Köhler, G.   126
Kulberg, A.   530, 531

l

Lac repressor   272

lactation   517
lactonization   146, 506
lanosterol   148, 400
large-scale production   399
leaving group   477
leucocytes   345
leukemic cell line   138
light-induced chemistry   350
limiting dilution   496
linear interaction energy (LIE)   74, 109
linear interaction method   92
lipid kinase   507, 545
localization   287
lock-and-key   7, 10, 19, 23, 47, 165, 195, 443, 447
long-range effect   27
Lpp-OmpA   273
lupus phage   560
lymphocyte proliferation   510
lymphoid cells   517
lysine residue   392

m

macrocyclic lactone   141
macrolactonization   138, 139
maltohexaose   553
maltooligosaccharides   553
maltopentoside hydrolysis   552
maltotriose   553
mammalian neonates   556
mammary gland   513
maturation   38
MBP   509
MD calculations   102
MD studies   75
medium effect   387
memory-like   445
mercury(II)   441
meso substrate   378
mesoporphyrin   4
metal mediated allostery   447
metal-activated catalytic antibodies   441
metalloporphyrin   44

mevalonolactone 146, 149, 395
MFT 508
MIC 268
mice
   129 G/X+ 499
   A/J 493
   Balb/c 493, 499
   C57 493
   Swiss Webster 493
Michael addition 408, 432
Michaelis complex 4, 7
Michaelis-Menten 219
Michaelis-Menten approximation 360
Michaelis-Menten kinetics 548
Michaelis-Menten model 358
microbicidal action 340
microenvironment 187, 209
microtubules 137
Middle Ages 120
Milstein, C. 126
minimum inhibitory concentration (MIC) 268
mitochondria 557
mitochondrial tRNA 537
Mitsunobu reaction 139
molecular dynamics (MD) 74, 80, 87, 92, 100, 102
molecular excited state 213
molecular modeling 532
molecular recognition 126
monoclonal antibodies 456
monoclonality 496
monooxygenase enzyme 404
morphology
   bacteria 340
MS 558
mucosal site 513, 517
multastriatin 406
multiple myeloma 559
multiple sclerosis 508
mutagenesis 262, 263
mutation rate 263
myelin 509, 555
myelin-proteolipid shell 509
myeloma cells 494, 499
myoclonic epilepsy with ragged-red fibers (MERRF) 557
myxobacteria 137

n

N-aryl-β-lactam 479
N-methylmesoporphyrin 4
N-oxide 400
N-oxide center 401
NAD(P)H oxidase 345
naïve antibody-phage display library 269
naïve library 267
naproxen 149
naproxen acid 373
naproxen ethyl ester 473
natural amidases 435
natural antibody diversity 250
natural evolution 262
natural hydrolases 437
natural philosophy 120
natural product synthesis 132, 143
natural products 132, 149
natural sciences 118, 120
natural selection 243
nervous system 508
neuraminic acid 547
neurite retraction 546
neuroblastoma cells 293
neuropathological mechanisms 509
neutrophil activation 346
neutrophils 345
4-nitro phenyl group 477
nitrobenzylesterase 228
nitrocellulose membrane 518
2-nitrophenyl 479
2-nitrophenyl-β-lactam 479
non-immunoglobulin 543
non-steroidal *anti*-inflammatory drug (NSAID) 148
Norrish type I fragmentation reaction 362

Norrish type II photochemical reaction 360
Norrish type II reaction 211, 223
NTPs 543
nucleic acid metabolism 507
nucleic acids 127
nucleophilic addition 137
nucleophilic attack 418
nucleophilic catalysis 430, 432
nucleophilic lysine 137
nucleophilic mechanism 433, 434
nucleophilic substitution 46

o

4,6-O-ethylidene-α-D-maltoheptoside (EPS) 552
oligo-peptidase B 428
oligodendrocyte glycoprotein (MOG) 509
oligosaccharide kinase 549
oligosaccharides 549
OmpT 274
oncological patients 552
organic synthesis 132
organophosphorus nerve agents 482
ortho-phosphate 518, 544
over-expression 149
4-oxalocrotonate tautomerase (4-OT) 264
oxetanes 360
oxidation 18, 336, 402, 546
oxidative decarboxylation 226
oxidative killing 178
oxidosqualene 148
2,3-oxidosqualene 400
oxidosqualene cyclase 147
2,3-oxidosqualene cyclase 191
oxime 226
oxocarbenium intermediate 383
oxocarbonium ion 133, 406
oxorhenium chelate 385
oxy-Cope reaction 22
oxy-Cope rearrangement 34, 38, 101, 164, 386, 474

oxyanion 459
oxyanion hole 49, 426, 459
oxyanion pK_a 426
oxyanion stabilization 157, 421
oxyanionic stabilization 419
oxygenation reaction 404
ozone 336

p

^{32}P-labeling 545
p-methylsulfonate 432
packing forces 444
papain 428, 500, 501
partial digestion 500
passive immunization 482
Paul, S. 508, 543
Pauling, L. 1, 122
PCM (Polarized Continuum Model) 77, 105
PDB 153, 154
PDEPT 295
penicillin 498
pepsin 500, 502
peptidase 507
peptide conjugates 493
pericyclic processes 30, 133
pericyclic reactions 385
pericyclic transition state 204, 406
periodate 18
peroxidation 506
peroxide 437, 446
pH dependence 480
pH-binding profile 426
pH-rate dependency 418
pH-rate profile 433
phage display 132, 244, 245, 250, 251, 253, 264, 329
phage display library 245, 248, 250, 251, 253, 258, 261, 269
phage lambda vector 247
phage library 246, 251
phage particle 255, 260, 261
phage vector 251
phage-displayed 222
phage-displayed mutant 266

phagemid 253, 259
phagemid vector 251
phagocyte action 531
phenyloxazolone antibody 50
pheromone 136, 142
philosopher 121
philosophy 120
phosphatase 542, 551
phosphate 472, 476
phosphate diester 458
phosphatidyl choline 546
phosphatidyl ethanolamine 546
phosphatidyl inositol 546
phosphatidyl serine 546
phosphinate 382, 421
phospholipids 515
phosphonamidate 46, 458
phosphonate 46, 156, 158, 159, 421
phosphonate analog 373
phosphonate diester 431
phosphonate ester 534
phosphonate monoester 458
phosphonate tetrahedral 440
phosphonoamidate 421, 433
phosphorothioate 476
phosphorus (V) 458
phosphoryl group transfer 383
phosphorylation 543
phosphothioate 421
phosphotriesterase (PTE) 272
photo-antenna 353
photocatalysts 350
photocatalytic antibody 357, 366
photochemical cleavage 358, 360
photochemical reactions 350, 354, 357, 361
photocontrol 351
photodimer 358
photodimerization 358
photoenzyme 350, 357, 366
photolyase 354, 355
photoprocesses 211
photoproduct 352
photoproduct lyase 350
photosensitization 211

phylogenetic tree 347
pIII-display format 255
ping-pong bi-bi mechanism 209
ping-pong mechanism 210, 432
pivaloyloxymethyl 237
pIX-display libraries 255
pK_a 426, 427, 429
plasmid 535
plasmid display 272
plasticity 443
platelet activation 546
PNP-α-glucopyranoside 552
point mutation 264
political changes 120
polyanionic peptide 259
polyarthritis 536
polyclonal antibodies 390, 470
polymerase 272
polysaccharide kinase 513
polysaccharides 549
polysome 270, 271
polyspecificity 10, 25, 27
POM 237
porphyrin 6, 43
    deformations 6
Poser, C. M. 558
post-natal period 515
pre-equilibrium 443, 445
pre-steady-state kinetics 433, 446
pregnancy 514, 515, 552
prephenate 386
prodrug 31, 58, 289, 319, 323
prodrug activation 284, 292
product distribution 187, 199, 212
product inhibition 380, 388, 391, 401, 432, 445
product selectivity 359
proficiency 72
progenitor cell-specific surface glycoprotein AN2 509
prolactin receptor 14
proline 177
prolyl isomerase 228
prolyl oligopeptidase 427, 428
prostaglandin $F_{2\alpha}$ 379

protease A-chymostatin 427
proteases 157, 509, 525
Protein Data Bank   see PDB
protein dynamics 492
protein G 135, 238, 499, 500, 502
protein kinase 507, 526, 542, 543, 545
protein kinase C 346
proteinuria 519
prothrombin 533
protochlorophyllide oxidoreductase 351
protonolysis 134
protoporphyrin IX 3
proximity effects 34, 386
pseudodiaxial conformation 386
PTE 272
purification
   natural abzymes 519
pyridoxal 145
pyrimidine dimer photocleavage 211
pyrimidine dimerization 211
pyrimidine dimers 352
pyrophosphatase 544

q
quantum mechanical calculations (QM) 88
quantum mechanical theozymes 110
quantum mechanics/molecular mechanics (QM/MM) 75

r
rabbit polyclonal antibodies 472
racemic ethyl phosphonate 473
radical anions 211
radioactive isotopes 554
random mutagenesis 263, 266, 274
random mutation 261
rate acceleration 430
rate ceiling 446
rational design 243
rational reengineering 60
reaction coordinate 169
reactive carbocation 172
reactive immunization 54, 137, 148, 175, 202, 209, 288, 304, 309, 330, 371, 393, 431, 465
recombinant DNA vector 244
recombinant expression 220
recombination 2
redox enzymes 402
reduction 402, 403
regioisomer 390, 397
regioselective catalysts 186
regioselectivity 188, 370, 390, 391, 397, 466
relaxed specificity 377
remote interactions 442
Renaissance 120
reperfusion 346
residue frequency 190
restriction enzymes 127
retro-aldol reaction 308, 325
retro-aldol-*retro*-Michael 288, 299
retro-aldolization 392
retro-aldolization process 395
retro-Diels-Alder reaction 236, 287, 444
retro-Michael addition 235
retroaldol 32, 56
retroaldol reaction 136, 139, 141–143, 145
retrosynthetic analysis 134, 138
reverse-phase HPLC 223
ribosomal display 270, 271
ribosomal RNA 517
ribozymes 63
RNA catalyst 63
RNA structure 557
RNase 507, 514, 515, 534, 535
Robinson annulation 56, 147
rtPCR 257, 270

s
S-(+)-naproxen 148

saframycin H 146, 395
salt precipitation 472
SAS-fraction 135
scFv 248, 251, 258, 267, 269, 271, 273, 275, 277, 279, 462
scFv library 250, 255, 271, 273, 275
scFv phage display library 270
scFv-displaying phages 252
scFv-presenting clone 251
Schiff base 175, 177, 432, 526
Schultz, P. 127
SCI-PCM (Self-Consistent Isodensity PCM) 77
scientific practice 118
Scolytus multistriatus 134
screening 279, 495
screening assay 30, 43
screening experiment 217
screening methods 492
SDS-PAGE 502, 545, 546
selection 243, 250, 252, 255, 262, 271, 275, 279
selection protocol 43
selection strategy 46
selective chemotherapy 290
selenoxide elimination 387
self-immolative 293
self-ligation 63
sepharose 539
sequence analysis 189, 190
serendipity 432
serine hydrolase 159, 419
serine protease 53, 532
serum 471
serum titer 493, 494
seryl/cysteyl hydrolases 418
sex pheromone 145
shape complementarity 165, 168, 169, 174
sheep polyclonal antibodies 471
shigatoxin 547
sigmatropic rearrangement 22, 34, 386
   chorismate rearrangement 34
   oxy-Cope rearrangement 36

[3,3]-sigmatropic rearrangement 386
single turnover 267
single-chain Fv *see* scFv
single-turnover 481
single-turnover screening 267
singlet oxygen 178, 336, 341
site-directed mutagenesis 42, 264, 270
SLE 507, 535, 554, 558
Smith, A. 120
S_N1 191
SNase 259
sodium periodate 404
sol-gel 149
solvation 37
solvation/desolvation effects 421
Solvay Conference 126
solvolysis 225
somatic hypermutation 51
somatic mutation 2, 12, 17–19, 27, 42, 250
Sorangium cellulosum 137
sphingomyelin 546, 547
splenic colony-forming units 510
staggered extension process (StEP) 269
steady-state 481
stem cell differentiation 519
StEP 270
stepwise acquisition 27
stepwise increase 21
stereoselectivity 370, 403, 466
Stork-Eschenmoser concept 199
Stork-Eschenmoser hypothesis 172
strain 386
strain theory 7
streptavidin-coated bead 260, 272
streptavidin-coated magnetic beads 271
Streptomyces hygroscopicuus 145
streptomycin 498
structural flexibility 8, 17, 445
structural isomerization 445
structural rigidity 485

structure-based engineering   168
structure-function relationship   62
subcloning   238, 495, 503
substrate analog   193
substrate collapse   202
substrate destabilization   45, 421, 429, 442
substrate docking   364
substrate inhibition   441
substrate specificity   376, 378, 447
substrate strain   174
substrate trap   256, 257, 259
subtilisin   428
subtilisin E   270
subtilisin E mutant   270
succinimide   382
suicide inhibitor   256
suicide-type reaction   262
sulfoxides   404
supernatant   499, 503
superoxide   530
superoxide dismutase   507, 530
syn-elimination   169, 206, 386, 407, 444
synthetic methodology   132

t

T-cell   509
tail vein   494
tandem cationic cyclization   197
Taq-polymerase   272
targeting device   294, 297
taxol   138
TEM-1 β-lactamase   268
terpene cyclase enzyme   194
terpene-like electrophilic cyclization   196
terpenoid   400
terpenoid cyclases   199
tertiary aldol   145, 395
tetrahedral intermediate   79
tetrahedral transition state   420
tetrahydrofurans   399
tetrahydrooxepanes   399
tetrahydropyrans   399

tetranitromethane   433
Tg   508
theoretical calculations   205
theozyme   89, 90, 92, 100, 105, 107
therapeutic agents   481
therapeutic potential   284
therapeutics   559
thermal denaturation   526
thermodynamic information   338
thermostability   272
thiamin   145
thioester   476
thioesterases   437
thrombin   533
thyroglobulin   508, 510, 533
thyroglobulin-hydrolyzing IgGs   552
thyroid gland   508
thyroid hormone   533, 558
thyrotropic hormone   558
thyroxin   558
TLC   546
topoisomerase   512
total synthesis   133
Toyopearl HW 60   521
transacylation   379
transesterification   208
transition state   7, 72, 80, 85, 87, 90, 92, 94, 96, 98, 102, 104, 107, 304, 305, 307, 309
transition state analog   210, 218, 421, 457, 513
   binding energy   429
   design   422, 429
   phosphonamidate   155
   phosphonate   155
transition state binding   492
transition state epitope   424
transition state stabilization   128, 454
transition state theory   31
transition structures   76
trapped substrate   259
triosephosphate isomerase   14, 24

tripeptide transition state analog 372
tRNA 517
TRNOE measurement 36
trypsin 261, 428
trypsin-leupeptin 427
tryptophan fluorescence 446
tryptophan indole 359
tubulin polymerization 138
tumor cell 290
turnover 361
type I aldolase 137
tyrosine 426, 480
tyrosyl pK_a 433

u
ultra-filtration 499, 500
umbelliferone 237
unimolecular rearrangements 163

v
V genes 27
Van der Waals interactions 17
vasoactive intestinal peptide 484, 532
Vibrio cholerae 547
VIP 508, 510
Vκi gene 25

w
WAM 74
Web Antibody Modeling 74
96-well plate 228
Wieland-Miescher ketone 147, 394, 398
Wittig olefination 141

x
X-ray crystallography 205
X-ray structure 374, 389, 541
X-ray structure determination 199, 204
xanthine oxidase 531
xenon 178

y
Yang cyclization 360

z
zinc hydrolases
  mechanism 420
zinc ion 441
zinc-protease 441